FORD
PICK-UPS
EXPEDITION, NAVIGATOR 1997-17
REPAIR MANUAL

Covers U.S. and Canadian models of
Ford F-150 ('97-'03), F-150 Heritage ('04) and
F-250 Pick-Ups ('97-'99); Ford Expedition ('97-'17)
and Lincoln Navigator ('98-'17);
2 and 4 wheel drive, gasoline engines

Does not include diesel engine, F-250HD, Super Duty, F-350 or information specific to Lightning or other supercharged models

by Jay Storer

PUBLISHED BY **HAYNES NORTH AMERICA, Inc.**

Printed in Malaysia
©2009, 2013, 2015 and 2017 Haynes North America, Inc.
ISBN-10: 1-62902-310-6
ISBN-13: 978-1-62092-310-8
Library of Congress Control Number: 2017956669

Haynes Publishing Group
Sparkford Nr Yeovil
Somerset BA22 7JJ England

Haynes North America, Inc
859 Lawrence Drive
Newbury Park
California 91320 USA

Chilton is a registered trademark of W.G. Nichols, Inc., and has been licensed to Haynes North America, Inc.

Contents

INTRODUCTORY PAGES

About this manual – 0-5
Introduction – 0-5
Vehicle identification numbers – 0-6
Recall information – 0-8
Buying parts – 0-9
Maintenance techniques, tools and working facilities – 0-18
Jacking and towing – 0-26
Booster battery (jump) starting – 0-27
Automotive chemicals and lubricants – 0-28
Safety first! – 0-29
Troubleshooting – 0-30

1 TUNE-UP AND ROUTINE MAINTENANCE – 1-1

2
4.2L V6 ENGINE – 2A-1
3.5L V6 ENGINE – 2B-1
V8 ENGINES – 2C-1
GENERAL ENGINE OVERHAUL PROCEDURES – 2D-1

3 COOLING, HEATING AND AIR CONDITIONING SYSTEMS – 3-1

4 FUEL AND EXHAUST SYSTEMS – 4-1

5 ENGINE ELECTRICAL SYSTEMS – 5-1

6 EMISSIONS AND ENGINE CONTROL SYSTEMS – 6-1

MANUAL TRANSMISSION – 7A-1
AUTOMATIC TRANSMISSION – 7B-1
TRANSFER CASE – 7C-1

7

CLUTCH AND DRIVELINE – 8-1

8

BRAKES – 9-1

9

SUSPENSION AND STEERING SYSTEMS – 10-1

10

BODY – 11-1

11

CHASSIS ELECTRICAL SYSTEM – 12-1
WIRING DIAGRAMS – 12-25

12

MASTER INDEX – IND-1

MASTER INDEX

Photographer, mechanic and author with 1997 Ford F-150

ACKNOWLEDGEMENTS

Wiring diagrams and certain illustrations originated exclusively for Haynes North America, Inc. by Valley Forge Technical Information Services. Technical writers who contributed to this project include Jeff Killingsworth, Mike Stubblefield, Jeff Kibler and Rob Maddox.

All rights reserved. No part of this book may be reproduced or transmitted in any form or by any means, electronic or mechanical, including photocopying, recording or by any information storage or retrieval system, without permission in writing from the copyright holder.

"Ford" and the Ford logo are registered trademarks of Ford Motor Company. Ford Motor Company is not a sponsor or affiliate of Haynes Publishing Group or Haynes North America, Inc. and is not a contributor to the content of this manual.

While every attempt is made to ensure that the information in this manual is correct, no liability can be accepted by the authors or publishers for loss, damage or injury caused by any errors in, or omissions from, the information given.

INTRODUCTION 0-5

About this manual

ITS PURPOSE

The purpose of this manual is to help you get the best value from your vehicle. It can do so in several ways. It can help you decide what work must be done, even if you choose to have it done by a dealer service department or a repair shop; it provides information and procedures for routine maintenance and servicing; and it offers diagnostic and repair procedures to follow when trouble occurs.

We hope you use the manual to tackle the work yourself. For many simpler jobs, doing it yourself may be quicker than arranging an appointment to get the vehicle into a shop and making the trips to leave it and pick it up. More importantly, a lot of money can be saved by avoiding the expense the shop must pass on to you to cover its labor and overhead costs. An added benefit is the sense of satisfaction and accomplishment that you feel after doing the job yourself.

USING THE MANUAL

The manual is divided into Chapters. Each Chapter is divided into numbered Sections. Each Section consists of consecutively numbered paragraphs.

At the beginning of each numbered Section you will be referred to any illustrations which apply to the procedures in that Section. The reference numbers used in illustration captions pinpoint the pertinent Section and the Step within that Section. That is, illustration 3.2 means the illustration refers to Section 3 and Step (or paragraph) 2 within that Section.

Procedures, once described in the text, are not normally repeated. When it's necessary to refer to another Chapter, the reference will be given as Chapter and Section number. Cross references given without use of the word "Chapter" apply to Sections and/or paragraphs in the same Chapter. For example, "see Section 8" means in the same Chapter.

References to the left or right side of the vehicle assume you are sitting in the driver's seat, facing forward.

Even though we have prepared this manual with extreme care, neither the publisher nor the author can accept responsibility for any errors in, or omissions from, the information given.

➡ NOTE

A *Note* provides information necessary to properly complete a procedure or information which will make the procedure easier to understand.

✳✳ CAUTION

A *Caution* provides a special procedure or special steps which must be taken while completing the procedure where the Caution is found. Not heeding a Caution can result in damage to the assembly being worked on.

✳✳ WARNING

A *Warning* provides a special procedure or special steps which must be taken while completing the procedure where the Warning is found. Not heeding a Warning can result in personal injury.

Introduction to the F-150, F-250, Expedition and Navigator

These models are available in pick-up and four-door sport utility body styles.

The available engines are either a fuel injected 4.2L V6, 4.6L V8, 5.4L V8 or a direct injection twin-turbocharged 3.5L V6. All models are equipped with the On Board Diagnostic Second-Generation (OBD-II) computerized engine management system that controls virtually every aspect of engine operation. OBD-II monitors emissions system components for signs of degradation and engine operation for any malfunction that could affect emissions, turning on the CHECK ENGINE light if any faults are detected.

Chassis layout is conventional, with the engine mounted at the front and the power being transmitted through either a 5-speed manual, 4-speed automatic transmission or 6-speed automatic transmission and a driveshaft to the solid rear axle. On 4WD models, a transfer case transmits power to a front differential by way of a driveshaft and then to the front wheels through independent driveaxles.

These models feature independent front suspension with torsion bars (4WD) or coil springs (2WD) and shock absorbers and solid axle with leaf springs and shock absorbers (pick-ups) or coil springs or air springs (sport utility models) at the rear. On 2003 and later Expedition and Navigator models, the rear suspension is independent via a solidly-mounted differential with driveaxles connecting the differential to the rear wheels. The wheel assemblies are located by upper and lower control arms and upper and lower trailing arms. Coil spring/shock units or optional air shocks provide suspension.

The brakes are disc on the front and either drum or disc on the rear wheels, with an Anti-Lock Brake System (ABS) standard on most models.

The power-assisted recirculating ball-type steering is mounted on the chassis frame rail to the left of the engine.

0-6 VEHICLE IDENTIFICATION NUMBERS

Vehicle Identification numbers

Modifications are a continuing and unpublicized process in vehicle manufacturing. Since spare parts lists and manuals are compiled on a numerical basis, the individual vehicle numbers are necessary to correctly identify the component required.

VEHICLE IDENTIFICATION NUMBER (VIN)

This very important identification number is stamped on a plate attached to the dashboard inside the windshield on the driver's side of the vehicle (see illustration). The VIN also appears on the Vehicle Certificate of Title and Registration. It contains information such as where and when the vehicle was manufactured, the model year and the body style.

VIN ENGINE AND MODEL YEAR CODES

Two particularly important pieces of information found in the VIN are the engine code and the model year code. Counting from the left, the engine code letter designation is the 8th digit and the model year code letter designation is the 10th digit.

The VIN is visible through the windshield on the driver's side

The vehicle certification label is affixed to the driver's side door pillar

On the models covered by this manual the engine codes are:

2	4.2L V6
W	4.6L (Romeo) V8
6	4.6L (Windsor) V8
L	5.4L V8, 2V
R	5.4L V8, 4V
5	5.4L V8, 3V
T	3.5L twin-turbo V6

On the models covered by this manual the model year codes are:

V	1997
W	1998
X	1999
0	2000
1	2001
2	2002
3	2003
4	2004
5	2005
6	2006
7	2007
8	2008
9	2009
A	2010
B	2011
C	2012
D	2013
E	2014
F	2015
G	2016
H	2017

VEHICLE CERTIFICATION LABEL

The Vehicle Certification Label is attached to the driver's side door pillar (see illustration). Information on this label includes the name of the manufacturer, the month and year of production, as well as information on the options with which it is equipped. This label is especially useful for matching the color and type of paint for repair work.

ENGINE IDENTIFICATION NUMBER

Labels containing the engine code, engine number and build date can be found on the valve cover (see illustration). The engine number is also stamped onto a machined pad on the external surface of the engine block.

The engine identification label is affixed to the valve cover

VEHICLE IDENTIFICATION NUMBERS/BUYING PARTS   0-7

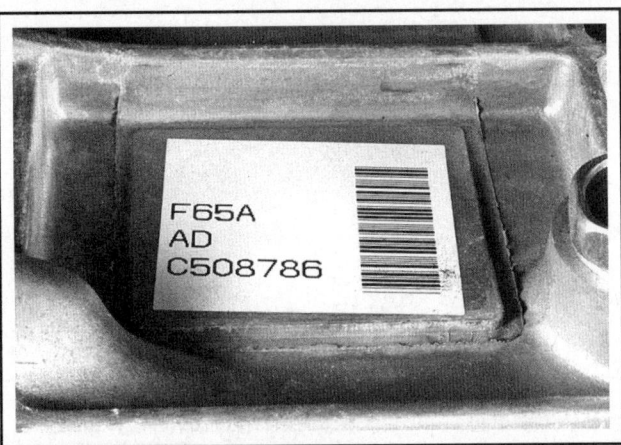

The manual transmission identification label is affixed to the passenger side of the transmission case

The transfer case identification tag is retained by a bolt at the rear of the transfer case

The differential identification tag is bolted to the differential cover

AUTOMATIC TRANSMISSION IDENTIFICATION NUMBER

The automatic transmission ID number is affixed to a label on the right side of the case.

MANUAL TRANSMISSION IDENTIFICATION NUMBER

The manual transmission ID number is affixed to a label on the right side of the case (see illustration).

TRANSFER CASE IDENTIFICATION NUMBER

The transfer case ID number is stamped on a tag which is bolted to the rear cover (see illustration).

DIFFERENTIAL IDENTIFICATION NUMBER

The differential ID number is stamped on a tag which is bolted to the differential cover (see illustration).

VEHICLE EMISSIONS CONTROL INFORMATION LABEL

This label is found in the engine compartment. See Chapter 6 for more information on this label.

0-8 RECALL INFORMATION

Recall information

Vehicle recalls are carried out by the manufacturer in the rare event of a possible safety-related defect. The vehicle's registered owner is contacted at the address on file at the Department of Motor Vehicles and given the details of the recall. Remedial work is carried out free of charge at a dealer service department.

If you are the new owner of a used vehicle which was subject to a recall and you want to be suvre that the work has been carried out, it's best to contact a dealer service department and ask about your individual vehicle - you'll need to furnish them your Vehicle Identification Number (VIN).

The table below is based on information provided by the National Highway Traffic Safety Administration (NHTSA), the body which oversees vehicle recalls in the United States. The recall database is updated constantly. For the latest information on vehicle recalls, check the NHTSA website at www.nhtsa.gov, www.safercar.gov, or call the NHTSA hotline at 1-888-327-4236.

Recall date	Recall campaign number	Model(s) affected	Concern
AUG 29, 1996	96V1610000	1997 F-150 and F-250	The certification labels on the some vehicles have incorrect rear tire inflation pressure designations. Incorrect air pressures could cause wear problems.
SEP 25, 1996	96V184000	1997 F-150 and F-250	On some models the retainer clip that holds the master cylinder pushrod to the brake pedal arm is missing. A loss of braking can occur, increasing the risk of an accident.
DEC 18, 1996	96V251000	1997 F-150	Operation of some models at highway speeds during extreme cold conditions can result in ice forming in the throttle body of the engine. This ice can cause the throttle plate to remain in open after the accelerator is released or the speed control is deactivated. The vehicle stopping distances would be increased possibly resulting in an accident.
DEC 18, 1996	96V252000	1997 Expedition	On some models, the rear axle track bar bracket can separate from the frame due to missing welds or inadequate weld penetration. If both brackets separate, the rear axle can move laterally until the tires contact either the frame or the wheelhouse resulting in either tire damage or a significant reduction in vehicle handling performance.

RECALL INFORMATION 0-9

Recall date	Recall campaign number	Model(s) affected	Concern
DEC 27, 1996	96V256000	1997 F-150	On some models, the seat belt anchorage attachments are missing or misinstalled. A loose fastener can eventually detach and would not properly restrain an occupant in the event of an accident.
SEP 09, 1997	97V147000	1997 F-150	On some models, the transmission bracket can separate from the shift cable assembly resulting in the inability to shift the transmission into the Park position. This can result in unintended vehicle movement if the parking brake is not set.
OCT 07, 1997	97V171000	1997, 1998 F-150, F-250 and Expedition, 1998 Navigator	On some models, the automatic transmission shift cable assembly was not fully attached to the steering column bracket; the shift cable can come out of the bracket. The driver would then not be able to shift into the Park position even though the gear shift selector would indicate "Park". This can result in unintended vehicle movement if the parking brake is not set.
OCT 29, 1997	97V187000	1997 F-250	On some models, the tires were damaged during mounting onto the wheels. The bead area of the tire was cut which could result in air loss. Sudden air loss could occur resulting in a loss of vehicle control.
JAN 21, 1998	98V007000	1997, 1998 F-150	On some models, the throttle is unable to return to idle due to ice forming in the throttle body when the temperature is below -10 degrees F.
FEB 10, 1998	98V028000	1998 Navigator	The text and/or graphics for the headlamp aiming instructions provided in the owner guides are not sufficiently clear. If the headlamp assemblies are replaced, customers may be confused by the aiming instructions, causing the headlamps to be improperly aimed.
MAY 13, 1998	98V095000	1997, 1998 F-150, F-250 and Expedition, 1998 Navigator	On some models, the lug nuts may not create sufficient clamp load, allowing wheel movement in relation to the hub/disc mounting surface. This can result in the loosening of lug nuts, stud fatigue failure, and the potential for a wheel separation from the vehicle, increasing the risk of a vehicle crash.

RECALL INFORMATION

Recall date	Recall campaign number	Model(s) affected	Concern
MAY 29, 1998	98V116000	1999 F-250	On some models, one or more of the four rivets on the lower steering shaft flex coupling may not have been crimped during the sub-assembly process. If a rivet is not crimped, the flex joint can rotate on the shaft and there would be a reduction in or loss of steering control.
AUG 19, 1998	98V194000	1999 F-150 and F-250	On some models, the fuel pressure regulator O-ring may have been damaged when the fuel pressure regulator was installed in the engine fuel rail. If the O-ring is damaged, fuel vapor or leakage could occur and could result in a fire.
AUG 24, 1998	98V192000	1999 F-150 and F-250	On some light duty trucks the speed control cable may not have fully seated into the accelerator bracket at time of installation. If the cable is not fully seated into this bracket, the cable could separate from the bracket. A loose cable can result in the throttle not fully returning to idle and cause an increase in stopping distance.
JUL 15, 1998	98V161000	1997, 1998 F-150	On some models, if the vehicle is overloaded, the rear leaf springs can be overstressed. A fatigue fracture of the spring can occur. The center leaf of the spring can fracture at the forward edge of the spring seat and, in some cases, contact the fuel tank. Fuel tank contact can damage the tank and result in fuel leakage. Fuel leakage in the presence of an ignition source can result in a fire.
NOV 17, 1998	98V296000	1998 Expedition, 1999 Navigator	Some models may have a missing or partially installed retainer clip that holds the master cylinder pushrod to the brake pedal arm. Increased brake stopping distance can occur, increasing the risk of a crash.
DEC 01, 1998	98V303000	1999 F-250	On some models, the airbag sensor was not calibrated properly, which may cause the airbags to deploy in lower speed impacts than design intent. Unexpected airbag deployment can result in personal injury.
DEC 07, 1998	98V312000	1999 Navigator	On some models, the fuel line assemblies may have been damaged during manufacturing, allowing leakage. Fuel leakage in the presence of an ignition source can result in a fire.
DEC 07, 1998	98V323000	1998 Expedition	On some models, the fuel line assemblies may have been damaged during manufacturing, allowing leakage. Fuel leakage in the presence of an ignition source can result in a fire.

RECALL INFORMATION 0-11

Recall date	Recall campaign number	Model(s) affected	Concern
MAR 01, 1999	99V039000	1999 F-150	On some models, the tire and rim identification information is incorrect on the certification labels. If incorrect tire information is given, underrated tires could be placed on the truck causing excessive tire wear or possible tire blowouts.
MAR 30, 1999	99V062001	1999 F-250	On some models, the cruise control cable can interfere with the speed control servo pulley and not allow the throttle to return to idle when disengaging the cruise control. If the cruise control is used and this condition is present, a stuck throttle could result, which could potentially result in a crash.
APR 27, 1999	99V093001	1999 F-250	On some models, six attachment bolts (M12 bolts) may have been damaged during a 30 mph front impact, allowing the bolts to shift or become detached, resulting in fuel spillage. Fuel leakage in the presence of an ignition source could result in a fire.
APR 27, 1999	99V093002	1999 F-250	On some models, contact between the fuel tank and a fuel tank locator bracket during a front impact test can result in a cut in the tank, resulting in fuel leakage. Fuel leakage in the presence of an ignition source could result in a fire.
APR 29, 1999	99V099000	1998 Expedition, 1999 Navigator	On some models, the clamp load can be lost on the wheel lugs due to insufficient wheel contact area with the hub. In some cases, the contact area can deform, resulting in a loss of lug nut torque. Loss of lug nut torque can cause vibration or separation of a wheel and tire from the vehicle.
AUG 12, 1999	99V219000	1999, 2000 F-250	On some models, the fuel tank front strap isolator may not be properly positioned, which could result in the strap cutting the tank during a 30 mph front impact, resulting in fuel spillage. Fuel leakage in the presence of an ignition source could result in a fire.
NOV 04, 1999	99V311000	1999, 2000 F-250	On some models, in a 30 mph frontal impact, a fuel tank vapor recovery valve could separate internally. Fuel leakage in the presence of an ignition source could result in a fire.
MAR 07, 2000	00V073000	1997, 1998, 1999, 2000 Expedition, and Navigator	On some models, the trailer hitch-to-frame bolts could loosen. The trailer hitch could then separate from the vehicle.

0-12 RECALL INFORMATION

Recall date	Recall campaign number	Model(s) affected	Concern
JUN 19, 2000	00V168000	1997, 1998 F-150 and Expedition	Some Imodels were imported into the U.S. erroneously (from Canada). The light given off from the daytime running lights is brighter than allowed by U.S. standards.
AUG 17, 2000	00V231000	1997 F-150 and F-250	On some models, the front fuel line assembly could have a hole rubbed through it. This condition could result in fuel leakage and could result in a fire.
AUG 18, 2000	00V228001	2000 F-150	On some models, the seat belt buckle bases were not properly heat treated. In the event of a crash, the occupant may not be properly restrained.
SEP 21, 2000	00V284000	1998, 1999, 2000 F-150, and F-250	On some bi-fuel CNG pickup trucks modified by CFI control systems, Inc., the label affixed to the fuel filler door by the vehicle modifier contains the inspection date and expiration date of the CNG tank. However, it does not contain the statement "See instructions on the fuel container for inspection and service life." These vehicles do not comply with the Federal requirements.
NOV 28, 2000	00V396000	2000, 2001 Expedition	On some models, the owner guides do not identify the locations of seating positions equipped with tether anchorages, and do not provide instructions for securing child seats to these tether anchorages. In the event of a crash, the child seat may not be properly attached increasing the risk of injury to the child
MAR 12, 2001	01V082000	2001 F-250 and Expedition	On some models, the inflator canister in the driver airbag module may have an inadequate weld near the igniter. In the event of a crash, the driver airbag may not deploy as intended, potentially resulting in reduced occupant protection, or a burn injury.
APR 05, 2001	01V114000	2001 F-150	On some models, the 35 gallon mid-ship fuel tank could develop a crack where it contacts the frame crossmember. If the fuel tank cracks, fuel leakage could occur. Fuel leakage in the presence of an ignition source could result in a fire.
MAY 22, 2001	01X001000	1997 F-150 and F-250, 1999, 2000, 2001 Expedition	Ford is replacing all Firestone Wilderness at 15, 16, and 17 inch tires mounted on Ford trucks and SUVs. The vehicles may have been originally equipped with Firestone wilderness A/T tires or may have had wilderness A/T tires installed during the Firestone recall. Both original equipment and replacement tires are affected. Should the tread separate at highway speeds, a vehicle crash could occur, possibly resulting in personal injury or death

RECALL INFORMATION O-13

Recall date	Recall campaign number	Model(s) affected	Concern
AUG 08, 2001	01V258000	2000 F-150, 2000, 2001 F-250, Expedition, and Navigator	On some models, a switch located in the plastic cover of the wiper motor gear case could malfunction and overheat, potentially resulting in loss of proper wiper function, or ignition of the plastic cover material. Loss of visibility while driving increases the risk of a crash
JUL 16, 2001	01V227001	2001 F-250, Expedition and Navigator	On some models, it is possible that the driver's and/or front passenger's outboard seat belt buckle may not fully latch. In the event of a crash, the restraint system may not provide adequate occupant protection, increasing the risk of personal injury to the seat occupant.
JAN 08, 2002	02V008000	1999 F-150	On some models, a typographical error occurred resulting in an incorrect tire size and pressure on the safety certification label. If the tires are inflated to the pressure indicated on the label, the customer will experience a harsher ride.
MAR 11, 2002	02V068000	1999 F-250	On some models equipped with manual transmissions, failure of the front parking brake cable input button could result in failure of the parking brake system to hold the vehicle stationary. This could result in unintended vehicle movement and a vehicle crash.
SEP 09, 2002	02V239000	2000 Expedition	On some 4x2 models, the rear tire pressure is incorrectly listed on the label for 4.2 vehicles. Customers may inflate their rear tires based on the incorrect tire pressure information on the certification label.
MAY 27, 2003	03V196000	1997 F-1500, F-250 and Expedition	On some models, if the intermediate shaft yoke separates from the steering gear input shaft, the steering system becomes disconnected. This could result in loss of steering control, potentially resulting in a vehicle crash.
SEP 22, 2003	03V394000	2003 F-150	On some models, the Pitman arm-to-steering gear retaining nut was not tightened to the required torque. The nut could back off completely and the Pitman arm could separate from the steering gear, resulting in a crash.
JAN 27, 2005	05V017000	2000 F-150, Expedition and Navigator	On some pickup trucks, the speed control deactivation switch may overheat, smoke, or burn. This condition could lead to a fire. Fires have occurred while the vehicles were parked with the ignition "OFF."

0-14 RECALL INFORMATION

Recall date	Recall campaign number	Model(s) affected	Concern
JUL 06, 2005	05V310000	2006 Navigator	On some vehicles, the front and rear tires on the driver's side may have been damaged at the center tread during production. Over time, the damage may be sufficient to allow belt corrosion, ultimately leading to a tread separation, which could result in a crash.
SEP 07, 2005	05V388000	1997, 1998, 1999, 2000, 2001, 2002 F-150, F-250, Expedition and 1998, 1999, 2000, 2001, 2002 Navigator	On some models equipped with speed control, the speed control deactivation switch may overheat, smoke, or burn. A fire at the switch could occur.
NOV 09, 2005	05V519000	2006 Navigator	Some models may have been built with an incorrectly manufactured automatic transmission parking pawl guide plate. This could cause the parking lock system not to engage, after the transmission is placed in the Park ("P") position. If the vehicle was parked on an incline, and the parking brake was not engaged, this could result in a vehicle rollaway condition.
NOV 09, 2005	05V520000	2006 Expedition and Navigator	On some models the windshield wiper motor may have been produced without grease being applied to the output shaft gear. Lack of grease on the output shaft gear may cause the gear to distort or fracture during operation resulting in a loss of wiper function, which could cause a crash due to impaired visibility.
AUG 01, 2006	06V286000	1997, 1998, 1999, 2000, 2001, 2002 F-250	On some models equipped with speed control, the speed control deactivation switch may overheat. A fire under the hood could occur.
MAR 06, 2007	07V078000	2003 F-150	On some models equipped with speed control, the speed control deactivation switch may overheat. A fire under the hood could occur.
AUG 03, 2007	07V336000	2001, 2002, 2003 F-150	On some models equipped with speed control, under certain conditions, the speed control deactivation switch can leak internally and then overheat, smoke, or burn. A fire under the hood could occur.
NOV 21, 2007	07V541000	2008 Expedition	On some models, the driver's airbag module may not have been properly assembled, preventing the airbag from properly deploying in the event of a crash, increasing the risk of injury.

RECALL INFORMATION   O-15

Recall date	Recall campaign number	Model(s) affected	Concern
FEB 04, 2008	08V051000	2001 F-150	Some models were previously repaired using a wiring harness that is not compatible with the vehicle circuit polarity. As a result, the fuse is located in the output circuit rather than in the intended input power feed circuit, and may not offer the intended protection in the event of an electrical short to ground.
FEB 05, 2008	08V057000	2007 Expedition and Navigator	Some models equipped with 5.4L engines may have fuel rail crossover hoses that contain weak areas. As a result, the hose may crack, causing fuel leaks, which, in the presence of an ignition source, could cause a fire.
FEB 05, 2008	08V058000	2007, 2008 Expedition and Navigator	On some models, the door handle housing embossment retaining the bottom of the interior door handle spring on all side doors may fracture during normal usage. In the event of a side impact crash, the interior door handle may cause the door latch to open increasing the risk of injury to a vehicle occupant.
APR 11, 2008	08V166000	2008 Expedition and Navigator	On some models equipped with a tire pressure monitoring system (TPMS) and the limo builder's package, the tire pressure was increased from 35 psi to 40 psi without a commensurate update to the TPMS calibrations. The TPMS indicator light will not illuminate when a tire pressure is 25 percent below the 40 psi value stated on the tire placard label occurs.
JUN 24, 2009	09V232000	2009 Expedition and Navigator	Some models may have improperly adjusted brake light switches, resulting in a delay of brake light illumination when the brake pedal is depressed. In instances of very light brake application, the brake lights may not illuminate at all. Delay or lack of brake light illumination may increase the risk of a crash.
OCT 13, 2009	09V399000	1997, 1999, 2000, 2001, 2002, 2003 F-250	On some models equipped with a Texas Instruments Speed Control Deactivation Switch (SCDS), the SCDS may leak internally and then overheat, smoke, or burn. A fire could occur with or without the engine running
JUL 13, 2011	11V352000	2002, 2003 F-250	On some models, the multi-function switch was shipped with a subcomponent (slider) that may experience deformation. This may cause a malfunction of the turn signal, tail lights, hazard warning signal flashers and/or brake lights. Non-functioning lights could increase the risk of a crash.
JUL 29, 2011	11V385000	1997, 1998, 1999, 2000, 2001, 2002, 2003 F-150, 1997, 1998, 1999 F-250	On some models, prolonged exposure to road de-icing chemicals may cause severe corrosion of the fuel tank straps which secure the tank to the vehicle. The corrosion may cause one or both straps to fail, allowing the fuel lines to separate from the tank, or in some cases, causing the tank to contact the ground. Either scenario may result in a fuel leak presenting a fire hazard.

RECALL INFORMATION

Recall date	Recall campaign number	Model(s) affected	Concern
FEB 02, 2012	12V034000	2012 Expedition and Navigator	Some vehicles manufactured from 11/19/11 through 12/2/11 may have inadequately welded head restraint supports on the second row seats. In the event of a crash, the head restraint may not provide the required strength, increasing the risk of injury.
MAY 03, 2012	12V190000	2011, 2012 Expedition and Navigator	Some models fail to comply with the Federal Motor Vehicle Safety Standards #102, "Transmission shift lever sequence, starter interlock and transmission braking effect," and #108, "Lamps, relfective devices and associated equipment." These vehicles may have a Transmission Range (TR) sensor that is improperly calibrated for Reverse, resulting in the vehicle not shifting into Reverse when the driver positions the shift lever to R, which can increase the risk of a crash.

BUYING PARTS 0-17

Buying parts

Replacement parts are available from many sources, which generally fall into one of two categories - authorized dealer parts departments and independent retail auto parts stores. Our advice concerning these parts is as follows:

Retail auto parts stores: Good auto parts stores will stock frequently needed components which wear out relatively fast, such as clutch components, exhaust systems, brake parts, tune-up parts, etc. These stores often supply new or reconditioned parts on an exchange basis, which can save a considerable amount of money. Discount auto parts stores are often very good places to buy materials and parts needed for general vehicle maintenance such as oil, grease, filters, spark plugs, belts, touch-up paint, bulbs, etc. They also usually sell tools and general accessories, have convenient hours, charge lower prices and can often be found not far from home.

Authorized dealer parts department: This is the best source for parts which are unique to the vehicle and not generally available elsewhere (such as major engine parts, transmission parts, trim pieces, etc.).

Warranty information: If the vehicle is still covered under warranty, be sure that any replacement parts purchased - regardless of the source - do not invalidate the warranty!

To be sure of obtaining the correct parts, have engine and chassis numbers available and, if possible, take the old parts along for positive identification.

Maintenance techniques, tools and working facilities

MAINTENANCE TECHNIQUES

There are a number of techniques involved in maintenance and repair that will be referred to throughout this manual. Application of these techniques will enable the home mechanic to be more efficient, better organized and capable of performing the various tasks properly, which will ensure that the repair job is thorough and complete.

Fasteners

Fasteners are nuts, bolts, studs and screws used to hold two or more parts together. There are a few things to keep in mind when working with fasteners. Almost all of them use a locking device of some type, either a lockwasher, locknut, locking tab or thread adhesive. All threaded fasteners should be clean and straight, with undamaged threads and undamaged corners on the hex head where the wrench fits. Develop the habit of replacing all damaged nuts and bolts with new ones. Special locknuts with nylon or fiber inserts can only be used once. If they are removed, they lose their locking ability and must be replaced with new ones.

Rusted nuts and bolts should be treated with a penetrating fluid to ease removal and prevent breakage. Some mechanics use turpentine in a spout-type oil can, which works quite well. After applying the rust penetrant, let it work for a few minutes before trying to loosen the nut or bolt. Badly rusted fasteners may have to be chiseled or sawed off or removed with a special nut breaker, available at tool stores.

If a bolt or stud breaks off in an assembly, it can be drilled and removed with a special tool commonly available for this purpose. Most automotive machine shops can perform this task, as well as other repair procedures, such as the repair of threaded holes that have been stripped out.

Flat washers and lockwashers, when removed from an assembly, should always be replaced exactly as removed. Replace any damaged washers with new ones. Never use a lockwasher on any soft metal surface (such as aluminum), thin sheet metal or plastic.

Fastener sizes

For a number of reasons, automobile manufacturers are making wider and wider use of metric fasteners. Therefore, it is important to be able to tell the difference between standard (sometimes called U.S. or SAE) and metric hardware, since they cannot be interchanged.

All bolts, whether standard or metric, are sized according to diameter, thread pitch and length. For example, a standard 1/2 - 13 x 1 bolt is 1/2 inch in diameter, has 13 threads per inch and is 1 inch long. An M12 - 1.75 x 25 metric bolt is 12 mm in diameter, has a thread pitch of 1.75 mm (the distance between threads) and is 25 mm long. The two bolts are nearly identical, and easily confused, but they are not interchangeable.

In addition to the differences in diameter, thread pitch and length, metric and standard bolts can also be distinguished by examining the bolt heads. To begin with, the distance across the flats on a standard bolt head is measured in inches, while the same dimension on a metric bolt is sized in millimeters (the same is true for nuts). As a result, a standard wrench should not be used on a metric bolt and a metric wrench should not be used on a standard bolt. Also, most standard bolts have slashes radiating out from the center of the head to denote the grade or strength of the bolt, which is an indication of the amount of torque that can be applied to it. The greater the number of slashes, the greater the strength of the bolt. Grades 0 through 5 are commonly used on automobiles. Metric bolts have a property class (grade) number, rather than a slash, molded into their heads to indicate bolt strength. In this case, the higher the number, the stronger the bolt. Property class numbers 8.8, 9.8 and 10.9 are commonly used on automobiles.

Strength markings can also be used to distinguish standard hex nuts from metric hex nuts. Many standard nuts have dots stamped into one side, while metric nuts are marked with a number. The greater the number of dots, or the higher the number, the greater the strength of the nut.

Metric studs are also marked on their ends according to property class (grade). Larger studs are numbered (the same as metric bolts), while smaller studs carry a geometric code to denote grade.

It should be noted that many fasteners, especially Grades 0 through 2, have no distinguishing marks on them. When such is the case, the only way to determine whether it is standard or metric is to measure the thread pitch or compare it to a known fastener of the same size.

Standard fasteners are often referred to as SAE, as opposed to metric. However, it should be noted that SAE technically refers to a non-metric fine thread fastener only. Coarse thread non-metric fasteners are referred to as USS sizes.

Since fasteners of the same size (both standard and metric) may have different strength ratings, be sure to reinstall any bolts, studs or nuts removed from your vehicle in their original locations. Also, when replacing a fastener with a new one, make sure that the new one has a strength rating equal to or greater than the original.

Tightening sequences and procedures

Most threaded fasteners should be tightened to a specific torque value (torque is the twisting force applied to a threaded component such as a nut or bolt). Overtightening the fastener can weaken it and cause it to break, while undertightening can cause it to eventually come loose. Bolts, screws and studs, depending on the material they are made of and their thread diameters, have specific torque values, many of which are noted in the Specifications at the end of each Chapter. Be sure to follow the torque recommendations closely. For fasteners not assigned a specific torque, a general torque value chart is presented here as a guide. These torque values are for dry (unlubricated) fasteners threaded into steel or cast iron (not aluminum). As was previously mentioned, the size and grade of a fastener determine the amount of torque that can safely be applied to it. The figures listed here are approximate for Grade 2 and Grade 3 fasteners. Higher grades can tolerate higher torque values.

Fasteners laid out in a pattern, such as cylinder head bolts, oil pan bolts, differential cover bolts, etc., must be loosened or tightened in sequence to avoid warping the component. This sequence will normally be shown in the appropriate Chapter. If a specific pattern is not given, the following procedures can be used to prevent warping.

Initially, the bolts or nuts should be assembled finger-tight only. Next, they should be tightened one full turn each, in a criss-cross or diagonal pattern. After each one has been tightened one full turn, return to the first one and tighten them all one-half turn, following the same pattern. Finally, tighten each of them one-quarter turn at a time until each fastener has been tightened to the proper torque. To loosen and remove the fasteners, the procedure would be reversed.

MAINTENANCE TECHNIQUES, TOOLS AND WORKING FACILITIES   0-19

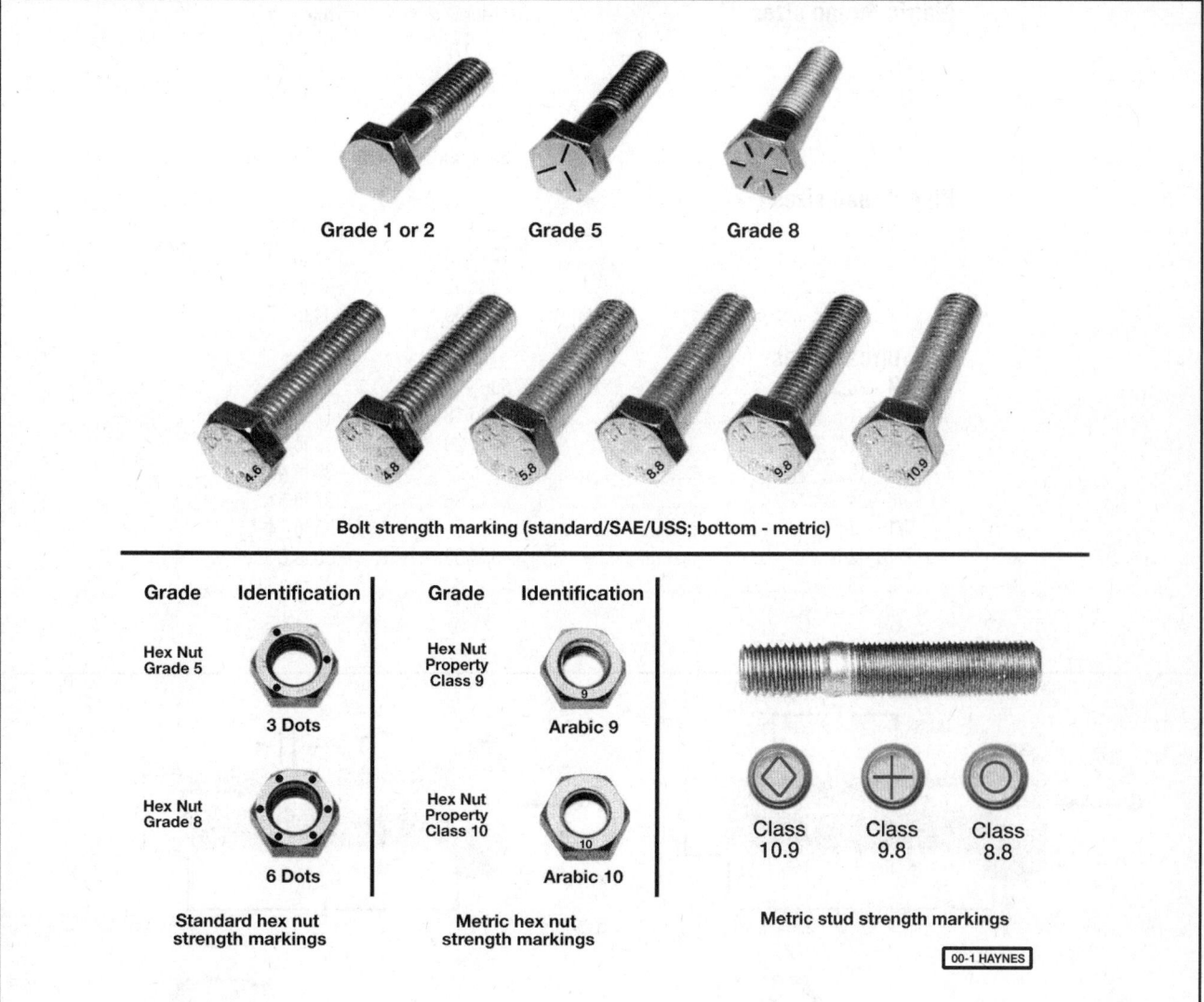

Component disassembly

Component disassembly should be done with care and purpose to help ensure that the parts go back together properly. Always keep track of the sequence in which parts are removed. Make note of special characteristics or marks on parts that can be installed more than one way, such as a grooved thrust washer on a shaft. It is a good idea to lay the disassembled parts out on a clean surface in the order that they were removed. It may also be helpful to make sketches or take instant photos of components before removal.

When removing fasteners from a component, keep track of their locations. Sometimes threading a bolt back in a part, or putting the washers and nut back on a stud, can prevent mix-ups later. If nuts and bolts cannot be returned to their original locations, they should be kept in a compartmented box or a series of small boxes. A cupcake or muffin tin is ideal for this purpose, since each cavity can hold the bolts and nuts from a particular area (i.e. oil pan bolts, valve cover bolts, engine mount bolts, etc.). A pan of this type is especially helpful when working on assemblies with very small parts, such as the carburetor, alternator, valve train or interior dash and trim pieces. The cavities can be marked with paint or tape to identify the contents.

Whenever wiring looms, harnesses or connectors are separated, it is a good idea to identify the two halves with numbered pieces of masking tape so they can be easily reconnected.

Gasket sealing surfaces

Throughout any vehicle, gaskets are used to seal the mating surfaces between two parts and keep lubricants, fluids, vacuum or pressure contained in an assembly.

Many times these gaskets are coated with a liquid or paste-type gasket sealing compound before assembly. Age, heat and pressure can sometimes cause the two parts to stick together so tightly that they are very difficult to separate. Often, the assembly can be loosened by striking it with a soft-face hammer near the mating surfaces. A regular hammer can be used if a block of wood is placed between the hammer and the part. Do not hammer on cast parts or parts that could be easily damaged. With any particularly stubborn part, always recheck to make sure that every fastener has been removed.

Avoid using a screwdriver or bar to pry apart an assembly, as they can easily mar the gasket sealing surfaces of the parts, which must remain smooth. If prying is absolutely necessary, use an old broom handle, but keep in mind that extra clean up will be necessary if the wood splinters.

After the parts are separated, the old gasket must be carefully scraped off and the gasket surfaces cleaned. Stubborn gasket material can be soaked with rust penetrant or treated with a special chemical to soften it so it can be easily scraped off.

Maintenance techniques, tools and working facilities

Metric thread sizes	Ft-lbs	Nm
M-6	6 to 9	9 to 12
M-8	14 to 21	19 to 28
M-10	28 to 40	38 to 54
M-12	50 to 71	68 to 96
M-14	80 to 140	109 to 154
Pipe thread sizes		
1/8	5 to 8	7 to 10
1/4	12 to 18	17 to 24
3/8	22 to 33	30 to 44
1/2	25 to 35	34 to 47
U.S. thread sizes		
1/4 - 20	6 to 9	9 to 12
5/16 - 18	12 to 18	17 to 24
5/16 - 24	14 to 20	19 to 27
3/8 - 16	22 to 32	30 to 43
3/8 - 24	27 to 38	37 to 51
7/16 - 14	40 to 55	55 to 74
7/16 - 20	40 to 60	55 to 81
1/2 - 13	55 to 80	75 to 108

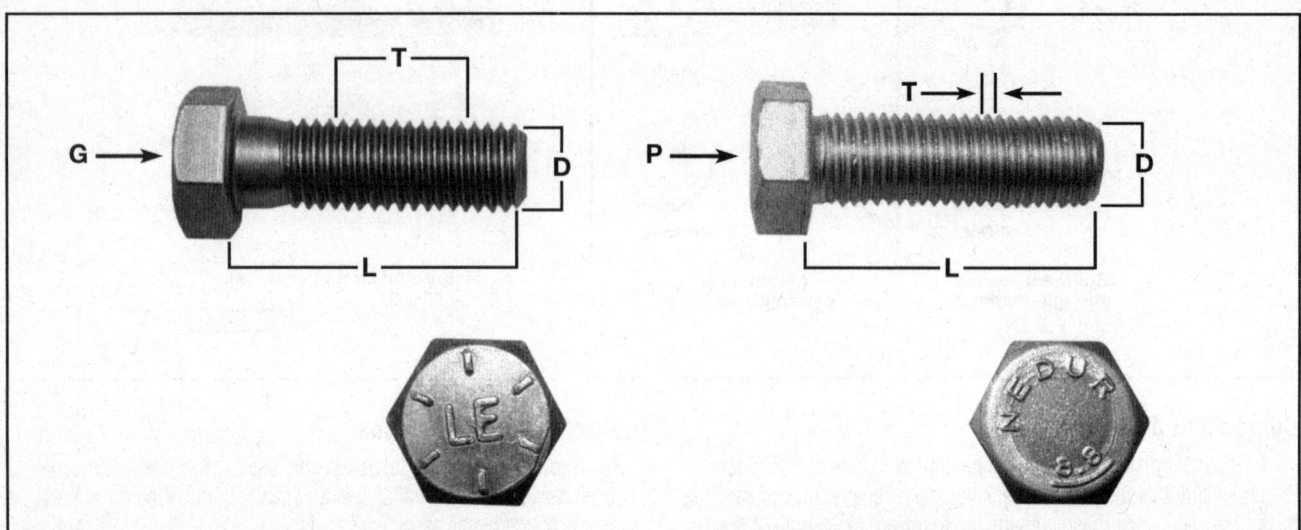

Standard (SAE and USS) bolt dimensions/grade marks

- G Grade marks (bolt strength)
- L Length (in inches)
- T Thread pitch (number of threads per inch)
- D Nominal diameter (in inches)

Metric bolt dimensions/grade marks

- P Property class (bolt strength)
- L Length (in millimeters)
- T Thread pitch (distance between threads in millimeters)
- D Diameter

✱✱ CAUTION:
Never use gasket removal solutions or caustic chemicals on plastic or other composite components.

A scraper can be fashioned from a piece of copper tubing by flattening and sharpening one end. Copper is recommended because it is usually softer than the surfaces to be scraped, which reduces the chance of gouging the part. Some gaskets can be removed with a wire brush, but regardless of the method used, the mating surfaces must be left clean and smooth. If for some reason the gasket surface is gouged, then a gasket sealer thick enough to fill scratches will have to be used during reassembly of the components. For most applications, a non-drying (or semi-drying) gasket sealer should be used.

MAINTENANCE TECHNIQUES, TOOLS AND WORKING FACILITIES   0-21

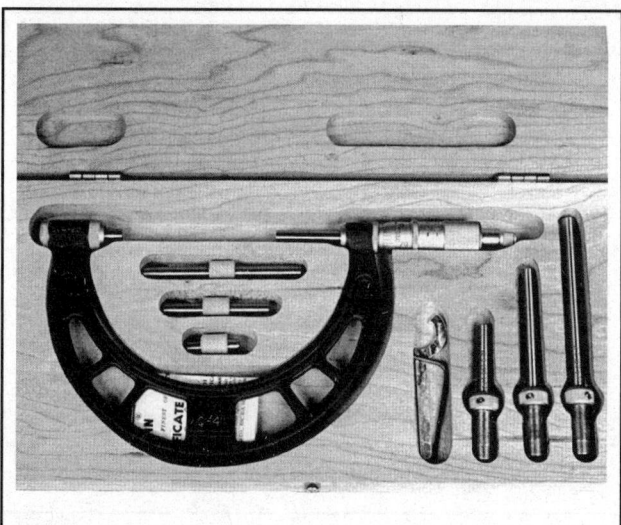

Micrometer set

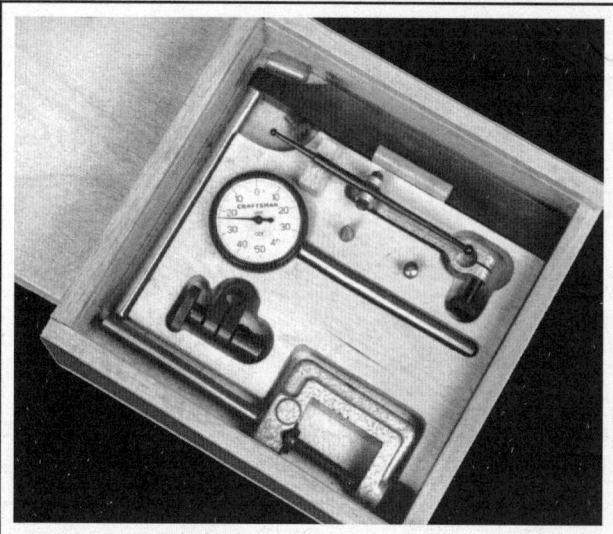

Dial indicator set

Hose removal tips

✱✱ WARNING:

If the vehicle is equipped with air conditioning, do not disconnect any of the A/C hoses without first having the system depressurized by a dealer service department or a service station.

Hose removal precautions closely parallel gasket removal precautions. Avoid scratching or gouging the surface that the hose mates against or the connection may leak. This is especially true for radiator hoses. Because of various chemical reactions, the rubber in hoses can bond itself to the metal spigot that the hose fits over. To remove a hose, first loosen the hose clamps that secure it to the spigot. Then, with slip-joint pliers, grab the hose at the clamp and rotate it around the spigot. Work it back and forth until it is completely free, then pull it off. Silicone or other lubricants will ease removal if they can be applied between the hose and the outside of the spigot. Apply the same lubricant to the inside of the hose and the outside of the spigot to simplify installation.

As a last resort (and if the hose is to be replaced with a new one anyway), the rubber can be slit with a knife and the hose peeled from the spigot. If this must be done, be careful that the metal connection is not damaged.

If a hose clamp is broken or damaged, do not reuse it. Wire-type clamps usually weaken with age, so it is a good idea to replace them with screw-type clamps whenever a hose is removed.

TOOLS

A selection of good tools is a basic requirement for anyone who plans to maintain and repair his or her own vehicle. For the owner who has few tools, the initial investment might seem high, but when compared to the spiraling costs of professional auto maintenance and repair, it is a wise one.

To help the owner decide which tools are needed to perform the tasks detailed in this manual, the following tool lists are offered: *Maintenance and minor repair*, *Repair/overhaul* and *Special*.

The newcomer to practical mechanics should start off with the *maintenance and minor repair* tool kit, which is adequate for the simpler jobs performed on a vehicle. Then, as confidence and experience grow, the owner can tackle more difficult tasks, buying additional tools as they are needed. Eventually the basic kit will be expanded into the *repair and overhaul* tool set. Over a period of time, the experienced do-it-yourselfer will assemble a tool set complete enough for most repair and overhaul procedures and will add tools from the special category when it is felt that the expense is justified by the frequency of use.

Maintenance and minor repair tool kit

The tools in this list should be considered the minimum required for performance of routine maintenance, servicing and minor repair work. We recommend the purchase of combination wrenches (box-end and open-end combined in one wrench). While more expensive than open end wrenches, they offer the advantages of both types of wrench.

Combination wrench set (1/4-inch to 1 inch or 6 mm to 19 mm)
Adjustable wrench, 8 inch
Spark plug wrench with rubber insert
Spark plug gap adjusting tool
Feeler gauge set
Brake bleeder wrench
Standard screwdriver (5/16-inch x 6 inch)
Phillips screwdriver (No. 2 x 6 inch)
Combination pliers - 6 inch
Hacksaw and assortment of blades
Tire pressure gauge
Grease gun
Oil can
Fine emery cloth
Wire brush
Battery post and cable cleaning tool
Oil filter wrench
Funnel (medium size)
Safety goggles
Jackstands (2)
Drain pan

➡**Note: If basic tune-ups are going to be part of routine maintenance, it will be necessary to purchase a good quality stroboscopic timing light and combination tachometer/dwell meter. Although they are included in the list of special tools, it is mentioned here because they are absolutely necessary for tuning most vehicles properly.**

0-22 MAINTENANCE TECHNIQUES, TOOLS AND WORKING FACILITIES

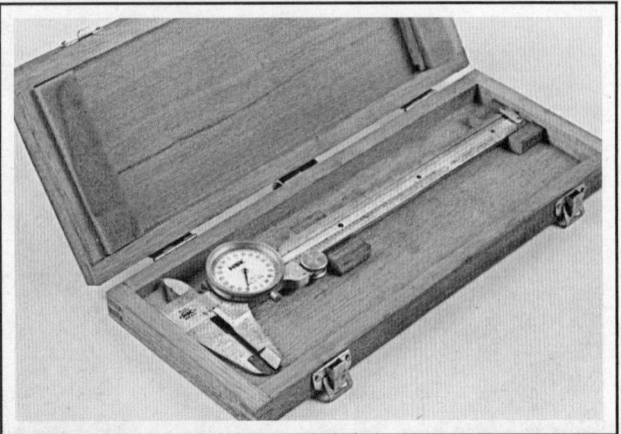

Dial caliper

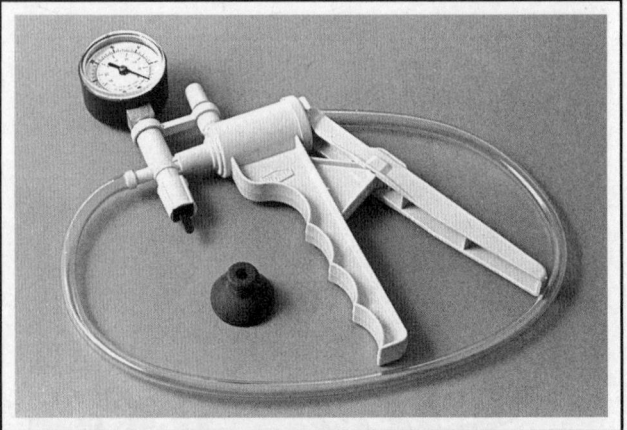

Hand-operated vacuum pump

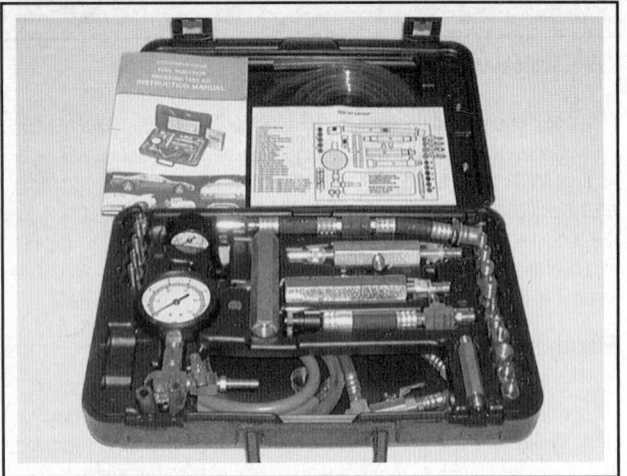

Fuel pressure gauge set

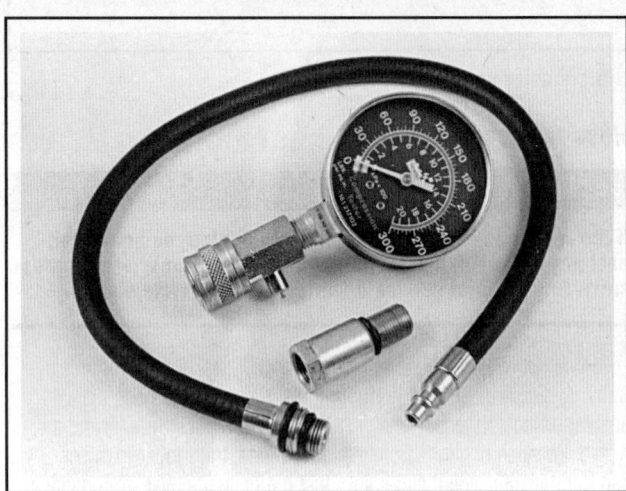

Compression gauge with spark plug hole adapter

Repair and overhaul tool set

These tools are essential for anyone who plans to perform major repairs and are in addition to those in the maintenance and minor repair tool kit. Included is a comprehensive set of sockets which, though expensive, are invaluable because of their versatility, especially when various extensions and drives are available. We recommend the 1/2-inch drive over the 3/8-inch drive. Although the larger drive is bulky and more expensive, it has the capacity of accepting a very wide range of large sockets. Ideally, however, the mechanic should have a 3/8-inch drive set and a 1/2-inch drive set.

Socket set(s)
Reversible ratchet
Extension - 10 inch
Universal joint
Torque wrench (same size drive as sockets)
Ball peen hammer - 8 ounce
Soft-face hammer (plastic/rubber)
Standard screwdriver (1/4-inch x 6 inch)
Standard screwdriver (stubby - 5/16-inch)
Phillips screwdriver (No. 3 x 8 inch)
Phillips screwdriver (stubby - No. 2)
Pliers - vise grip
Pliers - lineman's
Pliers - needle nose
Pliers - snap-ring (internal and external)

Cold chisel - 1/2-inch
Scribe
Scraper (made from flattened copper tubing)
Centerpunch
Pin punches (1/16, 1/8, 3/16-inch)
Steel rule/straightedge - 12 inch
Allen wrench set (1/8 to 3/8-inch or 4 mm to 10 mm)
A selection of files
Wire brush (large)
Jackstands (second set)
Jack (scissor or hydraulic type)

➡ **Note:** Another tool which is often useful is an electric drill with a chuck capacity of 3/8-inch and a set of good quality drill bits.

Special tools

The tools in this list include those which are not used regularly, are expensive to buy, or which need to be used in accordance with their manufacturer's instructions. Unless these tools will be used frequently, it is not very economical to purchase many of them. A consideration would be to split the cost and use between yourself and a friend or friends. In addition, most of these tools can be obtained from a tool rental shop on a temporary basis.

This list primarily contains only those tools and instruments widely available to the public, and not those special tools produced by the vehicle manufacturer for distribution to dealer service departments. Occasionally, references to the manufacturer's special tools are

MAINTENANCE TECHNIQUES, TOOLS AND WORKING FACILITIES 0-23

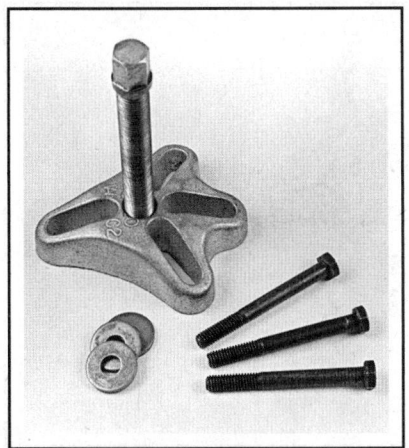
Damper/steering wheel puller

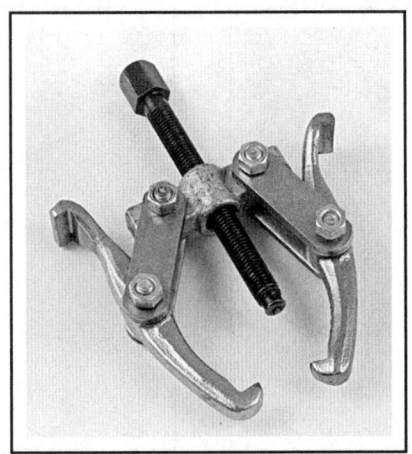
General purpose puller

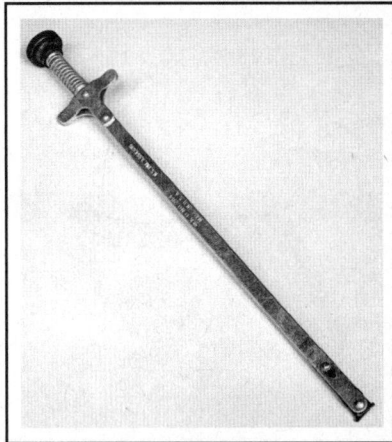
Hydraulic lifter removal tool

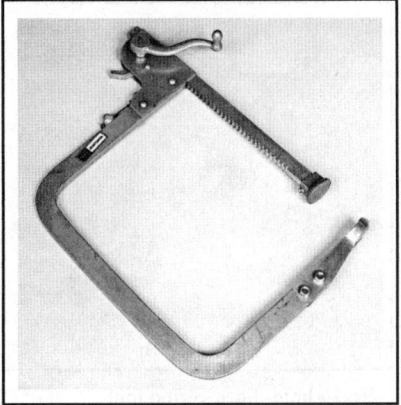
Valve spring compressor

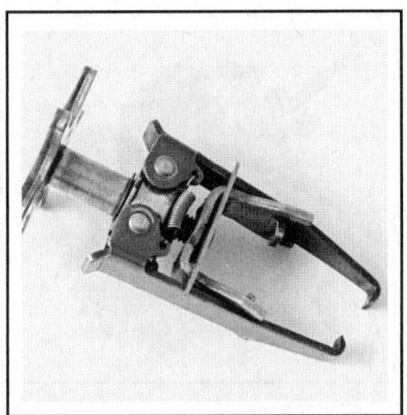
Valve spring compressor

Ridge reamer

included in the text of this manual. Generally, an alternative method of doing the job without the special tool is offered. However, sometimes there is no alternative to their use. Where this is the case, and the tool cannot be purchased or borrowed, the work should be turned over to the dealer service department or an automotive repair shop.

Valve spring compressor
Piston ring groove cleaning tool
Piston ring compressor
Piston ring installation tool
Cylinder compression gauge
Cylinder ridge reamer
Cylinder surfacing hone
Cylinder bore gauge
Micrometers and/or dial calipers
Hydraulic lifter removal tool
Balljoint separator
Universal-type puller
Impact screwdriver
Dial indicator set
Stroboscopic timing light (inductive pick-up)
Hand operated vacuum/pressure pump
Tachometer/dwell meter
Universal electrical multimeter
Cable hoist
Brake spring removal and installation tools
Floor jack

Buying tools

For the do-it-yourselfer who is just starting to get involved in vehicle maintenance and repair, there are a number of options available when purchasing tools. If maintenance and minor repair is the extent of the work to be done, the purchase of individual tools is satisfactory. If, on the other hand, extensive work is planned, it would be a good idea to purchase a modest tool set from one of the large retail chain stores. A set can usually be bought at a substantial savings over the individual tool prices, and they often come with a tool box. As additional tools are needed, add-on sets, individual tools and a larger tool box can be purchased to expand the tool selection. Building a tool set gradually allows the cost of the tools to be spread over a longer period of time and gives the mechanic the freedom to choose only those tools that will actually be used.

Tool stores will often be the only source of some of the special tools that are needed, but regardless of where tools are bought, try to avoid cheap ones, especially when buying screwdrivers and sockets, because they won't last very long. The expense involved in replacing cheap tools will eventually be greater than the initial cost of quality tools.

Care and maintenance of tools

Good tools are expensive, so it makes sense to treat them with respect. Keep them clean and in usable condition and store them properly when not in use. Always wipe off any dirt, grease or metal chips before putting them away. Never leave tools lying around in the work

0-24 MAINTENANCE TECHNIQUES, TOOLS AND WORKING FACILITIES

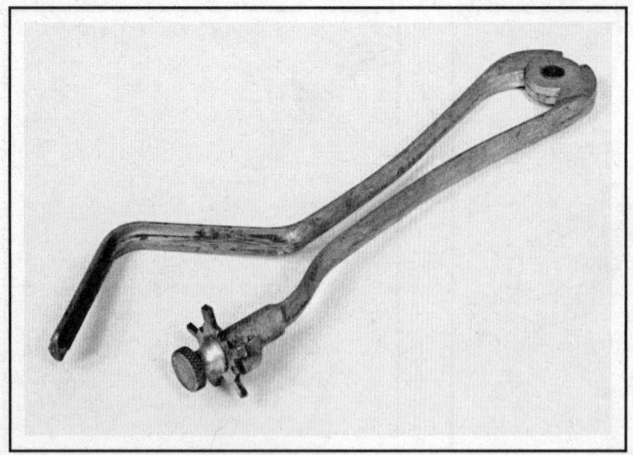
Piston ring groove cleaning tool

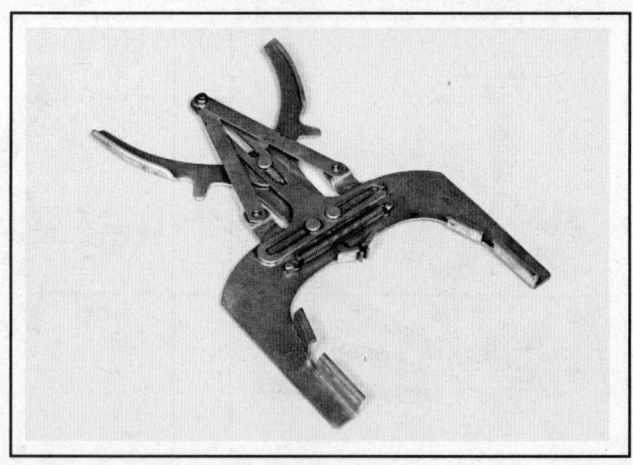
Ring removal/installation tool

Ring compressor

Cylinder hone

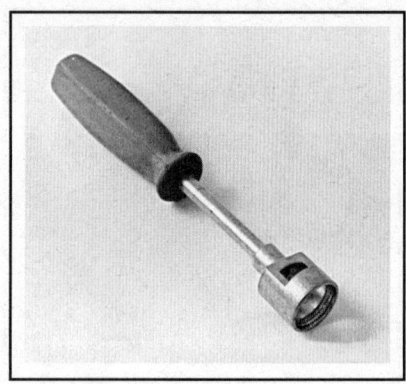
Brake hold-down spring tool

area. Upon completion of a job, always check closely under the hood for tools that may have been left there so they won't get lost during a test drive.

Some tools, such as screwdrivers, pliers, wrenches and sockets, can be hung on a panel mounted on the garage or workshop wall, while others should be kept in a tool box or tray. Measuring instruments, gauges, meters, etc. must be carefully stored where they cannot be damaged by weather or impact from other tools.

When tools are used with care and stored properly, they will last a very long time. Even with the best of care, though, tools will wear out if used frequently. When a tool is damaged or worn out, replace it. Subsequent jobs will be safer and more enjoyable if you do.

HOW TO REPAIR DAMAGED THREADS

Sometimes, the internal threads of a nut or bolt hole can become stripped, usually from overtightening. Stripping threads is an all-too-common occurrence, especially when working with aluminum parts, because aluminum is so soft that it easily strips out.

Usually, external or internal threads are only partially stripped. After they've been cleaned up with a tap or die, they'll still work. Sometimes, however, threads are badly damaged. When this happens, you've got three choices:

1) *Drill and tap the hole to the next suitable oversize and install a larger diameter bolt, screw or stud.*
2) *Drill and tap the hole to accept a threaded plug, then drill and tap the plug to the original screw size. You can also buy a plug already threaded to the original size. Then you simply drill a hole to the specified size, then run the threaded plug into the hole with a bolt and jam nut. Once the plug is fully seated, remove the jam nut and bolt.*
3) *The third method uses a patented thread repair kit like Heli-Coil or Slimsert. These easy-to-use kits are designed to repair damaged threads in straight-through holes and blind holes. Both are available as kits which can handle a variety of sizes and thread patterns. Drill the hole, then tap it with the special included tap. Install the Heli-Coil and the hole is back to its original diameter and thread pitch.*

Regardless of which method you use, be sure to proceed calmly and carefully. A little impatience or carelessness during one of these relatively simple procedures can ruin your whole day's work and cost you a bundle if you wreck an expensive part.

WORKING FACILITIES

Not to be overlooked when discussing tools is the workshop. If anything more than routine maintenance is to be carried out, some sort of suitable work area is essential.

It is understood, and appreciated, that many home mechanics do not have a good workshop or garage available, and end up removing an engine or doing major repairs outside. It is recommended, however, that the overhaul or repair be completed under the cover of a roof.

A clean, flat workbench or table of comfortable working height is an absolute necessity. The workbench should be equipped with a vise that

MAINTENANCE TECHNIQUES, TOOLS AND WORKING FACILITIES 0-25

Torque angle gauge

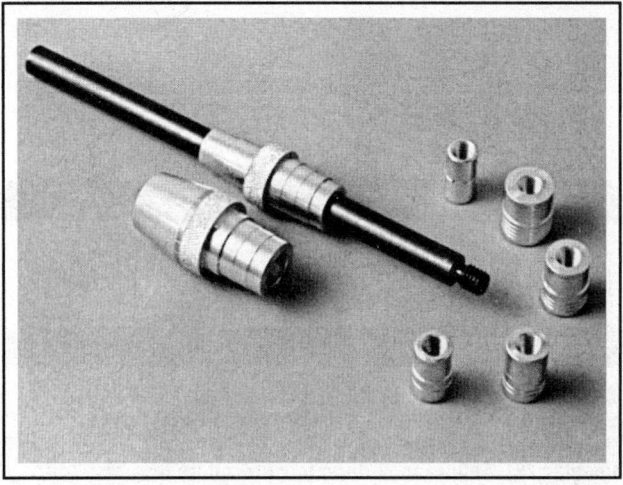

Clutch plate alignment tool

has a jaw opening of at least four inches.

As mentioned previously, some clean, dry storage space is also required for tools, as well as the lubricants, fluids, cleaning solvents, etc. which soon become necessary.

Sometimes waste oil and fluids, drained from the engine or cooling system during normal maintenance or repairs, present a disposal problem. To avoid pouring them on the ground or into a sewage system, pour the used fluids into large containers, seal them with caps and take them to an authorized disposal site or recycling center. Plastic jugs, such as old antifreeze containers, are ideal for this purpose.

Always keep a supply of old newspapers and clean rags available. Old towels are excellent for mopping up spills. Many mechanics use rolls of paper towels for most work because they are readily available and disposable. To help keep the area under the vehicle clean, a large cardboard box can be cut open and flattened to protect the garage or shop floor.

Whenever working over a painted surface, such as when leaning over a fender to service something under the hood, always cover it with an old blanket or bedspread to protect the finish. Vinyl covered pads, made especially for this purpose, are available at auto parts stores.

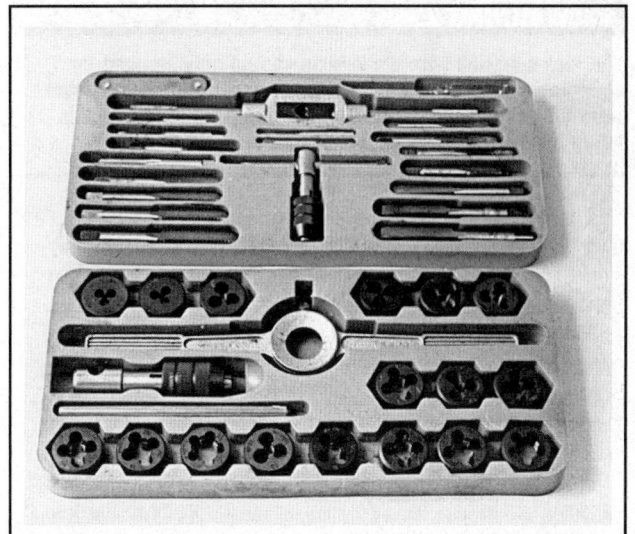

Tap and die set

0-26 JACKING AND TOWING

Jacking and towing

JACKING

✲✲ WARNING 1:

Some models covered by this manual are equipped with air suspension systems. Always disconnect electrical power to the suspension system before lifting or towing the vehicle (see Chapter 10). Failure to perform this procedure may result in unexpected shifting or movement of the vehicle which could cause personal injury.

✲✲ WARNING 2:

The jack supplied with the vehicle should only be used for changing a tire or placing jackstands under the frame. Never work under the vehicle or start the engine while this jack is being used as the only means of support.

The vehicle should be on level ground. Place the shift lever in Park, if you have an automatic, or Reverse if you have a manual transmission. Block the wheel diagonally opposite the wheel being changed. Set the parking brake.

Remove the spare tire and jack from stowage. Remove the wheel cover and trim ring (if so equipped) with the tapered end of the lug nut wrench by inserting and twisting the handle and then prying against the back of the wheel cover. Loosen the wheel lug nuts about 1/4-to-1/2 turn each.

Place the jack under the vehicle in the indicated position (see illustrations). Turn the jack handle clockwise until the tire clears the ground. Remove the lug nuts and pull the wheel off. Replace it with the spare.

Install the lug nuts with the beveled edges facing in. Tighten them snugly. Don't attempt to tighten them completely until the vehicle is lowered or it could slip off the jack. Turn the jack handle counterclockwise to lower the vehicle. Remove the jack and tighten the lug nuts in a diagonal pattern.

Install the cover (and trim ring, if used) and be sure it's snapped into place all the way around.

Stow the tire, jack and wrench. Unblock the wheels.

TOWING

We recommend these vehicles (except four-wheel drive models) be towed from the rear, with the rear wheels off the ground. If it's absolutely necessary, these vehicles can be towed from the front with the front wheels off the ground, provided that speeds don't exceed 35 mph and the distance is less than 50 miles; the transmission can be damaged if these mileage/speed limitations are exceeded. Vehicles with four-wheel drive must not be towed with all four wheels on the ground. They must only be towed with all four wheels off the ground.

Equipment specifically designed for towing should be used. It must be attached to the main structural members of the vehicle, not the bumpers or brackets.

Safety is a major consideration when towing and all applicable state and local laws must be obeyed. A safety chain must be used at all times.

The parking brake must be released and the transmission must be in Neutral. The steering must be unlocked (ignition switch in the Off position). Remember that power steering and power brakes won't work with the engine off.

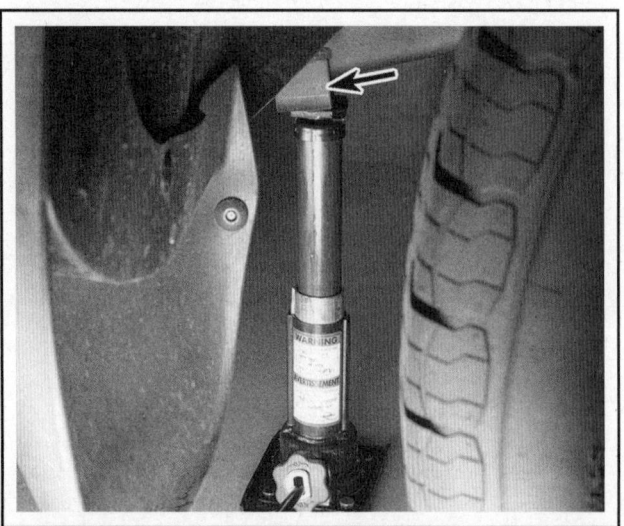

Front jacking location - position the jack under the welded bracket

Rear jacking locations - position the jack under the rear frame rail or (on pick-up models only) under the U-bolts on the rear leaf spring

JUMP STARTING 0-27

Booster battery (jump) starting

Observe the following precautions when using a booster battery to start a vehicle:

a) Before connecting the booster battery, make sure the ignition switch is in the Off position.
b) Turn off the lights, heater and other electrical loads.
c) Your eyes should be shielded. Safety goggles are a good idea.
d) Make sure the booster battery is the same voltage as the dead one in the vehicle.
e) The two vehicles MUST NOT TOUCH each other.
f) Make sure the transmission is in Park.
g) If the booster battery is not a maintenance-free type, remove the vent caps and lay a cloth over the vent holes.

Connect the red jumper cable to the positive (+) terminals of each battery.

Connect one end of the black cable to the negative (-) terminal of the booster battery. The other end of this cable should be connected to a good ground on the engine block (see illustration). Make sure the cable will not come into contact with the fan, drivebelts or other moving parts of the engine.

Start the engine using the booster battery, then, with the engine running at idle speed, disconnect the jumper cables in the reverse order of connection

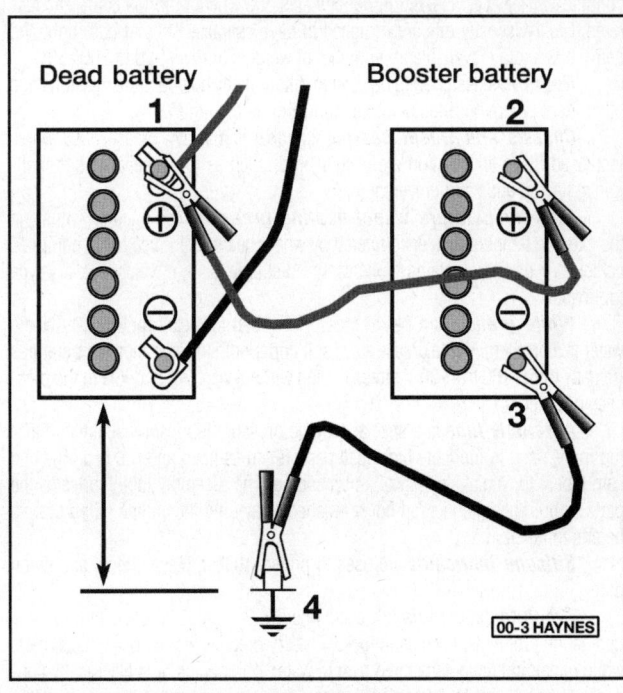

Make the booster battery cable connections in the numerical order shown (note that the negative cable of the booster battery is NOT attached to the negative terminal of the dead battery)

0-28 AUTOMOTIVE CHEMICALS AND LUBRICANTS

Automotive chemicals and lubricants

A number of automotive chemicals and lubricants are available for use during vehicle maintenance and repair. They include a wide variety of products ranging from cleaning solvents and degreasers to lubricants and protective sprays for rubber, plastic and vinyl.

CLEANERS

Carburetor cleaner and choke cleaner is a strong solvent for gum, varnish and carbon. Most carburetor cleaners leave a dry-type lubricant film which will not harden or gum up. Because of this film it is not recommended for use on electrical components.

Brake system cleaner is used to remove brake dust, grease and brake fluid from the brake system, where clean surfaces are absolutely necessary. It leaves no residue and often eliminates brake squeal caused by contaminants.

Electrical cleaner removes oxidation, corrosion and carbon deposits from electrical contacts, restoring full current flow. It can also be used to clean spark plugs, carburetor jets, voltage regulators and other parts where an oil-free surface is desired.

Demoisturants remove water and moisture from electrical components such as alternators, voltage regulators, electrical connectors and fuse blocks. They are non-conductive and non-corrosive.

Degreasers are heavy-duty solvents used to remove grease from the outside of the engine and from chassis components. They can be sprayed or brushed on and, depending on the type, are rinsed off either with water or solvent.

LUBRICANTS

Motor oil is the lubricant formulated for use in engines. It normally contains a wide variety of additives to prevent corrosion and reduce foaming and wear. Motor oil comes in various weights (viscosity ratings) from 0 to 50. The recommended weight of the oil depends on the season, temperature and the demands on the engine. Light oil is used in cold climates and under light load conditions. Heavy oil is used in hot climates and where high loads are encountered. Multi-viscosity oils are designed to have characteristics of both light and heavy oils and are available in a number of weights from 0W-20 to 20W-50.

Gear oil is designed to be used in differentials, manual transmissions and other areas where high-temperature lubrication is required.

Chassis and wheel bearing grease is a heavy grease used where increased loads and friction are encountered, such as for wheel bearings, ball-joints, tie-rod ends and universal joints.

High-temperature wheel bearing grease is designed to withstand the extreme temperatures encountered by wheel bearings in disc brake equipped vehicles. It usually contains molybdenum disulfide (moly), which is a dry-type lubricant.

White grease is a heavy grease for metal-to-metal applications where water is a problem. White grease stays soft under both low and high temperatures (usually from -100 to +190-degrees F), and will not wash off or dilute in the presence of water.

Assembly lube is a special extreme pressure lubricant, usually containing moly, used to lubricate high-load parts (such as main and rod bearings and cam lobes) for initial start-up of a new engine. The assembly lube lubricates the parts without being squeezed out or washed away until the engine oiling system begins to function.

Silicone lubricants are used to protect rubber, plastic, vinyl and nylon parts.

Graphite lubricants are used where oils cannot be used due to contamination problems, such as in locks. The dry graphite will lubricate metal parts while remaining uncontaminated by dirt, water, oil or acids. It is electrically conductive and will not foul electrical contacts in locks such as the ignition switch.

Moly penetrants loosen and lubricate frozen, rusted and corroded fasteners and prevent future rusting or freezing.

Heat-sink grease is a special electrically non-conductive grease that is used for mounting electronic ignition modules where it is essential that heat is transferred away from the module.

SEALANTS

RTV sealant is one of the most widely used gasket compounds. Made from silicone, RTV is air curing, it seals, bonds, waterproofs, fills surface irregularities, remains flexible, doesn't shrink, is relatively easy to remove, and is used as a supplementary sealer with almost all low and medium temperature gaskets.

Anaerobic sealant is much like RTV in that it can be used either to seal gaskets or to form gaskets by itself. It remains flexible, is solvent resistant and fills surface imperfections. The difference between an anaerobic sealant and an RTV-type sealant is in the curing. RTV cures when exposed to air, while an anaerobic sealant cures only in the absence of air. This means that an anaerobic sealant cures only after the assembly of parts, sealing them together.

Thread and pipe sealant is used for sealing hydraulic and pneumatic fittings and vacuum lines. It is usually made from a Teflon compound, and comes in a spray, a paint-on liquid and as a wrap-around tape.

CHEMICALS

Anti-seize compound prevents seizing, galling, cold welding, rust and corrosion in fasteners. High-temperature anti-seize, usually made with copper and graphite lubricants, is used for exhaust system and exhaust manifold bolts.

Anaerobic locking compounds are used to keep fasteners from vibrating or working loose and cure only after installation, in the absence of air. Medium strength locking compound is used for small nuts, bolts and screws that may be removed later. High-strength locking compound is for large nuts, bolts and studs which aren't removed on a regular basis.

Oil additives range from viscosity index improvers to chemical treatments that claim to reduce internal engine friction. It should be noted that most oil manufacturers caution against using additives with their oils.

Gas additives perform several functions, depending on their chemical makeup. They usually contain solvents that help dissolve gum and varnish that build up on carburetor, fuel injection and intake parts. They also serve to break down carbon deposits that form on the inside surfaces of the combustion chambers. Some additives contain upper cylinder lubricants for valves and piston rings, and others contain chemicals to remove condensation from the gas tank.

MISCELLANEOUS

Brake fluid is specially formulated hydraulic fluid that can withstand the heat and pressure encountered in brake systems. Care must be taken so this fluid does not come in contact with painted surfaces or plastics. An opened container should always be resealed to prevent contamination by water or dirt.

Weatherstrip adhesive is used to bond weatherstripping around doors, windows and trunk lids. It is sometimes used to attach trim pieces.

Undercoating is a petroleum-based, tar-like substance that is designed to protect metal surfaces on the underside of the vehicle from corrosion. It also acts as a sound-deadening agent by insulating the bottom of the vehicle.

Waxes and polishes are used to help protect painted and plated surfaces from the weather. Different types of paint may require the use of different types of wax and polish. Some polishes utilize a chemical or abrasive cleaner to help remove the top layer of oxidized (dull) paint on older vehicles. In recent years many non-wax polishes that contain a wide variety of chemicals such as polymers and silicones have been introduced. These non-wax polishes are usually easier to apply and last longer than conventional waxes and polishes.

SAFETY FIRST!

Safety first!

Regardless of how enthusiastic you may be about getting on with the job at hand, take the time to ensure that your safety is not jeopardized. A moment's lack of attention can result in an accident, as can failure to observe certain simple safety precautions. The possibility of an accident will always exist, and the following points should not be considered a comprehensive list of all dangers. Rather, they are intended to make you aware of the risks and to encourage a safety conscious approach to all work you carry out on your vehicle.

ESSENTIAL DOS AND DON'TS

DON'T rely on a jack when working under the vehicle. Always use approved jackstands to support the weight of the vehicle and place them under the recommended lift or support points.
DON'T attempt to loosen extremely tight fasteners (i.e. wheel lug nuts) while the vehicle is on a jack - it may fall.
DON'T start the engine without first making sure that the transmission is in Neutral (or Park where applicable) and the parking brake is set.
DON'T remove the radiator cap from a hot cooling system - let it cool or cover it with a cloth and release the pressure gradually.
DON'T attempt to drain the engine oil until you are sure it has cooled to the point that it will not burn you.
DON'T touch any part of the engine or exhaust system until it has cooled sufficiently to avoid burns.
DON'T siphon toxic liquids such as gasoline, antifreeze and brake fluid by mouth, or allow them to remain on your skin.
DON'T inhale brake lining dust - it is potentially hazardous (see Asbestos below).
DON'T allow spilled oil or grease to remain on the floor - wipe it up before someone slips on it.
DON'T use loose fitting wrenches or other tools which may slip and cause injury.
DON'T push on wrenches when loosening or tightening nuts or bolts. Always try to pull the wrench toward you. If the situation calls for pushing the wrench away, push with an open hand to avoid scraped knuckles if the wrench should slip.
DON'T attempt to lift a heavy component alone - get someone to help you.
DON'T rush or take unsafe shortcuts to finish a job.
DON'T allow children or animals in or around the vehicle while you are working on it.
DO wear eye protection when using power tools such as a drill, sander, bench grinder, etc. and when working under a vehicle.
DO keep loose clothing and long hair well out of the way of moving parts.
DO make sure that any hoist used has a safe working load rating adequate for the job.
DO get someone to check on you periodically when working alone on a vehicle.
DO carry out work in a logical sequence and make sure that everything is correctly assembled and tightened.
DO keep chemicals and fluids tightly capped and out of the reach of children and pets.
DO remember that your vehicle's safety affects that of yourself and others. If in doubt on any point, get professional advice.

STEERING, SUSPENSION AND BRAKES

These systems are essential to driving safety, so make sure you have a qualified shop or individual check your work. Also, compressed suspension springs can cause injury if released suddenly - be sure to use a spring compressor.

AIRBAGS

Airbags are explosive devices that can CAUSE injury if they deploy while you're working on the vehicle. Follow the manufacturer's instructions to disable the airbag whenever you're working in the vicinity of airbag components.

ASBESTOS

Certain friction, insulating, sealing, and other products - such as brake linings, brake bands, clutch linings, torque converters, gaskets, etc. - may contain asbestos or other hazardous friction material. Extreme care must be taken to avoid inhalation of dust from such products, since it is hazardous to health. If in doubt, assume that they do contain asbestos.

FIRE

Remember at all times that gasoline is highly flammable. Never smoke or have any kind of open flame around when working on a vehicle. But the risk does not end there. A spark caused by an electrical short circuit, by two metal surfaces contacting each other, or even by static electricity built up in your body under certain conditions, can ignite gasoline vapors, which in a confined space are highly explosive. Do not, under any circumstances, use gasoline for cleaning parts. Use an approved safety solvent.

Always disconnect the battery ground (-) cable at the battery before working on any part of the fuel system or electrical system. Never risk spilling fuel on a hot engine or exhaust component. It is strongly recommended that a fire extinguisher suitable for use on fuel and electrical fires be kept handy in the garage or workshop at all times. Never try to extinguish a fuel or electrical fire with water.

FUMES

Certain fumes are highly toxic and can quickly cause unconsciousness and even death if inhaled to any extent. Gasoline vapor falls into this category, as do the vapors from some cleaning solvents. Any draining or pouring of such volatile fluids should be done in a well ventilated area.

When using cleaning fluids and solvents, read the instructions on the container carefully. Never use materials from unmarked containers.

Never run the engine in an enclosed space, such as a garage. Exhaust fumes contain carbon monoxide, which is extremely poisonous. If you need to run the engine, always do so in the open air, or at least have the rear of the vehicle outside the work area.

THE BATTERY

Never create a spark or allow a bare light bulb near a battery. They normally give off a certain amount of hydrogen gas, which is highly explosive.

Always disconnect the battery ground (-) cable at the battery before working on the fuel or electrical systems.

If possible, loosen the filler caps or cover when charging the battery from an external source (this does not apply to sealed or maintenance-free batteries). Do not charge at an excessive rate or the battery may burst.

Take care when adding water to a non maintenance-free battery and when carrying a battery. The electrolyte, even when diluted, is very corrosive and should not be allowed to contact clothing or skin.

Always wear eye protection when cleaning the battery to prevent the caustic deposits from entering your eyes.

HOUSEHOLD CURRENT

When using an electric power tool, inspection light, etc., which operates on household current, always make sure that the tool is correctly connected to its plug and that, where necessary, it is properly grounded. Do not use such items in damp conditions and, again, do not create a spark or apply excessive heat in the vicinity of fuel or fuel vapor.

SECONDARY IGNITION SYSTEM VOLTAGE

A severe electric shock can result from touching certain parts of the ignition system (such as the spark plug wires) when the engine is running or being cranked, particularly if components are damp or the insulation is defective. In the case of an electronic ignition system, the secondary system voltage is much higher and could prove fatal.

HYDROFLUORIC ACID

This extremely corrosive acid is formed when certain types of synthetic rubber, found in some O-rings, oil seals, fuel hoses, etc. are exposed to temperatures above 750-degrees F (400-degrees C). The rubber changes into a charred or sticky substance containing the acid. *Once formed, the acid remains dangerous for years. If it gets onto the skin, it may be necessary to amputate the limb concerned.*

When dealing with a vehicle which has suffered a fire, or with components salvaged from such a vehicle, wear protective gloves and discard them after use.

0-30 TROUBLESHOOTING

Troubleshooting

CONTENTS

Section Symptom

Engine

1. Engine will not rotate when attempting to start
2. Engine rotates but will not start
3. Starter motor operates without turning engine
4. Engine hard to start when cold
5. Engine hard to start when hot
6. Starter motor noisy or engages roughly
7. Engine starts but stops immediately
8. Engine lopes while idling or idles erratically
9. Engine misses at idle speed
10. Excessively high idle speed
11. Battery will not hold a charge
12. Alternator light stays on
13. Alternator light fails to come on when key is turned on
14. Engine misses throughout driving speed range
15. Hesitation or stumble during acceleration
16. Engine stalls
17. Engine lacks power
18. Engine backfires
19. Engine surges while holding accelerator steady
20. Pinging or knocking engine sounds when engine is under load
21. Engine continues to run after being turned off
22. Low oil pressure
23. Excessive oil consumption
24. Excessive fuel consumption
25. Fuel odor
26. Miscellaneous engine noises

CHECK ENGINE light   See Chapter 6

Cooling system

27. Overheating
28. Overcooling
29. External coolant leakage
30. Internal coolant leakage
31. Abnormal coolant loss
32. Poor coolant circulation
33. Corrosion

Clutch

34. Fails to release (pedal pressed to the floor - shift lever does not move freely in and out of Reverse)
35. Clutch slips (engine speed increases with no increase in vehicle speed)
36. Grabbing (chattering) as clutch is engaged
37. Squeal or rumble with clutch fully engaged (pedal released)
38. Squeal or rumble with clutch fully disengaged (pedal depressed)
39. Clutch pedal stays on floor when disengaged

Manual transmission

40. Noisy in Neutral with engine running
41. Noisy in all gears

Section Symptom

42. Noisy in one particular gear
43. Slips out of high gear
44. Difficulty in engaging gears
45. Oil leakage

Automatic transmission

46. General shift mechanism problems
47. Transmission will not downshift with accelerator pedal pressed to the floor
48. Transmission slips, shifts rough, is noisy or has no drive in forward or reverse gears
49. Fluid leakage

Transfer case

50. Transfer case is difficult to shift into the desired range
51. Transfer case noisy in all gears
52. Noisy or jumps out of four-wheel drive Low range
53. Lubricant leaks from the vent or output shaft seals

Driveshaft

54. Oil leak at front of driveshaft
55. Knock or clunk when the transmission is under initial load (just after transmission is put into gear)
56. Metallic grinding sound consistent with vehicle speed
57. Vibration

Axles

58. Noise
59. Vibration
60. Oil leakage

Brakes

61. Vehicle pulls to one side during braking
62. Noise (high-pitched squeal with the brakes applied)
63. Excessive brake pedal travel
64. Brake pedal feels spongy when depressed
65. Excessive effort required to stop vehicle
66. Pedal travels to the floor with little resistance
67. Brake pedal pulsates during brake application

Suspension and steering systems

68. Vehicle pulls to one side
69. Shimmy, shake or vibration
70. Excessive pitching and/or rolling around corners or during braking
71. Excessively stiff steering
72. Excessive play in steering
73. Lack of power assistance
74. Excessive tire wear (not specific to one area)
75. Excessive tire wear on outside edge
76. Excessive tire wear on inside edge
77. Tire tread worn in one place

ENGINE

1 Engine will not rotate when attempting to start

1   Battery terminal connections loose or corroded. Check the cable terminals at the battery; tighten cable clamp and/or clean off corrosion as necessary (see Chapter 1).
2   Battery discharged or faulty. If the cable ends are clean and tight on the battery posts, turn the key to the On position and switch on the headlights or windshield wipers. If they won't run, the battery is discharged.
3   Automatic transmission not engaged in park (P) or Neutral (N).
4   Broken, loose or disconnected wires in the starting circuit. Inspect all wires and connectors at the battery, starter solenoid and ignition switch (on steering column).
5   Starter motor pinion jammed in driveplate ring gear. Remove starter (Chapter 5) and inspect pinion and driveplate (Chapter 2).
6   Starter solenoid faulty (Chapter 5).
7   Starter motor faulty (Chapter 5).
8   Ignition switch faulty (Chapter 12).
9   Engine seized. Try to turn the crankshaft with a large socket and breaker bar on the pulley bolt.
10 Starter relay faulty (Chapter 5)
11 Transmission Range (TR) sensor out of adjustment or defective (Chapter 6)

2 Engine rotates but will not start

1   Fuel tank empty.
2   Battery discharged (engine rotates slowly).
3   Battery terminal connections loose or corroded.
4   Fuel not reaching fuel injectors. Check for clogged fuel filter or lines and defective fuel pump. Also make sure the tank vent lines aren't clogged (Chapter 4).
5   Low cylinder compression. Check as described in Chapter 2C.
6   Water in fuel. Drain tank and fill with new fuel.
7   Defective ignition coil(s) (Chapter 5).
8   Dirty or clogged fuel injector(s) (Chapter 4).
9   Wet or damaged ignition components (Chapters 1 and 5).
10 Worn, faulty or incorrectly gapped spark plugs (Chapter 1).
11 Broken, loose or disconnected wires in the starting circuit (see previous Section).
12 Broken, loose or disconnected wires at the ignition coil or faulty coil (Chapter 5).
13 Timing chain failure or wear affecting valve timing (Chapter 2).
14 Fuel injection or engine control systems failure (Chapters 4 and 6).
15 Defective MAF sensor (Chapter 6)

3 Starter motor operates without turning engine

1   Starter pinion sticking. Remove the starter (Chapter 5) and inspect.
2   Starter pinion or driveplate teeth worn or broken. Remove the inspection cover and inspect.

4 Engine hard to start when cold

1   Battery discharged or low. Check as described in Chapter 1.
2   Fuel not reaching the fuel injectors. Check the fuel filter, lines and fuel pump (Chapters 1 and 4).
3   Defective spark plugs (Chapter 1).
4   Defective engine coolant temperature sensor (Chapter 6).
5   Fuel injection or engine control systems malfunction (Chapters 4 and 6).

5 Engine hard to start when hot

1   Air filter dirty (Chapter 1).
2   Bad engine ground connection.
3   Fuel injection or engine control systems malfunction (Chapters 4 and 6).

6 Starter motor noisy or engages roughly

1   Pinion or driveplate teeth worn or broken. Remove the inspection cover on the left side of the engine and inspect.
2   Starter motor mounting bolts loose or missing.

7 Engine starts but stops immediately

1   Loose or damaged wire harness connections at distributor, coil or alternator.
2   Intake manifold vacuum leaks. Make sure all mounting bolts/nuts are tight and all vacuum hoses connected to the manifold are attached properly and in good condition.
3   Insufficient fuel pressure (see Chapter 4).
4   Fuel injection or engine control systems malfunction (Chapters 4 and 6).

8 Engine lopes while idling or idles erratically

1   Vacuum leaks. Check mounting bolts at the intake manifold for tightness. Make sure that all vacuum hoses are connected and in good condition. Use a stethoscope or a length of fuel hose held against your ear to listen for vacuum leaks while the engine is running. A hissing sound will be heard. A soapy water solution will also detect leaks. Check the intake manifold gasket surfaces.
2   Leaking EGR valve or plugged PCV valve (see Chapters 1 and 6).
3   Air filter clogged (Chapter 1).
4   Fuel pump not delivering sufficient fuel (Chapter 4).
5   Leaking head gasket. Perform a cylinder compression check (Chapter 2).
6   Timing chain(s) worn (Chapter 2).
7   Camshaft lobes worn (Chapter 2).
8   Valves burned or otherwise leaking (Chapter 2).
9   Ignition timing out of adjustment (Chapter 5).
10 Ignition system not operating properly (Chapters 1 and 5).
11 Fuel injection or engine control systems malfunction (Chapters 4 and 6).

9 Engine misses at idle speed

1   Spark plugs faulty or not gapped properly (Chapter 1).
2   Faulty spark plug wires (Chapter 1).
3   Wet or damaged ignition components (Chapter 5).
4   Short circuits in ignition, coil or spark plug wires.
5   Sticking or faulty emissions systems (see Chapter 6).
6   Clogged fuel filter and/or foreign matter in fuel. Remove the fuel filter (Chapter 1) and inspect.
7   Vacuum leaks at intake manifold or hose connections. Check as described in Section
8   Low or uneven cylinder compression. Check as described in Chapter 2C.
9   Fuel injection or engine control systems malfunction (Chapters 4 and 6).

TROUBLESHOOTING

10 Excessively high idle speed

1 Sticking throttle linkage (Chapter 4).
2 Vacuum leaks at intake manifold or hose connections. Check as described in Section 8.
3 Fuel injection or engine control systems malfunction (Chapters 4 and 6).

11 Battery will not hold a charge

1 Alternator drivebelt defective or not adjusted properly (Chapter 1).
2 Battery cables loose or corroded (Chapter 1).
3 Alternator not charging properly (Chapter 5).
4 Loose, broken or faulty wires in the charging circuit (Chapter 5).
5 Short circuit causing a continuous drain on the battery.
6 Battery defective internally.

12 Alternator light stays on

1 Fault in alternator or charging circuit (Chapter 5).
2 Alternator drivebelt defective or not properly adjusted (Chapter 1).

13 Alternator light fails to come on when key is turned on

1 Faulty bulb (Chapter 12).
2 Defective alternator (Chapter 5).
3 Fault in the printed circuit, dash wiring or bulb holder (Chapter 12).

14 Engine misses throughout driving speed range

1 Fuel filter clogged and/or impurities in the fuel system. Check fuel filter (Chapter 1) or clean system (Chapter 4).
2 Faulty or incorrectly gapped spark plugs (Chapter 1).
3 Incorrect ignition timing (Chapter 5).
4 Defective spark plug wires (Chapter 1).
5 Emissions system components faulty (Chapter 6).
6 Low or uneven cylinder compression pressures. Check as described in Chapter 2C.
7 Weak or faulty ignition coil(s) (Chapter 5).
8 Weak or faulty ignition system (Chapter 5).
9 Vacuum leaks at intake manifold or vacuum hoses (see Section 8).
10 Dirty or clogged fuel injector(s) (Chapter 4).
11 Leaky EGR valve (Chapter 6).
12 Fuel injection or engine control systems malfunction (Chapters 4 and 6).

15 Hesitation or stumble during acceleration

1 Ignition system not operating properly (Chapter 5).
2 Dirty or clogged fuel injector(s) (Chapter 4).
3 Low fuel pressure. Check for proper operation of the fuel pump and for restrictions in the fuel filter and lines (Chapter 4).
4 Fuel injection or engine control systems malfunction (Chapters 4 and 6).

16 Engine stalls

1 Idle speed incorrect (Chapter 4).
2 Fuel filter clogged and/or water and impurities in the fuel system (Chapter 1).
3 Damaged or wet distributor cap and wires.
4 Emissions system components faulty (Chapter 6).
5 Faulty or incorrectly gapped spark plugs (Chapter 1). Also check the spark plug wires (Chapter 1).
6 Vacuum leak at the intake manifold or vacuum hoses. Check as described in Section 8.
7 Fuel injection or engine control systems malfunction (Chapters 4 and 6).

17 Engine lacks power

1 Incorrect ignition timing (Chapter 5).
2 Faulty or incorrectly gapped spark plugs (Chapter 1).
3 Air filter dirty (Chapter 1).
4 Faulty ignition coil(s) (Chapter 5).
5 Brakes binding (Chapters 1 and 10).
6 Automatic transmission fluid level incorrect, causing slippage (Chapter 1).
7 Fuel filter clogged and/or impurities in the fuel system (Chapters 1 and 4).
8 EGR system not functioning properly (Chapter 6).
9 Use of sub-standard fuel. Fill tank with proper octane fuel.
10 Low or uneven cylinder compression pressures. Check as described in Chapter 2.
11 Vacuum leak at intake manifold or vacuum hoses (check as described in Section 8).
12 Dirty or clogged fuel injector(s) (Chapters 1 and 4).
13 Fuel injection or engine control systems malfunction (Chapters 4 and 6).
14 Restricted exhaust system (Chapter 4).

18 Engine backfires

1 EGR system not functioning properly (Chapter 6).
2 Ignition timing incorrect (Chapter 5).
3 Damaged valve springs or sticking valves (Chapter 2).
4 Vacuum leak at the intake manifold or vacuum hoses (see Section 8).

19 Engine surges while holding accelerator steady

1 Vacuum leak at the intake manifold or vacuum hoses (see Section 8).
2 Restricted air filter (Chapter 1).
3 Fuel pump or pressure regulator defective (Chapter 4).
4 Fuel injection or engine control systems malfunction (Chapters 4 and 6).

20 Pinging or knocking engine sounds when engine is under load

1 Incorrect grade of fuel. Fill tank with fuel of the proper octane rating.
2 Ignition timing incorrect (Chapter 5).
3 Carbon build-up in combustion chambers. Remove cylinder head(s) and clean combustion chambers (Chapter 2).
4 Incorrect spark plugs (Chapter 1).
5 Fuel injection or engine control systems malfunction (Chapters 4 and 6).
6 Restricted exhaust system (Chapter 4).

TROUBLESHOOTING 0-33

21 Engine continues to run after being turned off

1 Idle speed too high (Chapter 4).
2 Ignition timing incorrect (Chapter 5).
3 Incorrect spark plug heat range (Chapter 1).
4 Vacuum leak at the intake manifold or vacuum hoses (see Section 8).
5 Carbon build-up in combustion chambers. Remove the cylinder head(s) and clean the combustion chambers (Chapter 2).
6 Valves sticking (Chapter 2).
7 EGR system not operating properly (Chapter 6).
8 Fuel injection or engine control systems malfunction (Chapters 4 and 6).
9 Check for causes of overheating (see Section 27).

22 Low oil pressure

1 Improper grade of oil.
2 Oil pump worn or damaged (Chapter 2).
3 Engine overheating (refer to Section 27).
4 Clogged oil filter (Chapter 1).
5 Clogged oil strainer (Chapter 2).
6 Oil pressure gauge not working properly (Chapter 2).

23 Excessive oil consumption

1 Loose oil drain plug.
2 Loose bolts or damaged oil pan gasket (Chapter 2).
3 Loose bolts or damaged front cover gasket (Chapter 2).
4 Front or rear crankshaft oil seal leaking (Chapter 2).
5 Loose bolts or damaged valve cover gasket (Chapter 2).
6 Loose oil filter (Chapter 1).
7 Loose or damaged oil pressure switch (Chapter 2).
8 Pistons and cylinders excessively worn (Chapter 2).
9 Piston rings not installed correctly on pistons (Chapter 2).
10 Worn or damaged piston rings (Chapter 2).
11 Intake and/or exhaust valve oil seals worn or damaged (Chapter 2).
12 Worn or damaged valves/guides (Chapter 2).
13 Faulty or incorrect PCV valve allowing too much crankcase airflow.
14 Leak at remote oil filter hose (Expedition/Navigator only).

24 Excessive fuel consumption

1 Dirty or clogged air filter element (Chapter 1).
2 Incorrect ignition timing (Chapter 5).
3 Incorrect idle speed (Chapter 4).
4 Low tire pressure or incorrect tire size (Chapter 10).
5 Inspect for binding brakes.
6 Fuel leakage. Check all connections, lines and components in the fuel system (Chapter 4).
7 Dirty or clogged fuel injectors (Chapter 4).
8 Fuel injection or engine control systems malfunction (Chapters 4 and 6).
9 Thermostat stuck open or not installed.
10 Improperly operating transmission.

25 Fuel odor

1 Fuel leakage. Check all connections, lines and components in the fuel system (Chapter 4).
2 Fuel tank overfilled. Fill only to automatic shut-off.
3 Charcoal canister filter in Evaporative Emissions Control system clogged (Chapter 1).
4 Vapor leaks from Evaporative Emissions Control system lines (Chapter 6).

26 Miscellaneous engine noises

1 A strong dull noise that becomes more rapid as the engine accelerates indicates worn or damaged crankshaft bearings or an unevenly worn crankshaft. To pinpoint the trouble spot, remove the spark plug wire from one plug at a time and crank the engine over. If the noise stops, the cylinder with the removed plug wire indicates the problem area. Replace the bearing and/or service or replace the crankshaft (Chapter 2).
2 A similar (yet slightly higher pitched) noise to the crankshaft knocking described in the previous paragraph, that becomes more rapid as the engine accelerates, indicates worn or damaged connecting rod bearings (Chapter 2). The procedure for locating the problem cylinder is the same as described in Paragraph 1.
3 An overlapping metallic noise that increases in intensity as the engine speed increases, yet diminishes as the engine warms up indicates abnormal piston and cylinder wear (Chapter 2). To locate the problem cylinder, use the procedure described in Paragraph 1.
4 A rapid clicking noise that becomes faster as the engine accelerates indicates a worn piston pin or piston pin hole. This sound will happen each time the piston hits the highest and lowest points in the stroke (Chapter 2). The procedure for locating the problem piston is described in Paragraph 1.
5 A metallic clicking noise coming from the water pump indicates worn or damaged water pump bearings or pump. Replace the water pump with a new one (Chapter 3).
6 A rapid tapping sound or clicking sound that becomes faster as the engine speed increases indicates "valve tapping." This can be identified by holding one end of a section of hose to your ear and placing the other end at different spots along the valve cover. The point where the sound is loudest indicates the problem valve. If the pushrod and rocker arm components are in good shape, you likely have a collapsed valve lifter. Changing the engine oil and adding a high viscosity oil treatment will sometimes cure a stuck lifter problem. If the problem persists, the lifters, pushrods and rocker arms must be removed for inspection (see Chapter 2).
7 A steady metallic rattling or rapping sound coming from the area of the timing chain cover indicates a worn, damaged or out-of-adjustment timing chain. Service or replace the chain and related components (Chapter 2).

COOLING SYSTEM

27 Overheating

1 Insufficient coolant in system (Chapter 1).
2 Drivebelt defective or not adjusted properly (Chapter 1).
3 Radiator core blocked or radiator grille dirty and restricted (Chapter 3).
4 Thermostat faulty (Chapter 3).
5 Cooling fan not functioning properly (Chapter 3).
6 Expansion tank cap not maintaining proper pressure. Have cap pressure tested by a gas station or repair shop.

0-34 TROUBLESHOOTING

7 Defective water pump (Chapter 3).
8 Improper grade of engine oil.
9 Inaccurate temperature gauge (Chapter 12).

28 Overcooling

1 Thermostat faulty (Chapter 3).
2 Inaccurate temperature gauge (Chapter 12).

29 External coolant leakage

1 Deteriorated or damaged hoses. Loose clamps at hose connections (Chapter 1).
2 Water pump seals defective. If this is the case, water will drip from the weep hole in the water pump body (Chapter 3).
3 Leakage from radiator core or header tank. This will require the radiator to be professionally repaired (see Chapter 3 for removal procedures).
4 Leakage from the coolant reservoir or degas bottle.
5 Engine drain plugs or water jacket freeze plugs leaking (see Chapters 1 and 2).
6 Leak from coolant temperature switch (Chapter 3).
7 Leak from damaged gaskets or small cracks (Chapter 2).
8 Leak from oil cooler or oil cooler adapter housing (Chapter 3).

30 Internal coolant leakage

➡ Note: *Internal coolant leaks can usually be detected by examining the oil. Check the dipstick and inside the rocker arm cover for water deposits and an oil consistency like that of a milkshake.*

1 Leaking cylinder head gasket. Have the system pressure tested or remove the cylinder head (Chapter 2) and inspect.
2 Cracked cylinder bore or cylinder head. Dismantle engine and inspect (Chapter 2).
3 Loose cylinder head bolts (tighten as described in Chapter 2).
4 Leakage from internal coolant pipe/hose (V8 engines) (accessible only with intake manifold removed) (Chapter 2B).

31 Abnormal coolant loss

1 Overfilling system (Chapter 1).
2 Coolant boiling away due to overheating (see causes in Section 27).
3 Internal or external leakage (see Sections 29 and 30).
4 Faulty expansion tank cap. Have the cap pressure tested.
5 Cooling system being pressurized by engine compression. This could be due to a cracked head or block or leaking head gasket(s). Have the system tested for the presence of combustion gas in the coolant at a shop.

32 Poor coolant circulation

1 Inoperative water pump. A quick test is to pinch the top radiator hose closed with your hand while the engine is idling, then release it. You may be able to feel a surge of coolant if the pump is working properly (Chapter 3).
2 Restriction in cooling system. Drain, flush and refill the system (Chapter 1). If necessary, remove the radiator (Chapter 3) and have it reverse flushed or professionally cleaned.
3 Loose water pump drivebelt (Chapter 1).
4 Thermostat sticking (Chapter 3).
5 Insufficient coolant (Chapter 1).

33 Corrosion

1 Excessive impurities in the water. Soft, clean water is recommended. Distilled or rainwater is satisfactory.
2 Insufficient antifreeze solution (refer to Chapter 1 for the proper ratio of water to antifreeze).
3 Infrequent flushing and draining of system. Regular flushing of the cooling system should be carried out at the specified intervals as described in Chapter 1.

CLUTCH

34 Fails to release (pedal pressed to the floor - shift lever does not move freely in and out of Reverse)

1 Leak in the clutch hydraulic system. Check the master cylinder, release cylinder and lines (Chapter 8).
2 Clutch plate warped or damaged (Chapter 8).

35 Clutch slips (engine speed increases with no increase in vehicle speed)

1 Clutch plate oil soaked or lining worn. Remove clutch (Chapter 8) and inspect.
2 Clutch plate not seated. It may take 30 or 40 normal starts for a new one to seat.
3 Pressure plate worn (Chapter 8).

36 Grabbing (chattering) as clutch is engaged

1 Oil on clutch plate lining. Remove (Chapter 8) and inspect. Correct any leakage source.
2 Worn or loose engine or transmission mounts. These units move slightly when the clutch is released. Inspect the mounts and bolts (Chapter 2).
3 Worn splines on clutch plate hub. Remove the clutch components (Chapter 8) and inspect.
4 Warped pressure plate or flywheel. Remove the clutch components and inspect.

37 Squeal or rumble with clutch fully engaged (pedal released)

Release bearing binding on transmission bearing retainer. Remove clutch components (Chapter 8) and check bearing. Remove any burrs or nicks; clean and relubricate bearing retainer before installing.

38 Squeal or rumble with clutch fully disengaged (pedal depressed)

1 Worn, defective or broken release bearing (Chapter 8).
2 Worn or broken pressure plate springs (or diaphragm fingers) (Chapter 8).

39 Clutch pedal stays on floor when disengaged

1 Linkage or release bearing binding. Inspect the linkage or remove the clutch components as necessary.
2 Make sure proper pedal stop (bumper) is installed.

TROUBLESHOOTING 0-35

MANUAL TRANSMISSION

➡ **Note:** All the following references are in Chapter 7A, unless noted.

40 Noisy in Neutral with engine running

1 Input shaft bearing worn.
2 Damaged main drive gear bearing.
3 Worn countershaft bearings.
4 Worn or damaged countershaft endplay shims.

41 Noisy in all gears

1 Any of the above causes, and/or:
2 Insufficient lubricant (see the checking procedures in Chapter 1).

42 Noisy in one particular gear

1 Worn, damaged or chipped gear teeth for that particular gear.
2 Worn or damaged synchronizer for that particular gear.

43 Slips out of high gear

1 Transmission loose on clutch housing.
2 Dirt between the transmission case and engine or misalignment of the transmission.

44 Difficulty in engaging gears

1 Clutch not releasing completely.
2 Loose, damaged or out-of-adjustment shift linkage. Make a thorough inspection, replacing parts as necessary.

45 Oil leakage

1 Excessive amount of lubricant in the transmission (see Chapter 1 for correct checking procedures). Drain lubricant as required.
2 Transmission oil seal or vehicle speed sensor O-ring in need of replacement.

AUTOMATIC TRANSMISSION

➡ **Note:** Due to the complexity of the automatic transmission, it's difficult for the home mechanic to properly diagnose and service this component. For problems other than the following, the vehicle should be taken to a dealer service department or a transmission shop.

46 General shift mechanism problems

1 Common problems which may be attributed to a misadjusted shift cable are:
   a) *Engine starting in gears other than Park or Neutral.*
   b) *Indicator on shifter pointing to a gear other than the one actually being selected.*
   c) *Vehicle moves when in Park.*
2 Refer to Chapter 7B to check the shift cable and or the Transmission Range (TR) sensor adjustment.

47 Transmission will not downshift with accelerator pedal pressed to the floor

Since these transmissions are electronically controlled, your dealer or a professional shop with the proper equipment will have to diagnose the probable cause.

48 Transmission slips, shifts rough, is noisy or has no drive in forward or reverse gears

1 There are many probable causes for the above problems, but the home mechanic should be concerned with only one possibility - fluid level.
2 Before taking the vehicle to a repair shop, check the level and condition of the fluid as described in Chapter 1. Correct fluid level as necessary or change the fluid and filter if needed. If the problem persists, have a professional diagnose the problem.

49 Fluid leakage

1 Automatic transmission fluid is a deep red color. Fluid leaks should not be confused with engine oil, which can easily be blown by air flow to the transmission.
2 To pinpoint a leak, first remove all built-up dirt and grime from around the transmission. Degreasing agents and/or steam cleaning will achieve this. With the underside clean, drive the vehicle at low speeds so air flow will not blow the leak far from its source. Raise the vehicle and determine where the leak is coming from. Common areas of leakage are:
   a) **Pan:** *Tighten the mounting bolts and/or replace the pan gasket as necessary (see Chapter 7B).*
   b) **Filler pipe:** *Replace the rubber seal where the pipe enters the transmission case.*
   c) **Transmission oil lines:** *Tighten the connectors where the lines enter the transmission case and/or replace the lines.*
   d) **Vent pipe:** *Transmission overfilled and/or water in fluid (see checking procedures, Chapter 1).*
   e) **Speedometer connector:** *Replace the O-ring where the speedometer sensor enters the transmission case (Chapter 7B).*

TRANSFER CASE

50 Transfer case is difficult to shift into the desired range

1 Speed may be too great to permit engagement. Stop the vehicle and shift into the desired range.
2 Shift linkage loose, bent or binding on a manual shift transfer case. Check the linkage for damage or wear and replace or lubricate as necessary (Chapter 7C).
3 Defective circuit and or range switch on electric shift transfer case (Chapter 7C).
4 If the vehicle has been driven on a paved surface for some time, the driveline torque can make shifting difficult. Stop and shift into two-wheel drive on paved or hard surfaces.
5 Insufficient or incorrect grade of lubricant. Drain and refill the transfer case with the specified lubricant. (Chapter 1).
6 Worn or damaged internal components. Disassembly and overhaul of the transfer case may be necessary (Chapter 7C).

0-36 TROUBLESHOOTING

51 Transfer case noisy in all gears

Insufficient or incorrect grade of lubricant. Drain and refill (Chapter 1).

52 Noisy or jumps out of four-wheel drive Low range

1 Transfer case not fully engaged. Stop the vehicle, shift into Neutral and then engage 4L.
2 Shift linkage loose, worn or binding. Tighten, repair or lubricate linkage as necessary.
3 Shift fork cracked, inserts worn or fork binding on the rail. See your dealer for a new or rebuilt unit.

53 Lubricant leaks from the vent or output shaft seals

1 Transfer case is overfilled. Drain to the proper level (Chapter 1).
2 Vent is clogged or jammed closed. Clear or replace the vent.
3 Output shaft seal incorrectly installed or damaged. Replace the seal and check contact surfaces for nicks and scoring.

DRIVESHAFT

54 Oil leak at seal end of driveshaft

Defective transmission or transfer case oil seal. See Chapter 7 for replacement procedures. While this is done, check the splined yoke for burrs or a rough condition which may be damaging the seal. Burrs can be removed with crocus cloth or a fine whetstone.

55 Knock or clunk when the transmission is under initial load (just after transmission is put into gear)

1 Loose or disconnected rear suspension components. Check all mounting bolts, nuts and bushings (see Chapter 10).
2 Loose driveshaft bolts. Inspect all bolts and nuts and tighten them to the specified torque.
3 Worn or damaged universal joint bearings. Check for wear (see Chapter 8).

56 Metallic grinding sound consistent with vehicle speed

Pronounced wear in the universal joint bearings. Check as described in Chapter 8.

57 Vibration

➡Note: Before assuming that the driveshaft is at fault, make sure the tires are perfectly balanced and perform the following test.

1 Install a tachometer inside the vehicle to monitor engine speed as the vehicle is driven. Drive the vehicle and note the engine speed at which the vibration (roughness) is most pronounced. Now shift the transmission to a different gear and bring the engine speed to the same point.
2 If the vibration occurs at the same engine speed (rpm) regardless of which gear the transmission is in, the driveshaft is NOT at fault since the driveshaft speed varies.
3 If the vibration decreases or is eliminated when the transmission is in a different gear at the same engine speed, refer to the following probable causes.
4 Bent or dented driveshaft. Inspect and replace as necessary (see Chapter 8).
5 Undercoating or built-up dirt, etc. on the driveshaft. Clean the shaft thoroughly and recheck.
6 Worn universal joint bearings. Remove and inspect (see Chapter 8).
7 Driveshaft and/or companion flange out of balance. Check for missing weights on the shaft. Remove the driveshaft (see Chapter 8) and reinstall 180-degrees from original position, then retest. Have the driveshaft professionally balanced if the problem persists.

AXLES

58 Noise

1 Road noise. No corrective procedures available.
2 Tire noise. Inspect tires and check tire pressures (Chapter 1).
3 Rear wheel bearings loose, worn or damaged (Chapter 8).

59 Vibration

See probable causes under Driveshaft. Proceed under the guidelines listed for the driveshaft. If the problem persists, check the rear wheel bearings by raising the rear of the vehicle and spinning the rear wheels by hand. Listen for evidence of rough (noisy) bearings. Remove and inspect (see Chapter 8).

60 Oil leakage

1 Pinion seal damaged (see Chapter 8).
2 Axleshaft oil seals damaged (see Chapter 8).
3 Differential inspection cover leaking. Tighten the bolts or replace the gasket as required (see Chapters 1 and 8).

BRAKES

➡Note: Before assuming that a brake problem exists, make sure that the tires are in good condition and inflated properly (see Chapter 1), that the front end alignment is correct and that the vehicle is not loaded with weight in an unequal manner.

61 Vehicle pulls to one side during braking

1 Defective, damaged or oil contaminated disc brake pads or shoes on one side. Inspect as described in Chapter 9.
2 Excessive wear of brake shoe or pad material or drum/disc on one side. Inspect and correct as necessary.
3 Loose or disconnected front suspension components. Inspect and tighten all bolts to the specified torque (Chapter 10).
4 Defective drum brake or caliper assembly. Remove the drum or caliper and inspect for a stuck piston or other damage (Chapter 9).

TROUBLESHOOTING 0-37

62 Noise (high-pitched squeal with the brakes applied)

1 Disc brake pads worn out. The noise comes from the wear sensor rubbing against the disc (does not apply to all vehicles) or the actual pad backing plate itself if the material is completely worn away. Replace the pads with new ones immediately (Chapter 9). If the pad material has worn completely away, the brake discs should be inspected for damage as described in Chapter 9.
2 Linings contaminated with dirt or grease. Replace pads or shoes.
3 Incorrect linings. Replace with correct linings.

63 Excessive brake pedal travel

1 Partial brake system failure. Inspect the entire system (Chapter 9) and correct as required.
2 Insufficient fluid in the master cylinder. Check (Chapter 1), add fluid and bleed the system if necessary (Chapter 9).
3 Rear brakes not adjusting properly. Make a series of starts and stops while the vehicle is in Reverse. If this does not correct the situation, remove the drums and inspect the self-adjusters (Chapter 9).
4 Problem with the anti-lock brake system (Chapter 9).

64 Brake pedal feels spongy when depressed

1 Air in the hydraulic lines. Bleed the brake system (Chapter 9).
2 Faulty flexible hoses. Inspect all system hoses and lines. Replace parts as necessary.
3 Master cylinder mounting bolts/nuts loose.
4 Master cylinder defective (Chapter 9).
5 Problem with the anti-lock brake system (Chapter 9).

65 Excessive effort required to stop vehicle

1 Power brake booster not operating properly (Chapter 9).
2 Excessively worn linings or pads. Inspect and replace if necessary (Chapter 9).
3 One or more caliper pistons or wheel cylinders seized or sticking. Inspect and rebuild as required (Chapter 9).
4 Brake linings or pads contaminated with oil or grease. Inspect and replace as required (Chapter 9).
5 New pads or shoes installed and not yet seated. It will take a while for the new material to seat against the drum (or disc).
6 Problem with the anti-lock brake system (Chapter 9).

66 Pedal travels to the floor with little resistance

1 Little or no fluid in the master cylinder reservoir caused by leaking wheel cylinder(s), leaking caliper piston(s), loose, damaged or disconnected brake lines. Inspect the entire system and correct as necessary.
2 Worn master cylinder seals (Chapter 9).
3 Problem with the anti-lock brake system (Chapter 9).

67 Brake pedal pulsates during brake application

1 Caliper improperly installed. Remove and inspect (Chapter 9).
2 Disc or drum defective. Remove (Chapter 9) and check for excessive lateral runout and parallelism. Have the disc or drum resurfaced or replace it with a new one.

SUSPENSION AND STEERING SYSTEMS

68 Vehicle pulls to one side

1 Tire pressures uneven (Chapter 1).
2 Defective tire (Chapter 1).
3 Excessive wear in suspension or steering components (Chapter 10).
4 Front end in need of alignment.
5 Front brakes dragging. Inspect the brakes as described in Chapter 9.

69 Shimmy, shake or vibration

1 Tire or wheel out-of-balance or out-of-round. Have professionally balanced.
2 Loose, worn or out-of-adjustment front wheel bearings (Chapter 1).
3 Shock absorbers and/or suspension components worn or damaged (Chapter 10).

70 Excessive pitching and/or rolling around corners or during braking

1 Defective shock absorbers. Replace as a set (Chapter 10).
2 Broken or weak springs and/or suspension components. Inspect as described in Chapter 10.

71 Excessively stiff steering

1 Lack of fluid in power steering fluid reservoir (Chapter 1).
2 Incorrect tire pressures (Chapter 1).
3 Lack of lubrication at steering joints (see Chapter 1).
4 Front end out of alignment.
5 Lack of power assistance (see Section 73).

72 Excessive play in steering

1 Loose front wheel bearings (Chapters 1 and 10).
2 Excessive wear in suspension or steering components (Chapter 10).
3 Steering gearbox damaged or out of adjustment (Chapter 10).

73 Lack of power assistance

1 Steering pump drivebelt faulty or not adjusted properly (Chapter 1).
2 Fluid level low (Chapter 1).
3 Hoses or lines restricted. Inspect and replace parts as necessary.
4 Air in power steering system. Bleed the system (Chapter 10).

74 Excessive tire wear (not specific to one area)

1 Incorrect tire pressures (Chapter 1).
2 Tires out-of-balance. Have professionally balanced.
3 Wheels damaged. Inspect and replace as necessary.
4 Suspension or steering components excessively worn (Chapter 10).

75 Excessive tire wear on outside edge

1 Inflation pressures incorrect (Chapter 1).
2 Excessive speed in turns.
3 Front end alignment incorrect. Have professionally aligned.
4 Suspension arm bent or twisted (Chapter 10).

0-38 TROUBLESHOOTING

76 Excessive tire wear on inside edge

1 Inflation pressures incorrect (Chapter 1).
2 Front end alignment incorrect (toe-out). Have professionally aligned.
3 Loose or damaged steering components (Chapter 10).

77 Tire tread worn in one place

1 Tires out-of-balance.
2 Damaged or buckled wheel. Inspect and replace if necessary.
3 Defective tire (Chapter 1).

Section

1. Maintenance schedule
2. Introduction
3. Tune-up general information
4. Fluid level checks
5. Tire and tire pressure checks
6. Power steering fluid level check
7. Automatic transmission fluid level check
8. Engine oil and filter change
9. Chassis lubrication
10. Battery check, maintenance and charging
11. Windshield wiper blade inspection and replacement
12. Tire rotation
13. Exhaust system check
14. Seat belt check
15. Underhood hose check and replacement
16. Cooling system check
17. Fuel system check
18. Fuel filter replacement
19. Steering and suspension check
20. Brake check
21. Manual transmission lubricant level check
22. Transfer case lubricant level check
23. Differential lubricant level check
24. Air filter check and replacement
25. Cooling system servicing (draining, flushing and refilling)
26. Automatic transmission fluid and filter change
27. Front wheel bearing check, repack and adjustment (2WD models)
28. Brake fluid change
29. Drivebelt check and replacement
30. Positive Crankcase Ventilation (PCV) valve check
31. Spark plug check and replacement
32. Ignition system component check and replacement
33. Manual transmission lubricant change
34. Transfer case lubricant change (4WD models)
35. Differential lubricant change

1
TUNE-UP AND ROUTINE MAINTENANCE

1-2 TUNE-UP AND ROUTINE MAINTENANCE

1 Ford F-150/F-250, Expedition and Lincoln Navigator maintenance schedule

The following maintenance intervals are based on the assumption that the vehicle owner will be doing the maintenance or service work, as opposed to having a dealer service department or other repair shop do the work. Although the time/mileage intervals are loosely based on factory recommendations, most have been shortened to ensure, for example, that such items as lubricants and fluids are checked/changed at intervals that promote maximum engine/driveline service life. Also, subject to the preference of the individual owner interested in keeping his or her vehicle in peak condition at all times, and with the vehicle's ultimate resale in mind, many of the maintenance procedures may be performed more often than recommended in the following schedule. We encourage such owner initiative.

When the vehicle is new it should be serviced initially by a factory authorized dealer service department to protect the factory warranty. In many cases the initial maintenance check is done at no cost to the owner (check with your dealer service department for more information).

EVERY 250 MILES OR WEEKLY, WHICHEVER COMES FIRST

Check the engine oil level (Section 4)
Check the engine coolant level (Section 4)
Check the windshield washer fluid level (Section 4)
Check the brake and the clutch fluid level (Section 4)
Check the tires and tire pressures (Section 5)

EVERY 3000 MILES OR 3 MONTHS, WHICHEVER COMES FIRST

All items listed above, plus:

Check the power steering fluid level (Section 6)
Check the automatic transmission fluid level (Section 7)
Change the engine oil and oil filter (Section 8)
Lubricate the chassis (Section 9)

EVERY 6000 MILES OR 6 MONTHS, WHICHEVER COMES FIRST

All items listed above, plus:

Check and service the battery (Section 10)
Inspect and replace, if necessary, the windshield wiper blades (Section 11)
Rotate the tires (Section 12)
Inspect the exhaust system (Section 13)
Check the seat belt operation (Section 14)

EVERY 15,000 MILES OR 12 MONTHS, WHICHEVER COMES FIRST

All items listed above, plus:

Inspect and replace, if necessary, all underhood hoses (Section 15)
Inspect the cooling system (Section 16)
Check the fuel system (Section 17)
Replace the fuel filter (Section 18)
Inspect the steering and suspension components (Section 19)
Inspect the brakes (Section 20)
Check the manual transmission lubricant level (Section 21)
Check the transfer case lubricant level (Section 22)
Check the rear axle (differential) lubricant level (Section 23)

EVERY 30,000 MILES OR 24 MONTHS, WHICHEVER COMES FIRST

Replace the air filter (Section 24)*
Service the cooling system (drain, flush and refill) (Section 25)
Change the automatic transmission fluid and filter (Section 26)**
Inspect and repack the front wheel bearings (2WD models) (Section 27)
Change the brake fluid (Section 28)

EVERY 60,000 MILES OR 48 MONTHS, WHICHEVER COMES FIRST

Check the engine drivebelt(s) (Section 29)
Check the PCV valve (Section 30)
Replace the spark plugs (Section 31)
Inspect and replace, if necessary, the spark plug wires (Section 32)
Change the manual transmission lubricant (Section 33)
Change the transfer case lubricant (Section 34)
Change the rear axle (differential) lubricant (Section 35)

* Replace more often if is the vehicle is driven in dusty areas
** If the vehicle is operated in continuous stop-and-go driving or in mountainous areas, change at 15,000 miles

TUNE-UP AND ROUTINE MAINTENANCE

Typical engine compartment components (V6 engine)

1. Battery
2. Ignition coil pack
3. Engine oil filler cap
4. Clutch master cylinder reservoir
5. Brake master cylinder reservoir
6. Engine compartment fuse box
7. Engine coolant expansion tank
8. Air filter housing
9. Power steering fluid reservoir
10. Engine oil dipstick
11. Upper radiator hose
12. Windshield washer fluid reservoir
13. Automatic transmission dipstick (not visible)

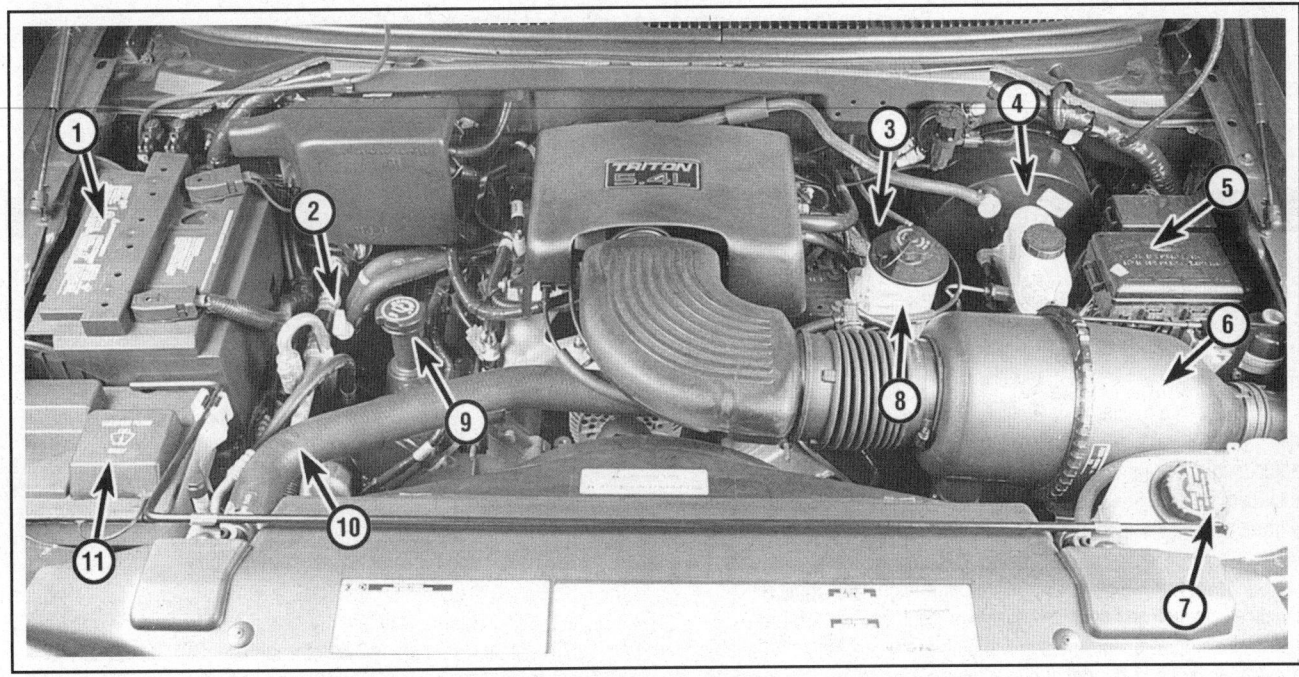

Typical engine compartment components (early 4.6L and 5.4L V8 engines)

1. Battery
2. Automatic transmission dipstick
3. Engine oil dipstick
4. Brake master cylinder reservoir
5. Engine compartment fuse box
6. Air filter housing
7. Coolant expansion tank
8. Power steering fluid reservoir
9. Engine oil filler cap
10. Upper radiator hose
11. Windshield washer fluid reservoir

1-4 TUNE-UP AND ROUTINE MAINTENANCE

Typical engine compartment underside components (2WD)

1. Stabilizer bar
2. Drivebelt
3. Oil filter
4. Shock absorber
5. Exhaust pipe
6. Engine oil drain plug
7. Lower control arm bushing
8. Steering linkage

2 Introduction

This Chapter is designed to help the home mechanic maintain the Ford F-150/F-250, the Ford Expedition and the Lincoln Navigator with the goals of maximum performance, economy, safety and reliability in mind.

Included is a master maintenance schedule, followed by procedures dealing specifically with each item on the schedule. Visual checks, adjustments, component replacement and other helpful items are included. Refer to the accompanying illustrations of the engine compartment and the underside of the vehicle for the locations of various components.

Servicing the vehicle in accordance with the mileage/time maintenance schedule and the step-by-step procedures will result in a planned maintenance program that should produce a long and reliable service life. Keep in mind that it is a comprehensive plan, so maintaining some items but not others at the specified intervals will not produce the same results.

As you service the vehicle, you will discover that many of the procedures can - and should - be grouped together because of the nature of the particular procedure you're performing or because of the close proximity of two otherwise unrelated components to one another.

For example, if the vehicle is raised for chassis lubrication, you should inspect the exhaust, suspension, steering and fuel systems while you're under the vehicle. When you're rotating the tires, it makes good sense to check the brakes since the wheels are already removed. Finally, let's suppose you have to borrow or rent a torque wrench. Even if you only need it to tighten the spark plugs, you might as well check the torque of as many critical fasteners as time allows.

The first step in this maintenance program is to prepare yourself before the actual work begins. Read through all the procedures you're planning to do, then gather up all the parts and tools needed. If it looks like you might run into problems during a particular job, seek advice from a mechanic or an experienced do-it-yourselfer.

TUNE-UP AND ROUTINE MAINTENANCE

Typical rear underside components (2WD)

1. Leaf spring
2. Differential cover
3. Rear shock absorber
4. Parking brake cable
5. Driveshaft
6. Gas tank

3 Tune-up general information

The term tune-up is used in this manual to represent a combination of individual operations rather than one specific procedure.

If, from the time the vehicle is new, the routine maintenance schedule is followed closely and frequent checks are made of fluid levels and high wear items, as suggested throughout this manual, the engine will be kept in relatively good running condition and the need for additional work will be minimized.

More likely than not, however, there will be times when the engine is running poorly due to lack of regular maintenance. This is even more likely if a used vehicle, which has not received regular and frequent maintenance checks, is purchased. In such cases, an engine tune-up will be needed outside of the regular routine maintenance intervals.

The first step in any tune-up or diagnostic procedure to help correct a poor running engine is a cylinder compression check. A compression check (see Chapter 2) will help determine the condition of internal engine components and should be used as a guide for tune-up and repair procedures. If, for instance, a compression check indicates serious internal engine wear, a conventional tune-up will not improve the performance of the engine and would be a waste of time and money. Because of its importance, the compression check should be done by someone with the right equipment and the knowledge to use it properly.

The following procedures are those most often needed to bring a generally poor running engine back into a proper state of tune.

MINOR TUNE-UP

Check all engine related fluids (Section 4)
Clean, inspect and test the battery (Section 10)
Check all underhood hoses (Section 15)
Check the cooling system (Section 16)
Check the fuel system (Section 17)
Check the air filter (Section 24)

MAJOR TUNE-UP

All items listed under Minor Tune-up, plus . .

Replace the fuel filter (Section 18)
Replace the air filter (Section 24)
Check the drivebelt (Section 29)
Replace the PCV valve (Section 30)
Replace the spark plugs (Section 31)
Replace the spark plug wires, if equipped (Section 32)
Check the charging system (Chapter 5)

1-6 TUNE-UP AND ROUTINE MAINTENANCE

4 Fluid level checks (every 250 miles or weekly)

1 Fluids are an essential part of the lubrication, cooling, brake and windshield washer systems. Because the fluids gradually become depleted and/or contaminated during normal operation of the vehicle, they must be periodically replenished. See *Recommended lubricants and fluids* at the end of this Chapter before adding fluid to any of the following components.

→Note: *The vehicle must be on level ground when fluid levels are checked.*

ENGINE OIL

▸ Refer to illustrations 4.2, 4.4, 4.6a and 4.6b

2 The oil level is checked with a dipstick, which is located on the left (driver's) side of the engine (see illustration). The dipstick extends through a metal tube down into the oil pan.

3 The oil level should be checked before the vehicle has been driven, or about 15 minutes after the engine has been shut off. If the oil is checked immediately after driving the vehicle, some of the oil will remain in the upper part of the engine, resulting in an inaccurate reading on the dipstick.

4 Pull the dipstick out of the tube and wipe all the oil from the end with a clean rag or paper towel. Insert the clean dipstick all the way back into the tube and pull it out again. Note the oil at the end of the dipstick. At its highest point, the level should be above the MIN mark, within the hatched marked section of the dipstick (see illustration).

5 It takes one quart of oil to raise the level from the MIN mark to the MAX mark on the dipstick. Do not allow the level to drop below the MIN mark or oil starvation may cause engine damage. Conversely, overfilling the engine (adding oil above the MAX mark) may cause oil fouled spark plugs, oil leaks or oil seal failures.

6 To add oil, remove the filler cap from the valve cover (see illustrations). After adding oil, wait a few minutes to allow the level to stabilize, then pull out the dipstick and check the level again. Add more oil if required. Install the filler cap and tighten it by hand only.

7 Checking the oil level is an important preventive maintenance step. A consistently low oil level indicates oil leakage through damaged seals, defective gaskets or past worn rings or valve guides. If the oil looks milky in color or has water droplets in it, the cylinder head gasket(s) may be blown or the head(s) or block may be cracked. The engine should be checked immediately. The condition of the oil should also be checked. Whenever you check the oil level, slide your thumb and index finger up the dipstick before wiping off the oil. If you see small dirt or metal particles clinging to the dipstick, the oil should be changed (see Section 8).

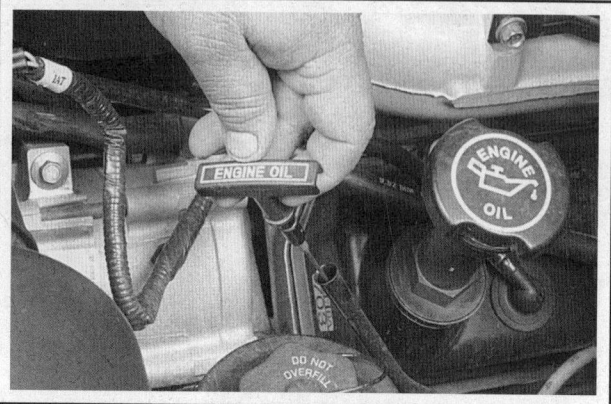

4.2 The engine oil dipstick is located on the driver's side of the engine

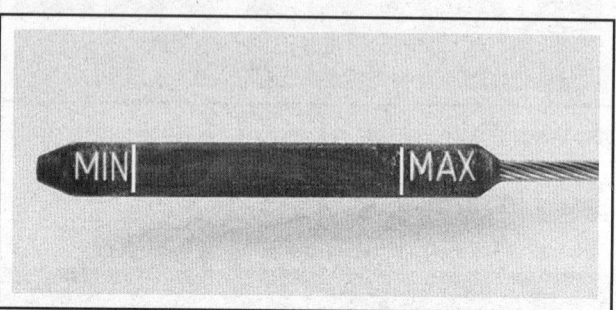

4.4 The oil level should be at or near the MAX area on the dipstick - if it isn't, add enough oil to bring the level to near the MAX mark (it takes one quart of oil to raise the level from the lower to the upper mark)

4.6a The oil filler cap on V6 engines is located on the left valve cover

4.6b The oil filler cap on V8 engines is located on the right valve cover

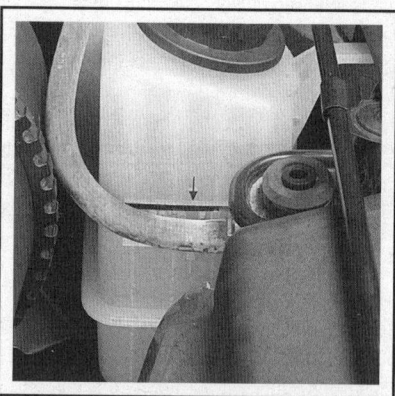

4.9 The coolant expansion tank is located at the front of the engine compartment on early models - keep the level near the arrow - DO NOT remove the cap until the engine has cooled completely

TUNE-UP AND ROUTINE MAINTENANCE

4.15a The brake fluid level should be kept between the MIN and MAX marks on the translucent plastic reservoir

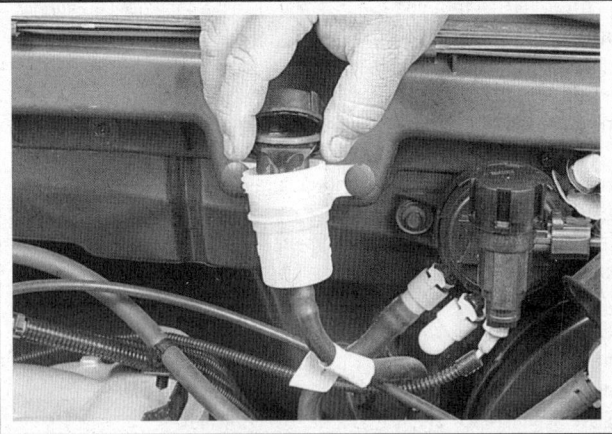

4.15b The clutch fluid reservoir is mounted on the firewall - add fluid until it reaches the FULL line

ENGINE COOLANT

♦ Refer to illustration 4.9

※※ WARNING:

Do not allow antifreeze to come in contact with your skin or painted surfaces of the vehicle. Flush contaminated areas immediately with plenty of water. Don't store new coolant or leave old coolant lying around where it's accessible to children or pets - they're attracted by its sweet smell. Ingestion of even a small amount of coolant can be fatal! Wipe up garage floor and drip pan spills immediately. Keep antifreeze containers covered and repair cooling system leaks as soon as they're noticed.

8 All vehicles covered by this manual are equipped with a pressurized coolant recovery system. A plastic expansion tank located at the front (early models) or rear (later models) of the engine compartment is connected by a hose to the radiator. As the engine heats up during operation, the expanding coolant fills the tank.

9 The coolant level in the tank should be checked regularly.

※※ WARNING:

Do not remove the expansion tank cap to check the coolant level when the engine is warm! The level in the tank varies with the temperature of the engine. When the engine is cold, the coolant level should be at or slightly above the FULL COLD mark on the reservoir. Once the engine has warmed up, the level should be at or near the FULL HOT mark. If it isn't, allow the engine to cool, then remove the cap from the tank and add a 50/50 mixture of ethylene glycol based antifreeze and water (see illustration).

10 Drive the vehicle and recheck the coolant level. Don't use rust inhibitors or additives. If only a small amount of coolant is required to bring the system up to the proper level, water can be used. However, repeated additions of water will dilute the antifreeze and water solution. In order to maintain the proper ratio of antifreeze and water, always top up the coolant level with the correct mixture. An empty plastic milk jug or bleach bottle makes an excellent container for mixing coolant.

11 If the coolant level drops consistently, there may be a leak in the system. Inspect the radiator, hoses, filler cap, drain plugs and water pump (see Section 16). If no leaks are noted, have the expansion tank cap pressure tested by a service station.

12 If you have to remove the expansion tank cap wait until the engine has cooled completely, then wrap a thick cloth around the cap and unscrew it slowly, stopping if you hear a hissing noise. If coolant or steam escapes, let the engine cool down longer, then remove the cap.

13 Check the condition of the coolant as well. It should be relatively clear. If it's brown or rust colored, the system should be drained, flushed and refilled. Even if the coolant appears to be normal, the corrosion inhibitors wear out, so it must be replaced at the specified intervals.

BRAKE AND CLUTCH FLUID

♦ Refer to illustrations 4.15a and 4.15b

14 The brake master cylinder is mounted on the front of the power booster unit in the engine compartment. The hydraulic clutch master cylinder used on manual transmission vehicles is located next to the brake master cylinder.

15 To check the fluid level of the brake and clutch master cylinders, simply look at the MAX and MIN marks on the reservoir (see illustrations). The level should be within the specified distance from the maximum fill line.

16 If the level is low, wipe the top of the reservoir cover with a clean rag to prevent contamination of the brake system before lifting the cover.

17 Add only the specified brake fluid to the brake and clutch reservoirs (refer to *Recommended lubricants and fluids* at the end of this Chapter or to your owner's manual). Mixing different types of brake fluid can damage the system. Fill the brake master cylinder reservoir only to the MAX line.

※※ WARNING:

Use caution when filling either reservoir - brake fluid can harm your eyes and damage painted surfaces. Do not use brake fluid that has been opened for more than one year or has been left open. Brake fluid absorbs moisture from the air. Excess moisture can cause a dangerous loss of braking.

1-8 TUNE-UP AND ROUTINE MAINTENANCE

18 While the reservoir cap is removed, inspect the master cylinder reservoir for contamination. If deposits, dirt particles or water droplets are present, the system should be drained and refilled.

19 After filling the reservoir to the proper level, make sure the lid is properly seated to prevent fluid leakage and/or system pressure loss.

20 The fluid in the brake master cylinder will drop slightly as the brake pads at each wheel wear down during normal operation. If either master cylinder requires repeated replenishing to keep it at the proper level, this is an indication of leakage in the brake or clutch system, which should be corrected immediately. If the brake system shows an indication of leakage check all brake lines and connections, along with the calipers, wheel cylinders and booster (see Section 20 for more information). If the hydraulic clutch system shows an indication of leakage check all clutch lines and connections, along with the clutch slave cylinder (see Chapter 8 for more information).

21 If, upon checking the brake or clutch master cylinder fluid level, you discover one or both reservoirs empty or nearly empty, the systems should be bled (see Chapter 9).

WINDSHIELD WASHER FLUID

♦ Refer to illustration 4.22

➡ **Note:** *The windshield washer fluid reservoir also supplies washer fluid for the liftgate wiper system on Expedition and Navigator models.*

22 Fluid for the windshield washer system is stored in a plastic reservoir located at the right front of the engine compartment (see illustration).

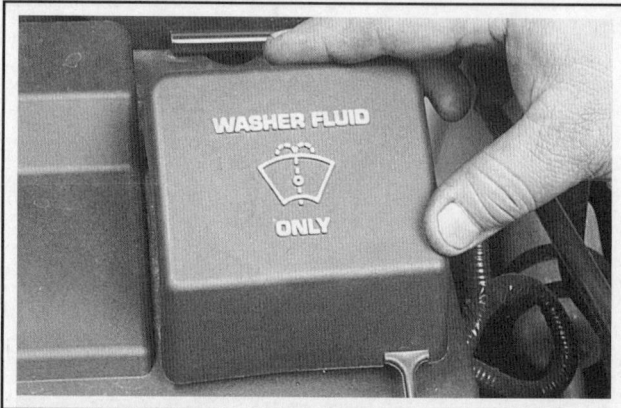

4.22 The windshield washer reservoir is located at the right front corner of the engine compartment next to the battery

23 In milder climates, plain water can be used in the reservoir, but it should be kept no more than 2/3 full to allow for expansion if the water freezes. In colder climates, use windshield washer system antifreeze, available at any auto parts store, to lower the freezing point of the fluid. Mix the antifreeze with water in accordance with the manufacturer's directions on the container.

✱✱ CAUTION:

Do not use cooling system antifreeze - it will damage the vehicle's paint.

5 Tire and tire pressure checks (every 250 miles or weekly)

♦ Refer to illustrations 5.2, 5.3, 5.4a, 5.4b and 5.8

1 Periodic inspection of the tires may spare you the inconvenience of being stranded with a flat tire. It can also provide you with vital information regarding possible problems in the steering and suspension systems before major damage occurs.

2 The original tires on this vehicle are equipped with 1/2-inch wide bands that will appear when tread depth reaches 1/16-inch, at which point they can be considered worn out. Tread wear can be monitored with a simple, inexpensive device known as a tread depth indicator (see illustration).

3 Note any abnormal tread wear (see illustration). Tread pattern irregularities such as cupping, flat spots and more wear on one side than the other are indications of front end alignment and/or balance problems. If any of these conditions are noted, take the vehicle to a tire shop or service station to correct the problem.

4 Look closely for cuts, punctures and embedded nails or tacks. Sometimes a tire will hold air pressure for a short time or leak down very slowly after a nail has embedded itself in the tread. If a slow leak persists, check the valve stem core to make sure it is tight (see illustration). Examine the tread for an object that may have embedded itself in the tire or for a "plug" that may have begun to leak (radial tire punctures are repaired with a plug that is installed in a puncture). If a puncture is suspected, it can be easily verified by spraying a solution of soapy water onto the puncture area (see illustration). The soapy solution will bubble if there is a leak. Unless the puncture is unusually large, a tire shop or service station can usually repair the tire.

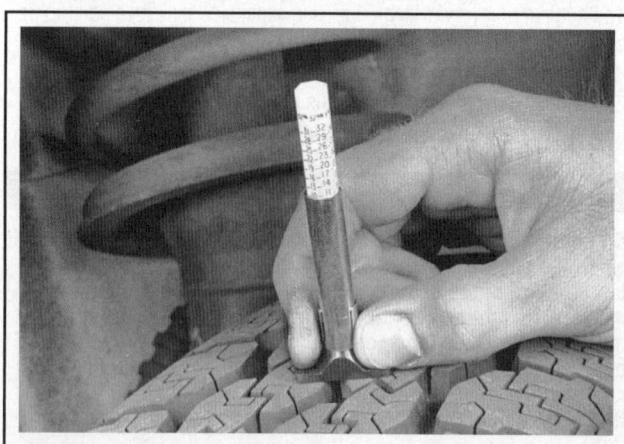

5.2 Use a tire tread depth indicator to monitor tire wear - they are available at auto parts stores and service stations and cost very little

5 Carefully inspect the inner sidewall of each tire for evidence of brake fluid leakage. If you see any, inspect the brakes immediately.

6 Correct air pressure adds miles to the life span of the tires, improves mileage and enhances overall ride quality. Tire pressure cannot be accurately estimated by looking at a tire, especially if it's a radial. A tire pressure gauge is essential. Keep an accurate gauge in the glove compartment. The pressure gauges attached to the nozzles of air

TUNE-UP AND ROUTINE MAINTENANCE 1-9

UNDERINFLATION

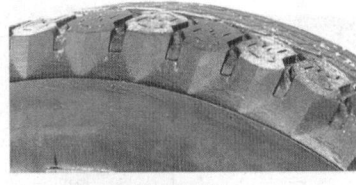
CUPPING

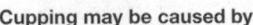
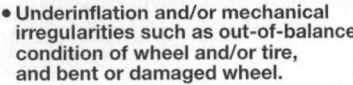
Cupping may be caused by:
- Underinflation and/or mechanical irregularities such as out-of-balance condition of wheel and/or tire, and bent or damaged wheel.
- Loose or worn steering tie-rod or steering idler arm.
- Loose, damaged or worn front suspension parts.

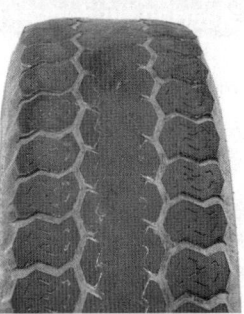
OVERINFLATION

INCORRECT TOE-IN OR EXTREME CAMBER

FEATHERING DUE TO MISALIGNMENT

5.3 This chart will help you determine the condition of the tires, the probable cause(s) of abnormal wear and the corrective action necessary

hoses at gas stations are often inaccurate.

7 Always check tire pressure when the tires are cold. Cold, in this case, means the vehicle has not been driven over a mile in the three hours preceding a tire pressure check. A pressure rise of four to eight pounds is not uncommon once the tires are warm.

8 Unscrew the valve cap protruding from the wheel or hubcap and push the gauge firmly onto the valve stem (see illustration). Note the reading on the gauge and compare the figure to the recommended tire pressure shown on the tire placard on the driver's side door. Be sure to reinstall the valve cap to keep dirt and moisture out of the valve stem mechanism. Check all four tires and, if necessary, add enough air to bring them up to the recommended pressure.

9 Don't forget to keep the spare tire inflated to the specified pressure (refer to your owner's manual or the decal attached to the right door pillar). Note that the pressure recommended for the temporary (mini) spare is higher than for the tires on the vehicle.

5.4a If a tire loses air on a steady basis, check the valve stem core first to make sure it's snug (special inexpensive wrenches are commonly available at auto parts stores)

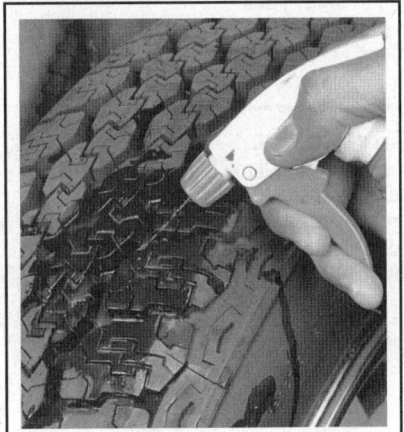

5.4b If the valve stem core is tight, raise the corner of the vehicle with the low tire and spray a soapy water solution onto the tread as the tire is turned slowly - leaks will cause small bubbles to appear

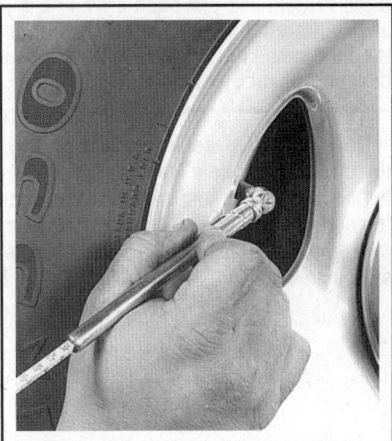

5.8 To extend the life of the tires, check the air pressure at least once a week with an accurate gauge (don't forget the spare!)

1-10 TUNE-UP AND ROUTINE MAINTENANCE

6 Power steering fluid level check (every 3000 miles or 3 months)

▶ Refer to illustration 6.3

1 Check the power steering fluid level periodically to avoid steering system problems, such as damage to the pump.

CAUTION:

DO NOT hold the steering wheel against either stop (extreme left or right turn) for more than five seconds. If you do, the power steering pump could be damaged.

2 The fluid reservoir for the power steering pump is mounted at the left front corner of the engine block.
3 The reservoir is translucent plastic and the fluid level can be checked visually (see illustration).
4 The fluid level should be kept between the MIN and MAX marks on the reservoir.
5 Add small amounts of fluid until the level is correct.

CAUTION:

Do not overfill the reservoir. If too much fluid is added, remove the excess with a clean syringe or suction pump.

6.3 The power steering fluid reservoir is translucent so the fluid level can be checked without removing the cap

6 If the reservoir requires frequent fluid additions, all power steering hoses, hose connections, the power steering pump and the steering gear assembly should be carefully checked for leaks.

7 Automatic transmission fluid level check (every 3000 miles or 3 months)

1 The automatic transmission fluid level should be carefully maintained. Low fluid level can lead to slipping or loss of drive, while overfilling can cause foaming and loss of fluid. Either condition can cause transmission damage.
2 Transmission fluid expands as it heats up. As a result, it's best to check the fluid level when the transmission is at normal operating temperature. We suggest driving the vehicle about 5 miles to warm it up.

CAUTION:

If the vehicle has just been driven for a long time at high speed or in city traffic in hot weather, or if it has been pulling a trailer, the fluid is too hot to check. Allow the transmission to cool down for about 30 minutes.

➡Note: On late model transmissions, the manufacturer states that the transmission fluid level is checked when it's within a specific temperature range and with the use a scan tool to monitor the temperature. Since this is not possible for most home mechanics, other methods of checking the fluid temperature can be utilized such as a cooking thermometer or digital laser thermometer.

3 Immediately after driving the vehicle, park it on a level surface with the parking brake set and the engine on. While the engine is idling, depress the brake pedal and move the selector lever through all the gear ranges, beginning and ending in Park.

MODELS WITH FILLER TUBE AND DIPSTICK (4R100, 4R70W, 4R70E-4R75E, E40D)

▶ Refer to illustrations 7.4 and 7.6

4 Locate the automatic transmission dipstick tube in the engine compartment (see illustration).

5 With the engine still idling, pull the dipstick from the tube, wipe it off with a clean rag, push it all the way back into the tube and withdraw it again, then note the fluid level.
6 If the transmission is COLD (room temperature with no warm-up), the level should be in the cold range on the dipstick (or between two holes in the dipstick); if it's HOT (within normal operating temperature in this case), the fluid level should be in the hot range (or in a cross-hatched area) on the dipstick (see illustration). If the level is low, add the specified automatic transmission fluid through the dipstick tube using a funnel to prevent spills.
7 Add just enough of the recommended fluid to fill the transmission to the proper level. It takes about one pint to raise the level from the low mark to the high mark when the fluid is hot, so add the fluid a

7.4 The automatic transmission dipstick is located on the passenger's side of the engine

TUNE-UP AND ROUTINE MAINTENANCE   1-11

little at a time and keep checking the level until it's correct.

8 The condition of the fluid should also be checked along with the level. If the fluid is black or a dark reddish-brown color, or if it smells burned, it should be changed (see Section 26). If you are in doubt about its condition, purchase some new fluid and compare the two for color and smell.

MODELS WITH SIDE CHECK/FILL PLUG (6HP26)

→ Note: These models are equipped with a horizontally mounted check/fill plug mounted in the side of the transmission case.

→ Note: The transmission fluid level check on these models can be performed every 30,000 miles or if a fluid leak is suspected. However, we recommend regular visual inspections for fluid leaks on vehicle models with these types of transmissions.

9 Secure the vehicle on level ground or place it on jackstands if necessary. The vehicle must be level.

10 With the transmission in Park and the engine idling, remove the fill/check plug.

✳✳ WARNING:

The transmission check/fill plug is located near the exhaust system. Be careful when removing the check/fill plug and when adding fluid to the transmission.

A slight amount of fluid dripping out is good. Check the temperature of the transmission fluid (see Step 2 in this Section). The manufacturer recommends that the transmission fluid temperature should be between 90 and 122 degrees F. If no transmission fluid came out when removing the fill/check plug and the transmission temperature is within range, add the specified automatic transmission fluid until fluid reaches the bottom of the hole and fluid starts to come out.

11 Install the check/fill plug and tighten it to the Specifications listed in this Chapter.

MODELS WITH DIPSTICK/FILL PLUG ASSEMBLY (6R75 AND 6R80)

▶ Refer to illustrations 7.15a and 7.15b

→ Note: These models are equipped with a dipstick/fill plug assembly. There is a plug portion and a dipstick portion. After unscrewing the plug from the transmission, the plug and dipstick are separated so that the dipstick portion can be reinserted back into fill hole to check the fluid level.

→ Note: The transmission fluid level check on these models can be performed every 30,000 miles or if a fluid leak is suspected. However, we recommend regular visual inspections for fluid leaks on vehicle models with these types of transmissions.

12 Secure the vehicle on level ground or place it on jackstands if necessary. The vehicle must be level.

13 With the transmission in Park and the engine idling, remove the dipstick/fill plug assembly.

✳✳ WARNING:

The transmission dipstick/fill plug assembly is located near the exhaust system. Be careful when removing the dipstick/fill plug assembly and when adding fluid to the transmission.

14 Check the temperature of the transmission fluid (see Step 2 in this Section). The manufacturer recommends that the transmission fluid temperature should be between 176 and 185 degrees F.

15 Separate the plug from the dipstick portion (see illustration), then insert the dipstick into the fill hole. The transmission fluid level must be at the upper level of the crosshatch mark (see illustration).

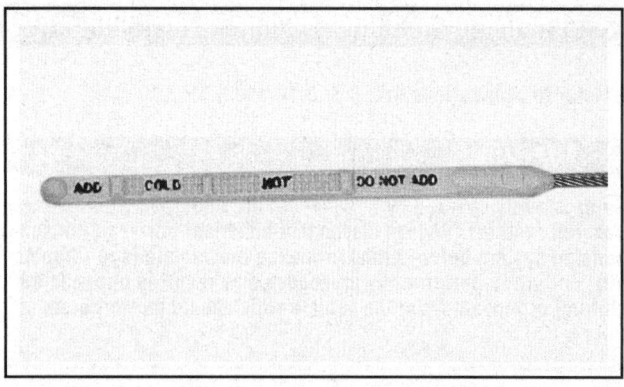

7.6 Check the fluid with the transmission at normal operating temperature - the level should be kept in the HOT range in the cross-hatched area (don't add fluid if the level is anywhere in the cross-hatched area)

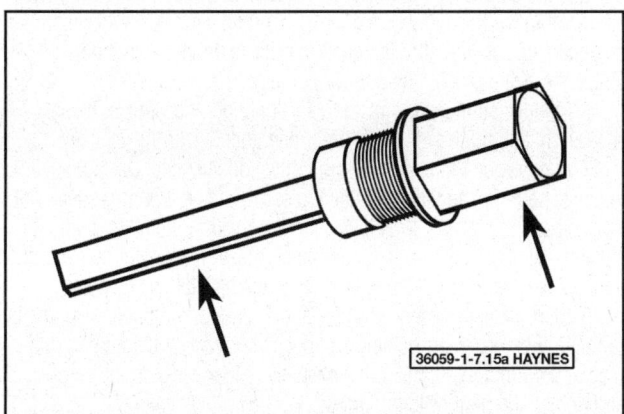

7.15a After the dipstick/fill plug assembly is removed from the transmission, separate the dipstick portion from the plug

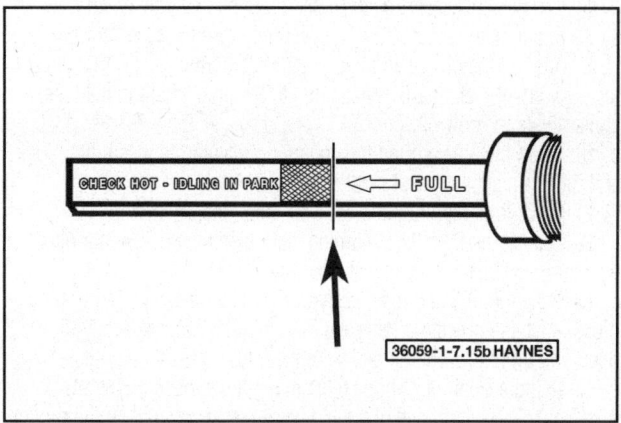

7.15b The transmission fluid level should be within the crosshatched area, but not above it

1-12 TUNE-UP AND ROUTINE MAINTENANCE

8 Engine oil and filter change (every 3000 miles or 3 months)

♦ Refer to illustrations 8.2, 8.7, 8.12 and 8.15

※ WARNING:

Some models covered by this manual are equipped with air suspension systems. Always disconnect electrical power to the suspension system before lifting or towing the vehicle (see Chapter 10). Failure to perform this procedure may result in unexpected shifting or movement of the vehicle which could cause personal injury.

1 Frequent oil changes are the most important preventive maintenance procedures that can be done by the home mechanic. As engine oil ages, in becomes diluted and contaminated, which leads to premature engine wear.

2 Make sure that you have all the necessary tools before you begin this procedure (see illustration). You should also have plenty of rags or newspapers handy for mopping up oil spills.

3 Access to the oil drain plug and filter will be improved if the vehicle can be lifted on a hoist, driven onto ramps or supported by jackstands.

※ WARNING:

Do not work under a vehicle supported only by a jack - always use jackstands!

4 If you haven't changed the oil on this vehicle before, get under it and locate the oil drain plug and the oil filter. The exhaust components will be warm as you work, so note how they are routed to avoid touching them when you are under the vehicle.

5 Start the engine and allow it to reach normal operating temperature - oil and sludge will flow out more easily when warm. If new oil, a filter or tools are needed, use the vehicle to go get them and warm up the engine/oil at the same time. Park on a level surface and shut off the engine when it's warmed up. Remove the oil filler cap from the valve cover.

6 Raise the vehicle and support it on jackstands. Make sure it is safely supported!

7 Being careful not to touch the hot exhaust components, position a drain pan under the plug in the bottom of the engine, then remove the plug (see illustration). It's a good idea to wear a rubber glove while unscrewing the plug the final few turns to avoid being scalded by hot oil.

8 It may be necessary to move the drain pan slightly as oil flow slows to a trickle. Inspect the old oil for the presence of metal particles.

9 After all the oil has drained, wipe off the drain plug with a clean rag. Any small metal particles clinging to the plug would immediately contaminate the new oil.

10 Clean the area around the drain plug opening, reinstall the plug and tighten it securely, but don't strip the threads.

11 Move the drain pan into position under the oil filter.

12 Loosen the oil filter by turning it counterclockwise with a filter wrench (see illustration). Any standard filter wrench will work.

13 Once the filter is loose, use your hands to unscrew it from the block. Just as the filter is detached from the block, immediately tilt the open end up to prevent the oil inside the filter from spilling out.

14 Using a clean rag, wipe off the mounting surface on the block. Also, make sure that none of the old gasket remains stuck to the mounting surface. It can be removed with a scraper if necessary.

15 Compare the old filter with the new one to make sure they are the

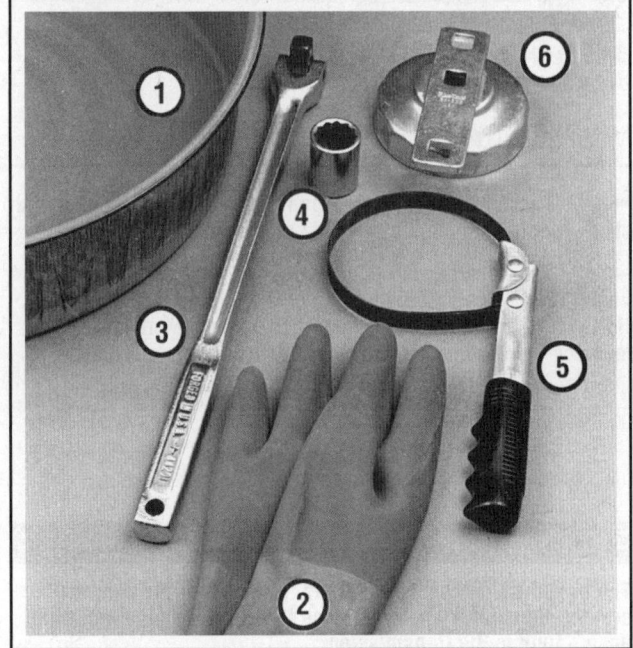

8.2 These tools are required when changing the engine oil and filter

1 **Drain pan** - It should be fairly shallow in depth, but wide to prevent spills
2 **Rubber gloves** - When removing the drain plug and filter, you will get oil on your hands (the gloves will prevent burns)
3 **Breaker bar** - Sometimes the oil drain plug is tight, and a long breaker bar is needed to loosen it
4 **Socket** - To be used with the breaker bar or a ratchet (must be the correct size to fit the drain plug - six-point preferred)
5 **Filter wrench** - This is a metal band-type wrench, which requires clearance around the filter to be effective
6 **Filter wrench** - This type fits on the bottom of the filter and can be turned with a ratchet or breaker bar (different-size wrenches are available for different types of filters)

same type. Smear some engine oil on the rubber gasket of the new filter and screw it into place (see illustration). Overtightening the filter will damage the gasket, so don't use a filter wrench. Most filter manufacturers recommend tightening the filter by hand only. Normally they should be tightened 3/4-turn after the gasket contacts the block, but be sure to follow the directions on the filter or container.

16 Remove all tools and materials from under the vehicle, being careful not to spill the oil in the drain pan, then lower the vehicle.

17 Add new oil to the engine through the oil filler cap. Use a funnel to prevent oil from spilling onto the top of the engine. Pour four quarts of fresh oil into the engine. Wait a few minutes to allow the oil to drain into the pan, then check the level on the dipstick (see Section 4 if necessary). If the oil level is in the OK range (hatched area), install the filler cap.

18 Start the engine and run it for about a minute. While the engine is running, look under the vehicle and check for leaks at the oil pan drain plug and around the oil filter. If either one is leaking, stop the engine and tighten the plug or filter slightly.

19 Wait a few minutes, then recheck the level on the dipstick. Add oil as necessary to bring the level into the OK range.

20 During the first few trips after an oil change, make it a point to

TUNE-UP AND ROUTINE MAINTENANCE

8.7 Use a proper size wrench or socket to remove the oil drain plug and avoid rounding it off

8.12 The oil filter is usually on very tight and will require a special oil filter wrench to remove it - DO NOT use the wrench to tighten the new filter

8.15 Lubricate the oil filter gasket with clean engine oil before installing the filter on the engine

check frequently for leaks and proper oil level.

21 The old oil drained from the engine cannot be reused in its present state and should be discarded. Oil reclamation centers, auto repair shops and gas stations will normally accept the oil, which can be recycled. After the oil has cooled, it can be drained into a container (plastic jugs, bottles, milk cartons, etc.) for transport to a disposal site.

OIL LIFE INDICATOR RESETTING (LATE MODELS ONLY)

Models with message centers

22 Press the SETUP button on the message center until it displays "OIL LIFE = XXX% HOLD RESET = NEW."

Models without message centers

➡ Note: On these models, all steps utilize the single button on the instrument cluster.

23 Press the instrument cluster button to scroll through the menu until "HOLD RESET FOR SETUP MENU" is displayed.
24 Press and hold the button until the display reads "HOLD RESET FOR SYSTEM CHECK."
25 Press the button once to display "OIL LIFE = XXX% HOLD RESET = NEW."

All models

26 Press and hold the RESET/instrument cluster button, releasing the button when the display reads "OIL LIFE SET TO 100%."
27 To exit the resetting procedure, press the SETUP button once.

9 Chassis lubrication (every 3000 miles or 3 months)

♦ Refer to illustrations 9.1, 9.2, 9.9, 9.10 and 9.11

✴✴ WARNING:

Some models covered by this manual are equipped with air suspension systems. Always disconnect electrical power to the suspension system before lifting or towing the vehicle (see Chapter 10). Failure to perform this procedure may result in unexpected shifting or movement of the vehicle which could cause personal injury.

1 Refer to *Recommended lubricants and fluids* at the end of this Chapter to obtain the necessary grease, etc. You will also need a grease gun (see illustration). Occasionally plugs will be installed rather than grease fittings. If so, grease fittings will have to be purchased and installed.

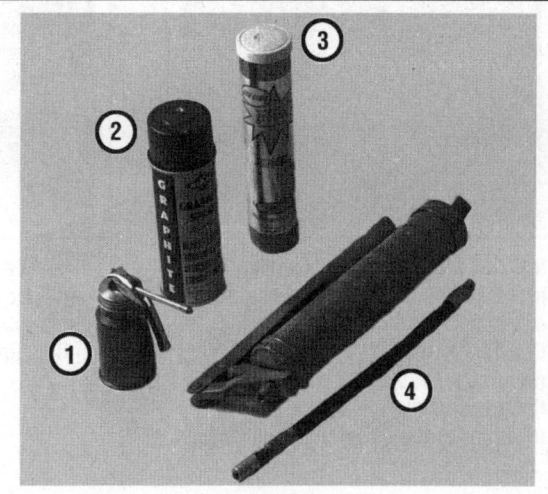

9.1 Materials required for chassis and body lubrication

1 **Engine oil** - Light engine oil in a can like this can be used for door and hood hinges
2 **Graphite spray** - Used to lubricate lock cylinders
3 **Grease** - Grease, in a variety of types and weights, is available for use in a grease gun. Check the Specifications for your requirements
4 **Grease gun** - A common grease gun, shown here with a detachable hose and nozzle, is needed for chassis lubrication. After use, clean it thoroughly!

9.2 Front suspension grease fitting locations (arrows) - upper balljoint (A) and lower balljoint (B) grease fittings not visible (2WD shown)

2 Look under the vehicle for grease fittings or plugs on the steering, suspension, and driveline components (see illustration). They are normally found on the balljoints, tie-rod ends and universal joints. If there are plugs, remove them and install grease fittings, which will thread into the component. An automotive parts store will be able to supply the correct fittings. Straight, as well as angled, fittings are available.

3 For easier access under the vehicle, raise it with a jack and place jackstands under the frame. Make sure it is safely supported by the stands. If the wheels are to be removed at this interval for tire rotation or brake inspection, loosen the lug nuts slightly while the vehicle is still on the ground.

4 Before beginning, force a little grease out of the nozzle to remove any dirt from the end of the gun. Wipe the nozzle clean with a rag.

5 With the grease gun and plenty of clean rags, crawl under the vehicle and begin lubricating the components.

6 Wipe the balljoint grease fitting clean and push the nozzle firmly over it. Squeeze the trigger on the grease gun to force grease into the component. The balljoints should be lubricated until the rubber seal is firm to the touch. Do not pump too much grease into the fittings as it could rupture the seal. For all other suspension and steering components, continue pumping grease into the fitting until it oozes out of the joint between the two components. If it escapes around the grease gun nozzle, the fitting is clogged or the nozzle is not completely seated on the fitting. Resecure the gun nozzle to the fitting and try again. If necessary, replace the fitting with a new one.

7 Wipe the excess grease from the components and the grease fitting. Repeat the procedure for the remaining fittings.

8 On models equipped with an automatic transmission lubricate the shift linkage with a little clean engine oil.

9 On 4WD models, lubricate the front driveshaft Constant Velocity (CV) joint, located at the transfer case end, using a special needle-like adapter on the grease gun (see illustration). Lubricate the transfer case shift mechanism contact surfaces with clean engine oil.

10 Lubricate the driveshaft slip-joints (see illustration).

➡**Note: It may be necessary to disassemble the driveshaft to lubricate the slip joint (see Chapter 8).**

11 Lubricate conventional universal joints until grease can be seen coming out of the contact points (see illustration).

12 While you are under the vehicle, clean and lubricate the parking brake cable along with the cable guides and levers. This can be done by smearing some chassis grease onto the cable and its related parts with your fingers.

13 Open the hood and smear a little chassis grease on the hood latch mechanism. Have an assistant pull the hood release lever from inside the vehicle as you lubricate the cable at the latch.

14 Lubricate all the hinges (door, hood, etc.) with engine oil to keep them in proper working order.

15 The key lock cylinders can be lubricated with spray-on graphite or silicone lubricant, which is available at auto parts stores.

16 Lubricate the door weatherstripping with silicone spray. This will reduce chafing and retard wear.

9.9 In addition to the conventional universal joints at each end of the driveshaft, 4WD models have a Constant Velocity (CV) joint which requires a needle-like adapter on the grease gun - the arrows show the locations of the CV joint grease fittings

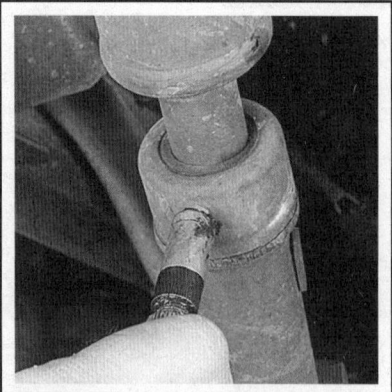

9.10 The slip joint grease fitting (if equipped) is located on the collar - pump grease into it until it comes out of the slip joint seal

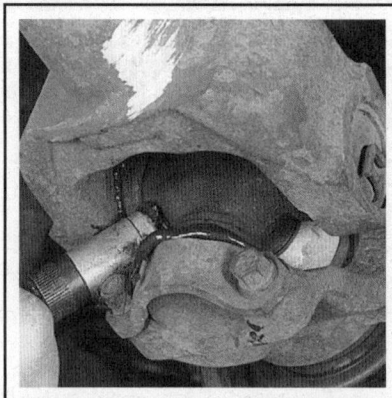

9.11 The universal joint grease fitting (if equipped) is located between the bearing caps on the main body of the joint - pump grease into it until it comes out of the contact surfaces

TUNE-UP AND ROUTINE MAINTENANCE

10 Battery check, maintenance and charging (every 6000 miles or 6 months)

▸ Refer to illustrations 10.1, 10.6a, 10.6b, 10.7a and 10.7b

> **※ WARNING:**
> Certain precautions must be followed when checking and servicing the battery. Hydrogen gas, which is highly flammable, is always present in the battery cells, so keep lighted tobacco and all other open flames and sparks away from the battery. The electrolyte inside the battery is actually dilute sulfuric acid, which will cause injury if splashed on your skin or in your eyes. It will also ruin clothes and painted surfaces. When removing the battery cables, always detach the negative cable first and hook it up last!

1 A routine preventive maintenance program for the battery in your vehicle is the only way to ensure quick and reliable starts. But before performing any battery maintenance, make sure that you have the proper equipment necessary to work safely around the battery (see illustration).

2 There are also several precautions that should be taken whenever battery maintenance is performed. Before servicing the battery, always turn the engine and all accessories off and disconnect the cable from the negative terminal of the battery.

3 The battery produces hydrogen gas, which is both flammable and explosive. Never create a spark, smoke or light a match around the battery. Always charge the battery in a ventilated area.

4 Electrolyte contains poisonous and corrosive sulfuric acid. Do not allow it to get in your eyes, on your skin or your clothes. Never ingest it. Wear protective safety glasses when working near the battery. Keep children away from the battery.

5 Note the external condition of the battery. If the positive terminal and cable clamp on your vehicle's battery is equipped with a rubber protector, make sure that it's not torn or damaged. It should completely cover the terminal. Look for any corroded or loose connections, cracks in the case or cover or loose hold-down clamps. Also check the entire length of each cable for cracks and frayed conductors.

6 If corrosion, which looks like white, fluffy deposits (see illustration) is evident, particularly around the terminals, the battery should be removed for cleaning. Loosen the cable clamp bolts with a wrench, being careful to remove the ground cable first, and slide them off the terminals (see illustration). Then disconnect the hold-down clamp bolt and nut, remove the clamp and lift the battery from the engine compartment.

7 Clean the cable clamps thoroughly with a battery brush or a

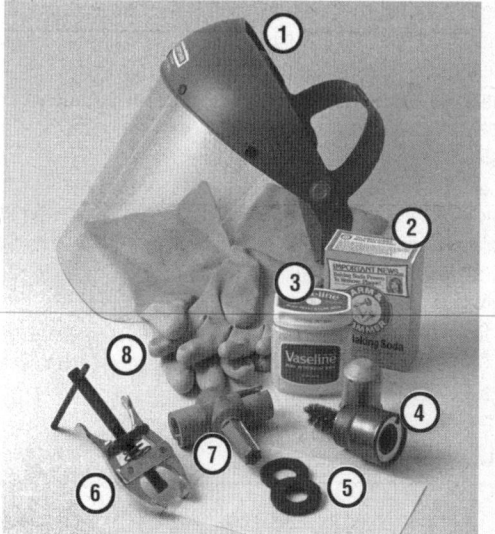

10.1 Tools and materials required for battery maintenance

1 *Face shield/safety goggles* - When removing corrosion with a brush, the acidic particles can easily fly up into your eyes
2 *Baking soda* - A solution of baking soda and water can be used to neutralize corrosion
3 *Petroleum jelly* - A layer of this on the battery posts will help prevent corrosion
4 *Battery post/cable cleaner* - This wire brush cleaning tool will remove all traces of corrosion from the battery posts and cable clamps
5 *Treated felt washers* - Placing one of these on each post, directly under the cable clamps, will help prevent corrosioan
6 *Puller* - Sometimes the cable clamps are very difficult to pull off the posts, even after the nut/bolt has been completely loosened. This tool pulls the clamp straight up and off the post without damage
7 *Battery post/cable cleaner* - Here is another cleaning tool which is a slightly different version of Number 4 above, but it does the same thing
8 *Rubber gloves* - Another safety item to consider when servicing the battery; remember that's acid inside the battery!

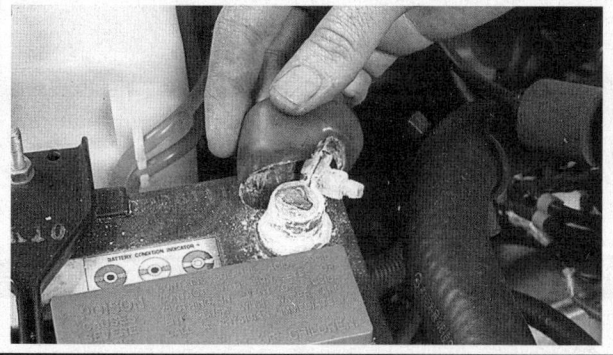

10.6a Battery terminal corrosion usually appears as light, fluffy powder

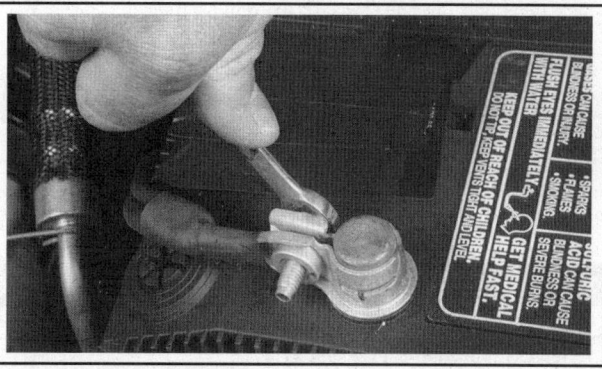

10.6b Removing the cable from a battery post with a wrench - sometimes special battery pliers are required for this procedure if corrosion has caused deterioration of the nut hex (always remove the ground cable first and hook it up last!)

1-16 TUNE-UP AND ROUTINE MAINTENANCE

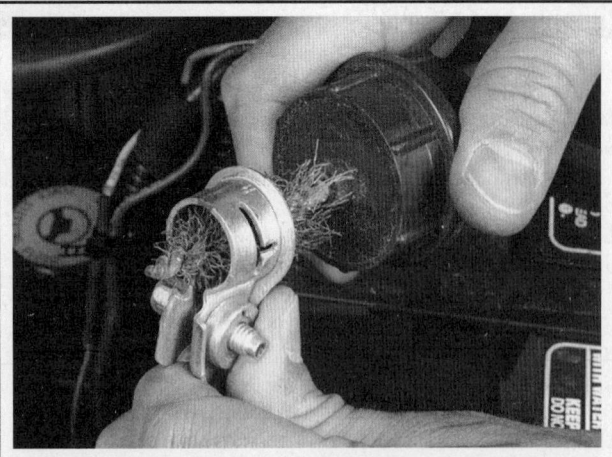

10.7a When cleaning the cable clamps, all corrosion must be removed (the inside of the clamp is tapered to match the taper on the post, so don't remove too much material)

10.7b Regardless of the type of tool used on the battery posts, a clean, shiny surface should be the result

terminal cleaner and a solution of warm water and baking soda (see illustration). Wash the terminals and the top of the battery case with the same solution but make sure that the solution doesn't get into the battery When cleaning the cables, terminals and battery top, wear safety goggles and rubber gloves to prevent any solution from coming in contact with your eyes or hands. Wear old clothes too - even diluted, sulfuric acid splashed onto clothes will burn holes in them. If the terminals have been extensively corroded, clean them up with a terminal cleaner (see illustration). Thoroughly wash all cleaned areas with plain water.

8 Make sure that the battery tray is in good condition and the hold-down clamp bolts are tight. If the battery is removed from the tray, make sure no parts remain in the bottom of the tray when the battery is reinstalled. When reinstalling the hold-down clamp bolts, do not overtighten them.

9 Information on removing and installing the battery can be found in Chapter 5. Information on jump starting can be found at the front of this manual.

CLEANING

10 Corrosion on the hold-down components, battery case and surrounding areas can be removed with a solution of water and baking soda. Thoroughly rinse all cleaned areas with plain water.

11 Any metal parts of the vehicle damaged by corrosion should be covered with a zinc-based primer, then painted.

CHARGING

❋❋ WARNING:

When batteries are being charged, hydrogen gas, which is very explosive and flammable, is produced. Do not smoke or allow open flames near a charging or a recently charged battery. Wear eye protection when near the battery during charging. Also, make sure the charger is unplugged before connecting or disconnecting the battery from the charger.

12 Slow-rate charging is the best way to restore a battery that's discharged to the point where it will not start the engine. It's also a good way to maintain the battery charge in a vehicle that's only driven a few miles between starts. Maintaining the battery charge is particularly important in the winter when the battery must work harder to start the engine and electrical accessories that drain the battery are in greater use.

13 It's best to use a one or two-amp battery charger (sometimes called a "trickle" charger). They are the safest and put the least strain on the battery. They are also the least expensive. For a faster charge, you can use a higher amperage charger, but don't use one rated more than 1/10th the amp/hour rating of the battery. Rapid boost charges that claim to restore the power of the battery in one to two hours are hardest on the battery and can damage batteries not in good condition. This type of charging should only be used in emergency situations.

14 The average time necessary to charge a battery should be listed in the instructions that come with the charger. As a general rule, a trickle charger will charge a battery in 12 to 16 hours.

11 Windshield wiper blade inspection and replacement (every 6000 miles or 6 months)

▸ Refer to illustrations 11.4 and 11.5

1 The windshield wiper and blade assembly should be inspected periodically for damage, loose components and cracked or worn blade elements.

2 Road film can build up on the wiper blades and affect their efficiency, so they should be washed regularly with a mild detergent solution.

3 If the wiper blade elements are cracked, worn or warped, or no longer clean adequately, they should be replaced with new ones.

TUNE-UP AND ROUTINE MAINTENANCE   1-17

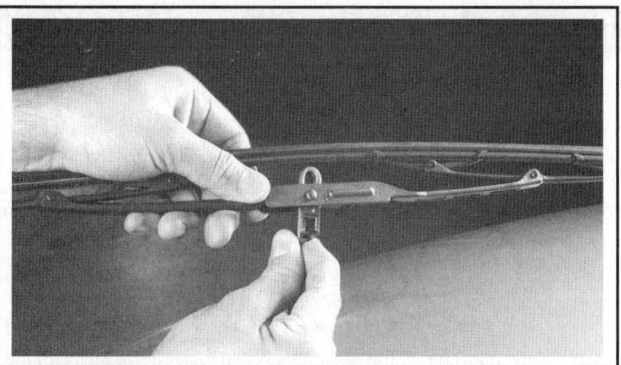

11.4 Press on the release tab and push the blade assembly down out of the hook in the arm

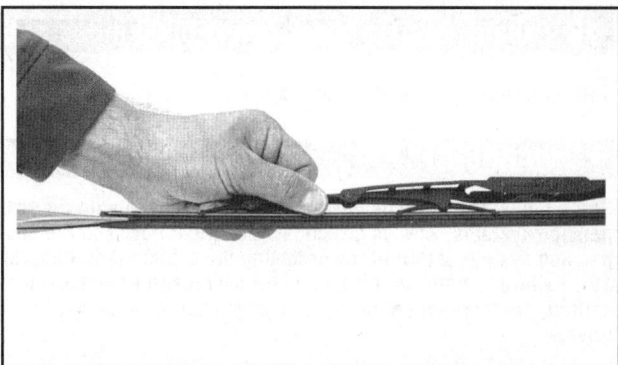

11.5 Use needle-nose pliers to compress the rubber element, then slide the element out - slide the new element in and lock the blade assembly fingers into the notches of the wiper element

4 Lift the arm assembly away from the glass for clearance, press on the release lever, then slide the wiper blade assembly out of the hook in the end of the arm (see illustration).

5 Use needle-nose pliers to compress the blade element, then slide the element out of the frame and discard it (see illustration).
6 Installation is the reverse of removal.

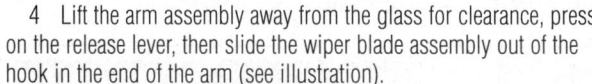

12 Tire rotation (every 6000 miles or 6 months)

♦ Refer to illustrations 12.2a, 12.2b and 12.2c

✶✶ WARNING:

Some models covered by this manual are equipped with air suspension systems. Always disconnect electrical power to the suspension system before lifting or towing the vehicle (see Chapter 10). Failure to perform this procedure may result in unexpected shifting or movement of the vehicle which could cause personal injury.

1 The tires should be rotated at the specified intervals and whenever uneven wear is noticed. Since the vehicle will be raised and the tires removed anyway, check the brakes also (see Section 20).

2 Radial tires must be rotated in a specific pattern (see illustrations). If your vehicle has a compact spare tire, don't include it in the rotation pattern.

3 Refer to the information in *Jacking and towing* at the front of this manual for the proper procedure to follow when raising the vehicle and changing a tire. If the brakes must be checked, don't apply the parking brake as stated.

4 The vehicle must be raised on a hoist or supported on jackstands to get all four wheels off the ground. Make sure the vehicle is safely supported!

5 After the rotation procedure is finished, check and adjust the tire pressures as necessary and be sure to check the lug nut tightness.

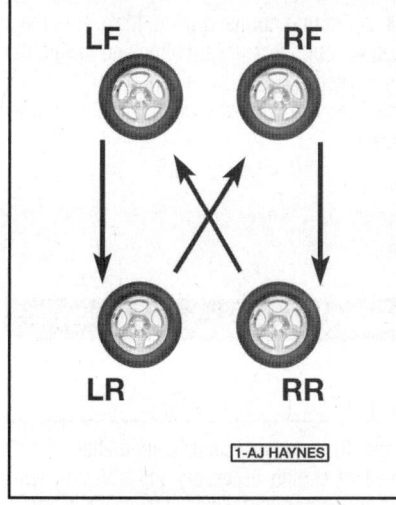

12.2a The recommended tire rotation pattern for early model *non-directional* tires

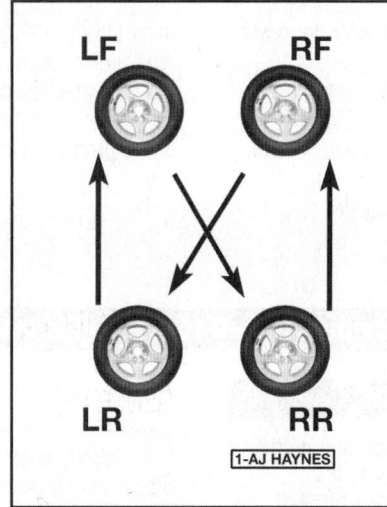

12.2b The recommended tire rotation pattern for later model *non-directional* tires

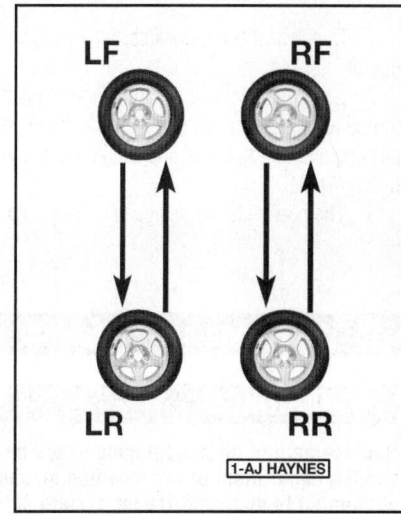

12.2c The recommended tire rotation pattern for *directional* tires

1-18 TUNE-UP AND ROUTINE MAINTENANCE

13 Exhaust system check (every 6000 miles or 6 months)

▶ Refer to illustrations 13.2a and 13.2b

WARNING:
Some models covered by this manual are equipped with air suspension systems. Always disconnect electrical power to the suspension system before lifting or towing the vehicle (see Chapter 10). Failure to perform this procedure may result in unexpected shifting or movement of the vehicle which could cause personal injury.

1 With the engine cold (at least three hours after the vehicle has been driven), check the complete exhaust system from the engine to the end of the tailpipe. Ideally, the inspection should be done with the vehicle on a hoist to permit unrestricted access. If a hoist isn't available, raise the vehicle and support it securely on jackstands.

2 Check the exhaust pipes and connections for evidence of leaks, severe corrosion and damage. Make sure that all brackets and hangers are in good condition and tight (see illustrations).

3 At the same time, inspect the underside of the body for holes, corrosion, open seams, etc. which may allow exhaust gases to enter the passenger compartment. Seal all body openings with silicone or body putty.

4 Rattles and other noises can often be traced to the exhaust system, especially the mounts and hangers. Try to move the pipes, muffler and catalytic converter. If the components can come in contact with the body or suspension parts, secure the exhaust system with new mounts.

5 Check the running condition of the engine by inspecting inside the end of the tailpipe. The exhaust deposits here are an indication of engine state-of-tune. If the pipe is black and sooty or coated with white deposits, the engine may need a tune-up, including a thorough fuel system inspection and adjustment.

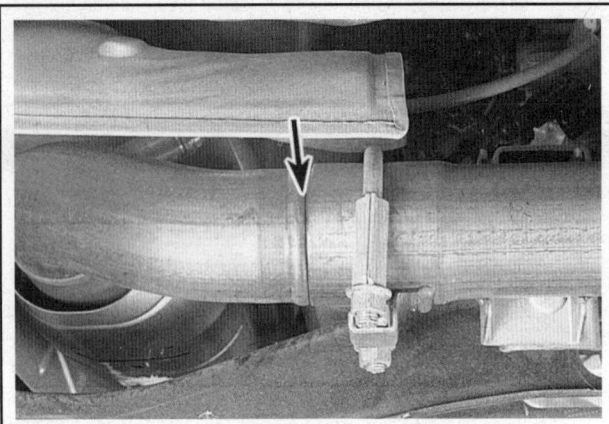

13.2a Check the connections for exhaust leaks - also check that the clamp retaining nuts are securely tightened

13.2b Check the exhaust system hangers for damage and cracks

14 Seat belt check (every 6000 miles or 6 months)

1 Check seat belts, buckles, latch plates and guide loops for obvious damage and signs of wear.

2 See if the seat belt reminder light comes on when the key is turned to the Run or Start position. A chime should also sound. On passive restraint systems, the shoulder belt should move into position in the A-pillar.

3 The seat belts are designed to lock up during a sudden stop or impact, yet allow free movement during normal driving. Make sure the retractors return the belt against your chest while driving and rewind the belt fully when the buckle is unlatched.

4 If any of the above checks reveal problems with the seat belt system, replace parts as necessary.

15 Underhood hose check and replacement (every 15,000 miles or 12 months)

WARNING:
Replacement of air conditioning hoses must be left to a dealer service department or air conditioning shop that has the equipment to depressurize the system safely. Never remove air conditioning components or hoses until the system has been depressurized.

GENERAL

1 High temperatures under the hood can cause deterioration of the rubber and plastic hoses used for engine, accessory and emission systems operation. Periodic inspection should be made for cracks, loose clamps, material hardening and leaks.

2 Information specific to the cooling system hoses can be found in Section 16

TUNE-UP AND ROUTINE MAINTENANCE   1-19

3 Most (but not all) hoses are secured to the fittings with clamps. Where clamps are used, check to be sure they haven't lost their tension, allowing the hose to leak. If clamps aren't used, make sure the hose has not expanded and/or hardened where it slips over the fitting, allowing it to leak.

PCV SYSTEM HOSE

4 To reduce hydrocarbon emissions, crankcase blow-by gas is vented through the PCV valve in the rocker arm cover to the intake manifold via a rubber hose on most models. The blow-by gases mix with incoming air in the intake manifold before being burned in the combustion chambers.

5 Check the PCV hose for cracks, leaks and other damage. Disconnect it from the valve cover and the intake manifold and check the inside for obstructions. If it's clogged, clean it out with solvent.

VACUUM HOSES

6 It's quite common for vacuum hoses, especially those in the emissions system, to be color coded or identified by colored stripes molded into them. Various systems require hoses with different wall thickness, collapse resistance and temperature resistance. When replacing hoses, be sure the new ones are made of the same material.

7 Often the only effective way to check a hose is to remove it completely from the vehicle. If more than one hose is removed, be sure to label the hoses and fittings to ensure correct installation.

8 When checking vacuum hoses, be sure to include any plastic T-fittings in the check. Inspect the fittings for cracks and the hose where it fits over each fitting for distortion, which could cause leakage.

9 A small piece of vacuum hose (1/4-inch inside diameter) can be used as a stethoscope to detect vacuum leaks. Hold one end of the hose to your ear and probe around vacuum hoses and fittings, listening for the "hissing" sound characteristic of a vacuum leak.

※※ WARNING:

When probing with the vacuum hose stethoscope, be careful not to come into contact with moving engine components such as drivebelts, the cooling fan, etc.

FUEL HOSE

※※ WARNING:

Gasoline is extremely flammable, so take extra precautions when you work on any part of the fuel system. Don't smoke or allow open flames or bare light bulbs near the work area, and don't work in a garage where a gas-type appliance (such as a water heater or clothes dryer) is present. Since gasoline is carcinogenic, wear latex gloves when there's a possibility of being exposed to fuel, and, if you spill any fuel on your skin, rinse it off immediately with soap and water. Mop up any spills immediately and do not store fuel-soaked rags where they could ignite. The fuel system is under constant pressure, so, if any fuel lines are to be disconnected, the fuel pressure in the system must be relieved first (see Chapter 4 for more information). When you perform any kind of work on the fuel system, wear safety glasses and have a Class B type fire extinguisher on hand.

10 The fuel lines are usually under pressure, so if any fuel lines are to be disconnected be prepared to catch spilled fuel.

※※ WARNING:

Your vehicle is equipped with fuel injection and you must relieve the fuel system pressure before servicing the fuel lines.

Refer to Chapter 4 for the fuel system pressure relief procedure.

11 Check all flexible fuel lines for deterioration and chafing. Check especially for cracks in areas where the hose bends and just before fittings, such as where a hose attaches to the fuel pump, fuel filter and fuel injection unit.

12 When replacing a hose, use only hose that is specifically designed for your fuel injection system.

13 Spring-type clamps are sometimes used on fuel return or vapor lines. These clamps often lose their tension over a period of time, and can be "sprung" during removal. Replace all spring-type clamps with screw clamps whenever a hose is replaced. Some fuel lines use spring-lock type couplings, which require a special tool to disconnect. See Chapter 4 for more information on these type of couplings.

METAL LINES

14 Sections of metal line are often used for fuel line between the fuel pump and the fuel injection unit. Check carefully to make sure the line isn't bent, crimped or cracked.

15 If a section of metal fuel line must be replaced, use seamless steel tubing only, since copper and aluminum tubing do not have the strength necessary to withstand vibration caused by the engine.

16 Check the metal brake lines where they enter the master cylinder and brake proportioning unit (if used) for cracks in the lines and loose fittings. Any sign of brake fluid leakage calls for an immediate thorough inspection of the brake system.

16 Cooling system check (every 15,000 miles or 12 months)

▶ Refer to illustrations 16.4a and 16.4b

1 Many major engine failures can be attributed to a faulty cooling system. The cooling system also plays an important role in prolonging transmission life because it cools the fluid.

2 The engine should be cold for the cooling system check, so perform the following procedure before the vehicle is driven for the day or after it has been shut off for at least three hours.

3 Remove the cap from the expansion tank. Clean the cap thoroughly, inside and out, with clean water. Also clean the filler neck on the expansion tank. The presence of rust or corrosion in the filler neck means the coolant should be changed (see Section 25). The coolant inside the radiator should be relatively clean and transparent. If it's rust colored, drain the system and refill it with new coolant.

4 Carefully check the radiator hoses and the smaller diameter heater

1-20 TUNE-UP AND ROUTINE MAINTENANCE

Check for a chafed area that could fail prematurely.

Check for a soft area indicating the hose has deteriorated inside.

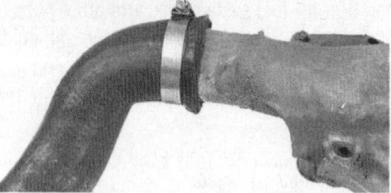

Overtightening the clamp on a hardened hose will damage the hose and cause a leak.

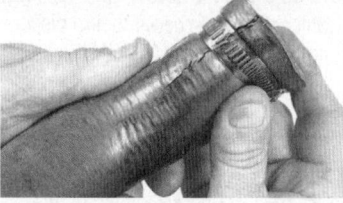

Check each hose for swelling and oil-soaked ends. Cracks and breaks can be located by squeezing the hose.

16.4a Hoses, like drivebelts, have a habit of failing at the worst possible time - to prevent the inconvenience of a blown radiator or heater hose, inspect them carefully as shown here

16.4b A leak in the V8 engine heater hose means the intake manifold will have to be removed for hose replacement - coolant coming out the back of the engine is the symptom

hoses (see illustration). Inspect each coolant hose along its entire length, replacing any hose which is cracked, swollen or deteriorated. Cracks will show up better if the hose is squeezed. Pay close attention to hose clamps that secure the hoses to cooling system components. Hose clamps can pinch and puncture hoses, resulting in coolant leaks. Some hoses are hidden from view so sometimes you'll have to trace a coolant leak. For example, on the 4.6L or 5.4L V8 engine the heater hose connects to the water pump under the intake manifold. If it leaks, coolant will run down the rear of the engine (see illustration).

5 Make sure that all hose connections are tight. A leak in the cooling system will usually show up as white or rust colored deposits on the area adjoining the leak. If wire-type clamps are used on the hoses, it may be a good idea to replace them with screw-type clamps.

6 Clean the front of the radiator and air conditioning condenser with compressed air, if available, or a soft brush. Remove all bugs, leaves, etc. embedded in the radiator fins. Be extremely careful not to damage the cooling fins or cut your fingers on them.

7 If the coolant level has been dropping consistently and no leaks are detectable, have the radiator cap and cooling system pressure checked at a service station.

17 Fuel system check (every 15,000 miles or 12 months)

※※ WARNING 1:

Gasoline is extremely flammable, so take extra precautions when you work on any part of the fuel system. Don't smoke or allow open flames or bare light bulbs near the work area, and don't work in a garage where a gas-type appliance (such as a water heater or clothes dryer) is present. Since gasoline is carcinogenic, wear latex gloves when there's a possibility of being exposed to fuel, and, if you spill any fuel on your skin, rinse it off immediately with soap and water. Mop up any spills immediately and do not store fuel-soaked rags where they could ignite. When you perform any kind of work on the fuel system, wear safety glasses and have a Class B type fire extinguisher on hand. The fuel system is under constant pressure, so, before any lines are disconnected, the fuel system pressure must be relieved. See Chapter 4.

※※ WARNING 2:

Some models covered by this manual are equipped with air suspension systems. Always disconnect electrical power to the suspension system before lifting or towing the vehicle (see Chapter 10). Failure to perform this procedure may result in unexpected shifting or movement of the vehicle which could cause personal injury.

1 If you smell gasoline while driving or after the vehicle has been sitting in the sun, inspect the fuel system immediately.

2 Remove the gas filler cap and inspect if for damage and corrosion. The gasket should have an unbroken sealing imprint. If the gasket is damaged or corroded, install a new cap.

3 Inspect the fuel feed and return lines for cracks. Make sure that

TUNE-UP AND ROUTINE MAINTENANCE 1-21

the connections between the fuel lines and the fuel injection system and between the fuel lines and the in-line fuel filter are tight.

※※ WARNING:

Your vehicle is fuel injected, so you must relieve the fuel system pressure before servicing fuel system components. The fuel system pressure relief procedure is outlined in Chapter 4.

4 Since some components of the fuel system - the fuel tank and part of the fuel feed and return lines, for example - are underneath the vehicle, they can be inspected more easily with the vehicle raised on a hoist. If that's not possible, raise the vehicle and support it on jackstands.

5 With the vehicle raised and safely supported, inspect the gas tank and filler neck for punctures, cracks and other damage. The connection between the filler neck and the tank is particularly critical. Sometimes a rubber filler neck will leak because of loose clamps or deteriorated rubber. Inspect all fuel tank mounting brackets and straps to be sure that the tank is securely attached to the vehicle.

※※ WARNING:

Do not, under any circumstances, try to repair a fuel tank (except rubber components). A welding torch or any open flame can easily cause fuel vapors inside the tank to explode.

6 Carefully check all rubber hoses and metal lines leading away from the fuel tank. Check for loose connections, deteriorated hoses, crimped lines and other damage. Repair or replace damaged sections as necessary (see Chapter 4).

18 Fuel filter replacement (every 15,000 miles or 12 months)

▸ Refer to illustration 18.3a and 18.3b

※※ WARNING 1:

Gasoline is extremely flammable, so take extra precautions when you work on any part of the fuel system. Don't smoke or allow open flames or bare light bulbs near the work area, and don't work in a garage where a gas-type appliance (such as a water heater or clothes dryer) is present. Since gasoline is carcinogenic, wear latex gloves when there's a possibility of being exposed to fuel, and, if you spill any fuel on your skin, rinse it off immediately with soap and water. Mop up any spills immediately and do not store fuel-soaked rags where they could ignite. When you perform any kind of work on the fuel system, wear safety glasses and have a Class B type fire extinguisher on hand.

※※ WARNING 2:

Some models covered by this manual are equipped with air suspension systems. Always disconnect electrical power to the suspension system before lifting or towing the vehicle (see Chapter 10). Failure to perform this procedure may result in unexpected shifting or movement of the vehicle which could cause personal injury.

➡ **Note:** This procedure requires a special fuel line disconnect tool which is available at most auto part stores.

➡ **Note:** This procedure does not apply to 2009 short wheel base models or 2010 and later models. On those models, the fuel filter is integral with the fuel pump module, and must be replaced as an assembly (see Chapter 4).

1 The fuel filter is mounted under the vehicle on the inside of the left frame rail. On later Expedition/Navigator models, there is a plastic protective cover over the fuel filter; twist the fasteners one-quarter turn counterclockwise to remove the cover.

2 Inspect the hose fittings at both ends of the filter to see if they're clean. If more than a light coating of dust is present, clean the fittings before proceeding.

3 Relieve the fuel system pressure (see Chapter 4). Removal of the fuel lines from the fuel filter is a two-stage procedure. First, detach the safety clips from the inlet and outlet lines. Second install the special fuel line disconnect tool onto the fuel filter inlet port, then push the disconnect tool into the fuel line coupling until it releases itself from the fuel filter (see illustrations). Repeat this procedure on the outlet port.

4 After both couplings have been released, grasp the fuel hoses, one at a time, and pull them straight off the filter. Be prepared for fuel spillage.

5 After the hoses have been detached, check the clips for damage and distortion. If they were damaged in any way during removal, new ones must be used when the hoses are reattached to the new filter (if new clips are packaged with the filter, be sure to use them in place of the originals).

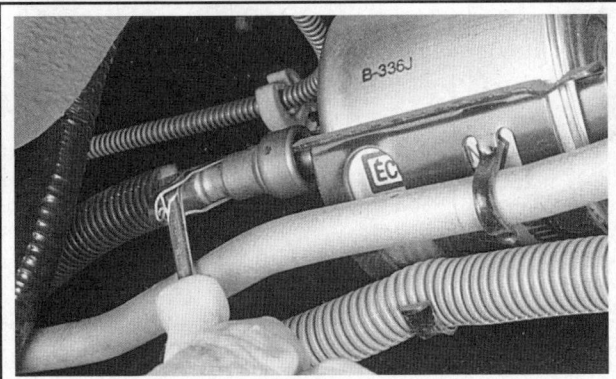

18.3a Detach the safety clips from the inlet and outlet ports of the fuel filter

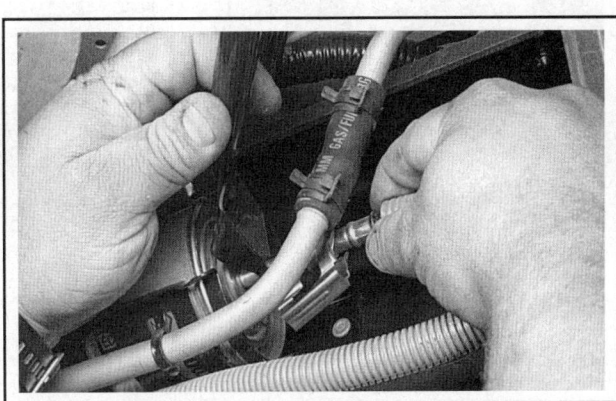

18.3b Install the fuel line disconnect tool onto the fuel filter port, then push the disconnect tool into the coupling until the fuel line releases itself

1-22 TUNE-UP AND ROUTINE MAINTENANCE

6 Use a screwdriver to loosen the fuel filter mounting clamp, while noting the direction the fuel filter is installed.

7 Remove the filter from the bracket and install the new fuel filter in the same direction.

8 Carefully push each hose onto the filter until it's seated against the collar on the fitting, then install the clips. Make sure the clips are securely attached to the hose fittings - if they come off, the hoses could back off the filter and a fire could result!

9 Start the engine and check for fuel leaks.

19 Steering and suspension check (every 15,000 miles or 12 months)

※※ WARNING:

Some models covered by this manual are equipped with air suspension systems. Always disconnect electrical power to the suspension system before lifting or towing the vehicle (see Chapter 10). Failure to perform this procedure may result in unexpected shifting or movement of the vehicle which could cause personal injury.

→Note: *The steering linkage and suspension components should be checked periodically. Worn or damaged suspension and steering linkage components can result in excessive and abnormal tire wear, poor ride quality and vehicle handling and reduced fuel economy. For detailed illustrations of the steering and suspension components, refer to Chapter 10.*

19.6 Check the shocks for leakage at the indicated area (arrow)

SHOCK ABSORBER CHECK

♦ **Refer to illustrations 19.6**

1 Park the vehicle on level ground, turn the engine off and set the parking brake. Check the tire pressures.

2 Push down at one corner of the vehicle, then release it while noting the movement of the body. It should stop moving and come to rest in a level position within one or two bounces.

3 If the vehicle continues to move up-and-down or if it fails to return to its original position, a worn or weak shock absorber is probably the reason.

4 Repeat the above check at each of the three remaining corners of the vehicle.

5 Raise the vehicle and support it securely on jackstands.

6 Check the shock absorbers for evidence of fluid leakage (see illustration). A light film of fluid is no cause for concern. Make sure that any fluid noted is from the shocks and not from some other source. If leakage is noted, replace the shocks as a set.

7 Check the shocks to be sure that they are securely mounted and undamaged. Check the upper mounts for damage and wear. If damage or wear is noted, replace the shocks as a set (front or rear).

8 If the shocks must be replaced, refer to Chapter 10 for the procedure.

STEERING AND SUSPENSION CHECK

♦ **Refer to illustrations 19.9a, 19.9b, 19.9c and 19.11**

9 Visually inspect the steering and suspension components for damage and distortion. Look for damaged seals, boots and bushings and leaks of any kind (see illustrations).

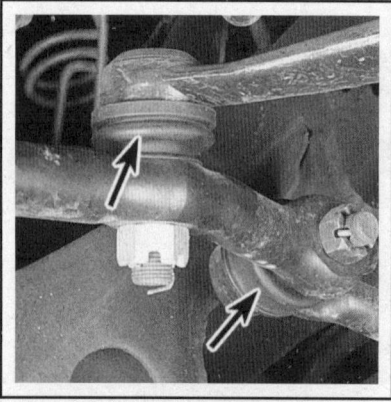

19.9a Inspect the steering and suspension components for torn grease seals

19.9b If equipped with 4WD, check the front driveaxle boots for cracks and/or leaking grease

19.9c Check the stabilizer bar link bushings (arrows) for deterioration at the front and the rear of the vehicle

TUNE-UP AND ROUTINE MAINTENANCE

10 Clean the lower end of the steering knuckle. Have an assistant grasp the lower edge of the tire and move the wheel in-and-out while you look for movement at the steering knuckle-to-control arm balljoint. If there is any movement the suspension balljoint(s) must be replaced.

11 Grasp each front tire at the front and rear edges, push in at the front, pull out at the rear and feel for play in the steering system components If any freeplay is noted, check the idler arm and the tie-rod ends for looseness (see illustration).

12 Additional steering and suspension system information and illustrations can be found in Chapter 10.

19.11 With the steering wheel locked and the vehicle raised, grasp the front tire as shown and try to move it back-and-forth - if any play is noted, check the idler arm and tie-rod ends for looseness

20 Brake check (every 15,000 miles or 12 months)

✱✱ WARNING 1:

The dust created by the brake system is harmful to your health. Never blow it out with compressed air and don't inhale any of it. An approved filtering mask should be worn when working on the brakes. Do not, under any circumstances, use petroleum-based solvents to clean brake parts. Use brake system cleaner only! Try to use non-asbestos replacement parts whenever possible.

✱✱ WARNING 2:

Some models covered by this manual are equipped with air suspension systems. Always disconnect electrical power to the suspension system before lifting or towing the vehicle (see Chapter 10). Failure to perform this procedure may result in unexpected shifting or movement of the vehicle which could cause personal injury.

➡Note: For detailed photographs of the brake system, refer to Chapter 9.

1 In addition to the specified intervals, the brakes should be inspected every time the wheels are removed or whenever a defect is suspected.

2 Any of the following symptoms could indicate a potential brake system defect: The vehicle pulls to one side when the brake pedal is depressed; the brakes make squealing or dragging noises when applied; brake pedal travel is excessive; the pedal pulsates; brake fluid leaks, usually onto the inside of the tire or wheel.

3 Loosen the wheel lug nuts.

4 Raise the vehicle and place it securely on jackstands.

5 Remove the wheels (see *Jacking and towing* at the front of this book, or your owner's manual, if necessary).

DISC BRAKES

♦ Refer to illustrations 20.6 and 20.11

6 There are two pads (an outer and an inner) in each caliper. The pads are visible through inspection holes in each caliper (see illustration).

7 Check the pad thickness by looking at each end of the caliper and through the inspection hole in the caliper body. If the lining material is less than the thickness listed in this Chapter's Specifications, replace the pads.

➡Note: Keep in mind that the lining material is riveted or bonded to a metal backing plate and the metal portion is not included in this measurement.

8 If it is difficult to determine the exact thickness of the remaining pad material by the above method, or if you are at all concerned about the condition of the pads, remove the caliper(s), then remove the pads from the calipers for further inspection (refer to Chapter 9).

9 Once the pads are removed from the calipers, clean them with brake cleaner and re-measure them with a ruler or a vernier caliper.

10 Measure the disc thickness with a micrometer to make sure that it still has service life remaining. If any disc is thinner than the specified minimum thickness, replace it (refer to Chapter 9). Even if the disc has service life remaining, check its condition. Look for scoring, gouging and burned spots. If these conditions exist, remove the disc and have it resurfaced (see Chapter 9).

11 Before installing the wheels, check all brake lines and hoses for damage, wear, deformation, cracks, corrosion, leakage, bends and

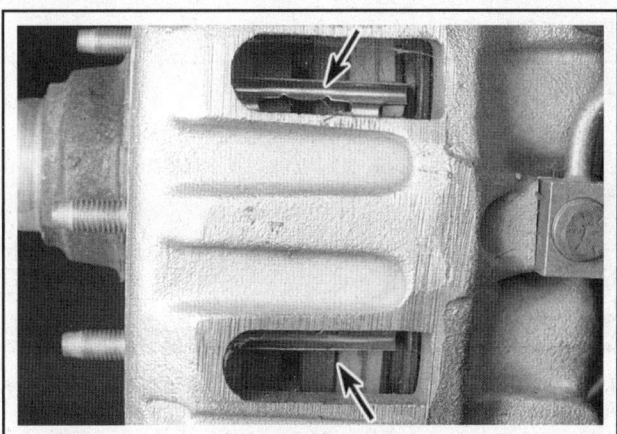

20.6 You will find an inspection hole(s) like this in each caliper - placing a ruler across the hole should enable you to determine the thickness of remaining pad material for the inner and outer pads

1-24 TUNE-UP AND ROUTINE MAINTENANCE

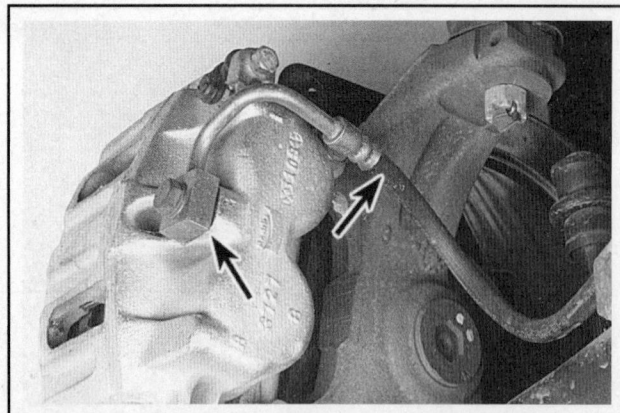

20.11 Check along the brake hoses and at each fitting for deterioration and cracks

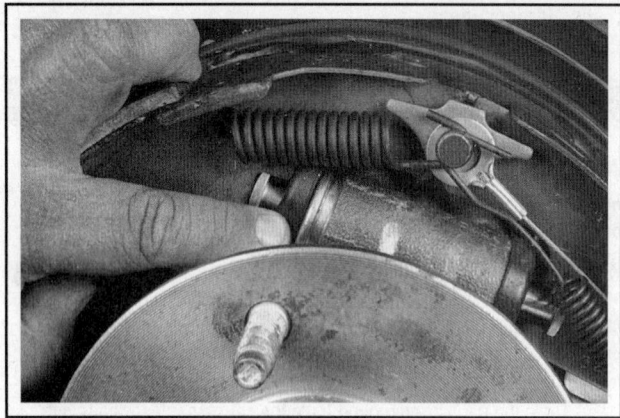

20.17 Check the wheel cylinder boots for leaking fluid indicating that the cylinder must be replaced or rebuilt

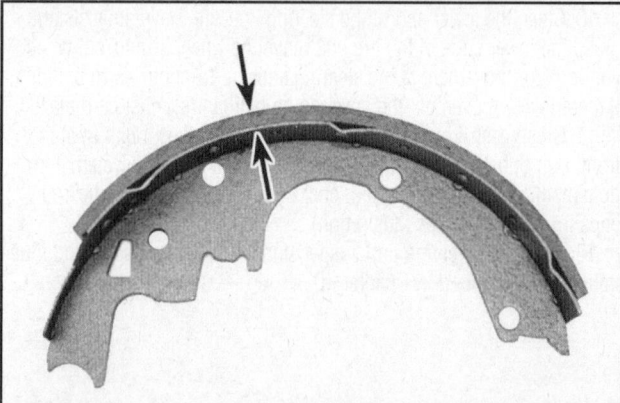

20.15 If the lining is bonded to the brake shoe, measure the lining thickness from the outer surface to the metal shoe, as shown here. If the lining is riveted to the shoe, measure from the lining outer surface to the rivet head

twists, particularly in the vicinity of the rubber hoses at the calipers (see illustration). Check the clamps for tightness and the connections for leakage. Make sure that all hoses and lines are clear of sharp edges, moving parts and the exhaust system. If any of the above conditions are noted, repair, reroute or replace the lines and/or fittings as necessary (see Chapter 9).

DRUM BRAKES

♦ Refer to illustrations 20.15 and 20.17

12 On rear drum brakes, make sure the parking brake is off then proceed to tap on the outside of the drum with a rubber mallet to loosen it.

13 Remove the brake drums.

14 With the drums removed, carefully clean the brake assembly with brake system cleaner.

WARNING:

Don't blow the dust out with compressed air and don't inhale any of it (it is harmful to your health).

15 Note the thickness of the lining material on both front and rear brake shoes. If the material has worn away to within 1/16-inch of the recessed rivets or metal backing, the shoes should be replaced (see illustration). The shoes should also be replaced if they're cracked, glazed (shiny areas), or covered with brake fluid.

16 Make sure all the brake assembly springs are connected and in good condition.

17 Check the brake components for signs of fluid leakage. With your finger or a small screwdriver, carefully pry back the rubber cups on the wheel cylinder located at the top of the brake shoes (see illustration). Any leakage here is an indication that the wheel cylinders should be overhauled immediately (see Chapter 9). Also, check all hoses and connections for signs of leakage.

18 Wipe the inside of the drum with a clean rag and denatured alcohol or brake cleaner. Again, be careful not to breathe the dangerous brake dust.

19 Check the inside of the drum for cracks, score marks, deep scratches and "hard spots" which will appear as small discolored areas. If imperfections cannot be removed with fine emery cloth, the drum must be taken to an automotive machine shop for resurfacing.

20 Repeat the procedure for the remaining wheel. If the inspection reveals that all parts are in good condition, reinstall the brake drums, install the wheels and lower the vehicle to the ground.

BRAKE BOOSTER CHECK

21 Sit in the driver's seat and perform the following sequence of tests.

22 With the brake fully depressed, start the engine - the pedal should move down a little when the engine starts.

23 With the engine running, depress the brake pedal several times - the travel distance should not change.

24 Depress the brake, stop the engine and hold the pedal in for about 30 seconds - the pedal should neither sink nor rise.

25 Restart the engine, run it for about a minute and turn it off. Then firmly depress the brake several times - the pedal travel should decrease with each application.

26 If your brakes do not operate as described, the brake booster has failed. Refer to Chapter 9 for the replacement procedure.

PARKING BRAKE

27 Vehicles equipped with rear drum brakes utilize a self-adjusting parking brake mechanism and do not require regular scheduled maintenance. Only vehicles equipped rear disc brakes require regular scheduled maintenance. For more detailed information on the parking brake assembly see Chapter 9.

TUNE-UP AND ROUTINE MAINTENANCE   1-25

21 Manual transmission lubricant level check (every 15,000 miles or 12 months)

▶ Refer to illustration 21.2

※※ WARNING:

Some models covered by this manual are equipped with air suspension systems. Always disconnect electrical power to the suspension system before lifting or towing the vehicle (see Chapter 10). Failure to perform this procedure may result in unexpected shifting or movement of the vehicle which could cause personal injury.

1 The manual transmission has a filler plug which must be removed to check the lubricant level. If the vehicle is raised to gain access to the plug, be sure to support it safely on jackstands - DO NOT crawl under a vehicle which is supported only by a jack! Be sure the vehicle is level or the check may be inaccurate.

2 Using a wrench, unscrew the plug from the transmission (see illustration) and use a finger to reach inside the housing to determine the lubricant level. The level should be at or near the bottom of the plug hole.

3 If it isn't, add the recommended lubricant through the plug hole with a pump or squeeze bottle.

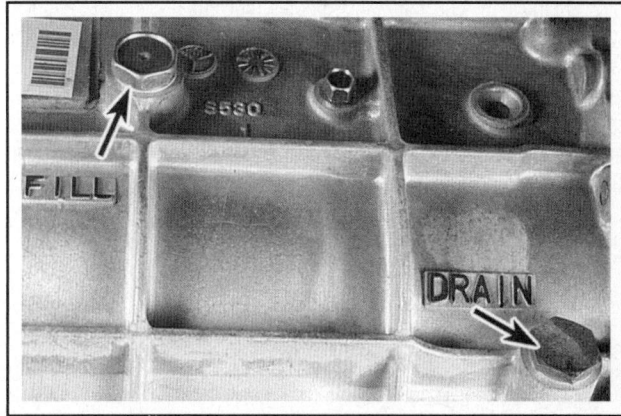

21.2 The manual transmission fill plug and drain plug are located on the side of the transmission case

4 Install and tighten the plug and check for leaks after the first few miles of driving.

22 Transfer case lubricant level check (every 15,000 miles or 12 months)

▶ Refer to illustration 22.1

※※ WARNING:

Some models covered by this manual are equipped with air suspension systems. Always disconnect electrical power to the suspension system before lifting or towing the vehicle (see Chapter 10). Failure to perform this procedure may result in unexpected shifting or movement of the vehicle which could cause personal injury.

1 The lubricant level is checked by removing a plug from the side of the case (see illustration). If the vehicle is raised to gain access to the plug, be sure to support it safely on jackstands - DO NOT crawl under the vehicle when it's supported only by a jack!

2 With the engine and transfer case cold, remove the plug. If lubricant immediately starts leaking out, thread the plug back into the case - the level is correct. If it doesn't, completely remove the plug and reach inside the hole with your little finger. The level should be even with the bottom of the plug hole.

3 If more lubricant is needed, use a syringe or small pump to add it through the opening.

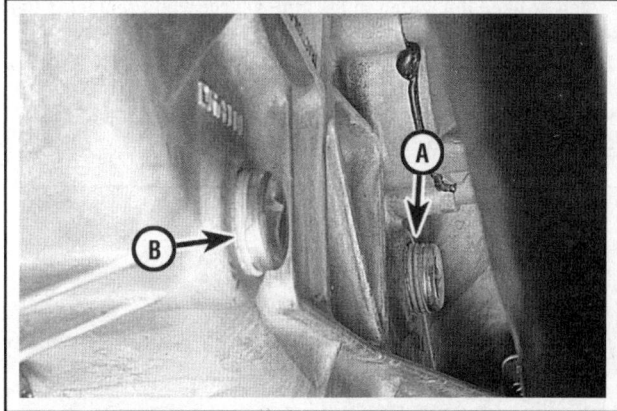

22.1 The transfer case fill plug (A) and drain plug (B) are located towards the rear of the transfer case

4 Thread the plug back into the case and tighten it securely. Drive the vehicle, then check for leaks around the plug. Install the rock guard.

23 Differential lubricant level check (every 15,000 miles or 12 months)

▶ Refer to illustration 23.2

※※ WARNING:

Some models covered by this manual are equipped with air suspension systems. Always disconnect electrical power to the suspension system before lifting or towing the vehicle (see Chapter 10). Failure to perform this procedure may result in unexpected shifting or movement of the vehicle which could cause personal injury.

1 The differential has a check/fill plug which must be removed to check the lubricant level. If the vehicle is raised to gain access to the

1-26 TUNE-UP AND ROUTINE MAINTENANCE

plug, be sure to support it safely on jackstands - DO NOT crawl under the vehicle when it's supported only by the jack!

2 Remove the check/fill plug from the differential (see illustration).

3 Use your little finger as a dipstick to make sure the lubricant level is even with the bottom of the plug hole. If not, use a syringe to add the recommended lubricant until it just starts to run out of the opening. On some models a tag is located in the area of the plug which gives information regarding lubricant type, particularly on models equipped with a Traction Lok differential.

4 Install the plug and tighten it securely.

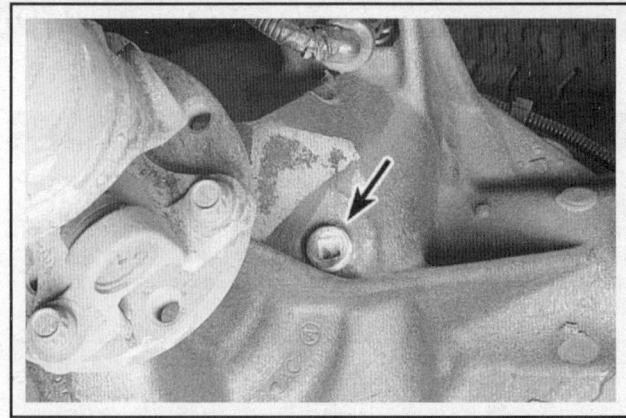

23.2 Use a 3/8-inch drive ratchet or breaker bar and an extension to remove the differential fill plug, then use your finger as a dipstick to check the lubricant level

24 Air filter check and replacement (every 30,000 miles or 24 months)

▸ Refer to illustration 24.1

※ CAUTION:

The MAF sensor is attached to the air filter housing. Do not stretch or bend the wiring to the MAF sensor while replacing the sensor.

➡ Note: The air filter is located inside a housing at the left (driver's) side of the engine compartment.

24.1 Release the clamp (A) then separate the housing halves to access the air filter

1 To remove the air filter on 2009 and earlier models, release the clamp that secures the two halves of the air filter housing together, then separate the cover halves and remove the air filter element (see illustration).

➡ Note: On 2007 through 2009 Expedition/Navigator models, reach under the connector and pull back the retaining clip, then disconnect the connector before pulling up the air filter housing.

2 On 2010 and later models, loosen the clamp securing the air outlet pipe to the air filter cover (see Chapter 4), then disconnect the Mass Air Flow (MAF) sensor electrical connector. Release the clamps that secure the two halves of the air cleaner housing together, slide the cover forward until the tabs at the rear of the housing are free, then separate the cover and remove the air filter element.

3 Inspect the outer surface of the filter element. If it is dirty, replace it. If it is only moderately dusty, it can be reused by blowing it clean from the back to the front surface with compressed air. Because it is a pleated paper type filter, it cannot be washed or oiled. If it cannot be cleaned satisfactorily with compressed air, discard and replace it. While the cover is off, be careful not to drop anything down into the housing.

※ CAUTION:

Never drive the vehicle with the air filter removed. Excessive engine wear could result and backfiring could even cause a fire under the hood.

4 Wipe out the inside of the air filter housing.
5 Place the new filter into the housing, making sure it seats properly.
6 The remainder of installation is the reverse of removal.

25 Cooling system servicing (draining, flushing and refilling) (every 30,000 miles or 24 months)

※ WARNING 1:

Do not allow antifreeze to come in contact with your skin or painted surfaces of the vehicle. Rinse off spills immediately with plenty of water. Antifreeze is highly toxic if ingested. Never leave antifreeze lying around in an open container or in puddles on the floor; children and pets are attracted by it's sweet smell and may drink it. Check with local authorities about disposing of used antifreeze. Many communities have collection centers which will see that antifreeze is disposed of safely.

TUNE-UP AND ROUTINE MAINTENANCE

25.4 The radiator drain fitting is located at the lower corner of the radiator

25.5 The block drain plugs are generally located about one to two inches above the oil pan - there is one on each side of the engine block

✱✱ WARNING 2:

Some models covered by this manual are equipped with air suspension systems. Always disconnect electrical power to the suspension system before lifting or towing the vehicle (see Chapter 10). Failure to perform this procedure may result in unexpected shifting or movement of the vehicle which could cause personal injury.

1 Periodically, the cooling system should be drained, flushed and refilled to replenish the antifreeze mixture and prevent formation of rust and corrosion, which can impair the performance of the cooling system and cause engine damage. When the cooling system is serviced, all hoses and the expansion tank cap should be checked and replaced if necessary.

DRAINING

▶ Refer to illustration 25.4 and 25.5

2 Apply the parking brake and block the wheels. If the vehicle has just been driven, wait several hours to allow the engine to cool down before beginning this procedure.

3 Once the engine is completely cool, remove the expansion tank cap.

4 Move a large container under the radiator drain to catch the coolant. Attach a 3/8-inch diameter hose to the drain fitting to direct the coolant into the container, then open the drain fitting (a pair of pliers may be required to turn it) (see illustration).

5 After the coolant stops flowing out of the radiator, move the container under the engine block drain plugs and allow the coolant in the block to drain (see illustration).

6 While the coolant is draining, check the condition of the radiator hoses, heater hoses and clamps (refer to Section 16 if necessary).

7 Replace any damaged clamps or hoses (refer to Chapter 3 for detailed replacement procedures).

FLUSHING

8 Once the system is completely drained, flush the radiator with fresh water from a garden hose until water runs clear at the drain. The flushing action of the water will remove sediments from the radiator but will not remove rust and scale from the engine and cooling tube surfaces.

9 These deposits can be removed by the chemical action of a cleaner available at auto parts stores. Follow the procedure outlined in the manufacturer's instructions. If the radiator is severely corroded, damaged or leaking, it should be removed (see Chapter 3) and taken to a radiator repair shop.

REFILLING

10 Close and tighten the radiator drain. Install and tighten the block drain plugs.

11 Place the heater temperature control in the maximum heat position.

12 Slowly add new coolant (a 50/50 mixture of water and antifreeze) to the expansion tank up to the Full Hot mark.

13 Leave the expansion tank cap off and run the engine in a well-ventilated area until the thermostat opens (coolant will begin flowing through the radiator and the upper radiator hose will become hot).

14 Turn the engine off and let it cool. Add more coolant mixture to bring the level back up to the Full Hot mark on the expansion tank.

15 Squeeze the upper radiator hose to expel air, then add more coolant mixture if necessary. Replace the expansion tank cap.

16 Place the heater temperature control and the blower motor speed control to their maximum setting.

17 Start the engine, allow it to reach normal operating temperature and check for leaks.

1-28 TUNE-UP AND ROUTINE MAINTENANCE

26 Automatic transmission fluid and filter change (every 30,000 miles or 24 months)

Refer to illustrations 26.7, 26.9, 26.10, 26.11 and 26.12

WARNING:
Some models covered by this manual are equipped with air suspension systems. Always disconnect electrical power to the suspension system before lifting or towing the vehicle (see Chapter 10). Failure to perform this procedure may result in unexpected shifting or movement of the vehicle which could cause personal injury.

Note: Some late models are equipped with transmissions that require less frequent fluid service as stated by the manufacturer. However, in order to prolong transmission life, regular transmission fluid changes are recommended.

1 At the specified intervals, the transmission fluid should be drained and replaced. Since the fluid will remain hot long after driving, perform this procedure only after the engine has cooled down completely.

2 Before beginning work, purchase the transmission fluid specified in *Recommended lubricants and fluids* in this Chapter's Specifications, a new filter and gasket.

Note: Some rubber gaskets are designed to be reused, but if you are unsure about the gasket's condition, we recommend replacement!

3 Other tools necessary for this job include jackstands to support the vehicle in a raised position, a drain pan capable of holding at least eight quarts, newspapers and clean rags.

4 Raise the vehicle and support it securely on jackstands.

WARNING:
On later Expedition/Navigator models, the exhaust pipe is close to the fill plug area, be careful not to burn yourself.

Note: If your vehicle is equipped with an E40D transmission, it may be necessary to disconnect the Transmission Range (TR) sensor connector (see Chapter 6).

5 With the drain pan in place, remove the drain plug (if equipped) and allow the fluid to run into the pan. If the transmission is equipped with a torque converter drain plug, remove the rubber cover and the plug to drain the torque converter. Install the drain plugs for the transmission and torque converter then tighten them to the torque listed in this Chapter's Specifications.

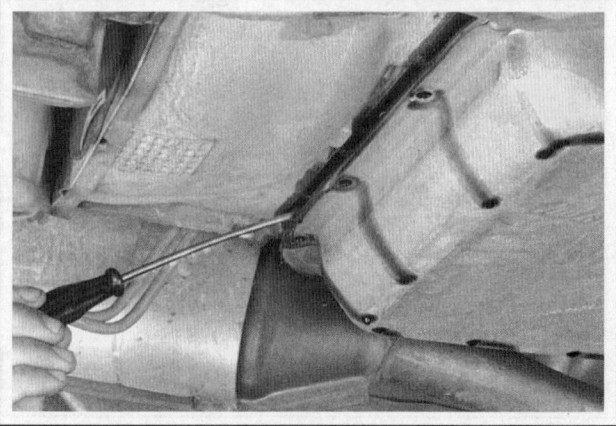

26.7 Pry the pan free of the gasket and allow the fluid to drain

6 On models without a drain plug, remove the bolts on the sides of the transmission pan, then loosen the front and rear pan bolts about a quarter-inch. Carefully pry around the pan's mating surface until the seal is completely broken and the loose pan is resting on the front and rear pan bolts. Hold the rear of the pan up against the transmission while removing the rear pan bolts, then slowly tilt the pan down and allow fluid to drain. Hold the pan up again and remove the front pan bolts. Detach the pan completely, being careful to not spill any remaining fluid.

7 On models with a drain plug, remove the pan bolts and carefully pry the transmission pan loose (see illustration). Don't damage the pan or transmission gasket surfaces or leaks could develop.

8 Carefully clean the mating surfaces of the transmission and pan to remove all traces of the old gasket and any sealant.

9 Clean the pan with solvent.

Note: Upon initial service, many transmissions will have a plastic plug lying in the bottom of the pan. This plug was used to keep contamination out of the transmission while on the assembly line. Discard the factory installed dust plug (see illustration).

10 Remove the old filter from the transmission (see illustration). If the filter seal did not come out with the filter, remove it from the transmission being careful not to gouge the seal bore in any way.

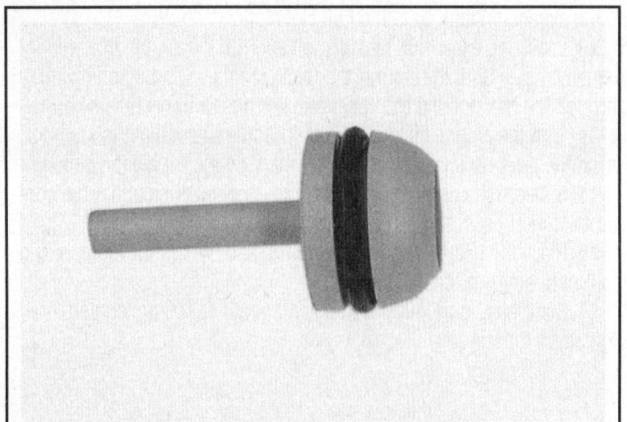

26.9 Discard the factory-installed dust plug. This plug will be in the pan if this is the first time the fluid has been drained

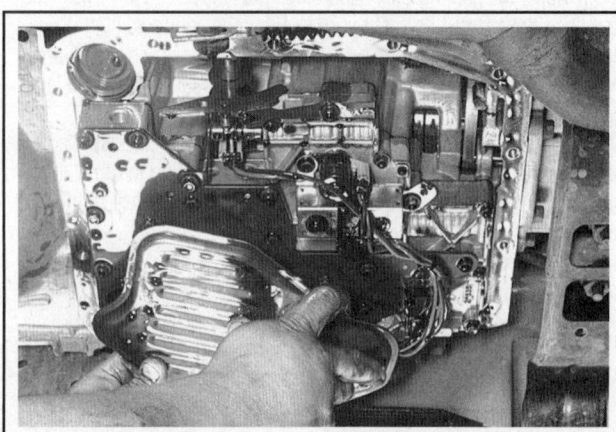

26.10 Pull straight down on the filter to remove it

TUNE-UP AND ROUTINE MAINTENANCE

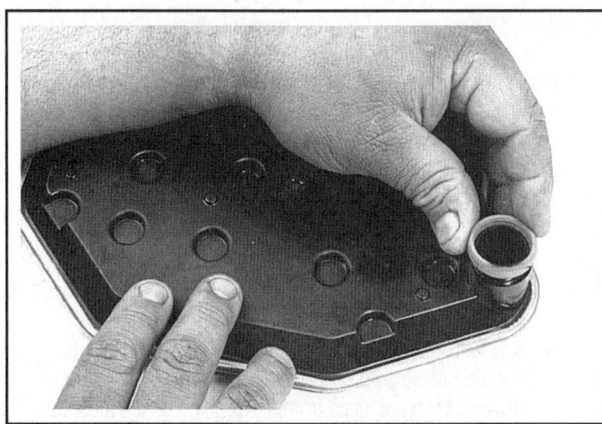

26.11 Install a new seal on the transmission filter

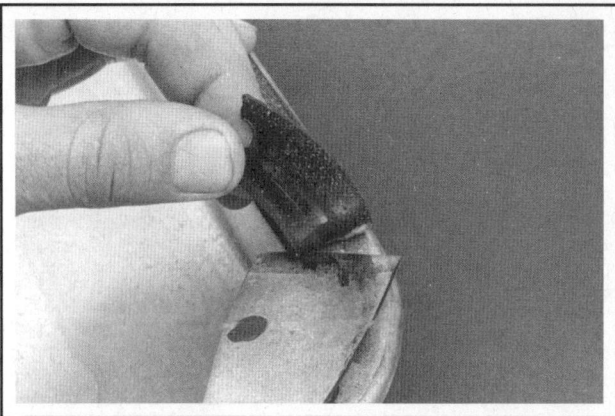

26.12 Be sure to clean all traces of the old gasket from the pan before installing a new one

11 Install a new seal and filter (see illustration).
12 Make sure the mating surfaces on the transmission and pan are clean, then install a new gasket (see illustration). Put the pan in place against the transmission and install the bolts. Working around the pan, tighten each bolt a little at a time until the final torque figure listed in this Chapter's Specifications is reached. Don't overtighten the bolts!
13 Lower the vehicle and add three and one-half quarts of automatic transmission fluid (see Section 7).

➡ Note: If the fluid was drained from the torque converter, add six quarts of transmission fluid.

14 With the transmission in Park and the parking brake set, run the engine at a fast idle, but don't race it.
15 Move the gear selector through each range and back to Park. Check the fluid level (see Section 7).
16 Check under the vehicle for leaks during the first few trips.

27 Front wheel bearing check, repack and adjustment (2WD models) (every 30,000 miles or 24 months)

✳ WARNING:

Some models covered by this manual are equipped with air suspension systems. Always disconnect electrical power to the suspension system before lifting or towing the vehicle (see Chapter 10). Failure to perform this procedure may result in unexpected shifting or movement of the vehicle which could cause personal injury.

CHECK AND REPACK

▸ Refer to illustrations 27.1, 27.3, 27.6, 27.7, 27.10, 27.11, 27.15 and 27.19

1 In most cases the front wheel bearings will not need servicing until the brake pads are changed. However, the bearings should be checked whenever the front of the vehicle is raised for any reason. Several items, including a torque wrench and special grease, are required for this procedure (see illustration).

27.1 Tools and materials needed for front wheel bearing maintenance

1 **Hammer** - A common hammer will do just fine
2 **Grease** - High-temperature grease that is formulated specially for front wheel bearings should be used
3 **Wood block** - If you have a scrap piece of 2x4, it can be used to drive the new seal into the hub
4 **Needle-nose pliers** - Used to straighten and remove the cotter pin in the spindle
5 **Torque wrench** - Used for pre-loading the bearing before adjustment
6 **Screwdriver** - Used to remove the seal from the hub (a long screwdriver is preferred)
7 **Socket/breaker bar** - Needed to loosen the nut on the spindle if it's extremely tight
8 **Brush** - Together with some clean solvent, this will be used to remove old grease from the hub and spindle

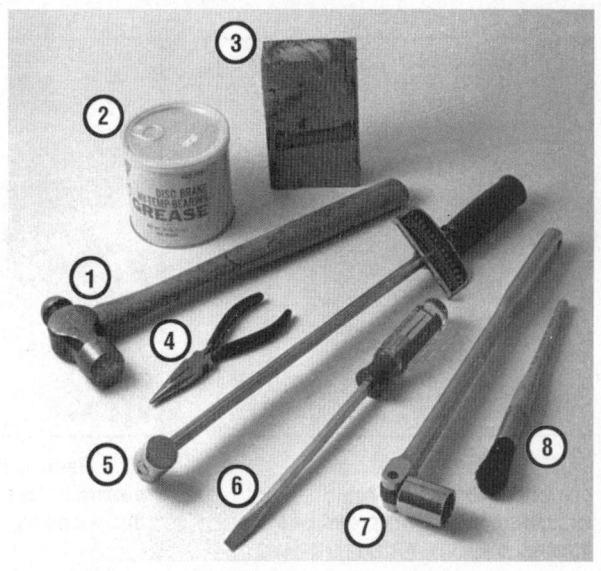

1-30 TUNE-UP AND ROUTINE MAINTENANCE

27.3 To check the wheel bearings, move the tire in and out as shown above - if there's any noticeable freeplay, the bearings should be checked and then repacked with grease or replaced if necessary

2 With the vehicle securely supported on jackstands, spin each wheel and check for noise, rolling resistance and freeplay.

3 Grasp the top of each tire with one hand and the bottom with the other (see illustration). Move the wheel in-and-out on the spindle. If there's any noticeable movement, the bearings should be checked and then repacked with grease or replaced if necessary.

4 Remove the wheel.

5 Remove the brake caliper (see Chapter 9) and hang it out of the way on a piece of wire. A wood block of the appropriate width can be slid between the brake pads to keep them separated, if necessary.

6 Dislodge the dust cap from the hub/disc assembly using a screwdriver or hammer and chisel (see illustration).

7 Straighten the bent ends of the cotter pin, then pull the cotter pin out of the nut lock (see illustration). Discard the cotter pin and use a new one during reassembly.

8 Remove the nut lock, nut and washer from the end of the spindle.

9 Pull the hub/disc assembly out slightly, then push it back into its original position. This should force the outer bearing off the spindle enough so it can be removed.

10 Pull the disc assembly off the spindle (see illustration).

11 Use a screwdriver or a seal puller tool to pry the seal out of the rear of the disc (see illustration). Note how the seal is installed.

12 Remove the inner wheel bearing from the disc.

13 Use solvent to remove all traces of the old grease from the bearings, hub and spindle. A small brush may prove helpful; however make sure no bristles from the brush embed themselves inside the bearing rollers. Allow the parts to air dry.

14 Carefully inspect the bearings for cracks, heat discoloration, worn rollers, etc. Check the bearing races inside the hub for wear and damage. If the bearing races are defective, the hubs should be taken to a machine shop with the facilities to remove the old races and press new ones in. Note that the bearings and races come as matched sets and old bearings should never be installed on new races.

15 Use high-temperature front wheel bearing grease to pack the bearings. Work the grease completely into the bearings, forcing it between the rollers, cone and cage from the back side (see illustration).

16 Apply a thin coat of grease to the spindle at the outer bearing seat, inner bearing seat, shoulder and seal seat.

17 Put a small quantity of grease inboard of each bearing race inside the hub. Using your finger, form a dam at these points to provide extra grease availability and to keep thinned grease from flowing out of the bearing.

18 Place the grease-packed inner bearing into the rear of the hub and put a little more grease outboard of the bearing.

19 Place a new seal over the inner bearing and tap the seal evenly into place until it's flush with the hub (see illustration).

20 Carefully place the hub assembly onto the spindle and push the grease-packed outer bearing into position.

ADJUSTMENT

21 Install the washer and spindle nut. Tighten the nut only slightly (no more than 12 ft-lbs of torque).

22 Spin the hub in a forward direction while tightening the spindle nut to approximately 30 ft-lbs to seat the bearings and remove any grease or burrs which could cause excessive bearing play later.

23 Loosen the spindle nut 1/4-turn, then using your hand (not a wrench of any kind), tighten the nut until it's snug. Install the nut lock and a new cotter pin through the hole in the spindle and the slots in the nut lock. If the slots in the nut lock don't line up with the hole in the spindle, reposition the nut lock until they do.

24 Check that the hub/disc assembly spins freely with no noticeable free play. If freeplay exists repeat Steps 22 and 23 until proper adjustment is obtained.

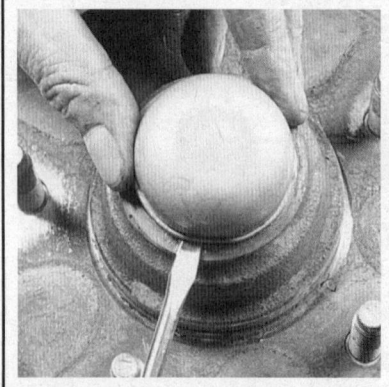

27.6 Dislodge the dust cap by working around the outer circumference with a screwdriver or a hammer and chisel

27.7 Remove the cotter pin and discard it - use a new one when the disc assembly is reinstalled

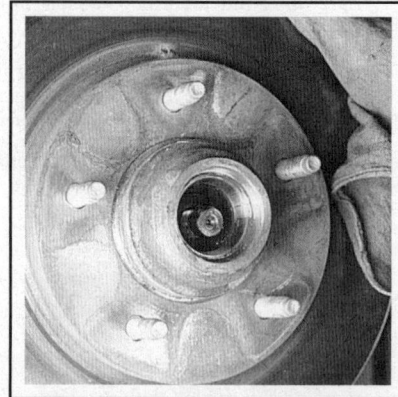

27.10 After the washer and outer wheel bearing have been dislodged, pull the disc off the spindle

TUNE-UP AND ROUTINE MAINTENANCE

27.11 Use a seal puller or large screwdriver to remove the inner grease seal - note the seal installed position

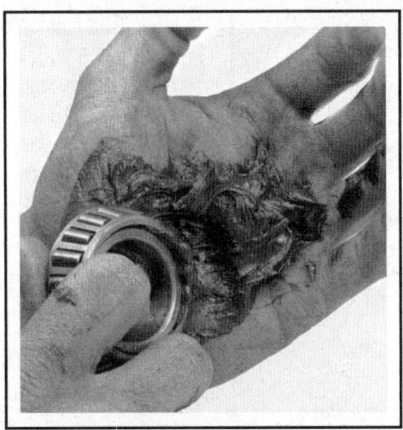

27.15 Work the grease completely into the bearing rollers

27.19 Gently tap the grease seal into place

25 Bend the ends of the cotter pin until they're flat against the nut. Cut off any extra length which could interfere with the dust cap.

26 Install the dust cap, lightly tapping it into place with a hammer.

27 Install brake caliper in the reverse order of removal (see Chapter 9).

28 Install the wheel on the hub and tighten the lug nuts.

29 Lower the vehicle and tighten the lug nuts to the torque listed in this Chapter's Specifications.

28 Brake fluid change (every 30,000 miles or 24 months)

✴✴ WARNING:

Brake fluid can harm your eyes and damage painted surfaces, so use extreme caution when handling or pouring it. Do not use brake fluid that has been standing open or is more than one year old. Brake fluid absorbs moisture from the air. Excess moisture can cause a dangerous loss of braking effectiveness.

1 At the specified intervals, the brake fluid should be drained and replaced. Since the brake fluid may drip or splash when pouring it, place plenty of rags around the master cylinder to protect any surrounding painted surfaces.

2 Before beginning work, purchase the specified brake fluid (see *Recommended lubricants and fluids* at the end of this Chapter).

3 Remove the cap from the master cylinder reservoir.

4 Using a hand suction pump or similar device, withdraw the fluid from the master cylinder reservoir.

5 Add new fluid to the master cylinder until it rises to the base of the filler neck.

6 Bleed the brake system as described in Chapter 9 at all four brakes until new and uncontaminated fluid is expelled from the bleeder screw. Be sure to maintain the fluid level in the master cylinder as you perform the bleeding process. If you allow the master cylinder to run dry, air will enter the system.

7 Refill the master cylinder with fluid and check the operation of the brakes. The pedal should feel solid when depressed, with no sponginess.

✴✴ WARNING:

Do not operate the vehicle if you are in doubt about the effectiveness of the brake system.

29 Drivebelt check and replacement (every 60,000 miles or 48 months)

▸ **Refer to illustrations 29.4, 29.5, 29.6, 29.8a, 29.8b and 29.8c**

1 The drivebelts are located at the front of the engine and play an important role in the overall operation of the vehicle and its components. Due to their function and material make-up, the drivebelts are prone to failure after a period of time and should be inspected and adjusted periodically to prevent major engine damage.

2 All models except for 3.5L models are equipped with a single automatically adjusted serpentine drivebelt, which is used to drive all of the accessory components such as the alternator, power steering pump, water pump and air conditioning compressor. 3.5L V6 models are equipped with two drivebelts - a main accessory serpentine belt that drives the alternator and water pump and is automatically adjusted, and an air conditioning compressor drivebelt that doesn't use a tensioner.

INSPECTION

3 With the engine off, open the hood and locate the drivebelt at the front of the engine. Using your fingers (and a flashlight, if necessary),

1-32   TUNE-UP AND ROUTINE MAINTENANCE

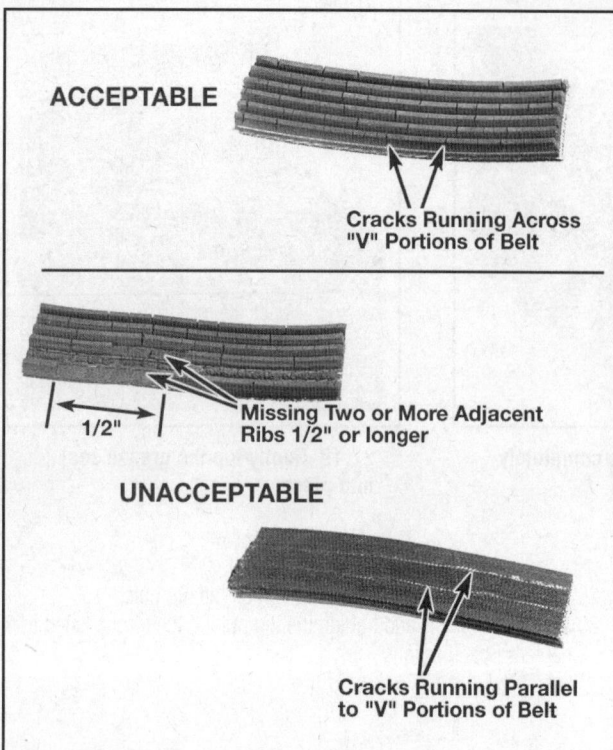

29.4 Small cracks in the underside of a V-ribbed belt are acceptable - lengthwise cracks, or missing pieces that cause the belt to make noise, are cause for replacement

move along the belts checking for cracks and separation of the belt plies. Also check for fraying and glazing, which gives the belt a shiny appearance. Both sides of each belt should be inspected, which means you will have to twist the belt to check the underside.

4 Check the ribs on the underside of the belt. They should all be the same depth, with none of the surface uneven (see illustration).

5 The tension of the belt is automatically adjusted by the belt tensioner and does not require any adjustments. Drivebelt wear can be checked visually by inspecting the wear indicator marks located on the side of the tensioner body. Locate the belt tensioner at the front of the engine on the right (passenger) side, adjacent to the lower crankshaft pulley, then find the tensioner operating marks (see illustration). If the indicator mark is outside the operating range, the belt should be replaced.

REPLACEMENT

Accessory drivebelt

6 To replace the belt, rotate the tensioner to relieve the tension on the belt (see illustration). Some models have a square hole in the tensioner arm that will accept a 1/2-inch drive breaker bar. On other models, place a wrench on the tensioner pulley bolt.

7 Remove the belt from the auxiliary components and carefully release the tensioner.

8 Route the new belt over the various pulleys, again rotating the tensioner to allow the belt to be installed, then release the belt tensioner. Make sure the belt fits properly into the pulley grooves - it must be completely engaged.

➡Note: Most models have a drivebelt routing decal on the upper radiator panel to help during drivebelt installation (see illustrations).

3.5L V6 air conditioning compressor drivebelt

9 Remove the accessory drivebelt (see Steps 6 and 7).
10 Remove the engine splash shield/under cover.
11 Using a pair of diagonal cutting pliers, cut the air conditioning belt to remove it.

➡Note: Some tool manufacturers make air conditioning compressor belt removal tools that can remove the belt without cutting it off. We only recommend using this type of tool if the belt if relatively new and will be reused. Always check the belt for damage after the removal process.

29.5 Belt wear indicator marks are located on the side of the tensioner body - when the belt reaches the maximum wear mark it must be replaced

29.6 Rotate the tensioner arm to relieve belt tension

TUNE-UP AND ROUTINE MAINTENANCE 1-33

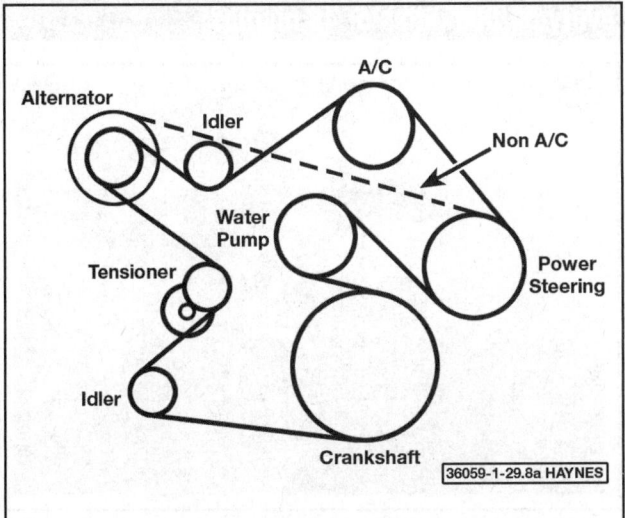

29.8a The routing schematic for the serpentine belt is usually found on the fan shroud (4.2L V6 engine shown)

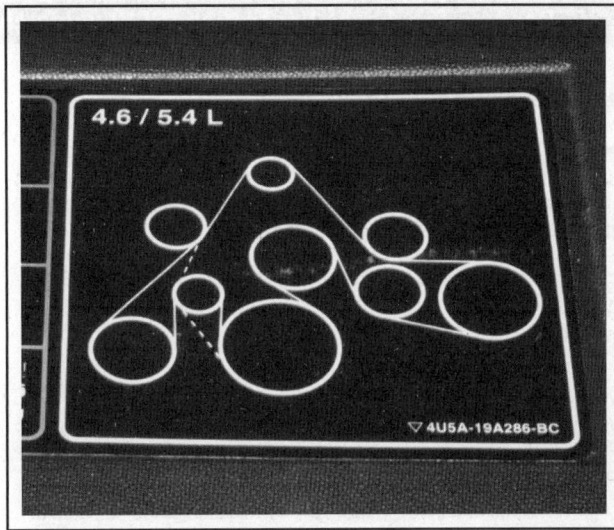

29.8b Drivebelt routing diagram on a late model V8 engine

12 Place one end of the new belt behind the crankshaft pulley, and the other end onto the compressor pulley, making sure the belt is fully seated in the pulley grooves.
13 Place part of the belt onto the top of the crankshaft pulley grooves, letting the rest of the belt toward the bottom of the crankshaft pulley hang/fold inward.
14 At this stage, the belt should be placed evenly onto the compressor pulley and part way onto the top of the crank pulley. While holding the belt in place at the top of the crank pulley, use a cable tie to hold the belt in place at the top crankshaft grooves.
15 Rotate the crankshaft pulley clockwise with a socket and ratchet, which should lift the belt onto the crankshaft pulley grooves. When the belt is in place, remove the cable tie.

❋❋ **WARNING:**

DO NOT crank the engine with the starter during this process! Rotate the engine by hand only.

16 The remainder of installation is the reverse of removal. Make sure the belt is even with the pulley grooves.

Tensioner replacement (2010 and later models)

17 Remove the drivebelt (see Steps 6 and 7).
18 Remove the tensioner mounting bolts and remove the tensioner.
19 Installation is the reverse of removal. See Step 8 to install the drivebelt.

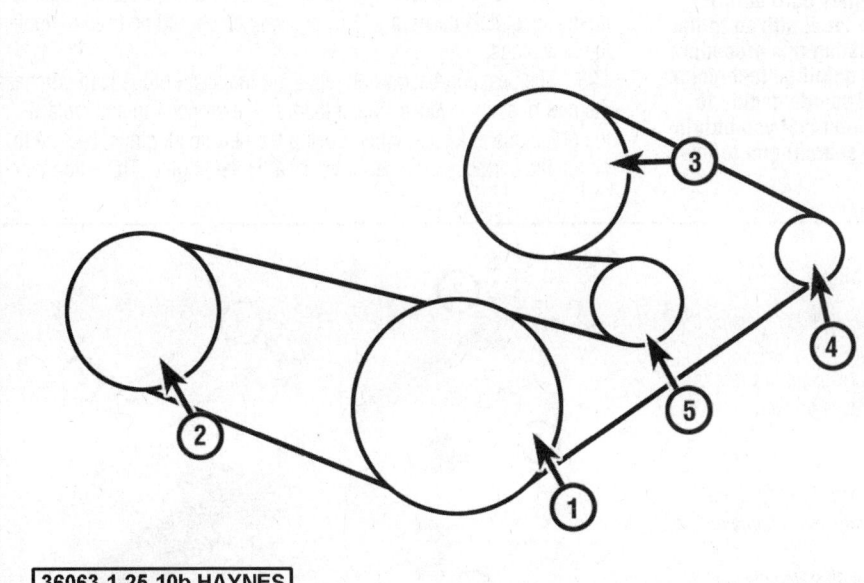

29.8c Drivebelt routing schematic for 3.5L V6 models

1 Crankshaft pulley
2 Air conditioning compressor
3 Water pump
4 Alternator
5 Tensioner and pulley

30 Positive Crankcase Ventilation (PCV) valve check (every 60,000 miles or 48 months)

◆ Refer to illustration 30.2

➟Note: To maintain efficient operation of the PCV system, clean the hoses and check the PCV valve at the intervals recommended in the maintenance schedule. For additional information on the PCV system, refer to Chapter 6.

1 The PCV valve is located on the right (2009 and earlier models and 2015 and later models) or the left (2010 through 2014 models) valve cover.

2 Start the engine and allow it to idle, then disconnect the PCV valve from the intake manifold at the front of the engine and feel for vacuum at the end of the valve (see illustration). If vacuum is felt, the PCV valve/system is working properly (see Chapter 6 for additional PCV system information).

3 If no vacuum is felt, remove the valve and check for vacuum at the hose. If vacuum is present at the hose but not at the valve, replace the valve. If no vacuum is felt at the hose, check for a plugged or cracked hose between the PCV valve and the intake plenum.

4 Check the rubber grommet in the valve cover for cracks and distortion. If it's damaged, replace it.

5 If the valve is clogged, the hose is also probably plugged. Remove the hose between the valve and the intake manifold and clean it with solvent.

30.2 With the engine running at idle, remove the PCV valve and verify that vacuum can be felt at the end of the valve

6 After cleaning the hose, inspect it for damage, wear and deterioration. Make sure it fits snugly on the fittings.

7 If necessary, install a new PCV valve. On 2010 and later models, the air intake resonator must be removed to access the PCV valve.

31 Spark plug check and replacement (every 60,000 miles or 48 months)

◆ Refer to illustrations 31.2, 31.5a, 31.5b, 31.6a, 31.6b, 31.8, 31.9, 31.10a and 31.10b

※※ CAUTION:

Ford has issued a Technical Service Bulletin (TSB 08-7-6) stating that attempting to remove the spark plugs is likely to result in the extended electrode of the plug breaking off in the cylinder head. This TSB pertains to 4.6L three-valve engines built before 11/30/07, and 5.4L and 6.8L three-valve engines built before 10/9/07. Engine build date can be found on a label affixed to the left valve cover. We highly recommend entrusting this procedure to a Ford dealer service department or other qualified technician equipped with the necessary special tools. If you do decide to attempt spark plug replacement, we recommend that you obtain the Ford TSB (it can be obtained via internet search) and follow the procedure exactly.

1 Vehicles equipped with 4.2L engines have the spark plugs located on the sides of the engine. Vehicles equipped with 4.6L and 5.4L engines have the spark plugs located at the top of the engine.

2 In most cases, the tools necessary for spark plug replacement include a spark plug socket which fits onto a ratchet (spark plug sockets are padded inside to prevent damage to the porcelain insulators on the new plugs), various extensions and a gap gauge to check and adjust the gaps on the new plugs (see illustration). A special plug wire removal tool is available for separating the wire boots from the spark plugs, but it isn't absolutely necessary. A torque wrench should be used to tighten the new plugs.

3 The best approach when replacing the spark plugs is to purchase the new ones in advance, adjust them to the proper gap and replace the plugs one at a time. When buying the new spark plugs, be sure to obtain the correct plug type for your particular engine. This information

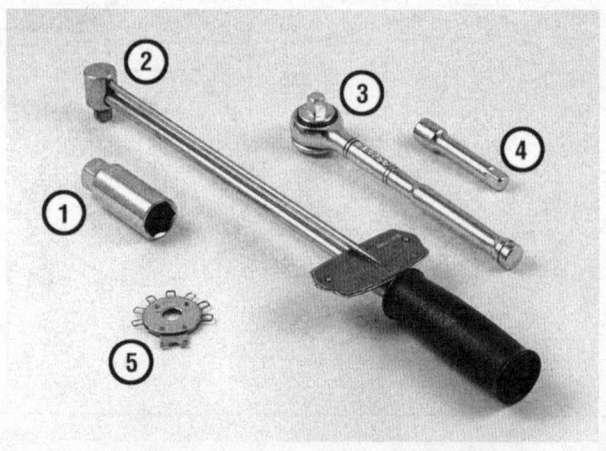

31.2 Tools required for changing spark plugs

1 **Spark plug socket** - This will have special padding inside to protect the spark plug's porcelain insulator
2 **Torque wrench** - Although not mandatory, using this tool is the best way to ensure the plugs are tightened properly
3 **Ratchet** - Standard hand tool to fit the spark plug socket
4 **Extension** - Depending on model and accessories, you may need special extensions and universal joints to reach one or more of the plugs
5 **Spark plug gap gauge** - This gauge for checking the gap comes in a variety of styles. Make sure the gap for your engine is included

TUNE-UP AND ROUTINE MAINTENANCE 1-35

31.5a Spark plug manufacturers recommend using a wire-type gauge when checking the gap - if the wire does not slide between the electrodes with a slight drag, adjustment is required

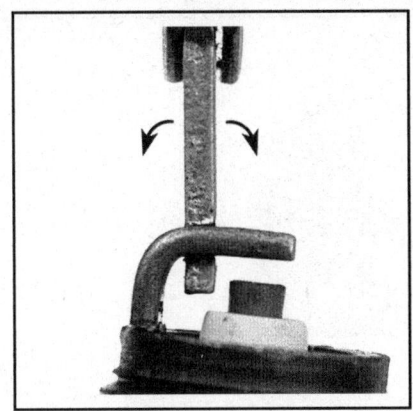

31.5b To change the gap, bend the side electrode only, as indicated by the arrows, and be very careful not to crack or chip the porcelain insulator surrounding the center electrode

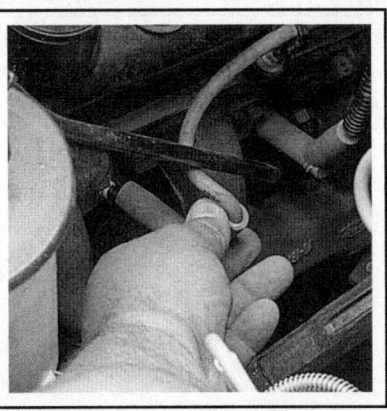

31.6a When removing the spark plug wires, pull only on the boot and twist it back-and-forth

can be found in the Specifications Section a the end of this Chapter, on the Emission Control Information label located under the hood or in the factory owner's manual. If differences exist between the plug specified on the emissions label, Specifications Section or in the owner's manual, assume that the emissions label is correct.

4 Allow the engine to cool completely before attempting to remove any of the plugs. Some engines are equipped with aluminum cylinder heads, which can be damaged if the spark plugs are removed when the engine is hot. While you are waiting for the engine to cool, check the new plugs for defects and adjust the gaps.

5 The gap is checked by inserting the proper thickness gauge between the electrodes at the tip of the plug (see illustration). The gap between the electrodes should be the same as the one specified on the Emissions Control Information label. The wire should just slide between the electrodes with a slight amount of drag. If the gap is incorrect, use the adjuster on the gauge body to bend the curved side electrode slightly until the specified gap is obtained (see illustration). If the side electrode is not exactly over the center electrode, bend it with the adjuster until it is. Check for cracks in the porcelain insulator (if any are found, the plug should not be used).

6 With the engine cool, remove the spark plug wire from one spark plug. Pull only on the boot at the end of the wire - do not pull on the wire (see illustration). A plug wire removal tool should be used if available. 5.4L V8 engines, 2000 and later 4.6L V8 and 8.5L V6 engines are equipped with individual coil packs which must be removed first to access the spark plugs (see illustration).

7 If compressed air is available, use it to blow any dirt or foreign material away from the spark plug hole. A common bicycle pump will also work. The idea here is to eliminate the possibility of debris falling into the cylinder as the spark plug is removed.

8 Place the spark plug socket over the plug and remove it from the engine by turning it in a counterclockwise direction (see illustration).

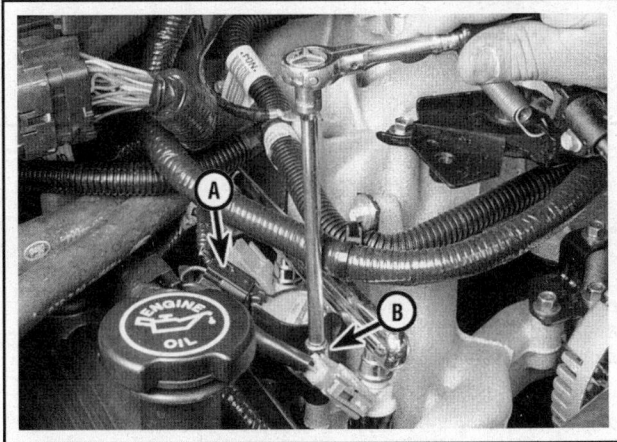

31.6b On 5.4L V8 engines and 2000 and later 4.6L V8 engines, the coil pack must be removed to access the spark plugs - disconnect the electrical connector (A) and remove the coil pack retaining screw (B) - pull straight up and out to remove the coil pack

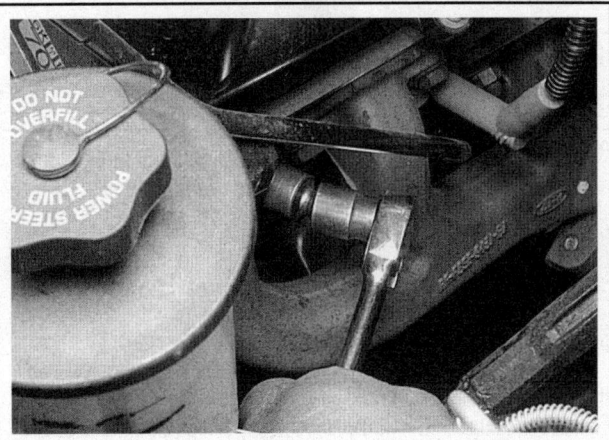

31.8 Use a spark plug socket wrench and extension to unscrew the spark plug

1-36 TUNE-UP AND ROUTINE MAINTENANCE

A normally worn spark plug should have light tan or gray deposits on the firing tip.

A carbon fouled plug, identified by soft, sooty, black deposits, may indicate an improperly tuned vehicle. Check the air cleaner, ignition components and engine control system.

An oil fouled spark plug indicates an engine with worn piston rings and/or bad valve seals allowing excessive oil to enter the chamber.

This spark plug has been left in the engine too long, as evidenced by the extreme gap- Plugs with such an extreme gap can cause misfiring and stumbling accompanied by a noticeable lack of power.

A physically damaged spark plug may be evidence of severe detonation in that cylinder. Watch that cylinder carefully between services, as a continued detonation will not only damage the plug, but could also damage the engine.

A bridged or almost bridged spark plug, identified by a build-up between the electrodes caused by excessive carbon or oil build-up on the plug.

31.9 Inspect the spark plug to determine engine running conditions

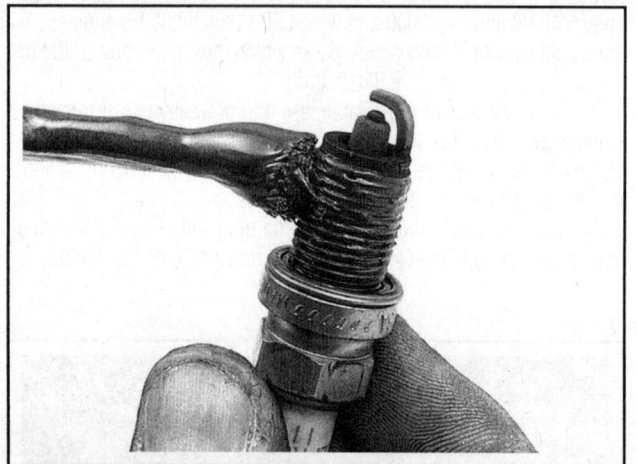

31.10a Apply a thin film of anti-seize compound to the spark plug threads to prevent damage to the cylinder head

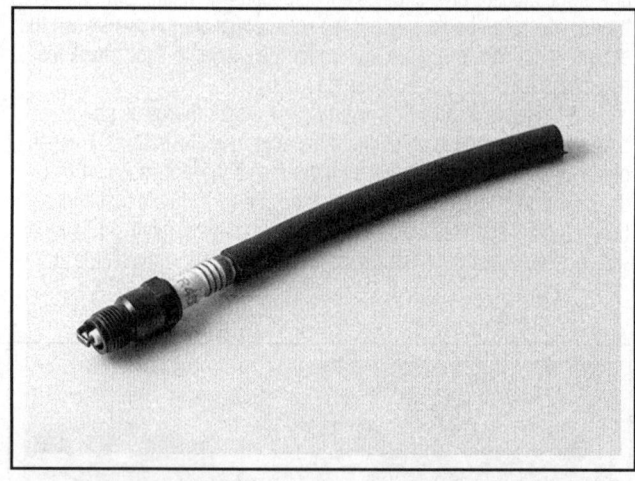

31.10b A length of rubber hose will save time and prevent damaged threads when installing the spark plugs

9   Compare the spark plug to those shown in this chart (see illustration) to get an indication of the general running condition of the engine.
10   Apply a small amount of anti-seize compound to the spark plug threads (see illustration). Install one of the new plugs into the hole until you can no longer turn it with your fingers, then tighten it with a torque wrench (if available) or the ratchet. It is a good idea to slip a short length of rubber hose over the end of the plug to use as a tool to thread it into place (see illustration). The hose will grip the plug well enough to turn it, but will start to slip if the plug begins to cross-thread in the hole - this will prevent damaged threads and the accompanying repair costs.

➥Note: On late 2008 and later Expedition/Navigator 5.4L 3V engines (identified by brown coil boots instead of black), the spark plugs require significantly less torque.

11   Before pushing the spark plug wire onto the end of the plug, inspect it following the procedures outlined in Section 32.
12   Attach the plug wire to the new spark plug, again using a twisting motion on the boot until it is seated on the spark plug.
13   Repeat the procedure for the remaining spark plugs, replacing them one at a time to prevent mixing up the spark plug wires.

TUNE-UP AND ROUTINE MAINTENANCE

32 Ignition system component check and replacement (every 60,000 miles or 48 months)

SPARK PLUG WIRES

▶ Refer to illustration 32.8

➥Note: Every time a spark plug wire is detached from a spark plug or the coil, silicone dielectric compound (a white grease available at auto parts stores) must be applied to the inside of each boot before reconnection. Use a small standard screwdriver to coat the entire inside surface of each boot with a thin layer of the compound.

1 The spark plug wires should be checked and, if necessary, replaced at the same time new spark plugs are installed.

2 The easiest way to identify bad wires is to make a visual check while the engine is running. In a dark, well-ventilated garage, start the engine and look at each plug wire. Be careful not to come into contact with any moving engine parts. If there is a break in the wire, you will see arcing or a small spark at the damaged area. If arcing is noticed, make a note to obtain new wires.

3 The spark plug wires should be inspected one at a time, beginning with the spark plug for the number one cylinder, (the cylinder closest to the radiator on the right bank), to prevent confusion. Clearly label each original plug wire with a piece of tape marked with the correct number. The plug wires must be reinstalled in the correct order to ensure proper engine operation.

4 Disconnect the plug wire from the first spark plug. A removal tool can be used, or you can grab the wire boot, twist it slightly and pull the wire free. Do not pull on the wire itself, only on the rubber boot.

5 Push the wire and boot back onto the end of the spark plug. It should fit snugly. If it doesn't, detach the wire and boot once more and use a pair of pliers to carefully crimp the metal connector inside the wire boot until it does.

6 Using a clean rag, wipe the entire length of the wire to remove built-up dirt and grease.

7 Once the wire is clean, check for burns, cracks and other damage. Do not bend the wire sharply or you might break the conductor.

8 Disconnect the wire from the coil pack. Pull only on the rubber

32.8 Remove each spark plug wire from the ignition coil pack(s) and at the spark plug - check for corrosion and a tight fit

boot. Check for corrosion and a tight fit (see illustration). Reinstall the wire.

9 Inspect each of the remaining spark plug wires, making sure that each one is securely fastened on each end.

10 If new spark plug wires are required, purchase a set for your specific engine model. Pre-cut wire sets with the boots already installed are available. Remove and replace the wires one at a time to avoid mix-ups in the firing order. Should a mix up occur refer to the Specifications at the end this Chapter.

IGNITION COIL PACKS

11 Clean the coil packs with a dampened cloth and dry them thoroughly.

12 Inspect each coil pack for cracks, damage and carbon tracking. If damage exists refer to Chapter 5 for the replacement procedure.

33 Manual transmission lubricant change (every 60,000 miles or 48 months)

❊❊ WARNING:

Some models covered by this manual are equipped with air suspension systems. Always disconnect electrical power to the suspension system before lifting or towing the vehicle (see Chapter 10). Failure to perform this procedure may result in unexpected shifting or movement of the vehicle which could cause personal injury.

1 Raise the vehicle and support it securely on jackstands.

2 Move a drain pan, rags, newspapers and wrenches under the transmission.

3 Remove the transmission drain plug at the bottom of the case and allow the lubricant to drain into the pan (see illustration 21.2).

4 After the lubricant has drained completely, reinstall the plug and tighten it securely.

5 Remove the fill plug from the side of the transmission case. Using a hand pump, syringe or funnel, fill the transmission with the specified lubricant until it is level with the lower edge of the filler hole. Reinstall the fill plug and tighten it securely.

6 Lower the vehicle.

7 Drive the vehicle for a short distance, then check the drain and fill plugs for leakage.

1-38 TUNE-UP AND ROUTINE MAINTENANCE

34 Transfer case lubricant change (4WD models) (every 60,000 miles or 48 months)

WARNING:
Some models covered by this manual are equipped with air suspension systems. Always disconnect electrical power to the suspension system before lifting or towing the vehicle (see Chapter 10). Failure to perform this procedure may result in unexpected shifting or movement of the vehicle which could cause personal injury.

1 Drive the vehicle for at least 15 minutes to warm the lubricant in the case. Perform this warm-up procedure with 4WD engaged, if possible. Use all gears, including Reverse, to ensure the lubricant is sufficiently warm to drain completely.

2 Raise the vehicle and support it securely on jackstands.
3 Remove the drain plug from the lower part of the case and allow the old lubricant to drain completely (see illustration 22.1).
4 After the lubricant has drained completely, reinstall the plug and tighten it securely
5 Remove the filler plug from the case
6 Fill the case with the specified lubricant until it is level with the lower edge of the filler hole.
7 Install the filler plug and tighten it securely.
8 Drive the vehicle for a short distance and recheck the lubricant level. In some instances a small amount of additional lubricant will have to be added.

35 Differential lubricant change (every 30,000 miles or 24 months)

WARNING:
Some models covered by this manual are equipped with air suspension systems. Always disconnect electrical power to the suspension system before lifting or towing the vehicle (see Chapter 10). Failure to perform this procedure may result in unexpected shifting or movement of the vehicle which could cause personal injury.

DRAIN

▶ Refer to illustration 35.6, 35.8a, 35.8b, 35.8c and 35.10

1 This procedure should be performed after the vehicle has been driven so the lubricant will be warm and therefore flow out of the differential more easily.
2 Raise the vehicle and support it securely on jackstands.
3 The easiest way to drain the differential(s) is to remove the lubricant through the filler plug hole with a suction pump. If the differential cover gasket is leaking, it will be necessary to remove the cover to drain the lubricant (which will also allow you to inspect the differential.

→ **Note:** If you're changing the front differential lubricant on a 4WD vehicle simply remove the drain plug to drain the lubricant.

Changing the lubricant with a suction pump

4 Remove the filler plug from the differential (see Section 23).
5 Insert the flexible hose.
6 Work the hose down to the bottom of the differential housing and pump the lubricant out (see illustration).

Changing rear differential lubricant by removing the cover

7 Move a drain pan, rags, newspapers and wrenches under the vehicle.
8 Remove the bolts on the lower half of the cover. Loosen the bolts on the upper half and use them to loosely retain the cover. Allow the oil to drain into the pan, then completely remove the cover (see illustrations).

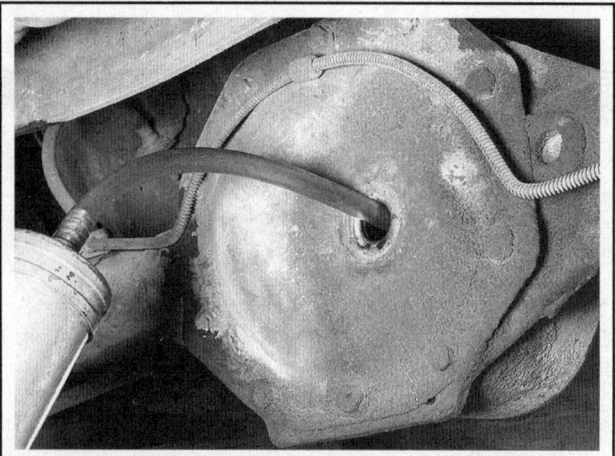

35.6 This is the easiest way to remove the differential lubricant - work the end of the hose to the bottom of the differential housing and draw out the old lubricant with a suction pump

35.8a Remove the bolts from the lower edge of the cover . . .

TUNE-UP AND ROUTINE MAINTENANCE   1-39

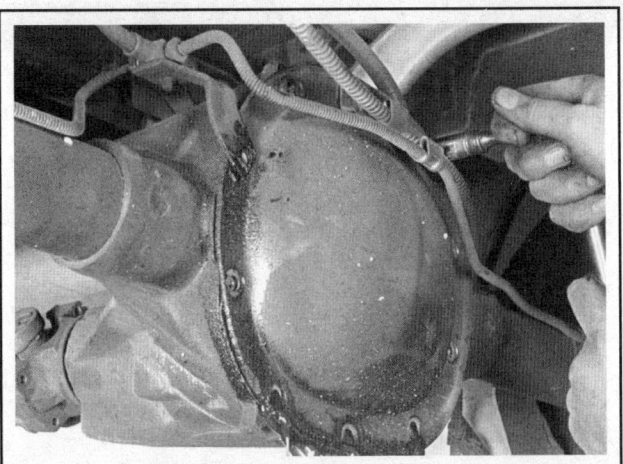

35.8b . . . then loosen the top bolts and let the lubricant drain

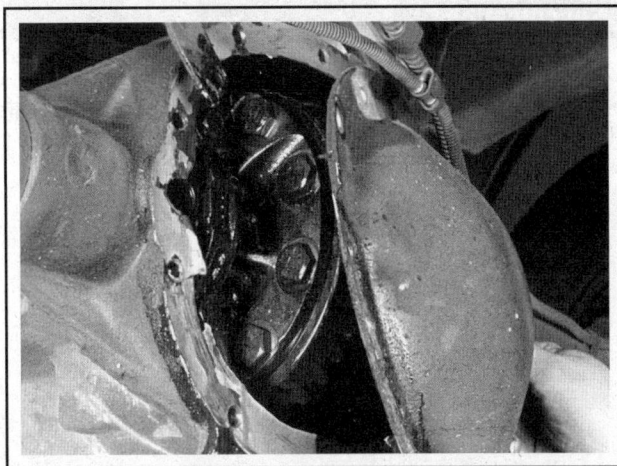

35.8c Once the lubricant has drained, remove the cover

35.10 Carefully scrape off the old material to ensure a leak-free seal

9 Using a lint-free rag, clean the inside of the cover and the accessible areas of the differential housing. As this is done, check for chipped gears and metal particles in the lubricant, indicating that the differential should be more thoroughly inspected and/or repaired.

10 Thoroughly clean the gasket mating surfaces of the differential housing and the cover plate. Use a gasket scraper or putty knife to remove all traces of the old gasket (see illustration).

11 Apply a thin layer of RTV sealant to the cover flange, then press a new gasket into position on the cover. Make sure the bolt holes align properly.

REFILL

12 Use a hand pump, syringe or funnel to fill the differential housing with the specified lubricant until it's level with the bottom of the filler plug hole.

13 Install the fill plug and tighten it securely.

1-40 TUNE-UP AND ROUTINE MAINTENANCE

Specifications

Recommended lubricants and fluids

→ Note: Listed here are manufacturer recommendations at the time this manual was written. Manufacturers occasionally upgrade their fluid and lubricant specifications, so check with your local auto parts store for current recommendations.

Engine oil	
Type	API "certified for gasoline engines"
Viscosity	
1999 and earlier models	5W-30
2000 and later models (non-turbocharged engines)	5W-20
2015 and later models (turbocharged engines)	5W-30 Synthetic blend (USA), 5W-30 premium oil (Canada)
Fuel	Unleaded gasoline, 87 octane or higher
Engine coolant*	
2009 and earlier models	50/50 mixture of Motorcraft Premium Engine Coolant (green colored) or Motorcraft Premium Gold Engine Coolant (yellow colored) and distilled water
2010 models	50/50 mixture of Motorcraft Premium Gold Engine Coolant (yellow colored) or Motorcraft Specialty Orange Engine Coolant with Bittering Agent and distilled water
2011 and later models	50/50 mixture of Motorcraft Specialty Orange Engine Coolant with Bittering Agent and distilled water
Brake fluid	DOT 3 heavy duty brake fluid
Clutch fluid	DOT 3 heavy duty brake fluid
Power steering fluid	
2005 and earlier models	MERCON automatic transmission fluid
2006 and later models	MERCON V automatic transmission fluid
Automatic transmission fluid	
1997 models	MERCON automatic transmission fluid
1998 and later models	
E4OD transmission	MERCON automatic transmission fluid
4R70W transmission	MERCON V automatic transmission fluid
4R100 transmission	MERCON automatic transmission fluid
4R75E transmission	MERCON V automatic transmission fluid
6HP26 transmission	MERCON SP automatic transmission fluid
4R70E - 4R75E transmissions	MERCON V automatic transmission fluid
6R75 transmission	MERCON SP automatic transmission fluid
6R80 transmission	MERCON ULV automatic transmission fluid
Manual transmission lubricant	MERCON automatic transmission fluid
Transfer case lubricant	
2007 and earlier models	MERCON automatic transmission fluid
2008 through 2011 models	Motorcraft transfer case fluid
2012 models	
4WD ESOF	Motorcraft transfer case fluid
Torque-on-demand	MERCON LV automatic transmission fluid
Chassis grease	SAE NLGI no. 2 chassis grease
Differential lubricant	
Front	
2009 and earlier	SAE 75W-90 GL-5 gear lubricant
2010 and later	SAE 80W-90 Premium rear axle lubricant
Rear	SAE 75W-140 GL-5 synthetic gear lubricant**/***

* **Caution:** Do not mix coolants of different colors. Doing so might damage the cooling system and/or the engine. The manufacturer specifies whether a green, yellow or orange colored coolant is to be used in these systems, depending on what was originally installed in the vehicle.

**Add 4 oz. of friction modifier (Ford part no. C8AZ-19B546-A) to 8.8 and 9.75 inch rear axles when lubricant is changed.

***Add 8 oz. of friction modifier (Ford part no. C8AZ-19B546-A) to 10.25 inch rear axles when lubricant is changed.

TUNE-UP AND ROUTINE MAINTENANCE

Capacities*

Engine oil (with filter)	
2004 and earlier models	6.0 quarts
2005 through 2014 models	7.0 quarts
2015 and later models	6.0 quarts
Cooling system	
F-150/F-250	
4.2L V6 engine	From 15.7 to 20.1 qts
4.6L and 5.4L V8 engines	From 17.9 to 23.9 qts
Expedition/Navigator	
3.5L V6 turbocharged engine	15.6 quarts
4.6L V8 engine	
Standard	18.0 qts
Heavy duty	19.9 qts
5.4L V8 engine	
2006 and earlier models	20.8 quarts
2007 through 2009 models	23.2 quarts
2010 and later models	
Standard	
With rear heat	19.0 quarts
Without rear heat	16.4 quarts
Heavy duty	
With rear heat	19.5 quarts
Without rear heat	16.9 quarts
Automatic transmission (from dry - see Section 26 for drain and refill fluid requirements)	
4R70W	13.9 qts
E4OD	
2WD	15.9 qts
4WD	16.4 qts
4R100 F-150	
2WD	17.1 qts
4WD	17.7 qts
4R100 Expedition	
2WD	15.9 qts
4WD	16.4 qts
4R70E - 4R75E	13.9 qts
6HP26	9.5 qts
6R75	11.0 qts
6R80	13.0 qts
Manual transmission	3.75 qts
Transfer case	
1997 models	4.1 quarts
1998 and later models	up to 2 quarts
2012 torque-on-demand models	1.5 quarts
Front differential	3.6 pts
Rear differential	
F-150	5.5 to 6.9 pts
F-250	7.8 pts
Expedition/Navigator	5.7 pts
2003 and earlier models	5.7 pints
2004 and later models	4.5 pints

All capacities approximate. Add as necessary to bring to appropriate level.

1-42 TUNE-UP AND ROUTINE MAINTENANCE

General

Radiator cap pressure rating	16 psi
Disc brake pad thickness (minimum)	1/8 inch
Drum brake shoe thickness (minimum)	1/16 inch

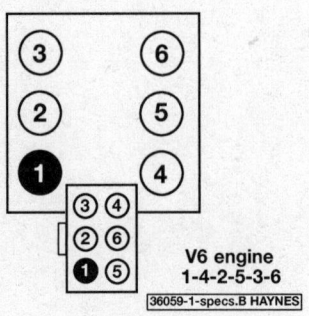

Cylinder location and coil terminal identification diagram - V6 engines

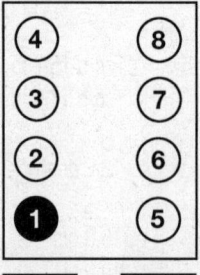

Cylinder location and coil terminal identification diagram - V8 engines (2003 and later engines are equipped with coil-on-plug ignition)

Ignition system

Spark plug type and gap
    1997
        4.2L V6 — Motorcraft AWSF-42EG or equivalent @ 0.054 inch
        4.6L V8 — Motorcraft AWSF-32PG or equivalent @ 0.054 inch
        5.4L V8 — Motorcraft AWSF-22EE or equivalent @ 0.054 inch
    1998 through 2000
        4.2L V6 — Motorcraft AGSF-34EE or equivalent @ 0.054 inch
        4.6L V8 — Motorcraft AWSF-32PP or equivalent @ 0.054 inch
        5.4L V8 — Motorcraft AWSF-22E or equivalent @ 0.054 inch
    2001 and later models
        4.2L V6 — Motorcraft AGSF-34EE or equivalent @ 0.054 inch
        4.6L V8 — Motorcraft AWSF-32P or equivalent @ 0.054 inch
        5.4L V8 — Motorcraft AWSF-22W or equivalent @ 0.054 inch
            2001 through 2003 — Motorcraft ASWF-22 or equivalent @ 0.045 inch
            2004 through 2006 — Motorcraft SP462 or equivalent @ 0.044 inch
            2007 through early 2008 — Motorcraft SP507 or equivalent @ 0.045 inch
            Late 2008 (brown coil boots) and later — Motorcraft SP509 or equivalent @ 0.041 inch
        3.5L V6 turbocharged engine — Motorcraft #12405 or equivalent @ 0.028 to 0.037 inch
Firing order
    V6 models — 1-4-2-5-3-6
    V8 models — 1-3-7-2-6-5-4-8

TUNE-UP AND ROUTINE MAINTENANCE

Torque specifications — Ft-lbs (unless otherwise indicated)

➡ **Note:** One foot-pound (ft-lb) of torque is equivalent to 12 inch-pounds (in-lbs) of torque. Torque values below approximately 15 foot-pounds are expressed in inch-pounds, because most foot-pound torque wrenches are not accurate at these smaller values.

Automatic transmission pan bolts	108 to 132 in-lbs
Automatic transmission pan drain plug	
4R100	18
6HP26	71 in-lbs
Automatic transmission side check/fill plug	
6HP26, 6R80, 6R75	26
Automatic transmission torque converter drain plug	
E4OD	22
4R100	19
4R70W	22
Wheel lug nuts	
12 mm wheel studs	100
14 mm wheel studs	150
Spark plugs	
2002 and earlier models	84 to 168 in-lbs
2003 and 2004 models	15
2005 through early 2008 models	25
Late 2008 (brown coil boots) and later models (non-turbo charged engines)	106 to 108 in-lbs
3.5L V6 turbocharged engine	133 in-lbs
Oil pan drain plug	
2009 and earlier	96 to 144 in-lbs
2010 through 2014	17
2015 and later	20
Manual transmission drain and fill plug	30 to 43
Transfer case fill and drain plug	
2002 and earlier models	80 to 203 in-lbs
2003 and later models	132 in-lbs
Front differential fill plug	15 to 22
Front differential cover bolts	20 to 27
Rear differential fill plug	15 to 30
Rear differential cover bolts	
2002 and earlier models	28 to 38
2003 and later models	24
Drive tensioner bolt(s)	18

1-44 TUNE-UP AND ROUTINE MAINTENANCE

Notes

2A
4.2L V6 ENGINE

Section

1 General information
2 Repair operations possible with the engine in the vehicle
3 Top Dead Center (TDC) for number one piston - locating
4 Valve covers - removal and installation
5 Rocker arms and pushrods - removal, inspection and installation
6 Valve springs, retainers and seals - replacement
7 Intake manifold - removal and installation
8 Exhaust manifolds - removal and installation
9 Cylinder heads - removal and installation
10 Timing chain cover - removal and installation
11 Timing chain and sprockets - inspection, removal and installation
12 Valve lifters - removal, inspection and installation
13 Camshaft, balance shaft and bearings - removal, inspection and installation
14 Oil pan - removal and installation
15 Oil pump - removal and installation
16 Crankshaft oil seals - replacement
17 Flywheel/driveplate - removal and installation
18 Engine mounts - check and replacement

Reference to other Chapters

Cylinder compression check - See Chapter 2D
Drivebelt check, adjustment and replacement - See Chapter 1
Engine - removal and installation - See Chapter 2D
Engine oil and filter change - See Chapter 1
Engine overhaul - general information - See Chapter 2D
Spark plug replacement - See Chapter 1
Water pump - removal and installation - See Chapter 3

2A-2 4.2L V6 ENGINE

1 General information

This Part of Chapter 2 is devoted to in-vehicle repair procedures for the 4.2L V6 engine. This engine design utilizes a cast-iron block with six cylinders arranged in a V-shape at a 90-degree angle between the two banks. The cylinder heads are also cast-iron and the block-mounted camshaft operates pushrods and rocker arms for valve actuation. The engine also features a balance shaft geared to the camshaft in the block.

All information concerning engine removal and installation and engine block and cylinder head overhaul can be found in Part D of this Chapter. The following repair procedures are based on the assumption that the engine is installed in the vehicle. If the engine has been removed from the vehicle and mounted on a stand, many of the steps outlined in this Part of Chapter 2 will not apply.

The Specifications included in this Part of Chapter 2 apply only to the procedures contained in this Part. Part D of Chapter 2 contains the Specifications necessary for cylinder head and engine block rebuilding.

2 Repair operations possible with the engine in the vehicle

Many major repair operations can be accomplished without removing the engine from the vehicle.

Clean the engine compartment and the exterior of the engine with some type of pressure washer before any work is done. It will make the job easier and help keep dirt out of the internal areas of the engine.

It may help to remove the hood to improve access to the engine as repairs are performed (refer to Chapter 11 if necessary).

If vacuum, exhaust, oil or coolant leaks develop, indicating a need for gasket or seal replacement, the repairs can generally be made with the engine in the vehicle. The intake and exhaust manifold gaskets, timing cover gasket, oil pan gasket, crankshaft oil seals and cylinder head gaskets are all accessible with the engine in place.

Exterior engine components, such as the intake and exhaust manifolds, the oil pan (and the oil pump), the water pump, the starter motor, the alternator, and the fuel system components can be removed for repair with the engine in place.

Since the cylinder heads can be removed without pulling the engine, valve component servicing can also be accomplished with the engine in the vehicle. Replacement of the timing chain and sprockets is also possible with the engine in the vehicle.

In extreme cases caused by a lack of necessary equipment, repair or replacement of piston rings, pistons, connecting rods and rod bearings is possible with the engine in the vehicle. However, this practice is not recommended because of the cleaning and preparation work that must be done to the components involved.

3 Top Dead Center (TDC) for number one piston - locating

▶ Refer to illustration 3.6

1 Top Dead Center (TDC) is the highest point in the cylinder that each piston reaches as it travels up-and-down when the crankshaft turns. Each piston reaches TDC on the compression stroke and again on the exhaust stroke, but TDC generally refers to piston position on the compression stroke. The timing marks on the vibration damper installed on the front of the crankshaft are referenced to the number one piston at TDC on the compression stroke.

2 Positioning the piston(s) at TDC is an essential part of procedures such as timing chain and sprocket replacement.

3 In order to bring any piston to TDC, the crankshaft must be turned using one of the methods outlined below. When looking at the front of the engine, normal crankshaft rotation is clockwise.

✱✱ WARNING:

Before beginning this procedure, be sure to place the transmission in Neutral and remove the ignition key.

a) The preferred method is to turn the crankshaft with a large socket and breaker bar attached to the vibration damper bolt threaded into the front of the crankshaft.
b) A remote starter switch, which may save some time, can also be used. Attach the switch leads to the S (switch) and B (battery) terminals on the starter motor. Once the piston is close to TDC, use a socket and breaker bar as described in the previous paragraph.
c) If an assistant is available to turn the ignition switch to the Start position in short bursts, you can get the piston close to TDC without a remote starter switch. Use a socket and breaker bar as described in Paragraph a) to complete the procedure.

4 Disable the ignition system by disconnecting the primary electrical connectors at the ignition coil pack/modules (see Chapter 5).

5 Remove the spark plugs and install a compression gauge in the number one cylinder. Turn the crankshaft clockwise with a socket and breaker bar as described above.

6 When the piston approaches TDC, compression will be noted on the compression gauge. Continue turning the crankshaft until the

3.6 Turn the crankshaft clockwise until the zero on the vibration damper scale is directly opposite the pointer

notch in the crankshaft damper is aligned with the TDC mark on the front cover (see illustration). At this point number one cylinder is at TDC on the compression stroke. If the marks aligned but there was no compression, the piston was on the exhaust stroke. Continue rotating the crankshaft one complete revolution (360-degrees).

7 After the number one piston has been positioned at TDC on the compression stroke, TDC for any of the remaining cylinders can be located by turning the crankshaft and following the firing order (refer to the Specifications). Divide the crankshaft pulley into three equal sections with chalk marks at three points, each indicating 120-degrees of crankshaft rotation. Rotating the engine 120-degrees past TDC #1 will put the engine at TDC for cylinder no. 4.

4 Valve covers - removal and installation

REMOVAL

▶ **Refer to illustrations 4.2 and 4.5**

1 Disconnect the cable from the negative battery terminal.
2 Note their locations, then detach the spark plug wire clips from the valve cover studs (see illustration).
3 Detach the spark plug wires from the plugs (see Chapter 1). Position the wires out of the way.
4 On some vehicles with cruise control, it may be necessary to disconnect the servo linkage at the throttle body and remove the servo bracket.
5 If you're removing the left (driver's side) valve cover, detach the oil fill cap and crankcase vent tube (see illustration).
6 If you're removing the right (passenger's side) valve cover, position the air cleaner duct out of the way (see Chapter 4) and remove the PCV valve.
7 Remove the valve cover bolts/nuts (see illustration 4.5), then detach the cover from the cylinder head.

➡**Note: If the cover is stuck to the cylinder head, bump one end with a wood block and a hammer to jar it loose. If that doesn't work, try to slip a flexible putty knife between the cylinder head and cover to break the gasket seal. Don't pry at the cover-to-cylinder head joint or damage to the sealing surfaces may occur (leading to oil leaks in the future). Some valve covers are made of plastic - be extra careful when tapping or pulling on them. On some models, the valve cover bolts stay with the valve cover. Do not attempt to remove them completely.**

INSTALLATION

8 The mating surfaces of each cylinder head and valve cover must be perfectly clean when the covers are installed. Use a gasket scraper to remove all traces of sealant and old gasket material, then clean the mating surfaces with lacquer thinner or acetone. If there's sealant or oil on the mating surfaces when the cover is installed, oil leaks may develop.
9 Clean the mounting bolt threads with a die to remove any corrosion and restore damaged threads. Make sure the threaded holes in the cylinder head are clean - run a tap into them to remove corrosion and restore damaged threads. Apply a small amount of light oil to the bolt threads.
10 The gaskets should be mated to the covers before the covers are installed. Make sure the tabs on the gaskets(s) engage in the slots in the cover(s). On engines that don't have gaskets, apply a 3/16-inch bead of RTV sealant to the cover flange, inside of the bolt holes.

4.2 Remove the spark plug wire clips

4.5 Detach the crankcase vent tube (A) - B indicates two of the valve cover bolts

11 Carefully position the cover on the cylinder head and install the bolts/nuts.
12 Tighten the bolts in three or four steps to the torque listed in this Chapter's Specifications. Plastic valve covers are easily damaged, so don't overtighten the bolts!
13 The remaining installation steps are the reverse of removal.
14 Start the engine and check carefully for oil leaks as the engine warms up.

5 Rocker arms and pushrods - removal, inspection and installation

REMOVAL

◆ **Refer to illustrations 5.2 and 5.4**

1 Refer to Section 4 and detach the valve cover(s) from the cylinder head(s).

2 Beginning at the front of one cylinder head, remove the rocker arm fulcrum bolts (see illustration). Store them separately in marked containers to ensure that they will be reinstalled in their original locations.

➥**Note: If the pushrods are the only items being removed, loosen each bolt just enough to allow the rocker arms to be rotated to the side so the pushrods can be lifted out.**

3 Lift off the rocker arms and fulcrums. Store them in the marked containers with the bolts (they must be reinstalled in their original locations).

4 Remove the pushrods and store them separately to make sure they don't get mixed up during installation (see illustration).

INSPECTION

◆ **Refer to illustration 5.7**

5 Check each rocker arm for wear, cracks and other damage, especially where the pushrods and valve stems contact the rocker arm faces.

6 Make sure the hole at the pushrod end of each rocker arm is open.

7 Check each rocker arm pivot area and fulcrum for wear, cracks and galling (see illustration). If the rocker arms are worn or damaged, replace them with new ones and use new fulcrums as well.

8 Inspect the pushrods for cracks and excessive wear at the ends. Roll each pushrod across a piece of plate glass to see if it's bent (if it wobbles, it's bent).

INSTALLATION

9 Lubricate the lower end of each pushrod with clean engine oil or moly-base grease and install them in their original locations. Make sure each pushrod seats completely in the lifter.

10 Apply moly-base grease to the ends of the valve stems and the upper ends of the pushrods before positioning the rocker arms and fulcrums.

11 Apply moly-base grease to the fulcrums to prevent damage to the mating surfaces before engine oil pressure builds up. Set the rocker arms and guides in place, then install the fulcrums and bolts. Tighten the rocker arm bolts to Specifications.

12 Install the valve covers.

13 The remainder of installation is the reverse of removal.

5.2 The rocker arm fulcrum bolts (arrow) may not have to be completely removed in all cases - loosen them several turns and see if the rocker arms can be pivoted out of the way to allow pushrod removal

5.4 A perforated cardboard box can be used to store the pushrods to ensure that they are reinstalled in their original locations - note the label indicating the front of the engine

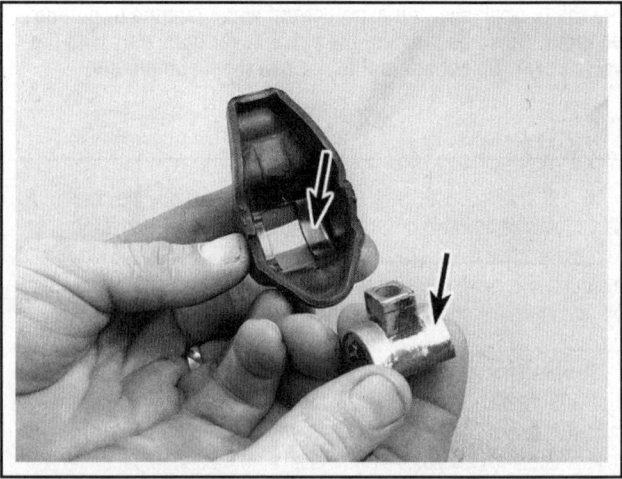

5.7 Check for wear on the rocker arm and fulcrum contact areas

4.2L V6 ENGINE 2A-5

6 Valve springs, retainers and seals - replacement

▸ Refer to illustrations 6.4, 6.8, 6.9 and 6.16

➡Note: Broken valve springs and defective valve stem seals can be replaced without removing the cylinder heads. Two special tools and a compressed air source are normally required to perform this operation, so read through this Section carefully and rent or buy the tools before beginning the job.

1 Refer to Section 4 and remove the valve cover from the affected cylinder head. If all of the valve stem seals are being replaced, remove both valve covers.

2 Remove the spark plug from the cylinder which has the defective component. If all of the valve stem seals are being replaced, all of the spark plugs should be removed.

3 Turn the crankshaft until the piston in the affected cylinder is at Top Dead Center on the compression stroke (refer to Section 3 for instructions). If you're replacing all of the valve stem seals, begin with cylinder number one and work on the valves for one cylinder at a time. Move from cylinder-to-cylinder following the firing order sequence (see this Chapter's Specifications).

4 Thread an adapter into the spark plug hole (see illustration) and connect an air hose from a compressed air source to it. Most auto parts stores can supply the air hose adapter.

➡Note: Many cylinder compression gauges utilize a screw-in fitting that may work with your air hose quick-disconnect fitting.

5 Remove the bolt, fulcrum and rocker arm for the valve with the defective part and pull out the pushrod. If all of the valve stem seals are being replaced, all of the rocker arms and pushrods should be removed (refer to Section 5).

6 Apply compressed air to the cylinder.

❊❊ WARNING:

The piston may be forced down by compressed air, causing the crankshaft to turn suddenly. If the wrench used when positioning the number one piston at TDC is still attached to the bolt in the crankshaft nose, it could cause damage or injury when the crankshaft moves.

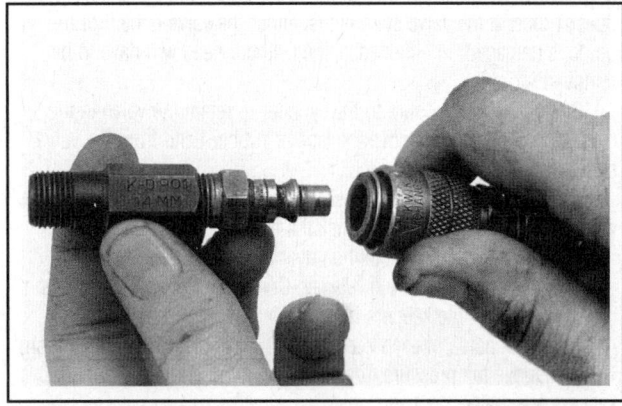

6.4 This is what the air hose adapter that threads into the spark plug hole looks like - they're commonly available from auto parts stores

7 The valves should be held in place by the air pressure.

8 Stuff shop rags into the cylinder head holes above and below the valves to prevent parts and tools from falling into the engine, then use a valve spring compressor to compress the spring. Remove the keepers with small needle-nose pliers or a magnet (see illustration).

➡Note: A couple of different types of tools are available for compressing the valve springs with the cylinder head in place. One type, shown here, grips the lower spring coils and presses on the retainer as the knob is turned, while the other type utilizes the rocker arm bolt for leverage. Both types work very well, although the lever type is usually less expensive.

9 Remove the spring keepers, retainer and spring, then remove the valve guide seal (see illustration).

➡Note: If air pressure fails to hold the valve in the closed position during this operation, the valve face or seat is probably damaged. If so, the cylinder head will have to be removed for additional repair operations.

10 Wrap a rubber band or tape around the top of the valve stem so the valve won't fall into the combustion chamber, then release the air pressure.

11 Inspect the valve stem for damage. Rotate the valve in the guide

6.8 Once the spring is depressed, the keepers can be removed with a small magnet or needle-nose pliers (a magnet is preferred to prevent dropping the keepers)

6.9 The seal can be pulled off the guide with a pair of pliers

2A-6 4.2L V6 ENGINE

and check the end for eccentric movement, which would indicate that the valve is bent.

12 Move the valve up-and-down in the guide and make sure it doesn't bind. If the valve stem binds, either the valve is bent or the guide is damaged. In either case, the cylinder head will have to be removed for repair.

13 Reapply air pressure to the cylinder to retain the valve in the closed position, then remove the tape or rubber band from the valve stem.

14 Lubricate the valve stem with engine oil and the valve stem tip with polyethylene grease, then install a new guide seal.

15 Install the spring in position over the valve.

16 Install the valve spring retainer. Compress the valve spring and carefully position the keepers in the groove. Apply a small dab of grease to the inside of each keeper to hold it in place (see illustration).

17 Remove the pressure from the spring tool and make sure the keepers are seated.

18 Disconnect the air hose and remove the adapter from the spark plug hole.

19 Refer to Section 5 and install the rocker arm(s) and pushrod(s).

20 Install the spark plug(s) and hook up the wire(s).

6.16 Apply a small dab of grease to each keeper as shown here before installation - it'll keep them in place on the valve stem as the spring is released

21 Refer to Section 4 and install the valve cover(s).

22 Start and run the engine, then check for oil leaks and unusual sounds coming from the valve cover area.

7 Intake manifold - removal and installation

REMOVAL

▶ Refer to illustrations 7.4, 7.5, 7.6, 7.7, 7.8 and 7.9

1 Relieve the fuel pressure and remove the air duct assembly (see Chapter 4).

2 Disconnect the negative cable from the battery.

3 Drain the cooling system (see Chapter 1).

4 Remove the upper intake plenum and throttle body (see Chapter 4). Disconnect the IMRC vacuum and electrical connectors, the fuel lines, fuel pressure regulator and injector connectors (see illustration).

5 Cover the air intake passages with a shop towel (see illustration). Disconnect the upper radiator hose and heater hoses from the intake manifold fittings.

6 Disconnect the EGR valve vacuum hose and the EGR tube fitting at the EGR valve (see illustration).

7.4 Disconnect the two electrical connectors on the IMRC solenoids at the back of the intake manifold

7.5 Cover the air intake with a shop towel to prevent debris from falling into the engine

7.6 Disconnect the lines and vacuum hose at the EGR valve

4.2L V6 ENGINE 2A-7

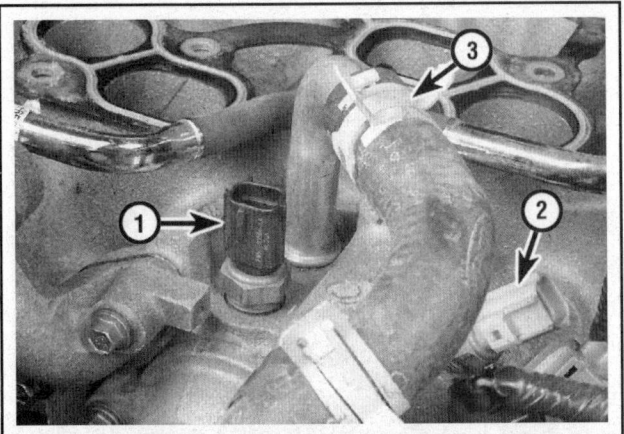

7.7 Label and disconnect the hoses and wiring

1 Engine Coolant Temperature sensor connector
2 Coolant temperature sending unit connector
3 Coolant hose

7 Label and disconnect the vacuum and emissions hoses and wire harness connectors attached to the intake manifold (see illustration).

8 Loosen the intake manifold mounting bolts in 1/4-turn increments until they can be removed by hand (see illustration). Not all bolts are easily noticed; there are 14 in all, six long ones and eight short ones. Keep track of which ones go where.

9 The intake manifold will probably be stuck to the cylinder heads and force may be required to break the gasket seal. A prybar can be positioned under the cast-in lug (see illustration) to pry up the front of the intake manifold, but make sure all bolts have been removed first!

※ CAUTION:

Don't pry between the engine block and intake manifold or the cylinder heads and intake manifold or damage to the gasket sealing surfaces may occur, leading to vacuum and oil leaks.

INSTALLATION

▸ **Refer to illustrations 7.10, 7.13, 7.14 and 7.17**

※ CAUTION:

The mating surfaces of the cylinder heads, engine block and manifold must be perfectly clean when the manifold is installed. Gasket removal solvents in aerosol cans are available at most auto parts stores and may be helpful when removing old gasket material that's stuck to the cylinder heads and manifold (since the manifold is made of aluminum, aggressive scraping can cause damage). Be sure to follow directions printed on the container.

10 Use a gasket scraper to remove all traces of sealant and old gasket material (see illustration), then clean the mating surfaces with lacquer thinner or acetone. If there's old sealant or oil on the mating surfaces when the manifold is installed, oil or vacuum leaks may develop. When working on the cylinder heads and block, cover the lifter valley with shop rags to keep debris out of the engine. Use a vacuum cleaner to remove any gasket material that falls into the intake ports in the cylinder heads.

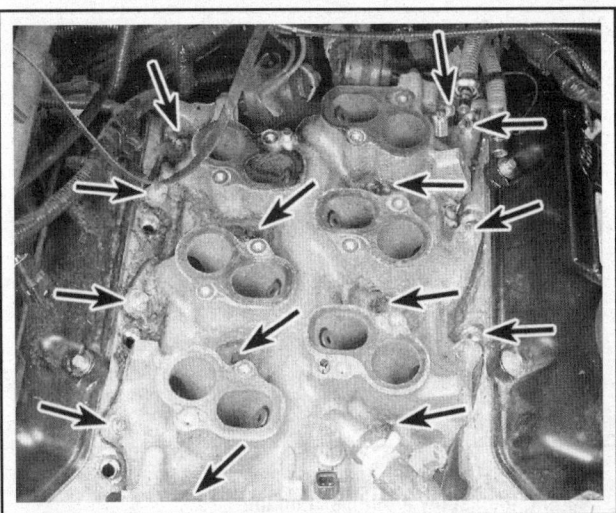

7.8 Remove the intake manifold mounting bolts

7.9 Pry against a casting protrusion to break the intake manifold loose

7.10 After covering the lifter valley, use a gasket scraper to remove all traces of sealant and old gasket material from the cylinder head and intake manifold mating surfaces

2A-8 4.2L V6 ENGINE

7.13 Install the rubber end seals over a thin bead of RTV sealant, then apply a thin bead on top

11 Use a tap of the correct size to chase the threads in the bolt holes, then use compressed air (if available) to remove the debris from the holes.

✴ WARNING:

Wear safety glasses or a face shield to protect your eyes when using compressed air! Remove excessive carbon deposits and corrosion from the exhaust and coolant passages in the cylinder heads and manifold.

12 Make a final inspection of the gasket surfaces before installing the manifold.

7.17 Intake manifold bolt tightening sequence

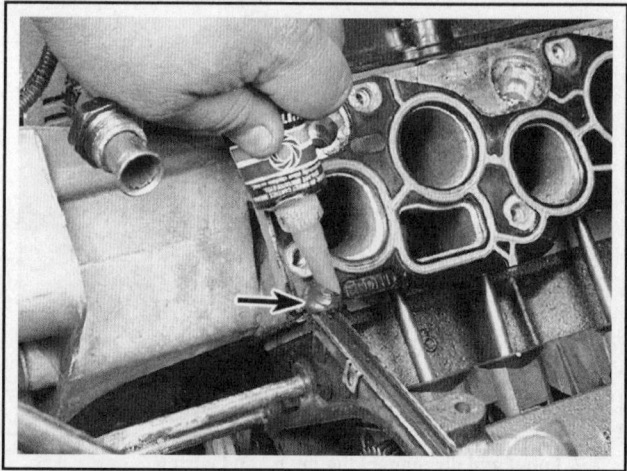

7.14 Apply a dab of RTV sealant to the corners where the engine block, cylinder heads and intake manifold converge, then position the side gaskets in place and apply an additional bead of RTV sealant where the end seals and intake manifold gaskets meet

13 Apply a 1/8-inch wide bead of RTV sealant to the front and rear of the block surface and install the rubber front and rear manifold seals (see illustration).

➡ **Note: This sealant sets up in 15 minutes. Do not take longer to install and tighten the manifold once the sealant is applied, or leaks may occur.**

14 Apply a small dab of RTV at the four corners where the side gaskets will fit against the end seals. Position the side gaskets on the cylinder heads, over the alignment studs. The upper side of each gasket will have a TOP or THIS SIDE UP label stamped into it to ensure correct installation (see illustration).

15 Make sure all intake port openings, coolant passage holes and bolt holes are aligned correctly.

16 Carefully set the manifold in place while the sealant is still wet.

✴ CAUTION:

Don't disturb the gaskets. Make sure the end seals haven't been disturbed.

17 Lightly oil the manifold bolts, install them and tighten to the torque listed in this Chapter's Specifications, following the recommended sequence (see illustration). Work up to the final torque in three steps.

18 The remaining installation steps are the reverse of removal. Start the engine and check carefully for oil and coolant leaks at the intake manifold joints.

4.2L V6 ENGINE   2A-9

8 Exhaust manifolds - removal and installation

REMOVAL

▶ Refer to illustrations 8.5, 8.7 and 8.8

1 Disconnect the negative battery cable from the battery.
2 Raise the vehicle and support it securely on jackstands.

➡ Note: On models with air suspension, turn off the air suspension switch before raising the vehicle.

3 Disconnect the oxygen sensor electrical connector (see Chapter 6).
4 Working under the vehicle, apply penetrating oil to the exhaust Y-pipe-to-manifold studs and nuts (they're usually rusty).
5 Remove the nuts holding the exhaust Y-pipe to the exhaust manifolds (see illustration). In extreme cases you may have to heat them with a propane or acetylene torch in order to loosen them.

Right (passenger's side) manifold

6 Disconnect the EGR tube from the manifold (see Chapter 6).

Left (driver's side) manifold

7 Remove the nut holding the oil dipstick tube support at the front of the manifold, then pull the dipstick tube up and out of the oil pan (see illustration).

Both manifolds

8 Remove the mounting bolts and separate the exhaust manifold(s) from the cylinder head (see illustration). Note the locations of the bolts and studs, and remove the old gaskets.

INSTALLATION

9 Check the exhaust manifold for cracks and make sure the bolt threads are clean and undamaged. The exhaust manifold and cylinder head mating surfaces must be clean before the manifolds are reinstalled - use a gasket scraper to remove all carbon deposits and old gasket material.

8.5 From below, remove the two exhaust Y-pipe-to-manifold nuts

➡ Note: If the exhaust manifold is being replaced with a new one, remove the oxygen sensor. Clean the threads of the oxygen sensor with a wire brush and coat the threads with high-temperature anti-seize compound before transferring the sensor to the new exhaust manifold.

10 Position the exhaust manifold and new gasket on the cylinder head and install the mounting bolts and studs.

➡ Note: Exhaust manifold warpage is possible on some V6 engines. If the manifold is warped slightly, install the pilot bolts first. Sometimes it's necessary to elongate (with a round file) some holes in the exhaust manifolds to start the bolts - but never file out the pilot bolt holes!

11 When tightening the mounting bolts, tighten the center pair first, the front pair, then the rear pair, and be sure to use a torque wrench. Tighten the bolts in three equal steps until the torque listed in this Chapter's Specifications is reached.
12 The remaining installation steps are the reverse of removal.
13 Start the engine and check for exhaust leaks.

8.7 Remove the nut and pull the oil dipstick tube out

8.8 Remove the exhaust manifold bolts and studs (left side shown, right side similar)

2A-10 4.2L V6 ENGINE

9 Cylinder heads - removal and installation

> **※ CAUTION:**
>
> The engine must be completely cool when the cylinder heads are removed. Failure to allow the engine to cool off could result in cylinder head warpage.

REMOVAL

1 Disconnect the cable from the negative battery terminal.
2 Remove the valve covers (see Section 4).
3 Remove the pushrods and rocker arms (see Section 5).
4 Remove the upper intake plenum (see Chapter 4) and the lower intake manifold (see Section 7).
5 Unbolt the exhaust manifold (see Section 8) for the head(s) being removed.

Left (driver's side) cylinder head

▶ Refer to illustration 9.8

6 Unbolt the power steering reservoir and tie it aside in an upright position. Leave the hoses connected (see Chapter 10).
7 Remove the drivebelts (see Chapter 1).
8 Remove the air-conditioning compressor/power steering pump bracket from the head and position the assembly out of the way (see illustration) (see Chapter 3). DO NOT disconnect the hoses!

Right cylinder head

▶ Refer to illustration 9.9

9 Disconnect the alternator wiring (see Chapter 5). Unbolt the alternator bracket (see illustration).

➡ **Note:** The bracket can be left in place, with just the upper bolt in place, loose enough to allow the steel bracket to be swung out of the way. For removal of the aluminum compressor/alternator mount, remove all the bolts/studs.

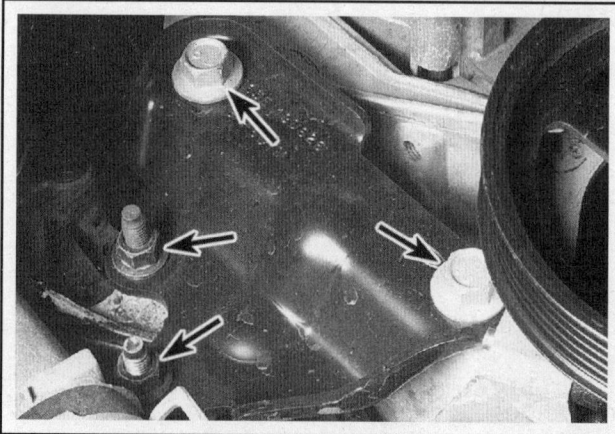

9.8 Remove the bolts and nuts (arrows) until the compressor and power steering pump steel bracket can be moved aside, away from the LH cylinder head

Both cylinder heads

▶ Refer to illustrations 9.10 and 9.11

10 Loosen the cylinder head bolts in 1/4-turn increments until they can be removed by hand. Work from bolt-to-bolt in a pattern that's the reverse of the tightening sequence.

➡ **Note 1:** Head bolts should not be reused. Remove the bolts and discard them - new bolts must be used when installing the cylinder head(s).

➡ **Note 2:** At the back of the cylinder heads, near the firewall, remove the bolts holding the wiring harness to each head (see illustration).

11 Lift the cylinder head(s) off the engine. If resistance is felt, DO NOT pry between the cylinder head and engine block as damage to the mating surfaces will result. To dislodge the cylinder head, place a wood block against the end of it and strike the wood block with a hammer or

9.9 Remove the bolts/nuts (arrows) holding the compressor/alternator bracket to the RH head idler pulley (shown removed)

9.10 Remove the two bolts (arrows) at the back of the heads holding the wiring harness (transmission removed for clarity)

4.2L V6 ENGINE 2A-11

9.11 Using a prybar, carefully lever it against a casting protrusion to lift the cylinder head and break the gasket seal

place a prybar against a casting protrusion (see illustration). Store the cylinder heads on blocks of wood to prevent damage to the gasket sealing surfaces.

> **CAUTION:**
> Do not slide the heads across the floor or workbench, the aluminum is easily gouged.

12 Cylinder head disassembly and inspection procedures are covered in detail in Chapter 2, Part D.

INSTALLATION

▸ **Refer to illustrations 9.14, 9.16 and 9.19**

13 The mating surfaces of the cylinder heads and engine block must be perfectly clean when the cylinder heads are installed. Use a gasket scraper to remove all traces of carbon and old gasket material, then clean the mating surfaces with lacquer thinner or acetone. If there's oil on the mating surfaces when the cylinder heads are installed, the gaskets may not seal correctly and leaks may develop. When working on the engine block, cover the lifter valley with shop rags to keep debris out of the engine. Use a vacuum cleaner to remove any debris that falls into the valley or intake ports.

14 Check the engine block and cylinder head mating surfaces for nicks, deep scratches and other damage. If damage is slight, it can be removed with a file - if it's excessive, machining may be the only alternative. Do not use a rotary abrasive tool or wire brush on the aluminum cylinder heads. Use a straightedge and feeler gauges to check for warpage (see illustration). If the warpage is beyond Specifications, have the head machined at an automotive machine shop.

15 Use a tap of the correct size to chase the threads in the cylinder head bolt holes. Dirt, corrosion, sealant and damaged threads will affect torque readings.

16 Position the new gasket(s) over the dowel pins in the engine block (see illustration). Make sure it's facing the right way.

17 Carefully position the cylinder head(s) on the engine block without disturbing the gasket(s).

18 Before installing the new cylinder head bolts, lightly oil the threads.

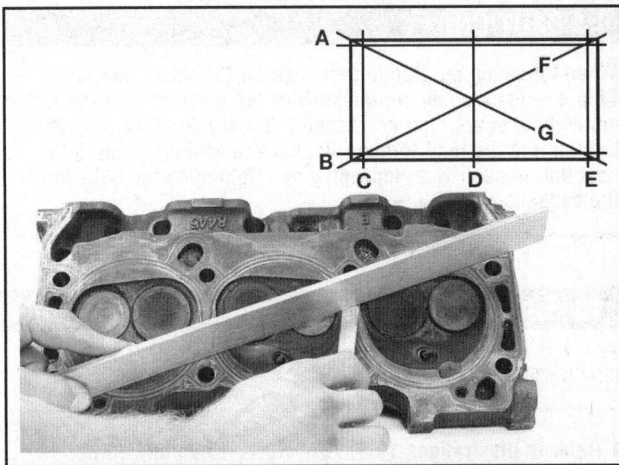

9.14 Check the cylinder head and block surfaces for flatness with a straightedge - if a feeler gauge in excess of the Specification fits under it, have the head machined

9.16 Locating dowels (arrows) are used to position the gaskets on the engine block - make sure the mark (circled) is correctly oriented

9.19 Cylinder head bolt tightening sequence

19 Install the bolts (long bolts inside, short bolts outside row) and tighten them finger tight. Follow the recommended sequence and tighten the bolts, in the recommended steps, to the torque listed in this Chapter's Specifications (see illustration).

2A-12 4.2L V6 ENGINE

✳✳ CAUTION:

When following the torque sequence for the V6 engine, at Step 4, do not loosen all the bolts at the same time or the gasket will not seal properly. Loosen the first bolt in the sequence, tighten it to the final torque (Steps 5 and 6) then go on to the next bolt in sequence, loosening and tightening the bolts until the sequence is completed.

20 The remaining installation steps are the reverse of removal.
21 Change the engine oil and filter (Chapter 1), then start the engine and check carefully for oil and coolant leaks.

10 Timing chain cover - removal and installation

REMOVAL

▶ Refer to illustrations 10.3, 10.7, 10.8, 10.9a and 10.9b

1 Refer to Chapter 1 and drain the cooling system. Refer to Chapter 3 and remove the fan, fan shroud and water pump. Disconnect the radiator and heater hoses.

✳✳ WARNING:

Some models covered by this manual are equipped with an air suspension system. Always disconnect the electrical power to the suspension system before lifting or towing (see Chapter 10). Failure to perform this procedure may result in unexpected shifting or movement of the vehicle, which could cause personal injury.

2 Drain the engine oil and remove the oil filter (see Chapter 1).
3 Remove the crankshaft pulley and vibration damper (see Section 16). Loosen the bolt on the belt idler and swing it aside (see illustration). It isn't necessary to remove the idler.
4 Unbolt and remove all accessory brackets attached to the timing chain cover. When unbolting the power steering pump (see Section 9), tie it aside with the hoses still connected. On air-conditioned models, remove the compressor/power steering pump assembly and bracket (see Section 9).
5 Position the number one piston at TDC on the compression stroke (Section 3), then disconnect the electrical connector at the camshaft position sensor and remove the camshaft position sensor (see Chapter 6). Also disconnect the knock sensor and crankshaft position sensor electrical connectors (see Chapter 6).
6 Remove the front oil pan bolts (see Section 14).
7 Remove the socket-head bolt (see illustration).
8 Disconnect the two heater hoses, then unbolt the heater outlet tube and pull it out (see illustration).

➡ Note: Pry the wiring harness clip out of the hole in the heater tube bracket.

10.3 Loosen the bolt on the idler assembly and swing the idler out of the way of the cover

10.7 Remove this Allen bolt (arrow) from the bottom left of the front cover (oil filter and pump shown removed)

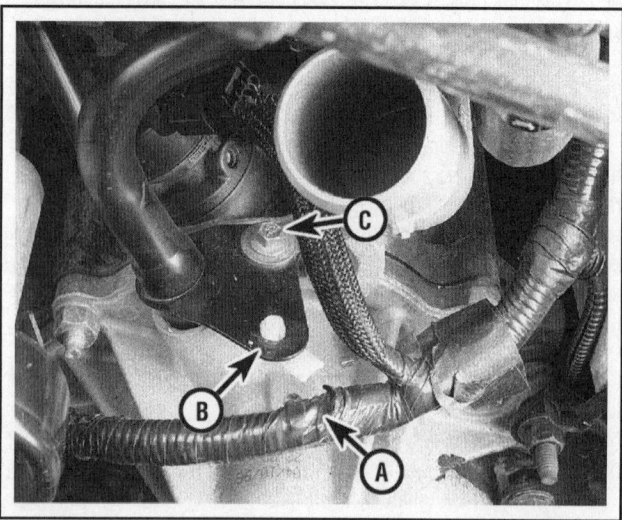

10.8 Pry the wiring harness clip (A) out of the heater tube mount (B), then remove the bolt (C) and pull the tube out

4.2L V6 ENGINE 2A-13

10.9a Gently tap the timing chain cover loose with a soft-face hammer

9 Remove the bolts and separate the timing chain cover from the engine block. If it's stuck, tap it gently with a soft-face hammer (see illustrations).

✻✻ CAUTION:

DO NOT use excessive force or you may crack the cover. If the cover is difficult to remove, double check to make sure all of the bolts have been removed.

INSTALLATION

▸ Refer to illustrations 10.10, 10.12 and 10.16

10 Use a gasket scraper to remove all traces of old gasket material and sealant from the cover, oil pan and engine block, then clean them with lacquer thinner or acetone (see illustration).

11 The oil pump is mounted in the timing chain cover and driven by the intermediate shaft, which is driven by the camshaft position sensor. See Section 15 for oil pump information.

12 Inspect the gear and intermediate shaft inside the front cover (see illustration).

13 While the cover is off the engine, it's a good idea to install a new crankshaft front seal (see Section 16).

14 Lubricate the front crankshaft oil seal lip with engine oil.

10.12 While the cover is off, inspect the camshaft position sensor driven gear (A) and the shaft (B)

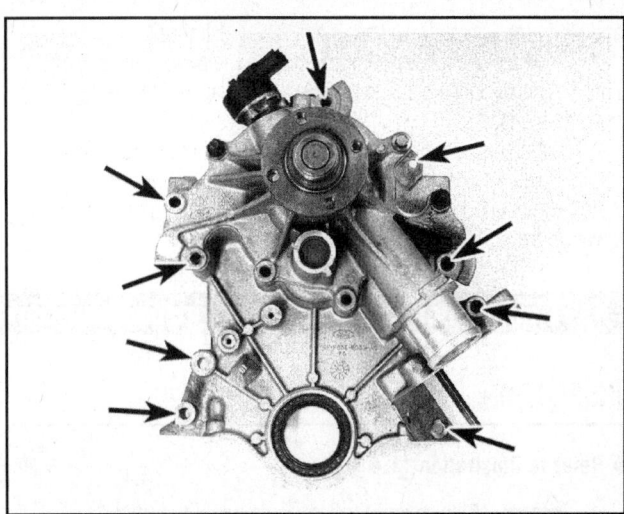

10.9b Timing chain cover bolt and stud locations (arrows)

10.10 Scrape the old gasket and sealant from the block and cover, then clean the surfaces thoroughly

15 Apply a small bead or RTV sealant along the oil pan-to-block joints.

16 Install the front cover gasket on the block and press it into place. Then again apply a small bead of RTV sealant to the block-to-pan joints

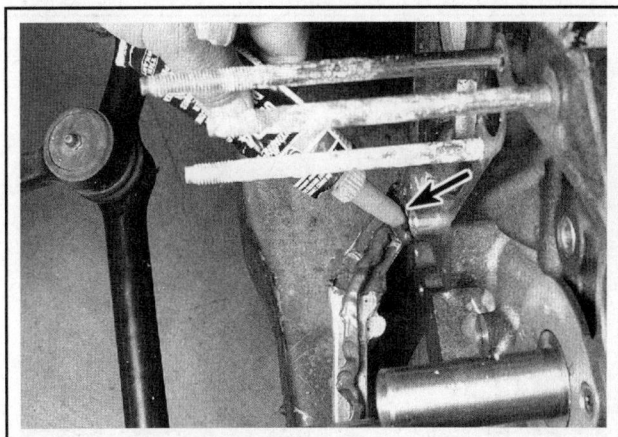

10.16 With the new cover gasket on the engine, apply a small bead of RTV sealant along the pan-to-cover flange, and a small bead on each side (arrow) where the pan, cover and block meet

2A-14 4.2L V6 ENGINE

and along the front edge of the oil pan (see illustration).

17 Slide the front cover onto the engine. The dowel pins will position it correctly. Don't damage the seal and make sure the gasket remains in place.

18 Install the bolts finger tight. Tighten them to the torque listed in this Chapter's Specifications only after the water pump has been installed (some of the water pump bolts also hold the timing chain cover in place).

→ Note: When installing the water pump, be sure to coat the threads of the water pump bolt with sealant (see illustration 10.9b).

19 Install the oil pan bolts (see Section 14).
20 Install the remaining parts in the reverse order of removal.
21 Add engine oil and coolant (see Chapter 1).
22 Run the engine and check for leaks.

11 Timing chain and sprockets - inspection, removal and installation

INSPECTION

▸ **Refer to illustration 11.4**

1 Disconnect the negative battery cable from the battery.
2 Refer to Section 3 and position the number one piston several degrees before TDC on the compression stroke.
3 Remove the right valve cover (see Section 4).
4 Attach a dial indicator to the cylinder head with the plunger in-line with and resting on the pushrod side of the rocker arm (see illustration).
5 Remove the timing chain cover (see Section 10).
6 Temporarily remove the timing chain and sprockets to remove the chain tensioner (see below).
7 Temporarily install the timing chain and sprockets without the tensioner and slip the timing chain cover and vibration damper in place to provide timing marks.
8 Turn the crankshaft clockwise until the number one piston is at TDC (see Section 3). This will take up the slack on the right side of the chain.
9 Zero the dial indicator.
10 Slowly turn the crankshaft counterclockwise until the slightest movement is seen on the dial indicator. Stop and note how far the number one piston has moved away from TDC by looking at the ignition timing marks.
11 If the mark has moved more than 6 degrees, install a new timing chain and sprockets.

11.4 Install a dial indicator to measure timing chain deflection - use a short length of vacuum hose to hold the plunger on the pushrod (rocker arm removed), or set the indicator on the pushrod end of the rocker arm

REMOVAL

▸ **Refer to illustrations 11.14a, 11.14b, 11.15, 11.16 and 11.17**

12 Position the number one piston at TDC on the compression stroke (see Section 3).
13 Remove the timing chain cover (see Section 10). Try to avoid turning the crankshaft during vibration damper removal.
14 Make sure the crankshaft, camshaft and balance shaft sprocket timing marks are aligned (see illustrations). If they aren't, install the vibration damper bolt and use it to turn the crankshaft clockwise until all marks are aligned.

11.14a Align the timing marks on the crankshaft and camshaft sprockets (arrows) before removing the sprockets from the shafts

11.14b When the marks are aligned for TDC, the balance shaft keyway (arrow) will point straight up

4.2L V6 ENGINE 2A-15

11.15 Remove the camshaft sprocket bolt without turning the engine

11.16 Compress the timing chain tensioner with a screwdriver and insert an Allen wrench as a retaining pin

11.17 Remove the camshaft sprocket and chain from the camshaft

➡ **Note:** When all marks are aligned, the crankshaft keyway is straight up, the camshaft keyway is straight down, and the balance shaft keyway is straight up.

15 Remove the camshaft sprocket mounting bolt and camshaft position sensor drive gear (see illustration).

✳✳ CAUTION:

Do not allow the crankshaft to be turned from TDC while removing the camshaft sprocket bolt.

16 Compress the timing chain tensioner with a screwdriver and insert a drill or Allen wrench as a retaining pin to hold it in the retracted position (see illustration).

17 Pull the camshaft sprocket/chain off the camshaft (see illustration).

➡ **Note:** The crankshaft sprocket will slide off with the chain and camshaft sprocket.

18 If you intend to remove the tensioner assembly, remove the two bolts holding it to the block.

INSTALLATION

▶ **Refer to illustrations 11.20 and 11.22**

19 Turn the crankshaft until the key is facing up (12 o'clock position).

20 Drape the chain over the camshaft sprocket and turn the sprocket until the timing mark faces down (6 o'clock position). Position the chain over the camshaft and crankshaft sprockets with their timing marks aligned, then slip both sprockets onto the camshaft and crankshaft (see illustration).

21 When correctly installed, a straight line should pass through the center of the balance shaft gear, the camshaft, the camshaft timing mark (in the 6 o'clock position), the crankshaft timing mark (in the 12 o'clock position) and the center of the crankshaft (see illustrations 11.14a and 11.14b). DO NOT proceed until the valve timing is correct!

22 Install the camshaft position sensor drive gear (see illustration).

23 Apply a non-hardening thread locking compound to the threads and install the camshaft sprocket bolt. Tighten the bolt to the torque listed in this Chapter's Specifications.

24 Remove the retaining pin from the timing chain tensioner. Reinstall the remaining parts in the reverse order of removal.

11.20 Assemble the chain and both sprockets so the marks (arrows) are aligned, then slip the assembly onto the crankshaft and camshaft

11.22 Align the keyway (arrow) on the camshaft position sensor drive gear with the Woodruff key on the camshaft

2A-16 4.2L V6 ENGINE

12 Valve lifters - removal, inspection and installation

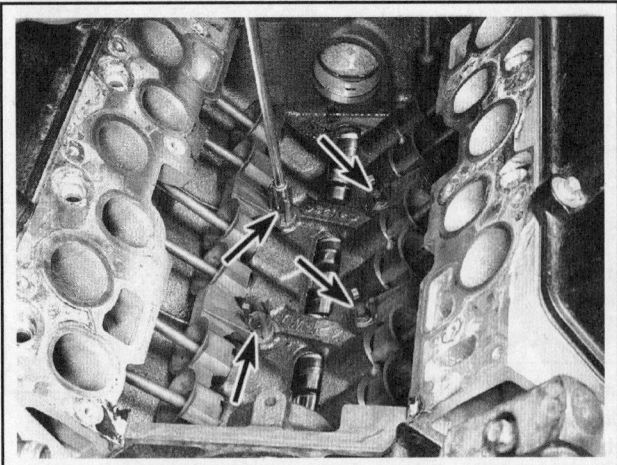

12.3a Remove the lifter guide retainer bolts (arrows)

12.3b Be sure to store the lifters in an organized manner to make sure they're reinstalled in their original locations

REMOVAL

▶ **Refer to illustrations 12.3a, 12.3b, 12.4a and 12.4b**

1 Remove the upper intake plenum (see Chapter 4) and lower intake manifold (see Section 7).

2 Remove the rocker arms and pushrods (see Section 5).

3 Before removing the lifters, arrange to store them in a clearly labeled box to ensure that they're reinstalled in their original locations. Remove the lifter guide plates (see illustrations).

4 There are several ways to extract the lifters from the bores. Special tools designed to grip and remove lifters are manufactured by many tool companies and are widely available (see illustration), but may not be needed in every case. On newer engines without a lot of varnish buildup, the lifters can often be removed with a small magnet (see illustration) or even with your fingers. A machinist's scribe with a bent end can be used to pull the lifters out by positioning the point under the retainer ring in the top of each lifter.

✳✳ CAUTION:

Don't use pliers to remove the lifters unless you intend to replace them with new ones (along with the camshaft). The pliers may damage the precision machined and hardened lifters, rendering them useless.

On engines with a lot of sludge and varnish, work the lifters up and down, using carburetor cleaner spray to loosen the deposits.

INSPECTION

▶ **Refer to illustrations 12.6 and 12.7**

5 Clean the lifters with solvent and dry them thoroughly without mixing them up.

6 Check each lifter wall and pushrod seat for scuffing, score marks and uneven wear. If the lifter walls are damaged or worn (which isn't very likely), inspect the lifter bores in the engine block as well. If the pushrod seats (see illustration) are worn, check the pushrod ends.

12.4a If the lifters are difficult to remove, you may have to remove them with a special puller . . .

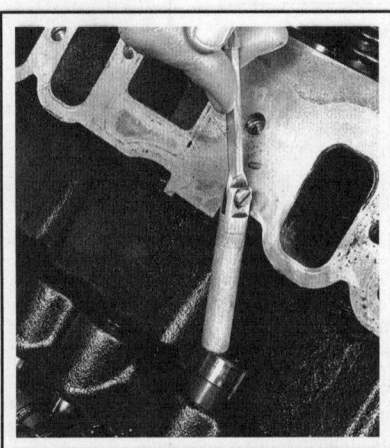

12.4b . . . or you may be able to remove the lifters with a magnet

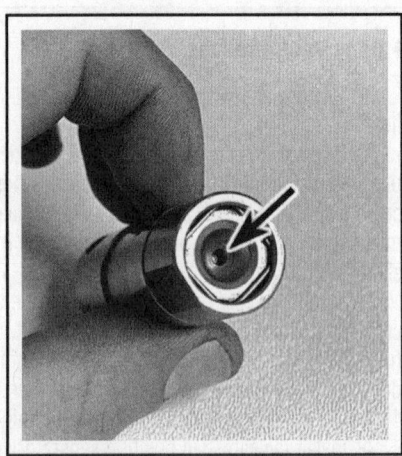

12.6 Inspect the pushrod seat (arrow) in the top of each lifter for wear

4.2L V6 ENGINE 2A-17

7 Check the roller carefully for wear and damage and make sure they turn freely without excessive play (see illustration).

8 If any lifters are found to be defective, they can be replaced with new ones without having to replace the camshaft (unlike conventional, non-roller lifters), but if the camshaft is being replaced due to high mileage, all the lifters should be replaced as well.

INSTALLATION

9 The original lifters, if they're being reinstalled, must be returned to their original locations. Coat them with moly-base grease or engine assembly lube.

10 Install the lifters in the bores.
11 Install the guide plates and guide retainer.
12 Install the pushrods and rocker arms.

12.7 The roller must turn freely - check for wear and excessive play as well

13 Install the intake manifold and valve covers.
14 Change the engine oil and filter (see Chapter 1).

13 Camshaft, balance shaft and bearings - removal, inspection and installation

CAMSHAFT LOBE LIFT CHECK

1 In order to determine the extent of cam lobe wear, the lobe lift should be checked prior to camshaft removal. Refer to Section 4 and remove the valve covers. The rocker arms must also be removed (Section 5), but leave the pushrods in place.

2 Position the number one piston at TDC on the compression stroke (see Section 3).

3 Beginning with the number one cylinder, mount a dial indicator on the engine and position the plunger in-line with and resting on the first rocker arm (see illustration 11.4).

4 Zero the dial indicator, then very slowly turn the crankshaft in the normal direction of rotation until the indicator needle stops and begins to move in the opposite direction. The point at which it stops indicates maximum cam lobe lift.

5 Record this figure for future reference, then reposition the piston at TDC on the compression stroke.

6 Move the dial indicator to the remaining number one cylinder pushrod and repeat the check. Be sure to record the results for each valve.

7 Repeat the check for the remaining valves. Since each piston must be at TDC on the compression stroke for this procedure, work from cylinder-to-cylinder following the firing order sequence (see Section 3).

8 After the check is complete, compare the results to the Specifications. If camshaft lobe lift is less than specified, cam lobe wear has occurred and a new camshaft should be installed.

REMOVAL

▸ Refer to illustrations 13.11, 13.12 and 13.13

9 Refer to the appropriate Sections and remove the pushrods, the valve lifters and the timing chain and camshaft sprocket. The radiator should be removed also (see Chapter 3). You also may have to remove the air conditioning condenser and the grille as well, but wait and see if the camshaft can be pulled out of the engine.

10 Check the camshaft end play with a dial indicator aligned with the front of the camshaft. Insert the camshaft sprocket bolt and use it to pull the camshaft fore and aft. If the play is greater than specified, replace the thrust plate with a new one when the camshaft is reinstalled.

11 Slide the larger balance shaft drive gear from the end of the camshaft (see illustration).

12 Remove the camshaft thrust plate bolts. A T-30 Torx bit is required. Remove the thrust plate and the spacer ring (see illustration).

13.11 Slide the balance shaft drive gear (arrow) from the camshaft

13.12 T-30 Torx screws (arrows) retain the camshaft thrust plate - remove the thrust plate and spacer (at center of plate)

2A-18 4.2L V6 ENGINE

13.13 Carefully guide the camshaft out of the engine block to avoid nicking the bearings with the lobes

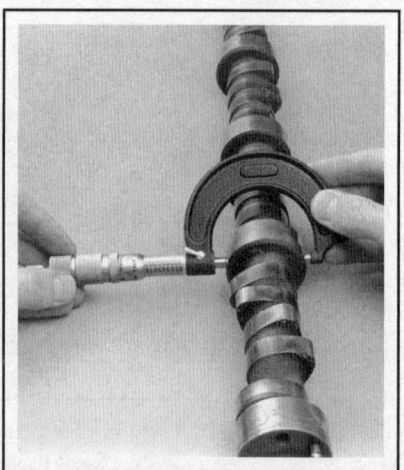

13.15 The camshaft bearing journal diameters are checked to pinpoint excessive wear and out-of-round conditions

13.17 Remove the balance shaft thrust plate bolts (arrows)

13 Carefully pull the camshaft out. Support the cam so the lobes don't nick or gouge the bearings as it's withdrawn (see illustration).

INSPECTION

▶ **Refer to illustration 13.15**

14 After the camshaft has been removed, clean it with solvent and dry it, then inspect the bearing journals for uneven wear, pitting and evidence of seizure. If the journals are damaged, the bearing inserts in the engine block are probably damaged as well. Both the camshaft and bearings will have to be replaced. Replacement of the camshaft bearings requires special tools and techniques which place it beyond the scope of the home mechanic. The engine block will have to be removed from the vehicle and taken to an automotive machine shop for this procedure.

15 Measure the bearing journals with a micrometer (see illustration) to determine whether they are excessively worn or out-of-round.

16 Inspect the camshaft lobes for heat discoloration, score marks, chipped areas, pitting and uneven wear. If the lobes are in good condition and if the lobe lift measurements are as specified, you can reuse the camshaft.

BALANCE SHAFT

▶ **Refer to illustrations 13.17 and 13.18**

17 Remove the bolts holding the balance shaft thrust plate (see illustration).

18 Carefully guide the balance shaft out of the block (see illustration).

➡ **Note:** There is very little to go wrong with the balance shaft or its bearings, which are pressed into the front and rear of the block like camshaft bearings. If abnormal wear is noticed on either bearings or the front and rear balance shaft journals, the bearings must be replaced at an automotive machine shop, and a new balance shaft installed. The balance shaft gear can be replaced without replacing the balance shaft, but requires a machine shop press to replace it or the thrust plate on the balance shaft.

19 When reinstalling the balance shaft, lubricate the front and rear journals and the back of the thrust plate with camshaft installation or engine assembly lube, and line the gear's keyway at the 12 o'clock position. Torque the thrust plate bolts to Specifications.

INSTALLATION

▶ **Refer to illustrations 13.20 and 13.21**

20 Lubricate the camshaft bearing journals and cam lobes with cam-

13.18 Guide the balance shaft out with one hand supporting the shaft inside the engine's valley

13.20 Apply camshaft installation lube to the camshaft lobes and journals prior to installation

4.2L V6 ENGINE 2A-19

shaft installation lube (see illustration).

21 Slide the camshaft into the engine. Support the cam near the engine block and be careful not to scrape or nick the bearings. Align the camshaft keyway straight down (6 o'clock position) and the balance shaft drive gear (on the camshaft) should align perfectly with the smaller gear on the balance shaft. Align their marks (see illustration).

22 Apply moly-base grease or engine assembly lube to both sides of the thrust plate, then position it on the engine block with the oil grooves in (against the engine block). Install the bolts and tighten them to the torque listed in this Chapter's Specifications.

23 Refer to the appropriate Sections and install the lifters, pushrods, rocker arms, timing chain/sprocket, timing chain cover and valve covers.

24 The remaining installation steps are the reverse of removal.

25 Before starting and running the engine, change the oil and install a new oil filter (see Chapter 1).

13.21 Align the marks (arrows) on the balance shaft drive gear on the camshaft (A) and the balance shaft driven gear (B) above it

14 Oil pan - removal and installation

REMOVAL

▸ Refer to illustrations 14.10 and 14.11

➡ **Note: On 2WD models, it is necessary to remove the engine (see Chapter 2D) to remove the oil pan.**

1 If the vehicle is equipped with air suspension, turn the switch OFF before raising the vehicle.

2 Disconnect the cable from the negative battery terminal.

3 Remove the oil dipstick.

4 Refer to Chapter 1 and drain the engine oil and remove the oil filter.

5 Raise the vehicle and place it securely on jackstands.

6 Remove the engine mount through-bolts (see Section 18).

7 Raise the engine about two inches with either an overhead support fixture or a crane attached to the lifting eyes on the top of the engine.

✳✳ CAUTION:

Do not use a jack under the oil pan itself, it is cast-aluminum and can be damaged easily.

8 Disconnect the electrical connector from the low oil level sensor located in the oil pan if so equipped.

9 On 4WD models, the front differential must be removed (see Chapter 8).

10 Remove the oil pan-to-transmission bolts (see illustration).

11 Remove the oil pan-to-engine bolts (see illustration).

12 Remove the oil pan. It may be necessary to break the seal of the pan with a thin putty knife, but do not pry between the pan and block or the pan's sealing edge could be gouged, leading to oil leaks later. Empty any residual oil from the oil pan, and clean it out with solvent.

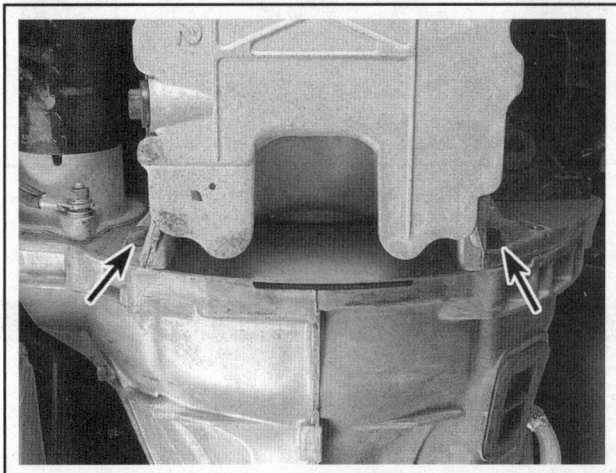

14.10 Remove the two oil pan-to-transmission bolts (arrows)

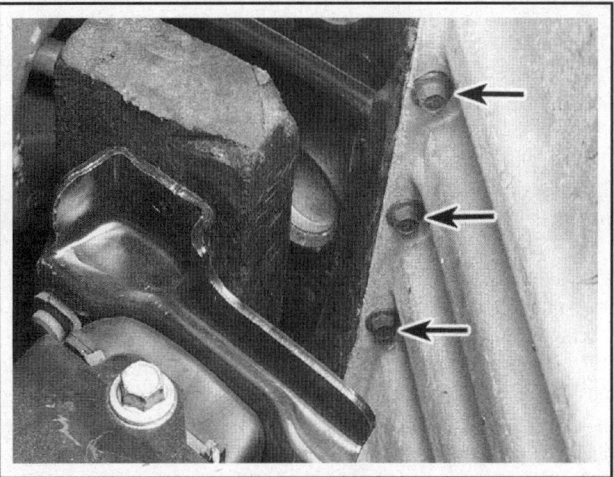

14.11 Remove the pan-to-engine bolts (three shown) - note block of wood between engine and mount

2A-20 4.2L V6 ENGINE

INSTALLATION

▶ **Refer to illustration 14.15**

13 Use a gasket scraper or putty knife to remove all traces of old gasket material and sealant from the pan and engine block. Remove the rubber seas at the rear main and clean any sealant residue with lacquer thinner.

14 Clean the mating surfaces with lacquer thinner or acetone. Make sure the bolt holes in the engine block are clean.

15 Use a dab of gasket adhesive to hold the rear rubber seals in the rear main cap, using a small screwdriver to force the new seal tightly into the groove (see illustration). Stick the side gaskets in place on the engine block with a few dabs of adhesive and tuck the front and rear tabs into the slots in the ends of the front and rear rubber seals. Apply a dab of RTV sealant to the four corners where the side gaskets meet the end seals.

16 Make two alignment dowels by cutting off the heads of two long bolts that fit the holes in the block. Slot the ends of the "studs" with a hacksaw and install one at the front of the block and one at the rear.

17 Apply a bead of RTV sealant on the block and front cover, and a small dab at each corner of the rubber rear seal.

※※ CAUTION:

The oil pan must be installed within 15 minutes or the sealant will "kick off" and must be cleaned off and new sealer applied.

18 Carefully position the pan against the engine block and install the bolts finger tight. When all of the pan-to-engine bolts are started, tighten the bolts to the torque listed in this Chapter's Specifications in two steps, removing the two alignment studs and replacing them with pan bolts. Start at the rear of the pan and work out toward the front. After the pan-to-engine bolts are tightened, install and tighten the pan-to-transmission bolts.

19 The remaining steps are the reverse of removal.

※※ CAUTION:

Don't forget to refill the engine with oil and install a new oil filter before starting it (see Chapter 1).

20 Start the engine and check carefully for oil leaks at the oil pan. Drive the vehicle and check again.

14.15 Press the new rear rubber seal onto the rear main cap and place a dab of RTV on each side (arrows)

15 Oil pump - removal and installation

▶ **Refer to illustrations 15.3, 15.11, 15.12a, 15.12b and 15.13**

➡ **Note: If there is insufficient oil pressure, see Chapter 2, Part C for oil pressure testing.**

1 The oil pump is mounted externally on the timing chain cover.
2 Detach the oil filter (see Chapter 1).
3 Detach the cover and O-ring, then remove the gears from the cavity in the timing chain cover (see illustration). Discard the O-ring.
4 Clean and inspect the oil pump cavity. If the oil pump gear pocket in the timing chain cover is damaged or worn, replace the timing chain cover.
5 Remove all traces of gasket material from the oil pump cover, then check it for warpage with a straightedge and feeler gauges. If it's warped more than 0.0016-inch, replace it with a new one.
6 To remove the pressure relief valve, first detach the timing chain cover from the engine (see Section 10). Drill a hole in the plug, then pry it out or remove it with a slide hammer and screw adapter. Remove the spring and valve from the bore.
7 Remove all metal chips from the bore and the valve, then check them carefully for wear, score marks and galling. If the bore is worn or damaged, a new timing chain cover will be required. The valve should fit in the bore with no noticeable side play or binding.
8 If the spring appears to be fatigued or collapsed, replace it with a new one.
9 Apply clean engine oil to the valve and install it in the bore, small end first. Insert the spring, then install a new plug. Carefully tap it in until it's 0.010-inch below the machined surface of the cover.
10 Intermediate shaft removal and installation is covered in Section 10.

15.3 After the filter is removed, the oil pump cover bolts are accessible (arrows)

4.2L V6 ENGINE 2A-21

11 The oil pump pickup tube is inside the oil pan. For access, remove the oil pan (see Section 14). Remove the pick-up tube nut and the two mounting bolts (see illustration). When reinstalling it, replace the gasket at the front.

12 Using feeler gauges, measure the clearance between the rotors and the pump housing, and the clearance over the rotors (see illustrations).

13 Installation is the reverse of removal.

> ※※ **CAUTION:**
> **Fill the oil pump with clean engine oil before installation.**

Install a new cover O-ring and tighten the bolts to the torque listed in this Chapter's Specifications in a criss-cross pattern (see illustration). Use a new pick-up tube gasket and tighten the mounting bolts securely.

15.11 To detach the oil pickup tube, remove the nut and bolts (arrows)

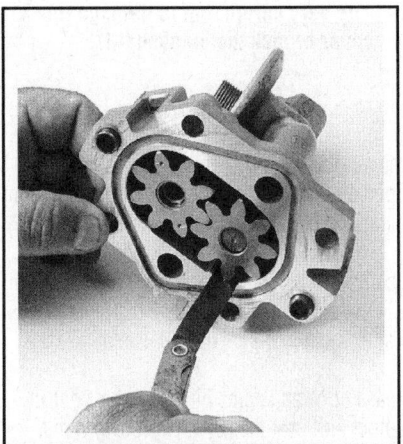
15.12a Measure the clearance between the rotor tips and the pump housing and compare to the Specifications

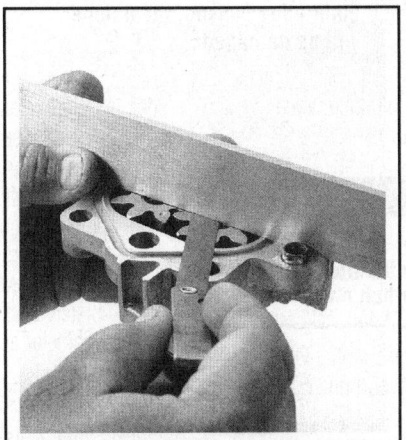
15.12b Put a straightedge across the pump body and measure the clearance above the gears

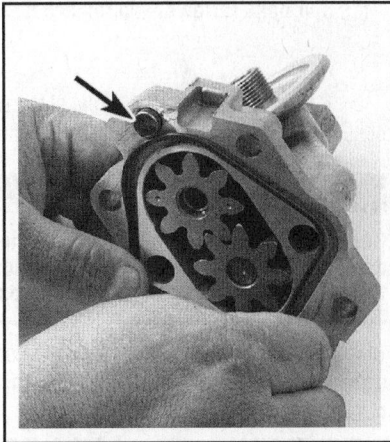
15.13 Install a new O-ring (arrow) lubricated with clean oil before reinstalling the oil pump

16 Crankshaft oil seals - replacement

FRONT SEAL - TIMING CHAIN COVER IN PLACE

▶ Refer to illustrations 16.4a, 16.4b, 16.5, 16.6, 16.8 and 16.10

1 Disconnect the cable from the negative battery terminal.
2 Remove the electric cooling fan/shroud assembly (see Chapter 3).
3 Remove the drivebelts (see Chapter 1).
4 Mark the crankshaft pulley and vibration damper so they can be reassembled in the same relative position. This is important, since the damper and pulley are initially balanced as a unit. Unbolt and remove the pulley (see illustrations).

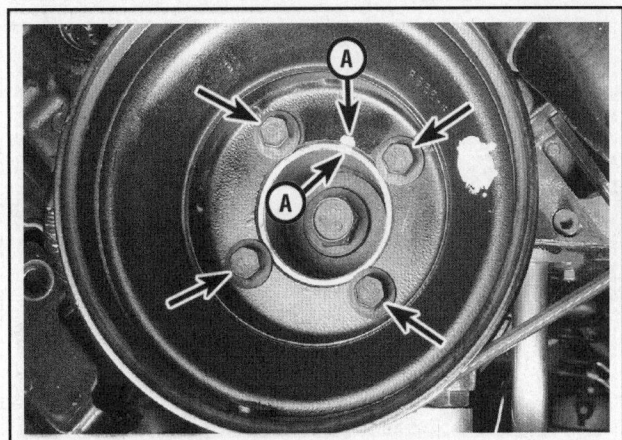
16.4a Mark the pulley and vibration damper (marks A) before removing the four bolts (arrows) - the large vibration damper bolt in the center is usually very tight, so use a six-point socket and a breaker bar to loosen it

2A-22 4.2L V6 ENGINE

16.4b While the damper is off, inspect the crankshaft position sensor ring; if it is damaged, remove the four bolts (arrows) and replace it

16.5 Use the recommended puller to remove the vibration damper - if a puller that applies force to the outer edge is used, the damper will be damaged!

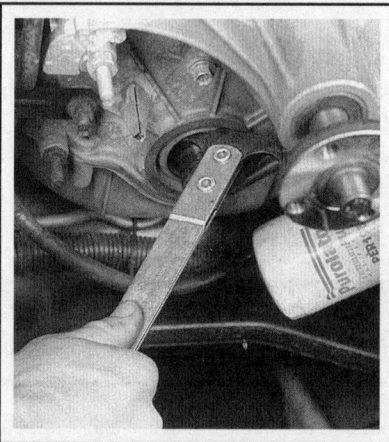

16.6 Use a screwdriver or seal removal tool (shown) to work the seal out of the timing chain cover - be very careful not to damage the cover or nick the crankshaft!

5 Remove the bolt from the front of the crankshaft, then use a puller to detach the vibration damper (see illustration).

※ CAUTION:

Don't use a puller with jaws that grip the outer edge of the damper. The puller must be the type shown in the illustration that utilizes bolts to apply force to the damper hub only.

Clean the crankshaft nose and the seal contact surface on the vibration damper with lacquer thinner or acetone. Leave the Woodruff key in place in the crankshaft keyway.

→Note: *The damper to crankshaft bolt is very tight. Have an assistant hold the flywheel from turning while removing the bolt, or hold the damper with a special strap-wrench designed for this purpose.*

6 Carefully remove the seal from the cover with a screwdriver or seal removal tool (see illustration). Be careful not to damage the cover or scratch the wall of the seal bore. If the engine has accumulated a lot of miles, apply penetrating oil to the seal-to-cover joint and allow it to soak in before attempting to remove the seal.

7 Check the seal bore and crankshaft, as well as the seal contact surface on the vibration damper for nicks and burrs. Position the new seal in the bore with the open end of the seal facing IN. A small amount of oil applied to the outer edge of the new seal will make installation easier - don't overdo it!

8 Drive the seal into the bore with a large socket and hammer until it's completely seated (see illustration). Select a socket that's the same outside diameter as the seal (a section of pipe can be used if a socket isn't available).

9 Apply moly-base grease or clean engine oil to the seal contact surface of the vibration damper and coat the keyway (groove) with a thin layer of RTV sealant.

10 Install the damper on the end of the crankshaft. The keyway in the damper bore must be aligned with the Woodruff key in the crankshaft nose. If the damper can't be seated by hand, tap it into place with a soft-face hammer (see illustration) or slip a large washer over the bolt, install the bolt and tighten it to press the damper into place. Remove the large washer, then install the bolt and tighten it to the torque listed in this Chapter's Specifications.

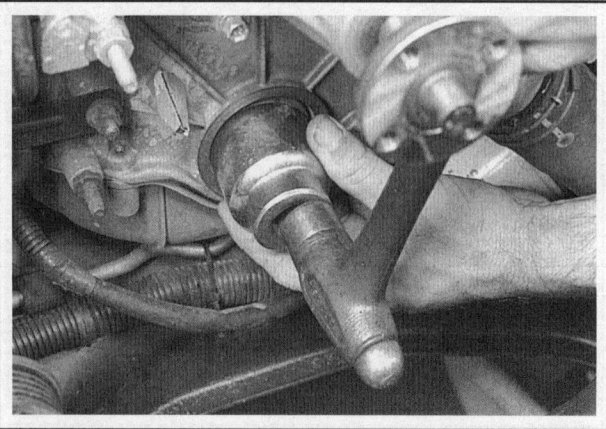

16.8 Clean the bore, then apply a small amount of oil to the outer edge of the new seal and drive it squarely into the opening with a large socket and a hammer - don't damage the seal in the process and make sure it's completely seated

16.10 A soft-face hammer can be used to tap the vibration damper onto the crankshaft - don't use a steel hammer!

4.2L V6 ENGINE 2A-23

16.15 The new seal can also be driven in carefully with the cover off the engine - be sure to support the cover from below with a wood block to prevent damage to the cover

16.17 If you're very careful not to damage the crankshaft or the seal bore, the rear seal can be pried out with a screwdriver - normally a special puller is used for this procedure

11 Install the remaining parts removed for access to the seal.
12 Start the engine and check for leaks at the seal-to-cover joint.

FRONT SEAL - TIMING CHAIN COVER REMOVED

▶ Refer to illustration 16.15

13 Use a punch or screwdriver and hammer to drive the seal out of the cover from the back side. Support the cover as close to the seal bore as possible. Be careful not to distort the cover or scratch the wall of the seal bore. If the engine has accumulated a lot of miles, apply penetrating oil to the seal-to-cover joint on each side and allow it to soak in before attempting to drive the seal out.

14 Clean the bore to remove any old seal material and corrosion. Support the cover on blocks of wood and position the new seal in the bore with the open end of the seal facing IN. A small amount of oil applied to the outer edge of the new seal will make installation easier - don't overdo it!

15 Drive the seal into the bore with a large socket and hammer until it's completely seated (see illustration). Select a socket that's the same outside diameter as the seal (a section of pipe can be used if a socket isn't available).

REAR MAIN SEAL

▶ Refer to illustration 16.17

16 Refer to Chapter 7 and remove the transmission, then detach the flywheel or driveplate and the rear cover plate from the engine (Section 17).

17 The old seal can be removed by prying it out with a screwdriver (see illustration) or by making one or two small holes in the seal flange with a sharp pick, then using a screw-in type slide-hammer puller. Be sure to note how far the seal is recessed into the bore before removing it; the new seal will have to be recessed an equal amount.

✳✳ CAUTION:

Be very careful not to scratch or otherwise damage the crankshaft or the bore in the housing or oil leaks could develop!

18 Clean the crankshaft and seal bore with lacquer thinner or acetone. Check the seal contact surface very carefully for scratches and nicks that could damage the new seal lip and cause oil leaks. If the crankshaft is damaged, the only alternative is a new or different crankshaft.

19 Make sure the bore is clean, then apply a thin coat of engine oil to the outer edge of the new seal. Apply moly-based grease to the seal lips. The seal must be pressed squarely into the bore, a special seal installation tool is highly recommended. Hammering it into place is not recommended. If you don't have access to the special tool, you may be able to tap the seal in with a large section of pipe and a hammer. If you must use this method, be very careful not to damage the seal or crankshaft! And work the seal lip carefully over the end of the crankshaft with a blunt tool such as the rounded end of a socket extension.

20 Reinstall the engine rear cover plate, the flywheel or driveplate and the transmission.

17 Flywheel/driveplate - removal and installation

▶ Refer to illustrations 17.3, 17.4a and 17.4b

1 Raise the vehicle and support it securely on jackstands, then refer to Chapter 7 and remove the transmission. If it's leaking, now would be a very good time to replace the front pump seal/O-ring (automatic transmission only).

✳✳ WARNING:

Some models covered by this manual are equipped with an air suspension system. Always disconnect the electrical power to the suspension system before lifting or towing (see Chapter 10). Failure to perform this procedure may result in unexpected shifting or movement of the vehicle, which could cause personal injury.

2A-24 4.2L V6 ENGINE

2 Remove the pressure plate and clutch disc (see Chapter 8) (manual transmission equipped vehicles). Now is a good time to check/replace the clutch components and pilot bearing.

3 Look for factory paint marks that indicate flywheel-to-crankshaft alignment. If they aren't there, use a center-punch or paint to make alignment marks on the flywheel/driveplate and crankshaft to ensure correct alignment during reinstallation crankshaft (see illustration).

4 Remove the bolts that secure the flywheel/driveplate to the crankshaft (see illustration). If the crankshaft turns, use a flywheel-holding tool or wedge a screwdriver through the starter opening to jam the flywheel.

➡Note: On manual-shift transmission flywheels, there are two extra threaded holes. Insert two bolts in the holes and tighten them evenly from side to side to force the flywheel off the crankshaft (see illustration).

5 Remove the flywheel/driveplate from the crankshaft. Since the flywheel is fairly heavy, be sure to support it while removing the last bolt.

✳✳ **WARNING:**

The flywheel is heavy and the ring gear teeth may be sharp, wear gloves to protect your hands.

6 Clean the flywheel with brake cleaner to remove grease and oil. Inspect the surface for cracks, rivet grooves, burned areas and score marks. Light scoring can be removed with emery cloth. Check for cracked and broken ring gear teeth. Lay the flywheel on a flat surface and use a straightedge to check for warpage.

7 Clean and inspect the mating surfaces of the flywheel/driveplate and the crankshaft. If the crankshaft rear seal is leaking, replace it

17.3 Make two matching marks (arrows) on the flywheel and pilot bearing rim for flywheel alignment

before reinstalling the flywheel/driveplate.

8 Position the flywheel/driveplate against the crankshaft. Be sure to align the marks made during removal. Note that some engines have an alignment dowel or staggered bolt holes to ensure correct installation. Before installing the bolts, apply Teflon thread sealant to the threads.

9 Use a flywheel-holding tool or wedge a screwdriver through the starter motor opening to keep the flywheel/driveplate from turning as you tighten the bolts to the torque listed in this Chapter's Specifications.

10 The remainder of installation is the reverse of the removal procedure.

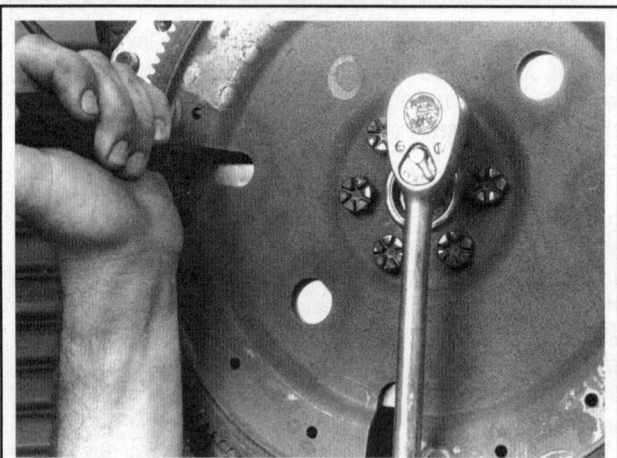

17.4a On automatic transmission driveplates, insert a prybar through a hole to keep the crankshaft from turning when loosening/tightening the bolts

17.4b On manual transmission flywheels, use two bolts (arrows) in the threaded holes to force the flywheel from the crankshaft

18 Engine mounts - check and replacement

1 Engine mounts seldom require attention, but broken or deteriorated mounts should be replaced immediately or the added strain placed on the driveline components may cause damage or wear.

CHECK

2 During the check, the engine must be raised slightly to remove the weight from the mounts.

3 Raise the vehicle and support it securely on jackstands, then position a jack under the engine oil pan. Place a large wood block between the jack head and the oil pan, then carefully raise the engine just enough to take the weight off the mounts.

✳ WARNING:

DO NOT place any part of your body under the engine when it's supported only by a jack!

4 Check the mounts to see if the rubber is cracked, hardened or separated from the metal plates. Sometimes the rubber will split right down the center. Rubber preservative should be applied to the mounts to slow deterioration.

5 Check for relative movement between the mount plates and the engine or frame (use a large screwdriver or pry bar to attempt to move the mounts). If movement is noted, lower the engine and tighten the mount fasteners.

REPLACEMENT

▶ **Refer to illustrations 18.9 and 18.11**

6 Disconnect the cable from the negative battery terminal.
7 Detach the air cleaner duct.
8 Support the engine from above, using an engine support fixture or overhead crane. Take a slight amount of weight off the engine.
9 Remove the engine mount throughbolts (see illustration).
10 Raise the engine high enough to clear the brackets. Do not force the engine up too high. If it touches anything before the mounts are free, remove the part for clearance.
11 Unbolt the mount from the engine block and remove it from the vehicle (see illustration).

➡**Note:** On vehicles equipped with self-locking nuts and bolts, replace them with new ones whenever they are disassembled. Prior to assembly, remove hardened residual adhesive from the engine block holes with an appropriate-size bottoming tap.

12 Attach the new mount to the engine block and install the bolts and stud/nuts in the appropriate locations. Tighten the fasteners securely.
13 Lower the engine into place. Install the through bolts and tighten them to Specifications.
14 Complete the installation by reinstalling all parts removed to gain access to the mounts.

18.9 Remove the engine mount throughbolts (A) - B indicates one of the mount-to-block bolts (left side shown here)

18.11 Remove the remaining bolts (arrows) holding the mount to the engine - front of left mount shown here

2A-26 4.2L V6 ENGINE

Specifications

General

Displacement	4.2 liters (256 cubic inches)
Cylinder numbers (front-to-rear)	
Left (driver's) side	4-5-6
Right side	1-2-3
Firing order	1-4-2-5-3-6

Camshaft and lifters

Lobe lift	
Intake	0.245 inch
Exhaust	0.259 inch
Lobe wear limit	0.005 inch
Endplay	0.001 to 0.006 inch
Journal diameter (all)	2.0515 to 2.0505 inches
Cam bearing inside diameter	2.0535 to 2.0525 inches
Journal-to-bearing (oil) clearance	0.001 to 0.0006 inch
Journal runout limit	0.002 inch
Journal out-of-round limit	0.001 inch

**V6 engine
1-4-2-5-3-6**

Cylinder and coil terminal locations - V6 engine

Oil pump

Gear backlash	0.008 to 0.0012 inch
Gear radial clearance	0.002 to 0.0049 inch
Gear height (housing to gear tops)	0.0004 to 0.0033 inch

Torque specifications   Ft-lbs (unless otherwise indicated)

→ **Note:** One foot-pound (ft-lb) of torque is equivalent to 12 inch-pounds (in-lbs) of torque. Torque values below approximately 15 ft-lbs are expressed in inch-pounds, because most foot-pound torque wrenches are not accurate at these smaller values.

→ **Note:** Refer to Part C for additional specifications.

Camshaft sprocket bolt	30 to 36
Camshaft thrust plate bolts	72 to 120 in-lbs
Balance shaft thrust plate bolts	72 to 120 in-lbs
Timing chain cover-to-block bolts	15 to 22
Water pump-to-timing chain cover bolts	15 to 22
Oil pan-to-block bolts	
Step 1	36 to 44 in-lbs
Step 2	80 to 106 in-lbs
Oil pan-to-transmission bolts	29 to 38
Oil pump and filter body to front cover	
8-mm bolts	15 to 22
6-mm bolts	71 to 97 in-lbs
Cylinder head bolts (oiled)	
Step 1	168 in-lbs
Step 2	29
Step 3	36
Step 4	Loosen 3 turns (DO NOT loosen all the bolts at the same time, from this point on work on one bolt at a time) (see Section 7)
Step 5	
Long bolts	30 to 36
Short bolts	15 to 22
Step 6 (all bolts)	Tighten an additional 175 to 185-degrees

4.2L V6 ENGINE 2A-27

Torque specifications (continued) — Ft-lbs (unless otherwise indicated)

Rocker arm fulcrum bolts	23 to 29
Lower intake manifold-to-cylinder head bolts	
Step 1	44 in-lbs
Step 2	89 in-lbs
Exhaust manifold bolts	15 to 22
Crankshaft pulley-to-vibration damper bolts	20 to 28
Valve cover bolts	71 to 102 in-lbs
Vibration damper bolt	103 to 117
Flywheel/driveplate mounting bolts	54 to 64

Notes

2B

3.5L V6 ENGINES

Section

1 General information
2 Repair operations possible with the engine in the vehicle
3 Valve clearance - check and adjustment
4 Valve covers - removal and installation
5 Intake manifold - removal and installation
6 Exhaust manifolds - removal and installation
7 Crankshaft pulley and crankshaft front oil seal - replacement
8 Cylinder heads - removal and installation
9 Engine front cover - removal and installation
10 Timing chains and sprockets - removal and installation
11 Camshafts and tappets - removal, inspection and installation
12 Oil pan - removal and installation
13 Oil pump - removal and installation
14 Driveplate - removal and installation
15 Rear main oil seal - replacement
16 Engine mounts - check and replacement

2B-2 3.5L V6 ENGINES

1 General Information

Chapter 2B is devoted to in-vehicle repair procedures for the 3.5L GTDI (Gasoline Turbocharged Direct Injection) 4-valve V6 engine. The engine utilizes a cast-aluminum block. The cylinders are arranged in a V-shape at a 60-degree angle between the two banks. The cylinder heads are cast-aluminum and the valve actuation is via two overhead camshafts in each cylinder head with four valves per cylinder. Information concerning engine removal and installation and engine overhaul can be found in Chapter 2D. The following repair procedures are based on the assumption that the engine is installed in the vehicle. If the engine has been removed from the vehicle and mounted on a stand, many of the steps outlined in this Chapter will not apply.

2 Repair operations possible with the engine in the vehicle

Many major repair operations can be accomplished without removing the engine from the vehicle.

Clean the engine compartment and the exterior of the engine with some type of pressure washer before any work is done. It will make the job easier and help keep dirt out of the internal areas of the engine.

If vacuum, exhaust, oil or coolant leaks develop, indicating a need for gasket or seal replacement, the repairs can generally be made with the engine in the vehicle. The intake and exhaust manifold gaskets, timing cover gasket, oil pan gasket, crankshaft oil seals and cylinder head gaskets are all accessible with the engine in place.

Exterior engine components, such as the intake and exhaust manifolds, the oil pan (and the oil pump), the water pump, the starter motor, the alternator, and the fuel system components can be removed for repair with the engine in place.

Since the cylinder heads can be removed without pulling the engine, valve component servicing can also be accomplished with the engine in the vehicle. Replacement of the timing chain and sprockets is also possible with the engine in the vehicle.

In extreme cases caused by a lack of necessary equipment, repair or replacement of piston rings, pistons, connecting rods and rod bearings is possible with the engine in the vehicle. However, this practice is not recommended because of the cleaning and preparation work that must be done to the components involved.

3 Valve clearance - check and adjustment

♦ Refer to illustrations 3.5 and 3.8

➡ **Note: The engine must be room temperature before checking the valve clearances.**

➡ **Note: Checking and, if necessary, adjusting the valve clearance is only necessary after replacement of the camshaft(s) or other valve-related parts, or if the valvetrain is making excessive noise.**

1 Disconnect the cable from the negative terminal of the battery (see Chapter 5).

2 Remove the spark plugs (see Chapter 1).

3 Remove the valve covers (see Section 4).

4 Using a socket and a breaker bar on the crankshaft pulley center bolt, rotate the engine until the cam lobes on the cylinder to be checked are pointing away from the tappets.

5 Measure the clearance of the indicated valves with a feeler gauge (see illustration). Record each measurement and compare your measurements with the desired valve clearance listed in this Chapter's Specifications. Note which are out of specification, as this data will be used later to determine the required replacement tappets.

6 Repeat Steps 4 and 5 until the clearances for all valves have been measured.

7 If a clearance is out of specification, the tappet must be replaced with a new tappet that has a different thickness head to correct the clearance. Refer to Section 11 and remove the camshafts to access the tappets.

8 Mark the lifters that are to be replaced, and record which valve they came from. Use a micrometer to measure the thickness of the head

3.5L V6 ENGINES

3.5 Measure the clearance for each valve with a feeler gauge of the specified thickness - if the clearance is correct, you should feel a slight drag on the gauge as you pull it out

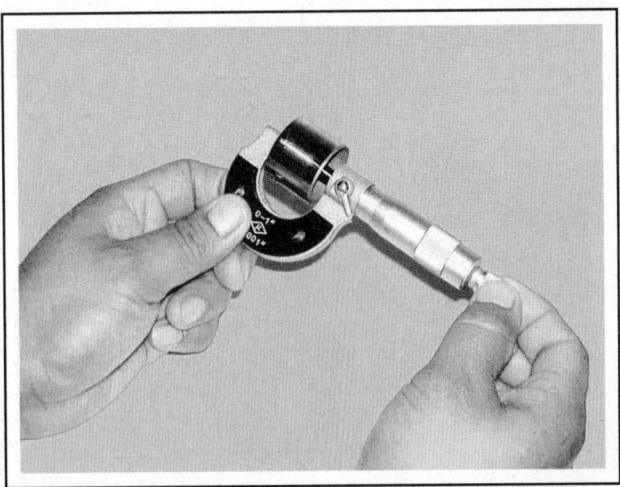

3.8 Use a micrometer to measure the thickness of the head of the tappet

of the lifter, making sure the measurement is precise and on the center projection on the underside of the lifter (see illustration).

9 To calculate the correct thickness for a replacement lifter that will place the valve clearance within the specified value, use the following formula:

$N = R + (M1 - M2)$
N = thickness of new tappet
R = measured valve clearance
$M1$ = thickness of old tappet
$M2$ = standard valve clearance

10 Lifters are marked on the underside as to their size. A marking of 3.310 on the underside of the lifter indicates a thickness value of 3.31 mm, or 0.13 inch.

11 Mark the new lifters as to their destination, lubricate them with engine assembly lube and install them. After replacing the lifters, refer to Section 11 to reinstall the camshafts and Section 10 to reinstall the timing chains, then re-check the valve clearances.

4 Valve covers - removal and installation

REMOVAL

1 Relieve the fuel system pressure (see Chapter 4).
2 Disconnect the cable from the negative battery terminal (see Chapter 5).
3 Remove the engine cover, if equipped.

Left (driver's side) valve cover

4 Remove the dipstick.
5 Disconnect the quick-connectors for the PCV hose then detach the PCV hose from the valve cover and manifold.
6 Remove the nuts securing the charge air cooler bracket from the cylinder head.
7 Disconnect the quick-connectors for the two turbocharger regulator tubes and set them aside then remove the inlet and outlet tube from the turbocharger (see Chapter 4).
8 Remove the high-pressure fuel pump (see Chapter 4).

Both valve covers

9 Remove the ignition coils from the cover to be removed (see Chapter 5).
10 Disconnect the two electrical connectors at the valve cover for the VCT solenoids (being careful not to twist the solenoids) and the electrical connectors from the camshaft position sensors (CMP).
11 Remove the valve cover fasteners, then separate the wiring/clips and remove the valve cover from the cylinder head. Wiggle the valve cover by hand to detach it from the cylinder head if it's stuck.

✲✲✲ CAUTION:

Do not use a hammer to knock the cover loose. Try to slip a flexible putty knife between the cylinder head and cover to break the gasket seal. Don't pry at the cover-to-cylinder head joint or damage to the sealing surfaces may occur (leading to oil leaks in the future).

INSTALLATION

▶ Refer to illustrations 4.13, 4.14, 4.15 and 4.17

12 The mating surfaces of each cylinder head and valve cover must be perfectly clean when the covers are installed. Remove all traces of sealant and old gasket material, then clean the mating surfaces with brake cleaner. If there's sealant or oil on the mating surfaces when the cover is installed, oil leaks may develop.

2B-4 3.5L V6 ENGINES

4.13 Check the spark plug tube seals, VCT solenoid seals and valve cover gasket. If they are in good condition they can be reused

4.14 Check the bolt seals - replace them if they're not in good condition

4.15 Apply RTV sealant to the seams where the front cover meets the cylinder head

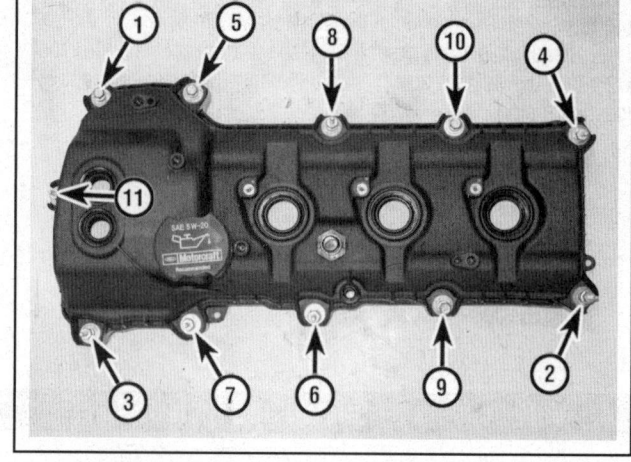

4.17 Valve cover bolt tightening sequence (left side shown, right side similar)

13 Inspect the spark plug seals, the seals for the VCT solenoids and the valve cover gasket (see illustration). Do not replace them unless necessary.

14 Make sure the bolt seals are in good condition, too (see illustration).

15 Apply a bead of RTV sealant to the joints where the engine front cover meets the cylinder head and install the valve cover within four minutes (see illustration).

16 Carefully position the cover on the cylinder head and install the bolts.

17 Tighten the bolts a little at a time, in the proper sequence (see illustration), to the torque listed this Chapter's Specifications.

18 The remaining installation steps are the reverse of removal.

19 Start the engine and check carefully for oil leaks as the engine warms up.

3.5L V6 ENGINES   2B-5

5 Intake manifold - removal and installation

REMOVAL

1 Remove the three bolts and the engine cover bracket from the intake manifold.
2 Loosen the large hose clamp and detach the charge air cooler hose from the throttle body.
3 Disconnect the electrical connector at the throttle body.
4 Disconnect the PCV electrical connector.
5 Disconnect the brake booster vacuum hose from the intake manifold, then disconnect the MAP sensor and the EVAP canister purge valve connectors.
6 Disconnect the turbocharger vacuum regulator hose from the intake manifold, then release the pushpin securing the turbocharger vacuum regulator to the intake manifold.
7 Remove the manifold mounting bolts and detach the manifold from the engine.

➡ **Note: Do not use a hammer, and once the manifold is off, cover the intake ports of the engine with clean shop rags.**

INSTALLATION

▶ **Refer to illustration 5.9**

8 Remove the old gaskets from the manifold. Clean the manifold and cylinder head mating surfaces.

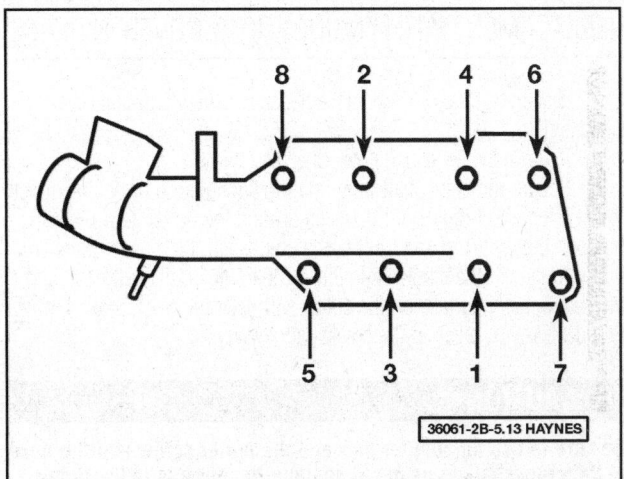

5.9 Intake manifold bolt tightening sequence

9 Install the new manifold gaskets to the grooves in the manifold. Install the manifold and bolts. Tighten the bolts in the indicated sequence to the torque listed in this Chapter's Specifications (see illustration).
10 The remainder of the installation is the reverse of removal.

6 Exhaust manifolds - removal and installation

▶ **Refer to illustration 6.10**

1 Disconnect the cable from the negative battery terminal (see Chapter 5).
2 Loosen the lug nuts on the manifold side you're working on. Raise the vehicle and support it securely on jackstands, then remove the wheel.
3 Also remove the inner fender splash shield (see Chapter 11).
4 Remove the turbocharger from the manifold (see Chapter 4).
5 As a precaution, disconnect the oxygen sensor electrical connectors (see Chapter 6).
6 Remove the manifold mounting nuts and detach the manifold from the cylinder head.
7 Remove the exhaust gaskets and clean the gasket surfaces on the manifold and cylinder heads.

INSTALLATION

➡ **Note: The manufacturer recommends removing the exhaust mounting studs from the cylinder heads and installing new studs.**

8 Check the exhaust manifold for cracks and make sure all the stud threads are clean and undamaged. The exhaust manifold and cylinder head mating surfaces must be clean before the manifolds are reinstalled.

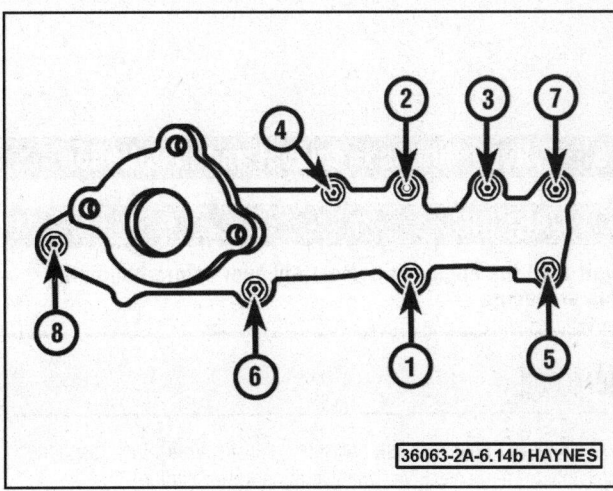

6.10 Engine manifold tightening sequence

9 Position the exhaust manifold and new gasket over the studs on the cylinder head and install the mounting nuts.
10 Tighten the nuts in sequence (see illustration) to the torque listed in this Chapter's Specifications is reached.
11 The remaining installation steps are the reverse of removal.
12 Start the engine and check for exhaust leaks.

2B-6 3.5L V6 ENGINES

7 Crankshaft pulley and crankshaft front oil seal - replacement

REPLACEMENT WITH ENGINE FRONT COVER IN PLACE

1 Disconnect the cable from the negative battery terminal (see Chapter 5).
2 Remove the drivebelt(s) (see Chapter 1).
3 Secure the crankshaft from rotating with a strap or chain wrench (wrap a length of rag or old drivebelt around the pulley to protect it). Remove the bolt from the front of the crankshaft, then use a three-jaw puller to detach the crankshaft pulley. Clean the crankshaft nose and the seal contact surface on the pulley with RTV remover. Leave the Woodruff key in place in the crankshaft keyway.

✳✳ CAUTION:

Be sure to use an adapter between the puller screw and the nose of the crankshaft so as not to damage the threads in the crankshaft. Also, don't use a puller with jaws that grip the outer edge of the damper. The puller must be the type that applies force to the damper hub only.

4 Carefully remove the seal from the timing chain cover with a screwdriver or seal removal tool. Be careful not to damage the cover or scratch the wall of the seal bore. If the engine has accumulated a lot of miles, apply penetrating oil to the seal-to-cover joint and allow it to soak in before attempting to remove the seal.
5 Check the seal bore and crankshaft, as well as the seal contact surface on the pulley for nicks and burrs. Position the new seal in the bore with the open end of the seal facing IN. A film of engine oil applied to the outer edge of the new seal will make installation easier.
6 Drive the seal into the bore with a seal driver or a large socket and hammer until it's completely seated. If you're using a socket, select one that's the same outside diameter as the seal.

7 Apply clean engine oil to the seal contact surface of the crankshaft pulley and coat the keyway (groove) with a thin layer of RTV sealant.
8 Install the pulley on the end of the crankshaft. The keyway in the pulley bore must be aligned with the Woodruff key in the crankshaft nose. If the pulley can't be seated by hand, tap it into place with a soft-face hammer or slip a large washer over the bolt, install the bolt and tighten it to press the pulley into place. Remove the large washer, then install the bolt and tighten it in Steps to the torque listed in this Chapter's Specifications.
9 Install the remaining parts removed for access to the seal.
10 Start the engine and check for leaks.

REPLACEMENT WITH ENGINE FRONT COVER REMOVED

11 Remove the engine front cover (see Section 9).
12 Use a punch or screwdriver and hammer to drive the seal out of the cover from the back side. Support the cover as close to the seal bore as possible, using blocks of wood. Be careful not to distort the cover or scratch the wall of the seal bore. If the engine has accumulated a lot of miles, apply penetrating oil to the seal-to-cover joint on each side and allow it to soak in before attempting to drive the seal out.
13 Clean the bore to remove any old seal material and corrosion. Support the cover on blocks of wood and position the new seal in the bore with the open end of the seal facing IN. A film of oil applied to the outer edge of the new seal will make installation easier.
14 Drive the seal into the bore with a seal driver or a large socket and hammer until it's completely seated. If you're using a socket, select one that's the same outside diameter as the seal.
15 Reinstall the engine front cover (see Section 9) and crankshaft pulley (see Step 8).
16 Start the engine and check for leaks.

8 Cylinder heads - removal and installation

✳✳ WARNING:

Wait until the engine is completely cool before beginning this procedure.

REMOVAL

1 Relieve the fuel system pressure (see Chapter 4), then disconnect the cable from the negative battery terminal (see Chapter 5).
2 Drain the cooling system (see Chapter 1).
3 Remove the drivebelt and the drivebelt tensioner (see Chapter 1).
4 Remove the intake manifold (see Section 5).
5 Remove the coolant crossover manifold bolts and crossover manifold at the rear of the cylinder heads.
6 Remove the exhaust manifold (see Section 6).
7 Remove the turbocharger(s) (see Chapter 4).
8 Disconnect the high-pressure fuel line(s) from the fuel rail and high-pressure fuel pump (see Chapter 4).

✳✳ CAUTION:

The manufacturer requires the fuel line from the fuel rail to the high-pressure fuel pump always be replaced once it has been removed.

9 Remove the valve covers (see Section 4).
10 Remove the front cover (see Section 9), timing chains and timing chain guides (see Section 10).
11 Remove the camshafts and tappets (see Section 11).
12 Remove the two Camshaft Position (CMP) sensors at the rear of each cylinder head and disconnect any other electrical connectors from the cylinder head and mark them with tape for correct reassembly.
13 Loosen the cylinder head bolts in 1/4-turn increments until they can be removed by hand. Work from bolt-to-bolt in a pattern that's the reverse of the tightening sequence (see illustration 8.22).

3.5L V6 ENGINES 2B-7

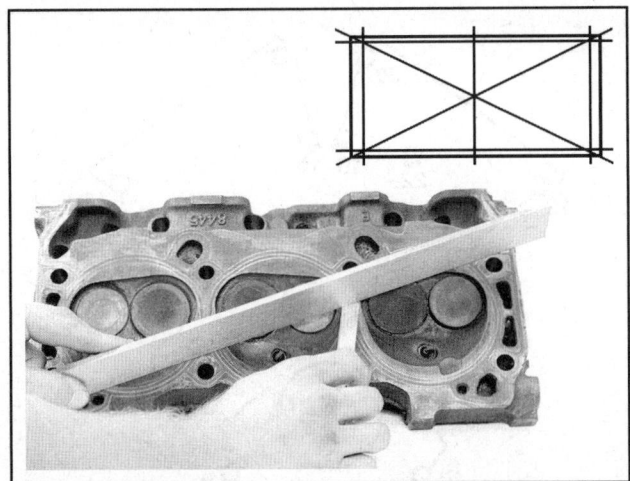

8.17 Check the cylinder head and block surfaces for flatness with a straightedge

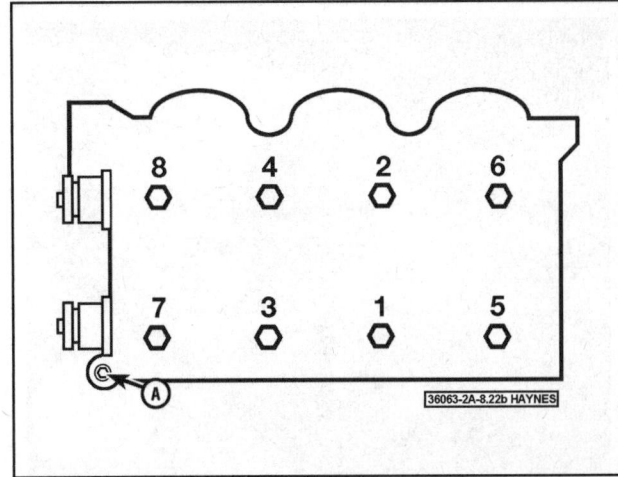

8.22 Cylinder head bolt tightening sequence ("A" is an M6 bolt)

CAUTION:

Remove the bolts and discard them - new bolts must be used when installing the cylinder head(s).

14 Lift the cylinder head(s) off the engine. If resistance is felt, DO NOT pry between the cylinder head and engine block as damage to the mating surfaces will result. To dislodge the cylinder head, place a wood block against the end of it and strike the wood block with a hammer or pry against a casting protrusion. Store the cylinder heads on blocks of wood to prevent damage to the gasket sealing surfaces.

15 Cylinder head disassembly and inspection procedures should be performed by a qualified automotive machine shop.

INSTALLATION

▸ **Refer to illustrations 8.17 and 8.22**

16 The mating surfaces of the cylinder heads and engine block must be perfectly clean when the cylinder heads are installed.

17 Check the engine block and cylinder head mating surfaces for nicks, deep scratches and other damage. If damage is slight, it can be removed with a file - if it's excessive, machining may be the only alternative. Use a straightedge and feeler gauges to check for warpage (see illustration). If the warpage is beyond Specifications, have the head machined at an automotive machine shop.

18 Use a tap of the correct size to chase the threads in the block head bolt holes. Dirt, corrosion, sealant and damaged threads will affect torque readings.

19 Position the new gasket(s) over the dowel pins in the engine block. Make sure it's facing the right way. If the cylinder head is to be replaced, a new secondary timing chain tensioner will be required.

20 Carefully position the cylinder head(s) on the engine block without disturbing the gasket(s).

21 Before installing the new cylinder head bolts, lightly oil the threads.

22 Install the main cylinder head bolts and tighten them finger-tight, then tighten the bolts, in the recommended sequence, to the torque listed in this Chapter's Specifications (see illustration).

23 Install and tighten the M6 bolt at the front of the cylinder head (see illustration 8.22, bolt A) to the torque listed in this Chapter's Specifications.

24 The remaining installation steps are the reverse of removal.

25 Change the engine oil and filter and refill the cooling system (see Chapter 1), then start the engine and check carefully for oil and coolant leaks.

9 Engine front cover - removal and installation

WARNING:

The engine must be completely cool before beginning this procedure.

REMOVAL

▸ **Refer to illustration 9.13**

1 Relieve the fuel system pressure (see Chapter 4).
2 Disconnect the cable from the negative terminal of the battery (see Chapter 5).
3 Drain the cooling system and engine oil, and remove the oil filter (see Chapter 1).
4 Remove the expansion tank (see Chapter 3).
5 Remove the air filter housing and inlet and outlet ducts (see Chapter 4).
6 Disconnect the upper and lower radiator hoses from the engine, then secure the hoses out of the way.
7 On turbocharged models, remove the charge air cooler inlet and outlet pipes (see Chapter 4).
8 Remove the turbocharger inlet and outlet tube (see Chapter 4).
9 Remove the engine oil cooler (see Chapter 3).
10 Remove the intake manifold (see Section 5).
11 Remove the valve covers (see Section 4).

2B-8 3.5L V6 ENGINES

9.13 Remove the drivebelt, the idler pulley (indicated), alternator and crankshaft pulley

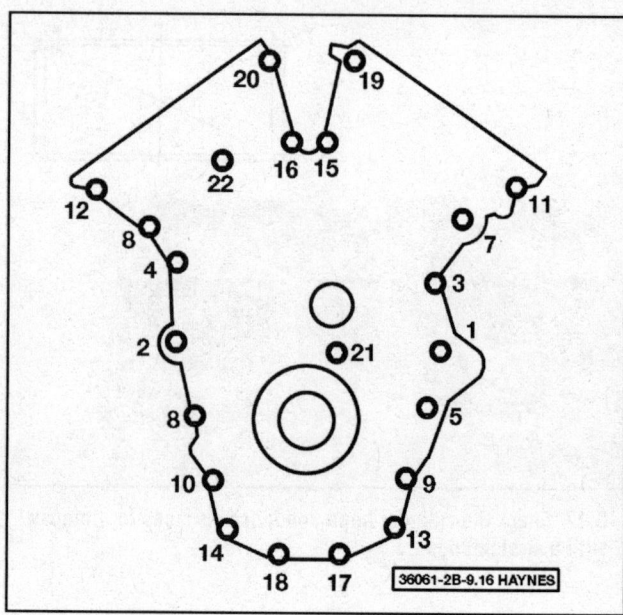

9.26 Engine front cover bolt tightening sequence

12 Remove the crankshaft pulley and crankshaft front oil seal (see Section 7).

13 Remove the drivebelt tensioner (see Chapter 1) and remove all accessory brackets attached to the engine front cover (see illustration).

14 Remove drivebelt idler pulley bolt and pulley.

15 Remove the alternator bolt and swing the alternator out of the way.

16 Remove the water pump (see Chapter 3).

17 Remove the air conditioning compressor mounting bolts without disconnecting the refrigerant lines (see Chapter 3).

18 Disconnect the oil control solenoid electrical connector then detach the wiring harness retainers and move the harness out of the way.

19 Remove the front cover bolts and separate the cover from the engine block.

※ CAUTION:

DO NOT use excessive force or you may crack the cover. If the cover is difficult to remove, double check to make sure all of the bolts have been removed.

➡ *Note: There are seven pry spots around the front cover. Work gently at each one until the cover comes off.*

INSTALLATION

♦ **Refer to illustration 9.26**

20 Clean the gasket mating surfaces with brake system cleaner.

21 While the cover is off the engine, it's a good idea to install a new crankshaft front seal (see Section 7).

22 Apply a 1/8-inch bead of High Performance RTV sealant to the mating surface of the cover. Be sure to also apply the sealant to the bosses on the interior of the cover where the bolts go through.

23 Apply a 3/16-inch bead of High Performance RTV sealant to the areas where the oil pan and cylinder heads meet the engine block.

24 Slide the front cover onto the engine. The dowel pins will position it correctly.

25 Install all the bolts finger-tight within four minutes of applying the High performance RTV.

26 Follow the correct sequence (see illustration), tighten the bolts to the torque listed in this Chapter's Specifications. All the bolts must be tightened within 35 minutes.

27 Install the remaining parts in the reverse order of removal.

28 Install a new oil filter, then add engine oil and coolant (see Chapter 1).

29 Run the engine and check for leaks.

3.5L V6 ENGINES 2B-9

10 Timing chains and sprockets - removal and installation

※※ **CAUTION:**

This engine is difficult to work on and requires some special tools. On any procedure involving the timing chains, the Steps must be read carefully and disassembly must proceed using the special tools, otherwise damage to the engine will result.

※※ **CAUTION:**

Because this is an "interference" engine design, if the timing chain has broken, there will be damage to the valves (and possibly the pistons) and will require removal of the cylinder heads.

※※ **CAUTION:**

The timing system is complex. Severe engine damage will occur if you make any mistakes. Do not attempt this procedure unless you are highly experienced with this type of repair. If you are at all unsure of your abilities, consult an expert. Double-check all your work and be sure everything is correct before you attempt to start the engine.

※※ **CAUTION:**

This model is equipped with Direct Injection (DI) and a high-pressure fuel pump. Fuel pressure in the high-pressure system on Direct Injection (DI) systems is under extremely high pressure. Be sure to correctly perform the fuel pressure relief procedure prior to servicing any of the high-pressure fuel system components to prevent injury (see Chapter 4, Section 2).

REMOVAL

▶ **Refer to illustration 10.5 and 10.10**

1 Disconnect the cable from the negative battery terminal (see Chapter 5).
2 Remove the valve covers (see Section 4).
3 Remove the engine front cover (see Section 9).
4 Turn the engine clockwise until the VCT units (camshaft sprockets) on each intake camshaft are positioned at 90-degrees (12 o'clock) in relation to the top surface of the cylinder heads.
5 Remove the bolts that secure the valvetrain oil tubes to each cylinder head (see illustration).
6 Install the special camshaft locking tools (303-1248), with the tools holding the flats on each camshaft.
7 If you can't see the timing marks on the timing chain, add a paint dot to align with the marks on the VCT units and the crankshaft timing sprocket.
8 Remove the two bolts and the primary timing chain tensioner arm, then unbolt and remove the two tensioners.
9 Remove the mounting bolts and the lower-left chain guide, then the lower right chain guide.
10 Remove the bolts securing the VCT solenoids to the cylinder heads (see illustration). You may have to twist or wiggle the solenoids to disengage them.

➡ **Note: The intake solenoids are white, the exhaust solenoids are black.**

11 Keep the VCT solenoids in clean plastic sandwich bags, marked with their destination.
12 Remove the primary timing chain.
13 The two camshafts on each cylinder head are jointly driven by a smaller secondary timing chain.

10.5 Remove three of the camshaft cap bolts to remove the valvetrain oil tubes

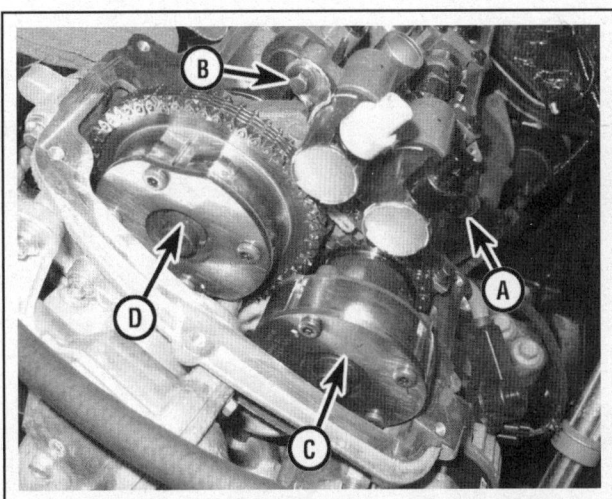

10.10 Remove the VCT oil control solenoid mounting bolts (A and B) - (C) is the exhaust VCT unit and (D) is the intake VCT unit

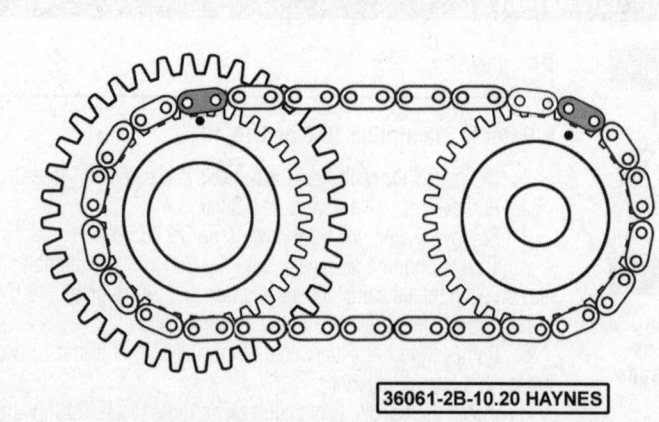

10.20 Align the colored links of the secondary timing chain with the marks on the back of the sprockets (VCT units)

14 Depress and lock the secondary chain tensioner on each bank, using the factory tool (303-1530) inserted in the large hole at the camshaft first bearing cap, or a length of threaded rod and nuts to hold the tensioner down, by pushing from the camshaft bearing cap.

15 With the camshafts still locked at TDC, remove the mounting bolts from the VCT units (camshaft sprockets), and remove the sprockets and secondary timing chains from the camshafts.

16 If the primary timing chain sprocket (at the center of the front of the block) is to be replaced, remove the nine bolts securing the plate that mounts the gear to the front of the block.

INSTALLATION

▸ Refer to illustration 10.20 and 10.22

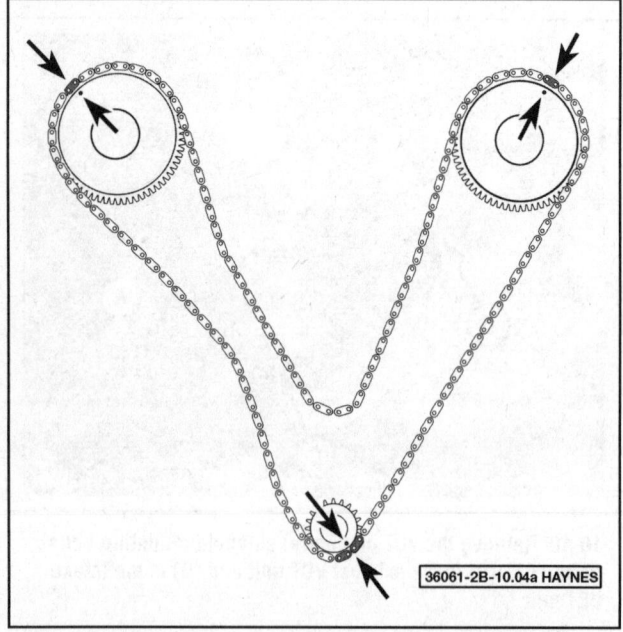

10.22 Align the colored links on the primary timing chain with the marks on the crankshaft and camshaft sprockets (VCT units)

17 If the primary chain gear and its plate were removed, clean the mounting surface and the block and install the sprocket and plate with a new gasket. Prepare the primary chain tensioners by pushing in the release button and compressing the plunger until a nail or large paper clip can be inserted to hold the plunger.

18 If the secondary timing chain tensioners were removed, they cannot be reused. New tensioners must be installed.

※ CAUTION:

Do not remove the plastic clip that is keeping the new tensioner compressed.

19 The tensioner is installed to the correct depth when you hear a "snap" sound. Install the tensioner shoe. The plastic clip can now be removed with pliers.

20 Align the colored links with the marks on the VCT units and install the chain on the backside of the VCT units (see illustration). Align the two VCT units with the dowel pins on the front of the camshafts.

21 New bolts must be used to install the VCT units. Tighten the bolts to the torque listed in this Chapter's Specifications.

22 When the secondary chains and VCT units are installed and aligned, install the primary timing chain, aligning the colored links with the marks on the sprockets (see illustration). When all components are installed and aligned, pull the pins holding back the two primary chain tensioners.

23 Reinstall the remaining parts in the reverse order of removal.

24 Carefully rotate the crankshaft by hand through at least two full revolutions (use a socket and breaker bar on the crankshaft pulley center bolt).

※ CAUTION:

If you feel any resistance, STOP! There is something wrong - most likely, valves are contacting the pistons. You must find the problem before proceeding.

25 Install a new oil filter, then add engine oil and coolant (see Chapter 1).

26 Run the engine and check for leaks.

3.5L V6 ENGINES   2B-11

11 Camshafts and tappets - removal, inspection and installation

※ CAUTION:

This engine is difficult to work on and require some special tools. On any procedure involving the timing chains, the Steps must be read carefully and disassembly must proceed using the special tools, otherwise damage to the engine will result.

※ CAUTION:

Because this is an "interference" engine design, if the timing chain has broken, there will be damage to the valves (and possibly the pistons) and will require removal of the cylinder heads.

※ CAUTION:

The timing system is complex. Severe engine damage will occur if you make any mistakes. Do not attempt this procedure unless you are highly experienced with this type of repair. If you are at all unsure of your abilities, consult an expert. Double-check all your work and be sure everything is correct before you attempt to start the engine.

※ CAUTION:

This model is equipped with Direct Injection (DI) and a high-pressure fuel pump. Fuel pressure in the high-pressure system on Direct Injection (DI) systems is under extremely high pressure. Be sure to correctly perform the fuel pressure relief procedure prior to servicing any of the high-pressure fuel system components to prevent injury (see Chapter 4, Section 2).

REMOVAL

➡ Note: If you're removing the camshafts from both sides, remove the right (passenger's) side first.

1 Before removing the camshafts, check the valve clearances (see Section 3).

2 Refer to Section 10 and remove the timings chain and camshaft VCT units also the tensioner holding tool and camshaft holding tools
3 When removing the right-side camshafts, remove the brake vacuum pump (see Chapter 9).
4 Loosen the camshaft caps and "mega-cap" mounting bolts, a little at a time, and remove the caps.

※ CAUTION:

Keep the caps in order and don't mix them up.

5 Remove the camshafts from the cylinder head.
6 With the camshafts removed, the tappets can be removed using a magnet.

※ CAUTION:

They must be stored in an egg carton or other divided and marked container, so they can be restored to their original locations during reassembly.

INSPECTION

▸ Refer to illustrations 11.9a and 11.9b

7 After the camshafts have been removed, clean them with solvent, then inspect the bearing journals for uneven wear and pitting. If the journals are damaged, the bearing saddles in the cylinder heads and caps are probably damaged as well. The head and camshaft caps will have to be replaced.
8 Measure the bearing journals with a micrometer and compare to the Specifications in this Chapter to determine whether they are excessively worn or out-of-round
9 Measure the camshaft lobe height and the base circle (see illustrations). The difference between the two measurements is the lobe lift (lobe height - base circle = lobe lift). Record this figure for future reference and repeat the check on the remaining camshaft lobes. Compare the results to the values listed in this Chapter's Specifications.

11.9a Lobe lift can be obtained by measuring camshaft lobe height . . .

11.9b . . . and by measuring the camshaft base circle - the difference between the two measurements equals lobe lift

2B-12 3.5L V6 ENGINES

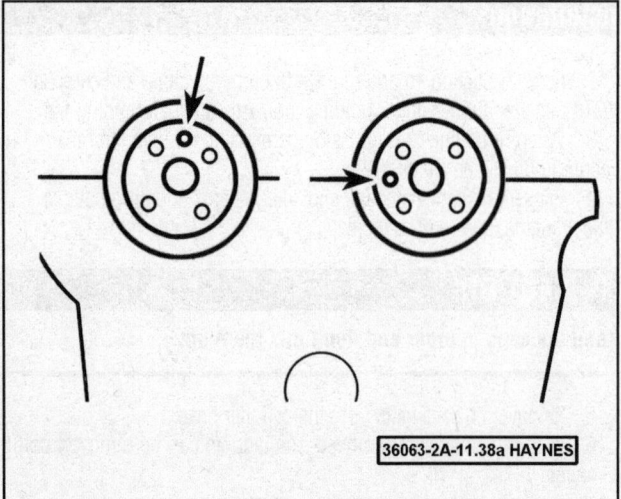

11.15a When installing the left (driver's) side cylinder head camshafts, position the dowel pins like this

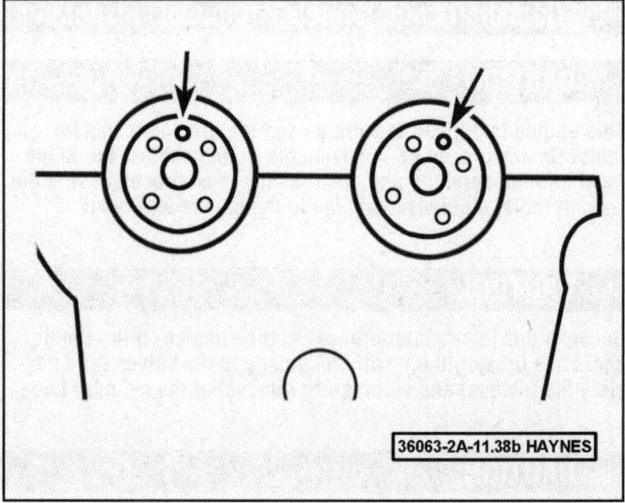

11.15b When installing the right (passenger's) side cylinder head camshafts, position the dowel pins like this

10 Inspect the camshaft lobes for heat discoloration, score marks, chipped areas, pitting and uneven wear. If the lobes are in good condition and if the lobe lift measurements are as specified, you can reuse the camshafts.

INSTALLATION

♦ **Refer to illustrations 11.15a, 11.15b, 11.16, 11.19a and 11.19b**

11 Rotate the crankshaft counterclockwise and position the keyway in the 9 o'clock position (this is the neutral position).

12 Lubricate the tappets with clean engine oil and reinstall them in their original locations. If any valve clearances were out-of-specification, replace the tappet(s) with new ones of the proper thickness to achieve the desired clearance.

13 At the front of each camshaft, there are two grooves to hold seals. Obtain new seals and install them in the camshaft grooves.

➡ **Note: The split where the ends of each seal come together must face UP (12 o'clock position) on the camshafts when they are installed.**

14 Lubricate the camshaft bearing journals and cam lobes with moly-based grease or camshaft installation lubricant.

15 Position the camshafts in the cylinder head in their neutral positions (see illustrations).

16 Install the camshaft bearing caps (in their original locations), and tighten the bolts, in sequence, to the torque listed in this Chapter's Specifications (see illustration).

17 Recheck the valve clearances (see Section 3). Since the crankshaft is in the neutral position, the camshafts can be turned without the valves contacting the pistons.

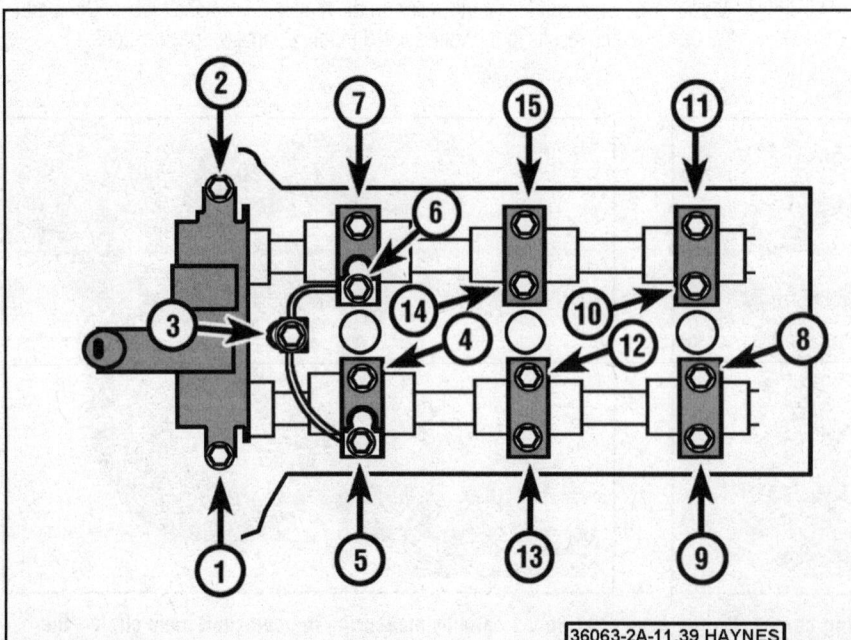

11.16 Camshaft bearing cap tightening sequence

3.5L V6 ENGINES 2B-13

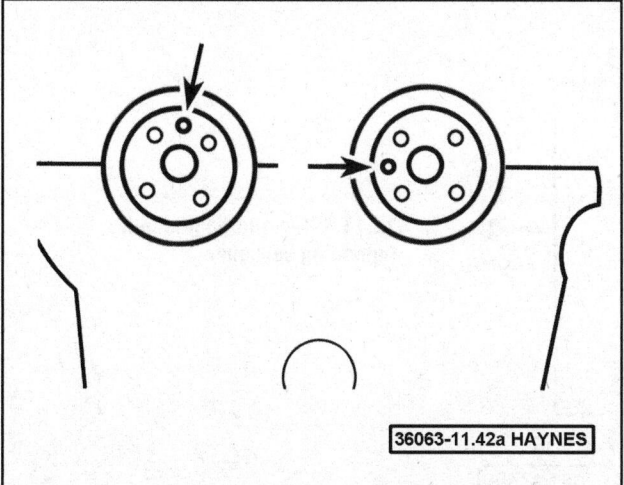

11.19a Rotate the left (driver's) side cylinder head camshafts so the dowel pins are positioned like this (TDC position), then install the camshaft holding tool

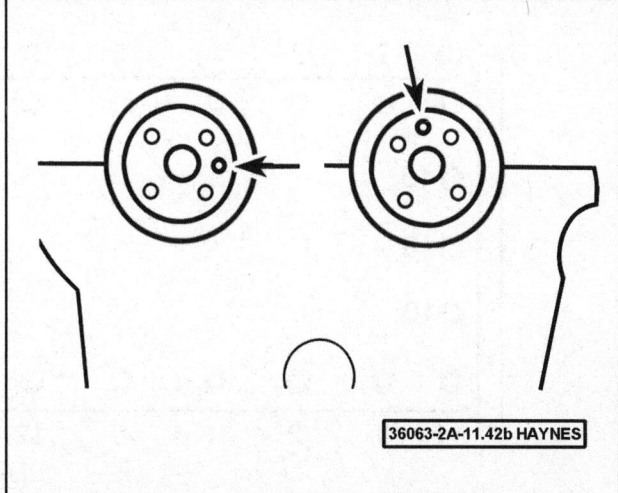

11.19b Rotate the right (passenger's) side cylinder head camshafts so the dowel pins are positioned like this (TDC position), then install the camshaft holding tool

18 Remove the camshaft cap bolts that secure the valvetrain oil tubes. Remove the oil tubes.

19 Rotate the camshafts to position the dowel pins in their proper positions for installing the VCT units and timing chains (see illustrations). Install the camshaft holding tools.

20 Rotate the crankshaft clockwise to position the keyway in the 11 o'clock position.

21 Install the VCT units (see Section 10).

22 Install the timing chains (see Section 10).

23 reinstall the valvetrain oil tubes and tighten the bearing cap bolts to the torque listed in this Chapter's Specifications.

24 The remaining installation steps are the reverse of removal.

25 Before starting and running the engine, refill the cooling system, change the oil and install new oil filter (see Chapter 1).

12 Oil pan - removal and installation

REMOVAL

1 Disconnect the cable from the negative battery terminal (see Chapter 5).

2 Remove the engine oil dipstick.

3 Raise the vehicle and support it securely on jackstands. Remove the under-vehicle splash shield by twisting the quarter-turn fasteners counterclockwise

4 On models with a skid-plate, remove the mounting bolts and the skid-plate.

5 Drain the engine oil and remove the oil filter (see Chapter 1).

6 On 4WD models, strap the front differential to a sturdy jack, remove the bushing bolts and lower the front axle enough to allow clearance for oil pan removal (see Chapter 8).

7 Remove the transmission cooler bracket fasteners then disconnect the transmission cooler from the transmission (see Chapter 7B).

8 Disconnect the oxygen-sensor harness clips and the starter motor harness clips from the oil pan. Position the harness out of the way.

9 Disconnect the alternator harness from the engine front cover.

10 On models with an air dam below the radiator, remove the two bolts and the air dam.

11 Remove the nuts/bolts securing the crossmember below the oil pan, and remove the crossmember.

12 Remove the front cover-to-oil pan bolts, the oil pan-to-transmission bolts and the oil pan-to-engine block bolts.

13 Loosen the following bolts 3/16-inch: the four upper bellhousing bolts, the bellhousing bolt above the starter motor, the two mounting bolts on the left side of the bellhousing, and the two nuts in the center of the transmission crossmember. Move the transmission rearward as allowed by the loosened fasteners.

14 There are two notches at the top rear corners of the oil pan where a large screwdriver can safely be used to pry the pan loose.

INSTALLATION

▶ **Refer to illustration 12.18**

✸✸ CAUTION:

Do not use abrasive discs or scrapers on either the engine or the pan mounting surfaces. The aluminum and plastic components can easily be gouged, leading to oil leakage.

15 Clean the surfaces of the pan, the engine, front cover and transmission of all traces of RTV sealant.

16 Apply a 7/32-inch (5.5 mm) bead of High Performance RTV sealant at the junctures of the rear main seal housing and the block, and also where the pan meets the front cover.

17 Apply a 1/8-inch (3 mm) bead of High Performance RTV sealant around the inner perimeter of the oil pan's mounting surface and install the oil pan. Install the four corner bolts within four minutes of applying the RTV sealant. Install and tighten the remaining bolts within an hour.

2B-14 3.5L V6 ENGINES

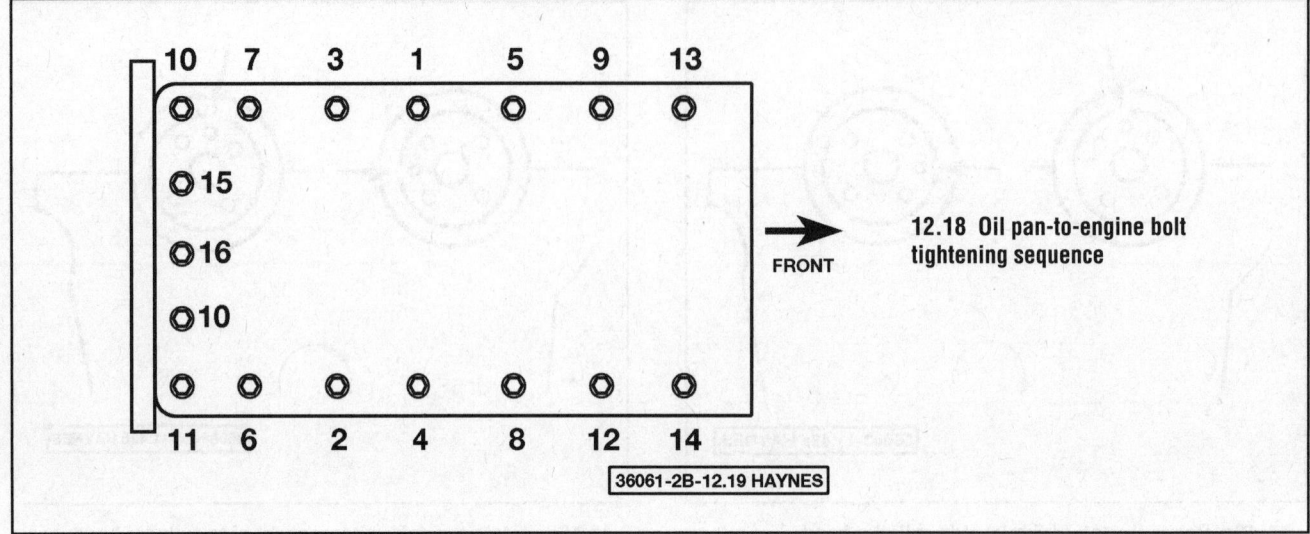

12.18 Oil pan-to-engine bolt tightening sequence

18 When all of the pan-to-engine bolts are started, tighten the bolts, in sequence, to the torque listed in this Chapter's Specifications (see illustration). After the pan-to-engine bolts are tightened, install and tighten the pan-to-transmission bolts, and all fasteners loosened in Steps 13 and 14.

✳ CAUTION:
You must work fast to secure the oil pan to the engine within 10 minutes of applying the RTV.

➥ **Note: All bolts must be fully tightened within 60 minutes of applying the High Performance RTV or oil leaks can occur.**

19 The remaining steps are the reverse of removal.

✳ CAUTION:
Don't forget to add engine oil and install a new oil filter (see Chapter 1), but wait at least 90 minutes before doing so.

20 Start the engine and check carefully for oil leaks at the oil pan.

13 Oil pump - removal and installation

✳ WARNING:
The engine must be completely cool before beginning this procedure.

1 Drain the oil and remove the oil filter (see Chapter 1).
2 Remove the oil pan (see Section 12).
3 Remove the engine front cover (see Section 9).
4 Remove the primary timing chain (see Section 10).
5 Slide the crankshaft sprocket from the crankshaft.
6 Remove the three fasteners and the oil pump pickup and screen assembly.
7 Remove the three oil pump mounting bolts and remove the oil pump.
8 Before installing the oil pump, pour a couple of tablespoons of engine oil into the pump the rotate the pump sprocket by hand to prime the pump.
9 Installation is the reverse of removal. After bolting the oil pump in place tighten the bolts to the torque listed in this Chapter's Specifications, reinstall the oil pump pickup tube and screen with a new O-ring.
10 Install a new oil filter, refill the engine with oil and fill the cooling system (see Chapter 1).

14 Driveplate - removal and installation

▶ Refer to illustration 14.7

1 Raise the vehicle and support it securely on jackstands, then refer to Chapter 7B and remove the transmission. If it's leaking, now would be a very good time to replace the front pump seal/O-ring. Look for factory paint marks that indicate driveplate-to-crankshaft alignment. If they aren't there, use a center-punch or paint to make alignment marks on the driveplate and crankshaft to ensure correct alignment during reinstallation.
2 Remove the bolts that secure the driveplate to the crankshaft. If the crankshaft turns, use a flywheel holding tool or wedge a screwdriver in the ring gear teeth to jam the driveplate.
3 Remove the driveplate from the crankshaft.
4 Check for cracked and broken ring gear teeth.

3.5L V6 ENGINES 2B-15

5 Clean and inspect the mating surfaces of the driveplate and the crankshaft. If the crankshaft rear seal is leaking, replace it before reinstalling the driveplate (see Section 15).

6 Position the driveplate against the crankshaft, but make sure the crankshaft sensor ring is positioned between the driveplate and the engine. Be sure to align the marks made during removal.

➡ **Note that some engines have an alignment dowel or staggered bolt holes to ensure correct installation.**

7 Tighten the bolts, in sequence (see illustration) to the torque listed in this Chapter's Specifications.

8 The remainder of installation is the reverse of the removal procedure.

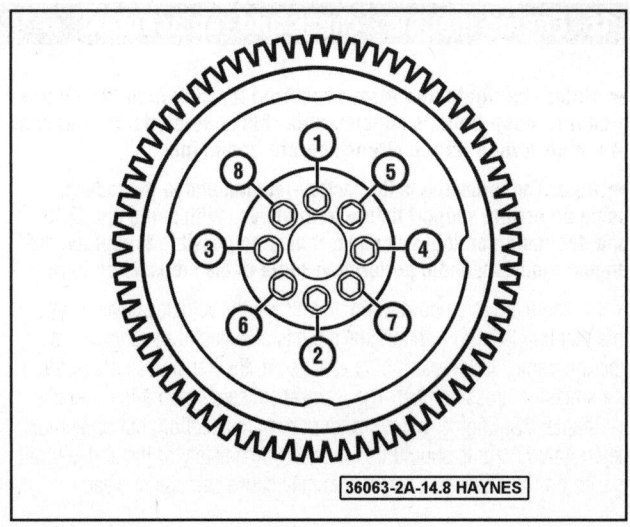

14.7 Driveplate bolt tightening sequence

15 Rear main oil seal - replacement

▶ Refer to illustration 15.10

➡ **Note: The crankshaft rear seal and retainer must be replaced as an assembly, the seal cannot be replaced separately.**

1 Remove the transmission (see Chapter 7B).

2 Remove the driveplate (see Section 14) then remove the crankshaft sensor ring from the end of the crankshaft.

3 Disconnect the crankshaft sensor electrical connector, then remove the crankshaft sensor mounting bolts and the sensor from the seal housing (see Chapter 6).

4 Remove the oil pan-to-seal housing bolts.

5 Remove the seal retainer mounting bolts and the retainer from the rear of the engine.

6 Use only a plastic gasket scraper or putty knife to remove all traces of old gasket material and sealant from the pan and engine block.

✳✳ CAUTION:

To prevent an oil leak after the new seal is installed, be very careful not to scratch or otherwise damage the crankshaft sealing surface or the engine block.

7 Remove all traces of RTV sealant. Apply a 3/16-inch bead of high performance RTV along the sealing surfaces of the retainer where the block and oil pan flanges meet.

8 Apply a film of clean engine oil to the lips of the new seal and around the crankshaft where the seal makes contact.

9 Install the retainer assembly to the engine at an angle, with the bottom of the retainer touching the oil pan, tilt the retainer up and onto the rear of the cylinder block then install all the bolts within 10 minutes of applying the sealant.

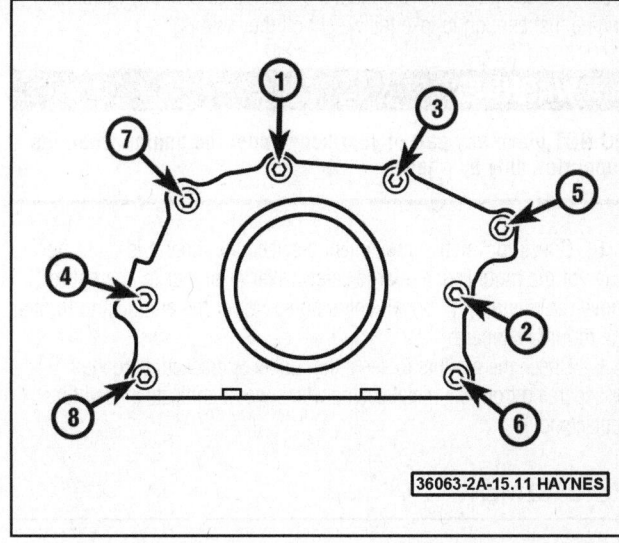

15.10 Crankshaft rear oil seal retainer bolt tightening sequence

10 Install the seal retainer-to-engine block bolts first, then tighten the bolts, in sequence (see illustration) to the torque listed in this Chapter's Specifications, then install the oil pan-to-seal retainer bolts and tighten the bolts to the torque listed in this Chapter's Specifications.

11 Reinstall the crankshaft sensor (see Chapter 6), the crankshaft sensor ring and the driveplate, making sure the crankshaft sensor ring is between the crankshaft and the driveplate.

12 Reinstall the transmission (see Chapter 7B).

16 Engine mounts - check and replacement

➡ **Note:** The tightening torque required for installing the engine mount through-bolts is considerable. Make sure you have access to a high-torque torque wrench before beginning.

➡ **Note:** The following is the factory-recommended procedure, using an engine support fixture from above. With some creativity and depending on tool selection, it may be possible to replace the engine mounts without performing some of the steps listed here.

1 There are three powertrain mounts on the vehicles covered by this manual; left and right engine mounts attached to the engine block and the frame, and a rear mount attached to the transmission and the transmission crossmember. The rear transmission mount is covered in Chapter 7B. Engine mounts seldom require attention, but broken or deteriorated mounts should be replaced immediately or the added strain placed on the driveline components may cause damage or wear.

CHECK

2 During the check, the engine must be raised slightly to remove the weight from the mounts.
3 Raise the vehicle and support it securely on jackstands.
4 Position a jack under the engine oil pan. Place a large wood block between the jack head and the oil pan, then carefully raise the engine just enough to take the weight off the mounts.

✻ WARNING:

DO NOT place any part of your body under the engine when it's supported only by a jack!

5 Check for relative movement between the inner and outer portions of the mount (use a large screwdriver or prybar to attempt to move the mounts). If movement is noted, lower the engine and tighten the mount fasteners.
6 Check the mounts to see if the rubber is cracked, hardened or separated from the metal casing which would indicate a need for replacement.

REPLACEMENT

✻ WARNING:

A high amount of torque must be employed to remove/install the engine mounts. Unless you have the proper tools, you may not be able to fully tighten the fasteners, which is an unsafe condition.

✻ CAUTION:

Use only hand tools to either remove or install the engine mount through-bolts.

7 Disconnect the cable from the negative terminal of the battery (see Chapter 5).
8 Raise the front of the vehicle and support it securely on jackstands.

Left-side mount

9 Remove the hood (see Chapter 11).
10 Remove the air filter housing (see Chapter 4).
11 Remove the charge air cooler intake and outlet pipes (see Chapter 4).
12 Remove the intake manifold (see Section 5).
13 Remove the right rear ignition coil (see Chapter 5).
14 Attach lifting eyes at the right-front and left-rear of the engine. Support the engine with a hoist or a lifting device such as an engine support fixture.

✻ CAUTION:

Raise the engine just enough to take the weight off the engine mounts.

15 Remove the under-vehicle splash shield.
16 Remove the skid plate, if equipped.
17 Remove the front exhaust Y pipe bolts from all three ends and lower the pipe.
18 Loosen, but don't remove, the nuts on the transmission mount crossmember (see Chapter 7B).
19 On 4WD models, place a floor jack under the front differential and chain or otherwise securely retain the differential to the jack. Raise the front axle slightly, and remove the differential mounting bushing bolts, then lower the differential for room to access the engine mounts. Remove the front driveshaft (see Chapter 8).
20 Remove the engine mount through-bolt. use only hand tools do not use impact tools to remove the mount nuts/bolts.
21 Remove the bracket-to-engine block bolts, then remove the bracket.
22 Remove the three bolts securing the mount insulator to the frame rail.

Right-side mount

23 Remove the hood (see Chapter 11).
24 Remove the air filter housing (see Chapter 4).
25 Remove the charge air cooler intake and outlet pipes (see Chapter 4).
26 Remove the intake manifold (see Section 5).
27 Remove the right rear ignition coil (see Chapter 5).
28 Attach lifting eyes at the right-front and left-rear of the engine. Support the engine with a hoist or a lifting device such as an engine support fixture

✻ CAUTION:

Raise the engine just enough to take the weight off the engine mounts.

29 Remove the under-vehicle splash shield and skid plate, if equipped.
30 Remove the starter (see Chapter 5).
31 Remove the drivebelt (see Chapter 1).
32 Disconnect the electrical connector to the coolant pump, then remove the coolant pump mounting bolts, without disconnecting the coolant hoses (see Chapter 3).
33 Disconnect the electrical connectors to the air conditioning compressor then remove the air conditioning compressor mounting bolts and secure the compressor out of the way without disconnecting the compressor lines (see Chapter 3).

3.5L V6 ENGINES 2B-17

34 Remove the front exhaust Y pipe bolts from all three ends and lower the pipe.
35 Loosen, but don't remove, the nuts on the transmission mount crossmember (see Chapter 7B).
36 On 4WD models, place a floor jack under the front differential and chain or otherwise securely retain the differential to the jack. Raise the front axle slightly, and remove the differential mounting bushing bolts, then lower the differential for room to access the engine mounts. Remove the front driveshaft (see Chapter 8).
37 Remove the engine mount through-bolt. use only hand tools do not use impact tools to remove the mount nuts/bolts.
38 Remove the bracket-to-engine block bolts, then remove the bracket.
39 Remove the three bolts securing the mount insulator to the frame rail.

All models

40 Install the new engine mount, tightening the bolts to the torque listed in this Chapter's Specifications.
41 Install the engine mount onto the engine support bracket and tighten the bolts securely. Then install the mount through-bolts, using thread locking compound on the threads.
42 The remainder of the installation is the reverse of removal. Remove the engine hoist and the jackstands and lower the vehicle.

✲✲ CAUTION:
Be sure to tighten the nuts at the transmission crossmember.

Specifications

General

Displacement	3.5 liters (214 cubic inches)
Cylinder numbers (front-to-rear)	
Left (driver's) side	4-5-6
Right side	1-2-3
Firing order	1-4-2-5-3-6
Cylinder head warpage limit	
Lengthwise	0.003 inch
Widthwise	0.002 inch

Cylinder locations

Camshaft

Lobe lift (intake and exhaust)	0.373 inch
Lobe wear limit	0.0024 inch
Endplay	
Standard	0.001 to 0.006 inch
Service limit	0.0074 inch
Journal diameter	
First journal	1.535 to 1.536 inches
Intermediate journals	1.021 to 1.022 inches
Journal-to-bearing (oil) clearance	0.0029 inch
Journal runout limit	0.0015 inch
Valve clearance	
Intake	0.006 to 0.010 inch
Exhaust	0.0142 to 0.0181 inch

Torque specifications Ft-lbs (unless otherwise indicated)

➡ **Note:** One foot-pound (ft-lb) of torque is equivalent to 12 inch-pounds (in-lbs) of torque. Torque values below approximately 15 ft-lbs are expressed in inch-pounds, since most foot-pound torque wrenches are not accurate at these smaller values.

Variable Camshaft Timing (VCT) unit (camshaft sprocket) bolt (new)	
Step 1	30
Step 2	Loosen one full turn
Step 3	18
Step 4	Tighten an additional 180-degrees

2B-18 3.5L V6 ENGINES

Torque specifications (continued) — Ft-lbs (unless otherwise indicated)

→ **Note:** One foot-pound (ft-lb) of torque is equivalent to 12 inch-pounds (in-lbs) of torque. Torque values below approximately 15 ft-lbs are expressed in inch-pounds, since most foot-pound torque wrenches are not accurate at these smaller values.

Camshaft cap bolts (see illustration 11.16)	
Step 1	71 in-lbs
Step 2	Tighten an additional 45-degrees
Step 3	Loosen bolts 8 through 11
Step 4	Repeat Steps 1 and 2 on bolts 8 through 11
Crankshaft pulley bolt	
Step 1	89
Step 2	Loosen one turn
Step 3	37
Step 4	Tighten an additional 90-degrees
Crankshaft rear seal retainer	89 in-lbs
Cylinder head bolts (new, oiled) (in sequence - see illustration 8.22)	
Step 1	177 in-lbs
Step 2	26
Step 3	Tighten an additional 90-degrees
Step 4	Tighten an additional 90-degrees
Step 5	Tighten an additional 45-degrees
M6 bolt at front of cylinder head	89 in-lbs
Drivebelt tensioner pulley mounting bolt(s)	18
Exhaust manifold nuts	
Step 1	162 in-lbs
Step 2	18
Exhaust manifold studs	106 in-lbs
Driveplate bolts	59
Intake manifold bolts (see illustration 5.9)	
Step 1	89 in-lbs
Step 2	Tighten an additional 45-degrees
Oil pan-to-block bolts (see illustration 12.18)	89 in-lbs
Oil pump screen nuts-to-engine block nuts	89 in-lbs
Oil pump mounting bolts	89 in-lbs
Engine front cover-to-block bolts (see illustration 9.26)	
Step 1, bolts 1 through 21	89 in-lbs
Step 2, bolts 1 through 20	18
Step 3, bolt 21	177 in-lbs
Step 4, bolt 21	Tighten an additional 90-degrees
Step 5, bolt 22	89 in-lbs
Step 6, bolt 22	Tighten an additional 45-degrees
Oil pan-to-timing chain cover bolts	89 in-lbs
Engine mount fasteners	
Through-bolts	258
Mount bracket-to-engine block bolts	57
Mount-to-frame bolts	129
Mount nuts	111
Timing chain tensioner bolts	96 in-lbs
Turbocharger-to-exhaust manifold bolts	24
Turbocharger bracket-to-block bolts	89 in-lbs
Turbocharger bracket-to-turbocharger bolt	159 in-lbs
Valve cover bolts	89 in-lbs
Variable Camshaft Timing oil control solenoid bolt	
Step 1	71 in-lbs
Step 2	Tighten an additional 20 degrees

Section

1. General information
2. Repair operations possible with the engine in the vehicle
3. Top Dead Center (TDC) for number one piston - locating
4. Valve covers - removal and installation
5. Crankshaft pulley - removal and installation
6. Timing chain cover - removal and installation
7. Timing chains, tensioners and sprockets - removal, inspection and installation
8. Rocker arms and valve lash adjusters - removal, inspection and installation
9. Camshaft(s) - removal, inspection and installation
10. Valve springs, retainers and seals - removal and installation
11. Intake manifold - removal and installation
12. Exhaust manifolds - removal and installation
13. Cylinder heads - removal and installation
14. Oil pan - removal and installation
15. Oil pump - removal and installation
16. Flywheel/driveplate - removal and installation
17. Crankshaft oil seals - replacement
18. Engine mounts - check and replacement

Reference to other Chapters
CHECK ENGINE light - See Chapter 6

2C
V8 ENGINES

2C-2 V8 ENGINES

1 General information

This Part of Chapter 2 is devoted to in-vehicle repair procedures for the 4.6L and 5.4L Single Overhead Cam (SOHC) engines. Both engines are of the same modular design; the 5.4L engine has a longer stroke for more displacement. Both engines have aluminum heads, iron blocks, a single camshaft for each cylinder head, and two valves per cylinder. All information concerning engine removal and installation and engine block and cylinder head overhaul can be found in Part D of this Chapter.

There are two versions of the 4.6L V8, depending on where it was produced; the Romeo (Romeo, Michigan plant) and the Windsor (Windsor, Ontario, Canada), with minor differences in some of the components and procedures. It's important when working on the 4.6L engine to know which version you have. The REP (Romeo Engine Plant) 4.6L has 11 bolts on the left valve cover, while the WEP (Windsor Engine Plant) 4.6L has 13 bolts on the left valve cover, and the 5.4L V8 has 14 bolts. The engines can also be identified from the VIN number of the vehicle. The eighth place in the VIN is the engine identification. See "Vehicle Identification Numbers" at the beginning of this manual for engine code information.

The following repair procedures are based on the assumption that the engine is installed in the vehicle. If the engine has been removed from the vehicle and mounted on a stand, many of the steps outlined in this Part of Chapter 2 will not apply.

The Specifications included in this Part of Chapter 2 apply only to the procedures contained in this Part. Part D of Chapter 2 contains the Specifications necessary for cylinder head and engine block rebuilding.

2 Repair operations possible with the engine in the vehicle

Many major repair operations can be accomplished without removing the engine from the vehicle.

If possible, clean the engine compartment and the exterior of the engine with some type of pressure washer before any work is started. It will make the job easier and help keep dirt out of the internal areas of the engine.

It may help to remove the hood to improve access to the engine as repairs are performed (refer to Chapter 11 if necessary).

If vacuum, exhaust, oil or coolant leaks develop, indicating a need for gasket or seal replacement, the repairs can generally be made with the engine in the vehicle. The intake and exhaust manifold gaskets, timing cover gasket, oil pan gasket, crankshaft oil seals and cylinder head gaskets are all accessible with the engine in place.

Exterior engine components, such as the intake and exhaust manifolds, the oil pan, the water pump, the starter motor, the alternator and the fuel system components can be removed for repair with the engine in place.

Since the cylinder heads can be removed without pulling the engine, valve component servicing can also be accomplished with the engine in the vehicle. Replacement of the timing chain and sprockets and oil pump is also possible with the engine in the vehicle.

In extreme cases caused by a lack of necessary equipment, repair or replacement of piston rings, pistons, connecting rods and rod bearings is also possible with the engine in the vehicle. However, this practice is not recommended because of the cleaning and preparation work that must be done to the components involved.

3 Top Dead Center (TDC) for number one piston - locating

3.1 When placing the engine at Top Dead Center (TDC), align the notch in the crankshaft pulley (arrow) with the TDC indicator on the timing chain cover

▶ **Refer to illustration 3.1**

Refer to Chapter 2, Part A for the TDC locating procedure, but use the illustration provided with this Section for the appropriate reference marks and the following exceptions:

a) *Disable the ignition system by disconnecting the primary electrical connectors at the ignition coil pack/modules (see Chapter 5).*
b) *Remove the spark plugs and install a compression gauge in the number one cylinder. Turn the crankshaft clockwise with a socket and breaker bar as described in Chapter 2.*
c) *When the piston approaches TDC, compression will be noted on the compression gauge. Continue turning the crankshaft until the notch in the crankshaft pulley is aligned with the TDC mark on the front cover (see illustration). At this point number one cylinder is at TDC on the compression stroke.*

V8 ENGINES 2C-3

4 Valve covers - removal and installation

REMOVAL

➡ **Note:** On 1998 and later V8 engines, the valve cover bolts are connected to the valve covers. Do not attempt to remove them from the covers, just loosen them until they are free of the cylinder head.

4.6L engine

Right valve cover

▶ Refer to illustration 4.3

1 Disconnect the cable from the negative battery terminal.
2 Disconnect the PCV hose and pull the wiring harness from the studs on the valve cover.
3 Loosen the valve cover bolts (do not try to remove them completely on a 1998 or later model) and remove the cover (see illustration). If it's stuck tap it with a hammer and block of wood.

Left valve cover

▶ Refer to illustration 4.6

4 Refer to Chapter 4 and remove the air cleaner assembly. Disconnect the cable from the negative battery terminal.
5 Remove the three bolts retaining the power steering fluid reservoir and bracket (see Chapter 10). Set the reservoir aside.
6 Remove the nut and brake booster vacuum hose bracket (see illustration).
7 Refer to Chapter 6 and disconnect the EGR pipe and the two feedback transducer hoses.
8 Loosen the valve cover bolts (do not try to remove them completely on a 1998 or later model) and remove the valve cover. If it's stuck tap it with a hammer and block of wood.

5.4L engine

Right valve cover

▶ Refer to illustrations 4.12 and 4.16

9 Refer to Chapter 4 and remove the air cleaner and the plastic shield over the throttle cable. Disconnect the cable from the negative battery terminal. On 2003 and later Expedition/Navigator models, have

4.3 Remove the bolts around the perimeter of the valve cover (arrows)

the refrigerant recovered before beginning the procedure. On models with two-piece intake manifolds, remove the upper intake plenum (see Section 11).

10 Remove the plastic shield over the battery cable junction box and disconnect the two cables (see Chapter 12).
11 Refer to Chapter 12 and remove the two mega-fuses.
12 At the junction block bracket, disconnect the two bulkhead connectors, the single-wire connector and the 16-pin connector, then remove the bolts and set the junction block bracket aside (see illustration). On later Expedition/Navigator models, disconnect the PCM harness connectors, then unbolt and set aside the PCM (see Chapter 6).
13 Drain the cooling system (see Chapter 1) and disconnect and set aside the two heater hoses from the heater core and the ends near the intake manifold (see Chapter 3).
14 From underneath, disconnect the electrical connectors at the air conditioning compressor (see Chapter 3) and the crankshaft position sensor (see Chapter 6).
15 Disconnect the electrical connectors on the right-hand fuel injectors (see Chapter 4) and the right hand ignition coil connectors (see Chapter 5).
16 Pull the harness away from the valve cover to allow enough room to get the cover out (see illustration).

4.6 Remove the nut (arrow) and set aside the bracket supporting the power brake booster vacuum hose

4.12 Loosen the bolts retaining the bulkhead connectors and unbolt the junction block bracket from the firewall (arrows)

4.16 Detach the wiring harness from the valve cover studs (arrow)

2C-4 V8 ENGINES

4.24 Apply a dab of RTV sealant to the mating joints (arrows) between the timing cover and the cylinder head before installing the valve cover

17 Loosen the valve cover bolts (do not try to remove them completely on a 1998 or later model) and remove the valve cover. If it's stuck tap it with a hammer and block of wood.

Left valve cover

18 Follow Steps 4 through 7 for the left valve cover on the 4.6L engine. On models with two-piece intake manifolds, remove the upper intake plenum (see Section 11).

19 Disconnect the electrical connectors on the left bank fuel injectors (see Chapter 4) and the left bank ignition coil connectors (see Chapter 5).

20 Pull the harness away from the valve cover to allow enough room to get the cover out (see illustration 4.16). Disconnect the PCV hose and unbolt the oil dipstick tube. On 2005 and later 5.4L 3V engines, disconnect the VCT (Variable Camshaft Timing) connector just ahead of the left valve cover.

21 Loosen the valve cover bolts (do not try to remove them completely on a 1998 or later model) and remove the valve cover. If it's stuck tap it with a hammer and block of wood.

INSTALLATION - ALL MODELS

▸ **Refer to illustration 4.24**

22 The mating surfaces of each cylinder head and valve cover must be perfectly clean when the valve covers are installed. Remove all traces of sealant, and clean the mating surfaces with lacquer thinner or acetone. If there's old sealant or oil on the mating surfaces when the valve cover is installed, oil leaks may develop.

23 The valve cover gaskets should be mated to the valve covers with gasket adhesive before the valve covers are installed. Make sure the gasket is pushed all the way into the groove in the valve cover.

24 At the mating joint (two spots per cylinder head) between the timing chain cover and cylinder head, apply a dab of RTV sealant before installing the valve cover (see illustration).

25 Carefully position the valve cover on the cylinder head and install the nuts and bolts.

➡ **Note: Install the covers within five minutes of applying the RTV sealant.**

26 Tighten the fasteners in two steps to the torque listed in this Chapter's Specifications. Wait two minutes between the first and the second round of tightening. On all except 4.6L REP models, tighten the bolts/nuts in a sequence starting in the center and working alternately toward each end of the valve cover. On 4.6L REP models, tighten the upper row of bolts first, then the lower row, from rear to front.

❊❊ CAUTION:

Be careful with the plastic valve covers, don't over tighten the bolts!

27 The remaining installation steps are the reverse of removal.
28 Start the engine and check for oil leaks as the engine warms up.

5 Crankshaft pulley - removal and installation

5.2 Have an assistant hold the crankshaft and use a breaker bar and socket to remove the crankshaft pulley bolt (arrow)

REMOVAL

▸ **Refer to illustrations 5.2 and 5.3**

1 If necessary for access, remove the cooling fan and shroud (see Chapter 3). Remove the accessory drivebelt (see Chapter 1). On 2006 through 2009 Expedition/Navigator models, remove the drivebelt splash shield.

2 Remove the flywheel inspection cover, if equipped (see Chapter 7A) and with the help of an assistant, wedge a large screwdriver into the starter ring gear teeth to prevent the crankshaft from turning. Remove the large center bolt from the crankshaft pulley with a breaker bar and socket (see illustration).

3 Using a suitable puller, detach the crankshaft pulley (see illustration). Leave the Woodruff key in place in the crankshaft keyway.

❊❊ CAUTION:

Don't use a puller with jaws that grip the outer edge of the crankshaft pulley. The puller must be the type shown in the illustration that utilizes bolts to apply force to the crankshaft pulley hub only.

V8 ENGINES 2C-5

5.3 Remove the crankshaft pulley with a puller that bolts to the crankshaft pulley hub

5.4 Inspect the crankshaft pulley for signs of damage or excessive wear

1 Oil seal surface   2 Woodruff keyway

INSTALLATION

♦ Refer to illustration 5.4

4 Lubricate the oil seal contact surface of the crankshaft pulley hub (see illustration) with moly-base grease or clean engine oil. Apply a dab of RTV sealant to the front end of the keyway in the crankshaft pulley before installation.

5 Install the crankshaft pulley on the end of the crankshaft. The keyway in the crankshaft pulley must be aligned with the Woodruff key in the crankshaft. If the crankshaft pulley cannot be seated by hand, slip the large washer over the bolt, install the bolt and tighten it to pull the crankshaft pulley into place. Now loosen the bolt one full turn, then tighten the bolt to the torque listed in this Chapter's Specifications.

6 The remaining installation steps are the reverse of removal.

7 Check the oil level. Run the engine and check for oil leaks.

6 Timing chain cover - removal and installation

6.9 Remove the bolts (arrows) and the drivebelt tensioner assembly from the right side

REMOVAL

♦ Refer to illustrations 6.9 and 6.11

1 Disconnect the cable from the negative battery terminal.

2 Drain the engine oil and remove the oil filter (see Chapter 1).

3 Remove the drive belt and the water pump pulley. Remove the crankshaft pulley (see Section 5).

4 Refer to Section 4 and remove both valve covers.

5 Drain the cooling system (see Chapter 1). Refer to Chapter 3 and remove the radiator and water pump.

6 From below, remove the four front oil pan bolts (see Section 14).

7 Disconnect the electrical connectors to the camshaft sensor and the crankshaft sensor (see Chapter 6). On 4.6L engines, disconnect and remove both ignition coil packs (one is attached to each side of the timing chain cover, see Chapter 5). Unbolt the ignition coil brackets.

8 Remove the bolts securing the power steering pump to the engine (see Chapter 10).

➡ Note: The front lower bolt on the power steering pump will not come all the way out.

Position the pump aside and secure it out of the way.

9 Unbolt and remove the drivebelt tensioner assembly from the right side of the timing cover (see illustration).

10 Remove the timing chain cover-to-engine block bolts and the four oil pan-to-timing chain cover bolts from underneath. Including the oil pan bolts, there are 19 bolts to be removed from the timing chain cover. Note the locations of studs and different length bolts so they can be reinstalled in their original locations.

2C-6 V8 ENGINES

6.11 Separate the timing chain cover from the engine, using a soft-faced hammer if necessary to break the gasket seal

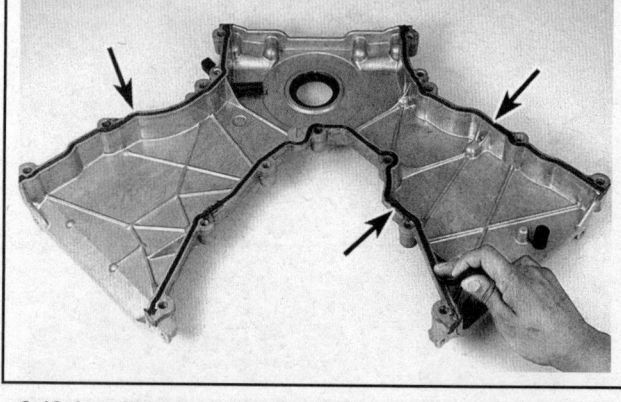

6.13 Install three new gaskets (arrows) into the grooves in the back of the timing chain cover

11 Separate the timing chain cover from the engine block (see illustration). If it's stuck, tap it gently with a soft-face hammer to break the gasket bond.

※※ CAUTION:

DO NOT use excessive force or you may crack the cover. If the cover is difficult to remove, make sure all of the bolts have been removed.

INSTALLATION

▸ Refer to illustrations 6.13, 6.14 and 6.16

12 Clean the mating surfaces of the timing chain cover, engine block and cylinder heads to remove all traces of old gasket material, oil and dirt. Final cleaning should be with lacquer thinner or acetone.

※※ WARNING:

Be careful when cleaning any of the aluminum components. Use of a metal scraper could cause scratches or gouges that could lead to an oil leak later.

13 Adhere the three new gaskets to the backside of the timing chain cover (see illustration).
14 Apply a 1/8-inch bead of RTV sealant to the junctions of the oil pan-to-engine block and the cylinder head-to-engine block (see illustration). Apply a small dab of RTV where the timing chain cover and engine block meet at the valve cover surface.
15 Lubricate the timing chains and the lip of the crankshaft front oil seal with clean engine oil.
16 Install the timing chain cover on the engine, within five minutes of applying the RTV sealant. Position the bottom/front edge of the timing chain cover flush with the front edge of the oil pan and "tilt" the top of the cover into place against the engine. Do not press the cover straight in against the engine or the sealant may be scraped off the front of the oil pan and cause a leak. Tighten the timing chain cover-to-engine block bolts in the recommended sequence (see illustration), to the torque and sequence listed in this Chapter's Specifications. Tighten the timing chain cover-to-oil pan bolts.
17 Install the remaining parts in the reverse order of removal.
18 Add the proper type and quantity of engine oil and coolant (see Chapter 1). Run the engine and check for leaks.

6.14 Apply a small bead of RTV sealant to the mating junctions of the oil pan-to-engine block and cylinder head-to-engine block

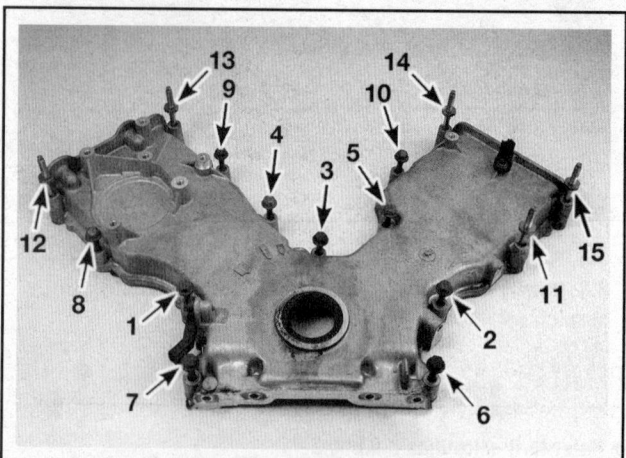

6.16 Timing chain cover bolt tightening sequence

V8 ENGINES 2C-7

7 Timing chains, tensioners and sprockets - removal, inspection and installation

⚠ CAUTION:

At no time, once the timing chain(s) have been removed, can the crankshaft or the camshafts be rotated. If moved, damage to the valves and/or pistons can occur. Special tools, available from your local dealer, are necessary to prevent the camshafts from moving when the timing chain is removed. Read through the entire procedure and obtain the necessary tools before proceeding.

➡ Note: Because this is an interference engine design, if the chain has broken, there will be damage to the valves and/or pistons and will require removal of the cylinder heads.

REMOVAL

▸ Refer to illustrations 7.3 and 7.4

⚠ CAUTION

The timing system is complex. Severe engine damage will occur if you make any mistakes. Do not attempt this procedure unless you are highly experienced with this type of repair. If you are at all unsure of your abilities, consult an expert. Double-check all your work and be sure everything is correct before you attempt to start the engine.

1 Disconnect the cable from the negative battery terminal.
2 Position the number one cylinder on TDC (see Section 3), and remove the spark plugs (see Chapter 1).
3 Remove the valve covers (see Section 4) and the timing chain cover (see Section 6). Two long timing chains connect the crankshaft to the camshafts (see illustration).
4 Remove the crankshaft position sensor toothed-wheel (see illustration) by sliding it off the end of the crankshaft nose. Note the stamped word "front" or "rear" on the wheel to be sure it's reinstalled in the correct direction.

5.4L 2V (1997 through 1999) and 4.6L Windsor (1997 through 2004)

▸ Refer to illustrations 7.5a and 7.5b

5 Install the camshaft and crankshaft retaining tools (see illustrations). Because of different methods of retaining the camshaft sprockets on the WEP and REP engines, different tools are required for each version.

7.4 The crankshaft sensor tooth wheel has a specific direction to be installed, look for the word "front" or "rear" stamped in it (arrow)

7.3 All V8 engines have two long primary timing chains with two main tensioners (arrows)

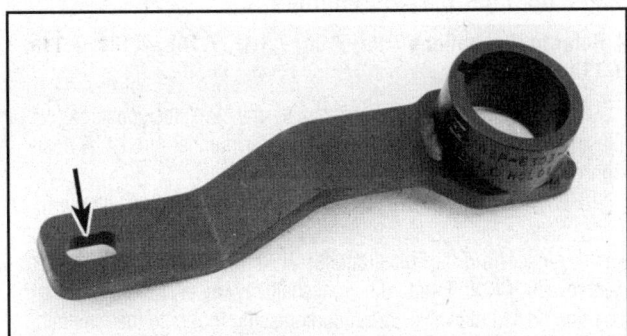

7.5a On Windsor-built engines, use this crankshaft positioning tool - the side hole (arrow) fits over the dowel on the right side of the engine

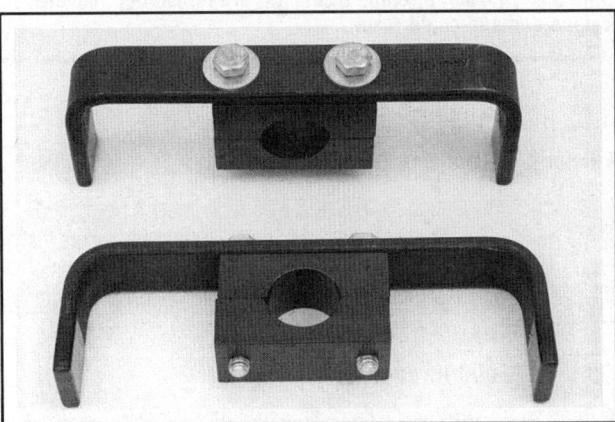

7.5b Each Windsor camshaft should be locked in TDC position with the camshaft holding tool, which bolts onto the camshaft between the second and third (from the front) camshaft retaining caps - the legs fit against the valve cover mounting surface on the cylinder head

2C-8 V8 ENGINES

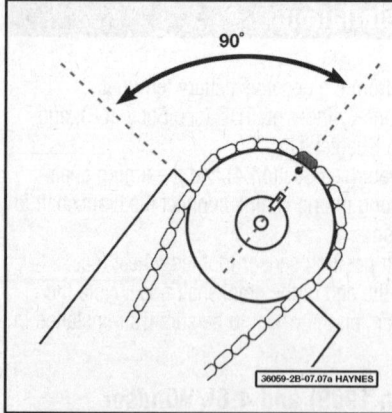

7.7a Both camshaft keyways on the Romeo-built engine must be perpendicular to the valve cover mounting surface of the heads

7.7b The camshaft positioning tool for the Romeo engine and adapters installs over the rear of the camshaft at TDC

7.8a On 5.4L 4V engines, the primary sprocket (A) drives the exhaust camshaft, which drives the secondary timing chain (B) and the intake camshaft (C)

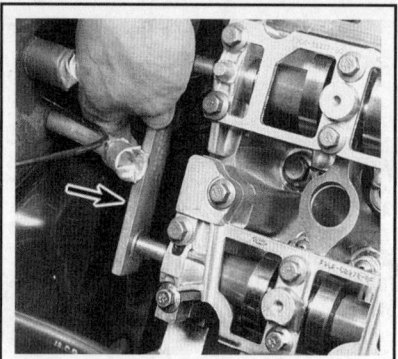

7.8b On 5.4L 4V engines, the camshaft locking tool, one for each cylinder head, fits into the D-slots on the rear of the camshafts to position the camshafts into TDC

7.8c After positioning the camshaft to TDC, install the special tool to lock the camshafts

6 The tools lock the camshafts in position to prevent any movement of the camshafts in either direction that could occur due to valve spring pressure, when the timing chains are removed.

※※ **CAUTION:**

The camshaft(s) MUST be retained exactly at TDC. If the valve timing is off when the timing chain(s) are reinstalled, severe engine damage could result.

4.6L Romeo (1997 through 2004)

▶ Refer to illustrations 7.7a and 7.7b

7 When positioning the engine at TDC for number 1 piston, look at the keyways on the camshaft sprockets. Use a square to insure that each keyway is at 90-degrees to the valve-cover-mounting surface of the cylinder head (see illustration). Then install the camshaft position tool on the rear of the camshafts (see illustration).

5.4L 4V (2000 through 2004)

▶ Refer to illustrations 7.8a, 7.8b and 7.8c

8 Install the appropriate camshaft retaining tools (see illustrations). The tool locks the camshaft(s) in position when the timing chains are removed. Be sure to lock the crankshaft in position with a special tool (see illustration 7.5a).

※※ **CAUTION:**

The camshaft(s) MUST be retained exactly at TDC. If the valve timing is off when the timing chain(s) are reinstalled, severe engine damage could result.

The retaining and positioning tools are installed into the rear of the camshaft and hold the cam in place by locking into the specially shaped hole in the rear of the camshaft. This will prevent any movement of the camshafts in either direction, due to valve spring pressure, when the timing chains are removed.

2005 and later 5.4L 3V engines

▶ Refer to illustrations 7.9a, 7.9b, 7.10a, 7.10b, 7.10c, 7.11a, 7.11b and 7.13

9 Set the groove in the crankshaft at the 12 o'clock position (see illustration). The camshaft lobes for cylinder number 1 must be positioned at a specific angle (see illustration).

10 Using a special valve spring depressing tool (see illustration), depress the valve springs and remove the specified rocker arms from each cylinder head (see illustrations). Next, rotate the crankshaft 180-degrees CLOCKWISE until the crankshaft keyway is in the 6 o'clock position. Install the valve spring compressing tool onto the remaining valve springs, compress the springs and remove the remaining rocker arms. Camshaft rocker arms MUST be reinstalled with the same camshaft lobe that they were removed from. Label and store all components to avoid confusion during reassembly.

V8 ENGINES 2C-9

7.9a On 2005 and later 5.4L engines, position the crankshaft with the keyway in the 12 o'clock position . . .

7.9b . . . and position the camshaft lobes for cylinder number one like this:

   A   Intake lobes   B   Exhaust lobe

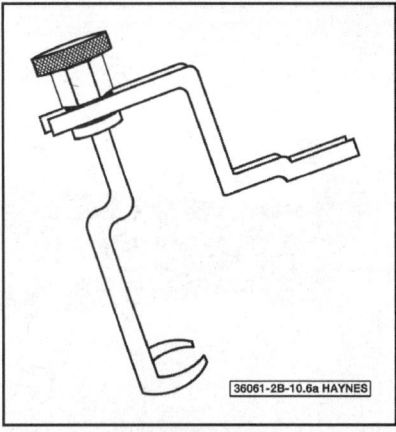

7.10a Use a special tool to compress the valve springs and remove the rocker arms (2005 and later 5.4L engine)

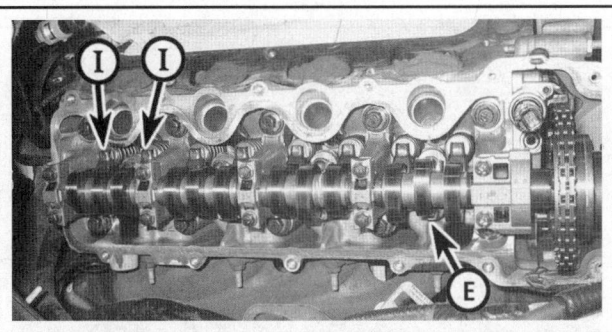

7.10b With the crankshaft keyway positioned at 12 o'clock (just past TDC for cylinder number one), the exhaust rocker arm can be removed from cylinder number 1, the intake rocker arms from cylinder number 4 . . .

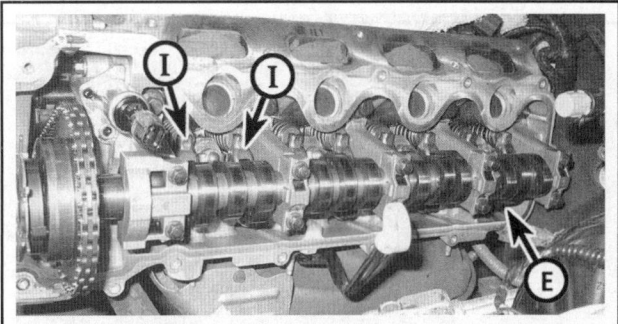

7.10c . . . the intake rocker arms from cylinder number 5 and the exhaust rocker arm from cylinder number 8 (2005 and later 5.4L engine)

7.11a Location of the left side timing chain tensioner mounting bolts (2005 and later 5.4L engine)

7.11b Location of the right side timing chain tensioner mounting bolts (2005 and later 5.4L engine)

11 Remove the tensioners and the left and right tensioner arms (see illustrations). At this point, the chains should have sufficient slack that you can lift them off the camshaft sprockets. Lift the right chain off the camshaft sprocket, then remove it from the crankshaft sprocket. Repeat on the left sprocket.

✴ CAUTION:

Be sure to note the locations of the timing marks on the sprockets and chains so you can reinstall them in the same relationship. Take photographs to ensure correct reassembly.

12 Slide both crankshaft sprockets off the crankshaft, noting how they're installed (the shoulders on the sprockets should face each other). Unbolt and remove the left and right chain guides.

13 The camshaft sprockets incorporate phasers that change camshaft timing during engine operation. If you need to remove the camshaft sprocket/phaser assemblies, lock them in position with a special tool. The tool bolts to the cylinder head (see illustration) and locks against the sprocket teeth. Remove the phaser/sprocket bolt and the phaser/sprocket assembly. Discard the old bolt and replace it with a new one on reassembly.

2C-10 V8 ENGINES

7.13 A special tool that is bolted to the cylinder head locks the camshaft sprocket in position (2005 and later 5.4L engine)

7.14a To remove the timing chain tensioner from the engine block, remove the two bolts from the tensioner (arrows) . . .

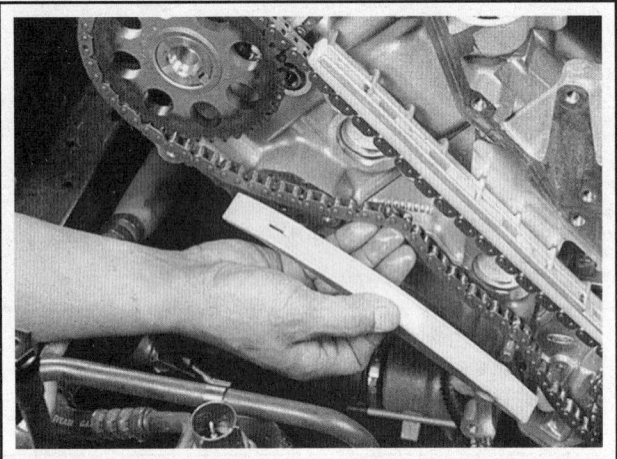

7.14b . . . and detach the guide assembly from the dowel at the opposite end (all except 2005 and later 5.4L engine)

7.16 The stationary chain guide is removed by removing the bolts (arrows) at the mount plate (all except 2005 and later 5.4L engine)

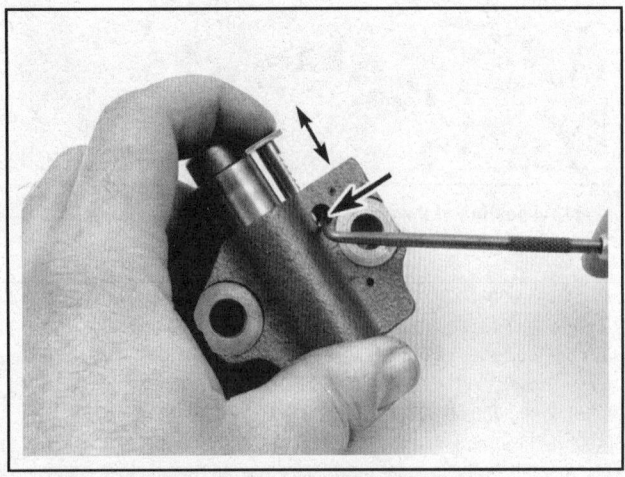

7.21a To fully retract the primary tensioner, release the plunger lock (arrow) and push the plunger into the tensioner body (earlier style tensioners)

7.21b Check the tensioner oil feed hole (arrow) to be sure it's not plugged by debris (earlier style tensioners)

V8 ENGINES 2C-11

All engines through 2004

▶ Refer to illustrations 7.14a, 7.14b and 7.16

14 Remove the right side timing chain tensioner (see illustrations).

15 Remove the right timing chain from the crankshaft and camshaft sprockets, by slipping the chain off the camshaft sprocket and pulling the crankshaft sprocket off with the chain.

16 Remove the stationary guide (see illustration).

➡ Note: The appearance of the chain guides is different between the WEP and REP engines, but the procedure and fasteners are the same.

17 Remove the left side timing chain tensioner. Lift the chain off the camshaft sprocket and remove the chain along with the crankshaft sprocket.

18 Remove the left side stationary guide.

19 If the camshaft sprockets are to be replaced, remove the camshaft sprocket bolt and large washer from the camshafts (REP engines), then pull the camshaft sprockets off. On WEP engines, use a two-bolt puller to remove the camshaft sprockets.

➡ Note: Note the location and direction of any spacers on the crankshaft or camshafts, but do not remove the spacers unless necessary.

INSPECTION

20 Inspect the individual sprocket teeth and keyways for wear and damage. Check the chain for cracked plates, pitted or worn rollers. Check the wear surface of the chain guides/tensioner arms for wear and damage. Replace any excessively worn or defective parts with new ones.

✽✽ CAUTION:

If excessive plastic material is missing from the chain guides/tensioner arms, the oil pan should be removed and cleaned of all debris (see Section 14). Check the oil pick-up tube and screen. Replace the assembly if it is clogged.

Early-style tensioners

▶ Refer to illustrations 7.21a and 7.21b

21 On earlier ratchet-type tensioners, check them for proper operation:

a) Release the plunger lock (see illustration) and make sure the piston moves freely.
b) Submerge the tensioner in a can of oil or solvent, remove from the fluid and depress the plunger to make sure the oil feed oil is not plugged (see illustration).

➡ Note: Also inspect the oil feed hole in the engine block to be certain it's not plugged.

Later-style tensioners

▶ Refer to illustrations 7.22a and 7.22b

22 On later-style tensioners, check as follows:

a) Check the condition of the tensioner seal (see illustration). Make sure the seal is intact and not broken, chipped or damaged.
b) Check the condition of the plunger. Depress the plunger to make sure it moves freely.
c) For installation, compress the tensioner in a vise and lock it in the retracted position using a special retainer clip (see illustration).

INSTALLATION

✽✽ CAUTION:

Before starting the engine, carefully rotate the crankshaft by hand through at least two full revolutions (use a socket and breaker bar on the crankshaft pulley center bolt). If you feel any resistance, STOP! There is something wrong - most likely, valves are contacting the pistons. You must find the problem before proceeding. Check your work and see if any updated repair information is available.

All engines through 2004

▶ Refer to illustration 7.24

23 Install the stationary chain guides, for both sides, and tighten the bolts to the torque listed in this Chapter's Specifications.

24 If removed, install the left crankshaft sprocket on the crankshaft with the shoulder of the sprocket facing forward. When the two sprockets are placed correctly the hubs will be facing each other and there will be the maximum space possible between the two sprockets (see illustration). For now, only install the inner (left chain) crankshaft sprocket, with its bevel forward.

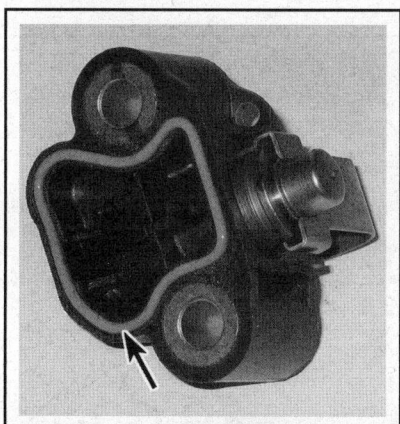

7.22a Check the condition of the tensioner seal (later-style tensioners)

7.22b Install a special tool (A) onto the plunger and tensioner side rail (B) to lock the plunger in the retracted position (later-style tensioners)

7.24 When both are installed, the two crankshaft sprockets should align like this; here the left sprocket is installed with its shoulder forward, and the right sprocket (in hand) is ready to install with its shoulder (arrow) toward the other sprocket

2C-12 V8 ENGINES

7.28a When installing the timing chain, align one of the bright links in the timing chain with the dimple on the camshaft sprocket (arrows) - sprocket should also align with the keyway in the camshaft (Windsor-built 4.6L and 5.4L engines through 2004)

7.28b Align the bright link at the lower end of the chain with the timing mark on the crankshaft sprocket (it should be at 6 o'clock), then bolt the primary camshaft sprocket to the camshaft - both chains are aligned here (Windsor-built 4.6L and 5.4L engines through 2004)

7.33 With the secondary chain and sprockets in place, be sure to reinstall the spacer, washer and bolt in the intake camshaft

7.34 When both are installed, the two crankshaft sprockets should align like this; here the left sprocket is installed bevel forward, and the right sprocket (in hand) is ready to install with the bevel toward the other sprocket

7.35a When installing the timing chain, align one of the bright links in the timing chain with the dimple on the primary camshaft sprocket - sprocket should also align with the keyway in the camshaft

7.35b Align the bright link at the lower end of the chain with the timing mark on the crankshaft sprocket (it should be at 6 o'clock) then bolt the primary sprocket to the camshaft - both primary chains are aligned here

7.41a Colored link on the timing chain aligned with the mark on the crankshaft sprocket (right timing chain shown, left timing chain similar) (2005 and later Expedition/Navigator 5.4L engine)

V8 ENGINES 2C-13

25 The two timing chains should each have two bright or colored links. The links separate the chain in two equal halves. If no colored links are present, lay the chain down, make a paint mark on a link, then count links and make another paint mark halfway around the chain. Refer to Step 15 and reinstall both stationary chain guides.

➡ **Note:** The longer bolts are the ones that hold the guide to the cylinder head, the shorter ones go to the engine block.

5.4L 2V (1997 through 1999) and 4.6L Windsor (1997 through 2004)

♦ Refer to illustrations 7.28a and 7.28b

26 Loosen the camshaft holding tools just enough to permit minor movement of the camshafts for alignment. Using a camshaft positioning tool, available at most auto parts stores, or a two-pin spanner that catches two of the camshaft sprocket holes, turn the left camshaft until the timing mark on the sprocket is at 12 o'clock (viewed from the front of the engine). Turn the right sprocket until its mark is at 11 o'clock, then retighten both camshaft holding tools.

27 Install the inner crankshaft sprocket with the shoulder facing out, if not already done (see illustration 7.24).

28 Install the left timing chain, aligning the bright link with the dimple on the camshaft sprocket (see illustration). Loop the timing chain under the crankshaft sprocket and align the bright link with the alignment mark on the crankshaft sprocket. The timing marks on the crankshaft sprocket should be in the 6 o'clock position (see illustration).

➡ **Note:** The slack side of the chain (towards the water pump) must be below the dowel pin on the block.

29 Install the right side chain and crank-shaft sprocket (with its shoulder facing in) as above, but its slack side (the bottom run of the chain) should be above the dowel on the block.

➡ **Note:** If necessary to achieve exact alignment of the chains and sprockets, move the camshaft sprockets slightly by loosening the holding tools.

4.6L Romeo (1997 through 2004)

30 Install the camshaft sprocket spacers (if removed before) and bolt on the camshaft sprockets, tightening the bolts to the torque listed in this Chapter's Specifications. Make sure the sprocket keyways are aligned with the camshaft keys before forcing the sprockets on or installing the bolt.

31 Install the left timing chain, aligning its colored link to the mark on the crankshaft sprocket and to the mark on the camshaft sprocket.

7.41b Align the L on the left camshaft sprocket (A) with the two colored links (B) on the timing chain (2005 and later Expedition/Navigator 5.4L engine)

32 Install the right timing chain and its crankshaft sprocket. When both chains are installed and all colored links and marks are aligned, the keyways of the camshaft sprockets should be at a 90-degree angle to the valve cover mounting surface of the cylinder heads (see illustration 7.24). Remember, the shoulders on the crankshaft sprockets must face each other.

5.4L 4V (2000 through 2004)

♦ Refer to illustrations 7.33, 7.34, 7.35a and 7.35b

33 If necessary, install the secondary timing chain with its camshaft sprockets and spacers onto the two camshafts (on each cylinder head), then install the tensioner (see Section 9). Do not remove the pins from the secondary tensioners until both secondary camshaft sprockets are in place. Make sure the spacer and washer are installed correctly onto the camshaft (see illustration).

34 If removed, install the left crankshaft sprocket on the crankshaft with the beveled hub of the sprocket facing forward. When the two sprockets are placed correctly, the hubs will be facing each other and there will be the maximum space possible between the two sprockets. The timing marks on each sprocket will also align (see illustration). For now, only install the inner (left chain) crankshaft sprocket, with its bevel forward.

35 Install the left timing chain on the camshaft sprocket, aligning the bright link with the dimple (see illustration). Loop the timing chain under the crankshaft sprocket and align the bright link with the alignment mark on the crankshaft sprocket. The timing marks on the crankshaft sprocket should be in the 6 o'clock position (see illustration).

➡ **Note:** If the plated links on your chain aren't noticeable, lay the chain down on the bench where the chain is exactly halved. Paint a bright mark on the two links at each end, then use those marks for alignment.

36 Install the right cam sprocket and chain on the camshaft. Install the camshaft sprocket bolt and tighten the bolt to the torque listed in this Chapter's Specifications. Install the right chain and sprocket in the same manner. Verify that all timing marks are in alignment, at each primary camshaft sprocket and both crankshaft sprockets.

37 Install the primary timing chain stationary guides, if removed earlier.

➡ **Note:** The longer bolts are the ones that hold the guide to the cylinder head, the shorter ones go to the engine block.

2005 and later 5.4L 3V

♦ Refer to illustrations 7.41a, 7.41b and 7.43

38 Install the camshafts (see Section 9) if they were removed.

✳✳ CAUTION:

DO NOT install the rocker arms at this time. The rocker arms will be installed as a final step when the timing chains and tensioners have been installed.

39 Using the special tool (see illustration 7.5a), position the crankshaft keyway in the 10 o'clock position. This is the TDC number 1 position.

40 The timing chains should have three bright (copper) or colored links on each chain. If no colored links are present, purchase timing chains equipped with colored links.

41 Loop the left side timing chain under the crankshaft sprocket and align the bright link with the alignment mark on the crankshaft sprocket (see illustration). Install the timing chain over the camshaft sprocket, aligning the two bright links with the L on the sprocket (see illustration).

➡ **Note:** The slack side of the chain should be below the tensioner arm dowel on the block.

2C-14 V8 ENGINES

42 Install the left timing chain tensioner guide. Before assembling the tensioner with the chain guide, compress the tensioner and lock it in this position with a retainer clip (see illustration 7.22b). Install the tensioner, tighten the bolts and remove the retainer clip to apply force against the tensioner chain guide so the tensioner fully extends against the chain guide and all slack is removed from the chain.

43 Loop the right side timing chain under the crankshaft sprocket and align the bright link with the alignment mark on the crankshaft sprocket (see illustration 7.41a). Install the timing chain over the camshaft sprocket, aligning the two bright links with the I or R on the camshaft sprocket (see illustration).

➥ Note: *The slack side of the chain should be above the tensioner arm dowel on the block.*

44 Install the right timing chain tensioner guide. Before assembling the tensioner with the chain guide, compress the tensioner and lock it in this position with a retainer clip (see illustration 7.22b). Install the tensioner, tighten the bolts securely and remove the retainer clip to apply force against the tensioner chain guide so the tensioner fully extends against the chain guide and all slack is removed from the chain.

45 Recheck all the timing marks to make sure they are still in alignment.

All engines through 2004

♦ Refer to illustration 7.46

46 The steps for installing the timing chain tensioners/guides are the same for both sides, either side can be done first. Before assembling the tensioner with the chain guide, compress the tensioner and lock it in this position with a straightened paper clip, Allen wrench or drill bit (see illustration).

47 Remove the slack from the chain by hand and install the moveable guide assembly to the engine block and install the tensioner in the retracted position. Tighten the bolts to the torque listed in this Chapter's Specifications. Repeat for the other chain.

48 Remove the paper clip and apply pressure against the tensioner chain guide so the tensioner fully extends against the chain guide and all slack is removed from the chain.

All engines

49 Recheck all the timing marks to make sure they are still in alignment.

50 Remove the camshaft positioning tools and retaining tools. On 2005 and later Expedition/Navigator 5.4L engines, oil and reinstall the removed rocker arms in their original locations.

51 Slowly rotate the crankshaft in the normal direction of rotation (clockwise) at least two revolutions and again bring the engine to TDC. If you feel any resistance, stop and find out why. Check all alignment marks to verify that everything is properly assembled.

52 The remainder of installation is the reverse of removal.

7.43 Align the "I" or "R" mark on the right camshaft sprocket (A) with the two colored links (B) on the timing chain (2005 and later Expedition/Navigator 5.4L engine)

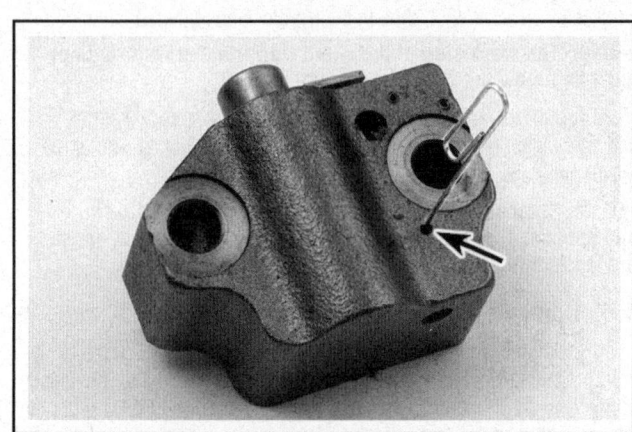

7.46 Lock the primary timing chain tensioner in the fully retracted position by placing a paper clip into the hole in the tensioner body (arrow) (all except 2005 and later 5.4L engine)

8 Rocker arms and valve lash adjusters - removal, inspection and installation

REMOVAL (ALL EXCEPT 2005 AND LATER 5.4L ENGINES)

♦ Refer to illustrations 8.3a, 8.3b and 8.4

1 There are two methods of removing the rocker arms and lash adjusters on these engines. The method recommended by the manufacturer accomplishes the removal of the camshaft roller followers without the removal of the camshaft(s) by using two special tools specified by the manufacturer: a valve spring spacer and a valve spring compressor, which are made specifically for the overhead cam V8 engine and are available at auto parts stores that carry special tools. The valve spring compressor uses the camshaft as a pivot point and, with a ratchet or bar attached, pushes down on the spring to release tension on the cam follower. The spring spacer keeps the spring from collapsing too far and hitting the valve stem seal. The alternative method requires the removal of the camshaft (see Section 9) in order to remove the cam followers. Either method will achieve the same results, but it is much easier using the special tools, if they can be located.

2 Remove the valve cover(s) (see Section 4). Because of the interference design of these modular engines, the pistons must be positioned off TDC before compressing the valve springs to remove the rockers arms or lash adjusters. For whatever cylinder you are removing the rockers arms from, remove the spark plug and insert a plastic pen (hold onto it) to see how far the piston is from the top of its travel. If necessary, turn the crankshaft until the pen indicates the pistons is down at least an inch or two from TDC.

V8 ENGINES 2C-15

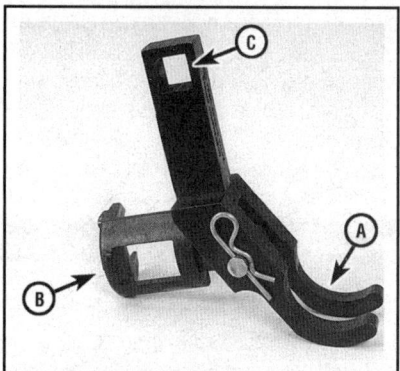

8.3a The special valve spring compressor hooks under the camshaft at (A), pushes on the valve spring retainer at (B), and is operated by a 1/2-inch drive breaker bar placed at (C)

8.3b Compress the valve spring until you can slip the rocker arm out - keep the rocker arms and lash adjusters matched to their original location

8.4 Pull the lash adjuster straight up and out of the cylinder head

3 Install valve spring compressor and compress the spring enough to remove the rocker arm (see illustrations). Camshaft rocker arms and hydraulic lash adjusters MUST be reinstalled with the same camshaft lobe that they were removed from. Label and store all components to avoid confusion during reassembly.

✳✳ CAUTION:

A valve spring spacer should be inserted into the coils of the spring before compressing it. If the spacer isn't in place between one of the valve spring coils, the spring can be compressed too far and the valve seal may be damaged.

4 Remove the hydraulic lash adjuster (see illustration). If there are many miles on the vehicle, the adjusters may have become varnished and difficult to remove. Apply a little penetrating oil around the lash adjuster to help loosen the varnish.

➡ **Note: Keep the rocker arm and lash adjuster for each valve together in a marked plastic sandwich bag.**

REMOVAL (2005 AND LATER 5.4L 3V ENGINES)

5 Remove the rocker arms, as described in Section 7 (start at Step 8).
6 Remove the camshafts (see Section 9).
7 Remove the lash adjusters, as described in Step 4 of this Section.

INSPECTION (ALL MODELS)

▸ **Refer to illustrations 8.8 and 8.10**

8 Inspect each adjuster carefully for signs of wear or damage. The areas of possible wear are the ball tip that contacts the cam follower and the sides of the adjuster that contact the bore in the cylinder head (see illustration). Since the lash adjusters frequently become clogged as mileage increases, we recommend replacing them if you're concerned about their condition or if the engine is exhibiting valve tapping noises.

9 A thin wire or paper clip can be placed in the oil hole to move the plunger and make sure it's not stuck.

➡ **Note: The lash adjuster must have no more than 1/16-inch of total plunger travel.**

It's recommended that if replacement of any of the adjusters is neces-

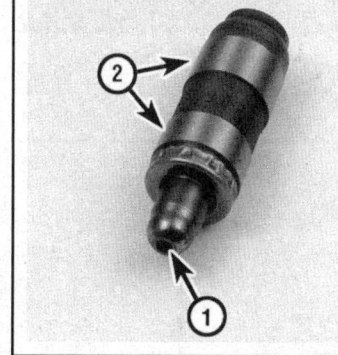

8.8 Inspect the lash adjuster for signs of excessive wear or damage, such as pitting, scoring or signs of overheating (bluing or discoloration), where the tip contacts the camshaft follower (1) and the side surfaces that contact the lifter bore in the cylinder head (2)

sary, that the entire set be replaced. This will avoid the need to repeat the repair procedure as the others require replacement in the future.

10 Inspect the rocker arms for signs of wear or damage. The areas of wear are the ball socket that contacts the lash adjuster and the roller where the follower contacts the camshaft (see illustration).

INSTALLATION (ALL MODELS)

11 Before installing the lash adjusters, bleed as much air as possible out of them. Stand the adjusters upright in a container of oil. Use a thin wire or paper clip to work the plunger up and down. This primes

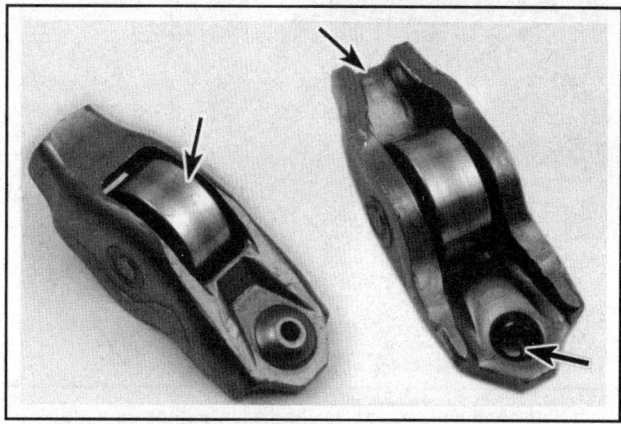

8.10 Check the roller surface (arrow) of the rocker arms and the areas where the valve stem and lash adjuster contact the rocker (arrows at right)

2C-16 V8 ENGINES

the adjuster and removes most of the air. Leave the adjusters in the oil until ready to install.

12 Lubricate the valve stem tip, rocker arm, and lash adjuster bore with clean engine oil.

13 Install the lash adjusters and, with the valve spring depressed as in Step 3, install each rocker arm.

14 The remainder of installation is the reverse of the removal procedure.

15 When re-starting the engine after replacing the adjusters, the adjusters will normally make some tapping noises, until all the air is bled from the lash adjusters. After the engine is warmed-up, raise the speed from idle to 3,000 rpm for one minute. Stop the engine and let it cool down. All of the noise should be gone when it is restarted.

9 Camshaft(s) - removal, inspection and installation

REMOVAL

▶ **Refer to illustrations 9.3a, 9.3b, 9.3c, 9.4 and 9.5**

1 Remove the valve covers (see Section 4), and the timing chain cover (see Section 6).

2 Remove the timing chains, camshaft sprockets and spacers (see Section 7).

❋❋ **CAUTION:**

Don't mix up the sprockets, they are marked, as RB (right bank) and LB (left bank), and must go back on the appropriate camshaft.

3 On 5.4L 4V engines, the secondary (smaller) timing chain has a tensioner between the two sprockets. Compress the tensioner with a C-clamp, then insert two pins to lock the tensioner in position (see illustrations). Remove the tensioner, and repeat for the opposite cylinder head.

❋❋ **CAUTION:**

The plunger is spring-loaded, do not remove the tensioner from the engine without pinning it, or it could drop out and be damaged.

Remove the camshaft sprocket bolt and large washer from the intake camshaft (5.4L-4V) and slip both camshaft sprockets (and their spacers) off with the secondary chain (see illustration).

9.3a Compress the secondary tensioner with a C-clamp . . .

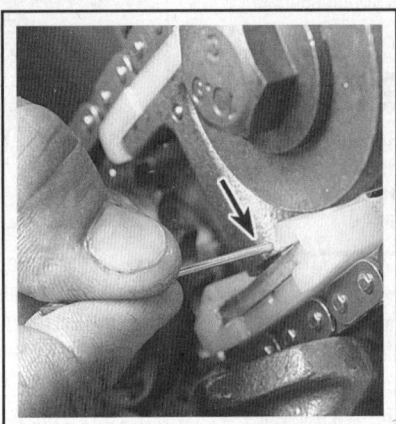

9.3b . . . then insert a drill bit or pin to hold the tensioner in the compressed position

9.3c With both camshaft sprocket bolts removed, slide the sprockets off with the secondary chain

9.4 Camshaft endplay can be checked by setting up a dial indicator off the front of the camshaft and prying the camshaft gently forward and back

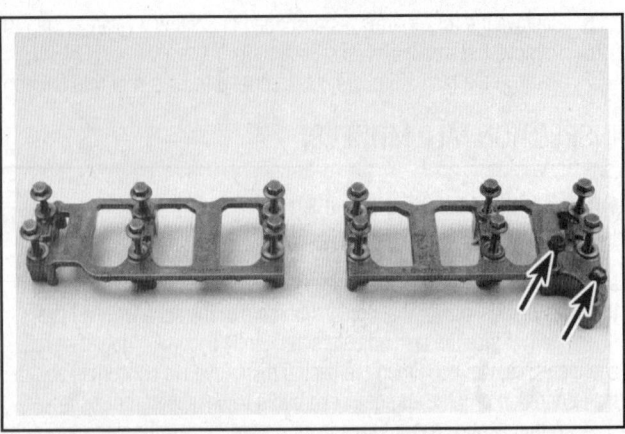

9.5 REP engines have two "camshaft cap clusters" (rather than individual bearing caps like the WEP engines), that hold the camshaft in place on the cylinder head. Note the position of the two different bolts (arrows)

V8 ENGINES 2C-17

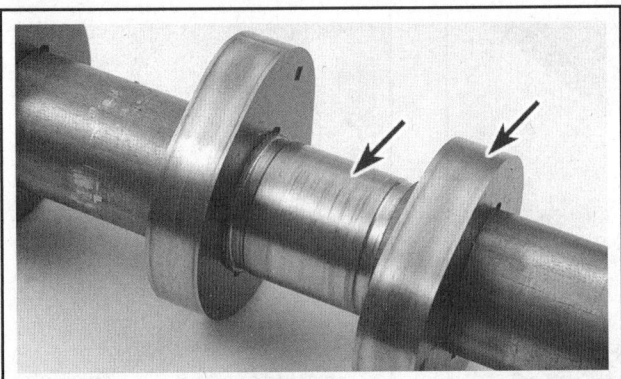

9.10a Areas to look for excessive wear or damage on the camshafts are the bearing surfaces and the camshaft lobes

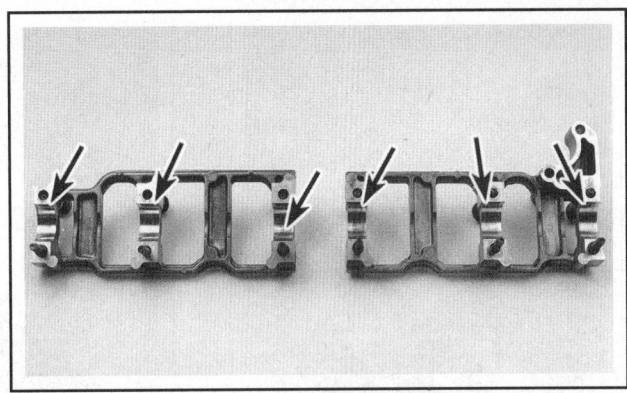

9.10b Inspect the bearing surfaces of the camshaft bearing caps (arrows) for signs of excessive wear, damage or overheating

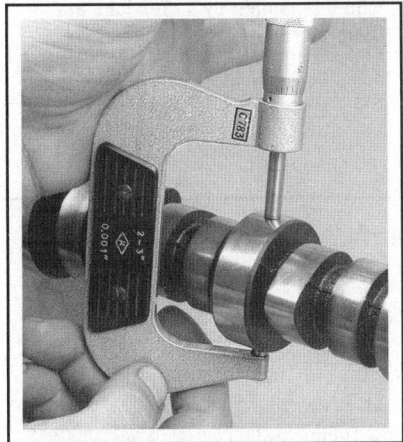

9.11a Measuring the camshaft bearing journal diameter

9.11b Measure the camshaft lobe at its greatest dimension . . .

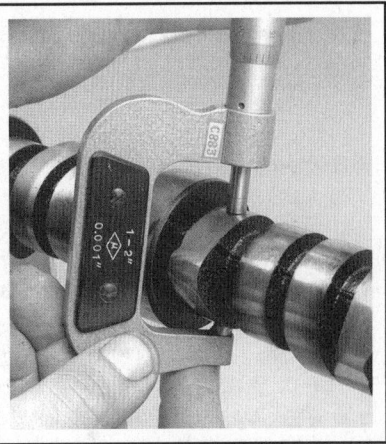

9.11c . . . and subtract the camshaft lobe diameter at its smallest dimension to obtain the lobe lift specification

4 Measure the thrust clearance (endplay) of the camshaft(s) with a dial indicator (see illustration). If the clearance is greater than the value listed in this Chapter's Specifications, replace the camshaft and/or the cylinder head.

5 There are two designs for holding the camshaft journals. On Windsor-built engines, and 2005 and later 5.4L engines, there are individual camshaft caps at each journal, while Romeo-built engines have two "camshaft cap clusters" for each camshaft. The configuration of the two cap clusters are different and must be placed in their original locations. Mark the camshaft cap clusters with a front and rear indication, for both the left and right cylinder heads.

➡ **Note:** *Two bolts used on one of the camshaft cap clusters are different than the others (see illustration), be sure they go back in the same locations on reassembly.*

6 Refer to Section 6 for removal of the front cover and Section 7 for removal of the timing chains.

➡ **Note:** *While the timing chains are off, the crankshaft must not be turned or valve-to-piston interference could result.*

7 It's IMPORTANT to loosen the bearing cap bolts only 1/4-turn at a time, following the reverse of the tightening sequence (see illustrations 9.17a, 9.17b, 9.17c and 9.17d), until they can be removed by hand.

8 Remove the caps and lift the camshaft off the cylinder head. You may have to tap lightly under the camshaft caps to jar them loose. Don't mix up the camshafts or any of the components. They must all go back on the same positions, and on the same cylinder head they were removed from.

9 Repeat this procedure for removal of the remaining camshaft.

INSPECTION

▸ **Refer to illustrations 9.10a, 9.10b, 9.11a, 9.11b, 9.11c, 9.12a, 9.12b and 9.14**

10 Visually examine the cam lobes and bearing journals for score marks, pitting, galling and evidence of overheating (blue, discolored areas). Look for flaking of the hardened surface of each lobe (see illustrations).

11 Using a micrometer, measure the diameter of each camshaft journal and the lift of each camshaft lobe (see illustrations). Compare the measurements for excessive variation (only compare the intake lobe measurements against each other, and the exhaust lobe measurements against each other). If the measurements for either the intake lobes or the exhaust lobes vary more than 0.004 inch, replace the camshaft.

12 Check the oil clearance for each camshaft journal as follows:

   a) Clean the bearing surfaces and the camshaft journals with lacquer thinner or acetone.
   b) Carefully lay the camshaft(s) in place in the cylinder head. Don't install the rocker arms or lash adjusters and don't use any lubrication.

2C-18 V8 ENGINES

9.12a Lay a strip of Plastigage on each of the camshaft journals

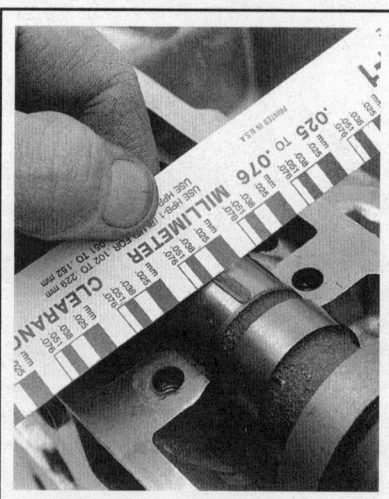

9.12b Compare the width of the crushed Plastigage to the scale on the envelope to determine the oil clearance

9.14 Oil is delivered to the timing chain tensioner by a feed tube and reservoir in the cylinder head

1 Tensioner oil feed tube
2 Reservoir

   c) Lay a strip of Plastigage on each journal (see illustration).
   d) Install the camshaft bearing caps.
   e) Tighten the cap bolts, a little at a time, to the torque listed in this Chapter's Specifications.

➡ **Note:** *Don't turn the camshaft while the Plastigage is in place.*

   f) Remove the bolts and detach the caps.
   g) Compare the width of the crushed Plastigage (at its widest point) to the scale on the Plastigage envelope (see illustration).
   h) If the clearance is greater than specified, and the diameter of any journal is less than specified, replace the camshaft. If the journal diameters are within specifications but the oil clearance is too great, the cylinder head is worn and must be replaced.

13 Scrape off the Plastigage with your fingernail or the edge of a credit card - don't scratch or nick the journals or bearing surfaces.

14 Finally, be sure to check the timing chain tensioner oil feed tube and reservoir before installing the cam caps (see illustration). it must be absolutely clean and free of all obstructions or it will affect the operation of the timing chain tensioner.

INSTALLATION

▶ **Refer to illustrations 9.17a, 9.17b, 9.17c and 9.17d**

15 If the lash adjusters and/or camshaft followers have been removed, install them in their original locations (see Section 8).

16 Apply moly-base grease or camshaft installation lube to the camshaft lobes and bearing journals, then install the camshaft(s).

17 Install the camshaft caps in the correct locations, and loosely install all the bolts. Refer to Section 7 to align the camshaft sprockets before tightening the cap bolts. Following the correct bolt-tightening sequence (see illustrations), tighten the bolts in 1/4-turn increments to the torque listed in this Chapter's Specifications.

18 Reinstall the camshaft positioning and retaining tools (see Section 7) to set the camshafts at TDC before reinstalling the timing chain(s).

19 Install the timing chain(s), tensioners and timing chain cover (see Section 7).

20 The remainder of installation is the reverse of the removal procedure.

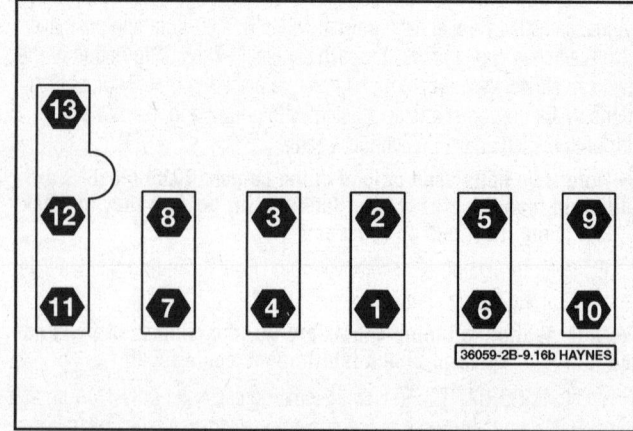

9.17a The camshaft cap cluster bolt tightening sequence, REP engines - notice that each cap is tightened separately and has its own sequence

9.17b Camshaft cap tightening sequence - WEP engines

V8 ENGINES 2C-19

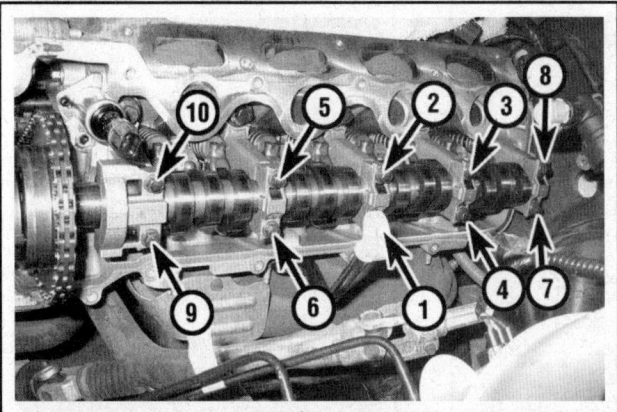

9.17c Camshaft cap bolt TIGHTENING sequence for 2005 and later 5.4L engines

9.17d Camshaft cap tightening sequence - 5.4L 4V engines

10 Valve springs, retainers and seals - removal and installation

Broken valve springs and/or defective valve stem seals can be replaced without removing the cylinder heads. There is a method described in Section 8, using a tool recommended by the manufacturer, which accomplishes the removal of the valve springs and seals without the removal of the camshafts. The alternative method uses a more commonly available tool, but will require the removal of the camshaft (see Section 9) in order to remove the valve spring. Either method will achieve the same results, but it is much easier using the manufacturer's recommended procedure, if the correct tools can be located.

In either repair procedure, a compressed air source is normally required to perform this operation, so read through this Section carefully and rent or buy the tools before beginning the job.

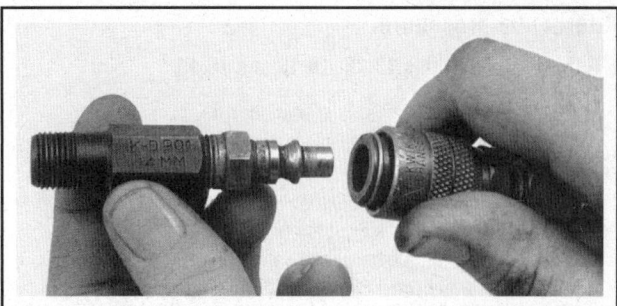

10.4 Thread the air hose adapter into the spark plug hole - they're commonly available from auto parts stores

REMOVAL

♦ **Refer to illustration 10.4**

1 Remove the valve cover (see Section 4).
2 Remove the spark plug from the cylinder with the defective component. If all of the valve stem seals are being replaced, remove all the spark plugs.
3 Turn the crankshaft until the piston in the affected cylinder is at Top Dead Center (TDC) on the compression stroke (see Section 3). If you're replacing all of the valve stem seals, begin with cylinder number one and work on the valves for one cylinder at a time. Move from cylinder-to-cylinder following the firing order sequence (see this Chapter's Specifications).
4 Thread an air hose adapter into the spark plug hole (see illustration) and connect an air hose from a compressed air source to it. Most auto parts stores can supply the air hose adapter.

➡ **Note: Many cylinder compression gauges utilize a screw-in fitting that may work with your air hose quick-disconnect fitting.**

5 Apply compressed air to the cylinder.

✲✲ WARNING:

The piston may be forced down by compressed air, causing the crankshaft to turn suddenly. If the wrench used when positioning the number one piston at TDC is still attached to the bolt in the crankshaft nose, it could cause damage or injury when the crankshaft moves.

6 The valves should be held in place by the air pressure.
7 Stuff shop rags into the cylinder head holes above and below the valves to prevent parts and tools from falling into the engine.

Method using special tools

8 Install a valve spacer, available at most auto parts stores (see illustration 8.4a).

✲✲ CAUTION:

If the spacer isn't in place between one of the valve coils, the spring can be compressed too far, possibly damaging the valve seal.

9 Compress the spring and remove the rocker arm (see Section 8). Rocker arms and hydraulic lash adjusters MUST be reinstalled with the same camshaft lobe that they were removed from. Label and store all components to avoid confusion during reassembly.
10 Keeping the spring compressed, remove the keepers with small needle-nose pliers or a magnet (see illustration 10.13). Remove the spring retainer and valve spring. Remove the valve stem seal (see illustration 10.15). If air pressure fails to hold the valve in the closed position during this operation, the valve face and/or seat is probably damaged. If so, the cylinder head will have to be removed for additional repair operations.

2C-20 V8 ENGINES

10.12 Installation of a valve spring compressor more commonly available at local automotive parts stores (note the camshaft must be removed for the use of this type of valve spring compressor)

10.13 Once the spring is compressed remove the keepers with a needle-nose pliers or a magnet, as shown here

10.15 Use pliers, of any type, to firmly grasp the old seal and pull it off the valve guide

Alternative procedure

♦ Refer to illustrations 10.12, 10.13 and 10.15

11 Remove the camshaft(s) (see Section 9).

12 Install the commonly available clamp-type valve spring compressor (see illustration).

13 Compress the spring and remove the keepers with small needle-nose pliers or a magnet (see illustration).

14 Remove the spring retainer and valve spring.

15 Remove the stem seal (see illustration). If air pressure fails to hold the valve in the closed position during this operation, the valve face and/or seat is probably damaged. If so, the cylinder head will have to be removed for additional repair operations.

INSTALLATION

♦ Refer to illustrations 10.20a, 10.20b and 10.22

16 Wrap a rubber band or tape around the top of the valve stem so the valve won't fall into the combustion chamber, then release the air pressure.

17 Inspect the valve stem for damage. Rotate the valve in the guide and check the end for eccentric movement, which would indicate that the valve is bent.

18 Move the valve up-and-down in the guide and make sure it doesn't bind. If the valve stem binds, either the valve is bent or the guide is damaged. In either case, the cylinder head will have to be removed for repair.

19 Reapply air pressure to the cylinder to retain the valve in the closed position, then remove the tape or rubber band from the valve stem.

20 Lubricate the valve stem with engine oil and install a new seal (see illustration). There is a special tool for the installation of the valve seal. If the tool isn't available, a socket that will fit over the seal and is deep enough to make contact with the seat (see illustration), can be used to carefully tap the new seal into place.

> ※※ **CAUTION:**
>
> The valve seal used on the V8 engines is a combination seal and spring seat. Never place a valve spring directly against the aluminum cylinder head (without a seal/spring seat) - the hardened spring would damage the cylinder head.

10.20a The valve stem seals combine a seal with the valve spring seat

10.20b There is special valve seal installation tool available, but a deep socket that fits over the seal can be used to gently tap the seal into place

10.22 Apply a small dab of grease to each keeper as shown here before installation - it'll hold them in place on the valve stem as the spring is released

21 Make sure the garter spring on the seal is still in place, and install the spring in position over the valve.

22 Install the valve spring retainer. Compress the valve spring and carefully position the keepers in the groove. Apply a small dab of grease to the inside of each keeper to hold it in place if necessary (see illustration).

23 Remove the pressure from the spring tool and make sure the keepers are seated.

24 Disconnect the air hose and remove the adapter from the spark plug hole.

25 If the camshaft(s) were removed, reinstall them at this time (see Section 9).

26 Install the spark plug(s) and connect the wire(s).

27 The remaining installation steps are the reverse of removal.

28 Start and run the engine, then check for oil leaks and unusual sounds coming from the valve cover area.

11 Intake manifold - removal and installation

REMOVAL

1 Disconnect the cable from the negative battery terminal. Relieve the fuel system pressure (see Chapter 4).

2 Drain the cooling system and remove the drivebelt (see Chapter 1).

3 Disconnect the radiator hose from the thermostat housing and remove the thermostat housing (see Chapter 3). The thermostat housing bolts also retain the intake manifold.

4 Remove the air cleaner and outlet tube (see Chapter 4).

5 Label and disconnect all vacuum lines connected to the intake manifold.

6 Remove the PCV and canister purge hoses from the valve covers (see Section 4).

7 Refer to Chapter 4 for the following procedures: disconnect the accelerator cable, speed control linkage (if so equipped), electrical connectors at the throttle body, fuel injectors, fuel rails, and disconnect and plug the fuel supply and return lines.

8 Refer to Chapter 5 and disconnect the ignition wire brackets, boots and wires and set them out of the way.

→**Note: Pull the spark plug wire separators from their mounts on the valve cover studs and lay the wires out of the way.**

9 Refer to Chapter 6 for the following procedures: disconnect the idle air control valve, differential pressure feedback transducer and hoses, engine vacuum regulator sensor and EGR valve-to-exhaust-manifold pipe.

10 Disconnect the alternator electrical connectors and remove the alternator (see Chapter 5).

11 Remove the bolts retaining the alternator bracket to the intake manifold (see Chapter 5).

12 Disconnect the ignition coils (see Chapter 5).

13 Disconnect the electrical connectors from the camshaft sensor, coolant temperature sending unit, air charge temperature sensor, throttle positioner and idle speed control solenoid (see Chapter 6). On 2005 and later Expedition/Navigator 5.4L engines, disconnect the PCV heater connector near the right valve cover.

14 On models through 2004, disconnect the heater hose at the manifold, then unbolt the heater hose tube from the back of the engine. On 2005 and later Expedition/Navigator 5.4L engines, there is a coolant crossover casting that bolts between the cylinder heads at the front. Disconnect the hoses, remove the crossover mounting bolts and remove the crossover.

15 Pull the engine wiring harness off the valve cover and intake studs and away from the intake manifold.

16 Unbolt the power steering reservoir bracket and set the pump aside (see Chapter 10).

17 Unbolt the bracket supporting the power brake booster hose at the left-rear of the engine (see illustration 4.6).

18 Loosen the intake manifold bolts and nuts in 1/4-turn increments, following the reverse order of the tightening sequence (see illustration 11.27), until they can be removed by hand.

19 Lift the intake manifold from the cylinder heads and disconnect the intake manifold tuning valve (see Chapter 4) from the lower plenum. The manifold may be stuck to the cylinder heads and force may be required to break the gasket seal. A prybar can be used to pry up the manifold, but make sure all bolts and nuts have been removed first!

✳✳ CAUTION:

Don't pry between the engine block and manifold or the cylinder heads, or damage to the gasket sealing surface may occur, leading to vacuum and oil leaks. Pry only at the manifold protrusion.

20 Remove the intake manifold gaskets and clean all traces of gasket or sealant material from the sealing surfaces of the cylinder heads and intake manifold.

INSTALLATION

♦ **Refer to illustrations 11.24, 11.25 and 11.27**

✳✳ CAUTION:

The mating surfaces of the cylinder heads, engine block and intake manifold must be perfectly clean. Gasket removal solvents in aerosol cans are available at most auto parts stores and may be helpful when removing old gasket material that's stuck to the cylinder heads and intake manifold. Since the cylinder heads are aluminum and the intake manifold is aluminum or plastic, aggressive scraping can cause damage! Be sure to follow directions printed on the container, and use only a plastic-tipped scraper, not a metal one.

21 If the intake manifold was disassembled from the lower plenum, reassemble it or if you are replacing it, transfer all components to the new intake manifold, including the IMRC components. Use electrically conductive sealant on the temperature sending unit threads. Use a new EGR valve gasket.

22 Remove the old gaskets, then clean the mating surfaces with lacquer thinner or acetone. If there's old sealant or oil on the mating surfaces when the intake manifold is installed, oil or vacuum leaks may develop.

23 When working on the cylinder heads and engine block, cover the open engine areas with shop rags to keep debris out of the engine. Use a vacuum cleaner to remove any gasket material that falls into the intake ports in the cylinder heads.

24 Use a tap of the correct size to chase the threads in the bolt holes, then use compressed air (if available) to remove the debris from the holes.

2C-22 V8 ENGINES

11.24 Check the condition of the water pump hose (arrow) and replace it if necessary - REP engine shown

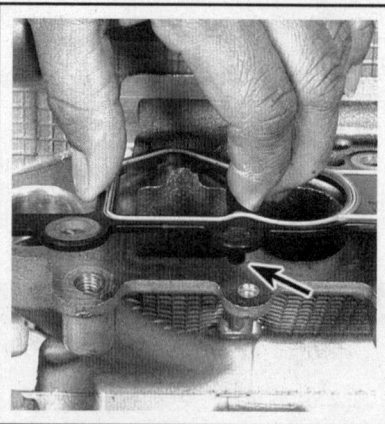

11.25 Install the new intake manifold gaskets to the cylinder head, aligning the plastic pins (arrow) to their holes in the cylinder head

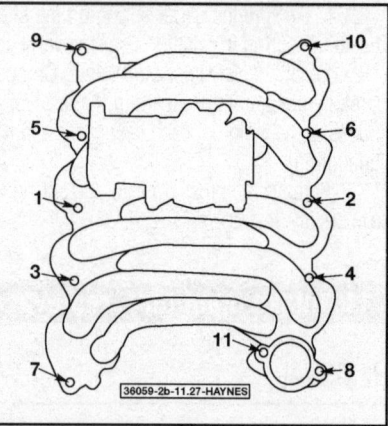

11.27 Intake manifold bolt tightening sequence

※ WARNING:
Wear safety glasses or a face shield to protect your eyes when using compressed air!

Remove excessive carbon deposits and corrosion from the exhaust and coolant passages in the cylinder heads and intake manifold.

➡ **Note:** If the vehicle has many miles on it, replace the water pump hose (normally hidden by the intake manifold) before replacing the manifold (see illustration). On REP engines, the hose is retained by a clamp, and a metal tube and nut are used on WEP engines. On WEP engines, replace the O-ring at the tube nut.

25 Install the gaskets on the cylinder heads (see illustration). Make sure all alignment tabs, intake port openings, coolant passage holes and bolt holes are aligned correctly. The gasket that goes on the cylinder head will have projecting plastic pins to align it with holes in the cylinder head.

26 Carefully set the intake manifold in place. Don't disturb the gaskets and don't move the manifold fore-and-aft after it contacts the gaskets on the engine block.

27 Install the intake manifold bolts and, following the recommended tightening sequence (see illustration), tighten them to the torque listed in this Chapter's Specifications. Replace the O-ring seal on the thermostat housing. Install the thermostat housing and tighten the bolts to the torque listed in Chapter 3 Specifications. On 2005 and later Expedition/Navigator 5.4L engines, install the coolant crossover with new gaskets and reconnect the hoses.

28 The remaining installation steps are the reverse of removal. Start the engine and check carefully for oil and coolant leaks.

12 Exhaust manifolds - removal and installation

REMOVAL

♦ **Refer to illustrations 12.5, 12.6a and 12.6b**

1 Disconnect the cable from the negative battery terminal. On 4WD models, remove the front driveshaft (see Chapter 8)
2 Remove the nut and move the power brake booster vacuum hose support bracket (see illustration 4.6).
3 Raise the vehicle and support it securely on jackstands.

※ WARNING:
Some models covered by this manual are equipped with an air suspension system. Always disconnect the electrical power to the suspension system before lifting or towing (see Chapter 10). Failure to perform this procedure may result in unexpected shifting or movement of the vehicle, which could cause personal injury. Remove the inner splash shield from the fenderwell.

4 Working under the vehicle, apply penetrating oil to the exhaust pipe-to-manifold studs and nuts (they're usually corroded or rusty). Also apply some to the EGR pipe fitting on the left exhaust manifold.

5 Remove the nuts retaining the exhaust pipes to the manifolds. On the left side, remove the nut and bracket securing the oil dipstick tube (see illustration). On 2005 and later Expedition/Navigator models, lock the steering column in the straight-ahead position and disconnect the

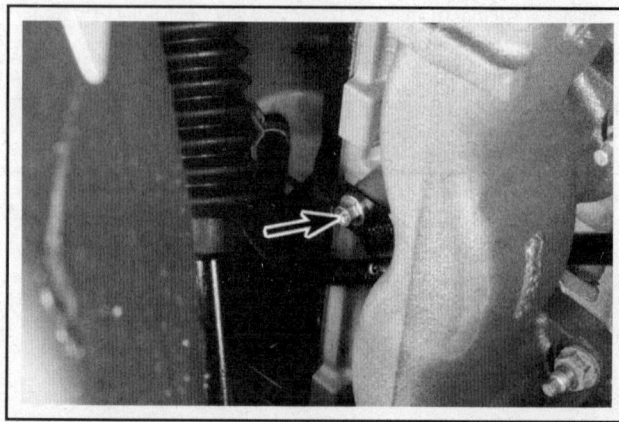

12.5 Remove the nut (arrow) securing the oil dipstick tube

V8 ENGINES 2C-23

lower steering shaft bolt (see Chapter 10) to allow moving the shaft aside for removal of the left exhaust manifold.

6 Remove the EGR pipe fitting from the left exhaust manifold. Remove the eight mounting nuts from each manifold (see illustrations).

➡ **Note: The exhaust manifold nuts are difficult to access. You'll need a flex-socket and various-length 3/8-inch drive or 1/4-inch drive extensions to reach all the bolts.**

INSTALLATION

7 Check the exhaust manifolds for cracks. Make sure the bolt threads are clean and undamaged. The exhaust manifold and cylinder head mating surfaces must be clean before the exhaust manifolds are reinstalled - use a gasket scraper to remove all carbon deposits.

8 Position a new gasket in place and slip the exhaust manifold over the studs on the cylinder head. Install the mounting nuts.

9 When tightening the mounting nuts, work from the rear to the front, alternating between top and bottom rows. Tighten the bolts in three equal steps to the torque listed in this Chapter's Specifications.

10 The remaining installation steps are the reverse of removal. When reconnecting the EGR tube to the left manifold, use a slight amount of anti-seize compound on the threads.

11 Start the engine and check for exhaust leaks.

12.6a Disconnect the EGR tube (arrow) from the left exhaust manifold

12.6b Remove the exhaust manifold nuts (three are shown here on the left manifold)

13 Cylinder heads - removal and installation

※ CAUTION:

The engine must be completely cool when the cylinder heads are removed. Failure to allow the engine to cool off could result in cylinder head warpage.

➡ **Note 1: Cylinder head removal is a difficult and time-consuming job requiring several special tools, read through the procedure and obtain the necessary tools before beginning.**

➡ **Note 2: On 2003 and later models, the manufacturer recommends that the engine be removed from the vehicle to perform this procedure (see Chapter 2D).**

REMOVAL

1 Disconnect the negative battery cable and drain the cooling system (see Chapter 1).

2 Remove the valve covers (see Section 4) and the intake manifold (see Section 11).

3 Remove the exhaust manifolds (see Section 12).

4 Remove the timing chain cover (see Section 6).

5 The cylinder heads can be removed with the camshafts, rocker arms and lash adjusters in place. Remove the timing chains, tensioners and guides (see Section 7).

※ CAUTION:

Use the required camshaft retaining fixtures to lock the camshafts and leave the tools in place.

6 If the cylinder head is to be completely overhauled, refer to Section 8 and remove the rocker arms, and Section 9 for removal of the camshafts.

7 Following the reverse of the tightening sequence (see illustration 13.18a), use a breaker bar to remove the cylinder head bolts. Loosen the bolts in sequence 1/4-turn at a time.

8 Use a pry bar at the corners of the cylinder head-to-engine block mating surface to break the cylinder head gasket seal. Do not pry between the cylinder head and engine block in the gasket sealing area.

9 Lift the cylinder head(s) off the engine. If resistance is felt, place a wood block against the end and strike the wood block with a hammer.

※ CAUTION:

The cylinder heads are aluminum, store them on wood blocks to prevent damage to the gasket sealing surfaces.

10 Remove the cylinder head gasket(s). Before removing, note which gasket goes on which side (they are different and cannot be interchanged).

2C-24 V8 ENGINES

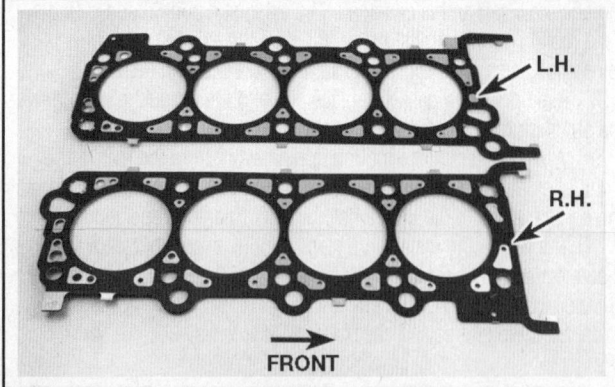

13.15 Identify the left and right cylinder head gaskets, the shapes are different and cannot be interchanged

11 Cylinder head disassembly and inspection procedures are covered in detail in Chapter 2, Part D.

INSTALLATION

▶ Refer to illustrations 13.15, 13.16, 13.18a and 13.18b

✳✳ CAUTION:

New cylinder head bolts must be used for reassembly. Failure to use new bolts may result in cylinder head gasket leakage and engine damage.

12 The mating surfaces of the cylinder heads and engine block must be perfectly clean when the cylinder heads are installed. Use a gasket scraper to remove all traces of carbon and old gasket material, then clean the mating surfaces with lacquer thinner or acetone. If there's oil on the mating surfaces when the cylinder heads are installed, the gaskets may not seal correctly and leaks may develop. When working on the engine block, cover the open areas of the engine with shop rags to keep debris out during repair and reassembly. Use a vacuum cleaner to remove any debris that falls into the cylinders.

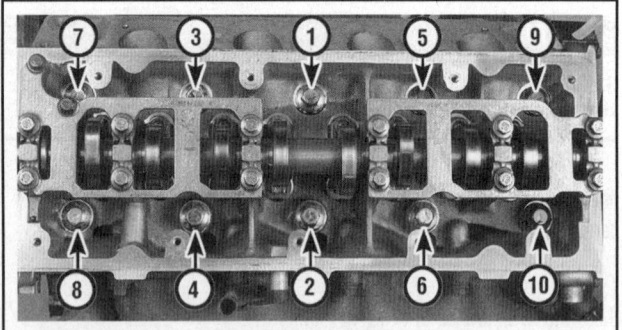

13.18a Cylinder head bolt-tightening sequence

13.18b Mark each cylinder head bolt with a paint stripe (arrows) and using a breaker bar and socket, tighten the bolts in sequence the additional 1/4-turn (90-degrees)

13.16 Position the gaskets on the correct cylinder banks, then push them down over the alignment dowels (arrows)

✳✳ CAUTION:

Do not use abrasive wheels or sharp metal scrapers on the heads or block surface, use a plastic scraper and chemical gasket remover, or the head gasket surfaces could have future leaks.

13 Check the engine block and cylinder head mating surfaces for nicks, deep scratches and other damage.

14 Use a tap of the correct size to chase the threads in the cylinder head bolt holes. Dirt, corrosion, sealant and damaged threads will affect torque readings.

15 Make sure the new gaskets are installed on the correct cylinder banks (see illustration). They are not interchangeable.

16 Position the new gasket(s) over the alignment dowels (see illustration) in the engine block.

17 If the cylinder heads are being installed with the camshafts in place, make sure the camshafts are back in their original TDC positions (see Sections 7 and 9) before placing the cylinder heads on the engine block.

✳✳ CAUTION:

If the camshafts aren't in the position described, damage may result to either the pistons and/or valve train parts.

18 Carefully position the cylinder heads on the engine block without disturbing the gaskets. Install NEW cylinder head bolts (the cylinder head bolts are torque-to-yield design and cannot be reused). Following the recommended sequence (see illustration), tighten the cylinder head bolts, in three steps, to the torque listed in this Chapter's Specifications.

➡ **Note:** *The method used for the cylinder head bolt tightening procedure is referred to as "torque-angle" or "torque-to-yield" method. Follow the procedure exactly.*

Tighten the bolts in the first step using a torque wrench, then use a breaker bar and a special torque-angle adapter (available at most auto parts stores) to tighten the bolts the required angle. If the adapter is not available, mark each bolt with a paint stripe to aid in the torque angle process (see illustration).

19 The remaining installation steps are the reverse of removal.

20 Change the engine oil and filter (see Chapter 1), then start the engine and check carefully for oil and coolant leaks.

V8 ENGINES 2C-25

14 Oil pan - removal and installation

REMOVAL

▶ Refer to illustrations 14.6, 14.13a and 14.13b

> ※※ **WARNING:**
>
> Some models covered by this manual are equipped with an air suspension system. Always disconnect the electrical power to the suspension system before lifting or towing (see Chapter 10). Failure to perform this procedure may result in unexpected shifting or movement of the vehicle, which could cause personal injury.

1 Disconnect the cable from the negative terminal of the battery.
2 On 2009 and earlier models, remove the air filter assembly (see Chapter 4).
3 Raise the vehicle and support it securely on jackstands.
4 Drain the engine oil and remove the oil filter (see Chapter 1). Remove the oil level dipstick.
5 Remove the skid plate, if equipped.
6 On 2005 and later Expedition/Navigator models, remove the front crossmember (see illustration).

7 On 2009 and earlier models, remove the engine cooling fan assembly and shroud.
8 On 4WD models, remove the front axle assembly (see Chapter 8).
9 On 2003 and earlier models, attach an engine support fixture to the engine lifting eyes adjacent to the exhaust manifolds.

➡ Note: Many equipment rental yards rent engine support fixtures.

10 On 2009 and earlier models, raise the engine slightly with the support fixture, then remove the engine mount through-bolts (see Section 18).
11 On 2009 and earlier models, raise the engine a little at a time with the support fixture, watching for interference with any engine compartment components. When the engine is raised about two inches, place short lengths of 2x4 lumber between the motor mounts and the crossmember for safety.

➡ Note: On 2005 through 2009 models, raising the engine is not necessary. Remove the four bolts/nuts and take out the front crossmember for clearance to remove the oil pan.

12 On 2010 and later models, disconnect the starter motor wiring harness, the oil pressure switch wiring harness and the transmission oil cooler tube, then move the tube and harnesses aside.
13 Remove the oil pan mounting bolts (see illustrations).
14 Carefully separate the oil pan from the engine block. Don't pry between the engine block and oil pan or damage to the sealing surfaces may result and oil leaks could develop. Instead, dislodge the oil pan with a large rubber mallet or a wood block and a hammer.

➡ Note: Use a new gasket on installation.

INSTALLATION

▶ Refer to illustrations 14.17 and 14.18

15 Use a gasket scraper or putty knife to remove all traces of old gasket material and sealant from the pan and engine block.

> ※※ **CAUTION:**
>
> Be careful not to gouge the oil pan or block, or oil leaks could develop later.

14.6 Crossmember mounting bolt locations (2005 and later Expedition/Navigator)

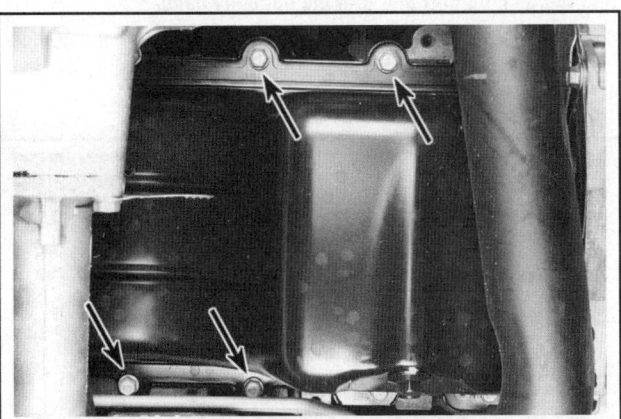

14.13a Remove the bolts from around the perimeter of the oil pan (four shown) and lower the pan to the frame crossmember

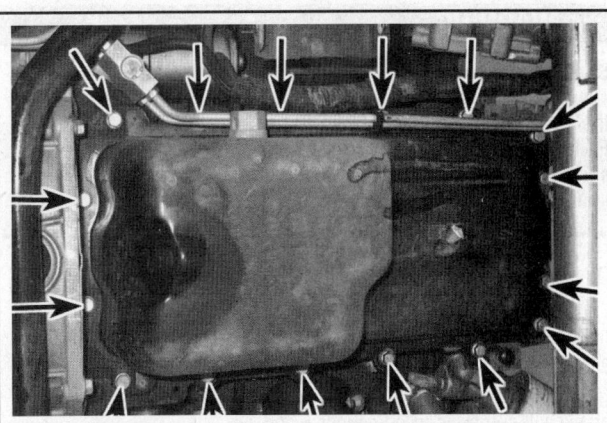

14.13b 2009 and later Expedition/Navigator oil pan bolt locations

2C-26 V8 ENGINES

14.17 Apply a bead of RTV sealant at the junctions of the front cover-to-engine block and the rear-seal retainer-to-engine block before installing the oil pan

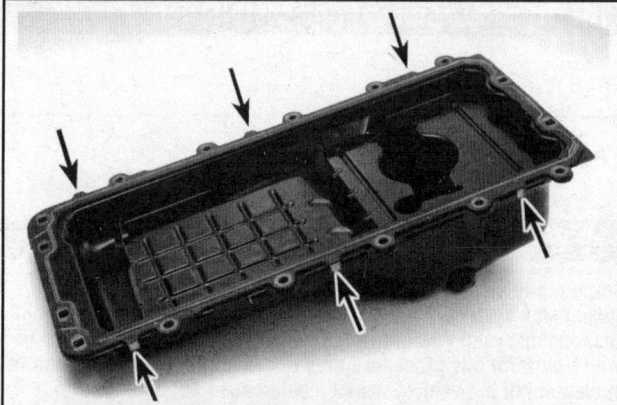

14.18 Place the gasket on the oil pan, the locating tabs (arrows) on each side of the gasket will keep the gasket aligned during installation

16 Clean the mating surfaces with lacquer thinner or acetone. Make sure the bolt holes in the engine block are clean.

17 Apply a bead of RTV sealant to the four corner seams where the rear seal retainer meets the engine block, and the front cover meets the engine block (see illustration).

→Note: You must work fast to secure the oil pan to the engine within 5 minutes of applying the RTV.

18 Carefully position the oil pan against the engine block and install the bolts finger tight. Make sure the gaskets haven't shifted, then tighten the bolts to the torque listed in this Chapter's Specifications (see illustration). Start at the center of the oil pan and work out toward the ends in a spiral pattern.

19 The remainder of installation is the reverse of removal.

⁂ CAUTION:

Allow at least 90 minutes for the RTV to dry, then refill the engine with oil before starting it (see Chapter 1).

20 Start the engine and check carefully for oil leaks at the oil pan. Drive the vehicle and check again.

15 Oil pump - removal and installation

→Note: The oil pump is available as a complete replacement unit only. No service parts or repair specifications are available from the manufacturer.

REMOVAL

▶ Refer to illustrations 15.2 and 15.4

1 Raise the vehicle and support it securely on jackstands.

⁂ WARNING:

Some models covered by this manual are equipped with an air suspension system. Always disconnect the electrical power to the suspension system before lifting or towing (see Chapter 10). Failure to perform this procedure may result in unexpected shifting or movement of the vehicle, which could cause personal injury.

Drain the engine oil (see Chapter 1).

2 Unbolt and lower the oil pan as described in Section 14. It's not necessary to completely remove the oil pan. Remove the two bolts that attach the oil pump pick-up tube to the oil pump (see illustration).

3 Remove the timing chain cover, timing chains, chain guides and crankshaft sprockets (see Sections 6 and 7).

4 Remove the four oil pump mounting bolts (see illustration) and separate the pump from the engine block. On 2005 and later Expedition/Navigator 5.4L engines, there are 3 mounting bolts.

INSTALLATION

▶ Refer to illustrations 15.5

5 Inspect the O-ring gasket on the pick-up tube (see illustration). If it's damaged, replace it.

6 Install the oil pump to the engine and tighten the bolts to the torque listed in this Chapter's Specifications.

→Note: Prime the oil pump prior to installation. Pour clean oil into the pick-up port and turn the pump by hand.

7 The remainder of installation is the reverse of removal procedure.

8 Fill the engine with the correct type and quantity of oil. Start the engine and check for leaks.

15.2 Remove the two bolts (arrows) retaining the pickup tube to the oil pump

V8 ENGINES 2C-27

15.4 Remove the four oil pump mounting bolts (arrows) and oil pump from the engine block

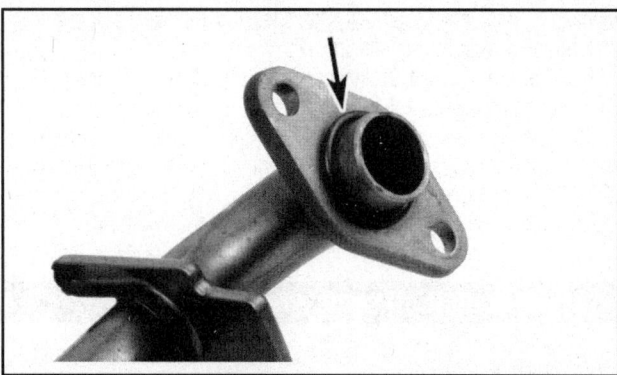

15.5 Before bolting the pickup tube back into the oil pump, inspect the O-ring (arrow) and replace it if necessary

16 Flywheel/driveplate - removal and installation

REMOVAL

◆ **Refer to illustration 16.4**

1 Disconnect the cable from the negative battery terminal.
2 Raise the vehicle and support it securely on jackstands.

> ❋❋ **WARNING:**
>
> Some models covered by this manual are equipped with an air suspension system. Always disconnect the electrical power to the suspension system before lifting or towing (see Chapter 10). Failure to perform this procedure may result in unexpected shifting or movement of the vehicle, which could cause personal injury.

3 Refer to Chapter 7 and remove the transmission.

➥ **Note: On models with automatic transmissions, the driveplate-to-torque converter bolts can be accessed for removal through the large rubber plug on the right rear of the engine block.**

4 Look for factory paint marks that indicate driveplate-to-crankshaft alignment. If they aren't there, scribe or paint marks on the driveplate and crankshaft to ensure correct alignment during reassembly (see illustration).

5 Remove the bolts that secure the flywheel/driveplate to the crankshaft. Use a flywheel retaining tool to keep the crankshaft from turning while loosening the flywheel/driveplate bolts.

➥ **Note: There are six flywheel retaining bolts on REP engines, and eight on WEP engines.**

6 Remove the flywheel/driveplate from the crankshaft. Be sure to support it while removing the last bolt.

> ❋❋ **WARNING:**
>
> The teeth on the flywheel may be sharp, and the manual transmission flywheel is heavy. Be sure to hold it with gloves or rags.

After the driveplate is removed, there is a sheetmetal reinforcement/mounting plate that is located between the engine block and the driveplate. It doesn't need to be removed, unless necessary.

INSTALLATION

◆ **Refer to illustration 16.7**

7 If removed, be sure the reinforcement plate is installed as shown (see illustration), so it is correctly positioned for the starter installation.
8 Clean and inspect the mating surfaces of the driveplate and the

16.4 Make an alignment mark (arrow), if not already on the driveplate, to enssure proper reassembly

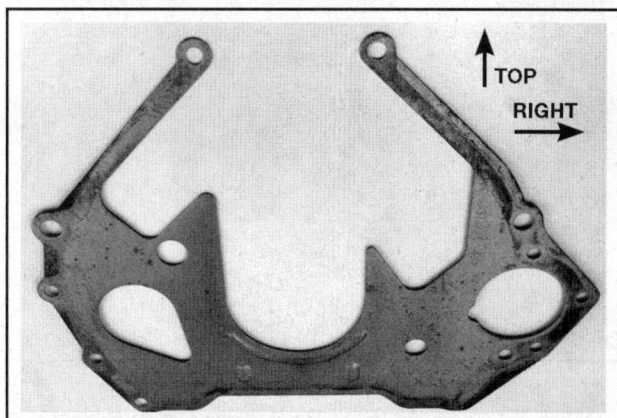

16.7 If the reinforcement plate is removed for any reason, it should be reinstalled in the direction shown here

2C-28 V8 ENGINES

crankshaft. If the crankshaft rear seal is leaking, replace it before reinstalling the driveplate (see Section 17).

9 Check for cracked, broken or missing ring gear teeth. If any of these conditions are found, replace the flywheel/driveplate.

10 Install the flywheel/driveplate to the engine, aligning the marks made during removal. Note that some engines have an alignment dowel or staggered bolt holes to ensure correct installation. Before installing the bolts, apply Teflon thread sealant to the threads.

11 Use a flywheel retaining tool to keep the driveplate from turning as you tighten the bolts to the torque listed in this Chapter's Specifications.

12 The remainder of installation is the reverse of the removal procedure.

17 Crankshaft oil seals - replacement

FRONT SEAL

▸ Refer to illustrations 17.4 and 17.6

1 Disconnect the negative battery cable.
2 Remove the drivebelt (see Chapter 1) and fan/shroud assembly (see Chapter 3).
3 Remove the crankshaft pulley (see Section 5).
4 Carefully remove the seal from the cover with a seal removal tool (see illustration). If a seal removal tool is not available, carefully use a screwdriver. If the timing cover is removed, use a chisel or small punch and hammer to drive the seal out of the cover from the back side. Support the cover as close to the seal bore as possible with two wood blocks. Be careful not to damage the cover or scratch the wall of the seal bore.
5 Check the seal bore and crankshaft, as well as the seal contact surface on the crankshaft pulley for nicks and burrs. Position the new seal in the bore with the open end of the seal facing IN. A small amount of engine oil applied to the outer edge of the new seal will make installation easier - but don't overdo it!
6 Drive the seal into the bore with a large socket and hammer until it's completely seated (see illustration). If the cover is removed, support the cover on wood blocks. Select a socket that's the same outside diameter as the seal (a section of pipe can be used if a socket isn't available).
7 Lubricate the lip of the seal with clean engine oil and install the crankshaft pulley on the end of the crankshaft. The keyway in the crankshaft pulley bore must be aligned with the Woodruff key in the crankshaft nose.

➡ Note: Before reinstalling the crankshaft pulley, apply as small dab of RTV sealant to the front end of the crankshaft key groove.

8 If the crankshaft pulley can't be seated by hand, tap it into place with a soft-face hammer, or install the bolt and washer and tighten it to press the crankshaft pulley into place.
9 Tighten the crankshaft pulley bolt to the torque listed in this Chapter's Specifications.
10 Install the drivebelt.
11 Install the remaining parts removed for access to the seal.
12 Start the engine and check for leaks.

REAR SEAL

▸ Refer to illustrations 17.15, 17.16, 17.17 and 17.18

13 Disconnect the cable from the negative battery terminal. Raise the vehicle and support it securely on jackstands.

✳✳ WARNING:

Some models covered by this manual are equipped with an air suspension system. Always disconnect the electrical power to the suspension system before lifting or towing (see Chapter 10). Failure to perform this procedure may result in unexpected shifting or movement of the vehicle, which could cause personal injury.

Refer to Chapter 7 and remove the transmission.

14 Remove the flywheel/driveplate and the rear cover plate from the engine (see Section 16).

17.4 Using a special seal removal tool or screwdriver, remove the crankshaft front oil seal, being very careful not to scratch the crankshaft during seal removal

17.6 There is special tool for installing the front oil seal into the timing chain cover, but if the tool is unavailable a large socket or section of tubing (the same diameter as the seal) can be used to drive the seal into place

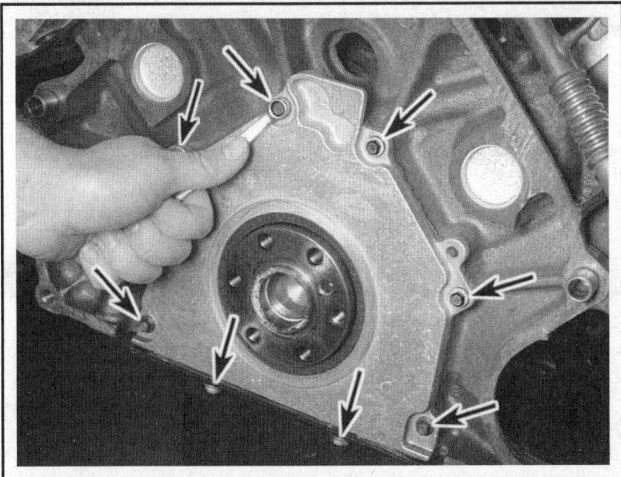

17.15 Remove the eight bolts (arrows) and separate the seal retainer from the engine block

17.16 Support the seal retainer on two wood blocks and drive out the old seal with a blunt punch and hammer

17.17 Support the seal retainer and drive the new seal into the housing with a wood block (be careful not to cock the seal in the bore while installing)

17.18 Inspect the seal contact surface on the crankshaft (arrow) for signs of excessive wear or grooves

15 Remove the bolts, detach the seal retainer (see illustration) and clean off all the old gasket and/or sealant material from both the engine block and the seal retainer.
16 Support the seal and retainer assembly on wood blocks and drive the old seal out from the back side with a drift punch and hammer (see illustration).
17 Drive the new seal into the retainer with a wood block (see illustration).
18 Clean the crankshaft and seal bore with lacquer thinner or acetone. Check the seal contact surface on the crankshaft very carefully for scratches or nicks that could damage the new seal lip and cause oil leaks (see illustration). If the crankshaft is damaged, the only alternative is a new or different crankshaft.
19 Lubricate the crankshaft seal journal and the lip of the new seal with engine oil.
20 Place a 1/16-inch wide bead of anaerobic sealant on either the engine block or the seal retainer.
21 Install the oil seal retainer by slowly and carefully pushing the seal onto the crankshaft. The seal lip is stiff, so work it onto the crankshaft with a smooth object such as the end of a socket extension as you push the retainer against the engine block.
22 Install the retainer bolts and tighten them, a little at a time using a criss-cross pattern, to the torque listed in this Chapter's Specifications.
23 Reinstall the engine rear cover plate, driveplate and the transmission.
24 The remaining steps are the reverse of removal.
25 Check the oil level and add if necessary, run the engine and check for oil leaks.

2C-30 V8 ENGINES

18 Engine mounts - check and replacement

CHECK

♦ Refer to illustration 18.4

1 Engine mounts seldom require attention, but broken or deteriorated mounts should be replaced immediately or the added strain placed on the driveline components may cause damage or wear.
2 During the check, the engine must be raised slightly to remove the weight from the mounts.
3 Raise the vehicle and support it securely on jackstands, then position a jack under the engine oil pan.

WARNING:

Some models covered by this manual are equipped with an air suspension system. Always disconnect the electrical power to the suspension system before lifting or towing (see Chapter 10). Failure to perform this procedure may result in unexpected shifting or movement of the vehicle, which could cause personal injury.

Place a large wood block between the jack head and the oil pan, then carefully raise the engine just enough to take the weight off the mounts.
4 Check the mounts to see if the rubber is cracked, hardened or separated from the metal plates (see illustration). Sometimes the rubber will split right down the center.
5 Check for relative movement between the mount plates and the engine or frame (use a large screwdriver or pry bar to attempt to move the mounts). If movement is noted, lower the engine and tighten the mount fasteners.
6 Rubber preservative should be applied to the mounts to slow deterioration.

REPLACEMENT

♦ Refer to illustrations 18.11 and 18.12

7 Disconnect the negative battery cable.
8 Refer to Chapter 3 and remove the engine cooling fan assembly and shroud.
9 Attach an engine support fixture to the engine lifting eyes adjacent to the exhaust manifolds.

➡Note: Many equipment rental yards rent the engine support fixture.

10 Raise the vehicle and support it securely on jackstands.

WARNING:

Some models covered by this manual are equipped with an air suspension system. Always disconnect the electrical power to the suspension system before lifting or towing (see Chapter 10). Failure to perform this procedure may result in unexpected shifting or movement of the vehicle, which could cause personal injury.

11 Take some of the weight off the engine mounts with the support fixture, then remove the engine mount through-bolts (see illustration).
12 Raise the engine a little at a time with the support fixture, watching for interference with any engine compartment components. When the engine is raised about two inches, remove the bolts retaining the mount to the engine block (see illustration).
13 Installation is the reverse of removal. Tighten the bolts to the torque listed in this Chapter's Specifications.

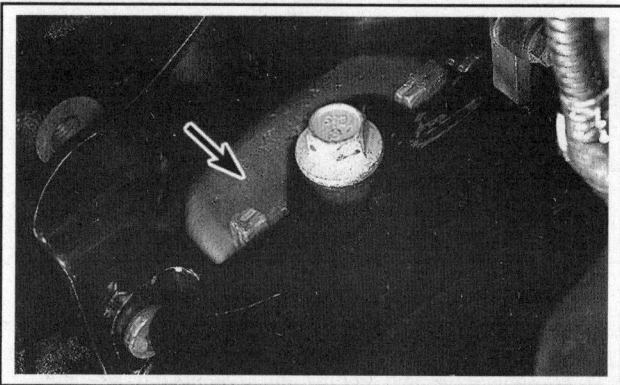

18.4 Inspect the engine mount components for cracked rubber insulators (arrow), missing bolts or cracked metal brackets

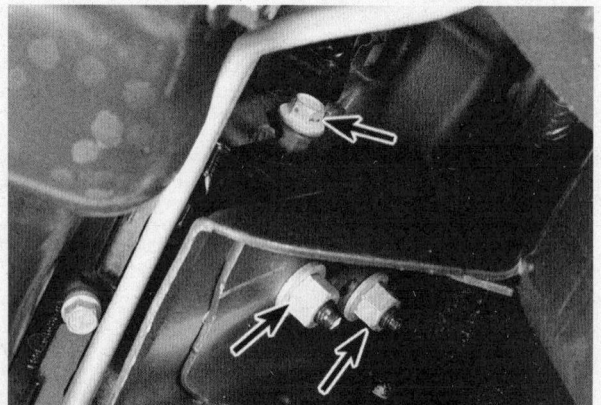

18.11 With the engine raised slightly, remove the mount through-bolt (upper arrow) - lower arrows indicate nuts to remove if the lower bracket is to be replaced

18.12 There are three bolts holding the mount to the block (two shown; there is one more at the other end of the mount)

V8 ENGINES 2C-31

Specifications

General

Displacement
  4.6L   4.6 liters (281 cubic inches)
  5.4L   5.4 liters (330 cubic inches)
Bore and stroke
  4.6L   3.554 X 3.546 inches
  5.4L   3.554 X 4.168 inches
Cylinder numbers (front to rear)
  Right side   1-2-3-4
  Left (driver's) side   5-6-7-8
Firing order   1-3-7-2-6-5-4-8

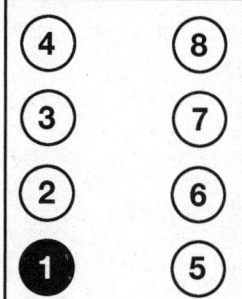

Camshaft

Lobe lift   See Section 9
Endplay
  4.6L   0.001 to 0.006 inch
  5.4L
    2004 and earlier   0.00106 to 0.00748 inch
    2005 through 2009   0.003 to 0.007 inch
    2010 and later   0.001 to 0.007 inch
Journal diameter
  4.6L   1.060 to 1.061 inches
  5.4L
    2004 and earlier   1.061 to 1.062 inches
    2005 and later   1.126 to 1.127 inches
Bearing inside diameter
  4.6L   1.062 to 1.063 inches
  5.4L
    2004 and earlier   1.063 to 1.064 inches
    2005 and later   1.128 to 1.129 inches
Journal-to-bearing (oil) clearance
  Standard   0.001 to 0.003 inch
  Service limit   0.005 inch maximum

V8 ENGINES

Torque specifications — Ft-lbs (unless otherwise indicated)

→ **Note:** One foot-pound (ft-lb) of torque is equivalent to 12 inch-pounds (in-lbs) of torque. Torque values below approximately 15 ft-lbs are expressed in inch-pounds, because most foot-pound torque wrenches are not accurate at these smaller values.

Front crossmember bolts	
2005 through 2009 models	75
2010 and later models	66
Camshaft sprocket bolts (REP engines)	81 to 95
Camshaft sprocket/phaser bolt (2005 and later 5.4L engine)	
Step 1	30
Step 2	Rotate an additional 1/4-turn (90-degrees)
Camshaft caps to cylinder head	71 to 106 in-lbs
Coolant crossover manifold bolts	89 in-lbs
Timing chain cover bolts	
2002 and earlier models	
Bolts 1 through 5	15 to 22
Bolts 6 through 15	30 to 41
2003 and 2004 models	
Step 1, bolts 1 through 5	18
Step 2, bolts 6 and 7	37
Step 3, bolts 11 through 15	18
2005 and later models	
Step 1, bolts 1 through 15	18
Step 2, bolts 6 and 7	35
Step 3, oil pan-to-cover bolts	18
Step 4, oil pan-to-cover bolts	Rotate an additional 60 degrees
Drivebelt idler and tensioner bolts	15 to 22
Cylinder head bolts	
Step 1	30
Step 2	Tighten an additional 90-degrees
Step 3	Tighten an additional 90-degrees
Crankshaft pulley-to-crankshaft bolt	
Step 1	66
Step 2	Loosen one full turn
Step 3	34 to 39
Step 4	Tighten an additional 90-degrees
Valve cover bolts	71 to 106 in-lbs
Oil pan-to-engine block bolts	
Step 1	18 in-lbs
Step 2	15
Step 3	Tighten an additional 60-degrees
Exhaust manifold-to-cylinder head nuts	17 to 20
Exhaust manifold-to-cylinder head studs	89 to 115 in-lbs
Intake manifold-to-cylinder head bolts	
2004 and earlier models	
Step 1	18 in-lbs
Step 2	15 to 22
2005 through 2009 models	18 in-lbs
2010 and later models	
Step 1	18 in-lbs
Step 2	89 in-lbs

Torque specifications (continued) — Ft-lbs (unless otherwise indicated)

Oil filter adapter bolts	15 to 22
Oil cooler-to-filter adapter nut	30 to 40
Oil pump-to-engine block mounting bolts	71 to 106 in-lbs
Oil pick-up tube-to-main bearing cap nut	15 to 22
Oil pick-up tube-to-oil pump bolts	71 to 106 in-lbs
Flywheel/driveplate mounting bolts	54 to 64
Rear crankshaft oil seal retainer bolts	89 in-lbs
Timing chain guides	71 to 107 in-lbs
Timing chain tensioners	15 to 20
Engine mount-to-engine block bolts	39 to 53
Engine mount throughbolts	
All models through 2006	50 to 68
2007 and later Expedition/Navigator models	259

Notes

2D GENERAL ENGINE OVERHAUL PROCEDURES

Section

1 General information
2 Engine overhaul - general information
3 Cylinder compression check
4 Vacuum gauge diagnostic checks
5 Engine removal - methods and precautions
6 Engine - removal and installation
7 Engine rebuilding alternatives
8 Engine overhaul - disassembly sequence
9 Cylinder head - disassembly
10 Cylinder head - cleaning and inspection
11 Valves - servicing
12 Cylinder head - reassembly
13 Pistons/connecting rods - removal
14 Crankshaft - removal
15 Engine block - cleaning
16 Engine block - inspection
17 Cylinder honing
18 Pistons/connecting rods - inspection
19 Crankshaft - inspection
20 Main and connecting rod bearings - inspection
21 Engine overhaul - reassembly sequence
22 Piston rings - installation
23 Crankshaft - installation and main bearing oil clearance check
24 Pistons/connecting rods - installation and rod bearing oil clearance check
25 Initial start-up and break-in after overhaul

2D-2 GENERAL ENGINE OVERHAUL PROCEDURES

1 General information

Included in this portion of Chapter 2 are the general overhaul procedures for the cylinder head(s) and internal engine components.

The information ranges from advice concerning preparation for an overhaul and the purchase of replacement parts to detailed, step-by-step procedures covering removal and installation of internal engine components and the inspection of parts.

The following Sections have been written based on the assumption that the engine has been removed from the vehicle. For information concerning in-vehicle engine repair, as well as removal and installation of the external components necessary for the overhaul, see Parts A, B or C of this Chapter and Section 7 of this Part.

The Specifications included in this Part are only those necessary for the inspection and overhaul procedures which follow. Refer to Parts A, B or C for additional Specifications.

2 Engine overhaul - general information

▶ Refer to illustrations 2.4a, 2.4b and 2.4c

It is not always easy to determine when, or if, an engine should be completely overhauled, as a number of factors must be considered.

High mileage is not necessarily an indication that an overhaul is needed, while low mileage does not preclude the need for an overhaul. Frequency of servicing is probably the most important consideration. An engine that has had regular and frequent oil and filter changes, as well as other required maintenance, will most likely give many thousands of miles of reliable service. Conversely, a neglected engine may require an overhaul very early in its life.

Excessive oil consumption is an indication that piston rings and/or valve guides are in need of attention. Make sure that oil leaks are not responsible before deciding that the rings and/or guides are bad. Test the cylinder compression (see Section 3) or have a leak down test performed by an experienced tune-up mechanic to determine the extent of the work required.

If the engine is making obvious knocking or rumbling noises, the connecting rod and/or main bearings are probably at fault. To accurately test oil pressure, temporarily connect a mechanical oil pressure gauge in place of the oil pressure sending unit (see illustrations). Compare the reading to the pressure listed in this Chapter's Specifications. If the pressure is extremely low, the bearings and/or oil pump are probably worn out.

Loss of power, rough running, excessive valve train noise and high fuel consumption rates may also point to the need for an overhaul, especially if they are all present at the same time. If a complete tune-up does not remedy the situation, major mechanical work is the only solution.

An engine overhaul involves restoring the internal parts to the specifications of a new engine. During an overhaul, the piston rings are replaced and the cylinder walls are reconditioned (rebored and/or honed). If a re-bore is done, new pistons are required. The main bearings, connecting rod bearings and camshaft bearings are generally replaced with new ones and, if necessary, the crankshaft may be reground to restore the journals. Generally, the valves are serviced as well, since they are usually in less-than-perfect condition at this point. While the engine is being overhauled, other components, such as the distributor, starter and alternator, can be rebuilt as well. The end result should be a like-new engine that will give many trouble free miles.

➡ **Note: Critical cooling system components such as the hoses, the drivebelts, the thermostat and the water pump MUST be replaced with new parts when an engine is overhauled. The radiator should be checked carefully to ensure that it isn't clogged or leaking. Some engine rebuilding shops will not honor their engine warranty unless you have had the radiator replaced or professionally cleaned. If in doubt, replace it with a new one. Also, we do not recommend overhauling the oil pump - always install a new one when an engine is rebuilt.**

Before beginning the engine overhaul, read through the entire procedure to familiarize yourself with the scope and requirements of the job. Overhauling an engine is not difficult, but it is time consuming. Plan on the vehicle being tied up for a minimum of two weeks, especially if parts must be taken to an automotive machine shop for repair or reconditioning. Check on availability of parts and make sure that any necessary

2.4a Remove the oil pressure sending unit and attach an oil pressure gauge - be sure the fittings you use have the same thread as the sender

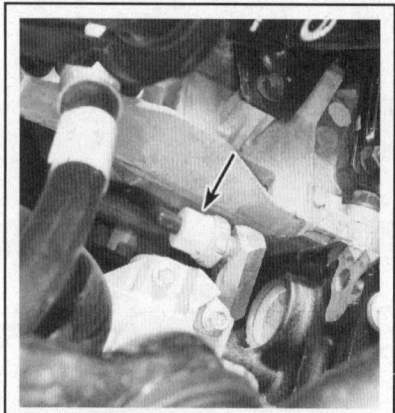

2.4b 4.2L V6 oil pressure sending unit, left front of block. On 3.5L V6 engines, the oil pressure sending unit is located near the lower left side of the engine on the oil filter adapter behind the alternator

2.4c The V8 oil pressure sending unit is located at the lower left (driver's side) corner of the engine, near the oil filter

GENERAL ENGINE OVERHAUL PROCEDURES

special tools and equipment are obtained in advance. Most work can be done with typical hand tools, although a number of precision measuring tools are required for inspecting parts to determine if they must be replaced. Often an automotive machine shop will handle the inspection of parts and offer advice concerning reconditioning and replacement.

➡ **Note: Always wait until the engine has been completely disassembled and all components, especially the engine block, have been inspected before deciding what service and repair operations must be performed by an automotive machine shop.**

Since the block's condition will be the major factor to consider when determining whether to overhaul the original engine or buy a rebuilt one, never purchase parts or have machine work done on other components until the block has been thoroughly inspected. As a general rule, time is the primary cost of an overhaul, so it does not pay to install worn or substandard parts.

As a final note, to ensure maximum life and minimum trouble from a rebuilt engine, everything must be assembled with care in a spotlessly clean environment.

3 Cylinder compression check

▸ **Refer to illustration 3.6**

1 A compression check will tell you what mechanical condition the upper end (pistons, rings, valves, head gaskets) of your engine is in. Specifically, it can tell you if the compression is down due to leakage caused by worn piston rings, defective valves and seats or a blown head gasket.

➡ **Note: The engine must be at normal operating temperature for this check and the battery must be fully charged.**

2 Begin by cleaning the area around the spark plugs before you remove them (compressed air works best for this). This will prevent dirt from getting into the cylinders as the compression check is being done.

3 Remove all of the spark plugs from the engine (see Chapter 1).

4 Block the throttle wide open.

5 Disable the ignition system by disconnecting the primary (low voltage) wires from the coil(s). Also disable the fuel pump (see Chapter 4, Section 2).

6 With the compression gauge in the number one spark plug hole, crank the engine over at least four compression strokes and watch the gauge (see illustration). The compression should build up quickly in a healthy engine. Low compression on the first stroke, followed by gradually increasing pressure on successive strokes, indicates worn piston rings. A low compression reading on the first stroke, which does not build up during successive strokes, indicates leaking valves or a blown head gasket (a cracked head could also be the cause). Record the highest gauge reading obtained.

7 Repeat the procedure for the remaining cylinders and compare the results to the Specifications.

8 Add some engine oil (about three squirts from a plunger-type oil can) to each cylinder, through the spark plug hole, and repeat the test.

9 If the compression increases after the oil is added, the piston rings are definitely worn. If the compression does not increase significantly, the leakage is occurring at the valves or head gasket. Leakage past the valves may be caused by burned valve seats and/or faces or warped, cracked or bent valves.

3.6 A compression gauge with a threaded fitting for the spark plug hole is preferred over the type that requires hand pressure to maintain the seal - be sure to open the throttle valve as far as possible during the compression check

10 If two adjacent cylinders have equally low compression, there is a strong possibility that the head gasket between them is blown. The appearance of coolant in the combustion chambers or the crankcase would verify this condition.

11 If the compression is unusually high, the combustion chambers are probably coated with carbon deposits. If that is the case, the cylinder heads should be removed and decarbonized.

12 If compression is way down or varies greatly between cylinders, it would be a good idea to have a leak-down test performed by an automotive repair shop. This test will pinpoint exactly where the leakage is occurring and how severe it is.

4 Vacuum gauge diagnostic checks

▸ **Refer to illustration 4.4**

1 A vacuum gauge provides valuable information about what is going on in the engine at a low-cost. You can check for worn rings or cylinder walls, leaking head or intake manifold gaskets, incorrect carburetor adjustments, restricted exhaust, stuck or burned valves, weak valve springs, improper ignition or valve timing and ignition problems.

2 Unfortunately, vacuum gauge readings are easy to misinterpret, so they should be used in conjunction with other tests to confirm the diagnosis.

3 Both the absolute readings and the rate of needle movement are important for accurate interpretation. Most gauges measure vacuum in inches of mercury (in-Hg). The following references to vacuum assume the diagnosis is being performed at sea level. As elevation increases (or atmospheric pressure decreases), the reading will decrease. For every 1,000 foot increase in elevation above approximately 2000 feet, the gauge readings will decrease about one inch of mercury.

2D-4 GENERAL ENGINE OVERHAUL PROCEDURES

4 Connect the vacuum gauge directly to intake manifold vacuum, not to ported (throttle body) vacuum (see illustration). Be sure no hoses are left disconnected during the test or false readings will result.

5 Before you begin the test, allow the engine to warm up completely. Block the wheels and set the parking brake. With the transmission in Park, start the engine and allow it to run at normal idle speed.

※ WARNING:

Carefully inspect the fan blades for cracks or damage before starting the engine. Keep your hands and the vacuum gauge clear of the fan and do not stand in front of the vehicle or in line with the fan when the engine is running.

6 Read the vacuum gauge; an average, healthy engine should normally produce about 17 to 22 inches of vacuum with a fairly steady needle. Refer to the following vacuum gauge readings and what they indicate about the engine's condition:

7 A low steady reading usually indicates a leaking gasket between the intake manifold and cylinder head(s) or throttle body, a leaky vacuum hose, late ignition timing or incorrect camshaft timing. Check ignition timing with a timing light and eliminate all other possible causes, utilizing the tests provided in this Chapter before you remove the timing chain cover to check the timing marks.

8 If the reading is three to eight inches below normal and it fluctuates at that low reading, suspect an intake manifold gasket leak at an intake port or a faulty fuel injector.

9 If the needle has regular drops of about two-to-four inches at a steady rate, the valves are probably leaking. Perform a compression check or leak-down test to confirm this.

10 An irregular drop or down-flick of the needle can be caused by a sticking valve or an ignition misfire. Perform a compression check or leak-down test and read the spark plugs.

11 A rapid vibration of about four in.-Hg vibration at idle combined with exhaust smoke indicates worn valve guides. Perform a leak-down test to confirm this. If the rapid vibration occurs with an increase in engine speed, check for a leaking intake manifold gasket or head gasket, weak valve springs, burned valves or ignition misfire.

4.4 A simple vacuum gauge can make a very useful diagnosis of an engine's condition

12 A slight fluctuation, say one inch up and down, may mean ignition problems. Check all the usual tune-up items and, if necessary, run the engine on an ignition analyzer.

13 If there is a large fluctuation, perform a compression or leak-down test to look for a weak or dead cylinder or a blown head gasket.

14 If the needle moves slowly through a wide range, check for a clogged PCV system, incorrect idle fuel mixture, carburetor/throttle body or intake manifold gasket leaks.

15 Check for a slow return after revving the engine by quickly snapping the throttle open until the engine reaches about 2,500 rpm and let it shut. Normally the reading should drop to near zero, rise above normal idle reading (about 5 in.-Hg over) and then return to the previous idle reading. If the vacuum returns slowly and doesn't peak when the throttle is snapped shut, the rings may be worn. If there is a long delay, look for a restricted exhaust system (often the muffler or catalytic converter). An easy way to check this is to temporarily disconnect the exhaust ahead of the suspected part and redo the test.

5 Engine removal - methods and precautions

If you have decided that an engine must be removed for overhaul or major repair work, several preliminary steps should be taken.

Locating a suitable work area is extremely important. A shop is, of course, the most desirable place to work. Adequate work space, along with storage space for the vehicle, will be needed. If a shop or garage is not available, at the very least a flat, level, clean work surface made of concrete or asphalt is required.

Cleaning the engine compartment and engine before beginning the removal procedure will help keep tools clean and organized.

An engine hoist or A-frame will be needed. Make sure that the equipment is rated in excess of the combined weight of the engine and its accessories. Safety is of primary importance, considering the potential hazards involved in lifting the engine out of the vehicle.

※ WARNING:

Some models covered by this manual are equipped with an air suspension system. Always disconnect the electrical power to the suspension system before lifting or towing (see Chapter 10). Failure to perform this procedure may result in unexpected shifting or movement of the vehicle, which could cause personal injury.

If the engine is being removed by a novice, a helper should be available. Advice and aid from someone more experienced would also be helpful. There are many instances when one person cannot simultaneously perform all of the operations required when lifting the engine out of the vehicle.

GENERAL ENGINE OVERHAUL PROCEDURES   2D-5

Plan the operation ahead of time. Ar-range for or obtain all of the tools and equipment you will need prior to beginning the job. Some of the equipment necessary to perform engine removal and installation safely and with relative ease are (in addition to an engine hoist) a heavy duty floor jack, complete sets of wrenches and sockets as described in the front of this manual, wooden blocks and plenty of rags and cleaning solvent for mopping up spilled oil, coolant and gasoline. If the hoist is to be rented, make sure that you arrange for it in advance and perform beforehand all of the operations possible without it. This will save you money and time.

Plan for the vehicle to be out of use for a considerable amount of time. A machine shop will be required to perform some of the work which the do-it-yourselfer cannot accomplish due to a lack of special equipment. These shops often have a busy schedule, so it would be wise to consult them before removing the engine in order to accurately estimate the amount of time required to rebuild or repair components that may need work.

Always use extreme caution when removing and installing the engine. Serious injury can result from careless actions. Plan ahead, take your time and a job of this nature, although major, can be accomplished successfully.

6 Engine - removal and installation

※※ WARNING 1:

The air conditioning system is under high pressure. DO NOT loosen any fittings or remove any components until after the system has been discharged. Air conditioning refrigerant should be properly discharged into an EPA-approved container at a dealer service department or an automotive air conditioning repair facility. Always wear eye protection when disconnecting air conditioning system fittings.

※※ WARNING 2:

Gasoline is extremely flammable, so take extra precautions when you work on any part of the fuel system. Don't smoke or allow open flames or bare light bulbs near the work area, and don't work in a garage where a gas-type appliance (such as a water heater or a clothes dryer) is present. Since gasoline is carcinogenic, wear latex gloves when there's a possibility of being exposed to fuel, and, if you spill any fuel on your skin, rinse it off immediately with soap and water. Mop up any spills immediately and do not store fuel-soaked rags where they could ignite. The fuel system is under constant pressure, so, if any fuel lines are to be disconnected, the fuel pressure in the system must be relieved first (see Chapter 4 for more information). When you perform any kind of work on the fuel system, wear safety glasses and have a Class B type fire extinguisher on hand.

REMOVAL

▶ Refer to illustrations 6.6, 6.26, 6.29 and 6.30

※※ WARNING:

Some models covered by this manual are equipped with an air suspension system. Always disconnect the electrical power to the suspension system before lifting or towing (see Chapter 10). Failure to perform this procedure may result in unexpected shifting or movement of the vehicle, which could cause personal injury.

1 Relieve the fuel system pressure (see Chapter 4).
2 Disconnect the negative cable from the battery.
3 Cover the fenders and cowl and remove the hood (see Chapter 11). Special pads are available to protect the fenders, but an old bedspread or blanket will also work.
4 Remove the air filter assembly (see Chapter 4).
5 Drain the cooling system (see Chapter 1).

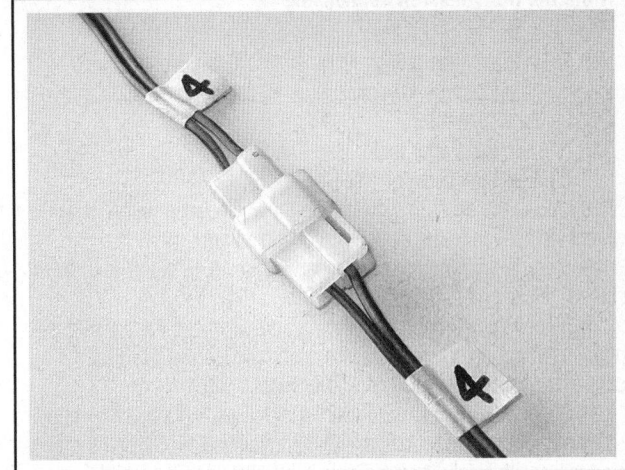

6.6 Label each wire before unplugging the connector

6 Label the vacuum lines, emissions system hoses, electrical connectors, ground straps and fuel lines that would interfere with engine removal, to ensure correct reinstallation, then detach them. Pieces of masking tape with numbers or letters written on them work well (see illustration). If there's any possibility of confusion, make a sketch of the engine compartment and clearly label the lines, hoses and wires.
7 Label and detach all coolant hoses from the engine.
8 Remove the cooling fan/shroud and radiator (see Chapter 3).
9 Remove the drivebelt(s) (see Chapter 1).
10 Disconnect the accelerator cable and Throttle Valve (TV) linkage/speed control cable from the engine (see Chapter 4). Remove the air cleaner assembly.
11 Unbolt the power steering pump (see Chapter 10). Leave the lines/hoses attached and make sure the pump is kept in an upright position in the engine compartment (use wire or rope to restrain it out of the way).
12 On air conditioned models, unbolt the compressor (see Chapter 3) and set it aside. Do not disconnect the hoses.
13 Refer to Part A, B or C and remove the intake manifold. On 4.6L engines, disconnect and remove the ignition coils (Refer to Part B).
14 Disconnect the main wiring harness plugs in the engine compartment and position the harness out of the way. There are many connectors and harnesses. Take your time, making sure all attachment points are identified and the wiring is carefully set aside.

2D-6 GENERAL ENGINE OVERHAUL PROCEDURES

6.26 Use a prybar or a large screwdriver and pry the engine from the transmission bellhousing

15 Drain the engine oil (see Chapter 1) and remove the filter. On models with oil coolers, disconnect the oil cooler hoses and remove the oil cooler (see Chapter 3).

16 Remove the starter motor (see Chapter 5).

17 Remove the alternator (see Chapter 5).

18 Unbolt the exhaust system from the engine (see Part A, B or C, whichever applies to the engine being worked on).

19 If you're working on a turbocharged model, remove the turbocharger intake and outlet tubes (see Chapter 4).

20 If you're working on a model with an automatic transmission, remove the torque converter access cover and remove the torque converter-to-driveplate fasteners (see Chapter 7B).

21 Support the transmission with a jack. Position a block of wood between them to prevent damage to the transmission. Special transmission jacks with safety chains are available - use one if possible.

22 Attach an engine sling or a length of sturdy chain to the engine and then to the engine hoist.

➡ Note: Some engines have lifting brackets already installed on them, usually on the exhaust manifold studs. Other engines do not have them, but they are available separately from Ford dealer parts departments.

23 Roll the hoist into position and connect the sling or chain to it. Take up the slack in the sling or chain, but don't lift the engine.

※※ **WARNING:**

DO NOT place any part of your body under the engine when it's supported only by a hoist or other lifting device.

24 Remove the transmission-to-engine block bolts (see Chapter 7).

25 Remove the engine mount through-bolts from both sides (see Part A, B or C, depending which engine is being worked on).

26 Recheck to be sure nothing is still connecting the engine to the transmission or vehicle. Disconnect anything still remaining.

27 Raise the engine slightly. Carefully work it forward to separate it from the transmission (see illustration). If you're working on a vehicle with an automatic transmission, be sure the torque converter stays in the transmission (clamp a pair of vise-grips to the transmission housing to keep the converter from sliding out). Slowly raise the engine out of the engine compartment. Check carefully to make sure nothing is hanging up.

28 Remove the driveplate (see Part A, B or C of this Chapter).

29 Mount the engine on an engine stand (see illustration).

30 Once the engine is removed, support the transmission with a piece of pipe that crosses from side to side to hold the transmission as the floor jack is removed (see illustration).

INSTALLATION

31 Check the engine and transmission mounts. If they're worn or damaged, replace them.

32 Carefully lower the engine into the engine compartment - make sure the engine mounts line up.

33 Guide the torque converter into the crankshaft following the procedure outlined in Chapter 7B. Don't pull the converter away from the transmission; let the engine move back against the transmission.

34 Install the transmission-to-engine bolts and tighten them securely.

※※ **CAUTION:**

DO NOT use the bolts to force the transmission and engine together!

35 Reinstall the remaining components in the reverse order of removal.

36 Add coolant and oil as needed. Run the engine and check for leaks and proper operation of all accessories, then install the hood and test drive the vehicle.

6.29 Use long, high-strength bolts (arrows) to hold the engine block on the engine stand - make sure they are tight before lowering the hoist and placing the entire weight of the engine on the stand

6.30 Use a piece of pipe to support the transmission once the engine has been removed, then remove the floor jack that supported the transmission during engine removal

GENERAL ENGINE OVERHAUL PROCEDURES 2D-7

7 Engine rebuilding alternatives

The do-it-yourselfer is faced with a number of options when performing an engine overhaul. The decision to replace the engine block, piston/connecting rod assemblies and crankshaft depends on a number of factors, with the number one consideration being the condition of the block. Other considerations are cost, access to machine shop facilities, parts availability, time required to complete the project and the extent of prior mechanical experience on the part of the do-it-yourselfer.

Some of the rebuilding alternatives include:

Individual parts - If the inspection procedures reveal that the engine block and most engine components are in reusable condition, purchasing individual parts may be the most economical alternative. The block, crankshaft and piston/connecting rod assemblies should all be inspected carefully. Even if the block shows little wear, the cylinder bores should be surface honed.

Crankshaft kit - This rebuild package consists of a reground crankshaft and a matched set of pistons and connecting rods. The pistons will already be installed on the connecting rods. Piston rings and the necessary bearings will be included in the kit. These kits are commonly available for standard cylinder bores, as well as for engine blocks which have been bored to a regular oversize.

Short block - A short block consists of an engine block with renewed crankshaft and piston/connecting rod assemblies already installed. All new bearings are incorporated and all clearances will be correct. The existing cylinder heads, camshaft(s), valve train components and external parts can be bolted to the short block with little or no machine shop work necessary.

Long block - A long block consists of a short block plus an oil pump, oil pan, cylinder heads, valve covers, camshaft(s) and valve train components, timing sprockets, timing chain and timing cover. All components are installed with new bearings, seals and gaskets incorporated throughout. The installation of manifolds and external parts is all that is necessary.

Give careful thought to which alternative is best for you and discuss the situation with local automotive machine shops, auto parts dealers or parts store countermen before ordering or purchasing replacement parts.

8 Engine overhaul - disassembly sequence

✱✱ CAUTION:
The cylinder head bolts on all engines in the vehicles covered by this manual are "torque-to-yield" bolts and are NOT reusable. A predetermined stretch of the bolt gives the even clamping load needed to seal the cylinders properly. Once removed they must be replaced.

1 It's much easier to disassemble and work on the engine if it's mounted on a portable engine stand. A stand can often be rented quite cheaply from an equipment rental yard. Before the engine is mounted on a stand, the flywheel/driveplate and rear main seal retainer plate (V8) should be removed from the engine.

2 If a stand isn't available, it's possible to disassemble the engine with it blocked up on the floor. Be extra careful not to tip or drop the engine when working without a stand.

3 If you're going to obtain a rebuilt engine, all external components must come off first, to be transferred to the replacement engine, just as they will if you're doing a complete engine overhaul yourself. These include:

Alternator and brackets
Emissions control components
Camshaft position sensor
Spark plug wires and spark plugs
Thermostat and housing cover
Water pump
EFI components
Intake/exhaust manifolds
Oil filter (replace)
Engine mounts
Driveplate
Engine rear plate
Crankshaft damper

➡ **Note: When removing the external components from the engine, pay close attention to details that may be helpful or important during installation. Note the installed position of gaskets, seals, spacers, pins, brackets, washers, bolts and other small items.**

4 If you're obtaining a short block, which consists of the engine block, crankshaft, pistons and connecting rods all assembled, then the cylinder head(s), oil pan and oil pump will have to be removed as well. See Engine rebuilding alternatives for additional information regarding the different possibilities to be considered.

5 If you're planning a complete overhaul, the engine must be disassembled and the components removed in the following order:

Driveplate
Valve covers
Intake manifold
Exhaust manifolds
Rocker arms and pushrods (4.2L V6)
Camshafts and followers (V8)
Valve lifters (V6) or hydraulic lash adjusters (V8)
Vibration damper
Timing chain cover
Timing chain(s), sprockets, guides and tensioners
Camshaft(s)
Cylinder heads
Oil pan
Oil pump (replace)
Piston/connecting rod assemblies
Crankshaft and main bearings (replace)

2D-8 GENERAL ENGINE OVERHAUL PROCEDURES

6 Before beginning the disassembly and overhaul procedures, make sure the following items are available. Also, refer to Engine overhaul - reassembly sequence for a list of tools and materials needed for engine reassembly.

Common hand tools
Small cardboard boxes and plastic bags for storing parts
Gasket scraper
Ridge reamer
Vibration damper puller
Micrometers
Telescoping gauges
Dial indicator set
Valve spring compressor
Cylinder surfacing hone
Piston ring groove cleaning tool
Electric drill motor
Tap and die set
Wire brushes
Oil gallery brushes
Cleaning solvent

9 Cylinder head - disassembly

▶ Refer to illustrations 9.1, 9.2 and 9.3

➡ **Note:** *New and rebuilt cylinder heads are commonly available for most engines at dealerships and auto parts stores. Due to the fact that some specialized tools are necessary for the disassembly and inspection procedures, and replacement parts may not be readily available, it may be more practical and economical for the home mechanic to purchase replacement head(s) rather than taking the time to disassemble, inspect and recondition the original(s). However, most machine shops will not have a core set of heads unless the subject vehicle has been on the market for close to five years, so on a newer vehicle, chances are you will have to wait until the rebuilding work is done to your heads.*

1 Cylinder head disassembly involves removal of the intake and exhaust valves and related components. If they're still in place, remove the rocker arm nuts, pivot balls and rocker arms from the cylinder head studs. Label the parts or store them separately (see illustration) so they can be reinstalled in their original locations and in the same valve guides they are removed from.

2 Compress the springs on the first valve with a spring compressor and remove the keepers (see illustration). Carefully release the valve spring compressor and remove the retainer, sleeve (if used), the spring and the spring seat (if used).

3 Pull the valve out of the head, then remove the oil seal from the guide. If the valve binds in the guide (won't pull through), push it back into the head and deburr the area around the keeper groove with a fine file or whetstone (see illustration).

4 Repeat the procedure for the remaining valves. Remember to keep all the parts for each valve together so they can be reinstalled in the same locations.

5 Once the valves and related components have been removed and stored in an organized manner, the head should be thoroughly cleaned and inspected. If a complete engine overhaul is being done, finish the engine disassembly procedures before beginning the cylinder head cleaning and inspection process.

9.1 A small plastic bag, with an appropriate label, can be used to store the valve train components so they can be kept together and reinstalled in the original position

9.2 Use a valve spring compressor to compress the spring, then remove the keepers from the valve stem with needle-nose pliers or a magnet, as shown here on a 4.6L engine

9.3 If the valve won't pull through the guide, deburr the edge of the stem end and the area around the top of the keeper groove with a file or whetstone

GENERAL ENGINE OVERHAUL PROCEDURES 2D-9

10 Cylinder head - cleaning and inspection

1 Thorough cleaning of the cylinder head(s) and related valve train components, followed by a detailed inspection, will enable you to decide how much valve service work must be done during the engine overhaul.

➡ **Note: If the engine was severely overheated, the cylinder head is probably warped (see Step 12).**

CLEANING

2 Scrape all traces of old gasket material and sealing compound off the head gasket, intake manifold and exhaust manifold sealing surfaces. Be very careful not to gouge the cylinder head. Special gasket removal solvents that soften gaskets and make removal much easier are available at auto parts stores.

3 Remove all built up scale from the coolant passages.

4 Run a stiff wire brush through the various holes to remove deposits that may have formed in them.

5 Run an appropriate-size tap into each of the threaded holes to remove corrosion and thread sealant that may be present. If compressed air is available, use it to clear the holes of debris produced by this operation.

⚠ WARNING:
Wear eye protection when using compressed air!

6 Clean the exhaust manifold stud threads, if equipped.

7 Clean the cylinder head with solvent and dry it thoroughly. Compressed air will speed the drying process and ensure that all holes and recessed areas are clean.

➡ **Note: Decarbonizing chemicals are available and may prove very useful when cleaning cylinder heads and valve train components. They are very caustic and should be used with caution. Be sure to follow the instructions on the container.**

8 Clean the valvetrain components with solvent and dry them thoroughly (don't mix them up during the cleaning process).

➡ **Note: Compressed air will speed the drying process and can be used to clean out the oil passages.**

9 Clean all the valve springs, spring seats, keepers and retainers with solvent and dry them thoroughly. Work the components from one valve at a time to avoid mixing up the parts.

10 Scrape off any heavy deposits that may have formed on the valves, then use a motorized wire brush to remove deposits from the valve heads and stems. Again, make sure the valves don't get mixed up.

INSPECTION

➡ **Note: Be sure to perform all of the following inspection procedures before concluding that machine shop work is required. Make a list of the items that need attention.**

Cylinder head
▸ **Refer to illustrations 10.12 and 10.14**

11 Inspect the head very carefully for cracks, evidence of coolant leakage and other damage. If cracks are found, check with an automotive machine shop concerning repair. If repair isn't possible, a new cylinder head should be obtained.

12 Using a straightedge and feeler gauge, check the head gasket mating surface for warpage (see illustration). Check the head both straight across and corner-to-corner. If the warpage exceeds the limit listed in this Chapter's Specifications, it can be resurfaced at an automotive machine shop.

➡ **Note: If the 5.0L cylinder heads are resurfaced, the intake manifold flanges will also require machining.**

13 Examine the valve seats in each of the combustion chambers. If they're pitted, cracked or burned, the head will require valve service that's beyond the scope of the home mechanic.

14 Check the valve stem-to-guide clearance by measuring the lateral movement of the valve stem with a dial indicator attached securely to the head (see illustration). The valve must be in the guide and approximately 1/16-inch off the seat. The total valve stem movement indicated by the gauge needle must be divided by two to obtain the actual clearance. After this is done, if there's still some doubt regarding the condition of the valve guides they should be checked by an automotive machine shop (the cost should be minimal).

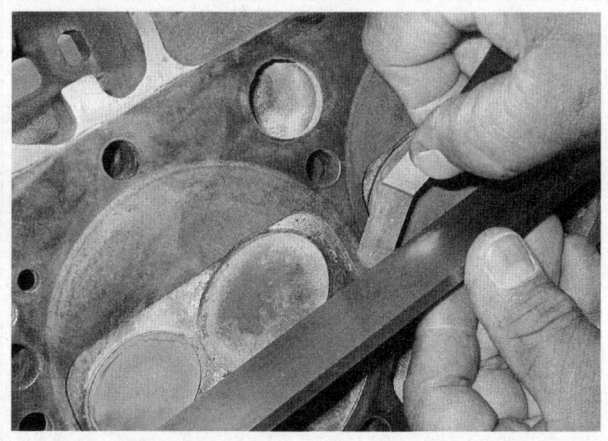

10.12 Check the cylinder head gasket surface for warpage by trying to slip a feeler gauge under the straightedge (see this Chapter's Specifications for the maximum warpage allowed and use a feeler gauge of that thickness)

10.14 Lay the head on its edge, pull each valve out about 1/8 inch, set up a dial indicator with the probe touching the valve stem, wiggle the valve and measure its movement

2D-10 GENERAL ENGINE OVERHAUL PROCEDURES

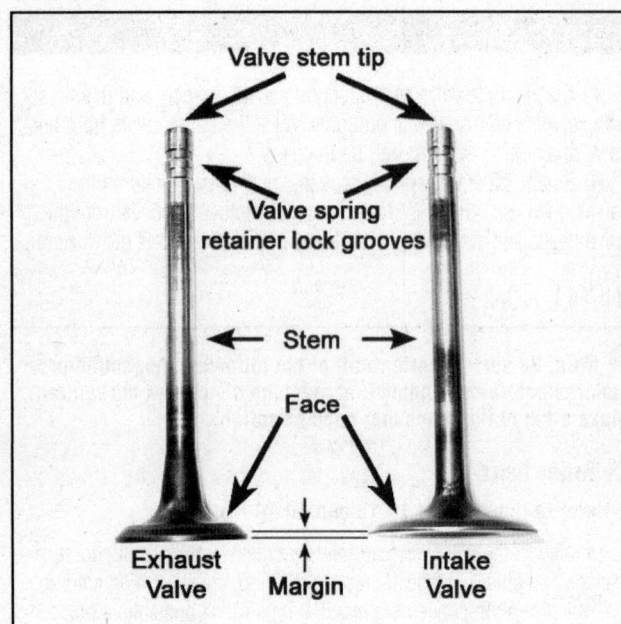

10.15 Check for valve wear at the points shown here

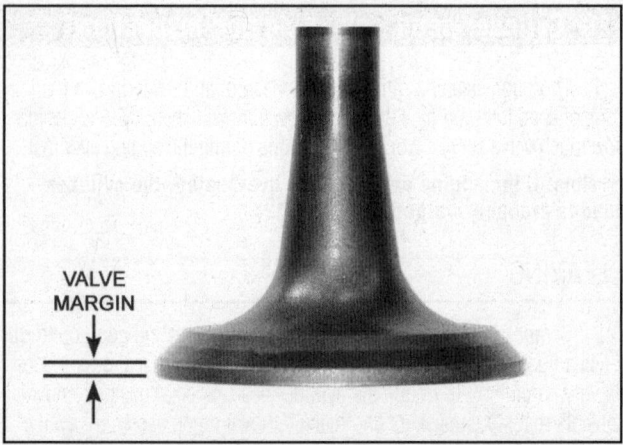

10.16 The margin width on each valve must be as specified (if no margin exists, the valve cannot be reused)

Valves

▶ Refer to illustrations 10.15 and 10.16

15 Carefully inspect each valve face for uneven wear, deformation, cracks, pits and burned areas (see illustration). Check the valve stem for scuffing and galling and the neck for cracks. Rotate the valve and check for any obvious indication that it's bent. Look for pits and excessive wear on the end of the stem. The presence of any of these conditions indicates the need for valve service by an automotive machine shop.

16 Measure the margin width on each valve (see illustration). Any valve with a margin narrower than 1/32-inch will have to be replaced with a new one.

Valve components

▶ Refer to illustrations 10.17 and 10.18

17 Check each valve spring for wear (on the ends) and pits. Measure the free length and compare it to the Specifications (see illustration).

Any springs that are shorter than specified have sagged and should not be reused. The tension of all springs should be checked with a special fixture before deciding that they're suitable for use in a rebuilt engine (take the springs to an automotive machine shop for this check).

18 Stand each spring on a flat surface and check it for squareness (see illustration). If any of the springs are distorted or sagged, replace all of them with new parts.

19 Check the spring retainers and keepers for obvious wear and cracks. Any questionable parts should be replaced with new ones, as extensive damage will occur if they fail during engine operation.

Rocker arm components

▶ Refer to illustrations 10.22a, 10.22b, 10.22c and 10.22d

20 Clean all the parts thoroughly. Make sure all oil passages are open.

21 Check the rocker arm faces (the areas that contact the pushrod ends and valve stems) for pits, wear, galling, score marks and rough spots.

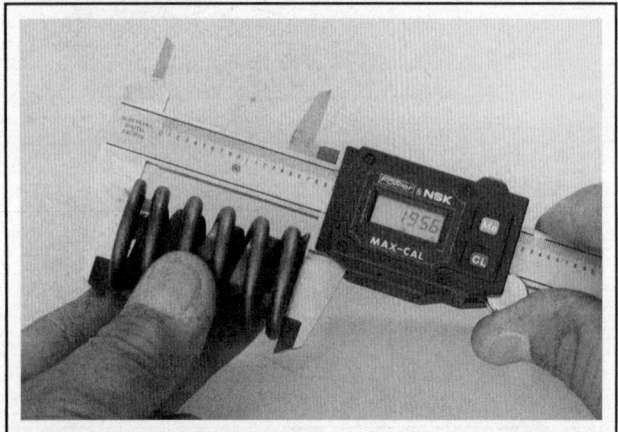

10.17 Measure the free length of each valve spring with a dial or vernier caliper

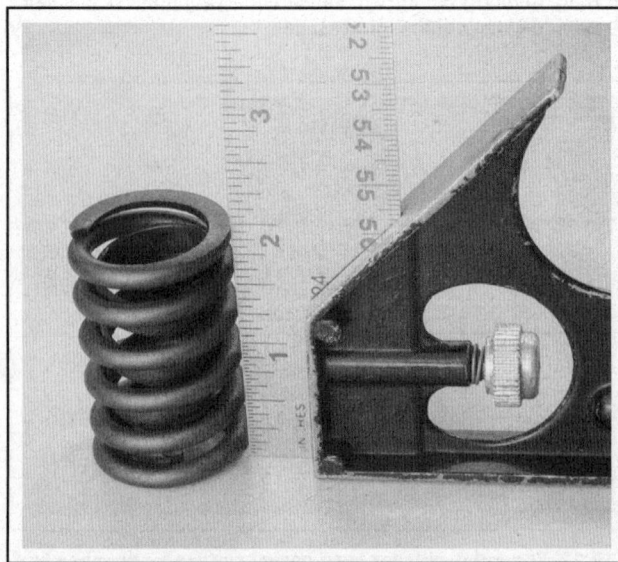

10.18 Check each valve spring for squareness; if it is bent it should be replaced

GENERAL ENGINE OVERHAUL PROCEDURES 2D-11

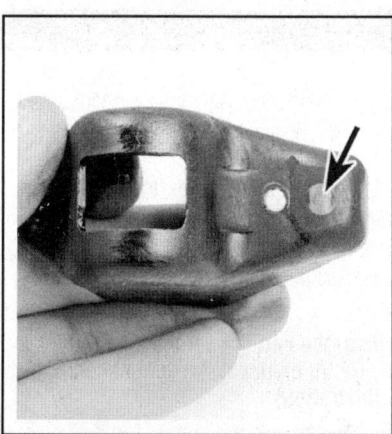

10.22a On V6 engines, check each rocker arm where the valve stem contacts it (arrow) . . .

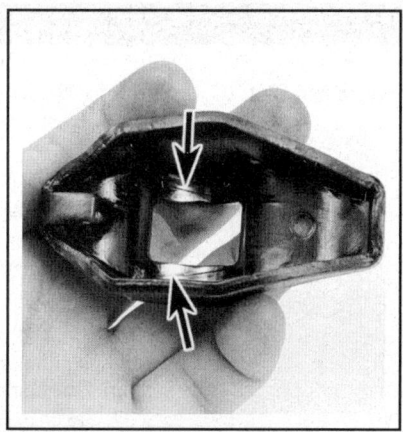

10.22b . . . the fulcrum seats (arrows) in the top of the rocker arm . . .

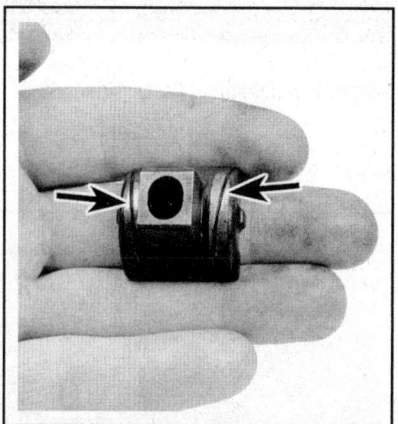

10.22c . . . and the fulcrums themselves for wear and galling (arrows)

22 Check the rocker arm pivot contact areas and fulcrums. Look for cracks in each rocker arm and bolt (see illustrations).

23 Inspect the pushrod ends for scuffing and excessive wear. On 4.2L V6 engines, roll each pushrod on a flat surface, like a piece of plate glass, to determine if it's bent.

24 Check the rocker arm studs in the cylinder heads on 4.2L V6 engines for damaged threads and secure installation.

25 Any damaged or excessively worn parts must be replaced with new ones.

26 If the inspection process indicates that the valve components are in generally poor condition and worn beyond the limits specified, which is usually the case in an engine that's being overhauled, reassemble the valves in the cylinder head and refer to Section 11 for valve servicing recommendations.

27 Clean all the parts thoroughly. Make sure all oil passages are open.

28 Any damaged or excessively worn parts must be replaced with new ones.

29 If the inspection process indicates that the valve components are in generally poor condition and worn beyond the limits specified, which is usually the case in an engine that's being overhauled, reassemble the valves in the cylinder head and refer to Section 11 for valve servicing recommendations.

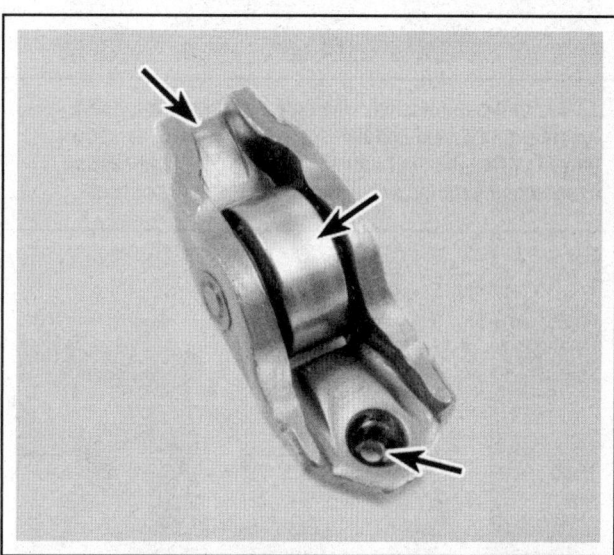

10.22d On 4.6L and 5.4L V8 engines, check the rocker arms for wear (arrows) at the valve stem end, roller and the pocket that contacts the lash adjuster

11 Valves - servicing

1 Because of the complex nature of the job and the special tools and equipment needed, servicing of the valves, the valve seats and the valve guides, commonly known as a valve job, should be done by a professional.

2 The home mechanic can remove and disassemble the head, do the initial cleaning and inspection, then reassemble and deliver it to a dealer service department or an automotive machine shop for the actual service work. Doing the inspection will enable you to see what condition the head and valve train components are in and will ensure that you know what work and new parts are required when dealing with an automotive machine shop.

3 The dealer service department or automotive machine shop will remove the valves and springs, recondition or replace the valves and valve seats, recondition the valve guides, check and replace the valve springs, spring retainers or rotators and keepers (as necessary), replace the valve seals with new ones, reassemble the valve components and make sure the installed spring height is correct. The cylinder head gasket surface will also be resurfaced if it's warped.

4 After the valve job has been performed by a professional, the head will be in like-new condition. When the head is returned, be sure to clean it again before installation on the engine to remove any metal particles and abrasive grit that may still be present from the valve service or head resurfacing operations. Use compressed air, if available, to blow out all the oil holes and passages.

2D-12 GENERAL ENGINE OVERHAUL PROCEDURES

12 Cylinder head - reassembly

Refer to illustrations 12.3a, 12.3b, 12.3c, 12.7a, 12.7b and 12.9

1 Regardless of whether or not the head was sent to an automotive repair shop for valve servicing, make sure it's clean before beginning reassembly.

2 If the head was sent out for valve servicing, the valves and related components will already be in place. Begin the reassembly procedure with Step 9.

3 On all engines, lubricate and install the valves, then install new seals on each of the valve guides. Using a hammer and deep socket, gently tap each seal into place until it's seated on the guide (see illustrations). Don't twist or cock the seals during installation or they will not seat properly on the valve stems.

➡ **Note: On 4.2L V6 engines, the spring seat must be in place before installing the seal. On V8 engines, the seal and spring seat are one-piece (see illustration).**

4 On V8 engines, reinstall the valve lifters.
5 Beginning at one end of the head, lubricate and install the first valve. Apply moly-base grease or clean engine oil to the valve stem.
6 Set the valve spring and retainer in place.
7 Apply a small dab of grease to each keeper to hold it in place (see illustration). Compress the springs with a valve spring compressor and carefully install the keepers in the upper groove (see illustration), then slowly release the compressor and make sure the keepers seat properly.

12.3a On V6 models with the type of seal shown, use a hammer and a seal installer (or a deep socket, as shown here) to drive the seal onto the valve guide/head casting boss (make sure the valve spring seat is in place first)

12.3b Installing a valve stem seal on a V8 engine - the socket must contact the flange (spring seat) of the seal

12.3c 4.6L and 5.4L V8 engines use valve stem seals that are a valve spring seat and seal combined into one part - make sure replacement parts are the same as the ones removed earlier

12.7a Apply a small dab of grease to each keeper as shown here before installation - it'll hold them in place on the valve stem as the spring is released

12.7b Compress the springs with a valve spring compressor and position the keepers in the upper groove, then slowly release the compressor and make sure the keepers seat properly

GENERAL ENGINE OVERHAUL PROCEDURES 2D-13

➡ **Note:** When the camshafts are not on the cylinder head on V8 engines, only the valve spring compressor type shown can be used; the special compressor uses the camshafts for leverage and is used during disassembly only (see Part C of this Chapter).

8 Repeat the procedure for the remaining valves. Be sure to return the components to their original locations - don't mix them up!

9 Check the installed valve spring height with a ruler graduated in 1/32-inch increments or a dial caliper. If the head was sent out for service work, the installed height should be correct (but don't automatically assume that it is). The measurement is taken from the top of each spring seat or shim(s) to the bottom of the retainer (see illustration). If the height is greater than the figure listed in this Chapter's Specifications, shims can be added under the springs to correct it.

✱✱ CAUTION:

Don't, under any circumstances, shim the springs to the point where the installed height is less than specified.

12.9 Valve spring installed height is the distance from the spring seat on the head to the bottom of the spring retainer

10 Apply moly-base grease to the rocker arm faces and the fulcrums, then install the rocker arms and fulcrums on the cylinder head studs.

13 Pistons/connecting rods - removal

◆ Refer to illustrations 13.1, 13.3, 13.4, 13.6a, 13.6b and 13.8

➡ **Note:** Prior to removing the piston/connecting rod assemblies, remove the cylinder head(s), the oil pan and the oil pump (on 5.4L engines) by referring to the appropriate Sections in Chapter 2, Part A, B or C, depending which engine is being overhauled. On some models, there will be a sheetmetal oil baffle plate attached with nuts to the main cap studs. Remove the nuts and the baffle.

1 Use your fingernail to feel if a ridge has formed at the upper limit of ring travel (about 1/4-inch down from the top of each cylinder). If carbon deposits or cylinder wear have produced ridges, they must be completely removed with a special tool (see illustration). Follow the manufacturer's instructions provided with the tool. Failure to remove the ridges before attempting to remove the piston/connecting rod assemblies may result in piston breakage.

➡ **Note:** Do not let the tool cut into the ring travel area more than 1/32-inch.

2 After the cylinder ridges have been removed, turn the engine upside-down so the crankshaft is facing up.

➡ **Note:** To access the main bearing caps and piston rod caps on 3.5L V6 engines, remove the engine block main bearing cap support brace (see Section 14).

3 Before the connecting rods are removed, check the endplay with a dial indicator or with feeler gauges (see illustration). Slide them between the first connecting rod and the crankshaft throw until the play is removed. The endplay is equal to the thickness of the feeler gauge(s). If the endplay exceeds the service limit, new connecting rods will be required. If new rods (or a new crankshaft) are installed, the endplay may fall under the specified minimum (if it does, the rods will have to be machined to restore it - consult an automotive machine shop for advice if necessary). Repeat the procedure for the remaining connecting rods.

13.1 A ridge reamer is required to remove the ridge from the top of each cylinder - do this before removing the pistons!

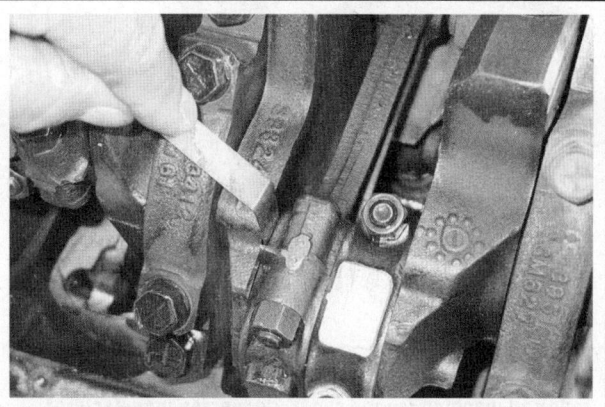

13.3 Check the connecting rod side clearance (endplay) with a dial indicator or a feeler gauge

2D-14 GENERAL ENGINE OVERHAUL PROCEDURES

13.4 Mark the rod bearing caps in order from the front of the engine to the rear (numbers can be used or use one mark for the front cap, two for the second one and so on)

13.6a Remove the rod cap and bearing insert together

4 Check the connecting rods and caps for identification marks. If they aren't plainly marked, use a small center-punch, number stamping die (see illustration), or scribe, to make the appropriate number of indentations, or marks, on each rod and cap (1, 2, 3, etc., depending on the engine type and cylinder they're associated with).

5 Loosen each of the connecting rod cap bolts 1/2-turn at a time until they can be removed by hand.

6 Remove the connecting rod cap and bearing insert (see illustration). Don't drop the bearing insert out of the cap.

➡ **Note:** The V6 and V8 engines covered by this manual use a new method, referred to as "fractured cap method," to more accurately match the rod cap to the connecting rod. The mating line of the rod and cap (see illustration) is made by breaking ("fracturing") the cap from the rod. This is supposed to ensure a perfect match once reassembled with its corresponding rod. This also means that no caps are interchangeable.

7 To protect the rod journals of the crankshaft during disassembly or assembly, take two rod cap bolts and cut the heads off, then slot that end of the "studs" with a hacksaw. Cover the non-threaded portion of these studs with rubber tubing and screw the studs into the rod to act as guides, so the journal isn't scratched as the piston/rod assembly is withdrawn or installed (see illustration 24.9b).

➡ **Note: These rod bolts are torque-to-yield design, so NEW bolts will be used during assembly. Discard the rest of the old rod bolts.**

8 Remove the bearing insert and push the connecting rod/piston assembly out through the top of the engine. Use a wooden or plastic hammer handle to push on the upper bearing surface in the connecting rod (see illustration). If resistance is felt, double-check to make sure that all of the ridge was removed from the cylinder.

9 Repeat the procedure for the remaining cylinders.

10 After removal, reassemble the connecting rod caps and bearing inserts in their respective connecting rods and install the cap nuts finger tight. Leaving the old bearing inserts in place until reassembly will help prevent the connecting rod bearing surfaces from being accidentally nicked or gouged.

11 Don't separate the pistons from the connecting rods (see Section 18 for additional information).

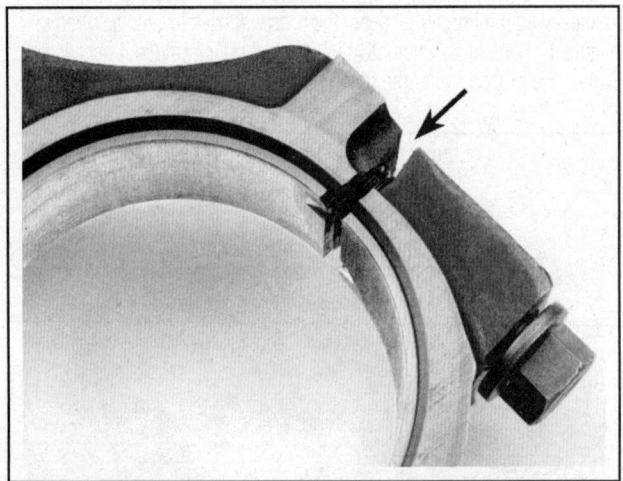

13.6b The new method used to manufacture connecting rods and machine the rod cap is unique; the manufacturer "fractures" (breaks) the cap from the rod to give a perfect match upon reassembly

13.8 Use a hammer handle to drive the piston and connecting rod assembly down and out of the cylinder block, being very careful not to nick the crankshaft on the way out

GENERAL ENGINE OVERHAUL PROCEDURES   2D-15

14 Crankshaft - removal

▶ Refer to illustrations 14.1, 14.3 and 14.4

➡ **Note:** The crankshaft can be removed only after the engine has been removed from the vehicle. It's assumed that the driveplate, crankshaft pulley, timing chains, oil pan, oil pump body, main cap support brace (3.5L engines) and piston/connecting rod assemblies have already been removed. The rear main oil seal retainer must be unbolted and separated from the block before proceeding with crankshaft removal.

1 Before the crankshaft removal procedure is started, check the endplay. Mount a dial indicator with the stem in line with the crankshaft and just touching the end of the crankshaft (see illustration).

2 Push the crankshaft all the way to the rear and zero the dial indicator. Next, pry the crankshaft to the front as far as possible and check the reading on the dial indicator. The distance that it moves is the endplay. If it's greater than the limit listed in this Chapter's Specifications, check the crankshaft thrust surfaces for wear. If no wear is evident, new main bearings should correct the endplay.

3 If a dial indicator isn't available, feeler gauges can be used. Gently pry or push the crankshaft all the way to the front of the engine. Slip feeler gauges between the crankshaft and the front face of the thrust main bearing to determine the clearance (see illustration).

4 Check the main bearing caps to see if they're marked to indicate their locations (see illustration). They should be numbered consecutively from the front of the engine to the rear. If they aren't, mark them with number stamping dies or a center-punch. Main bearing caps generally have a cast-in arrow, which points to the front of the engine.

V6 ENGINES

➡ **Note:** The thrust bearing on the 4.2L V6 engine is in the number three main bearing cap. The thrust bearing on the 3.5L V6 engine is installed at the number 4 crankshaft saddle and bearing cap. Both engines have an upper and lower thrust bearing shell.

5 On 3.5L engines, remove the main bearing cap support brace bolts, in reverse order of installation sequence (see illustration 10.38) and remove the support brace.

6 Loosen the main bearing cap bolts/studs 1/4-turn at a time each, until they can be removed by hand. Note if any stud bolts are used and make sure they're returned to their original locations when the crankshaft is reinstalled.

➡ **Note:** On some V6 engines, there are two main cap support braces, each spanning two main caps, and attached with nuts to the main cap studs. Remove the braces.

14.1 Checking crankshaft endplay with a dial indicator

14.3 Checking crankshaft endplay with a feeler gauge

V8 ENGINES

▶ Refer to illustrations 14.7, 14.8 and 14.9

7 The V8 engines have a more complex crankshaft removal and assembly procedure than V6 engines because of the number of bolts used to fasten the main caps to the cylinder block. On 4.6L REP models,

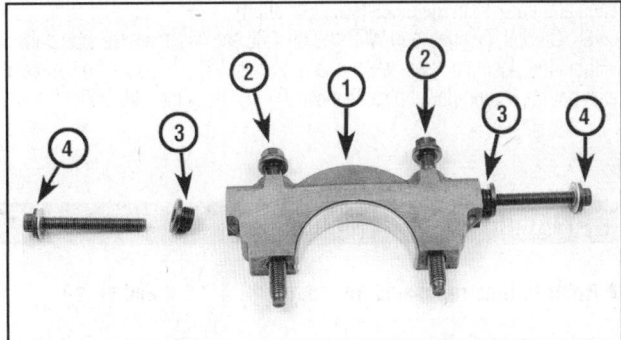

14.7 4.6L REP engines use main bearing caps that are fastened to the block through the use of a set of bolts, screws and specific adjustment procedures

   1 Main bearing cap   3 Jack screws (left handed thread)
   2 Main bearing bolts   4 Side bolts

14.4 The main bearing caps are usually marked to indicate their locations. They should be numbered consecutively from the front of the engine to the rear

2D-16 GENERAL ENGINE OVERHAUL PROCEDURES

14.8 Remove the side bolts - V8 models

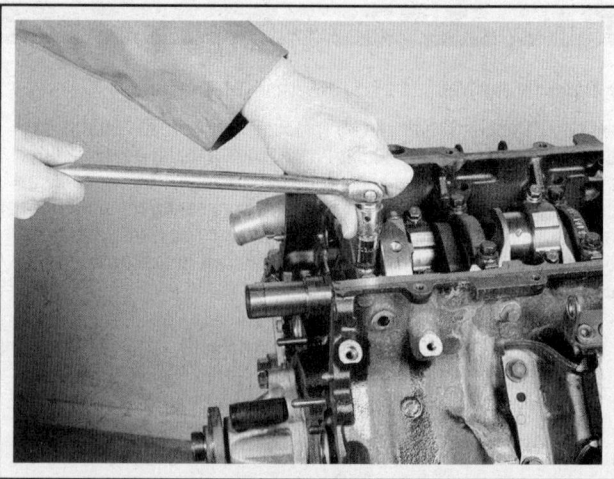

14.9 Following the reverse order of the tightening sequence (see Section 23), remove the main bearing cap bolts, loosening them 1/4 turn at a time until they can be removed by hand

9 Remove the main bearing cap bolts (see illustration). On WEP 4.6L and 5.4L engines, also pull the dowel pins from the main caps. **Caution:** *The main bearing cap bolts (top and side) are "torque-to-yield" bolts and are NOT reusable. A pre-determined stretch of the bolt, calculated by the manufacturer, gives the added rigidity required with this cylinder block. Once removed they must be replaced. The jack screws on REP engines are reusable.*

ALL ENGINES

▶ Refer to illustration 14.10

10 Gently tap the caps with a soft-face hammer, then separate them from the engine block. If necessary, use the bolts as levers to remove the caps. Try not to drop the bearing inserts if they come out with the caps. All main caps should have an arrow cast in to indicate the front of the engine, and a number to indicate which position they have on the block. On 4.6L engines, the number is stamped in on the left side of the cap (see illustration).

11 Carefully lift the crankshaft out of the engine. It may be a good idea to have an assistant available, since the crankshaft is quite heavy. With the bearing inserts in place in the engine block and main bearing caps, return the caps to their respective locations on the engine block and tighten the bolts finger tight.

14.10 Main caps should be numbered (arrow) and have an indicator (circle) of the direction facing the front of the engine

there are two main cap bolts, two jack screws and two side bolts on each of the main caps. It is extremely important to follow the removal and the installation procedure to ensure correct assembly and operation. Loosen and remove the main cap side bolts, then back off the jack screws into the caps and away from the block (see illustration).

8 On 5.4L engines, and WEP-design 4.6L engines, there are side bolts without the jacking screws, and the caps are securely positioned on the block with both bolts and dowel pins. Remove the side bolts (see illustration).

15 Engine block - cleaning

▶ Refer to illustrations 15.1a, 15.1b, 15.4, 15.8 and 15.10

❋❋ CAUTION:
The core plugs (also known as freeze plugs or soft plugs) may be difficult or impossible to retrieve if they're driven into the block coolant passages.

1 Using the wide end of a punch (see illustration) tap in on the outer edge of the core plug to turn the plug sideways in the bore. Then, using a pair of pliers, pull the core plug from the engine block (see illustration). Don't worry about the condition of the old core plugs as they are being removed because they will be replaced on reassembly with new plugs.

2 Using a gasket scraper, remove all traces of gasket material from the engine block. Be very careful not to nick or gouge the gasket sealing surfaces.

3 Remove the main bearing caps and separate the bearing inserts from

GENERAL ENGINE OVERHAUL PROCEDURES

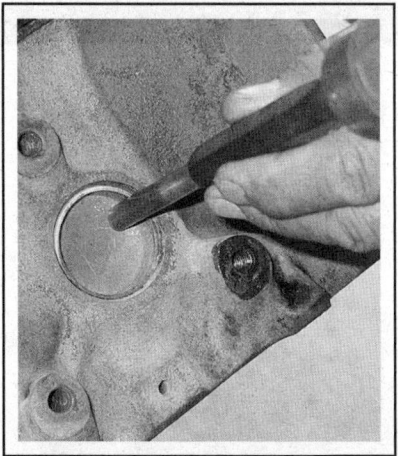

15.1a A hammer and a large punch can be used to knock the core plugs sideways in their bores

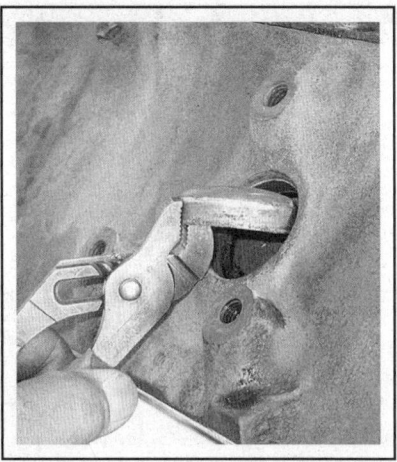

15.1b Pull the core plugs from the block with pliers

15.4 Remove the oil gallery plugs at the front and rear of the block (arrows indicate two at the rear of a typical engine)

the caps and the engine block (see Section 14). Tag the bearings, indicating which cylinder they were removed from and whether they were in the cap or the block, then set them aside.

4 Remove all of the threaded oil gallery plugs from the block (see illustration). The plugs are usually very tight - they may have to be drilled out and the holes re-tapped. Use new plugs when the engine is reassembled.

5 If the engine is extremely dirty it should be taken to an automotive machine shop to be steam cleaned or hot tanked.

6 After the block is returned, clean all oil holes and oil galleries one more time. Brushes specifically designed for this purpose are available at most auto parts stores. Flush the passages with warm water until the water runs clear, dry the block thoroughly and wipe all machined surfaces with a light, rust preventive oil. If you have access to compressed air, use it to speed the drying process and to blow out all the oil holes and galleries.

※※ WARNING:
Wear eye protection when using compressed air!

7 If the block isn't extremely dirty or sludged up, you can do an adequate cleaning job with hot soapy water and a stiff brush. Take plenty of time and do a thorough job. Regardless of the cleaning method used, be sure to clean all oil holes and galleries very thoroughly, dry the block completely and coat all machined surfaces with light oil.

8 The threaded holes in the block must be clean to ensure accurate torque readings during reassembly. Run the proper size tap into each of the holes to remove rust, corrosion, thread sealant or sludge and restore damaged threads (see illustration). If possible, use compressed air to clear the holes of debris produced by this operation.

9 Reinstall the main bearing caps and tighten all bolts finger tight.

10 After coating the sealing surfaces of the new core plugs with Permatex no. 2 sealant, install them in the engine block (see illustration). Make sure they're driven in straight and seated properly or leakage could result. Special tools are available for this purpose, but a large socket, with an outside diameter that will just slip into the core plug, a 1/2-inch drive extension and a hammer will work just as well.

11 Apply non-hardening sealant (such as Permatex no. 2 or Teflon pipe sealant) to the new oil gallery plugs and thread them into the holes in the block. Make sure they're tightened securely.

12 If the engine isn't going to be reassembled right away, cover it with a large plastic trash bag to keep it clean.

15.8 All bolt holes in the block - particularly the main bearing cap and head bolt holes - should be cleaned and restored with a tap (be sure to remove debris from the holes after this is done)

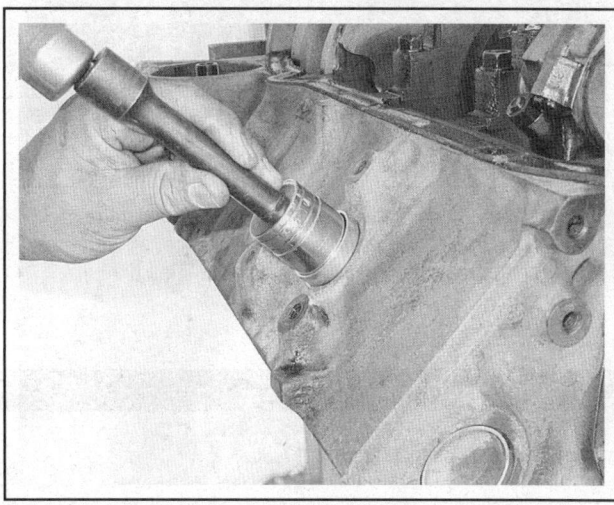

15.10 A large socket on an extension can be used to drive the new core plugs into the bores

2D-18 GENERAL ENGINE OVERHAUL PROCEDURES

16 Engine block - inspection

▶ Refer to illustrations 16.4a, 16.4b and 16.4c

1 Before the block is inspected, it should be cleaned as described in Section 15.

2 Visually check the block for cracks, rust and corrosion. Look for stripped threads in the threaded holes. It's also a good idea to have the block checked for hidden cracks by an automotive machine shop that has the special equipment to do this type of work. If defects are found, have the block repaired, if possible, or replaced.

3 Check the cylinder bores for scuffing and scoring.

4 Check the cylinders for taper and out-of-round conditions as follows (see illustrations):

5 Measure the diameter of each cylinder at the top (just under the ridge area), center and bottom of the cylinder bore, parallel to the crankshaft axis.

6 Next measure each cylinder's diameter at the same three locations perpendicular to the crankshaft axis.

7 The taper of the cylinder is the difference between the bore diameter at the top of the cylinder and the diameter at the bottom. The out-of-round specification of the cylinder bore is the difference between the parallel and perpendicular readings. Compare your results to those listed in this Chapter's Specifications.

8 Repeat the procedure for the remaining pistons and cylinders.

9 If the cylinder walls are badly scuffed or scored, or if they're out-of-round or tapered beyond the limits given in this Chapter's Specifications, have the engine block rebored and honed at an automotive machine shop. If a rebore is done, oversize pistons and rings will be required.

10 If the cylinders are in reasonably good condition and not worn to the outside of the limits, and if the piston-to-cylinder clearances can be maintained properly, then they don't have to be rebored. Honing is all that's necessary (see Section 17).

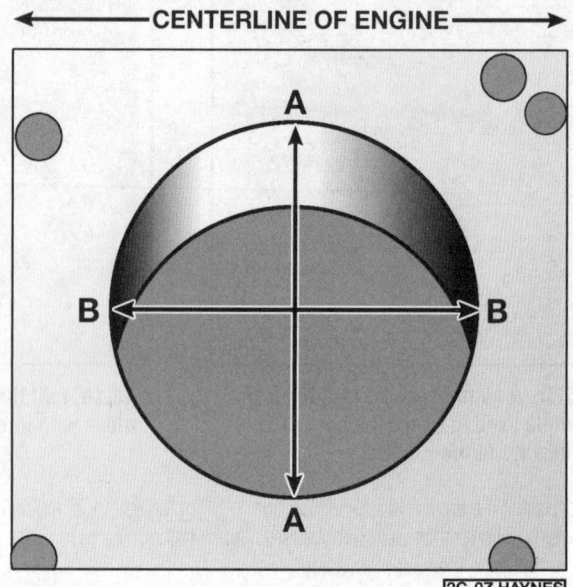

16.4a Measure the diameter of each cylinder at a right angle to the engine centerline (A), and parallel to engine centerline (B) - out-of-round is the difference between A and B; taper is the difference between A and B at the top of the cylinder and A and B at the bottom of the cylinder

16.4b The ability to "feel" when the telescoping gauge is at the correct point will be developed over time, so work slowly and repeat the check until you're satisfied the bore measurement is accurate

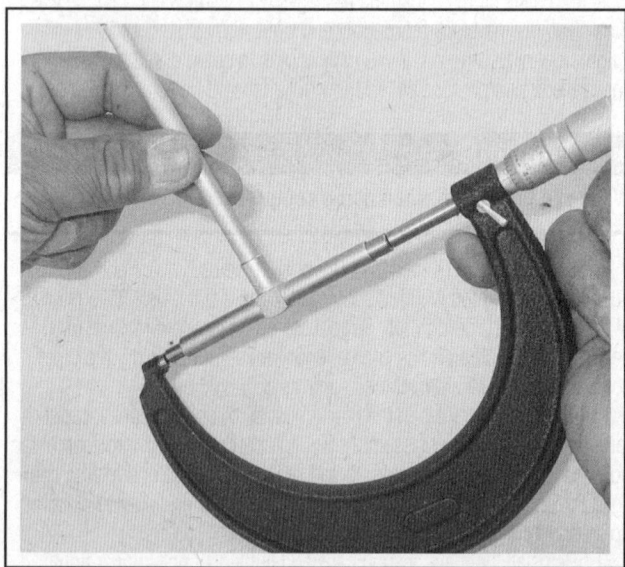

16.4c The gauge is then measured with a micrometer to determine the bore size

17 Cylinder honing

▶ Refer to illustrations 17.3a and 17.3b

1 Prior to engine reassembly, the cylinder bores must be honed so the new piston rings will seat correctly and provide the best possible combustion chamber seal.

➡ Note: If you don't have the tools or don't want to tackle the honing operation, most automotive machine shops will do it for a reasonable fee.

2 Before honing the cylinders, install the main bearing caps and tighten the bolts to the torque listed in this Chapter's Specifications.

GENERAL ENGINE OVERHAUL PROCEDURES   2D-19

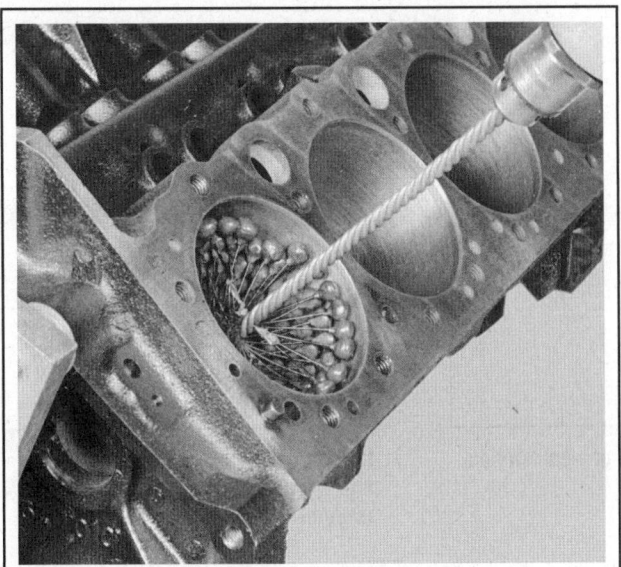

17.3a A "bottle brush" hone is the easiest type of hone to use

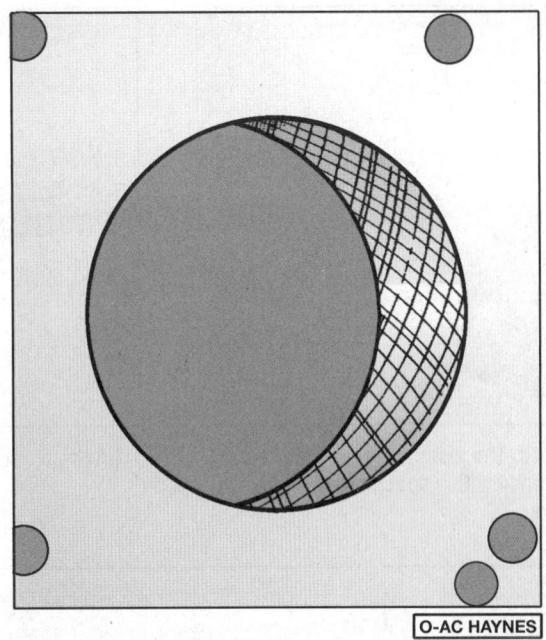

17.3b The cylinder hone should leave a smooth, crosshatch pattern with the lines intersecting at approximately a 60-degree angle

Make sure you use only the original main cap bolts, not the new ones for final assembly.

3   Two types of cylinder hones are commonly available - the flex hone or "bottle brush" type and the more traditional surfacing hone with spring-loaded stones. Both will do the job, but for the less experienced mechanic the "bottle brush" hone will probably be easier to use. You'll also need some kerosene or honing oil, rags and a 1/2-inch electric drill motor. Proceed as follows:

  a) Mount the hone in the drill motor, compress the stones and slip it into the first cylinder (see illustration). Be sure to wear safety goggles or a face shield!
  b) Lubricate the cylinder with plenty of honing oil, turn on the drill and move the hone up-and-down in the cylinder at a pace that will produce a fine crosshatch pattern on the cylinder walls. Ideally, the crosshatch lines should intersect at approximately a 60-degree angle (see illustration). Be sure to use plenty of lubricant and don't take off any more material than is absolutely necessary to produce the desired finish.

➡ Note: Piston ring manufacturers may specify a smaller crosshatch angle than the traditional 60-degrees - read and follow any instructions included with the new rings.

  c) Don't withdraw the hone from the cylinder while it's running. Instead, shut off the drill and continue moving the hone up-and-down in the cylinder until it comes to a complete stop, then compress the stones and withdraw the hone. If you're using a "bottle brush" type hone, stop the drill motor, then turn the chuck in the normal direction of rotation while withdrawing the hone from the cylinder.
  d) Wipe the oil out of the cylinder and repeat the procedure for the remaining cylinders.

4   After the honing job is complete, chamfer the top edges of the cylinder bores with a small file so the rings won't catch when the pistons are installed. Be very careful not to nick the cylinder walls with the end of the file.

5   The entire engine block must be washed again very thoroughly with warm, soapy water to remove all traces of the abrasive grit produced during the honing operation.

➡ **Note: The bores can be considered clean when a lint-free white cloth - dampened with clean engine oil - used to wipe them out doesn't pick-up any more honing residue, which will show up as gray areas on the cloth. Be sure to run a brush through all oil holes and galleries and flush them with running water.**

6   After rinsing, dry the block and apply a coat of light rust preventive oil to all machined surfaces. Wrap the block in a plastic trash bag to keep it clean and set it aside until reassembly.

18 Pistons/connecting rods - inspection

▶ Refer to illustrations 18.4a, 18.4b, 18.10 and 18.11

1   Before the inspection process can be carried out, the piston/connecting rod assemblies must be cleaned and the original piston rings removed from the pistons.

➡ **Note: Always use new piston rings when the engine is reassembled.**

2   Using a piston ring installation tool, carefully remove the rings from the pistons (see illustration 22.11). Be careful not to nick or gouge the pistons in the process.

3   Scrape all traces of carbon from the top of the piston. A hand-held wire brush or a piece of fine emery cloth can be used once the majority of the deposits have been scraped away. Do not, under any circumstances, use a wire brush mounted in a drill motor or caustic chemicals to remove deposits from the pistons. The piston material is soft and may be eroded away.

2D-20 GENERAL ENGINE OVERHAUL PROCEDURES

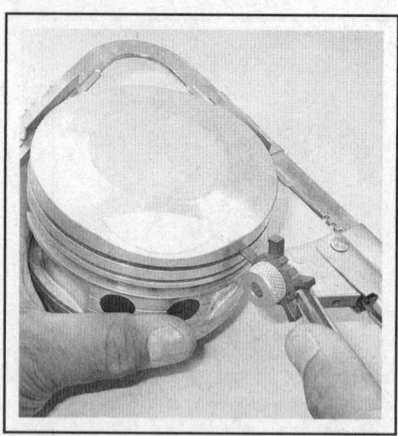

18.4a The piston ring grooves can be cleaned with a special tool, as shown here . . .

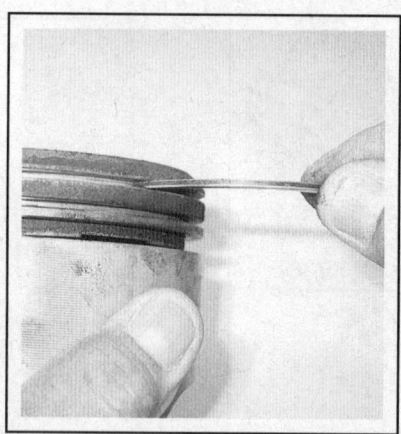

18.4b . . . or a section of a broken ring

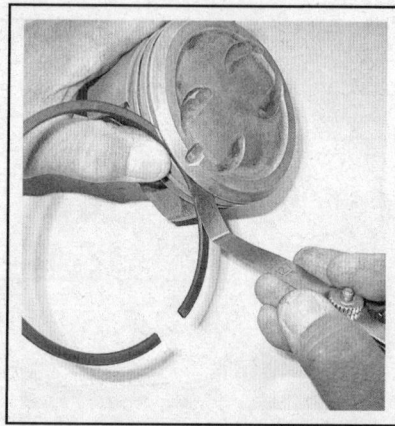

18.10 Check the ring side clearance with a feeler gauge at several points around the groove

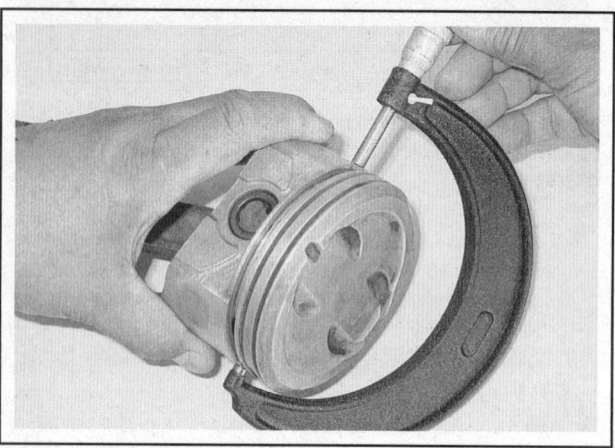

18.11 Measure the piston diameter at a 90-degree angle to the piston pin and in line with it

4 Use a piston ring groove cleaning tool to remove carbon deposits from the ring grooves. If a tool isn't available, a piece broken off the old ring will do the job (see illustrations). Be very careful to remove only the carbon deposits - don't remove any metal and do not nick or scratch the sides of the ring grooves.

5 Once the deposits have been removed, clean the piston/rod assemblies with solvent and dry them with compressed air (if available). Make sure the oil return holes in the back sides of the ring grooves are clear.

6 If the pistons and cylinder walls aren't damaged or worn excessively, and if the engine block is not rebored, new pistons won't be necessary. Normal piston wear appears as even vertical wear on the piston thrust surfaces and slight looseness of the top ring in its groove. New piston rings, however, should always be used when an engine is rebuilt.

7 Carefully inspect each piston for cracks around the skirt, at the pin bosses and at the ring lands.

8 Look for scoring and scuffing on the thrust faces of the skirt holes in the piston crown and burned areas at the edge of the crown. If the skirt is scored or scuffed, the engine may have been suffering from overheating and/or abnormal combustion, which caused excessively high operating temperatures. The cooling and lubrication systems should be checked thoroughly. A hole in the piston crown is an indication that abnormal combustion (pre-ignition) was occurring. Burned areas at the edge of the piston crown are usually evidence of spark knock (detonation). If any of the above problems exist, the causes must be corrected or the damage will occur again. The causes may include intake air leaks, incorrect fuel/air mixture, incorrect ignition timing and EGR system malfunctions.

9 Corrosion of the piston, in the form of small pits, indicates that coolant is leaking into the combustion chamber and/or the crankcase. Again, the cause must be corrected or the problem may persist in the rebuilt engine.

10 Measure the piston ring side clearance by laying a new piston ring in each ring groove and slipping a feeler gauge in beside it (see illustration). Check the clearance at three or four locations around each groove. Be sure to use the correct ring for each groove - they are different. If the side clearance is greater than the figure listed in this Chapter's Specifications, new pistons will have to be used.

11 Check the piston-to-bore clearance by measuring the bore (see Section 16) and the piston diameter. Make sure the pistons and bores are correctly matched. Measure the piston across the skirt, at a 90-degree angle to, and in line with, the piston pin (see illustration). Subtract the piston diameter from the bore diameter to obtain the clearance. If it's greater than specified, the block will have to be rebored and new pistons and rings installed.

12 Check the piston-to-rod clearance by twisting the piston and rod in opposite directions. Any noticeable play indicates excessive wear, which must be corrected. The piston/connecting rod assemblies should be taken to an automotive machine shop to have the pistons and rods resized and new pins installed.

13 If the pistons must be removed from the connecting rods for any reason, they should be taken to an automotive machine shop. While they are there have the connecting rods checked for bend and twist, since automotive machine shops have special equipment for this purpose.

➡ **Note: Unless new pistons and/or connecting rods must be installed, do not disassemble the pistons and connecting rods.**

14 Check the connecting rods for cracks and other damage. Temporarily remove the rod caps, lift out the old bearing inserts, wipe the rod and cap bearing surfaces clean and inspect them for nicks, gouges and scratches. After checking the rods, replace the old bearings, slip the caps into place and tighten the nuts finger tight.

➡ **Note: If the engine is being rebuilt because of a connecting rod knock, be sure to install new rods.**

GENERAL ENGINE OVERHAUL PROCEDURES 2D-21

19 Crankshaft - inspection

▶ Refer to illustrations 19.1, 19.2, 19.5 and 19.7

1 Remove all burrs from the crankshaft oil holes with a stone, file or scraper (see illustration).

2 Clean the crankshaft with solvent and dry it with compressed air (if available). Be sure to clean the oil holes with a stiff brush (see illustration) and flush them with solvent.

3 Check the main and connecting rod bearing journals for uneven wear, scoring, pits and cracks.

4 Check the rest of the crankshaft for cracks and other damage. It should be magnafluxed to reveal hidden cracks - an automotive machine shop will handle the procedure.

5 Using a micrometer, measure the diameter of the main and connecting rod journals and compare the results to this Chapter's Specifications (see illustration). By measuring the diameter at a number of points around each journal's circumference, you'll be able to determine whether or not the journal is out-of-round. Take the measurement at each end of the journal, near the crank throws, to determine if the journal is tapered.

6 If the crankshaft journals are damaged, tapered, out-of-round or worn beyond the limits given in the Specifications, have the crankshaft reground by an automotive machine shop. Be sure to use the correct size bearing inserts if the crankshaft is reconditioned.

7 Check the oil seal journals at each end of the crankshaft for wear and damage. If the seal has worn a groove in the journal, or if it's nicked or scratched (see illustration), the new seal may leak when the engine is reassembled. In some cases, an automotive machine shop may be able to repair the journal by pressing on a thin sleeve. If repair isn't feasible, a new or different crankshaft should be installed.

8 Refer to Section 20 and examine the main and rod bearing inserts.

19.1 The oil holes should be chamfered so sharp edges don't gouge or scratch the new bearings

19.2 Use a wire or stiff plastic bristle brush to clean the oil passages in the crankshaft

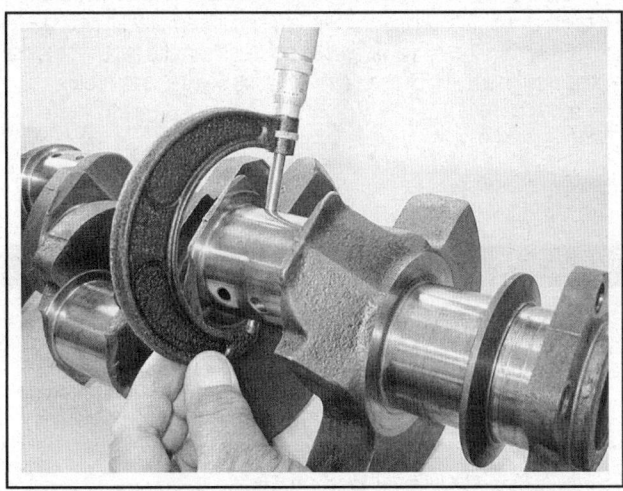

19.5 Measure the diameter of each crankshaft journal at several points to detect taper and out-of-round conditions

19.7 If the seals have worn grooves in the crankshaft journals, or if the seal contact surfaces are nicked or scratched, the new seals will leak

2D-22 GENERAL ENGINE OVERHAUL PROCEDURES

20 Main and connecting rod bearings - inspection

▸ **Refer to illustration 20.1**

1 Even though the main and connecting rod bearings should be replaced with new ones during the engine overhaul, the old bearings should be retained for close examination, as they may reveal valuable information about the condition of the engine (see illustration).

2 Bearing failure occurs because of lack of lubrication, the presence of dirt or other foreign particles, overloading the engine and corrosion. Regardless of the cause of bearing failure, it must be corrected before the engine is reassembled to prevent it from happening again.

3 When examining the bearings, remove them from the engine block, the main bearing caps, the connecting rods and the rod caps and lay them out on a clean surface in the same general position as their location in the engine. This will enable you to match any bearing problems with the corresponding crankshaft journal.

4 Dirt and other foreign particles get into the engine in a variety of ways. It may be left in the engine during assembly, or it may pass through filters or the PCV system. It may get into the oil, and from there into the bearings. Metal chips from machining operations and normal engine wear are often present. Abrasives are sometimes left in engine components after reconditioning, especially when parts are not thoroughly cleaned using the proper cleaning methods. Whatever the source, these foreign objects often end up embedded in the soft bearing material and are easily recognized. Large particles will not embed in the bearing and will score or gouge the bearing and journal. The best prevention for this cause of bearing failure is to clean all parts thoroughly and keep everything spotlessly clean during engine assembly. Frequent and regular engine oil and filter changes are also recommended.

5 Lack of lubrication (or lubrication breakdown) has a number of interrelated causes. Excessive heat (which thins the oil), overloading (which squeezes the oil from the bearing face) and oil leakage or throw off (from excessive bearing clearances, worn oil pump or high engine speeds) all contribute to lubrication breakdown. Blocked oil passages, which usually are the result of misaligned oil holes in a bearing shell, will also oil starve a bearing and destroy it. When lack of lubrication is the cause of bearing failure, the bearing material is wiped or extruded from the steel backing of the bearing. Temperatures may increase to the point where the steel backing turns blue from overheating.

6 Driving habits can have a definite effect on bearing life. Low speed operation in too high a gear (lugging the engine) puts very high loads on bearings, which tends to squeeze out the oil film. These loads cause the

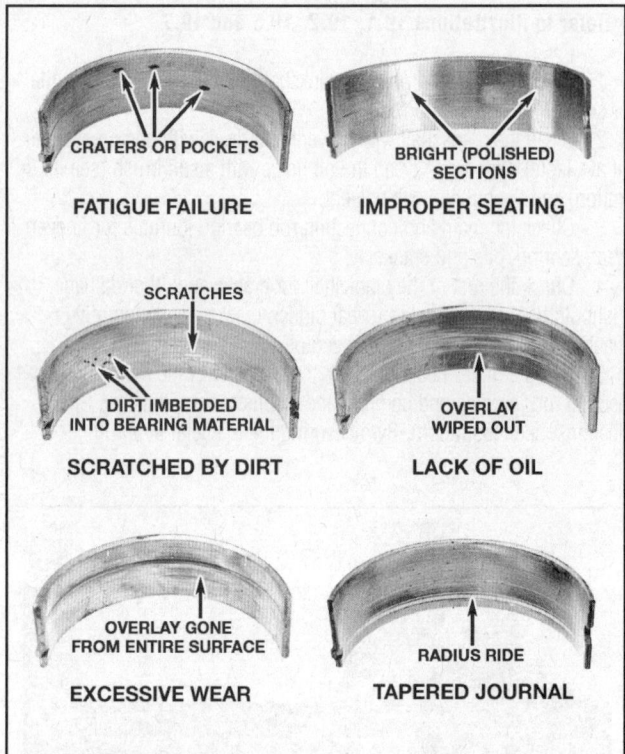

20.1 Typical bearing failures

bearings to flex, which produces fine cracks in the bearing face (fatigue failure). Eventually the bearing material will loosen in pieces and tear away from the steel backing. Short trip driving leads to corrosion of bearings because insufficient engine heat is produced to drive off the condensed water and corrosive gases. These products collect in the engine oil, forming acid and sludge. As the oil is carried to the engine bearings, the acid attacks and corrodes the bearing material.

7 Incorrect bearing installation during engine assembly will lead to bearing failure as well. Tight fitting bearings leave insufficient bearing oil clearance and will result in oil starvation. Dirt or foreign particles trapped behind a bearing insert result in high spots on the bearing which lead to failure.

21 Engine overhaul - reassembly sequence

1 Before beginning engine reassembly, make sure you have all the necessary new parts (including new head bolts, rod bolts and main cap bolts on V8 engines), gaskets and seals as well as the following items on hand:

Common hand tools
A 1/2-inch drive torque wrench
A 3/8-inch drive torque wrench (inch-lb. measurement)
Piston ring installation tool
Piston ring compressor
Vibration damper installation tool
Connecting rod guide bolts
Plastigage
Feeler gauges
A fine-tooth file
New engine oil
Engine assembly lube or moly-base grease
Gasket sealant
Thread-locking compound

2 In order to save time and avoid problems, engine reassembly must be done in the following general order:

GENERAL ENGINE OVERHAUL PROCEDURES 2D-23

V6 ENGINES

New camshaft bearings (recommended to be done by an automotive machine shop)
Piston rings
Crankshaft and main bearings
Piston/connecting rod assemblies
Oil pump
Oil pan
Camshaft
Valve lifters
Timing chain and sprockets
Timing chain cover
Cylinder heads
Rocker arms and pushrods
Intake and exhaust manifolds
Valve covers
Rear main seal
Driveplate

V8 ENGINES

Piston rings
Crankshaft and main bearings
Piston/connecting rod assemblies
Oil pump
Rear main seal/retainer plate
Oil pan
Cylinder heads
Valve lifters
Rocker arms
Camshaft(s)
Camshaft cap assemblies
Timing chains and sprockets
Timing chain guides and tensioners
Timing chain cover
Intake and exhaust manifolds
Valve covers
Driveplate

22 Piston rings - installation

♦ **Refer to illustrations 22.3, 22.4, 22.8a, 22.8b and 22.11**

1 Before installing the new piston rings, the ring end gaps must be checked. It's assumed that the piston ring side clearance has been checked and verified correct (see Section 18).

2 Lay out the piston/connecting rod assemblies and the new ring sets so the ring sets will be matched with the same piston and cylinder during the end gap measurement and engine assembly.

3 Insert the top (number one) ring into the first cylinder and square it up with the cylinder walls by pushing it in with the top of the piston (see illustration). The ring should be near the bottom of the cylinder, at the lower limit of ring travel.

4 To measure the end gap, slip feeler gauges between the ends of the ring until a gauge equal to the gap width is found (see illustra-

tion). The feeler gauge should slide between the ring ends with a slight amount of drag. Compare the measurement to this Chapter's Specifications. If the gap is larger or smaller than specified, double-check to make sure you have the correct rings before proceeding. If there is any doubt contact the parts store where the rings were purchased, to verify that the correct ring set is being used.

5 Excess end gap isn't critical unless it's greater than 0.040-inch. Again, double-check to make sure you have the correct rings for your engine.

6 Repeat the procedure for each ring that will be installed in the first cylinder and for each ring in the remaining cylinders. Remember to keep rings, pistons and cylinders matched up.

7 Once the ring end gaps have been checked/corrected, the rings can be installed on the pistons.

22.3 When checking piston ring end gap, the ring must be square in the cylinder bore (this is done by pushing the ring down with the top of a piston as shown)

22.4 With the ring square in the cylinder, measure the end gap with a feeler gauge

2D-24 GENERAL ENGINE OVERHAUL PROCEDURES

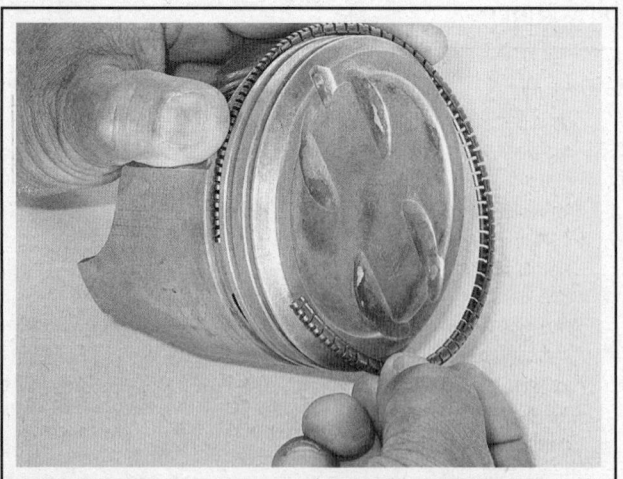

22.8a Installing the spacer/expander in the oil control ring groove . . .

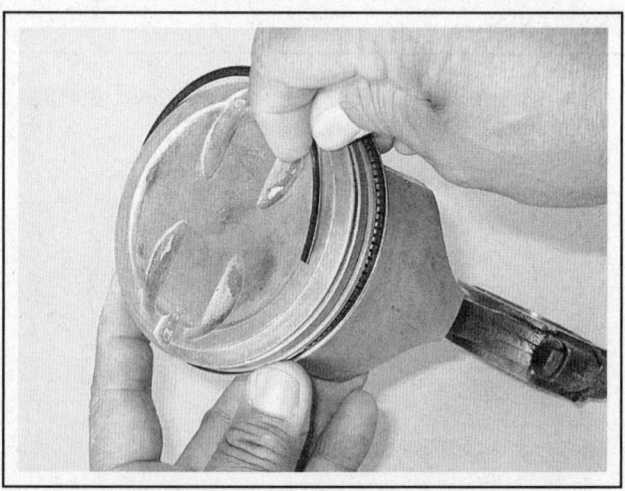

22.8b . . . followed by the side rails - DO NOT use a piston ring installation tool when installing the oil ring side rails

8 The oil control ring (lowest one on the piston) is usually installed first. It's composed of three separate components. Slip the spacer/expander into the groove (see illustration). Next, install the lower side rail. Don't use a piston ring installation tool on the oil ring side rails, as they may be damaged. Instead, place one end of the side rail into the groove between the spacer/expander and the ring land, hold it firmly in place and slide a finger around the piston while pushing the rail into the groove (see illustration). Next, install the upper side rail in the same manner.

9 After the three oil ring components have been installed, check to make sure that both the upper and lower side rails can be turned smoothly in the ring groove.

10 The number two (middle) ring is installed next. It's usually stamped with a mark which must face up, toward the top of the piston.

➡ **Note: Always follow the instructions printed on the ring package or box - different manufacturers may require different approaches. Do not mix up the top and middle rings, as they have different cross sections.**

11 Use a piston ring installation tool and make sure the identification mark is facing the top of the piston, then slip the ring into the middle groove on the piston (see illustration). Don't expand the ring any more than necessary to slide it over the piston.

12 Install the number one (top) ring in the same manner. Make sure the mark is facing up. Be careful not to confuse the number one and number two rings.

13 Repeat the procedure for the remaining pistons and rings.

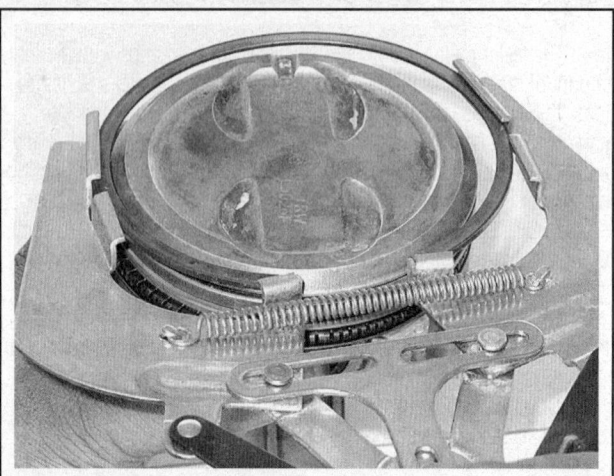

22.11 Installing the compressor rings with a ring expander - the mark (arrow) must face up

23 Crankshaft - installation and main bearing oil clearance check

⚠ CAUTION:

The main bearing cap bolts (and side bolts) on the V8 engines are all "torque-to-yield" bolts and are NOT reusable. A pre-determined stretch of the bolt, calculated by the manufacturer, gives the added rigidity required with this cylinder block. Once removed they must be replaced with new bolts. The main cap side bolts and jack screws are reusable. During clearance checks using Plastigage, use the old bolts and torque to Specifications, but use only new bolts for final assembly.

1 Crankshaft installation is the first step in engine reassembly. It's assumed at this point that the engine block and crankshaft have been cleaned, inspected and repaired or reconditioned.

2 Position the engine with the bottom facing up.

3 Remove the main bearing cap bolts/studs and lift out the caps. Lay them out in the proper order to ensure correct installation.

4 If they're still in place, remove the original bearing inserts from the block and the main bearing caps. Wipe the bearing surfaces of the block and caps with a clean, lint-free cloth. They must be kept spotlessly clean.

MAIN BEARING OIL CLEARANCE CHECK

▶ **Refer to illustrations 23.11, 23.12 and 23.15**

5 Clean the back sides of the new main bearing inserts and lay one in each main bearing saddle in the block. If one of the bearing inserts

GENERAL ENGINE OVERHAUL PROCEDURES 2D-25

23.11 Lay the Plastigage strips on the main bearing journals, parallel to the crankshaft centerline

23.12 On V8 engines tap the main caps down with a brass or plastic hammer before installing any top or side bolts - the cap must be square to the block before tapping it down

from each set has a large groove in it, make sure the grooved insert is installed in the block. Lay the other bearing from each set in the corresponding main bearing cap. Make sure the tab on the bearing insert fits into the recess in the block or cap.

> ※※ **CAUTION:**
>
> **The oil holes in the block must line up with the oil holes in the bearing insert. Do not hammer the bearing into place and don't nick or gouge the bearing faces. No lubrication should be used at this time.**

6 The flanged thrust bearing must be installed in the third cap and saddle on V6 engines, or the fifth cap and saddle on V8 engines.

7 Clean the faces of the bearings in the block and the crankshaft main bearing journals with a clean, lint-free cloth.

8 Check or clean the oil holes in the crankshaft, as any dirt here can go only one way - straight through the new bearings.

9 Once you're certain the crankshaft is clean, carefully lay it in position in the main bearings.

23.15 Compare the width of the crushed Plastigage to the scale on the envelope to determine the main bearing oil clearance (always take the measurement at the widest point of the Plastigage); be sure to use the correct scale - standard and metric ones are included

10 Before the crankshaft can be permanently installed, the main bearing oil clearance must be checked.

11 Cut several pieces of the appropriate-size Plastigage (they must be slightly shorter than the width of the main bearings) and place one piece on each crankshaft main bearing journal, parallel with the journal axis (see illustration).

12 Clean the faces of the bearings in the caps and install the caps in their respective positions (don't mix them up) with the arrows pointing toward the front of the engine (see Section 14). Don't disturb the Plastigage.

➡ **Note: On all V8 engines, the caps should be seated with a brass hammer or a dead-blow plastic mallet before installing any main cap bolts (see illustration). On models with jacking screws, make sure the screws are well into the caps and clear of the block.**

13 Starting with the center main and working out toward the ends, tighten the main bearing cap bolts/studs, in three steps, to the torque listed in this Chapter's Specifications.

➡ **Note 1: On V8 engines it is not necessary to tighten the jack screws or install the side bolts for Plastigage measurement purposes. Don't rotate the crankshaft at any time during this operation.**

➡ **Note 2: 3.5L V6 engines are equipped with side bolts and a main cap support brace, but it isn't necessary to install them for the oil clearance check.**

14 Remove the bolts/studs and carefully lift off the main bearing caps. Keep them in order. Don't disturb the Plastigage or rotate the crankshaft. If any of the main bearing caps are difficult to remove, tap them gently from side-to-side with a soft-face hammer to loosen them.

15 Compare the width of the crushed Plastigage on each journal to the scale printed on the Plastigage envelope to obtain the main bearing oil clearance (see illustration). Check the Specifications to make sure it's correct.

16 If the clearance is not as specified, the bearing inserts may be the wrong size (which means different ones will be required). Before deciding that different inserts are needed, make sure that no dirt or oil was between the bearing inserts and the caps or block when the clearance was measured. If the Plastigage was wider at one end than the other, the journal may be tapered (refer to Section 19).

17 Carefully scrape all traces of the Plastigage material off the main bearing journals and/or the bearing faces. Use your fingernail or the edge of a credit card - don't nick or scratch the bearing faces.

ENGINE BEARING ANALYSIS

Debris

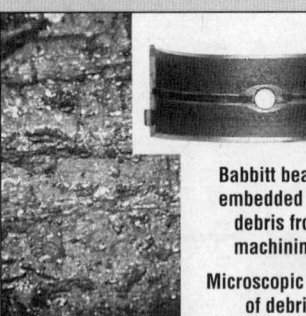

Babbitt bearing embedded with debris from machinings

Microscopic detail of debris

Microscopic detail of gouges

Overplated copper alloy bearing gouged by cast iron debris

Aluminum bearing embedded with glass beads

Microscopic detail of glass beads

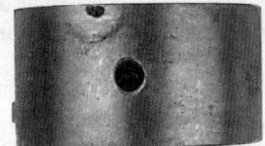

Damaged lining caused by dirt left on the bearing back

Misassembly

Result of a lower half assembled as an upper - blocking the oil flow

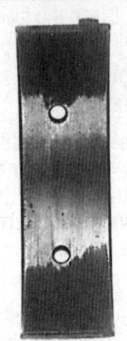

Excessive oil clearance is indicated by a short contact arc

Polished and oil-stained backs are a result of a poor fit in the housing bore

Result of a wrong, reversed, or shifted cap

Overloading

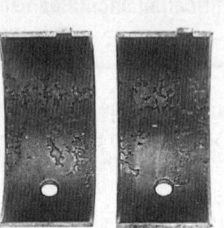

Damage from excessive idling which resulted in an oil film unable to support the load imposed

Damaged upper connecting rod bearings caused by engine lugging; the lower main bearings (not shown) were similarly affected

The damage shown in these upper and lower connecting rod bearings was caused by engine operation at a higher-than-rated speed under load

Misalignment

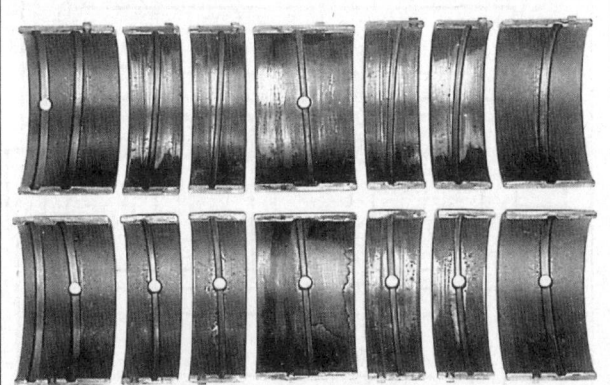

A warped crankshaft caused this pattern of severe wear in the center, diminishing toward the ends

A poorly finished crankshaft caused the equally spaced scoring shown

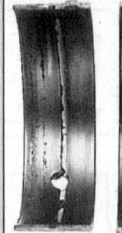

A tapered housing bore caused the damage along one edge of this pair

A bent connecting rod led to the damage in the "V" pattern

Lubrication

Result of dry start: The bearings on the left, farthest from the oil pump, show more damage

Corrosion

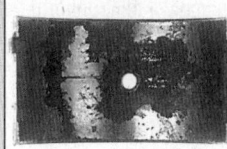

Corrosion is an acid attack on the bearing lining generally caused by inadequate maintenance, extremely hot or cold operation, or inferior oils or fuels

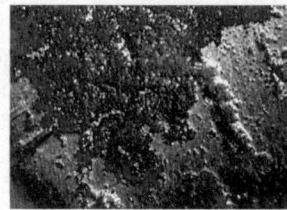

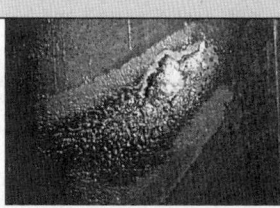

Example of cavitation - a surface erosion caused by pressure changes in the oil film

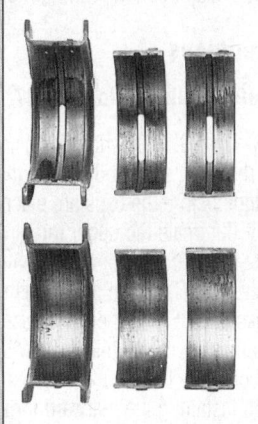

Result of a low oil supply or oil starvation

Severe wear as a result of inadequate oil clearance

Damage from excessive thrust or insufficient axial clearance

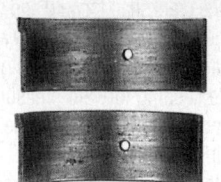

Bearing affected by oil dilution caused by excessive blow-by or a rich mixture

© 1986 Federal-Mogul Corporation
Copy and photographs courtesy of Federal Mogul Corporation

2D-28 GENERAL ENGINE OVERHAUL PROCEDURES

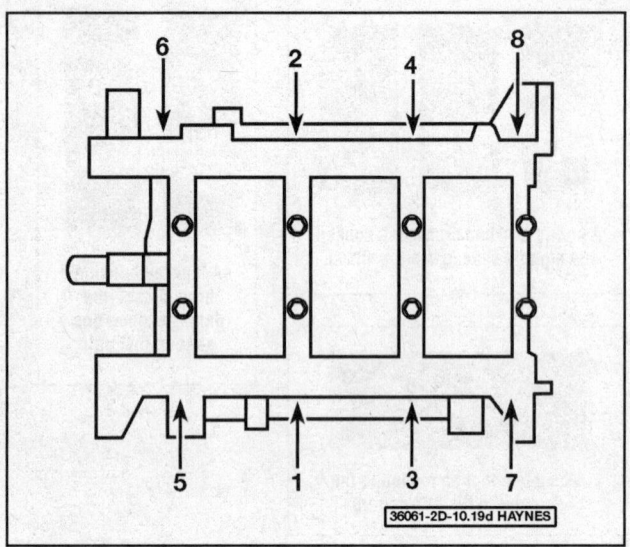

23.30a Main cap bolt tightening sequence - 3.5L V6 engines

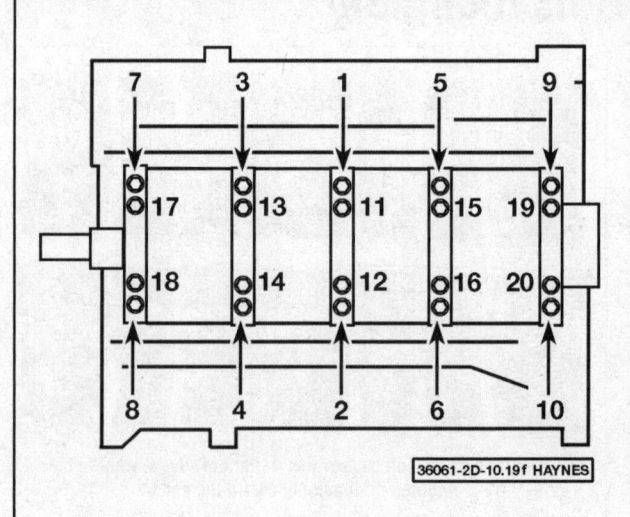

23.30b Side bolt tightening sequence - 3.5L V6 engines

FINAL CRANKSHAFT INSTALLATION

18 Carefully lift the crankshaft out of the engine.
19 Clean the bearing faces in the block, then apply a thin, uniform layer of moly-base grease or engine assembly lube to each of the bearing surfaces. Be sure to coat the thrust faces as well as the journal face of the thrust bearing.
20 Make sure the crankshaft journals are clean, then lay the crankshaft back in place in the block.
21 Clean the faces of the bearings in the caps, then apply lubricant to them.
22 Install the caps in their respective positions with the arrows pointing toward the front of the engine.

➡ Note: *On V6 engines, apply a thin bead of RTV sealant at the side joints between the rear main cap and the block. Install the cap and tighten the bolts (see Steps 23 and 24) within four minutes of applying the sealant.*

V6 engines

4.2L engines

23 Install the main cap bolts.
24 On 4.2L V6 engines, tighten all, except the thrust bearing cap bolts (number 3), to the torque listed in this Chapter's Specifications (work from the center out and approach the final torque in three steps). Tighten the thrust bearing cap bolts finger tight.
25 Pry the crankshaft forward and while holding pressure on the crankshaft, pry the thrust bearing cap backward. Forcing these two in opposite directions, against each other, will align the thrust bearing surfaces.
26 While keeping forward pressure on the crankshaft, re-tighten ALL main bearing cap bolts to the torque listed in this Chapter's Specifications, starting with cap #3, then #2, #4 and #1.
27 Rotate the crankshaft a number of times by hand to check for any obvious binding.
28 The final step is to check the crankshaft endplay with a feeler gauge or a dial indicator (see Section 13) The endplay should be correct if the crankshaft thrust faces aren't worn or damaged and new bearings have been installed.
29 Install the rear main oil seal (see Chapter 2, Part A).

3.5L engines

◆ Refer to illustration 23.30a, 23.30b and 23.34

30 Apply clean engine oil to all bolt threads prior to installation, then install all bolts finger-tight. Tighten the bearing cap assembly bolts in the sequence shown (see illustrations) progressing in steps, to the torque listed in this Chapter's Specifications.
31 Apply a bead of RTV sealant to the engine block on the rear main bearing cap parting line. Be sure the main bearing cap is installed within four minutes after the RTV sealant is applied.
32 Recheck the crankshaft endplay with a feeler gauge or a dial indicator. The endplay should be correct if the crankshaft thrust faces aren't worn or damaged and if new bearings have been installed.
33 Rotate the crankshaft a number of times by hand to check for any obvious binding. It should rotate with a running torque of 50 in-lbs or less. If the running torque is too high, correct the problem at this time.
34 Install the main cap support brace and tighten the bolts in sequence (see illustration) to the torque listed in this Chapter's Specifications.
35 Install the new rear main oil seal (see Chapter 2B).

V8 engines

◆ Refer to illustrations 23.37, 23.39, 23.42, 23.43a and 23.43b

36 On 4.6L REP engines, install the jack screws (if removed earlier) into the main caps and bottom them lightly against the caps, this must be done before the caps are placed into the block.
37 Lubricate the upper thrust washer with moly-base grease and rotate it into the block while prying rearward on the crankshaft (see illustration). The side of the washer with the oil grooves must face the crankshaft. REP 4.6L engines use one thrust washer at the back of the last crank journal, while 4.6L WEP and 5.4L engines use two thrust washers, one behind the last journal and one ahead of the last journal. When installing the rearward thrust washer, the crankshaft should be pried rearward. After installing the rearward thrust washer (all V8s), pry the crankshaft forward on 4.6L WEP and 5.4L engines and install the lubricated front thrust washer by rotating it into place.
38 Place the main caps on their correct journals and tap the caps into place with a brass or soft-face hammer.

GENERAL ENGINE OVERHAUL PROCEDURES 2D-29

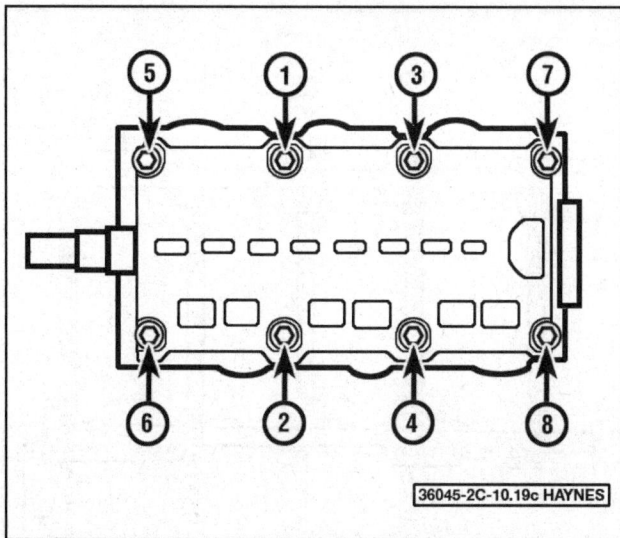

23.34 Main cap support brace bolt tightening sequence - 3.5L V6 engines

23.37 Roll the lubricated crankshaft thrust washer into place in front of the last crankshaft journal on V8 engines - the grooved side of the thrust washer must face the crankshaft (away from the main bearing saddle)

✲✲ CAUTION:

All main bearing caps MUST be tapped into position prior to tightening. Failure to do so may result in improper torque.

✲✲ CAUTION:

Once the crankshaft is pushed fully forward, to seat the thrust bearing, leave the screwdriver in position so that pressure stays placed on the crankshaft until after all main bearing cap bolts have been tightened.

39 Install the NEW main cap bolts and tighten them to 10 to 12 ft-lbs in the recommended sequence (see illustration). On 4.6L WEP engines and 5.4L engines, install the dowel pins in the outboard holes of the caps with their flat sides toward the center of the engine.

40 Push the crankshaft forward using a screwdriver or prybar to seat the thrust bearing.

41 Tighten the main bearing cap bolts in two steps in the sequence shown (see illustration 23.39) and to the torque and angle indicated in the Specifications listed at the beginning of this Chapter.

42 On REP 4.6L engines, tighten all jack screws against the block (left-hand threads) in two steps, in the sequence shown (see illustration), to the Specifications listed at the beginning of this Chapter.

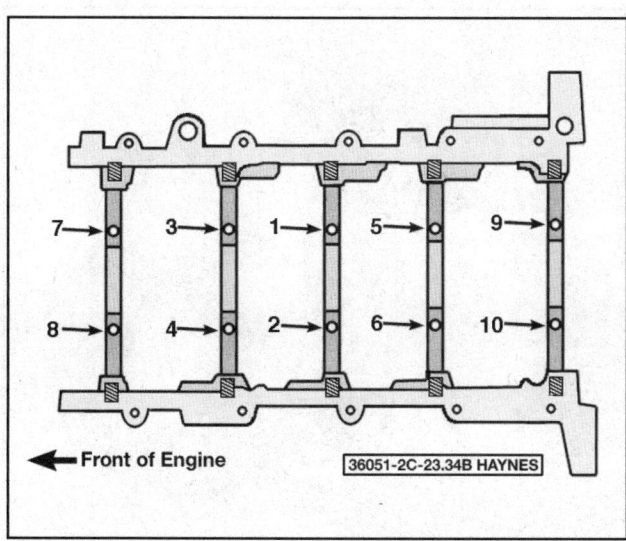

23.39 Main cap bolt tightening sequence - V8 engines

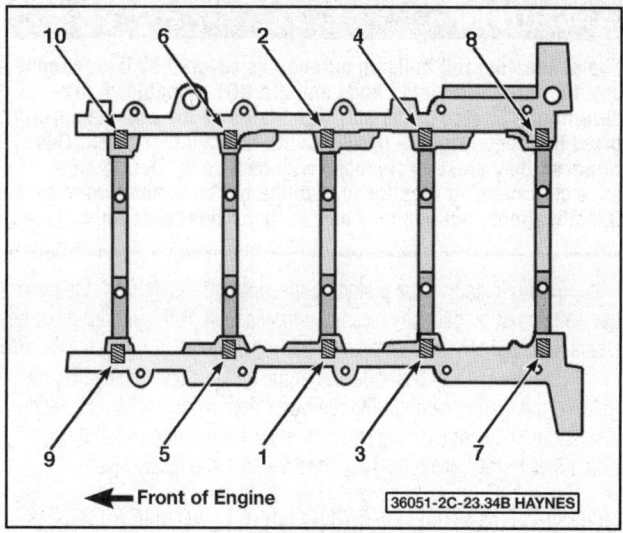

23.42 On REP 4.6L engines, tighten the ten jacking screws in this sequence

2D-30 GENERAL ENGINE OVERHAUL PROCEDURES

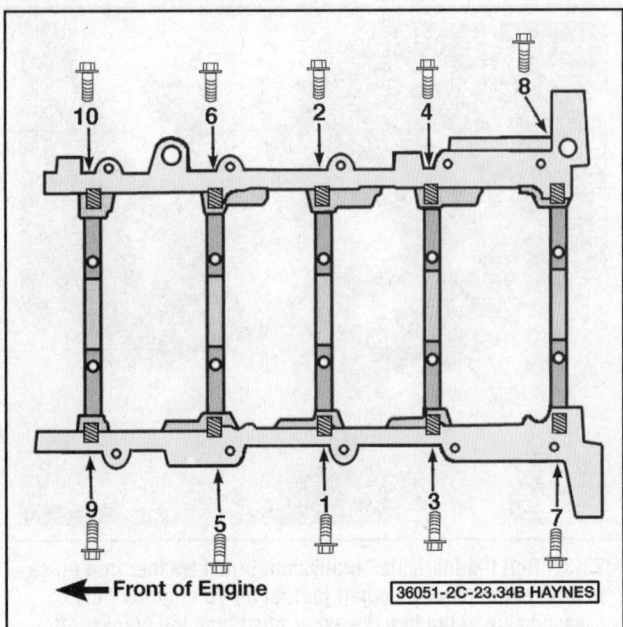

23.43a Tighten the side bolts in this sequence in two steps (2009 and earlier V8 engines)

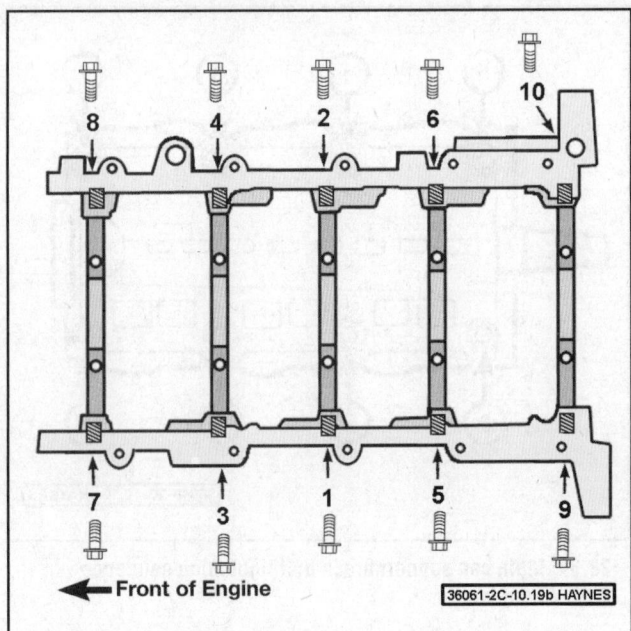

23.43b Main bearing cap side bolt tightening sequence (2010 and later models)

43 Tighten all side bolts in two steps, in the sequence shown (see illustrations), to the Specifications listed in this Chapter.

44 Check crankshaft endplay again and verify that it is correct (see Section 14).

45 Rotate the crankshaft a number of times by hand to check for any obvious binding.

46 Install the rear main oil seal (see Chapter 2, Part C).

24 Pistons/connecting rods - installation and rod bearing oil clearance check

※ CAUTION:

The connecting rod bolts on all engines covered by this manual are all "torque-to-yield" bolts and are NOT reusable. A pre-determined stretch of the bolt, calculated by the manufacturer, gives the added rigidity required with this cylinder block. Once removed they must be replaced with new bolts. During clearance checks using Plastigage, use the old bolts and torque to Specifications, but use only new bolts for final assembly.

1 Before installing the piston/connecting rod assemblies, the cylinder walls must be perfectly clean, the top edge of each cylinder must be chamfered, and the crankshaft must be in place.

2 Remove the cap from the end of the number one connecting rod (refer to the marks made during removal). Remove the original bearing inserts and wipe the bearing surfaces of the connecting rod and cap with a clean, lint-free cloth. They must be kept spotlessly clean.

CONNECTING ROD BEARING OIL CLEARANCE CHECK

▶ **Refer to illustrations 24.3, 24.5, 24.8, 24.9a, 24.9b, 24.11, 24.13 and 24.17**

3 Clean the back side of the new upper bearing insert, then lay it in

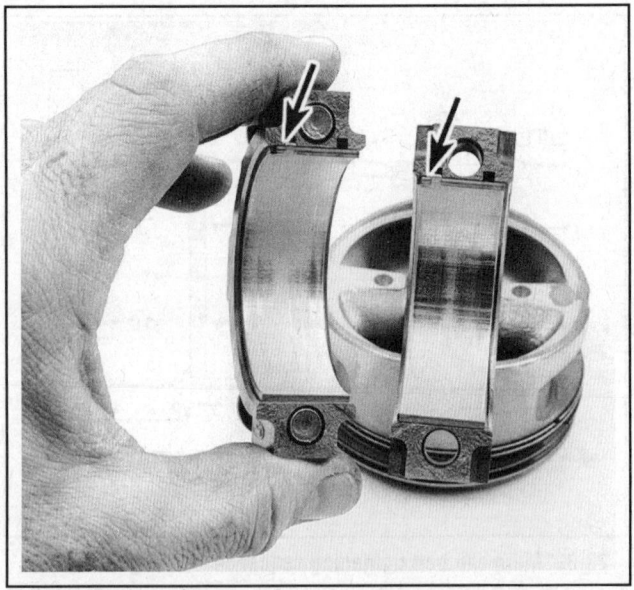

24.3 Insert the connecting rod bearing halves, making sure the bearing tabs are seated in the notches in the rod and cap

GENERAL ENGINE OVERHAUL PROCEDURES 2D-31

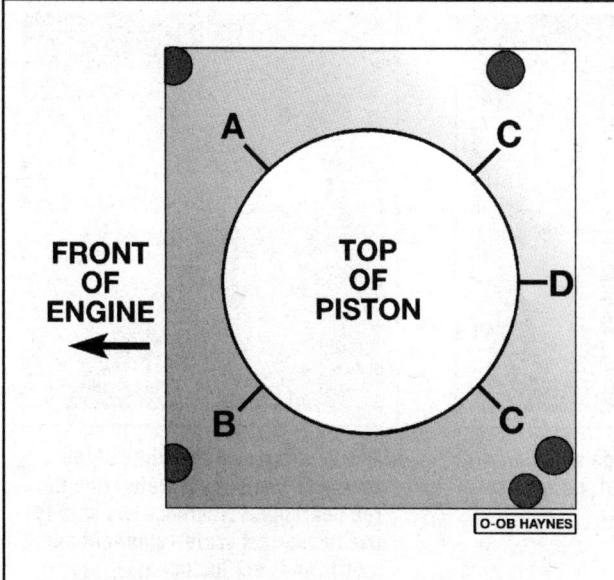

24.5 Ring end gap positions - Align the oil ring spacer gap at D, the oil ring side rails at C (one inch either side of the pin centerline), and the compression rings at A and B, one inch either side of the pin centerline

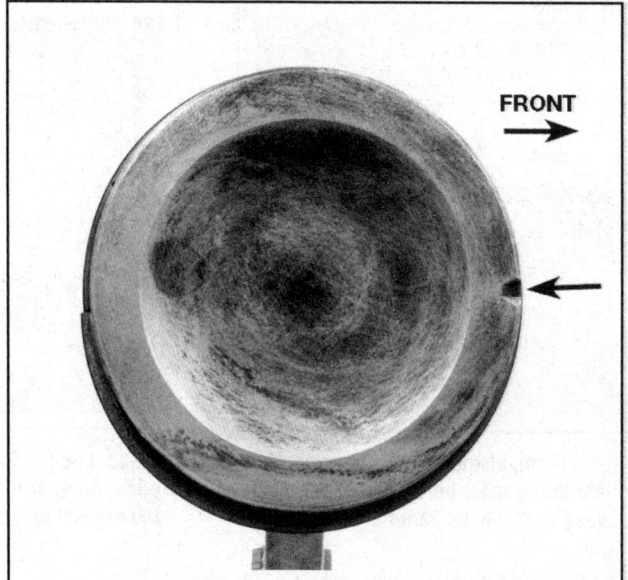

24.8 Turn the piston when installing it to make sure the mark/notch in the piston faces the front of the engine as they are installed

place in the connecting rod (see illustration). Make sure the tab on the bearing fits into the recess in the rod. Don't hammer the bearing insert into place and be very careful not to nick or gouge the bearing face. Don't lubricate the bearing at this time.

4   Clean the back side of the other bearing insert and install it in the rod cap. Again, make sure the tab on the bearing fits into the recess in the cap, and don't apply any lubricant. It's critically important that the mating surfaces of the bearing and connecting rod are perfectly clean and oil free when they're assembled.

5   Position the piston ring gaps at intervals around the piston (see illustration).

6   Lubricate the piston and rings with clean engine oil and attach a piston ring compressor to the piston. Leave the skirt protruding about 1/4-inch to guide the piston into the cylinder. The rings must be compressed until they're flush with the piston.

7   Rotate the crankshaft until the number one connecting rod journal is at BDC (bottom dead center) and apply a coat of engine oil to the cylinder walls.

8   With the arrow or notch on top of the piston (see illustration) facing the front of the engine, gently insert the piston/connecting rod assembly into the number one cylinder bore and rest the bottom edge of the ring compressor on the engine block.

9   The rod bolts are pressed lightly into the rod caps. They must be replaced with new bolts, so tap out two of the old ones (see illustration). Take these and cut the heads off, slip rubber hose over them, and screw them into the connecting rod to use as alignment devices during piston/rod installation (see illustration).

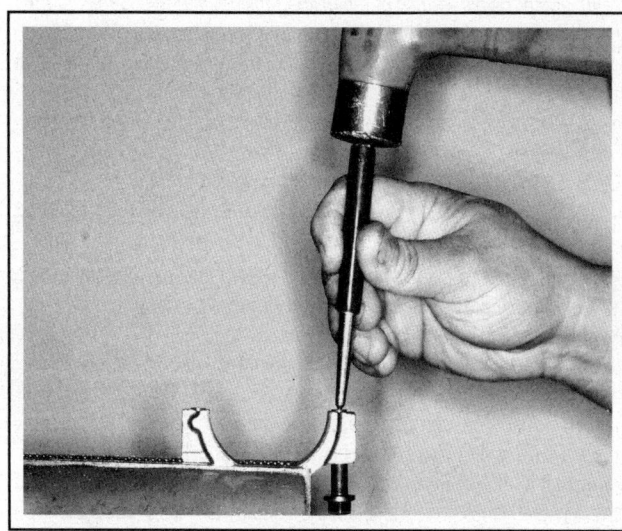

24.9a Hold the rod cap lightly in a vise while driving out the old rod bolts - it takes only a little tap to drive them out; then drive new bolts in

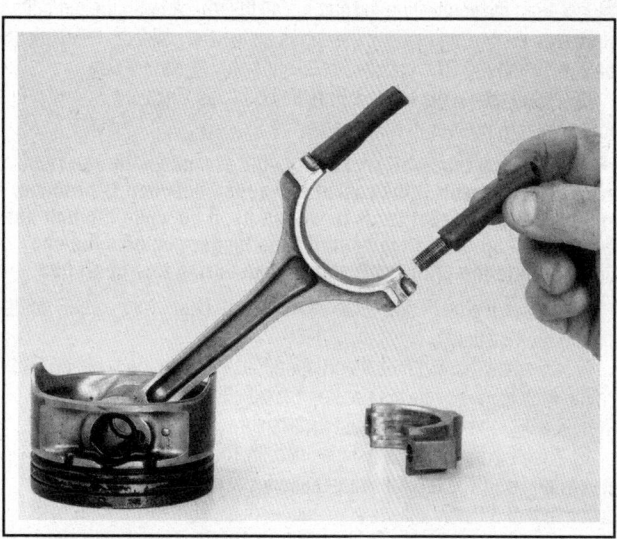

24.9b Two old rod bolts can be used (with the heads cut off) as rod installation guides with rubber hose over them

2D-32 GENERAL ENGINE OVERHAUL PROCEDURES

24.11 The piston can be driven gently into the cylinder bore with the end of a wooden or plastic hammer handle

24.13 Lay the Plastigage strips on each rod bearing journal, parallel to the crankshaft centerline

24.17 Measuring the width of the crushed Plastigage to determine the rod bearing oil clearance (be sure to use the correct scale - standard and metric ones are included)

10 Tap the top edge of the ring compressor to make sure it's contacting the block around its entire circumference.

11 Gently tap on the top of the piston with the end of a wooden hammer handle (see illustration) while guiding the end of the connecting rod into place on the crankshaft journal. The piston rings may try to pop out of the ring compressor just before entering the cylinder bore, so keep some downward pressure on the ring compressor. Work slowly, and if any resistance is felt as the piston enters the cylinder, stop immediately. Find out what's hanging up and fix it before proceeding. Do not, for any reason, force the piston into the cylinder - you might break a ring and/or the piston.

12 Once the piston/connecting rod assembly is installed, the connecting rod bearing oil clearance must be checked before the rod cap is permanently bolted in place.

13 Cut a piece of the appropriate-size Plastigage slightly shorter than the width of the connecting rod bearing and lay it in place on the number one connecting rod journal, parallel with the journal axis (see illustration).

14 Clean the connecting rod cap bearing face, remove the protective "studs" from the connecting rod and install the rod cap with the old bolts. Make sure the mating mark on the cap is on the same side as the mark on the connecting rod. The fractured-style caps only fit on one way.

15 Tighten the bolts to the torque listed in this Chapter's Specifications, working up to it in three steps.

➡ Note: Use a thin-wall socket to avoid erroneous torque readings that can result if the socket is wedged between the rod cap and bolt. If the socket tends to wedge itself between the bolt and the cap, lift up on it slightly until it no longer contacts the cap. Do not rotate the crankshaft at any time during this operation.

16 Loosen the bolts and detach the rod cap, being very careful not to disturb the Plastigage.

17 Compare the width of the crushed Plastigage to the scale printed on the Plastigage envelope to obtain the oil clearance (see illustration). Compare it to the Specifications to make sure the clearance is correct.

18 If the clearance is not as specified, the bearing inserts may be the wrong size (which means different ones will be required). Before deciding that different inserts are needed, make sure that no dirt or oil was between the bearing inserts and the connecting rod or cap when the clearance was measured. Also, recheck the journal diameter. If the Plastigage was wider at one end than the other, the journal may be tapered (refer to Section 19).

FINAL CONNECTING ROD INSTALLATION

19 Carefully scrape all traces of the Plastigage material off the rod journal and/or bearing face. Be very careful not to scratch the bearing - use your fingernail or the edge of a credit card.

20 Make sure the bearing faces are perfectly clean, then apply a uniform layer of clean moly-base grease or engine assembly lube to both of them. You'll have to push the piston into the cylinder to expose the face of the bearing insert in the connecting rod - be sure to slip the protective hoses over the rod bolts first.

21 Slide the connecting rod back into place on the journal, remove the protective "studs", install the rod cap (this time with the new bolts) and tighten the bolts to the torque listed in this Chapter's Specifications. Again, work up to the torque in three steps.

➡ Note: Install new rod bolts into the rod caps (see Step 9).

22 Repeat the entire procedure for the remaining pistons/connecting rods.
23 The important points to remember are:
   a) *Keep the back sides of the bearing inserts and the insides of the connecting rods and caps perfectly clean when assembling them.*
   b) *Make sure you have the correct piston/rod assembly for each cylinder.*
   c) *The notches or mark on the piston must face the FRONT of the engine.*
   d) *Lubricate the cylinder walls with clean oil.*
   e) *Lubricate the bearing faces when installing the rod caps after the oil clearance has been checked.*

24 After all the piston/connecting rod assemblies have been properly installed, rotate the crankshaft a number of times by hand to check for any obvious binding.

25 As a final step, the connecting rod endplay must be checked. Refer to Section 13 for this procedure.

26 Compare the measured endplay to the Specifications to make sure it's correct. If it was correct before disassembly and the original crankshaft and rods were reinstalled, it should still be right. If new rods or a new crankshaft were installed, the endplay may be inadequate. If so, the rods will have to be removed and taken to an automotive machine shop for resizing.

27 On 3.5L engines, install the main bearing cap support brace (see Section 14).

GENERAL ENGINE OVERHAUL PROCEDURES 2D-33

25 Initial start-up and break-in after overhaul

WARNING:

Have a fire extinguisher handy when starting the engine for the first time.

1 Once the engine has been installed in the vehicle, double-check the engine oil and coolant levels.

2 With the spark plugs out of the engine and the ignition system disabled (see Chap-ter 1), crank the engine until oil pressure registers on the gauge or the light goes out.

3 Install the spark plugs, hook up the plug wires and restore the ignition system functions.

4 Start the engine. It may take a few moments for the fuel system to build up pressure, but the engine should start without a great deal of effort.

➡ **Note: If backfiring occurs through the throttle body, recheck the valve timing and ignition timing.**

5 After the engine starts, it should be allowed to warm up to normal operating temperature. While the engine is warming up, make a thorough check for fuel, oil and coolant leaks.

6 Shut the engine off and recheck the engine oil and coolant levels.

7 Drive the vehicle to an area with no traffic, accelerate from 30 to 50 mph, then allow the vehicle to slow to 30 mph with the throttle closed. Repeat the procedure 10 or 12 times. This will load the piston rings and cause them to seat properly against the cylinder walls. Check again for oil and coolant leaks.

8 Drive the vehicle gently for the first 500 miles (no sustained high speeds) and keep a constant check on the oil level. It is not unusual for an engine to use oil during the break-in period.

9 At approximately 500 to 600 miles, change the oil and filter.

10 For the next few hundred miles, drive the vehicle normally. Do not pamper it or abuse it.

11 After 2000 miles, change the oil and filter again and consider the engine broken in.

GLOSSARY

B

Backlash - The amount of play between two parts. Usually refers to how much one gear can be moved back and forth without moving the gear with which it's meshed.

Bearing Caps - The caps held in place by nuts or bolts which, in turn, hold the bearing surface. This space is for lubricating oil to enter.

Bearing clearance - The amount of space left between shaft and bearing surface. This space is for lubricating oil to enter.

Bearing crush - The additional height which is purposely manufactured into each bearing half to ensure complete contact of the bearing back with the housing bore when the engine is assembled.

Bearing knock - The noise created by movement of a part in a loose or worn bearing.

Blueprinting - Dismantling an engine and reassembling it to EXACT specifications.

Bore - An engine cylinder, or any cylindrical hole; also used to describe the process of enlarging or accurately refinishing a hole with a cutting tool, as to bore an engine cylinder. The bore size is the diameter of the hole.

Boring - Renewing the cylinders by cutting them out to a specified size. A boring bar is used to make the cut.

Bottom end - A term which refers collectively to the engine block, crankshaft, main bearings and the big ends of the connecting rods.

Break-in - The period of operation between installation of new or rebuilt parts and time in which parts are worn to the correct fit. Driving at reduced and varying speed for a specified mileage to permit parts to wear to the correct fit.

Bushing - A one-piece sleeve placed in a bore to serve as a bearing surface for shaft, piston pin, etc. Usually replaceable.

C

Camshaft - The shaft in the engine, on which a series of lobes are located for operating the valve mechanisms. The camshaft is driven by gears or sprockets and a timing chain. Usually referred to simply as the cam.

Carbon - Hard, or soft, black deposits found in combustion chamber, on plugs, under rings, on and under valve heads.

Cast iron - An alloy of iron and more than two percent carbon, used for engine blocks and heads because it's relatively inexpensive and easy to mold into complex shapes.

Chamfer - To bevel across (or a bevel on) the sharp edge of an object.

Chase - To repair damaged threads with a tap or die.

Combustion chamber - The space between the piston and the cylinder head, with the piston at top dead center, in which air-fuel mixture is burned.

Compression ratio - The relationship between cylinder volume (clearance volume) when the piston is at top dead center and cylinder volume when the piston is at bottom dead center.

Connecting rod - The rod that connects the crank on the crankshaft with the piston. Sometimes called a con rod.

Connecting rod cap - The part of the connecting rod assembly that attaches the rod to the crankpin.

Core plug - Soft metal plug used to plug the casting holes for the coolant passages in the block.

Crankcase - The lower part of the engine in which the crankshaft rotates; includes the lower section of the cylinder block and the oil pan.

Crank kit - A reground or reconditioned crankshaft and new main and connecting rod bearings.

Crankpin - The part of a crankshaft to which a connecting rod is attached.

Crankshaft - The main rotating member, or shaft, running the length of the crankcase, with offset throws to which the connecting rods are attached; changes the reciprocating motion of the pistons into rotating motion.

Cylinder sleeve - A replaceable sleeve, or liner, pressed into the cylinder block to form the cylinder bore.

D

Deburring - Removing the burrs (rough edges or areas) from a bearing.

Deglazer - A tool, rotated by an electric motor, used to remove glaze from cylinder walls so a new set of rings will seat.

E

Endplay - The amount of lengthwise movement between two parts. As applied to a crankshaft, the distance that the crankshaft can move forward and back in the cylinder block.

F

Face - A machinist's term that refers to removing metal from the end of a shaft or the face of a larger part, such as a flywheel.

Fatigue - A breakdown of material through a large number of loading and unloading cycles. The first signs are cracks followed shortly by breaks.

Feeler gauge - A thin strip of hardened steel, ground to an exact thickness, used to check clearances between parts.

Free height - The unloaded length or height of a spring.

Freeplay - The looseness in a linkage, or an assembly of parts, between the initial application of force and actual movement. Usually perceived as slop or slight delay.

Freeze plug - See Core plug.

G

Gallery - A large passage in the block that forms a reservoir for engine oil pressure.

Glaze - The very smooth, glassy finish that develops on cylinder walls while an engine is in service.

H

Heli-Coil - A rethreading device used when threads are worn or damaged. The device is installed in a retapped hole to reduce the thread size to the original size.

I

Installed height - The spring's measured length or height, as installed on the cylinder head. Installed height is measured from the spring seat to the underside of the spring retainer.

J

Journal - The surface of a rotating shaft which turns in a bearing.

K

Keeper - The split lock that holds the valve spring retainer in position on the valve stem.

Key - A small piece of metal inserted into matching grooves machined into two parts fitted together - such as a gear pressed onto a shaft - which prevents slippage between the two parts.

Knock - The heavy metallic engine sound, produced in the combustion chamber as a result of abnormal combustion - usually detonation. Knock is usually caused by a loose or worn bearing. Also referred to as detonation, pinging and spark knock. Connecting rod or main bearing knocks are created by too much oil clearance or insufficient lubrication.

L

Lands - The portions of metal between the piston ring grooves.

Lapping the valves - Grinding a valve face and its seat together with lapping compound.

Lash - The amount of free motion in a gear train, between gears, or in a mechanical assembly, that occurs before movement can begin. Usually refers to the lash in a valve train.

Lifter - The part that rides against the cam to transfer motion to the rest of the valve train.

M

Machining - The process of using a machine to remove metal from a metal part.

Main bearings - The plain, or babbitt, bearings that support the crankshaft.

Main bearing caps - The cast iron caps, bolted to the bottom of the block, that support the main bearings.

O

O.D. - Outside diameter.

Oil gallery - A pipe or drilled passageway in the engine used to carry engine oil from one area to another.

Oil ring - The lower ring, or rings, of a piston; designed to prevent excessive amounts of oil from working up the cylinder walls and into the combustion chamber. Also called an oil-control ring.

Oil seal - A seal which keeps oil from leaking out of a compartment. Usually refers to a dynamic seal around a rotating shaft or other moving part.

O-ring - A type of sealing ring made of a special rubberlike material; in use, the O-ring is compressed into a groove to provide the sealing action.

Overhaul - To completely disassemble a unit, clean and inspect all parts, reassemble it with the original or new parts and make all adjustments necessary for proper operation.

P

Pilot bearing - A small bearing installed in the center of the flywheel (or the rear end of the crankshaft) to support the front end of the input shaft of the transmission.

Pip mark - A little dot or indentation which indicates the top side of a compression ring.

Piston - The cylindrical part, attached to the connecting rod, that moves up and down in the cylinder as the crankshaft rotates. When the fuel charge is fired, the piston transfers the force of the explosion to the connecting rod, then to the crankshaft.

Piston pin (or wrist pin) - The cylindrical and usually hollow steel pin that passes through the piston. The piston pin fastens the piston to the upper end of the connecting rod.

Piston ring - The split ring fitted to the groove in a piston. The ring contacts the sides of the ring groove and also rubs against the cylinder wall, thus sealing space between piston and wall. There are two types of rings: Compression rings seal the compression pressure in the combustion chamber; oil rings scrape excessive oil off the cylinder wall.

Piston ring groove - The slots or grooves cut in piston heads to hold piston rings in position.

Piston skirt - The portion of the piston below the rings and the piston pin hole.

Plastigage - A thin strip of plastic thread, available in different sizes, used for measuring clearances. For example, a strip of plastigage is laid across a bearing journal and mashed as parts are assembled. Then parts are disassembled and the width of the strip is measured to determine clearance between journal and bearing. Commonly used to measure crankshaft main-bearing and connecting rod bearing clearances.

Press-fit - A tight fit between two parts that requires pressure to force the parts together. Also referred to as drive, or force, fit.

Prussian blue - A blue pigment; in solution, useful in determining the area of contact between two surfaces. Prussian blue is commonly used to determine the width and location of the contact area between the valve face and the valve seat.

R

Race (bearing) - The inner or outer ring that provides a contact surface for balls or rollers in bearing.

Ream - To size, enlarge or smooth a hole by using a round cutting tool with fluted edges.

Ring job - The process of reconditioning the cylinders and installing new rings.

Runout - Wobble. The amount a shaft rotates out-of-true.

S

Saddle - The upper main bearing seat.

Scored - Scratched or grooved, as a cylinder wall may be scored by abrasive particles moved up and down by the piston rings.

Scuffing - A type of wear in which there's a transfer of material between parts moving against each other; shows up as pits or grooves in the mating surfaces.

Seat - The surface upon which another part rests or seats. For example, the valve seat is the matched surface upon which the valve face rests. Also used to refer to wearing into a good fit; for example, piston rings seat after a few miles of driving.

Short block - An engine block complete with crankshaft and piston and, usually, camshaft assemblies.

Static balance - The balance of an object while it's stationary.

Step - The wear on the lower portion of a ring land caused by excessive side and back-clearance. The height of the step indicates the ring's extra side clearance and the length of the step projecting from the back wall of the groove represents the ring's back clearance.

Stroke - The distance the piston moves when traveling from top dead center to bottom dead center, or from bottom dead center to top dead center.

Stud - A metal rod with threads on both ends.

T

Tang - A lip on the end of a plain bearing used to align the bearing during assembly.

Tap - To cut threads in a hole. Also refers to the fluted tool used to cut threads.

Taper - A gradual reduction in the width of a shaft or hole; in an engine cylinder, taper usually takes the form of uneven wear, more pronounced at the top than at the bottom.

Throws - The offset portions of the crankshaft to which the connecting rods are affixed.

Thrust bearing - The main bearing that has thrust faces to prevent excessive endplay, or forward and backward movement of the crankshaft.

Thrust washer - A bronze or hardened steel washer placed between two moving parts. The washer prevents longitudinal movement and provides a bearing surface for thrust surfaces of parts.

Tolerance - The amount of variation permitted from an exact size of measurement. Actual amount from smallest acceptable dimension to largest acceptable dimension.

U

Umbrella - An oil deflector placed near the valve tip to throw oil from the valve stem area.

Undercut - A machined groove below the normal surface.

Undersize bearings - Smaller diameter bearings used with re-ground crankshaft journals.

V

Valve grinding - Refacing a valve in a valve-refacing machine.

Valve train - The valve-operating mechanism of an engine; includes all components from the camshaft to the valve.

Vibration damper - A cylindrical weight attached to the front of the crankshaft to minimize torsional vibration (the twist-untwist actions of the crankshaft caused by cylinder firing impulses). Also called a harmonic balancer.

W

Water jacket - The spaces around the cylinders, between the inner and outer shells of the cylinder block or head, through which coolant circulates.

Web - A supporting structure across a cavity.

Woodruff key - A key with a radiused backside (viewed from the side).

2D-36 GENERAL ENGINE OVERHAUL PROCEDURES

Specifications

General

Oil pressure
- 3.5L V6 (engine hot at 1500 rpm) — 30 psi
- 4.2L V6 (engine hot at 2500 rpm) — 40 to 60 psi
- 4.6L V8 (engine hot at 1500 rpm) — 20 to 45 psi
- 5.4L V8 (engine hot at 1500 rpm) — 40 to 70 psi

Cylinder head warpage limit — 0.003 (in any 6 inches)/0.006 inch overall
Compression pressure — Lowest reading cylinder must be within 15 psi of highest reading cylinder (100 psi minimum)

3.5L V6 engines

Valves and related components

Seat angle — 44.5 to 45.5-degrees
Seat width
- Intake — 0.051 to 0.063 inch
- Exhaust — 0.067 to 0.079 inch

Stem diameter
- Intake — 0.2157 to 0.2164 inch
- Exhaust — 0.2151 to 0.2159 inch

Stem-to-guide clearance
- Intake — 0.0008 to 0.0027 inch
- Exhaust — 0.0014 to 0.0033 inch

Valve spring
- Free Length (approximately) — 2.17 inches
- Installed height — 1.45 inches

Crankshaft and connecting rods

Connecting rod journal
- Diameter — 2.204 to 2.205 inches
- Out-of-round limit — 0.00023 inch
- Taper limit — 0.00015 inch
- Bearing oil clearance — 0.0007 to 0.0021 inch
- Connecting rod side clearance — 0.0068 to 0.0167 inch

Main journal
- Diameter — 2.657 inches
- Out-of-round limit — 0.00023 inch
- Taper — 0.00015 inch

Main bearing oil clearance — 0.0010 to 0.0016 inch
Crankshaft maximum end play — 0.0039 to 0.0114 in

Cylinder bore

Diameter — 3.641 to 3.642 inches
Out-of-round limit — 0.003 inch
Taper limit — 0.0004 inch (per 1.000 inch)

Piston and rings

Piston diameter — 3.6407 to 3.6413 inch
Piston-to-cylinder bore clearance limit — 0.0003 to 0.0017 inch

Piston ring end gap
- Top compression ring — 0.0484 to 0.0492 inch
- Second compression ring — 0.0602 to 0.0610 inch
- Oil ring — 0.0996 to 0.1003 inch

Piston ring-to-groove clearance
(upper/lower compression rings) — 0.0014 to 0.0031 inch

GENERAL ENGINE OVERHAUL PROCEDURES 2D-37

Torque specifications   Ft-lbs (unless otherwise indicated)

➡ **Note:** Refer to Chapter 2, Part B for additional torque specifications.
➡ **Note:** One foot-pound (ft-lb) of torque is equivalent to 12 inch-pounds (in-lbs) of torque. Torque values below approximately 15 ft-lbs are expressed in inch-pounds, because most foot-pound torque wrenches are not accurate at these smaller values.

Main bearing cap bolts* (tighten the bolts in the order listed here)
  Vertical cap bolts
    Step 1   24
    Step 2   Tighten an additional 135-degrees
  Side bolts
    Step 1, side bolts (new)   33
    Step 2, side bolts   Tighten an additional 90-degrees
Main bearing cap support brace bolts*
  Step 1   18
  Step 2   Tighten an additional 180-degrees
Connecting rod bearing cap bolts*
  Step 1   17
  Step 2   32
  Step 3   Tighten an additional 90-degrees

Use new bolts

4.2L V6 engines

Valves and related components

Seat angle   44.5-degrees
Seat width   0.060 to 0.080 inch
Minimum valve margin width   1/32 inch
Stem diameter
  Intake   0.3423 to 0.3415 inch
  Exhaust   0.3418 to 0.3410 inch
Stem-to-guide clearance
  Intake   0.0008 to 0.0027 inch
  Exhaust   0.0018 to 0.0037 inch
Valve spring
  Free length   1.99 inches
  Installed height   1.566 to 1.637 inches
Valve lifter
  Lifter-to-bore clearance
    Standard   0.0007 to 0.0027 inch
    Service limit   0.005 inch

Crankshaft and connecting rods

Connecting rod journal
  Diameter   2.3103 to 2.3111 inches
  Out-of-round limit   0.0012 inch
  Taper limit   0.0006 inch per inch
  Bearing oil clearance   0.0010 to 0.0027 inch
Connecting rod side clearance (endplay)   0.0006 to 0.0177 inch
Main journal
  Diameter*   2.5190 to 2.5198 inches
  Out of round limit   0.0003 inch
  Taper limit   0.0006 inch per inch
Main bearing oil clearance   0.0010 to 0.0025 inch
Crankshaft endplay   0.0002 to 0.0078 inch

The crankshaft journals can't be machined more than 0.010 inch under the standard dimension.

2D-38 GENERAL ENGINE OVERHAUL PROCEDURES

4.2L V6 engines (continued)

Cylinder bore
Diameter	3.8139 inches
Out-of-round limit	0.002 inch
Taper limit	0.002 inch

Pistons and rings
Piston diameter	3.8103 inches
Piston-to-bore clearance limit	0.0007 to 0.0017 inch
Piston ring end gap	
Top compression ring	0.009 to 0.016 inch
Second compression ring	0.039 to 0.064 inch
Oil ring	0.0059 to 0.0064 inch
Piston ring side clearance	0.0011 to 0.0031 inch

Torque specifications   Ft-lbs (unless otherwise indicated)

→ **Note:** Refer to Chapter 2, Part A for additional torque specifications.

→ **Note:** One foot-pound (ft-lb) of torque is equivalent to 12 inch-pounds (in-lbs) of torque. Torque values below approximately 15 ft-lbs are expressed in inch-pounds, because most foot-pound torque wrenches are not accurate at these smaller values.

Main bearing cap stud/bolts	
Mains 1, 2 and 3	85 to 91
Rear main	81 to 88
Connecting rod cap	
Step 1	29
Step 2	Rotate an additional 90 degrees

V8 engines

Cylinder bore
Diameter, 4.6L and 5.4L	3.554 inches
Out-of-round	
Standard	0.0006 inch
Service limit	0.0008 inch
Taper	0.0002 inch maximum

Valves and related components
Valve arrangement (front to rear)	
2009 and earlier	
Left cylinder head	E-I-E-I-E-I-E-I
Right cylinder head	I-E-I-E-I-E-I-E
2010 and later (left and right cylinder heads)	I-E-I-I-E-I-I-E-I-I-E-I
Intake valve	
Seat angle	45 degrees
Seat width	
2004 and earlier models	0.0748 to 0.0827 inch
2005 and later 5.4L models	0.0470 to 0.0550 inch
Seat runout limit	0.0010 inch maximum (total indicator reading)
Stem diameter, standard	
2004 and earlier models	0.2746 to 0.2750 inch
2005 and later 5.4L models	0.2350 to 0.2360 inch
Valve stem-to-guide clearance	
2004 and earlier models	0.0008 to 0.0027 inch
2005 and later 5.4L models	0.0010 to 0.0020 inch
Valve face angle	45.5 degrees
Valve face runout limit	0.002 inch maximum

GENERAL ENGINE OVERHAUL PROCEDURES 2D-39

Valves and related components (continued)

    Exhaust valve
        Seat angle   45 degrees
        Seat width
            2009 and earlier   0.0748 to 0.0827 inch
            2010 and later   0.055 to 0.063 inch
        Seat runout limit   0.0010 inch maximum (total indicator reading)
        Stem diameter, standard
            2004 and earlier models   0.2740 to 0.2736 inch
            2005 and later 5.4L models   0.2340 to 0.2350 inch
        Valve stem-to-guide clearance
            2004 and earlier models   0.0018 to 0.0037 inch
            2005 and later 5.4L models   0.0030 to 0.0040 inch
        Valve face angle   45.5 degrees
        Valve face runout limit   0.0020 inch maximum
    Valve spring
        Free length
            4.6L   1.951 inches
            5.4L, 2004 and earlier   1.978 inches
            5.4L 3V, 2005 and later   2.194 inches
        Out-of-square limit   2 degrees maximum
        Installed height
            2004 and earlier models   1.576 inches
            2005 and later 5.4L models   1.660 inches
        Pressure, intake and exhaust
            4.6L
                Valve open   132 lbs at 1.104 inches
                Valve closed   55 lbs at 1.576 inches
            5.4L
                Valve open   142 to 157 lbs at 1.104 inches
                Valve closed   61 to 68 lbs at 1.576 inches
        Valve spring pressure service limit   10 percent pressure loss at 1.104 inches
    Hydraulic lash adjuster (lifter)
        Diameter (standard)
            2004 and earlier models   0.6304 to 0.6299 inch
            2005 and later 5.4L models   0.6294 to 0.6299 inch
        Lifter-to-bore clearance
            Standard   0.0007 to 0.0027 inch
            Service limit   0.0006 inch maximum
        Collapsed lifter gap - desired
            2004 and earlier models   0.0335 to 0.0177 inch
            2005 and later 5.4L models   0.0180 to 0.0330 inch
        Rocker arm ratio (roller cam followers)   1.75:1

Crankshaft and connecting rods

    Crankshaft
        Endplay
            4.6L   0.0051 to 0.012 inch
            5.4L   0.0029 to 0.0148 inch
        Runout to rear face of block (standard)   0.002 inch

2D-40 GENERAL ENGINE OVERHAUL PROCEDURES

V8 engines (continued)

Crankshaft and connecting rods (continued)

Connecting rods
    Connecting rod journal
        Diameter
            2004 and earlier models   2.0877 to 2.0883 inches
            2005 and later 5.4L models   2.0859 to 2.0867 inches
        Bearing oil clearance
            4.6L   0.0011 to 0.0027 inch
            5.4L   0.0010 to 0.0025 inch
    Connecting rod side clearance (endplay)
        Standard
            2004 and earlier models   0.0006 to 0.0177 inch
            2005 and later 5.4L models   0.0049 to 0.0187 inch
        Service limit   0.0197 inch
Main bearing journal
    Diameter
        2009 and earlier models   2.650 to 2.657 inches
        2010 and later models   2.6568 to 2.6576 inches
    Bearing oil clearance
        2004 and earlier models   0.0011 to 0.0027 inch
        2005 and later 5.4L models   0.0009 to 0.0019 inch

Pistons and rings

Piston diameter (2004 and earlier models)
    Coded red 1   3.5529 to 3.5537 inches
    Coded blue 2   3.5534 to 3.5542 inches
    Coded yellow 3   3.5539 to 3.5547 inches
Piston diameter (2005 and later models)
    Grade 1
        2005 through 2009   3.5498 to 3.5505 inches
        2010 and later   3.5498 to 3.5502 inches
    Grade 2   3.5502 to 3.5506 inches
    Grade 3   3.5506 to 3.5510 inches
Piston-to-bore clearance limit
    4.6L   0.0005 to 0.0010 inch
    5.4L
        2009 and earlier   0.0002 to 0.0010 inch
        2010 and later   0.0010 to 0.0018 inch
Piston ring end gap
    4.6L
        Compression rings   0.010 to 0.020 inch
        Oil ring   0.006 to 0.026 inch
    5.4L
        2004 and earlier models
            Top compression ring   0.005 to 0.011 inch
            Second compression ring   0.010 to 0.016 inch
            Oil ring   0.006 to 0.026 inch
        2005 and later models
            Top compression ring   0.0060 to 0.0120 inch
            Second compression ring   0.0098 to 0.0197 inch
            Oil ring   0.0059 to 0.0256 inch

GENERAL ENGINE OVERHAUL PROCEDURES 2D-41

Pistons and rings (continued)

Piston ring side clearance	
Compression ring (top)	
2004 and earlier	0.0016 to 0.0031 inch
2005 and later 5.4L engine	0.0008 to 0.0031 inch
Compression ring (second)	0.0012 to 0.0028 inch
2004 and earlier	0.0012 to 0.0031 inch
2005 and later 5.4L engine	0.0012 to 0.0028 inch
Service limit	0.006 inch maximum
Oil ring	Snug fit

Torque specifications
Ft-lbs (unless otherwise indicated)

➡ **Note:** Refer to Chapter 2, Part B for additional torque specifications.

➡ **Note:** One foot-pound (ft-lb) of torque is equivalent to 12 inch-pounds (in-lbs) of torque. Torque values below approximately 15 ft-lbs are expressed in inch-pounds, because most foot-pound torque wrenches are not accurate at these smaller values.

Main bearing cap bolts (tighten first)	
First step	30
Second step	Rotate and additional 90-degrees
Main bearing cap - jack screws (REP engines only, tighten second)	
First step	44 in-lbs
Second step	89 in-lbs
Main bearing cap - side bolts (tighten third)	
REP engines	
First step	89 in-lbs
Second step	14 to 17
WEP engines	
First step	22
Second step	Rotate and additional 90-degrees
Connecting rod cap nuts	
First step	29 to 33
Second step	Rotate an additional 90 to 120-degrees

Notes

Section

1   General Information
2   Antifreeze - general information
3   Thermostat - check and replacement
4   Engine cooling fan and fan clutch - check, removal and installation
5   Radiator and degas bottle - removal and installation
6   Coolant temperature sending unit - check and replacement
7   Engine oil cooler - removal and installation
8   Water pump - check
9   Water pump - removal and installation
10 Heater and air conditioning blower motor and circuit - check
11 Heater and air conditioning blower motor - removal and installation
12 Heater and air conditioning control assembly - removal and installation
13 Heater core - removal and installation
14 Air conditioning and heating system - check and maintenance
15 Air conditioning accumulator/drier - removal and installation
16 Air conditioning compressor - removal and installation
17 Air conditioning condenser - removal and installation
18 Air conditioning evaporator - removal and installation

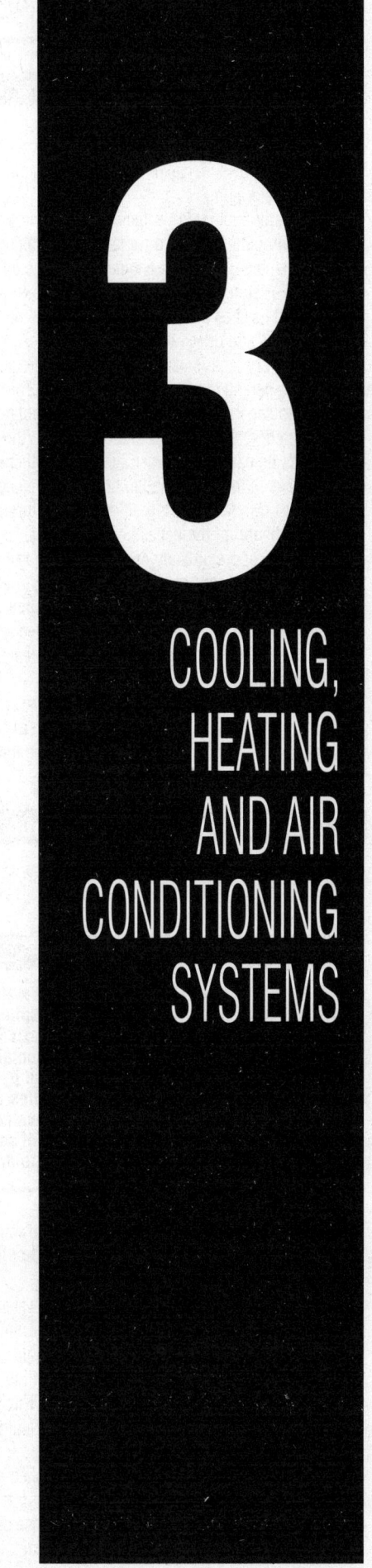

3
COOLING, HEATING AND AIR CONDITIONING SYSTEMS

3-2 COOLING, HEATING AND AIR CONDITIONING SYSTEMS

1 General information

The cooling system consists of a radiator and coolant reserve system, a radiator pressure cap, a thermostat, a cooling fan, and a pulley/belt-driven water pump.

The radiator cooling fan is mounted on the front of the water pump. The fan incorporates a fluid-drive fan clutch, which saves horsepower and reduces noise. When the engine is cold, the fluid in the clutch offers little resistance and allows the fan to freewheel. As the engine heats up and reaches a predetermined temperature, the fluid in the clutch thickens and drives the fan.

The recovery tank is called a "degas" bottle, and it functions somewhat differently than traditional recovery (or overflow) tanks. Designed to separate any trapped air in the coolant, it is pressurized by the radiator and has a pressure cap on top (the radiator on these models has no cap at all). When the engine's thermostat is closed, no coolant flows in the degas bottle, but when the engine is fully warmed up, coolant flows from the top of the radiator through a small hose that enters the top of the degas bottle. There, air separates and coolant falls to the approximately one quart coolant reserve in the bottle, which is fed to the cooling system through a larger hose connected to the lower radiator hose. Unlike traditional coolant recovery tanks, the cap on the degas bottle should never be opened when the engine is warm, since there is a danger of injury from steam or scalding coolant (see the **Warning** in Section 5).

Coolant in the left side of the radiator circulates through the lower radiator hose to the water pump, where it is forced through the water passages in the cylinder block. The coolant then travels up into the cylinder head, circulates around the combustion chambers and valve seats, travels out of the cylinder head past the open thermostat into the upper radiator hose and back into the radiator.

When the engine is cold, the thermostat restricts the circulation of coolant to the engine. When the minimum operating temperature is reached, the thermostat begins to open, allowing coolant to return to the radiator.

Automatic transmission-equipped models have a cooler element incorporated into the radiator to cool the transmission fluid.

The heating system works by directing air through the heater core, which is like a small radiator mounted behind the dash. Hot engine coolant heats the core, over which air passes to the interior of the vehicle by a system of ducts. Temperature is controlled by mixing heated air with fresh air, using a system of flapper doors in the ducts, and a heater bower motor. On Expedition and Navigator models, a second blower motor is located in the center console to direct airflow to the rear seat area.

Air conditioning is an optional accessory, consisting of an evaporator core located under the dash, a condenser in front of the radiator, an accumulator/drier in the engine compartment and a belt-driven compressor mounted at the front of the engine.

An option on models of the Navigator and Expedition is a rear-mounted heating/air-conditioning system. In this system, the driver has controls on the dashboard for the front system, plus an overhead control panel that operates the rear system. There is also an overhead control panel at the rear of the vehicle that controls only the rear heat/air. A second heater core, air-conditioning evaporator core and blower motor are located at the rear of the vehicle behind the left-rear interior panel.

2 Antifreeze - general information

▶ Refer to illustration 2.6

※ WARNING:

Do not allow antifreeze to come in contact with your skin or painted surfaces of the vehicle. Rinse off spills immediately with plenty of water. Antifreeze is highly toxic if ingested. Never leave antifreeze lying around in an open container or in puddles on the floor; children and pets are attracted by it's sweet smell and may drink it. Check with local authorities about disposing of used anti-freeze. Many communities have collection centers which will see that antifreeze is disposed of safely. Never dump used antifreeze on the ground or pour it into drains.

→ **Note: Non-toxic antifreeze is now manufactured and available at local auto parts stores, but even this type should be disposed of properly.**

The cooling system should be filled with a water/ethylene glycol based antifreeze solution which will prevent freezing down to at least -20-degrees F (even lower in cold climates). It also provides protection against corrosion and increases the coolant boiling point. The engines in the covered vehicles have aluminum heads. The manufacturer recommends that only coolant designated as safe for aluminum engine components be used.

The cooling system should be drained, flushed and refilled at least every other year (see Chapter 1). The use of antifreeze solutions for periods of longer than two years is likely to cause damage and encourage the formation of rust and scale in the system.

Before adding antifreeze to the system, check all hose connections. Antifreeze can leak through very minute openings.

The exact mixture of antifreeze to water which you should use depends on the relative weather conditions. The mixture should contain at least 50-percent antifreeze, but should never contain more than 70-percent anti-freeze. Consult the mixture ratio chart on the container before adding coolant. Hydrometers are available at most auto parts stores to test the coolant (see illustration). Use antifreeze which meets manufacturer specifications for engines with aluminum heads

2.6 An inexpensive hydrometer can be used to test the condition of your coolant

COOLING, HEATING AND AIR CONDITIONING SYSTEMS

3 Thermostat - check and replacement

※ WARNING:
The engine must be completely cool when this procedure is performed.

➡ **Note:** Don't drive the vehicle without a thermostat! The computer may stay in open loop and emissions and fuel economy will suffer.

CHECK

1 Before condemning the thermostat, check the coolant level, drivebelt tension and temperature gauge (or light) operation.

2 If the engine takes a long time to warm up, the thermostat is probably stuck open. Replace the thermostat.

3 If the engine runs hot, check the temperature of the upper radiator hose. If the hose isn't hot, the thermostat is probably stuck shut. Replace the thermostat.

4 If the upper radiator hose is hot, it means the coolant is circulating and the thermostat is open. Refer to the Troubleshooting Section at the front of this manual for the cause of overheating.

5 If an engine has been overheated, you may find damage such as leaking head gaskets, scuffed pistons and warped or cracked cylinder heads.

REPLACEMENT

♦ **Refer to illustrations 3.8a, 3.8b and 3.10**

6 Drain coolant from the radiator until the coolant level is below the thermostat housing (see Chapter 1).

7 Disconnect the upper radiator hose from the thermostat housing.

8 Remove the bolts and lift the cover off (see illustrations). Remove the O-ring seal. On V8 models, loosen the bolts on the power steering reservoir bracket enough to allow the thermostat housing to be withdrawn (one bracket bolt threads into the top of the thermostat housing).

9 Note how it's installed, then remove the thermostat. Be sure to use a replacement thermostat with the correct opening temperature (see this Chapter's Specifications).

10 On early 4.2L V6 engines, use a scraper or putty knife to remove all traces of old gasket material and sealant from the mating surfaces. Later models have an O-ring (see illustration). Make sure no gasket material falls into the coolant passages; it is a good idea to stuff a rag in the passage. Wipe the mating surfaces with a rag saturated with lacquer thinner or acetone.

11 Install the thermostat and make sure the correct end faces out - the spring end is directed toward the engine.

➡ **Note:** On some models, the thermostat housing has a notch that a tab on the thermostat fits into, which automatically locates the thermostat correctly.

12 On early 4.2L V6 models that use a conventional paper gasket, apply a thin coat of RTV sealant to both sides of the new gasket and position it on the engine side, over the thermostat, and make sure the gasket holes line up with the bolt holes in the housing.

➡ **Note:** No RTV sealant should be used on the O-ring seal on V8 engines.

3.8a Thermostat mounting bolt locations (V6 engine)

3.8b On V8 models, remove the upper bolt (upper arrow) and loosen the power steering reservoir bolts, then remove the two thermostat housing bolts (lower arrows)

3.10 When installing the thermostat, pay special attention to the direction in which it's placed in the engine; the spring side will go into the intake manifold (V6 shown) - arrow indicates the new O-ring

3-4 COOLING, HEATING AND AIR CONDITIONING SYSTEMS

13 On models that use an O-ring seal, install the new O-ring into the intake manifold or onto the thermostat housing.

14 Carefully position the cover and install the bolts. Tighten them to the torque listed in this Chapter's Specifications - do not overtighten them or the cover may be cracked or distorted.

15 Reattach the radiator hose to the cover and tighten the clamp - now may be a good time to check and replace the hoses and clamps (see Chapter 1).

16 Refer to Chapter 1 and refill the system, then run the engine and check carefully for leaks.

17 Repeat Steps 1 through 4 to be sure the repairs corrected the previous problem(s).

4 Engine cooling fan and clutch - check, removal and installation

※※ WARNING 1:

The models covered by this manual are equipped with Supplemental Restraint Systems (SRS), more commonly known as airbags. Always disconnect the negative battery cable, then the positive battery cable and wait two minutes before working in the vicinity of the impact sensors, steering column or instrument panel to avoid the possibility of accidental deployment of the airbag, which could cause personal injury (see Chapter 12). Do not use any electrical test equipment on any of the airbag system wires or tamper with them in any way.

※※ WARNING 2:

Keep hands, tools and clothing away from the fan when the engine is running. To avoid injury or damage DO NOT operate the engine with a damaged fan. Do not attempt to repair fan blades - replace a damaged fan with a new one.

CHECK

※※ WARNING:

In order to check the fan clutch, the engine will need to be at operating temperature, so while going through checks prior to Step 6 be careful that the engine is NOT started while the checks are being completed. Severe personal injury can result!

1 Symptoms of failure of the fan clutch are continuous noisy operation, looseness, vibration and evidence of silicone fluid leaks. On 2007 and later Expedition/Navigator models, if you have access to a scan tool, check for codes related to the electronic fan clutch system.

2 Rock the fan back and forth by hand to check for excessive bearing play.

3 With the engine cold, turn the blades by hand. The fan should turn freely.

4 Visually inspect for substantial fluid leakage from the fan clutch assembly, a deformed bi-metal spring or grease leakage from the cooling fan bearing. If any of these conditions exist, replace the fan clutch.

5 When the engine is warmed up, turn off the ignition switch. Turn the fan by hand. Some resistance should be felt. If the fan turns easily, replace the fan clutch.

REMOVAL AND INSTALLATION

2009 and earlier models

▶ Refer to illustrations 4.9, 4.12, 4.13 and 4.14

6 Disconnect the battery cable at the negative battery terminal.

7 Remove the air cleaner assembly (see Chapter 4). On 2007 and later Expedition/Navigator models, remove the screws securing the underhood fuse/relay box to the fan shroud and set the relay box aside.

4.9 Remove the screws (two shown) and remove the plastic air deflector over the radiator

8 Drain the cooling system (see Chapter 1).

9 Remove the screws and remove the plastic air deflector over the radiator (see illustration).

10 Remove the upper radiator hose.

11 Remove the degas bottle from the left side of the radiator and disconnect the hose connected to the radiator (see Section 5).

12 A special two-part fan wrench set, obtainable at most auto parts stores, is required to remove the cooling fan assembly. The clutch attaches to the drive hub with a large nut. Hold the water pump pulley with the clutch-holding tool and an extension, while turning the clutch nut with the longer tool (see illustration).

➡ **Note:** On some models, the fan clutch hub nut is left-hand thread (turn clockwise to loosen). Refer to the belt routing label on the fan shroud, if the hub nut is left-hand thread it should be stated as such on the label.

4.12 Two special wrenches are used - the shorter one holds the water pump pulley, while the longer one loosens the fan assembly hub nut

COOLING, HEATING AND AIR CONDITIONING SYSTEMS   3-5

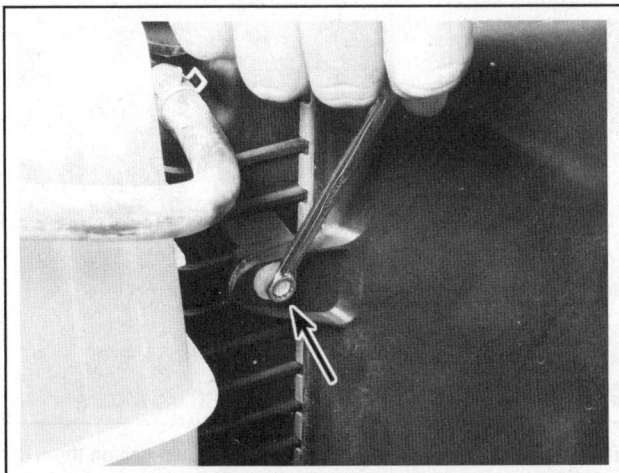

4.13 Remove the two fan shroud screws (left side shown) and pull the shroud out with the fan assembly

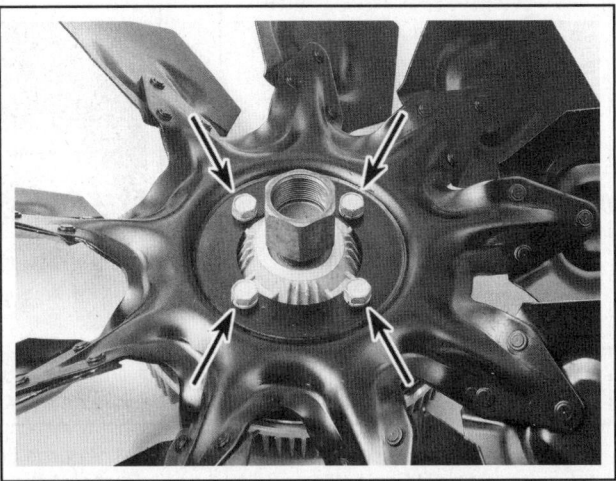

4.14 Remove the four bolts (arrows) and separate the fan from the fan clutch

13 Unbolt the fan shroud (see illustration) and lift the fan assembly and shroud up and out of the engine compartment together.

14 The fan clutch can be unbolted from the fan blade assembly for replacement (see illustration).

✱✱ CAUTION:

To prevent silicone fluid from draining from the clutch assembly into the fan drive bearing and ruining the lubricant, DON'T place the drive unit in a position with the rear of the shaft pointing down. Store the fan in its upright position if possible.

15 Installation is the reverse of removal.

2010 and later models

16 Disconnect the cable from the negative battery terminal (see Chapter 5).

17 Remove the air intake duct (see Chapter 4).

18 Detach the retainer for the power steering fluid cooler hose from the fan motor and shroud assembly.

19 Detach the retainer from the power steering fluid reservoir stud bolt.

20 Detach the wiring harness retainer from the left side of the fan motor and shroud assembly.

21 Detach the retainers for the alternator battery harness from the top of the cooling motor and shroud assembly.

22 Detach the wiring harness retainer from the right side of the fan motor and shroud assembly.

23 Disconnect the electrical connectors from both cooling fan motors.

24 Detach the cooling fan motor harness retainers.

25 Remove the battery positive cable nut and disconnect the battery positive cable terminals, then remove the battery junction box bracket bolts.

26 Remove the power steering fluid reservoir stud bolt and set the reservoir aside.

27 Remove the bolts that secure the cooling fan motor and shroud assembly and lift the assembly out of the engine compartment.

28 If you're only removing the fan shroud assembly to access some other component(s), no further disassembly is necessary. If you're replacing a fan or motor, proceed to the next step.

29 Unbolt the motor that you're replacing from the shroud.

30 Remove the fastener that secures the fan to the motor.

31 Installation is the reverse of removal.

5 Radiator and degas bottle - removal and installation

✱✱ WARNING 1:

The engine must be completely cool when this procedure is performed.

✱✱ WARNING 2:

The models covered by this manual are equipped with Supplemental Restraint Systems (SRS), more commonly known as airbags. Always disconnect the negative battery cable, then the positive battery cable and wait two minutes before working in the vicinity of the impact sensors, steering column or instrument panel to avoid the possibility of accidental deployment of the airbag, which could cause personal injury (see Chapter 12). Do not use any electrical test equipment on any of the airbag system wires or tamper with them in any way.

DEGAS BOTTLE (2009 AND EARLIER MODELS)

▶ Refer to illustrations 5.2 and 5.3

1 Disconnect the cable from the negative battery terminal.

3-6 COOLING, HEATING AND AIR CONDITIONING SYSTEMS

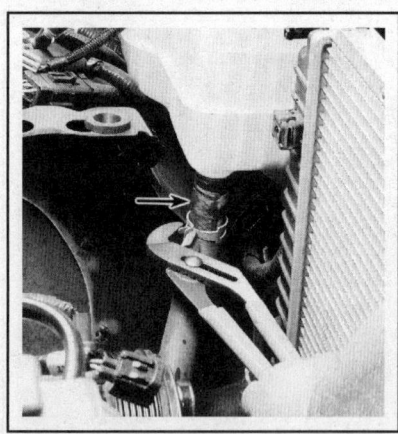

5.2 Disconnect the hose (arrow) from the bottom of the degas bottle

5.3 Remove the two bolts and overflow hose and pull the degas bottle straight up and out

5.8 Use a flare-nut wrench on the line fittings and a back-up wrench on the radiator fittings to prevent damage to the transmission cooler lines or radiator when disconnecting them from the radiator

2 Drain the cooling system as described in Chapter 1, then disconnect the lower hose from the degas bottle (see illustration).

3 Disconnect the overflow hose, then remove the bolts and pull the degas bottle straight up (see illustration).

4 Prior to installation make sure the degas bottle is clean and free of debris which could be drawn into the radiator (wash the inside with soapy water and a long brush if necessary).

5 Installation is the reverse of removal.

COOLANT EXPANSION TANK (2010 AND LATER MODELS)

6 The coolant expansion tank is an integral part of the air filter lower housing. See Chapter 4, Section 9, for the replacement procedure.

RADIATOR

→ Note: On 2007 and later models, have the refrigerant recovered before beginning this procedure.

2006 and earlier models

▸ Refer to illustrations 5.8 and 5.10

7 Remove the cooling fan and shroud assembly (see Section 4). Remove the degas bottle (see Steps 1 through 3).

8 On models with automatic transmission, detach the cooler lines from the radiator (see illustration) - be careful not to damage the lines or fittings. Plug the ends of the disconnected lines to prevent leakage and stop dirt from entering the system. Have a drip pan ready to catch spills.

9 Disconnect the upper and lower radiator hoses.

10 Remove the mounting bolts and remove the two upper radiator mounts (see illustration), and the mounts that hold the lug wrench and jack handle. Carefully lift the radiator out of the vehicle.

2007 through 2009 models

11 Remove the cooling fan/shroud assembly (see Section 4) and the two end panels that are secured by pushpins. Remove the degas bottle (see Steps 1 through 3).

12 If equipped with an automatic transmission, detach the cooler lines from the radiator (see illustration 5.8) - be careful not to damage

5.10 Remove the two bolts and mounts holding the top of the radiator (right mount shown)

the lines or fittings. Plug the ends of the disconnected lines to prevent leakage and stop dirt from entering the system. Have a drip pan ready to catch spills.

13 Disconnect the upper and lower radiator hoses.

14 Remove both headlight housings (see Chapter 12) and, on 2008 and later models, remove the front bumper cover.

15 Disconnect and plug the refrigerant lines at the condenser.

16 Disconnect the horn wiring at the side of the radiator and remove the horns.

17 Remove the bolts on the sides of the radiator. Push the radiator assembly toward the engine and push on the two lower condenser-to-radiator clips until the radiator is separated from the condenser.

18 Carefully lift the radiator out of the vehicle.

2010 and later models

19 Drain the cooling system (see Chapter 1). If the coolant is relatively new and in good condition, save it and reuse it.

20 Remove the headlight housing assemblies (see Chapter 11).

→ Note: On 2015 and later Navigator models, remove the front bumper cover (see Chapter 11).

21 Remove the cooling fan(s) and shroud (see Section 4).

COOLING, HEATING AND AIR CONDITIONING SYSTEMS 3-7

22 Disconnect the upper, lower and overflow radiator hoses. Loosen the hose clamps by squeezing the ends together. Hose clamp pliers work best, but regular pliers will work also.

→ **Note:** *The lower radiator hose is attached to the radiator with a spring clip. Separate the hose from the radiator by prying the clip up and remove the hose. If any hose is stuck, grasp it near the end with a pair of adjustable pliers and twist it to break the seal, then pull it off. If any hose is old or deteriorated, cut it off and install a new one.*

23 Remove the radiator mounting bolts, then remove the coolant expansion tank mounting bolts. Position the expansion tank aside.

24 Remove the horn assembly (see Chapter 11).

25 Remove the remaining fasteners between the condenser and the air deflectors. Disconnect the two transmission fluid cooler hoses from the radiator.

26 Pull the condenser assembly rearward to separate it from the radiator. Lift the condenser until all mounting brackets are detached from the radiator.

27 Carefully lift the radiator from the engine compartment. Don't spill coolant on the vehicle or scratch the paint. The rubber insulators that help secure the bottom of the radiator may stick to the radiator when it's removed. They will need to be returned to their original positions during installation.

All models

▶ Refer to illustration 5.30

28 Remove bugs and dirt from the radiator with compressed air and a soft brush. Don't bend the cooling fins. Inspect the radiator for leaks and damage. If it needs repair, have a radiator shop or a dealer service department do the work.

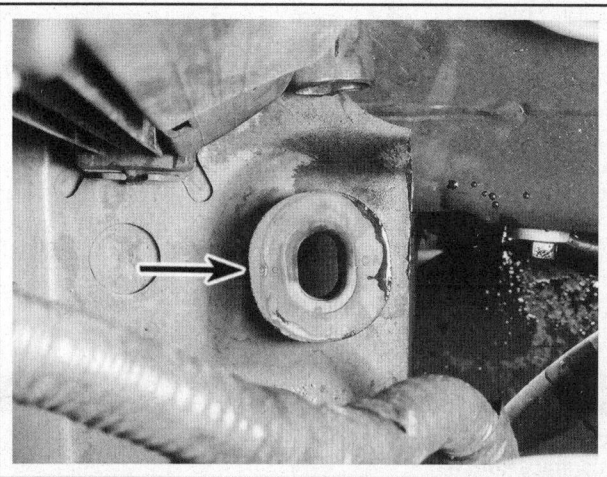

5.30 Make sure the radiator lower insulators (view is from above, looking down on right-side insulator) are in place before installing the radiator - if they're beginning to deteriorate, now is the time to replace them

29 Prior to installation of the radiator, replace any damaged hose clamps and radiator hoses.

30 Installation is the reverse of removal. When installing the radiator on earlier models, make sure it seats properly in the lower saddles and that the rubber mounts are intact (see illustration).

31 After installation, fill the system with the proper mixture of anti-freeze, and also check the automatic transmission fluid level.

6 Coolant temperature sending unit - check and replacement

CHECK

▶ Refer to illustrations 6.1a and 6.1b

→ **Note:** *On 2005 and later models with 5.4L engines and 2015 and later models with 3.5L engines, a Cylinder Head Temperature (CHT) sensor is used instead of a coolant temperature sending unit (see Chapter 6).*

1 The coolant temperature indicator system is composed of a temperature gauge mounted in the dash and a coolant temperature sending unit mounted on the engine (see illustrations). Some vehicles have more than one sending unit, but only one is used for the indicator system and the other is used to send engine temperature information to the computer.

2 If an overheating indication occurs, check the coolant level in the

6.1a The coolant temperature sending unit (A) on V6 models is on the intake manifold - (B) indicates the coolant temperature sensor for the computer

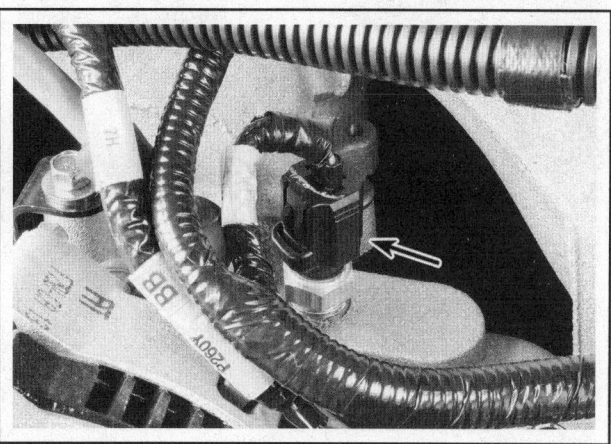

6.1b On early V8 engines, the coolant temperature sending unit is at the left front of the intake manifold

3-8 COOLING, HEATING AND AIR CONDITIONING SYSTEMS

system. Make sure the wiring between the gauge and the sending unit is secure and all fuses are intact.

3   To test the temperature sender, check that it reads in the cold range when the engine is cold. Disconnect the electrical connector from the sender and attach a jumper wire between the two pins of the connector. With the key ON, the gauge should now swing to full hot. If it doesn't, the problem is in the circuit from the sender to the instrument panel. If the gauge does swing to full hot when testing, but doesn't when the connector is in place and the engine is hot, replace the coolant temperature sending unit.

REPLACEMENT

※※ WARNING:
Wait until the engine is completely cool before beginning this procedure.

4   Prepare the new sending unit by wrapping its threads with Teflon tape or applying sealer. Remove the pressure cap from the degas bottle to release any pressure that may remain in the system, then reinstall it.

5   Disconnect the electrical connector and unscrew the sensor from the engine. Install the replacement as quickly as possible to minimize coolant loss. There will be some coolant loss as the unit is removed, so be prepared to catch it.

※※ CAUTION:
The sending unit is made of metal and plastic and is fragile. Use care not to crack the unit when removing it.

6   Check the coolant level after the replacement unit has been installed (see Chapter 1).

7   Engine oil cooler - removal and installation

REMOVAL

▶ Refer to illustrations 7.6 and 7.8

※※ WARNING:
Wait until the engine is completely cool before beginning this procedure.

3.5L V6 engines

1   Raise the vehicle and support it securely on jackstands.
2   Drain the engine coolant (see Chapter 1).
3   Remove the oil filter (see Chapter 1).
4   Disconnect the two coolant hoses from the oil cooler.
5   The oil cooler is attached to the engine by the same large threaded tube that the oil filter is screwed onto. Remove the oil cooler threaded tube and remove the oil cooler assembly.

V8 engines

※※ WARNING:
The engine should be completely cool for this procedure.

6   Some V8 models have an engine oil cooler (see illustration). Coolant flows through the cooler from the block and back to the block. Note: On 4WD models, the oil cooler is mounted out in front of the engine, while on 2WD models it is alongside the block at the left-rear, near the block-to-oil pan juncture.

7   To replace the oil cooler, refer to Chapter 1 for removal of the oil filter and draining of the cooling system.

8   Disconnect the oil and coolant hoses from the oil cooler (see illustration).

9   On 2WD models, use a 1/2-inch Allen wrench inside the threaded adapter holding the oil cooler housing to the oil filter adapter housing. On 4WD models, a single bolt retains the cooler to its bracket, and two bolts hold the oil line assembly on the cooler.

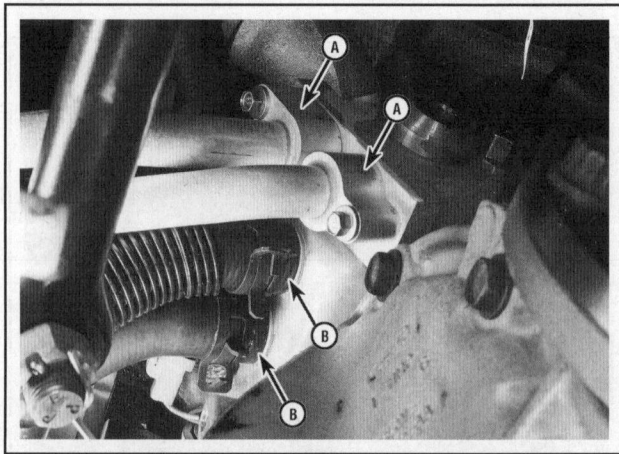

7.6 On some V8 models, the oil cooler is mounted at the lower left side of the engine block, with a hose to the remote oil filter (up front under the bumper)

7.8 The oil lines (A) are attached to the cooler with bolts, while the coolant hoses (B) are attached with clamps

COOLING, HEATING AND AIR CONDITIONING SYSTEMS

10 Pull the oil cooler from the adapter on the block.

11 If only the oil cooler is to be replaced, this is as far as you need to disassemble components. If the oil cooler adapter is to be removed, remove the bolts holding it to the block and remove it.

INSTALLATION

3.5L V6 engines

12 Installation is the reverse of removal. Be sure to use a new O-ring gasket between the oil cooler and the engine.

13 Refill the cooling system and check the engine oil level (see Chapter 1).

V8 engines

14 Clean the block and back of the oil cooler adapter of any old gasket material. Install the adapter to the block with a new gasket and tighten the bolts to the torque listed in this Chapter's Specifications.

15 Clean the coolant sealing surfaces of the adapter and oil cooler, and install new O-rings. Clean the threads of the threaded tube and apply non-hardening thread-locking compound to the end that goes into the oil cooler adapter. Insert it through the oil cooler and tighten it to this Chapter's Specifications.

16 The remainder of installation is the reverse of removal. Install a new oil filter, refill and bleed the cooling system (see Chapter 1), and run the engine to check for oil or coolant leaks.

8 Water pump - check

▶ Refer to illustration 8.2

WARNING:

The models covered by this manual are equipped with Supplemental Restraint Systems (SRS), more commonly known as airbags. Always disconnect the negative battery cable, then the positive battery cable and wait two minutes before working in the vicinity of the impact sensors, steering column or instrument panel to avoid the possibility of accidental deployment of the airbag, which could cause personal injury (see Chapter 12). Do not use any electrical test equipment on any of the airbag system wires or tamper with them in any way.

1 Water pump failure can cause overheating and serious damage to the engine. There are three ways to check the operation of the water pump while it's installed on the engine. If any one of the following quick checks indicates water pump problems, it should be replaced immediately.

2 A seal protects the water pump impeller shaft bearing from contamination by engine coolant. If this seal fails, a weep hole in the water pump snout will leak coolant (an inspection mirror can be used to look at the underside of the pump if the hole isn't on top). If the weep hole is leaking, shaft bearing failure will follow (see illustration). Replace the water pump immediately.

➡ **Note:** A small amount of gray discoloration is normal. A wet area or heavy brown deposits indicate the pump seal has failed.

3 Besides contamination by coolant after a seal failure, the water pump impeller shaft bearing can also prematurely wear out. If a noise is

8.2 If there's coolant leaking from the weep hole, the water pump must be replaced

coming from the water pump during engine operation, the shaft bearing has failed - replace the water pump immediately.

➡ **Note:** Do not confuse drivebelt noise with bearing noise. Loose or glazed drivebelts may emit a high-pitched squealing noise.

4 To identify excessive bearing wear before the bearing actually fails, grasp the water pump pulley (with the drivebelt removed) and try to force it up-and-down or from side-to-side. If the pulley can be moved either horizontally or vertically, the bearing is nearing the end of its service life. Replace the water pump.

9 Water pump - removal and installation

WARNING:

The models covered by this manual are equipped with Supplemental Restraint Systems (SRS), more commonly known as airbags. Always disconnect the negative battery cable, then the positive battery cable and wait two minutes before working in the vicinity of the impact sensors, steering column or instrument panel to avoid the possibility of accidental deployment of the airbag, which could cause personal injury (see Chapter 12). Do not use any electrical test equipment on any of the airbag system wires or tamper with them in any way.

REMOVAL

▶ Refer to illustrations 9.4, 9.6a and 9.6b

WARNING:

Wait until the engine is completely cool before starting this procedure.

1 Disconnect the cable from the negative battery terminal.
2 With the engine cold, drain the cooling system (see Chapter 1).

3-10 COOLING, HEATING AND AIR CONDITIONING SYSTEMS

All models except 3.5L V6 engines

3 Remove the fan shroud and the fan assembly (see Section 4).

➡ **Note:** This step may not be necessary on 2010 and later models; only remove the fan and shroud if it is necessary for working room.

4 Remove the drivebelt(s) (see Chapter 1) and remove the water pump pulley (see illustration).

➡ **Note:** It's helpful to loosen the pulley bolts/nuts while the belt is still in place. It helps hold the pulley from turning.

5 On V6 engines, disconnect the upper radiator hose from the water pump. Also remove the bolt and detach the heater water outlet tube from the top of the pump.

6 Remove the water pump retaining bolts and remove the water pump (see illustrations). Take note of the installed positions of the various length bolts and studs.

➡ **Note:** If the water pump sticks, dislodge it with a soft-face hammer or a hammer and a block of wood.

3.5L V6 engines

7 Remove the expansion tank mounting bolts and move the tank out of the way without disconnecting the hoses.

8 Disconnect the four hoses to the thermostat housing, then remove the housing bolts and housing from the front cover.

9 If the water pump pulley bolts aren't too tight, you might be able to loosen them now, using the drivebelt to hold the pulley. If that doesn't work, you'll have to hold the pulley with a strap wrench after you remove the drivebelt.

10 Remove the drivebelt (see Chapter 1).

➡ **Note:** Cover the air conditioning compressor drivebelt with kitchen plastic wrap. If coolant gets on the belt it can ruin it.

11 Remove the pulley bolts and remove the pulley. If you were unable to loosen the water pump pulley bolts with the drivebelt installed, use a strap wrench to hold the pulley as you loosen the bolts.

12 Remove the hose retaining clip and disconnect the quick-connect hose from the bottom of the expansion tank and heater hose at the engine.

13 Remove the water pump retaining bolts and remove the water pump. Take note of the installed positions of the various bolts.

➡ **Note:** If the water pump sticks, dislodge it with a soft-face hammer or a hammer and a block of wood.

9.4 While the drivebelt is still in place, remove the four bolts holding the water pump pulley

14 Before installation, remove and clean the mating surface on the front cover (and water pump, if the same pump is to be installed).

15 Lubricate a new O-ring seal with clean antifreeze and install it to the water pump also install a new gasket.

16 Install the water pump or water pump and tighten the bolts by hand evenly, then tighten the bolts to the torque listed in this Chapter's Specifications.

INSTALLATION

▸ **Refer to illustrations 9.18 and 9.19**

17 Before installation, remove and clean all gasket or sealant material from the water pump and cylinder block.

18 If you're working on a V8 engine, inspect the O-ring and sealing surface of water pump housing in the block for dirt and/or debris (see illustration). Clean them thoroughly before reassembly.

19 If you're working on a V8 engine, lubricate a new O-ring seal with clean antifreeze and install it to the water pump (see illustration).

20 If you're working on a V6 engine, coat a new gasket with a thin film of RTV sealant and install it on the pump.

21 Install the water pump and tighten the bolts to the torque listed in this Chapter's Specifications.

9.6a Water pump retaining bolt locations - V6 engine

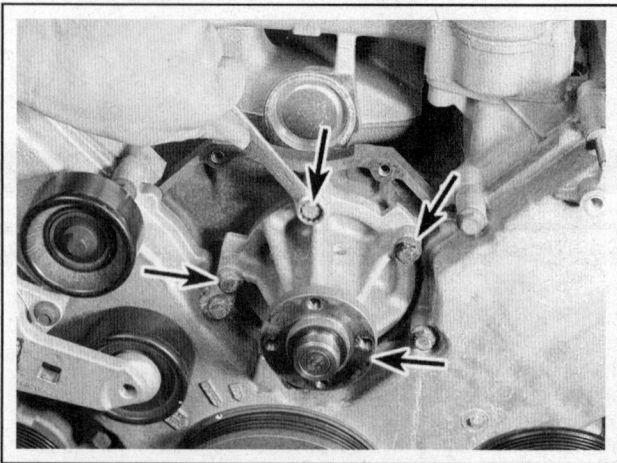

9.6b Water pump retaining bolts - V8 engines

COOLING, HEATING AND AIR CONDITIONING SYSTEMS   3-11

9.18 Inspect the sealing surface (arrow) in the pump cavity for dirt or signs of pitting (V8 engines)

9.19 On V8 engines, install a new O-ring seal on the water pump

All models

22 The remainder of the installation is the reverse of the disassembly sequence.

23 Fill the cooling system with the proper coolant mixture (see Chapter 1).

24 Start the engine and make sure there are no leaks. Check the level frequently during the first few weeks of operation to ensure there are no leaks and that the level in the system is stable.

10 Heater and air conditioning blower motor and circuit - check

▶ Refer to illustrations 10.5a, 10.5b, 10.6, 10.7 and 10.9

※ WARNING:

The models covered by this manual are equipped with Supplemental Restraint Systems (SRS), more commonly known as airbags. Always disconnect the negative battery cable, then the positive battery cable and wait two minutes before working in the vicinity of the impact sensors, steering column or instrument panel to avoid the possibility of accidental deployment of the airbag, which could cause personal injury (see Chapter 12). Do not use any electrical test equipment on any of the airbag system wires or tamper with them in any way.

➡ Note: The blower motor is switched on the ground-side of the circuit.

➡ Note: 2008 and later models are equipped with a HVAC module and a speed control module that also controls power to the blower motor and blower resistor. Both modules can only be diagnosed by using special Ford diagnostic equipment.

1 Check the fuse and all connections in the circuit for looseness and corrosion.

2 Make sure the battery is fully charged.

3 With the transmission in Park, the parking brake securely set, turn the ignition switch to the On position. It isn't necessary to start the vehicle.

4 Switch the heater controls to FLOOR and the blower speed to HI. Listen at the ducts to hear if the blower is operating. If it is, then switch the blower speed to LO and listen again. Try all the speeds.

5 The blower motor resistor assembly is located on the blower motor housing under the right side of the dash (see illustrations). There are three resistor elements mounted on the resistor board to provide

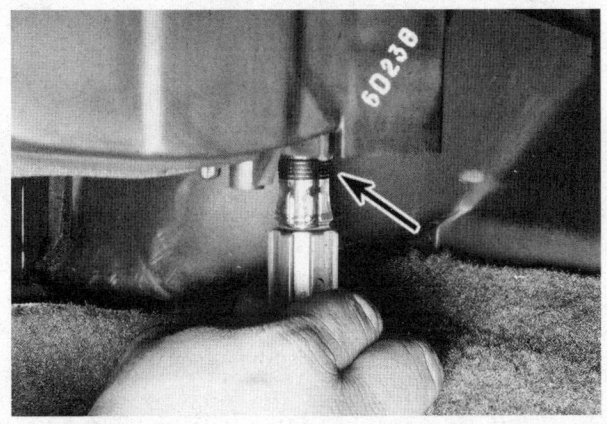

10.5a The blower motor resistor is located on the blower motor housing (arrow) - disconnect the wiring connector and remove the two screws

10.5b The thermal limiter (arrow) protects the components from excessive heat - check the thermal limiter for damage

3-12 COOLING, HEATING AND AIR CONDITIONING SYSTEMS

10.6 Test the terminals of the blower resistor with an ohmmeter for continuity

10.7 Backprobe the battery feed wire at the blower motor connector (arrow) - there should be voltage with the mode switch in any position other than Off and the ignition key On

low and medium blower speeds (HI bypasses the resistor). The blower operates continuously, anytime the ignition switch is On and the mode switch is in any position other than Off. A thermal limiter resistor is integrated into the circuits to prevent heat damage to the components. If the thermal limiter circuit has been opened as a result of excessive heat, it should be replaced only with the identical replacement part.

➡ **Note: Do not replace your blower resistor with a resistor that does not incorporate the thermal limiter.**

6 With the resistor removed from the vehicle, visually check the limiter for damage, indicated by the material melting out between the contacts of the limiter. Check the resistor block for continuity between terminals (see illustration). There should be continuity between terminals 2 and 3 with a resistance of approximately 8 ohms or less; continuity between terminals 2 and 4 with a resistance of approximately 1.8 to 2.0 ohms, and continuity between terminals 1 and 4 with total resistance of approximately 2.3 to 2.5 ohms. If any of the resistor elements do not pass the tests, replace the blower resistor.

7 Locate the electrical connector at the blower motor. Backprobe the brown/orange wire terminal; there should be at least 10 volts with the mode switch in any position other than Off and the ignition switch On (see illustration). If not, there is a problem in the circuit from the fuse panel to the heater/air conditioning control panel, or from the control panel to the blower.

8 If there is voltage at the feed wire, but the blower does not operate, backprobe the orange/black wire and connect it to a known good chassis ground with a jumper wire. If the blower now operates there is a problem in the ground circuit (which consists of the resistor, the blower motor switch and related wiring). If it still doesn't operate, replace the blower motor.

9 If the blower operates, but not at all speeds and you have already checked the blower resistor, refer to Section 12 and remove the heater/air conditioning control panel. Disconnect the electrical connector from the back of the blower speed switch and test the terminals for continuity (see illustration). In the Lo position, there should be no continuity between any terminals; in Medium Lo position, there should be continuity between terminals 2 and 3; in Medium Hi position there should be continuity between terminals 2, 3 and 4, and in HI position, there should be continuity between terminals 1, 2 and 4. If the continuity is not as described, replace the blower speed switch.

10 Locate the blower motor relay, in the relay box under the center dash panel (see Chapter 12). There are five pins on the back. Connect an ohmmeter to terminal 85, there should be resistance greater than 5 ohms between that pin and all others.

11 Energize the relay with jumper wires (battery voltage) applied to terminals 30 and 85. Connect a voltmeter to chassis ground and probe pin 87A with the positive lead of the voltmeter. Voltage should be greater than 10 volts.

12 With the jumper wires still in place, attach another jumper from pin 86 to chassis ground. Use the voltmeter as in Step 11 but check for 10 volts or better voltage at pin 87. If the relay fails any of these tests, replace the relay.

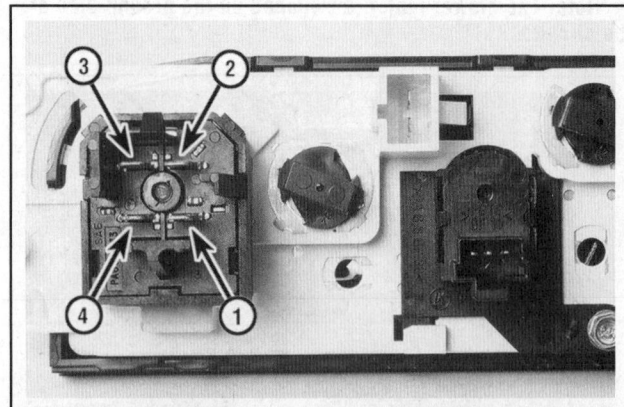

10.9 Check the blower speed switch for continuity

COOLING, HEATING AND AIR CONDITIONING SYSTEMS 3-13

11 Heater and air conditioning blower motor - removal and installation

WARNING:
The models covered by this manual are equipped with Supplemental Restraint Systems (SRS), more commonly known as airbags. Always disconnect the negative battery cable, then the positive battery cable and wait two minutes before working in the vicinity of the impact sensors, steering column or instrument panel to avoid the possibility of accidental deployment of the airbag, which could cause personal injury (see Chapter 12). Do not use any electrical test equipment on any of the airbag system wires or tamper with them in any way.

2009 AND EARLIER MODELS

Standard blower motor

▸ **Refer to illustrations 11.3 and 11.4**

1 Disconnect the blower motor electrical connector from the motor (see illustration 10.7).
2 Remove the three blower motor cover screws. Separate the blower motor cover from the blower motor by pushing the two plastic tabs.
3 Remove the three blower motor mounting screws (see illustration), and lower the blower motor carefully out of the housing.
4 If the blower motor is being replaced, the fan wheel should be transferred to the new motor at this time. It is attached to the blower motor shaft with a push nut. Grasp the nut with pliers and pull it off or pry it off with a small screwdriver, being careful not to crack the push nut or the fan (see illustration). To reinstall the nut, simply push it on to the shaft.
5 The remainder of the installation is the reverse of removal.

Console blower motor

6 Expedition and Navigator models have a second blower motor mounted in the center console. Refer to Chapter 11 for removal of the console.
7 Remove the screws, and take off the two sheetmetal braces from under the console.
8 Remove the large blower duct from the console. Unbolt the upper and lower duct covers from the blower motor assembly.
9 Disconnect the electrical connector from the blower motor and remove the two bolts to remove the blower motor.
10 Installation is the reverse of removal.

2010 AND LATER MODELS

11 Remove the passenger's side kick panel (if equipped) and the junction box cover on the A-pillar.
12 Disconnect the blower motor electrical connector.
13 Remove the blower motor mounting screws and remove the blower motor.
   a) Move the carpet under the blower motor aside.
   b) Rotate the blower so the vent tube is facing toward the front passenger's seat.
   c) Press in on the panel surrounding the blower until it is released.
14 Installation is the reverse of removal.

OPTIONAL REAR HEAT/AIR BLOWER (SOME EXPEDITION/NAVIGATOR MODELS)

▸ **Refer to illustration 11.16**

15 Some models of Expedition and Navigator vehicles have an auxiliary heating/air conditioning system located in the rear of the vehicle, behind the left-rear interior trim panel. On later Expedition/Navigator models, it is located behind the RH quarter-trim panel.
16 With the trim panel removed (see Chapter 11), there is access to remove/test the auxiliary blower motor (see illustration).

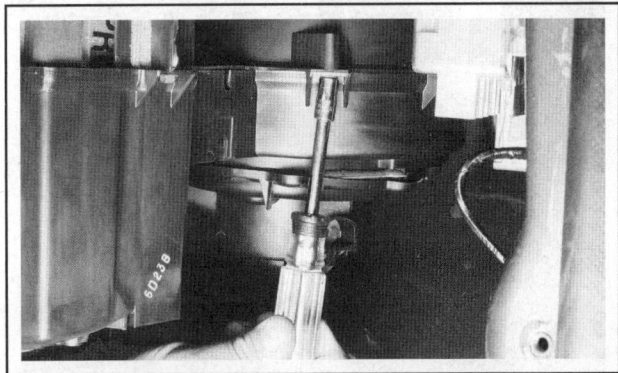

11.3 Remove the three blower cover screws and three motor mounting screws

11.4 Pry off the retaining clip retaining the blower fan to the blower motor shaft

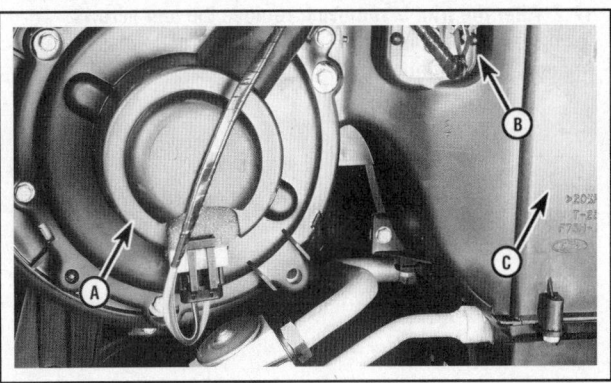

11.16 On some Navigator/Expedition models, there is a rear-mounted auxiliary blower motor (A) as part of the rear heat/air-conditioning system - B is the resistor for this rear blower, C is the housing containing the rear heater core and evaporator core

3-14 COOLING, HEATING AND AIR CONDITIONING SYSTEMS

12 Heater and air conditioning control assembly - removal and installation

♦ Refer to illustrations 12.2 and 12.3

⁕ WARNING:
The models covered by this manual are equipped with Supplemental Restraint Systems (SRS), more commonly known as airbags. Always disconnect the negative battery cable, then the positive battery cable and wait two minutes before working in the vicinity of the impact sensors, steering column or instrument panel to avoid the possibility of accidental deployment of the airbag, which could cause personal injury (see Chapter 12). Do not use any electrical test equipment on any of the airbag system wires or tamper with them in any way.

1 Refer to Chapter 11 for removal of the center dash bezel, around the control assembly and radio.

➡ **Note:** On 2006 and later models, the climate control assembly is mounted to the center trim panel. Once the trim panel is removed, the control assembly may be removed from the backside of the trim panel (see Chapter 11).

2 Remove the four screws retaining the control assembly to the instrument panel (see illustration).

➡ **Note:** The control assembly for the optional Electronic Automatic Temperature Control is a module which contains its own microprocessor (computer). Some models with EATC have extra function switches located in the steering wheel, in addition to the controls on the dashboard. The EATC system has its own self-diagnostic capability (see Section 14).

3 Pull the control assembly out of the instrument panel and disconnect the electrical connectors and vacuum harness (see illustration).

➡ **Note:** When disconnecting the vacuum lines. be careful to avoid cracking the plastic connectors and causing a vacuum leak (possibly internal within the control head).

4 Refer to Section 10 for electrical checks of the blower motor speed switch. The speed switch, function selector and blend-control switch can all be removed from the control head (manual air conditioning) by depressing plastic tabs at the back of the control head and pulling the switches off. The knobs will fall off the front side of the controls as each switch is removed.

5 An option on models of the Navigator and Expedition is a rear-mounted heating/air-conditioning system. In this system, the driver has controls on the dashboard for the front system, plus an overhead control panel that operates the rear system. There is also an overhead control panel at the rear of the vehicle that controls only the rear heat/air.

6 Installation is the reverse of the removal procedure.

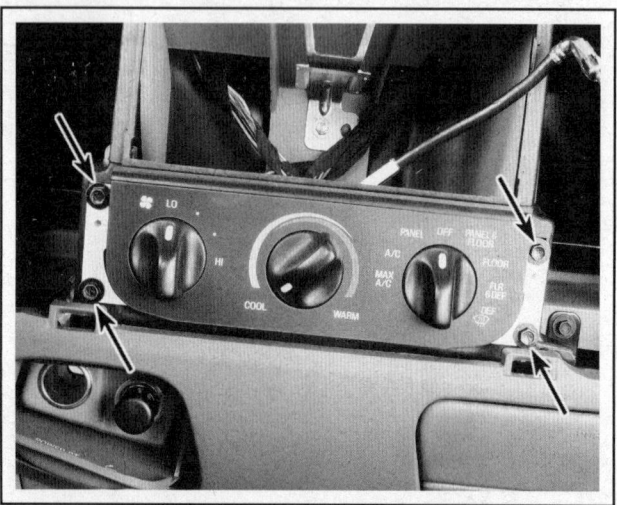

12.2 After the trim is removed, remove the four screws (arrows) and pull the control assembly out of the dash (manual air conditioning)

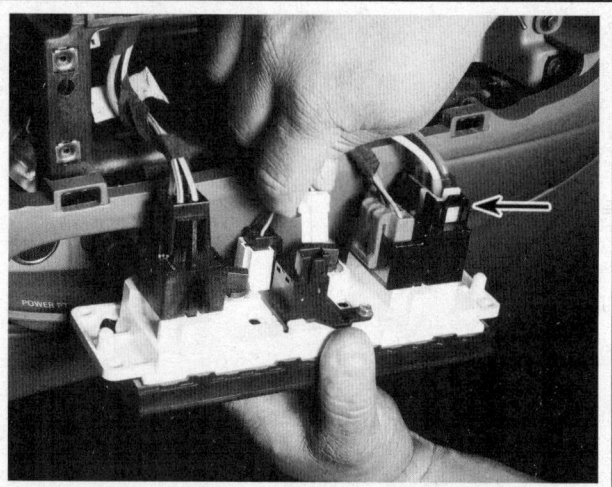

12.3 Disconnect the electrical connectors by gently prying up the clips and pulling the connectors off - arrow indicates the vacuum connection

COOLING, HEATING AND AIR CONDITIONING SYSTEMS   3-15

13 Heater core - removal and installation

▶ Refer to illustrations 13.3, 13.5, 13.6, 13.7a, 13.7b, 13.7c and 13.8

※※ WARNING 1:

The models covered by this manual are equipped with Supplemental Restraint Systems (SRS), more commonly known as airbags. Always disconnect the negative battery cable, then the positive battery cable and wait two minutes before working in the vicinity of the impact sensors, steering column or instrument panel to avoid the possibility of accidental deployment of the airbag, which could cause personal injury (see Chapter 12). Do not use any electrical test equipment on any of the airbag system wires or tamper with them in any way.

※※ WARNING 2:

The air conditioning system is under high pressure. DO NOT loosen any fittings or remove any components until after the system has been discharged. Air conditioning refrigerant should be properly discharged into an EPA-approved container at a dealer service department or an automotive air conditioning repair facility. Always wear eye protection when disconnecting air conditioning system fittings.

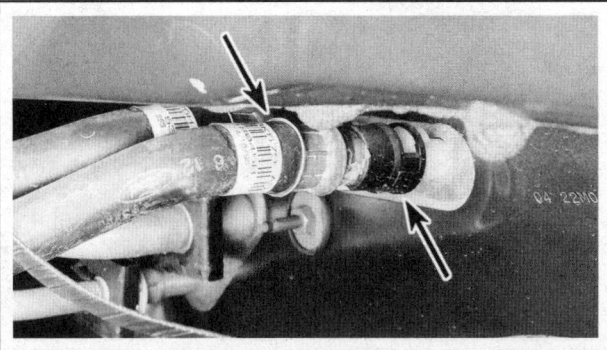

13.3 Squeeze the plastic tabs together and disconnect the heater hoses from the heater core inlet and outlet pipes (arrows) at the right side of the firewall

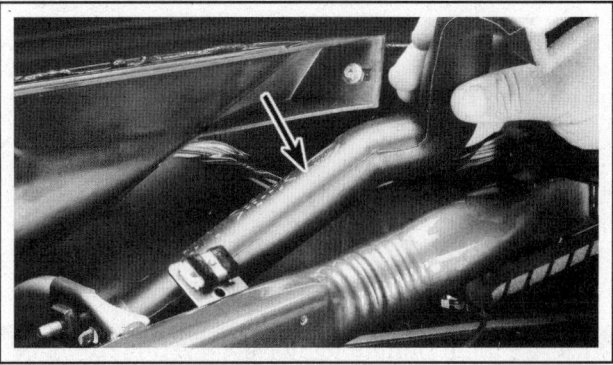

13.5 Remove the demister tube (arrow)

1 Take the vehicle to a dealer service department or automotive air conditioning shop and have the air conditioning system discharged. If equipped with adjustable pedals, put them into the fully forward position.

2 Disconnect the cable from the negative battery terminal. Drain the cooling system (see Chapter 1). Remove the instrument panel (see Chapter 11).

3 Disconnect the heater hoses from the heater core inlet and outlet tubes at the firewall (see illustration).

➡ Note: These fittings are quick-disconnect type. To remove, squeeze the two plastic tabs toward the hose and pull the hoses off the tubes from the heater core. On 2010 and later models, disconnect the Thermostatic Expansion Valve (TXV) and discard the gasket seals; new ones must be used on installation. If equipped with satellite radio, remove the antenna cable from brackets on the heater core housing.

4 Disconnect and plug the evaporator lines at the firewall (see Section 18). Plug the heater core tubes to avoid spilling any coolant during removal. Cap the evaporator lines to prevent the entry of dirt and moisture.

5 Disconnect the main vacuum line and remove the plenum demister tube (see illustration).

6 Remove the two screws holding the bracket to the top of the heater core housing (see illustration).

➡ Note: The number and location of the heating/cooling module mounting screws varies with year and model. Make sure you have located and removed all fasteners before attempting to pull out the module.

7 Remove the screws and the heater core cover, then pull out the

13.6 Remove the two screws and the bracket (arrow) over the heater core housing

13.7a Remove the screws (arrows indicate the front screws), then pull off the cover of the heater core/evaporator core housing

3-16 COOLING, HEATING AND AIR CONDITIONING SYSTEMS

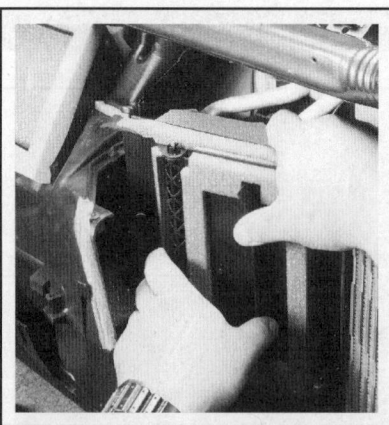

13.7b Pull the blend-door assembly out carefully to expose the heater core

13.7c Carefully pull the heater core (arrow) from the case

13.8 Reinstall the foam sealing material after installing the heater core in the case

heater core, being careful not to tear the foam sealing material (see illustrations).

8 When reinstalling the heater core in the housing, make sure the original foam sealing material is intact and in place (see illustration).

9 The remainder of the installation is the reverse of removal. Install new retainers and O-rings on the hoses that connect to the heater core at the firewall.

10 Fill the cooling system (see Chapter 1). Run the engine and check for coolant leaks. Have the air conditioning system charged and check for proper operation of the system.

11 On some models of Navigator/Expedition vehicles, there is an optional heat/air-conditioning system located behind the left-rear interior trim panel (behind the left-rear wheel housing), or right side panel on later Expedition/Navigator models. The assembly includes a blower, heater core and evaporator core (see illustration 11.12). Removal and installation procedures are similar to the procedures for the standard heater core up front, although the rear unit is much more accessible.

14 Air conditioning and heating system - check and maintenance

☼☼ WARNING:

The air conditioning system is under high pressure. DO NOT loosen any fittings or remove any components until after the system has been discharged. Air conditioning refrigerant should be properly discharged into an EPA-approved container at a dealer service department or an automotive air conditioning repair facility. Always wear eye protection when disconnecting air conditioning system fittings.

1 The following maintenance steps should be performed on a regular basis to ensure that the air conditioner continues to operate at peak efficiency.
   a) *Check the tension of the drivebelt and adjust if necessary (see Chapter 1).*
   b) *Check the condition of the hoses. Look for cracks, hardening and deterioration.*

☼☼ WARNING:

Do not replace air conditioning hoses until the system has been discharged by a dealer or air conditioning shop.

   c) *Check the fins of the condenser for leaves, bugs and other foreign material. A soft brush and compressed air can be used to remove them.*
   d) *Check the wire harness for correct routing, broken wires, damaged insulation, etc. Make sure the harness connections are clean and tight.*
   e) *Maintain the correct refrigerant charge.*

2 The system should be run for about 10 minutes at least once a month. This is particularly important during the winter months because long-term non-use can cause hardening of the internal seals.

3 Because of the complexity of the air conditioning system and the special equipment required to effectively work on it, accurate troubleshooting of the system should be left to a professional technician. One probable cause for poor cooling that can be determined by the home mechanic is low refrigerant charge. Should the system lose its cooling ability, the following procedure will help you pinpoint the cause.

CHECK

▶ **Refer to illustration 14.7**

4 Warm the engine up to normal operating temperature.

5 Place the air conditioning temperature selector at the coldest setting and put the blower at the highest setting. Open the doors (to make sure the air conditioning system doesn't cycle off as soon as it cools the passenger compartment).

6 After the system reaches operating temperature, feel the two pipes connected to the evaporator at the firewall.

7 The pipe (thinner tubing) leading from the condenser outlet to the evaporator should be cold, and the evaporator outlet line (the thicker tubing that leads back to the compressor) should be slightly colder (3 to 10-degrees F). If the evaporator outlet is considerably warmer than the inlet, the system needs a charge. Insert a thermometer in the center air distribution duct (see illustration) while operating

COOLING, HEATING AND AIR CONDITIONING SYSTEMS

14.7 Place an accurate thermometer in the center dash vent, turn the air conditioning on and check the output temperature

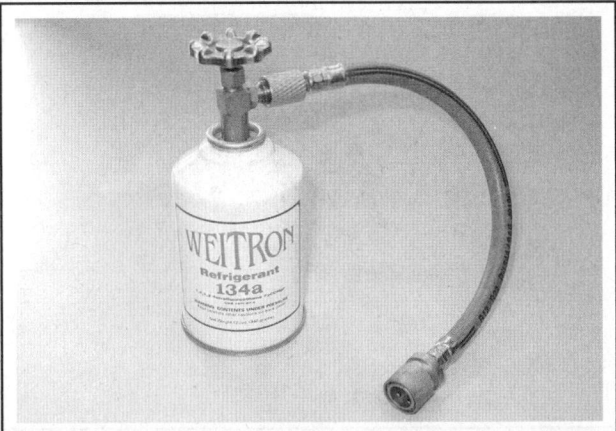

14.8 A basic charging kit for R-134a systems is available at most auto parts stores - it must say R-134a (not R-12) and so must the can of refrigerant

the air conditioning system - the temperature of the output air should be 35 to 40-degrees F below the ambient air temperature (down to approximately 40-degrees F). If the ambient (outside) air temperature is very high, say 110-degrees F, the duct air temperature may be as high as 60-degrees F, but generally the air conditioning is 35 to 40-degrees F cooler than the ambient air. If the air isn't as cold as it used to be, the system probably needs a charge. Further inspection or testing of the system is beyond the scope of the home mechanic and should be left to a professional.

ADDING REFRIGERANT

▶ Refer to illustrations 14.8 and 14.11

➡ Note: All models covered by this manual use refrigerant R-134a. When recharging or replacing air conditioning components, use only refrigerant, refrigerant oil and seals compatible with this system. The seals and compressor oil used with older, conventional R-12 refrigerant are not compatible with the components in this system.

8 Buy an automotive charging kit at an auto parts store. A charging kit includes a can of R-134a refrigerant, a tap valve and a short section of hose that can be attached between the tap valve and the system low side service valve (see illustration). Because one can of refrigerant may not be sufficient to bring the system charge up to the proper level, it's a good idea to buy a couple of additional cans. Try to find at least one can that contains red refrigerant dye. If the system is leaking, the red dye will leak out with the refrigerant and help you pinpoint the location of the leak.

➡ Note: New Ford vehicles are shipped with a leak-detect dye already in place. If your system has never been discharged, you can spot any refrigerant leaks with an ultraviolet spotlight, which causes leaks to glow greenish-yellow. The dye is said to be good for 500 hours of air conditioner operation, after which you can have more dye injected at your Ford dealer.

9 Connect the charging kit by following the manufacturer's instructions.

10 Back off the valve handle on the charging kit and screw the kit onto the refrigerant can, making sure first that the O-ring or rubber seal inside the threaded portion of the kit is in place.

※※ WARNING:

Wear protective eye wear when dealing with pressurized refrigerant cans.

11 Remove the dust cap from the low-side charging port and attach the quick-connect fitting on the kit hose (see illustration).

※※ WARNING:

DO NOT hook the charging kit hose to the system high side! The fittings on the charging kit are designed to fit *only* on the low side of the system.

12 Warm the engine to normal operating temperature and turn on the air conditioning. Keep the charging kit hose away from the fan and other moving parts.

13 Turn the valve handle on the kit until the stem pierces the can, then back the handle out to release the refrigerant. You should be able to hear the rush of gas. Add refrigerant to the low side of the system until both the outlet and the evaporator inlet pipe feel about the same temperature. Allow stabilization time between each addition.

14.11 Add R-134a only to the low-side port (arrow) - the procedure is easier if you wrap the can with a warm, wet towel to prevent icing

3-18 COOLING, HEATING AND AIR CONDITIONING SYSTEMS

14.21 Check the evaporator drain tube (arrow) for blockage that could lead to mildew on the core - this view is from below (transmission removed for clarity) - you can locate the drain by hand from above, behind the engine

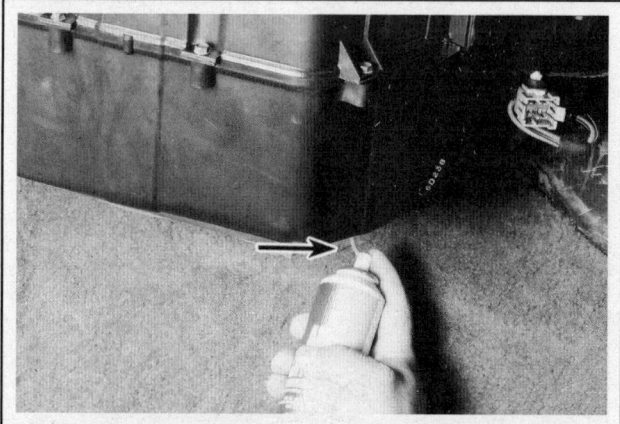

14.25 Remove the blower resistor and spray disinfectant through the hole (arrow) onto the core to destroy mildew that causes air conditioning odors

❊❊ WARNING:

Never add more than two cans of refrigerant to the system. The can may tend to frost up, slowing the procedure. Wet a shop towel with hot water and wrap it around the bottom of the can to keep it from frosting.

14 Put your thermometer back in the center register and check that the output air is getting colder.

15 When the can is empty, turn the valve handle to the closed position and release the connection from the low-side port. Replace the dust cap.

16 Remove the charging kit from the can and store the kit for future use with the piercing valve in the UP position, to prevent inadvertently piercing the can on the next use.

HEATING SYSTEMS

▶ Refer to illustration 14.21

17 If the air coming out of the heater vents isn't hot, the problem could stem from any of the following causes:

a) The thermostat is stuck open, preventing the engine coolant from warming up enough to carry heat to the heater core. Replace the thermostat (see Section 3).
b) A heater hose is blocked, preventing the flow of coolant through the heater core. Feel both heater hoses at the firewall. They should be hot. If one of them is cold, there is an obstruction in one of the hoses or in the heater core, or the heater control valve is shut. Detach the hoses and backflush the heater core with a water hose. If the heater core is clear but circulation is impeded, remove the two hoses and flush them out with a water hose.
c) If flushing fails to remove the blockage from the heater core, the core must be replaced (see Section 13).

18 If the blower motor speed does not correspond to the setting selected on the blower switch, the problem could be a bad fuse, circuit, control panel or blower resistor (see Section 10).

19 If there isn't any air coming out of the vents:

a) Turn the ignition ON and activate the fan control. Place your ear at the heating/air conditioning register (vent) and listen. Most motors are audible. Can you hear the motor running?
b) If you can't (and have already verified that the blower switch and the blower motor resistor are good), the blower motor itself is probably bad (see Section 11).

20 If the carpet under the heater core is damp, or if antifreeze vapor or steam is coming through the vents, the heater core is leaking. Remove it (see Section 13) and install a new unit (most radiator shops will not repair a leaking heater core).

21 Inspect the drain hose from the heater/evaporator assembly at the right-center of the firewall, make sure it is not clogged (see illustration). If there is a humid mist coming from the system ducts, this hose may be plugged with leaves or road debris.

ELIMINATING AIR CONDITIONING ODORS

▶ Refer to illustration 14.25

22 Unpleasant odors that often develop in air conditioning systems are caused by the growth of a fungus, usually on the surface of the evaporator core. The warm, humid environment there is a perfect breeding ground for mildew to develop.

23 The evaporator core on most vehicles is difficult to access, and factory dealerships have a lengthy, expensive process for eliminating the fungus by opening up the evaporator case and using a powerful disinfectant and rinse on the core until the fungus is gone. You can service your own system at home, but it takes something much stronger than basic household germ-killers or deodorizers.

24 Aerosol disinfectants for automotive air conditioning systems are available in most auto parts stores, but remember when shopping for them that the most effective treatments are also the most expensive. The basic procedure for using these sprays is to start by running the system in the RECIRC mode for ten minutes with the blower on its highest speed. Use the highest heat mode to dry out the system and keep the compressor from engaging by disconnecting the wiring connector at the compressor (see Section 16).

25 The disinfectant can usually comes with a long spray hose. Remove the blower motor resistor (see Section 10), point the nozzle inside the hole and spray, according to the manufacturer's recommendations (see illustration). Try to cover the whole surface of the evaporator core, by aiming the spray up, down and sideways. Follow the manu-

COOLING, HEATING AND AIR CONDITIONING SYSTEMS 3-19

facturer's recommendations for the length of spray and waiting time between applications.

26 Once the evaporator has been cleaned, the best way to prevent the mildew from coming back again is to make sure your evaporator housing drain tube is clear (see illustration 14.21).

ELECTRONIC AUTOMATIC TEMPERATURE CONTROL

27 Expedition and Navigator models may have an optional EATC system to control the temperature inside the vehicle. You set the desired temperature on the control panel and the EATC control module (computer) blends the right amount of cooled or heater air to maintain this cabin temperature. The speed of the blower motor in this system is electronically-controlled by the microprocessor.

28 Most repairs or diagnostics of the EATC are beyond the scope of the home mechanic, but there is an on-board diagnostics function in the EATC computer that will display trouble codes relating to the climate-control system. The codes can indicate what area, if any, is malfunctioning.

➡ **Note: On later Expedition/Navigator models, the system does not have the self-test capability. Use a scan tool to search for trouble codes related to the heating/air conditioning system.**

29 To begin the self-test function (vehicle interior at normal temperature), push the FLOOR and OFF buttons at the same time, followed within two seconds by pushing the AUTOMATIC button. During the half-minute test, any hard or intermittent Diagnostic Trouble Codes (DTCs) will appear on the control head where the temperature setting is usually displayed.

30 The test mode can be exited in two ways. If you push the blue button, the display will turn off, but keep the codes in memory. If you push the DEFROST button instead, the computer will erase the codes and then turn off the display.

➡ **Note: When the second method of exiting is used, a code 888 (meaning no codes are in memory) should show before the display turns off.**

EATC TROUBLE CODES

(Hard) Self-test faults	(Intermittent) Run-time faults	Problem
024	022	Blend door short
	025	Blend door failure
031		A/C in-car temperature sensor - open circuit
030		A/C in-car temperature sensor - short circuit
041	043	A/C ambient temperature sensor - open circuit
040	042	A/C ambient temperature sensor - short circuit
050	052	A/C solar radiation sensor circuit - shorted

15 Air conditioning accumulator/drier - removal and installation

REMOVAL

◆ Refer to illustrations 15.3, 15.4a, 15.4b, 15.4c and 15.4d

> ※ **WARNING:**
> The air conditioning system is under high pressure. DO NOT loosen any fittings or remove any components until after the system has been discharged. Air conditioning refrigerant should be properly discharged into an EPA-approved container at a dealer service department or an automotive air conditioning repair facility. Always wear eye protection when disconnecting air conditioning system fittings.

1 The accumulator/drier stores refrigerant and removes moisture from the system. When any major air conditioning component (compressor, condenser, evaporator) is replaced, or the system has been apart and exposed to air for any length of time, the accumulator/drier must be replaced.

2 Take the vehicle to a dealer service department or automotive air conditioning shop and have the air conditioning system discharged.

Disconnect the cable at the negative battery terminal.

2007 and earlier models

3 Disconnect the electrical connector at the compressor clutch cycling switch on top of the accumulator/drier (see illustration).

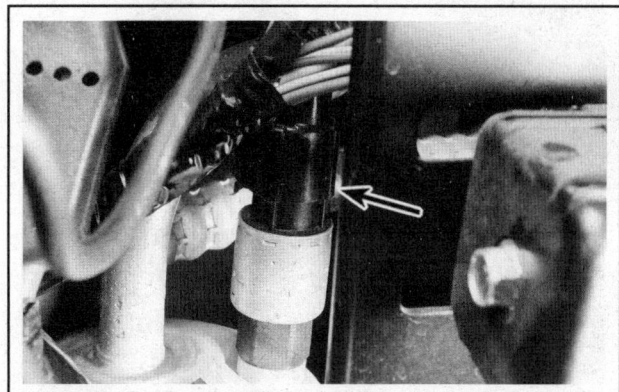

15.3 Disconnect the electrical connector (arrow) at the compressor clutch cycling switch

3-20 COOLING, HEATING AND AIR CONDITIONING SYSTEMS

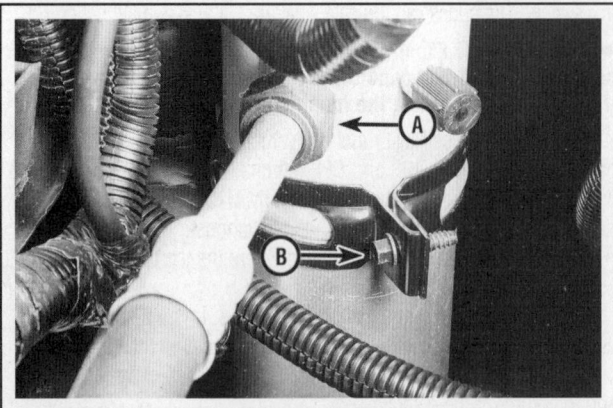

15.4a Disconnect this line (A) to the accumulator/drier using a backup wrench on the fitting at the drier - B is the mounting clamp bolt

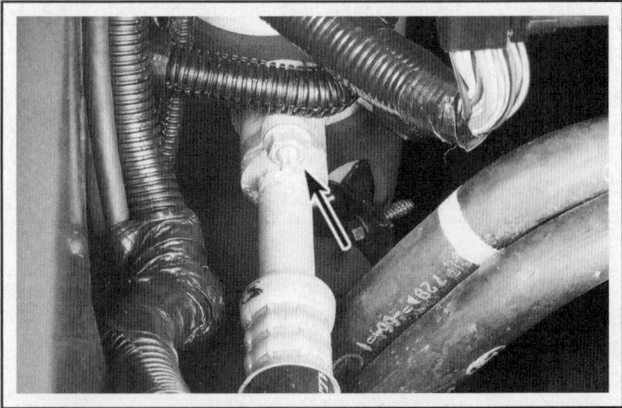

15.4b On some models, the line to the accumulator/drier is bolted to a flange (arrow) instead of a threaded fitting

4 Disconnect the refrigerant inlet and outlet lines (see illustrations). Some models have threaded fittings - others have spring lock couplings which require special tools to disconnect. On these kinds of fittings, remove the metal clips first, then use spring-lock coupling tools to disconnect the two lines (one of which is connected to the accumulator/drier) from the evaporator core tubes at the firewall. To disconnect a fitting, close the two halves of the tool over the connection and push the tool towards the garter spring. This expands the spring to release its hold. While the spring is expanded and tool is still in place, pull in opposite directions on the two lines to separate the connection. Cap or plug the open lines immediately.

➡ **Note:** : Special spring lock coupling tools are required to release the connectors used on the refrigerant lines throughout the air conditioning system, and are available at most auto parts stores in a set.

5 Remove the nut from the mounting bracket and slide the accumulator/drier assembly up and out of the mounting bracket (see illustration 15.4a).

2008 and later models

6 Remove the condenser (see Section 17).
7 Remove the threaded plug on the bottom of the condenser on the left end.
8 Using needle-nose pliers, pull the desiccant bag out of the receiver-drier.

INSTALLATION

2007 and earlier models

9 If you are replacing the accumulator/drier, drain the refrigerant oil from the old accumulator/drier. Add the same amount plus two ounces of clean refrigerant oil to the new accumulator. This will maintain the correct oil level in the system after the repairs are completed.

➡ **Note:** The manufacturer recommends that to properly drain all of the oil from the old accumulator/drier for an accurate measurement, you should drill two half-inch holes in the bottom of the accumulator/drier.

10 Place the new accumulator/drier into position, tighten the mounting bracket screw lightly, still allowing the accumulator drier to be turned to align the line connections.

11 Lubricate the O-rings using clean refrigerant oil and reconnect the inlet and outlet lines. Now tighten the clamp bolt securely and reconnect the electrical connector.

2008 and later models

12 Insert a new dessicant bag into the receiver-drier.
13 Install the plug and tighten it securely.

All models

14 Connect the cable to the negative terminal of the battery.
15 Have the system evacuated, recharged and leak tested by a dealer service department or an air conditioning repair facility.

15.4c Pull off the clip . . .

15.4d . . . and use the spring-lock coupling tool to separate the connection

COOLING, HEATING AND AIR CONDITIONING SYSTEMS 3-21

16 Air conditioning compressor - removal and installation

> ※ **WARNING:**
> The air conditioning system is under high pressure. DO NOT loosen any fittings or remove any components until after the system has been discharged. Air conditioning refrigerant should be properly discharged into an EPA-approved container at a dealer service department or an automotive air conditioning repair facility. Always wear eye protection when disconnecting air conditioning system fittings.

➡ **Note:** On some models, special spring-lock coupling tools are required to release the connectors used on the refrigerant lines throughout the air conditioning system. There are different special tools for each line size; these tools can usually be found at local auto parts stores, often in a set. See Section 15 for tool description and use.

REMOVAL

♦ Refer to illustrations 16.2 and 16.5

> ※ **CAUTION:**
> Whenever a compressor is replaced, it will be necessary to replace the accumulator/drier and the orifice tube.

1 Take the vehicle to a dealer service department or automotive air conditioning shop and have the air conditioning system discharged. Disconnect the cable from the negative battery terminal.
2 Remove the accessory drivebelt(s) (see Chapter 1). If you're working on an Expedition or Navigator model, raise the vehicle and support it securely on jackstands. On air-suspension-equipped vehicles, turn the switch OFF before jacking the vehicle up. Remove the plastic air deflector beneath the radiator (Expedition and Navigator models only) (see illustration).
3 Remove the bolt and disconnect the refrigerant lines from the compressor.

➡ **Note:** On 3.5L models, remove the turbocharger air inlet and outlet tubes (see Chapter 4).

4 Disconnect the electrical connection at the compressor clutch.
5 Remove the compressor mounting bolts (see illustration).
6 Remove the compressor from the mounting location. Drain and measure the refrigerant oil from the compressor.

INSTALLATION

♦ Refer to illustrations 16.9 and 16.10

7 If the compressor is being replaced, drain any shipping oil that may be in the new compressor.
8 If the amount of refrigerant oil drained from the old compressor was 3 to 5 ounces, add new oil in that amount plus an extra ounce to the new compressor. If the amount drained was more than 5 ounces, add that amount of new oil, and if the drained amount was less than three ounces, add three ounces of new oil to the new compressor.
9 Installation procedures are the reverse of those for removal. When installing the fitting block, use new O-rings and lubricate them with clean refrigerant oil (see illustration).

16.2 For access to the compressor on Expedition and Navigator models, remove this plastic panel (arrow) below the radiator

16.5 Remove the mounting bolts (arrows) and remove the air conditioning compressor from the engine compartment

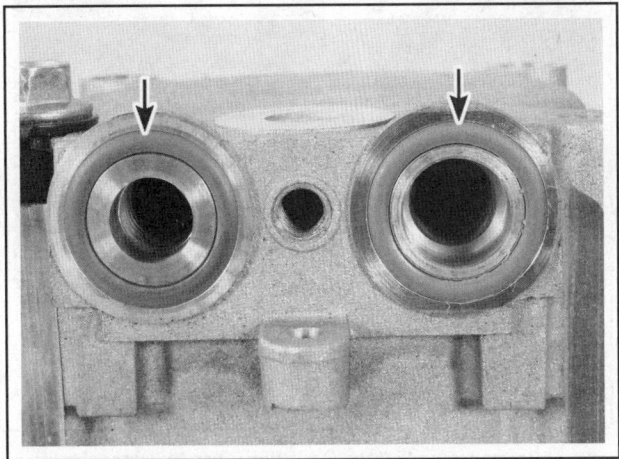

16.9 Use new O-rings (arrows), lubricated with refrigerant oil, when reattaching the line block to the compressor

3-22 COOLING, HEATING AND AIR CONDITIONING SYSTEMS

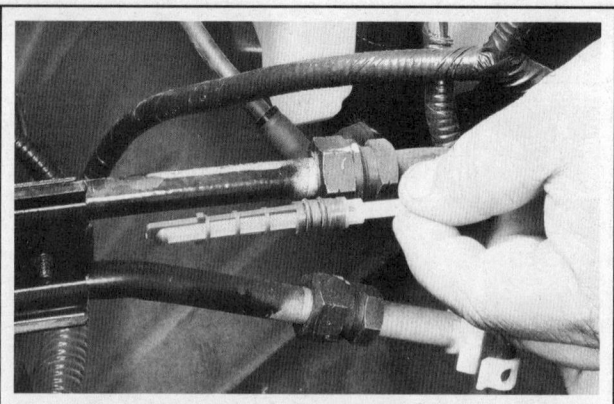

16.10 When the air conditioning compressor is replaced, the orifice tube must be replaced, too (2007 and earlier models)

10 On 2007 and earlier models, replace the orifice tube. The orifice is a plastic tube with a filter inside. On some models it is located in the high pressure line from the condenser (see illustration). On other models it's located in the evaporator inlet line. Recessed inside the refrigerant pipe slightly, it is difficult to remove without a small tool that is inserted and expanded until it grabs the orifice tube so that it can be pulled out.

➡ **Note: While it is sometimes possible to remove an orifice tube with a pair of pliers, this practice is not recommended since it can result in a broken orifice tube stuck inside the pipe.**

The same tool, available at auto parts stores, can be used to insert a new orifice tube. Make sure the tube is inserted in the same direction as the old one.

11 After the compressor is installed, have the system evacuated, recharged and leak tested by a dealer service department or an air conditioning repair facility.

17 Air conditioning condenser - removal and installation

※※ **WARNING 1:**

The models covered by this manual are equipped with Supplemental Restraint Systems (SRS), more commonly known as airbags. Always disconnect the negative battery cable, then the positive battery cable and wait two minutes before working in the vicinity of the impact sensors, steering column or instrument panel to avoid the possibility of accidental deployment of the airbag, which could cause personal injury (see Chapter 12). Do not use any electrical test equipment on any of the airbag system wires or tamper with them in any way.

※※ **WARNING 2:**

The air conditioning system is under high pressure. DO NOT loosen any fittings or remove any components until after the system has been discharged. Air conditioning refrigerant should be properly discharged into an EPA-approved container at a dealer service department or an automotive air conditioning repair facility. Always wear eye protection when disconnecting air conditioning system fittings.

➡ **Note: On some models, special spring-lock coupling tools are required to release the connectors used on the refrigerant lines throughout the air conditioning system. There are different special tools for each line size; these tools can usually be found at local auto parts stores, often in a set. See Section 15 for tool description and use.**

REMOVAL

▸ Refer to illustrations 17.5 and 17.6

※※ **CAUTION:**

Whenever a condenser is replaced, it will be necessary to replace the accumulator/drier (see Section 15).

1 Take the vehicle to a dealer service department or automotive air conditioning shop and have the air conditioning system discharged.
2 Disconnect the battery cables and remove the battery and battery tray (refer to Chapter 5).
3 Refer to Section 4 and remove the cooling fan and shroud assembly.
4 Refer to Section 5 and remove the degas bottle and radiator.
5 On 2010 and later models remove the headlight housings (see Chapter 12).
6 Disconnect the two fittings holding the refrigerant lines to the right side of the condenser (see illustration).

➡ **Note: Some models use threaded fittings; others use flange-type fittings secured by a bolt.**

7 Remove the two condenser mounting brackets bolted to the radiator support (see illustration). Carefully lift the condenser out of the bottom cradle supports. On 2007 and later Expedition/Navigator models, remove the six bolts securing the condenser/radiator unit to the body, then remove the condenser.

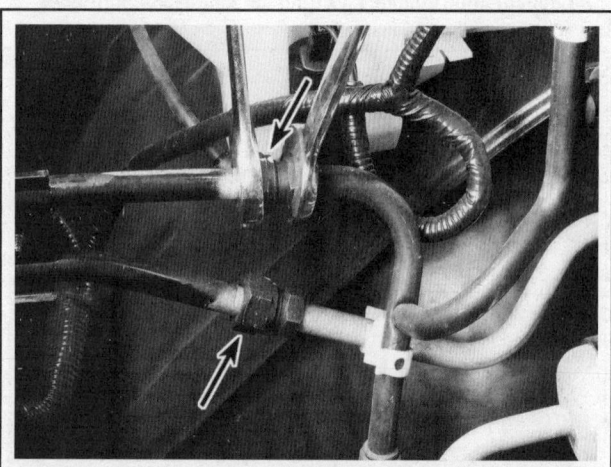

17.5 On models with threaded fittings, use two wrenches to disconnect the fittings retaining the refrigerant lines to the condenser tubes

COOLING, HEATING AND AIR CONDITIONING SYSTEMS   3-23

INSTALLATION

8 If the condenser is being replaced with a new one, transfer the brackets and mounts from the old unit to the new one.

9 When replacing the condenser add one ounce of new refrigerant oil to the condenser before reassembly. This will maintain the correct oil level in the system after the repairs are completed.

10 Before installation, check the bracket assemblies and mounts for excessive wear or damage. Replace them if necessary.

11 The installation procedures are the reverse of those for removal. When connecting the hose and fittings, use new O-rings and lubricate them with clean refrigerant oil (where applicable).

12 After the condenser is installed have the system evacuated, recharged and leak tested by a dealer service department or an air conditioning repair facility.

17.6 The condenser is mounted to the radiator support with two brackets (arrow indicates left-side bracket) - remove them and pull the condenser up and out of the vehicle

18 Air conditioning evaporator - removal and installation

▶ Refer to illustration 18.1

※ WARNING 1:

The models covered by this manual are equipped with Supplemental Restraint Systems (SRS), more commonly known as airbags. Always disconnect the negative battery cable, then the positive battery cable and wait two minutes before working in the vicinity of the impact sensors, steering column or instrument panel to avoid the possibility of accidental deployment of the airbag, which could cause personal injury (see Chapter 12). Do not use any electrical test equipment on any of the airbag system wires or tamper with them in any way.

※ WARNING 2:

The air conditioning system is under high pressure. DO NOT loosen any fittings or remove any components until after the system has been discharged. Air conditioning refrigerant should be properly discharged into an EPA-approved container at a dealer service department or an automotive air conditioning repair facility. Always eye wear protection when disconnecting air conditioning system fittings.

➡ Note: On some models, special spring-lock coupling tools are required to release the connectors used on the refrigerant lines throughout the air conditioning system. There are different special tools for each line size; these tools can usually be found at local auto parts stores, often in a set. See Section 15 for tool description and use.

1 The evaporator core is located inside the same housing as the heater core (see illustration).

➡ Note: Before replacing an evaporator core, determine for certain that the core is leaking by having a leak test performed with special equipment at dealer service department or automotive air conditioning repair facility.

2 Refer to Section 13 for removal of the cover to the heater/evaporator housing. This procedure requires the removal of the instrument panel (see Chapter 11).

➡ Note: Whenever the evaporator core is replaced with a new one, the accumulator/drier will also have to be replaced (see Section 15). Also remove the accumulator/drier mounting bracket from the firewall.

3 Disconnect the refrigerant lines at the firewall (see Section 15).

4 Pull the evaporator core from the heater/evaporator housing.

5 Installation is the reverse of the removal process. Add three ounces of new refrigerant oil to the accumulator/drier inlet tube when a new evaporator core is installed. Also, before the lines are reconnected, it's a good idea to replace the orifice tube (see Section 16).

6 On some models of Navigator/Expedition vehicles, there is an optional heat/air-conditioning system located behind the left-rear interior trim panel (behind the left-rear wheel housing). The assembly includes a blower, heater core and evaporator core (see illustration 11.12). Removal and installation procedures are similar to the procedures for the standard evaporator core up front, although the rear unit is much more accessible.

7 Have the system evacuated, recharged and leak tested by the dealer service department or an air conditioning repair facility.

18.1 The evaporator core (arrow) is inside the same case as the heater core

3-24 COOLING, HEATING AND AIR CONDITIONING SYSTEMS

Specifications

General

Cooling system capacity	See Chapter 1
Coolant type	See Chapter 1
Thermostat	
Opening temperature	188 to 195 degrees
Fully open temperature	208 to 215 degrees
Radiator cap pressure rating	See Chapter 1
Refrigerant type	R-134a
Refrigerant capacity	
2003 and earlier models	
Without auxiliary air conditioning	34 to 37 ounces
With auxiliary air conditioning	62 ounces
2004 through 2007 models	
Without auxiliary air conditioning	40 ounces
With auxiliary air conditioning	58 ounces
2008 and later models	
Without auxiliary air conditioning	22 ounces
With auxiliary air conditioning	33 ounces

Torque specifications   Ft-lbs (unless otherwise indicated)

➡ **Note:** One foot-pound (ft-lb) of torque is equivalent to 12 inch-pounds (in-lbs) of torque. Torque values below approximately 15 ft-lbs are expressed in inch-pounds, since most foot-pound torque wrenches are not accurate at these smaller values.

Thermostat housing bolts	
3.5L V6 engine	
Step 1	71 in-lbs
Step 2	Tighten an additional 45-degrees
4.2L V6 engine	71 to 103 in-lbs
Thermostat housing cover bolts (3.5L engines)	89 in-lbs
V8 engines	
2004 and earlier models	15 to 22
2005 and later models	89 in-lbs
Water pump-to-engine bolts	
3.5L V6 engine	
Step 1	Hand tighten
Step 2	89 in-lbs
Step 3	Tighten an additional 45-degrees
4.2L V6 engine	
Nuts	53 to 71 in-lbs
Bolts	15 to 22
V8 engines	15 to 22
Water pump pulley to hub	15 to 22
Transmission oil line fitting-to-radiator	12 to 18
Engine oil cooler insert fastener	41 to 44
Oil cooler adapter to block bolts	15 to 22
Fan shroud-to-radiator	
2009 and earlier	71 to 89 in-lbs
2010 and later	62 in-lbs
Fan assembly-to-fan clutch bolts	156 in-lbs
Fan clutch-to-water pump nut	41

4 FUEL AND EXHAUST SYSTEMS

Section

1. General information
2. Fuel pressure relief procedure
3. Fuel pump/fuel pressure - check
4. Fuel lines and fittings - general information
5. Fuel tank - removal and installation
6. Fuel tank - cleaning and repair
7. Fuel pump - removal and installation
8. Fuel level sending unit - check and replacement
9. Air fliter housing - removal and installation
10. Accelerator cable - removal, installation and adjustment
11. Fuel injection system - general information
12. Fuel injection system - check
13. Sequential Electronic Fuel Injection (SEFI) system (2014 and earlier models) - component check and replacement
14. Idle Air Control (IAC) valve (2004 and earlier models) - check, removal and adjustment
15. Intake Air Systems
16. Exhaust system servicing - general information
17. High-pressure fuel pump (3.5L V6 engine) - removal and installation
18. Fuel Pump Driver Module (FPDM) - replacement
19. Fuel rails and injectors (3.5L V6 engine) - removal and installation
20. Turbochargers (3.5L V6 engine) - removal and installation
21. Wastegate control actuator (3.5L V6 engine) - removal and installation
22. Intercooler (3.5L V6 engine) - removal and installation

1 General information

The fuel system consists of a fuel tank, an electric fuel pump (located in the fuel tank), a fuel pump relay, the fuel rail and fuel injectors, an air cleaner assembly and a throttle body unit. All models are equipped with a Sequential Electronic Fuel Injection (SEFI) system.

SEQUENTIAL ELECTRONIC FUEL INJECTION (SEFI) SYSTEM

Sequential Electronic Fuel Injection uses timed impulses to inject the fuel directly into the intake port of each cylinder according to its firing order. The injectors are controlled by the Powertrain Control Module (PCM). The PCM monitors various engine parameters and delivers the exact amount of fuel required into the intake ports. The throttle body serves only to control the amount of air passing into the system. Because each cylinder is equipped with its own injector, much better control of the fuel/air mixture ratio is possible.

FUEL PUMP AND LINES

Fuel is circulated from the fuel tank to the fuel injection system, and back to the fuel tank, through a pair of metal lines running along the underside of the vehicle. An electric fuel pump and fuel level sending unit is located inside the fuel tank. A vapor return system routes all vapors back to the fuel tank through a separate return line.

The fuel pump relay is equipped with a primary and secondary voltage circuit. The primary circuit is controlled by the PCM and the secondary circuit is linked directly to battery voltage from the ignition switch. With the ignition switch ON (engine not running), the PCM will ground the relay for one second. During cranking, the PCM grounds the fuel pump relay as long as the camshaft position sensor (CMP) sends its position signal (see Chapter 6). If there are no reference pulses, the fuel pump will shut off after two or three seconds.

The fuel pressure regulator is mounted in different locations depending on the year and engine. 1997 through 2004 models mount the fuel pressure regulator on the fuel rail. Some 2001 through 2004 and all 2005 and later models are equipped with the Returnless fuel system.

Return fuel system (1997 through 2004)

Fuel is circulated from the fuel tank to the fuel injection system and back to the fuel tank through a pair of metal lines running along the underside of the vehicle. An electric fuel pump is located inside the fuel tank. A vapor return system routes all vapors back to the fuel tank through a separate return line. The fuel pump will operate as long as the engine is running or cranking and the PCM is receiving ignition reference pulses from the electronic ignition system.

Returnless fuel system (2001 and later)

Fuel is circulated from the fuel tank to the fuel injection system through a metal line running along the underside of the vehicle. An electric fuel pump/fuel level sending unit is located inside the fuel tank. The fuel pump/fuel level sending unit assembly consists of the pump, the fuel level sending unit, an inlet filter (sometimes referred to as a *sock* or *strainer*), a check valve to maintain pressure after the pump is shut off and a pressure relief valve to protect the pump from over-pressurization in the event of a blocked fuel line.

Mechanical returnless fuel system (MRFS) - 2001 through 2004 5.4L Expedition models

This system incorporates the fuel pressure regulator as an integral component of the fuel pump/fuel level sending unit that is located in the fuel tank. This updated fuel supply system does not return fuel to the fuel tank; instead, the returnless fuel system bleeds off excess fuel directly at the fuel pressure regulator. Refer to Section 7 for the fuel pump and fuel pressure regulator replacement procedure.

Electronic returnless fuel system (ERFS) - 2005 and later models

What sets this fuel system apart from conventional in-tank pumps is its variable speed capability. The PCM controls fuel pressure by controlling the speed (rpm) of the pump. The PCM alters the fuel pressure by controlling the duty cycle of the Fuel Pump Driver Module (FPDM), which in turn controls the speed of the fuel pump by modulating the voltage to the fuel pump.

DIRECT INJECTION (DI) SYSTEM (3.5L V6 MODELS)

The Direct Injection (DI) system consists of the fuel tank, a 2-speed electric fuel pump/fuel level sending unit module mounted inside the tank, the fuel pressure regulator (integral with the fuel pump module), a fuel pump flow control module, the high-pressure fuel pump, the fuel rail, the fuel injectors, and the metal and flexible fuel lines that connect the various components of the DI system.

INERTIA SWITCH

The inertia switch (located behind passenger side kick panel) will disable the fuel pump circuit in the event of collision. The inertia switch is a cylindrical magnet with a steel ball that will release (breakaway) and trip a shutdown lever when the vehicle inertia reaches a certain peak value.

EXHAUST SYSTEM

The exhaust system includes an exhaust manifold, diverter pipes fitted with upstream (before catalytic converter) and downstream (after catalytic converter) oxygen sensors, a catalytic converter and a muffler.

The catalytic converter is an emission control device added to the exhaust system to reduce pollutants. A single-bed converter is used in combination with a three-way (reduction) catalyst. Refer to Chapter 6 for more information regarding the catalytic converter.

FUEL AND EXHAUST SYSTEMS 4-3

2 Fuel pressure relief procedure

※ WARNING:

Gasoline is extremely flammable, so take extra precautions when you work on any part of the fuel system. Don't smoke or allow open flames or bare light bulbs near the work area, and don't work in a garage where a gas-type appliance (such as a water heater or a clothes dryer) is present. Since gasoline is carcinogenic, wear latex gloves when there's a possibility of being exposed to fuel, and, if you spill any fuel on your skin, rinse it off immediately with soap and water. Mop up any spills immediately and do not store fuel-soaked rags where they could ignite. The fuel system is under constant pressure, so, if any fuel lines are to be disconnected, the fuel pressure in the system must be relieved first. When you perform any kind of work on the fuel system, wear safety glasses and have a Class B type fire extinguisher on hand.

→ **Note:** After the fuel pressure has been relieved, it's a good idea to lay a shop towel over any fuel connection to be disassembled, to absorb the residual fuel that may leak out when servicing the fuel system.

All except 3.5L engine

1 There are two methods for relieving the fuel system pressure; the easiest and most accessible is using a special fuel pressure gauge with a bleed-off valve. This special tool can be purchased at an automotive parts or tool company. In the event the tool is not available, locate the inertia switch and disable the fuel pump.

FUEL PRESSURE GAUGE BLEEDING METHOD

▸ Refer to illustration 2.3

2 Locate the fuel pressure test port on the fuel rail and install the fuel pressure gauge onto the Schrader valve.

3 Direct the bleed-off hose into a metal cup or suitable container for gasoline storage (see illustration).

4 Turn the valve and allow the excess fuel to bleed into the container.

5 Close the valve, remove the fuel pressure gauge and cap the test port.

INERTIA SWITCH METHOD

▸ Refer to illustrations 2.7 and 2.9

6 The fuel pump switch - sometimes called the "inertia switch" - which shuts off fuel to the engine in the event of a collision, affords a simple and convenient means by which fuel pressure can be relieved before servicing fuel injection components. The switch is located behind the passenger's side kick panel.

7 Unplug the inertia switch electrical connector (see illustration).

8 Start the engine and allow it to run until it stops. This should take only a few seconds.

9 The fuel system pressure is now relieved. When you're finished working on the fuel system, simply plug the electrical connector back into the switch. If the inertia switch was "popped" (activated) during this procedure, push the reset button on the top of the switch (see illustration).

3.5L V6 ENGINE

10 Unplug the Fuel Pump Driver Module (FPDM) electrical connector.

→ **Note: The FPDM is located on the left rear frame rail, above the spare tire.**

11 Start the engine and allow it to run until it stops. This should take only a few seconds.

※ WARNING:

The fuel pressure is relieved on the low-pressure side of the system, but pressure will still remain in the high-pressure side of the system. Before loosening any connections on the high-pressure side of the system, wait at least two hours.

12 The fuel system pressure is now relieved. Disconnect the cable from the negative terminal of the battery before performing any work on the fuel system.

13 When you're finished working on the fuel system, simply reconnect the electrical connector to the FPDM.

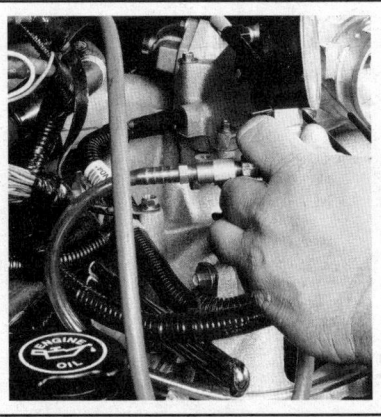

2.3 Install a special fuel pressure gauge onto the test port connector and bleed the fuel into a suitable container

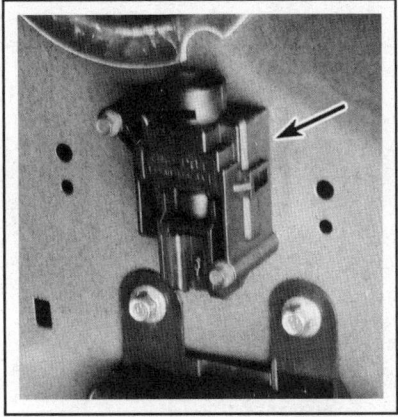

2.7 The inertia switch (arrow) is located behind the passenger side kick panel. Disconnect the electrical connector to disable the fuel pump

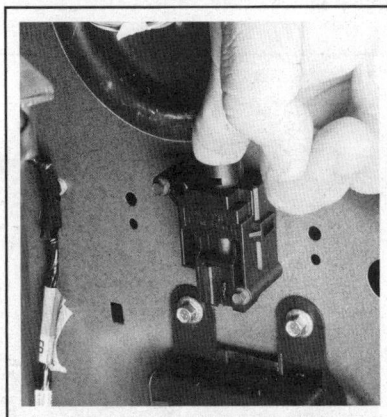

2.9 If necessary, push the reset button after connecting the inertia switch to energize the fuel pump

4-4 FUEL AND EXHAUST SYSTEMS

3 Fuel pump/fuel pressure - check

✻✻ WARNING:

Gasoline is extremely flammable, so take extra precautions when you work on any part of the fuel system. See the *Warning* in Section 2.

➡ Note 1: To perform the fuel pressure test, you will need to obtain a fuel pressure gauge and adapter set (fuel line fittings).

➡ Note 2: The fuel pump will operate as long as the engine is cranking or running and the PCM is receiving ignition reference pulses from the electronic ignition system. If there are no reference pulses, the fuel pump will shut off after two or three seconds.

➡ Note 3: After the fuel pressure has been relieved, it's a good idea to lay a shop towel over any fuel connection to be disassembled, to absorb the residual fuel that may leak out when servicing the fuel system.

PRELIMINARY INSPECTION

▸ Refer to illustrations 3.2 and 3.3

1 Should the fuel system fail to deliver the proper amount of fuel, or any fuel at all, inspect it as follows. Remove the fuel filler cap. Have an assistant turn the ignition key to the On position (engine not running) while you listen at the fuel filler opening. You should hear a whirring sound that lasts for a couple of seconds.

2 If you don't hear anything, check the fuel pump fuse (see Chapter 12). If the fuse is blown, replace it and see if it blows again (see illustration). If it does, trace the fuel pump circuit for a short. Refer to the wiring diagrams at the end of Chapter 12 for additional wiring schematics.

3 Check for battery voltage to the fuel pump relay connector and the PCM relay connector (see illustration). If there is battery voltage present, have the relay(s) tested at a dealer service department or other qualified automotive repair shop.

➡ Note 1: The inertia switch is an electrical device wired into the fuel pump circuit that will shut down power to the fuel pump in an accident. Be sure to check that the inertia switch is activated and in working order if the fuel pump is not receiving the proper voltage (see Section 2).

3.3 Remove the fuel pump relay and check for battery voltage to the relay with the ignition key ON (engine not running)

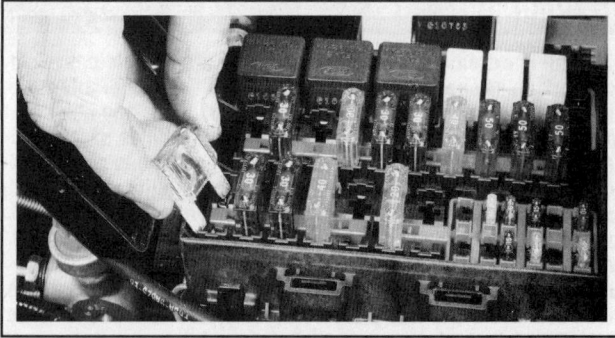

3.2 Remove the fuel pump fuse and make sure it is not blown

➡ Note 2: The fuel pump relay is equipped with a primary and secondary voltage circuit. The primary circuit is controlled by the PCM and the secondary circuit is linked directly to battery voltage from the ignition switch. With the ignition switch ON (engine not running), the PCM will ground the relay for one second. During cranking, the PCM grounds the fuel pump relay as long as the camshaft position sensor (CMP) sends its position signal (see Chapter 6). If there are no reference pulses, the fuel pump will shut off after two or three seconds. Refer to the wiring schematics at the end of Chapter 12 for additional information on the wiring color designations for the fuel pump relay.

4 If there is no voltage present, check the fuse(s) and the wiring circuit for the fuel pump relay and/or PCM power relay (see Chapter 12). If voltage is present, check for battery voltage at the fuel pump harness connector located near the fuel tank. If voltage is reaching the fuel pump, remove the fuel pump and have it checked by a dealer service department or other qualified automotive repair facility.

OPERATING PRESSURE CHECK

▸ Refer to illustration 3.6

➡ Note: Before proceeding, obtain a fuel pressure gauge capable of measuring fuel pressure well above the specified operating range of the fuel system you're going to test. You will also need fittings suitable for connecting the gauge onto the fuel rail (early models) or into the fuel system between the fuel delivery line and the fuel rail (late models).

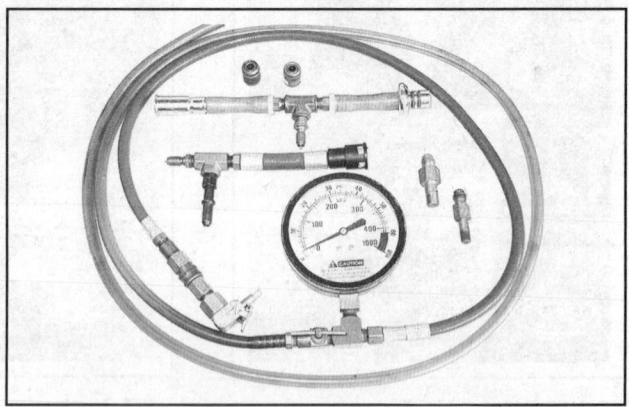

3.6 A typical fuel pressure gauge with hoses and fittings suitable for tee-ing in to the fuel system between the fuel delivery line and the fuel rail

FUEL AND EXHAUST SYSTEMS   4-5

3.7 If you don't have the correct adapter, it is possible to remove the Schrader valve from the fitting and install a standard fuel pressure gauge, using a hose clamp

3.9 Detach the vacuum line from the fuel pressure regulator and verify vacuum is present when the engine is running

3.12 Connect a hand-held vacuum pump to the fuel pressure regulator and read fuel pressure with vacuum applied. Pressure should decrease as vacuum is increased

5 Relieve the fuel system pressure (see Section 2).
6 In addition to a fuel pressure gauge capable of reading fuel pressure up to 70 psi, you'll need a hose and an adapter suitable for connecting the gauge onto the fuel rail (early models) or into the fuel system between the fuel delivery line and the fuel rail (late models) (see illustration).

Return fuel system (1997 through 2004)

▶ **Refer to illustrations 3.7, 3.9 and 3.12**

7 Remove the cap from the fuel pressure test port and attach a fuel pressure gauge (see illustration). If you don't have the correct adapter for the test port, remove the Schrader valve and connect the gauge hose to the fitting, using a hose clamp.
8 Start the engine.
9 Check the fuel pressure at idle. Compare your readings with the values listed in this Chapter's Specifications. Disconnect the vacuum hose from the fuel pressure regulator and watch the fuel pressure gauge - the fuel pressure should jump up considerably as soon as the hose is disconnected (see illustration). If it doesn't, check for a vacuum signal to the fuel pressure regulator (see Step 14).
10 If the fuel pressure is low, pinch the fuel return line shut and watch the gauge. If the pressure doesn't rise, the fuel pump is defective or there is a restriction in the fuel feed line. If the pressure rises sharply, replace the pressure regulator.

➡ **Note:** *If the vehicle is equipped with a nylon fuel return line (or fuel lines made up of steel or other rigid material), it will be necessary to install a special fuel testing harness between the fuel rail and the return line. This can be made up from compatible fuel line connectors (available at a dealer parts department and some auto parts stores), fuel hose and hose clamps.*

11 If the fuel pressure is too high, turn the engine off. Disconnect the fuel return line and blow through it to check for a blockage. If there is no blockage, replace the fuel pressure regulator.
12 Hook up a hand-held vacuum pump to the port on the fuel pressure regulator (see illustration).
13 Read the fuel pressure gauge with vacuum applied to the fuel pressure regulator and also with no vacuum applied. The fuel pressure should decrease as vacuum increases (and increase as vacuum decreases).

14 Connect a vacuum gauge to the pressure regulator vacuum hose. Start the engine and check for vacuum. If there isn't vacuum present, check for a clogged hose or vacuum port. If the amount of vacuum is adequate, replace the fuel pressure regulator.
15 Turn the ignition switch to OFF, wait five minutes and recheck the pressure on the gauge. Compare the reading with the hold pressure listed in this Chapter's Specifications. If the hold pressure is less than specified:

   a) *The fuel lines may be leaking.*
   b) *The fuel pressure regulator may be allowing the fuel pressure to bleed through to the return line.*
   c) *A fuel injector (or injectors) may be leaking.*
   d) *The fuel pump may be defective.*

Returnless fuel system (2001 and later)

➡ **Note:** *Some models are equipped with a fuel pressure test port mounted onto the fuel rail. Follow the fuel pressure gauge installation procedure in Step 7 but follow the fuel pressure testing procedure for the returnless fuel systems.*

16 Disconnect the quick-connect fitting at the connection between the fuel delivery hose and the fuel rail (see Section 4).
17 Tee into the fuel pressure gauge between the fuel delivery hose and the fuel rail. Note that some models may be equipped with a Schrader valve located on the fuel rail that allows easy attachment of the fuel pressure gauge (see illustration 3.6).
18 Turn off all the accessories, then start the engine and let it idle. The fuel pressure should be within the operating range listed in this Chapter's Specifications. If the pressure reading is within the specified range, the system is operating correctly.
19 If the fuel pressure is higher than specified, then the pump, the Fuel Pump Driver Module (FPDM), the Powertrain Control Module (PCM) or the circuit connecting these components is probably defective. Checking this circuit is beyond the scope of the home mechanic, so have the circuit checked by a professional.
20 If the fuel pressure is lower than specified, inspect the fuel delivery lines and hoses for an obstruction or a kink. Also inspect all fuel delivery line and hose quick-connect fittings for leaks. Replace the fuel filter (see Chapter 1) and re-check the pressure. If the lines, hoses, connections and the fuel filter are all in good shape, remove the fuel pump/fuel level sensor assembly (see Section 6) and inspect the fuel pump

4-6 FUEL AND EXHAUST SYSTEMS

inlet strainer for restrictions. If everything else is okay, replace the fuel pump (see Section 6).

21 Turn the ignition switch to OFF, wait five minutes and recheck the pressure on the gauge. Compare the reading with the hold pressure listed in this Chapter's Specifications. If the hold pressure is less than specified:

a) The fuel delivery line or a quick-connect fitting might be leaking.
b) A fuel injector (or injectors) may be leaking.
c) The fuel pump might be defective.

22 After the testing is complete, relieve the fuel pressure (see Section 2), remove the fuel pressure gauge and reconnect the fuel delivery line to the fuel rail (see Section 4).

3.5L V6 engine

23 Disconnect the quick-connect fitting at the connection between the fuel delivery hose and the high-pressure fuel pump. Tee in the fuel pressure gauge between the fuel delivery hose and the high-pressure fuel pump.

➡ **Note:** *The fuel line quick-connect fitting on the high-pressure fuel pump is at the rear of the right-side cylinder head.*

24 Turn the Key On, Engine Off (KOEO) and check for fuel leaks from the fuel pressure test adapter fittings.

25 Turn off all the accessories, then start the engine and let it idle. The fuel pressure should be within the operating range listed in this Chapter's Specifications. If the pressure reading is within the specified range, the system is operating correctly.

26 If the fuel pressure is higher than specified, then the pump, the Fuel Pump Driver Module (FPDM), the Powertrain Control Module (PCM) or the circuit connecting these components is probably defective. But checking this circuit is beyond the scope of the home mechanic, so have the circuit checked by a professional.

27 If the fuel pressure is lower than specified, inspect the fuel delivery lines and hoses for an obstruction or a kink. Also inspect all fuel delivery line and hose quick-connect fittings for leaks. If the lines, hoses, and connections are all in good shape, remove the fuel pump/fuel level sensor assembly (see Section 7) and inspect the fuel pump inlet strainer for restrictions. If everything else is okay, replace the fuel pump.

28 Turn the ignition switch to Off, wait five minutes and recheck the pressure on the gauge. Compare the reading with the hold pressure listed in this Chapter's Specifications. If the hold pressure is less than specified:

a) The fuel delivery line or a quick-connect fitting might be leaking.
b) A fuel injector (or injectors) may be leaking.
c) The fuel pump might be defective.

29 After the testing is complete, relieve the fuel pressure (see Section 2), remove the fuel pressure gauge and reconnect the fuel delivery line to the fuel rail.

4 Fuel lines and fittings - general information

⁂ WARNING:

Gasoline is extremely flammable, so take extra precautions when you work on any part of the fuel system. See the *Warning* in Section 2.

PUSH-CONNECT FITTINGS - DISASSEMBLY AND REASSEMBLY

1 The manufacturer uses two different push-connect fitting designs. Fittings used with 3/8 and 5/16-inch diameter lines have a "hairpin" type clip; fittings used with 1/4-inch diameter lines have a "duck bill" type clip. The procedure used for releasing each type of fitting is different. The clips should be replaced whenever a connector is disassembled.

2 Disconnect all push-connect fittings from fuel system components such as the fuel filter, the fuel charging assembly, the fuel tank, etc. before removing the assembly.

3/8 and 5/16-inch fittings (hairpin clip)

▶ **Refer to illustration 4.5**

3 Inspect the internal portion of the fitting for accumulations of dirt. If more than a light coating of dust is present, clean the fitting before disassembly.

4 Some adhesion between the seals in the fitting and the line will occur over a period of time. Twist the fitting on the line, then push and pull the fitting until it moves freely.

5 Remove the hairpin clip from the fitting by bending the shipping tab down until it clears the body. Then, using nothing but your hands, spread each leg about 1/8-inch to disengage the body and push the legs through the fitting. Remember, don't use any tools to perform this part of the procedure. Finally, pull lightly on the triangular end of the clip and work it clear of the line and fitting (see illustration).

6 Grasp the fitting and hose and pull it straight off the line.

7 Do not reuse the original clip in the fitting. A new clip must be used.

8 Before reinstalling the fitting on the line, wipe the line end with a clean cloth. Inspect the inside of the fitting to ensure that it's free of dirt and/or obstructions.

9 To reinstall the fitting on the line, align them and push the fitting into place. When the fitting is engaged, a definite click will be heard. Pull on the fitting to ensure that it's completely engaged. To install the new clip, insert it into any two adjacent openings in the fitting with the triangular portion of the clip pointing away from the fitting opening. Using your index finger, push the clip in until the legs are locked on the outside of the fitting.

4.5 A hairpin clip type push-connect fitting

FUEL AND EXHAUST SYSTEMS 4-7

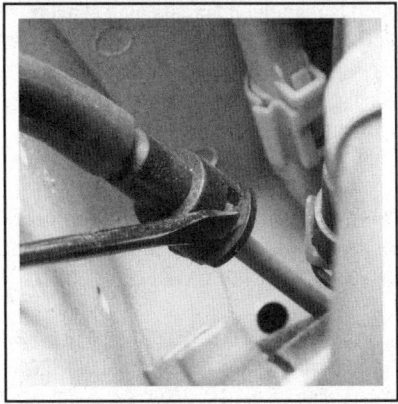

4.10 A push-connect fitting with a duck bill clip

4.15a If the spring lock couplings are equipped with safety clips, pry them off with a small screwdriver

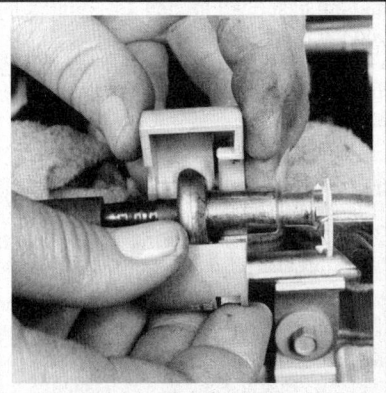

4.15b Open the spring-loaded halves of the spring lock coupling tool and place it in position around the coupling, then close it

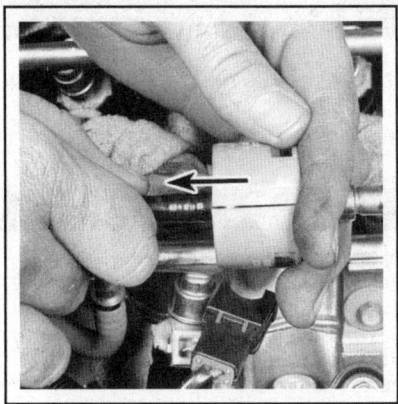

4.15c To disconnect the coupling, push the tool into the cage opening to expand the garter spring and release the female fitting, then pull the male and female fittings apart

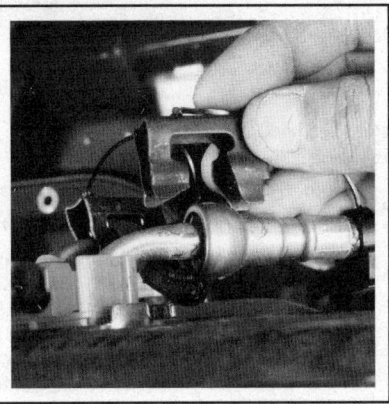

4.16 Remove the safety clamp

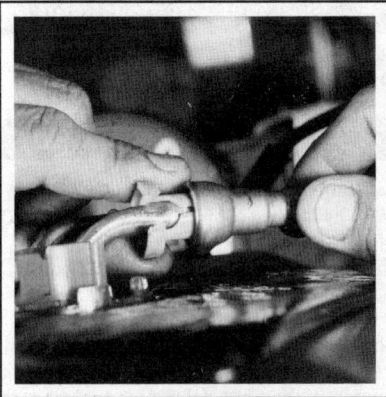

4.17 Duck bill clip fitting disassembly using the special tool

1/4-inch fittings (duck bill clip)

▶ Refer to illustration 4.10

10 The duck bill clip type fitting consists of a body, spacers, O-rings and the retaining clip (see illustration). The clip holds the fitting securely in place on the line. One of the two following methods must be used to disconnect this type of fitting.

11 Before attempting to disconnect the fitting, check the visible internal portion of the fitting for accumulations of dirt. If more than a light coating of dust is evident, clean the fitting before disassembly.

12 Some adhesion between the seals in the fitting and line will occur over a period of time. Twist the fitting on the line, then push and pull the fitting until it moves freely.

SPRING LOCK COUPLINGS - DISASSEMBLY AND REASSEMBLY

Type I

▶ Refer to illustrations 4.15a, 4.15b and 4.15c

13 The fuel supply and return lines used on SEFI engines utilize spring lock couplings at the engine fuel rail end instead of plastic push-connect fittings. The male end of the spring lock coupling, which is girded by two O-rings, is inserted into a female flared end engine fitting. The coupling is secured by a garter spring which prevents disengagement by gripping the flared end of the female fitting. A cup-tether assembly provides additional security.

14 To disconnect the 1/2-inch spring lock coupling supply fitting, you will need to obtain a spring lock coupling tool, available at most auto parts sales stores. Be aware that 1/2- inch and 3/8-inch fittings require different tools

15 Study the accompanying illustrations carefully before detaching either spring lock coupling fitting (see illustrations).

Type II

16 Remove the safety clamp from the fuel line (see illustration).

17 The preferred method used to disconnect the fitting requires a special tool. To disengage the line from the fitting, align the slot in the push-connect disassembly tool, available at most auto parts stores, with either tab on the clip (90-degrees from the slots on the side of the fitting) and insert the tool (see illustration). This disengages the duck bill from the line.

➡ **Note:** Some fuel lines have a secondary bead which aligns with the outer surface of the clip. The bead can make tool insertion difficult. If necessary, use the alternative disassembly

4-8 FUEL AND EXHAUST SYSTEMS

method described in Step 19. Holding the tool and the line with one hand, pull the fitting off.

➡ **Note:** *Only moderate effort is necessary if the clip is properly disengaged. The use of anything other than your hands should not be required.*

18 After disassembly, inspect and clean the line sealing surface. Also inspect the inside of the fitting and the line for any internal parts that may have been dislodged from the fitting. Any loose internal parts should be immediately reinstalled (use the line to insert the parts).

19 The alternative disassembly procedure requires a pair of small adjustable pliers. The pliers must have a jaw width of 3/16-inch or less.

20 Align the jaws of the pliers with the openings in the side of the fitting and compress the portion of the retaining clip that engages the body. This disengages the retaining clip from the body (often one side of the clip will disengage before the other - both sides must be disengaged).

21 Pull the fitting off the line.

➡ **Note:** *Only moderate effort is required if the retaining clip has been properly disengaged. Do not use any tools for this procedure.*

22 Once the fitting is removed from the line end, check the fitting and line for any internal parts that may have been dislodged from the fitting. Any loose internal parts should be immediately reinstalled (use the line to insert the parts).

23 The retaining clip will remain on the line. Disengage the clip from the line bead to remove it. Do not reuse the retaining clip - install a new one!

24 Before reinstalling the fitting, wipe the line end with a clean cloth. Check the inside of the fitting to make sure that it's free of dirt and/or obstructions.

25 To reinstall the fitting, align it with the line and push it into place. When the fitting is engaged, a definite click will be heard. Pull on the fitting to ensure that it's fully engaged.

26 Install the new replacement clip by inserting one of the serrated edges on the duck bill portion into one of the openings. Push on the other side until the clip snaps into place.

5 Fuel tank - removal and installation

▸ Refer to illustration 5.6 and 5.10

※ WARNING:

Gasoline is extremely flammable, so take extra precautions when you work on any part of the fuel system. See the *Warning* in Section 2.

➡ **Note:** *Don't begin this procedure until the gauge indicates that the tank is empty or nearly empty. If the tank must be removed when it's full (for example, if the fuel pump malfunctions), siphon any remaining fuel from the tank prior to removal.*

1 Unless the vehicle has been driven far enough to completely empty the tank, it's a good idea to siphon the residual fuel out before removing the tank from the vehicle.

※ WARNING:

DO NOT start the siphoning action by mouth! Use a siphoning kit (available at most auto parts stores).

2 Relieve the fuel pressure (see Section 2).
3 Detach the cable from the negative terminal of the battery.
4 Raise the vehicle and support it securely on jackstands.

※ WARNING:

Some models covered by this manual are equipped with self leveling suspension systems. Always disconnect electrical power to the suspension system before lifting or towing the vehicle (see Chapter 10). Failure to perform this procedure may result in unexpected shifting or movement of the vehicle which could cause personal injury.

5 Remove the fuel tank skid plate mounting bolts and lower the assembly (if equipped).

6 Remove the fuel tank filler hose and vapor hose from the fuel filler neck and the fuel tank (see illustration) and slide the assembly from the vehicle.

7 Disconnect the fuel lines (see Section 4) and EVAP vapor lines. On 2010 and later models, remove the Fuel Tank Pressure (FTP) sensor, then remove the stabilizer bar (see Chapter 10).

5.6 Remove the fuel tank filler hose and vent hose from the filler neck assembly

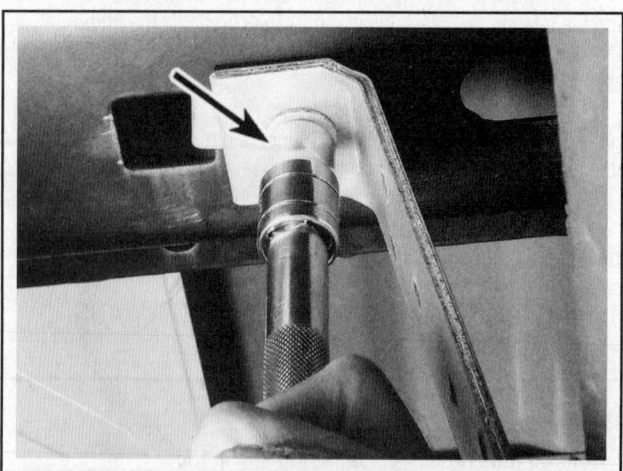

5.10 Remove the bolts (arrow) from the fuel tank straps

FUEL AND EXHAUST SYSTEMS   4-9

8 Disconnect the electric fuel pump and sending unit electrical connector with a screwdriver (see illustration 8.3).

9 On 2008 and later models, remove the rear stabilizer bar bracket nuts and allow the bar to hang down out of the way.

10 Place a floor jack under the tank and position a block of wood between the jack pad and the tank. Raise the jack until it's supporting the tank.

✳ CAUTION:

Take care when using a jack under the fuel tank. The fuel pump module could be damaged if the jack pushes up the sheetmetal directly below the module.

➡ **Note:** On Expedition/Navigator models with the 33.5 gallon fuel tank, before placing the jack under the fuel tank, remove the center tank support strap bolts and strap.

11 Remove the bolts that retain the fuel tank mounting straps (see illustration). The straps are hinged at the other end so you can swing them out of the way.

➡ **Note:** Expedition/Navigator models use a large bracket that straddles the entire fuel tank. The mounting bolts and bracket must be removed as a complete unit before the fuel tank can be lowered. The assembly will be lowered with the fuel tank because it will be positioned between the fuel tank and the lifting jack.

12 Lower the tank far enough to unplug any vapor lines or wire harness brackets that may be difficult to reach when the fuel tank is in the vehicle.

13 Slowly lower the jack while steadying the tank. Remove the tank from the vehicle.

14 If you're replacing the tank, or having it cleaned or repaired, refer to Section 6.

15 Refer to Section 7 or 8 to remove and install the fuel pump or sending unit.

16 Installation is the reverse of removal. Clean engine oil can be used as an assembly aid when pushing the fuel filler neck back into the tank.

17 Make sure the fuel tank heat shields are assembled correctly onto the fuel tank before reinstalling the tank in the vehicle.

18 Carefully angle the fuel tank filler neck into the filler pipe assembly and lift the tank into place.

6   Fuel tank - cleaning and repair

1 The fuel tanks installed in the vehicles covered by this manual are made of plastic and are not repairable.

2 If the fuel tank is removed from the vehicle, it should not be placed in an area where sparks or open flames could ignite the fumes coming out of the tank. Be especially careful inside a garage where a natural gas-type appliance is located, because the pilot light could cause an explosion.

7   Fuel pump - removal and installation

✳ WARNING:

Gasoline is extremely flammable, so take extra precautions when you work on any part of the fuel system. See the *Warning* in Section 2.

1 Unless the vehicle has been driven far enough to completely empty the tank, it's a good idea to siphon out the residual fuel before removing the fuel pump from the vehicle.

✳ WARNING:

DO NOT start the siphoning action by mouth! Use a siphoning kit (available at most auto parts stores).

2 Relieve the fuel pressure (refer to Section 2).
3 Detach the cable from the negative terminal of the battery.
4 Raise the vehicle and support it securely on jackstands.

✳ WARNING:

Some models covered by this manual are equipped with self leveling suspension systems. Always disconnect electrical power to the suspension system before lifting or towing the vehicle (see Chapter 10). Failure to perform this procedure may result in unexpected shifting or movement of the vehicle which could cause personal injury.

5 Remove the fuel tank from the vehicle (see Section 5).

➡ **Note:** A special fuel pump line removal tool, available at most auto parts stores, may be required to disconnect the fuel lines from the fuel pump.

2009 AND EARLIER MODELS

▸ **Refer to illustrations 7.6, 7.7, 7.8, 7.10, 7.11 and 7.12**

6 Use paint or a marking pen to highlight the alignment marks that are scribed into the fuel pump assembly and the fuel tank (see illustration).

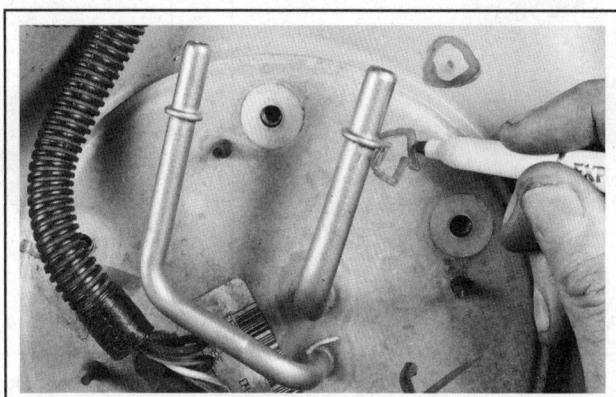

7.6 Using paint or a marker, highlight the alignment marks on the fuel pump assembly

4-10 FUEL AND EXHAUST SYSTEMS

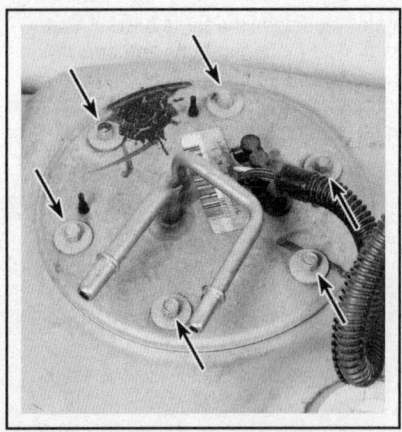

7.7 Remove the mounting bolts (arrows) from the fuel pump assembly

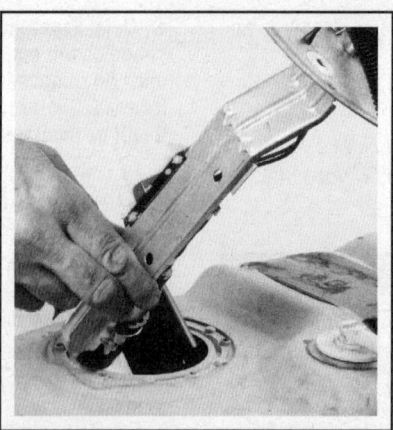

7.8 Carefully angle the fuel pump out of the fuel tank without damaging the fuel strainer

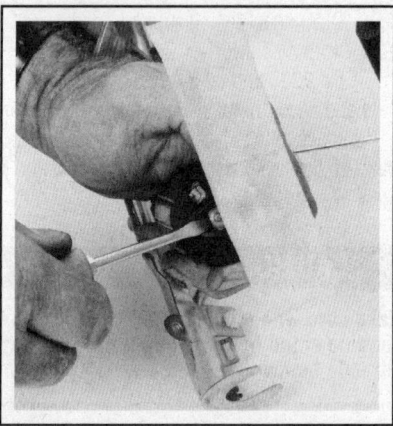

7.10 Remove the C-clip from the base of the fuel pump then separate the strainer from the fuel pump

7 Remove the mounting bolts (see illustration) from the fuel pump assembly.

➡ Note: Expedition/Navigator models use a threaded retainer. Use channel locks or a special tool, available at most auto parts sales stores, to remove the fuel pump assembly from the tank.

8 Carefully pull the fuel pump assembly from the tank (see illustration).

9 Remove the old seal ring and discard it.

10 If you're planning to reinstall the original fuel pump unit, remove the strainer (see illustration) by prying it off with a screwdriver, wash it in clean solvent, then push it back into place on the pump. If you're installing a new pump/sending unit, the assembly will include a new strainer.

11 To separate the fuel pump from the assembly, remove the clamp and disconnect the electrical connector from the fuel pump (see illustration).

12 Remove the fuel pump mounting clamp bolt (see illustration).

13 Clean the fuel pump mounting flange and the tank mounting surface and seal ring groove.

14 Installation is the reverse of removal. Apply a thin coat of heavy grease to the new seal ring to hold it in place during assembly.

2010 AND LATER MODELS

▸ Refer to illustration 7.16

➡ Note: On these models, the fuel pump module does not have separately serviceable parts. If any of it's components need to be replaced, you must replace the entire fuel pump module.

15 Remove the fuel pump module cover, if equipped. Disconnect the fuel supply tube and FTP sensor quick-connect couplings from the top of the pump module.

16 Using a lock ring wrench, or a hammer and a brass punch (do NOT use a steel punch!), loosen the fuel pump module lock ring (see illustration). When the lock ring is loose enough, fully unscrew it.

17 Remove the fuel pump module from the tank, being careful not to damage the fuel level sensor.

➡ Note: If the module sticks in the flange, it may be necessary to insert a screwdriver into the hole in the bottom of the pump module, and press the screwdriver inward until the module comes loose.

18 Remove and discard the O-ring seal.

19 Clean the fuel pump mounting flange and the tank mounting surface, particularly the area where the O-ring seal is installed.

20 Installation is the reverse of removal. Install a new O-ring seal.

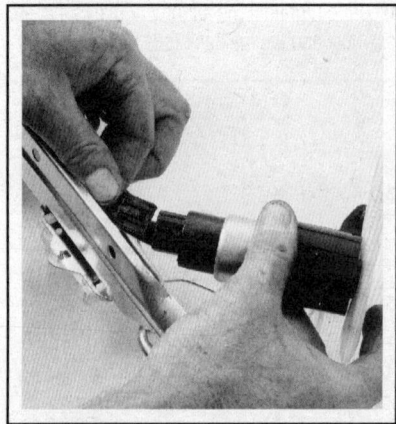

7.11 Disconnect the fuel pump electrical connector from the fuel pump

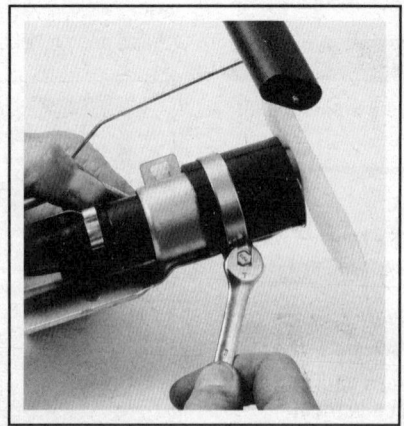

7.12 Loosen the fuel pump mounting clamp bolt

7.16 If you don't have the special wrench, loosen the lock ring with a hammer and brass punch (don't use a steel punch, which might cause sparks)

FUEL AND EXHAUST SYSTEMS 4-11

8 Fuel level sending unit - check and replacement

➡ **Note:** This procedure only applies to 2009 and earlier models. On 2010 and later models, the fuel level sensor is an integral part of the fuel pump module; if it fails, it must be replaced as a unit (see Section 7).

CHECK

▶ Refer to illustrations 8.3 and 8.5

1 Raise the vehicle and support it securely on jackstands.

✴ WARNING:

Some models covered by this manual are equipped with self leveling suspension systems. Always disconnect electrical power to the suspension system before lifting or towing the vehicle (see Chapter 10). Failure to perform this procedure may result in unexpected shifting or movement of the vehicle which could cause personal injury.

2 Disconnect the electrical connector for the fuel level sending unit.
3 Position the ohmmeter probes into the electrical connector and check the resistance (see illustration). Use the 200-ohm scale on the ohmmeter.
4 With the fuel tank completely full, the resistance should be about 160.0 ohms. With the fuel tank nearly empty, the resistance of the sending unit should be about 15.0 ohms.
5 If the readings are incorrect, replace the sending unit.

➡ **Note:** A more accurate check of the sending unit can be made by removing it from the fuel tank and checking its resistance while manually operating the float arm (see illustration).

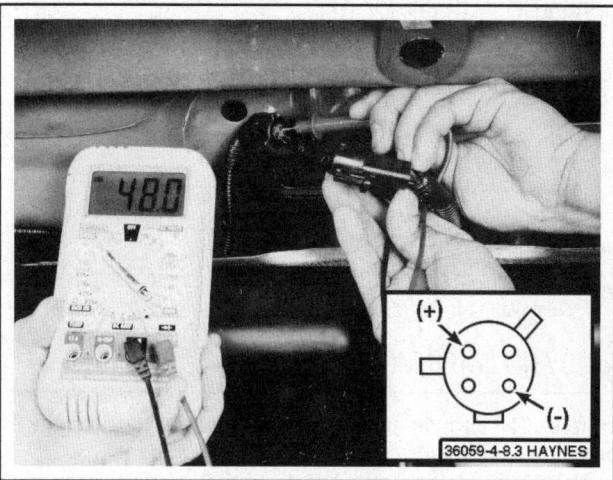

8.3 Using an ohmmeter, probe the terminals of the fuel sending unit connector to check the resistance

REPLACEMENT

▶ Refer to illustration 8.9

6 Remove the fuel tank (see Section 5).
7 Remove the fuel pump assembly mounting screws (pickups) or threaded retainer (Expedition/Navigator) (see Section 7).
8 Carefully angle the fuel pump/fuel level sending unit out of the opening without damaging the fuel level float located at the bottom of the assembly.
9 Remove the mounting bolt (see illustration) and harness connectors.
10 Installation is the reverse of removal.
11 Be sure to install a new rubber gasket.

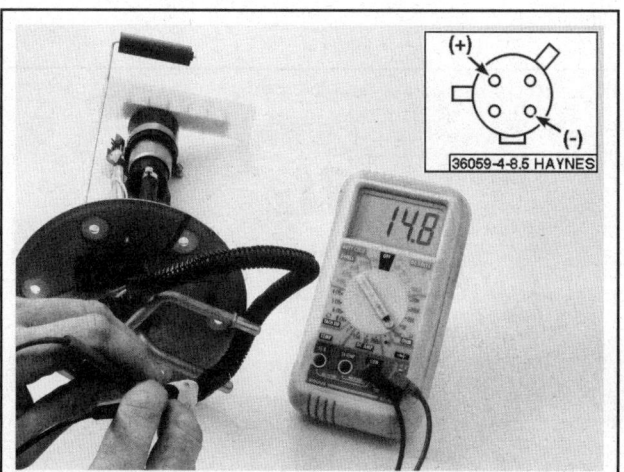

8.5 A more accurate check of the fuel level sending unit can be performed with the assembly on the bench. Connect the ohmmeter probes to the connector and check the resistance of the sending unit with the float positioned on "empty" and "full". Check for a smooth change in resistance as the float is moved between the positions

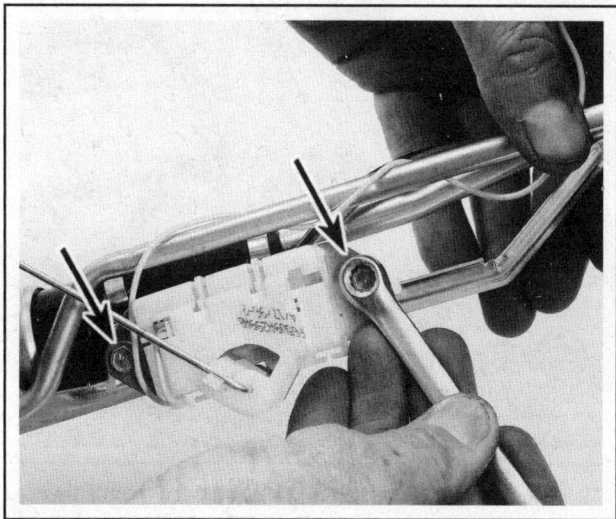

8.9 Remove the fuel level sending unit bolts (arrows) from the fuel pump assembly frame

4-12 FUEL AND EXHAUST SYSTEMS

9 Air filter housing - removal and installation

2009 AND EARLIER MODELS

1 Detach the cable from the negative terminal of the battery.
2 Remove the accelerator control protection shield.
3 Disconnect the IAT and MAF sensors (see Chapter 6), then disconnect the PCV hose and the IAC inlet hose.
4 Remove the air filter outlet tube mounting bolts.
5 Remove the air filter clamp from the assembly and remove the air filter element (see Chapter 1).
6 Lift the open end of the air filter housing assembly and remove it.
7 Installation is the reverse of removal.

2010 AND LATER MODELS

➡ Note: The air filter housing on these models is an integral part of the coolant expansion tank.

8 Drain the cooling system until the expansion tank is empty, or siphon the coolant from the expansion tank.
9 Remove the air filter element (see Chapter 1).
10 Remove the hood bumper and mounting bracket.
11 Disconnect the expansion tank overflow hose from the radiator.
12 Remove the expansion tank/air filter lower housing mounting bolts and lift the tank out of the engine compartment.
13 Installation is the reverse of removal.
14 Refill the cooling system with the proper type and concentration of antifreeze (see Chapter 1).

Air filter outlet duct

15 On Navigator models, remove the engine cover.
16 Loosen the clamps at each end of the assembly and remove the outlet pipe/resonator from the engine compartment.
17 Installation is the reverse of removal. Make sure the outlet pipe assembly is sealed securely and seated flush with the throttle body and air filter housing.

10 Accelerator cable - removal, installation and adjustment

REMOVAL

▶ Refer to illustrations 10.2, 10.3, 10.4 and 10.5

➡ Note: 2005 and later Expedition/Navigator models feature drive-by-wire electronic throttle control (ETC). The throttle body is operated electronically by the PCM and driver input from the Accelerator Pedal Position Switch. There is no cable at the throttle body.

1 Remove the accelerator control protection shield.
2 Detach the cruise control cable end from the throttle lever (see illustration).
3 Detach the accelerator cable from the throttle lever (see illustration).
4 Separate the accelerator cable from the cable bracket (see illustration).
5 Pull the cable end out from the accelerator pedal recess in the driver's compartment (see illustration).
6 Disconnect any cable clips or brackets securing the accelerator cable.
7 Remove the cable through the firewall from the engine compartment.

10.2 Remove the cruise control cable from the throttle lever

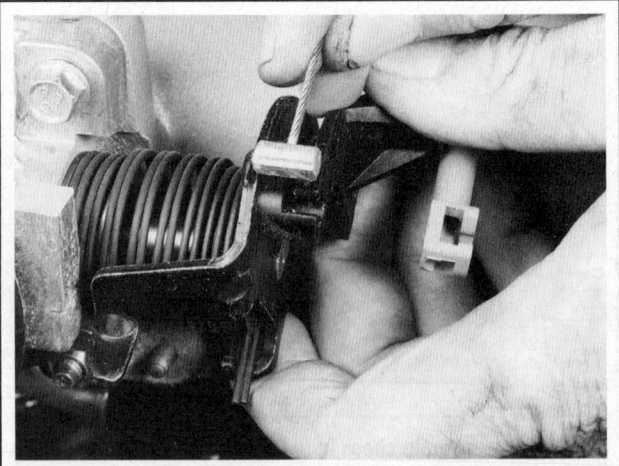

10.3 Rotate the accelerator cable until the slot in the throttle lever aligns with the cable and slide it out of the bracket

FUEL AND EXHAUST SYSTEMS   4-13

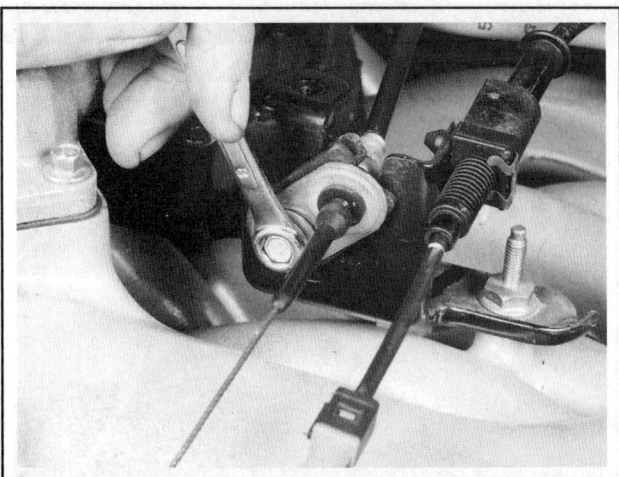

10.4 Unbolt the accelerator cable from its bracket

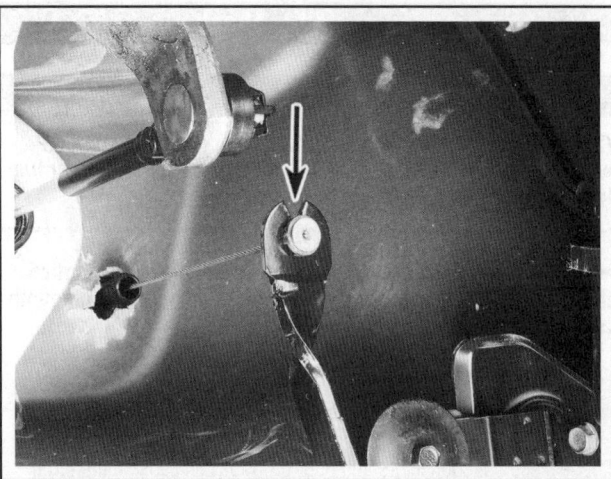

10.5 Working under the dash, pull the cable end from the accelerator pedal recess and lift it through the slot (arrow)

INSTALLATION

8 Installation is the reverse of removal. Be sure the cable is routed correctly and the grommet seats completely in the firewall.

9 If necessary, at the engine compartment side of the firewall, apply sealant around the accelerator cable to prevent water from entering the passenger compartment.

ADJUSTMENT

10 Measure the freeplay by firmly gripping the cable and pressing it down from the cable housing. There should be a slight amount of cable freeplay.

11 If there is no cable freeplay or the cable is binding and not allowing the throttle lever to completely close, replace the cable.

11 Fuel injection system - general information

SEQUENTIAL ELECTRONIC FUEL INJECTION (SEFI)

The Sequential Electronic Fuel Injection (SEFI) system is a multi-point fuel injection system. On the SEFI system, fuel is metered into each intake port in sequence with the engine firing order in accordance with engine demand through one injector per cylinder mounted on a tuned intake manifold. The intake manifold incorporates an air intake plenum to aid in air flow and distribution. Each engine uses a slightly different plenum design and fuel rail arrangement. The V6 engine uses a one-piece plenum mounted with twelve bolts. The air intake plenum bolts to the top of the intake manifold which sits directly in the middle of the engine block. The V8 engines incorporate an upper intake manifold with the air intake plenum mounted underneath. The throttle body is mounted to the upper intake manifold but the IMT valve and assembly is mounted below onto the air intake plenum (see Section 15 for additional information).

The V6 engine is equipped with the Intake Manifold Runner Control (IMRC) system. The IMRC system controls the air intake charge by opening or closing the butterfly valve on the secondary intake valve directly at the intake manifold. By closing the butterfly to the secondary intake valves under 3,000 rpm, low end driveability is improved. Above 3,000 rpm the butterfly valves open to increase high-end performance. The butterfly valves are controlled by the IMRC actuator and cable assembly.

4.6L and 5.4L engines are equipped with the Intake Manifold Tuning (IMT) system. The IMT system controls the air intake charge by opening or closing the Intake Manifold Tuning Valve located in the center of the air intake manifold, directly below the air intake plenum. By closing the IMTV, the dual plenum design sends the air intake charge through a single corridor into the intake system of the engine. Above 3,000 rpm, the IMTV is opened, allowing the intake pulses to blend together at the intake manifold thereby creating a more efficient air/fuel intake charge

for the additional rpm and engine load.

The Sequential Electronic Fuel Injection system incorporates an on-board Electronic Engine Control (EEC-V) computer that accepts inputs from various engine sensors to compute the required fuel flow rate necessary to maintain a prescribed air/fuel ratio throughout the entire engine operational range. The computer then outputs a command to the fuel injectors to meter the approximate quantity of fuel. The system automatically senses and compensates for changes in altitude, load and speed.

➡ **Note: The computer terminology has changed from Electronic Control Module (ECM) to the Powertrain Control Module (PCM) due to standardization of the Self Diagnosis system within the automotive industry.**

The fuel delivery systems include an electric in-tank fuel pump which forces pressurized fuel through a series of metal and plastic lines and an inline fuel filter/reservoir to the fuel charging manifold assembly. The SEFI system uses a single high-pressure pump mounted inside the tank.

The fuel rail assembly incorporates an electrically actuated fuel injector directly above each intake port. When energized, the injectors spray a metered quantity of fuel into the intake air stream.

A constant fuel pressure drop is maintained across the injector nozzles by a pressure regulator. The regulator is positioned downstream from the fuel injectors. Excess fuel passes through the regulator and returns to the fuel tank through a fuel return line.

On the SEFI system, each injector is energized once every other crankshaft revolution in sequence with engine firing order. The period of time that the injectors are energized (known as "on time" or "pulse width") is controlled by the PCM. Air entering the engine is sensed by speed, pressure and temperature sensors. The outputs of these sensors are processed by the PCM. The computer determines the needed injector pulse width and outputs a command to the injector to meter the exact quantity of fuel.

4-14 FUEL AND EXHAUST SYSTEMS

12 Fuel injection system - check

※ WARNING:

Gasoline is extremely flammable, so take extra precautions when you work on any part of the fuel system. See the *Warning* in Section 2.

→ Note: *The following procedure is based on the assumption that the fuel pump is working and the fuel pressure is adequate (see Section 3).*

PRELIMINARY CHECKS

1 Check all electrical connectors that are related to the system. Loose electrical connectors and poor grounds can cause many problems that resemble more serious malfunctions.

2 Check to see that the battery is fully charged, as the control unit and sensors depend on an accurate supply voltage in order to properly meter the fuel.

3 Check the air filter element - a dirty or partially blocked filter will severely impede performance and economy (see Chapter 1).

4 If a blown fuse is found, replace it and see if it blows again. If it does, search for a grounded wire in the harness to the fuel pump (see Chapter 12).

SYSTEM CHECKS

▸ **Refer to illustrations 12.7, 12.8 and 12.9**

5 Check the condition of the vacuum hoses connected to the intake manifold.

6 Remove the air intake duct from the throttle body and check for dirt, carbon or other residue build-up in the throttle body, particularly around the throttle plate.

※ CAUTION:

The throttle body on these models is coated with a sludge-resistant material designed to protect the bore and throttle plate. Do not attempt to clean the interior of the throttle body with carburetor or other spray cleaners. This throttle body is designed to resist sludge accumulation and cleaning may impair the performance of the engine.

12.7 Use a stethoscope or screwdriver to determine if the injectors are working properly - they should make a steady clicking sound that rises and falls as engine speed changes

7 With the engine running, place an automotive stethoscope against each injector, one at a time, and listen for a clicking sound, indicating operation (see illustration). If you don't have a stethoscope, you can place the tip of a long screwdriver against the injector and listen through the handle.

8 If an injector isn't functioning (not clicking), purchase a special injector test light (sometimes called a "noid" light) and install it into the injector electrical connector (see illustration). Start the engine and check to see if the noid light flashes. If it does, the injector is receiving proper voltage. If it doesn't flash, further diagnosis should be performed by a dealer service department or other properly equipped repair facility.

9 With the engine OFF and the fuel injector electrical connectors disconnected, measure the resistance of each injector (see illustration). Check the Specifications listed in this Chapter for the correct injector resistance.

10 The remainder of the system checks can be found in Section 13 and Chapter 6.

12.8 Install the fuel injector test light (or "noid light") into the fuel injector electrical connector and confirm that it blinks when the engine is cranked or running

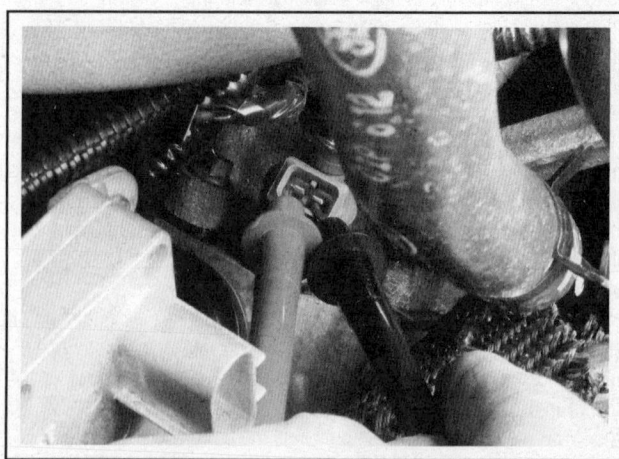

12.9 Measure the resistance of each injector. It should be within Specifications

FUEL AND EXHAUST SYSTEMS 4-15

13 Sequential Electronic Fuel Injection (SEFI) system (2014 and earlier models) - component check and replacement

AIR INTAKE PLENUM

➜ **Note:** Refer to Chapter 2B for the complete removal and installation procedure for the intake manifold on V8 models.

V6 models (2000 and earlier models)

Removal

▶ **Refer to illustrations 13.5, 13.8, 13.9 and 13.11**

1 Detach the cable from the negative terminal of the battery.
2 Remove the accelerator control protection shield.
3 Unplug the electrical connectors at the IAC valve, throttle position sensor (TPS) and EGR position sensor (see Chapter 6).
4 Detach the accelerator cable (see Section 10) and cruise control cable (if equipped) from the throttle body assembly.
5 Remove the accelerator cable bracket from the air intake plenum (see illustration) and position the bracket and cables out of the way.
6 Clearly label, then detach, the vacuum lines from the air intake plenum, the EGR valve and the fuel pressure regulator.
7 Disconnect the EGR valve tube from the air intake plenum (see Chapter 6).
8 Remove the PCV tube from the air intake plenum and the PCV valve on the valve cover. Remove the EVR bracket (see Chapter 6) from the air intake plenum. Disconnect the plenum brace (see illustration).
9 Remove the air intake plenum mounting bolts (see illustration).
10 Remove the upper air intake plenum and throttle body as an assembly from the lower intake manifold.

Installation

11 Be sure to clean and inspect the mounting faces of the lower intake manifold (see Chapter 2A) and the air intake plenum before positioning the new gasket(s) onto the lower intake mounting face. The use of alignment studs may be helpful. Install the air intake plenum and throttle body assembly onto the lower intake manifold. Ensure the gasket remains in place (if alignment studs are not used). Install the air intake plenum retaining bolts and tighten them to the torque listed in this Chapter's Specifications (see illustration).

13.5 Remove the accelerator cable bracket mounting nuts (arrows) and lift the assembly from the plenum

13.8 Remove the mounting brace bolt (arrow) and separate the brace from the air intake plenum

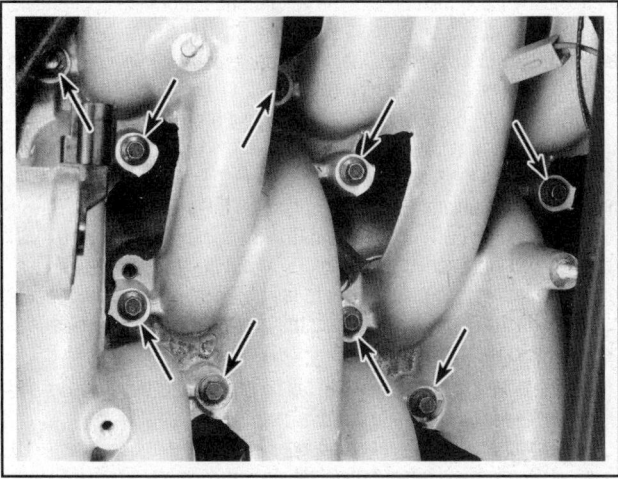

13.9 Remove the air intake plenum mounting bolts (V6 engine shown; not all bolts visible in this photo)

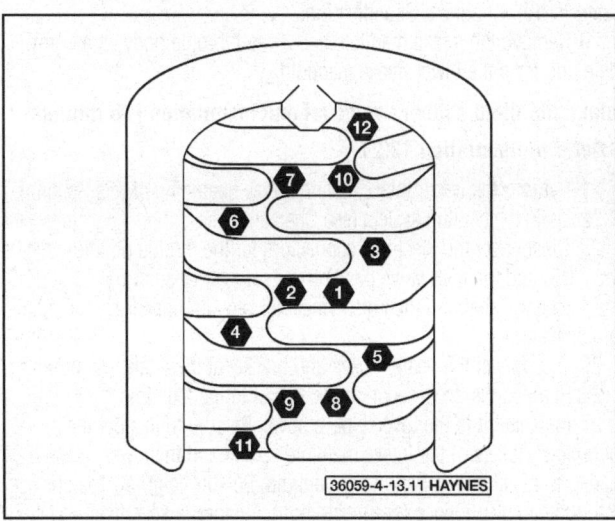

13.11 Air intake plenum tightening sequence - 4.2L V6 engine

4-16 FUEL AND EXHAUST SYSTEMS

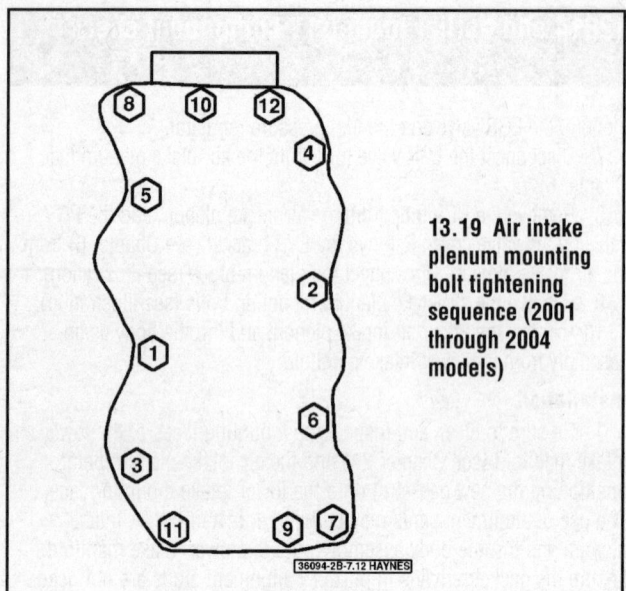

13.19 Air intake plenum mounting bolt tightening sequence (2001 through 2004 models)

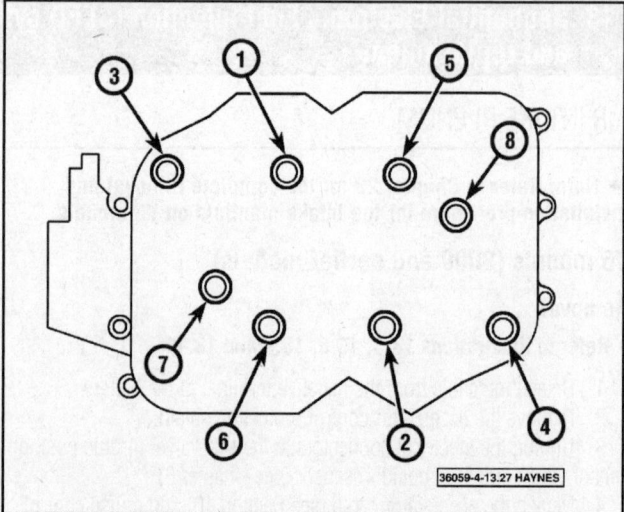

13.27 Intake manifold spacer mounting bolt tightening sequence

2001 through 2004 models

Removal and installation

▶ Refer to illustration 13.19

12 Remove the air cleaner assembly.
13 Remove the accelerator cable (see Section 10) and cruise control cable (if equipped) from the throttle body.
14 Disconnect the electrical connector for the idle air control (IAC) valve.
15 Remove the (IAC) valve (see Chapter 6).
16 Clearly label and then disconnect the vacuum lines from the upper intake manifold.
17 Remove the PCV tube from the upper intake manifold and the valve cover (see Chapter 6).
18 Remove the air intake plenum mounting bolts.
19 Installation is the reverse of removal. Be sure to clean and inspect the mounting faces of the intake manifold spacer and the air intake plenum before positioning the gaskets onto the intake manifold spacer mounting face. Be sure to follow the correct bolt tightening sequence (see illustration) and tighten the bolts to the torque listed in this Chapter's Specifications.
20 Remove the intake manifold spacer and throttle body as a complete unit from the lower intake manifold.

Intake manifold spacer - removal and installation (V6 models)

▶ Refer to illustration 13.27

21 Remove the upper intake manifold (see beginning of this Section).
22 Drain the cooling system (see Chapter 1).
23 Disconnect the electrical connectors for the throttle position sensor (TPS) and the EGR valve position sensor (see Chapter 6).
24 Clearly label and then disconnect all vacuum lines from the manifold spacer.
25 Disconnect the heater hoses and bracket at the intake manifold.
26 Remove the bolts securing the spacer to the manifold.
27 Installation is the reverse of removal. Be sure to inspect the mounting gaskets of the intake manifold spacer and the lower intake manifold. Install the upper intake manifold. Be sure to follow the correct bolt tightening sequence (see illustration) and torque specification listed at the beginning of this Chapter.

V8 models

Removal

▶ Refer to illustration 13.43

28 Detach the cable from the negative terminal of the battery.
29 Drain the cooling system (see Chapter 1).
30 Remove the accelerator control protection shield.
31 Remove the air cleaner intake and air cleaner outlet tube (see Section 9).
32 Unplug the electrical connectors at the IAC valve, throttle position sensor (TPS) and EGR valve (see Chapter 6).
33 On 2004 and earlier models, detach the accelerator cable (see Section 10) and cruise control cable (if equipped) from the throttle body assembly. Remove the accelerator cable bracket from the intake manifold and position the bracket and cables out of the way. On 2005 and later 5.4L engines, disconnect the electronic throttle body electrical connectors.
34 Disconnect the fuel feed and return lines from the fuel rail (see illustrations 13.56a and 13.56b). Refer to Section 4 for additional information concerning fuel line disconnecting.
35 Clearly label, then detach, the vacuum lines from the upper intake manifold; the vapor management hose (see Chapter 6), the engine vacuum regulator, PCV valve, the power brake booster, the fuel pressure regulator and the vacuum reservoir hose.
36 Remove the brake booster vacuum hose bracket nut and assembly from the upper intake manifold.
37 Disconnect the harness connectors from the DPFE, the EVR, the engine vacuum regulator (EVR) sensor and the IAC valve (see Chapter 6).
38 Remove the EGR tube on the 4.6L V8 engine (see Chapter 6).
39 On 2004 and earlier 5.4L engines, it will be necessary to remove several components to gain access to the upper intake manifold mounting bolts. On 2005 and later models, the intake manifold is one-piece (see Chapter 2B).

   a) *Disconnect the right bank and left bank fuel injector connectors and be sure to label them to avoid reassembly mistakes.*
   b) *Disconnect the coolant temperature sensor connector (see Chapter 6).*
   c) *Disconnect and remove the eight ignition coils (see Chapter 5).*
   d) *Remove the alternator (see Chapter 5).*
   e) *Remove the throttle body from the intake manifold.*

FUEL AND EXHAUST SYSTEMS 4-17

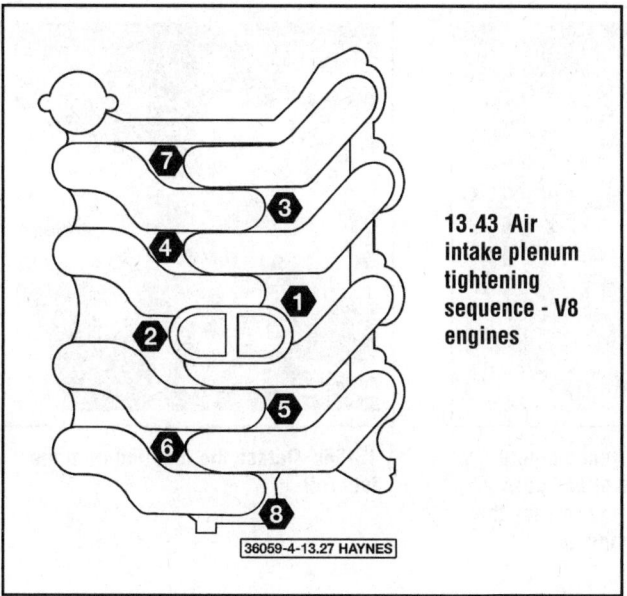

13.43 Air intake plenum tightening sequence - V8 engines

13.48 Throttle body mounting bolts (arrows)

f) Remove the heater hoses from the intake manifold (see Chapter 3).
g) Remove the thermostat housing (see Chapter 3).

40 Remove the upper intake manifold (see Chapter 2). Disconnect the IMT valve connector.

41 Working on a bench, remove the air intake plenum bolts from the upper intake manifold. There are eight bolts retaining the assembly.

42 Remove the Intake Manifold Tuning valve (IMT) if necessary. Be sure to use a new gasket for reassembly.

Installation

43 Be sure to clean and inspect the mounting faces of the upper intake manifold and the cylinder head (see Chapter 2A) and the air intake plenum before positioning the new gasket(s) onto the mounting face(s). The use of alignment studs may be helpful. Install the air intake plenum and IMT valve assembly onto the upper intake manifold. Ensure the gasket remains in place (if alignment studs are not used). Install the air intake plenum retaining bolts and tighten them to the torque listed in this Chapter's Specifications (see illustration). Installation is otherwise the reverse of removal.

THROTTLE BODY

✳✳ CAUTION:

The throttle body is coated with a sludge-resistant material designed to protect the bore and throttle plate. Do not attempt to clean the interior of the throttle body. The throttle body is designed to resist sludge accumulation and cleaning may impair the performance of the engine.

Removal

▶ Refer to illustration 13.48

44 Detach the cable from the negative terminal of the battery.
45 Detach the throttle position sensor (TPS) and Idle Air Control (IAC) valve electrical connectors.

46 On 2004 and earlier models, disconnect the accelerator cable (see Section 10) and the Throttle Valve (TV) cable from the throttle body. On 2005 and later models, disconnect the electronic throttle control connectors.

47 If equipped, remove the PCV vent closure hose at the throttle body (see Chapter 6).

48 Remove the four throttle body mounting nuts (see illustration).
49 Remove and discard the throttle body gasket.

Installation

50 Clean the gasket mating surfaces. If scraping is necessary, be careful not to damage the gasket surfaces or allow material to drop into the manifold. Installation is the reverse of removal. Be sure to tighten the throttle body mounting nuts to the torque listed in this Chapter's Specifications.

THROTTLE POSITION (TP) SENSOR

51 Refer to Chapter 6 for the check and replacement procedures for the TP sensor.

IDLE AIR CONTROL (IAC) VALVE

52 Refer to Section 14 for the check and replacement procedures for the IAC valve.

FUEL RAIL ASSEMBLY

▶ Refer to illustrations 13.56a, 13.56b, 13.56c, 13.57a and 13.57b

Removal

53 Relieve the fuel pressure (see Section 2).
54 Detach the cable from the negative terminal of the battery.
55 On V6 models, remove the air intake plenum assembly (see Steps 1 through 23).

4-18 FUEL AND EXHAUST SYSTEMS

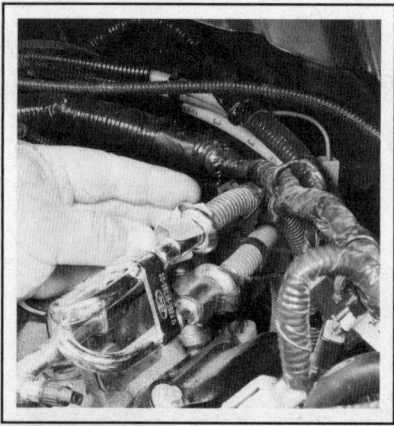

13.56a Press down to release the safety clamp from the fuel line connectors

13.56b Install the correct diameter spring lock coupling tool and push away from the fuel rail to release the internal locking mechanism

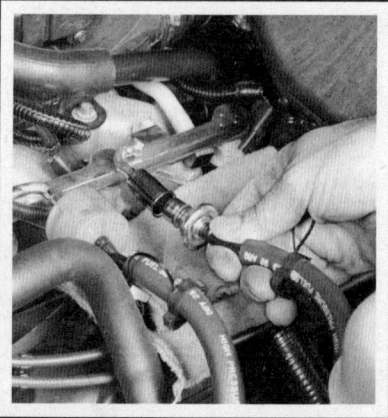

13.56c Detach the fuel line from the fuel rail

56 Using the special spring lock coupling tool, disconnect the fuel feed and return lines from the fuel rail assembly (see illustrations).

➡ Note: Refer to Section 4 for additional information on disconnecting fuel lines.

57 On V6 models, disconnect the fuel injector connectors and remove the fuel rail retaining bolts (two on each side) (see illustrations).

58 On V8 models, it will be necessary to remove several components to gain access to the fuel rail mounting bolts.

a) Disconnect the fuel injector connectors.
b) Remove the brake booster bracket and tube.
c) Remove the PCV hose (see Chapter 6).
d) Disconnect the EGR components (see Chapter 6).
e) Disconnect the vapor management valve (see Chapter 6).
f) On V8 engine models equipped with an ignition coil over each spark plug, disconnect the ignition coil harness connectors and the throttle body connectors (see Chapter 5).
g) Remove the right hand and left hand fuel rail mounting bolts.

59 Carefully remove the fuel rail with the fuel injectors attached as an assembly.

60 Use a rocking, side-to-side motion while lifting to remove the injectors from the fuel rail.

Installation

➡ Note: It's a good idea to replace the injector O-rings whenever the fuel rail is removed.

61 Ensure that the injector caps are clean and free of contamination.

62 Place the fuel rail over each of the injectors and seat the injectors into the fuel rail. Ensure that the injectors are well seated in the fuel rail assembly. Note: It may be easier to seat the injectors in the fuel rail and then seat the entire assembly in the lower intake manifold.

63 Secure the fuel rail assembly with the four retaining bolts and tighten them to the torque listed in this Chapter's Specifications.

64 The remainder of installation is the reverse of removal.

FUEL PRESSURE REGULATOR (RETURN FUEL SYSTEMS)

Check

➡ Note: This procedure assumes the fuel filter is in good condition and there are no leaks within the entire fuel rail and hoses network from the engine to the fuel tank.

65 Refer to the fuel system pressure checks in Section 3.

Replacement

▶ Refer to illustration 13.69

66 Relieve the fuel pressure from the system (see Section 2). Disconnect the cable from the negative terminal of the battery.

67 Clean any dirt from around the fuel pressure regulator.

68 Detach the vacuum hose from the fuel pressure regulator.

13.57a Fuel rail mounting bolt locations (arrows) - V6 engine

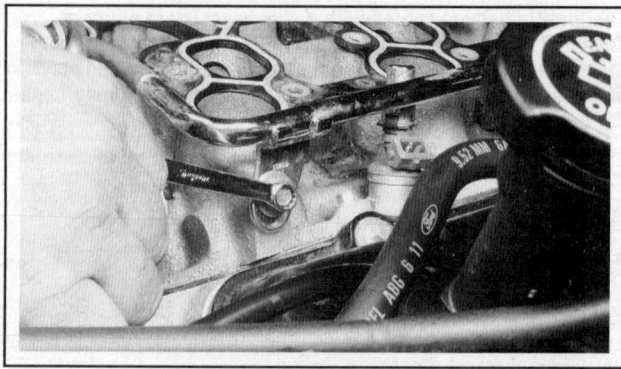

13.57b The fuel rail is mounted with bolts on each side of the air intake plenum - V6 engine

FUEL AND EXHAUST SYSTEMS   4-19

69 Remove the two bolts retaining the fuel pressure regulator (see illustration) and detach the regulator from the fuel rail.

70 Install new O-rings on the pressure regulator and lubricate them with a light coat of oil.

71 Installation is the reverse of removal. Tighten the pressure regulator mounting bolts securely.

FUEL INJECTOR

Removal

▶ Refer to illustrations 13.75, 13.78a and 13.78b

72 Relieve the system fuel pressure (see Section 2).

73 On V6 models, remove the air intake plenum assembly (see Steps 1 through 23).

74 Remove the fuel rail assembly (see above).

75 Disconnect the harness electrical connectors from the individual injectors as required (see illustration).

76 Carefully lift the assembly to gain access to the injectors.

77 Grasping the injector body, pull up while gently rocking the injector from side-to-side.

78 Inspect the injector O-rings (two per injector) for signs of deterioration (see illustrations). Replace as required.

➡ Note: As long as you have the fuel rail off, it's a good idea to replace all of the O-rings.

79 Inspect the injector plastic hat (covering the injector pintle) and washer for signs of deterioration. Replace as required. If the hat is missing, look for it in the intake manifold.

Installation

80 Lubricate the new O-rings with light grade oil and install two on each injector.

※ CAUTION:

Do not use silicone grease. It will clog the injectors.

81 Using a light twisting motion, install the injector(s).
82 The remainder of installation is the reverse of removal.

13.69 Remove the fuel pressure regulator bolts

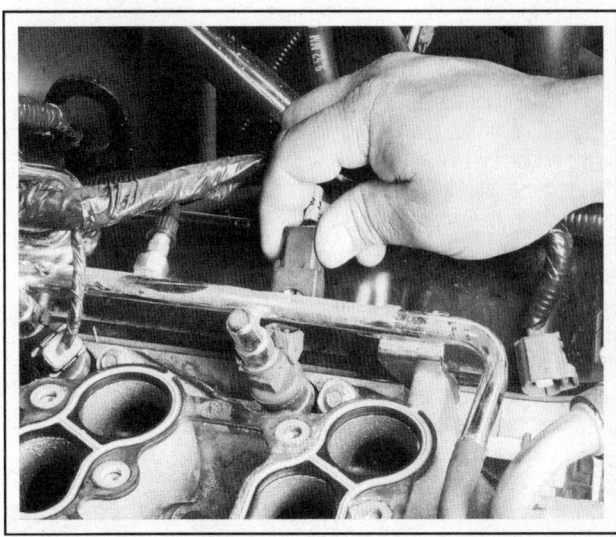

13.75 Disconnect the harness connector from each injector

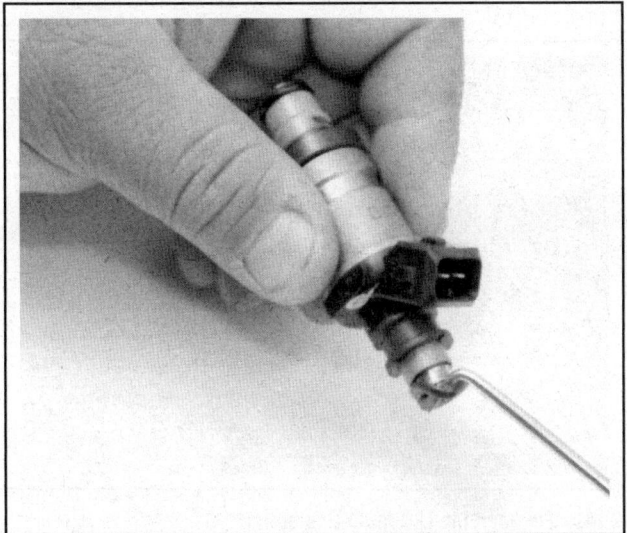

13.78a Remove the O-ring from the top of the fuel injector . . .

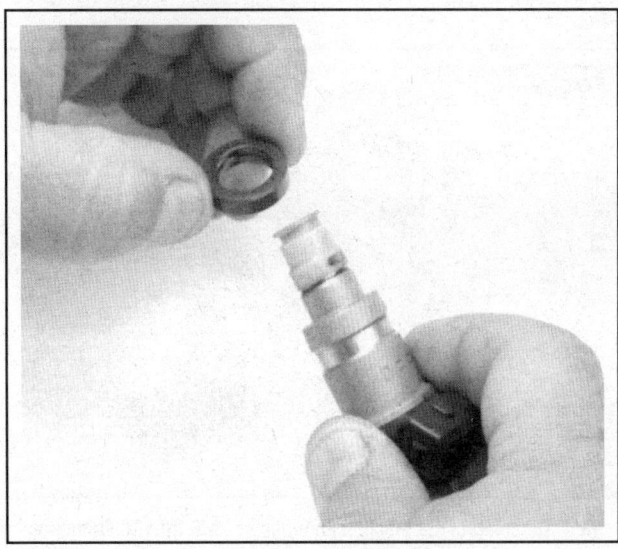

13.78b . . . then remove the lower O-ring from the injector

4-20 FUEL AND EXHAUST SYSTEMS

14 Idle Air Control (IAC) valve - check, removal and adjustment (2004 and earlier models)

CAUTION:

The throttle body on these models is coated with a sludge resistant material designed to protect the bore and throttle plate. Do not attempt to clean the interior of the throttle body. The throttle body is designed to resist sludge accumulation and cleaning may impair the performance of the engine.

→ Note: The minimum idle speed is pre-set at the factory and is not adjustable under normal circumstances. If the idle fluctuates, stalls, idles high or speeds out of control, follow these quick checks to determine if the IAC valve is damaged. Because idle problems involve possible air leaks, fuel injector problems, malfunctioning TPS, PCM problems, etc. have the IAC valve and system diagnosed by a dealer service department or other qualified repair facility.

CHECK

▶ Refer to illustration 14.2

1 The Idle Air Control (IAC) valve controls the amount of air that bypasses the throttle valve, which controls the engine idle speed. This output actuator is mounted on the throttle body and is controlled by voltage pulses sent from the PCM (computer). The IAC valve within the body moves in or out, allowing more or less intake air into the system. To increase idle speed, the PCM extends the IAC valve from the seat and allows more air to bypass the throttle bore. To decrease idle speed, the PCM retracts the IAC valve towards the seat, reducing the air flow.

2 To check the system, first check for the voltage signal from the PCM. Turn the ignition key On (engine not running) and with a voltmeter, probe the wires of the IAC valve electrical connector (harness side). It should be approximately 10.5 to 12.5 volts (see illustration). This indicates that the IAC valve is receiving the proper signal from the PCM.

3 If the IAC valve is receiving proper voltage, check the condition of the valve itself. Measure the resistance across the terminals on the IAC valve. There should be 6.0 to 13.0 ohms. If the resistance is incorrect, replace the IAC valve.

4 Check the IAC valve for an internal short circuit. Measure resistance from either terminal to the IAC body. There should be 10,000 ohms or greater. If less, the internal circuitry is grounding against the case; replace the IAC valve.

5 Next, remove the valve (proceed to Step 7) and check the pintle for excessive carbon deposits. If necessary, clean it with a soft rag. Also clean the valve housing to remove any deposits.

ADJUSTMENT

6 The idle speed is not adjustable. This procedure requires a special SCAN tool to extract working parameters from the EEC-V system while it is running. Have the procedure performed by a dealer service department or other qualified repair shop.

REMOVAL

▶ Refer to illustration 14.8

7 Unplug the electrical connector from the IAC valve.
8 Remove the two valve attaching screws and withdraw the valve assembly from the throttle body (see illustration).
9 Check the condition of the O-ring. If it's hardened or deteriorated, replace it.
10 Clean the sealing surface and the bore of the throttle body assembly with a shop rag or soft cloth to ensure a good seal.

CAUTION:

The IAC valve itself is an electrical component and must not be soaked in any liquid cleaner, as damage may result.

INSTALLATION

11 Position the new O-ring on the IAC valve. Lubricate the O-ring with a light film of engine oil.
12 Install the IAC valve and tighten the screws securely.
13 Plug in the electrical connector at the IAC valve assembly.

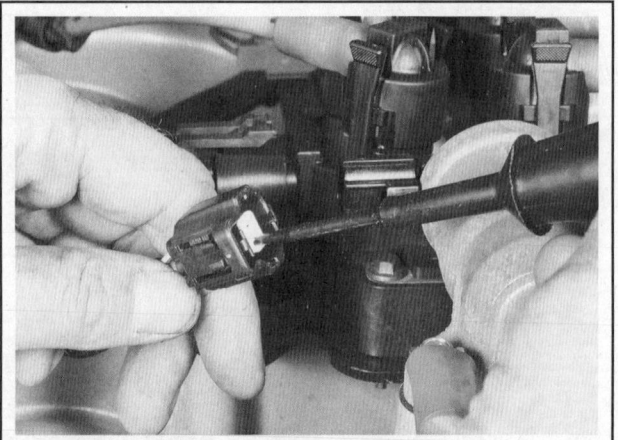

14.2 Disconnect the electrical connector from the IAC valve and check for voltage with the ignition key ON (engine not running)

14.8 Remove the IAC mounting bolts

FUEL AND EXHAUST SYSTEMS 4-21

15 Intake Air Systems

INTAKE MANIFOLD RUNNER CONTROL (IMRC) SYSTEM (4.2L V6 MODELS)

General information

1 The IMRC system controls the air intake charge by opening or closing the butterfly valve on the secondary intake valve directly at the intake manifold. By closing the butterfly to the secondary intake valves under 3,000 rpm, low end driveability is improved. Yet above 3,000 rpm the butterfly valves open to increase high-end performance. The butterfly valves are controlled by the IMRC actuator and cable assembly.

2 The IMRC system is difficult to check and requires a special SCAN tool to access the PCM for information and operating conditions. Have the system diagnosed by a dealer service department or a qualified repair shop.

Replacement

▶ Refer to illustrations 15.4

3 Remove the air intake plenum (see Section 13).
4 Remove the bolts that retain the IMRC actuator assembly to the cylinder head/intake manifold area (see illustration).
5 Disconnect the actuator cable from the lever on the intake manifold runner.
6 Installation is the reverse of removal.

INTAKE MANIFOLD TUNING (IMT) SYSTEM (2005 AND EARLIER V8 MODELS)

General information

7 The IMT system controls the air intake charge by opening or closing the Intake Manifold Tuning Valve located in the center of the air intake manifold, directly below the air intake plenum. By closing the IMTV, the dual plenum design sends the air intake charge through a single corridor into the intake system of the engine. Above 3,000 rpm, the IMTV is opened, allowing the intake pulses to blend together at the intake manifold thereby creating a more efficient air/fuel intake charge for the additional rpm and engine load.

8 The IMT system is difficult to check and requires a special SCAN tool to access the PCM for information and operating conditions. Have the system diagnosed by a dealer service department or other qualified repair shop.

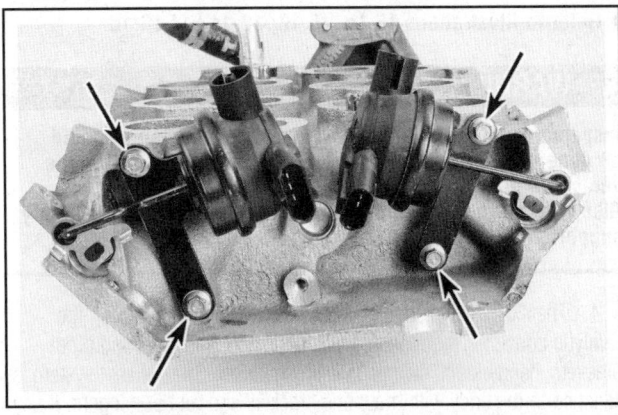

15.4 Remove the IMRC actuator mounting bolts (arrows)

Replacement

9 Remove the upper intake manifold from the engine (see Chapter 2) and then remove the air intake plenum from the intake manifold (see Section 13).
10 Remove the bolts that retain the IMT valve to the lower air intake plenum.
11 Installation is the reverse of removal.

CHARGE MOTION CONTROL VALVE (CMCV) SYSTEM (2006 THROUGH 2014 MODELS)

▶ Refer to illustrations 15.14a and 15.14b

General Information

12 This system controls the air intake charge by opening and closing the valves located inside the intake manifold, changing the path of airflow through the manifold, increasing torque at lower rpm and greatly improving the upper rpm power.

Replacement

13 Remove the intake manifold (see Chapter 2B).
14 Disconnect the CMCV rods from the back of the intake manifold (see illustrations).
15 Remove the mounting fasteners and remove the CMCV from the manifold.
16 Installation is the reverse of removal.

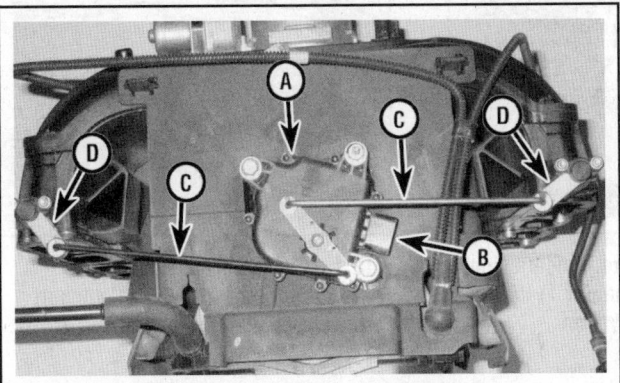

15.14a CMCV actuator assembly details

A Actuator
B Electrical connector
C Actuator rods
D Actuator arms

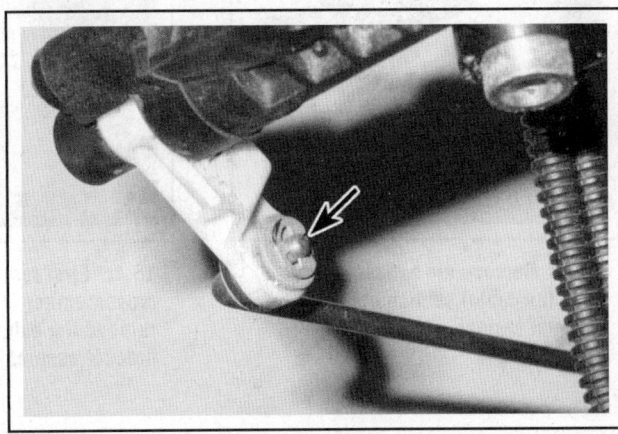

15.14b Each end of the CMCV actuator rods is retained to its crank arm on the intake manifold by a small E-clip

4-22 FUEL AND EXHAUST SYSTEMS

16 Exhaust system servicing - general information

▶ Refer to illustrations 16.1a, 16.1b, 16.1c and 16.1d

WARNING:

Inspection and repair of exhaust system components should be done only after enough time has elapsed after driving the vehicle to allow the system components to cool completely. Also, when working under the vehicle, make sure it is securely supported on jackstands.

1 The exhaust system consists of the exhaust manifold(s), the catalytic converter, the muffler, the tailpipe and all connecting pipes, brackets, hangers and clamps (see illustrations). The exhaust system is attached to the body with mounting brackets and rubber hangers. If any of the parts are improperly installed, excessive noise and vibration will be transmitted to the body.

2 Conduct regular inspections of the exhaust system to keep it safe and quiet. Look for any damaged or bent parts, open seams, holes, loose connections, excessive corrosion or other defects which could allow exhaust fumes to enter the vehicle. Deteriorated exhaust system components should not be repaired; they should be replaced with new parts.

3 If the exhaust system components are extremely corroded or rusted together, welding equipment will probably be required to remove them. The convenient way to accomplish this is to have a muffler repair shop remove the corroded sections with a cutting torch. If, however, you want to save money by doing it yourself (and you don't have a welding outfit with a cutting torch), simply cut off the old components with a hacksaw. If you have compressed air, special pneumatic cutting chisels can also be used. If you do decide to tackle the job at home, be sure to wear safety goggles to protect your eyes from metal chips and work gloves to protect your hands.

4 Here are some simple guidelines to follow when repairing the exhaust system:

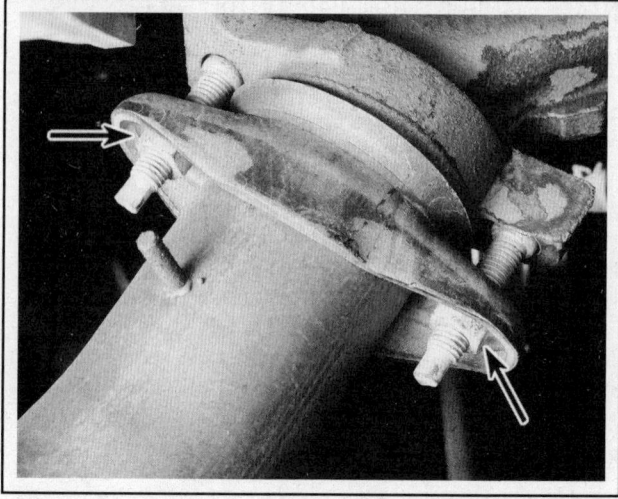

16.1a Be sure to apply penetrating lubricant to the exhaust flange nuts (arrows) before attempting to remove them

a) Work from the back to the front when removing exhaust system components.
b) Apply penetrating oil to the exhaust system component fasteners to make them easier to remove.
c) Use new gaskets, hangers and clamps when installing exhaust systems components.
d) Apply anti-seize compound to the threads of all exhaust system fasteners during reassembly.
e) Be sure to allow sufficient clearance between newly installed parts and all points on the underbody to avoid overheating the floor pan and possibly damaging the interior carpet and insulation. Pay particularly close attention to the catalytic converter and heat shield.

16.1b Remove the exhaust clamp from the center section of the exhaust system

16.1c First remove the transmission crossmember (see Chapter 7) then remove the exhaust system bracket assembly

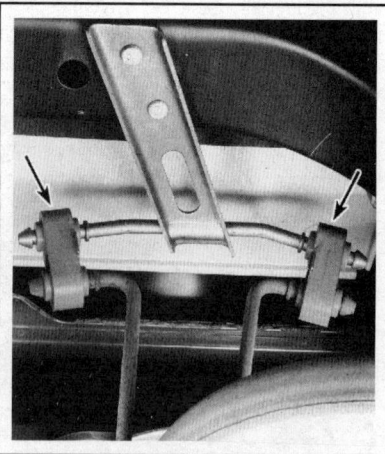

16.1d Check the condition of the rubber mounts (arrows) that support the weight of the center section

FUEL AND EXHAUST SYSTEMS 4-23

17 High-pressure fuel pump (3.5L V6 engine) - removal and installation

※ WARNING:

Fuel pressure in the high-pressure system on Direct Injection (DI) systems is under extremely high pressure. Be sure to correctly perform the fuel pressure release procedure prior to servicing any of the high-pressure fuel system components to prevent injury (see Section 2).

※ WARNING:

The fuel delivery system is made up of a low-pressure system and a high-pressure system. Once the pressure on the low-pressure side of the system has been relieved, wait at least two hours before loosening any fuel line fittings in the engine compartment.

➡ **Note:** Only 3.5L V6 models are equipped with Direct Injection (DI) and a high-pressure fuel pump.

➡ **Note:** The high-pressure fuel pump is located on the top of the passenger (right) side valve cover.

REMOVAL

1 Remove the two nuts at the front of the engine cover. Lift up at the front and unhook the retainers at the rear of the cover to remove.

2 Relieve the fuel system pressure (see Section 2), then wait at least two hours before proceeding.

3 Disconnect the cable from the negative terminal of the battery (see Chapter 5).

4 Remove the intake manifold (see Chapter 2B).

➡ **Note:** Raising the vehicle and working through the fenderwell makes access to certain parts of this procedure easier.

5 Locate the high-pressure fuel pump and disconnect the electrical connector.

6 Disconnect the fuel supply hose quick-connect fitting at the high-pressure fuel pump.

7 Remove the bolt attaching the fuel feed pipe to the intake manifold. Loosen the fuel feed line fitting at the high-pressure fuel pump. In this case, the fuel feed pipe can be reconnected and does not require replacement.

8 Disconnect the hose retainer from the high-pressure fuel pump.

9 Alternately loosen two nuts attaching the high-pressure fuel pump to the cylinder head, one complete turn at a time, until removed

10 Carefully pull the high-pressure fuel pump out of the cylinder head.

11 If necessary, remove the high-pressure fuel pump tappet (note installation orientation during removal).

➡ **Note:** Note the notch on the high-pressure fuel pump tappet and the groove in the high-pressure fuel pump bore.

INSTALLATION

12 Installation is reverse of removal noting the following points:
a) Rotate the engine CLOCKWISE ONLY until the high-pressure fuel pump drive lobe is at its lowest point.
b) Coat the bore with clean engine oil and align the notch on the high-pressure fuel pump tappet with the groove in the high-pressure fuel pump bore during installation.
c) Use a new O-ring on the high-pressure fuel pump. Coat the O-ring in clean engine oil.
d) Alternately tighten each of the high-pressure fuel pump nuts one complete turn until seated, then tighten them to the torque listed in this Chapter's Specifications.
e) Install a new fuel feed pipe.
f) Tighten all fasteners to the torque values listed in this Chapter's Specifications.
g) Turn the ignition ON and check for leaks before starting the engine.

18 Fuel Pump Driver Module (FPDM) - replacement

▸ Refer to illustration 18.1

➡ **Note:** The FPDM is located on the left rear frame rail, above the spare tire.

1 Locate the FPDM and disconnect the electrical connector (see illustration).

2 Remove the FPDM mounting bolts and detach the FPDM from the frame.

3 Installation is reverse of removal. No programming is necessary.

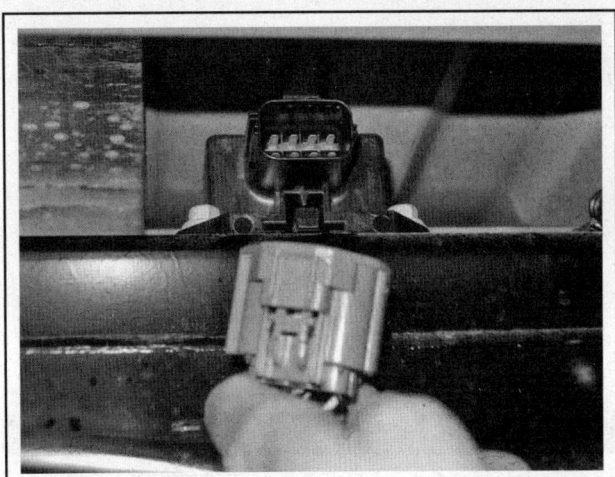

18.1 The FPDM is located on top of the left rear frame rail

4-24 FUEL AND EXHAUST SYSTEMS

19 Fuel rails and injectors (3.5L V6 engine) - removal and installation

WARNING:

Gasoline is extremely flammable, so take extra precautions when you work on any part of the fuel system. See the *Warning* Section 2.

WARNING:

The fuel delivery system on 3.5L turbocharged models is made up of a low-pressure system and a high-pressure system. Once the pressure on the low-pressure side of the system has been relieved, wait at least two hours before loosening any fuel line fittings in the engine compartment.

WARNING:

Wait until the engine is completely cool before beginning this procedure.

FUEL RAIL

1 Relieve the fuel pressure (see Section 2).
2 Disconnect the cable from the negative terminal of the battery (see Chapter 5).

REMOVAL

3 Remove the engine cover and high-pressure fuel pump cover.
4 Remove the intake manifold (see Chapter 2B).
5 Use compressed air to clean the area around the fuel injectors and fuel rails to prevent debris from entering the fuel system or engine.
6 Drain the cooling system (see Chapter 1).
7 Remove the three bolts and position the coolant crossover out of the way. Discard the O-rings for the crossover and install new ones during installation.
8 Remove the coolant pipe running between the cylinder banks by firmly pulling the pipe towards the rear of the engine to disengage the front coupling. Discard the O-rings for the coolant pipe ends and install new ones during installation.
9 Remove and discard the fuel feed line between the high-pressure fuel pump and the fuel rails. Remove the bolt attaching the fuel feed pipe to the intake manifold. The feed line is not to be re-used; obtain a NEW feed line for installation.
10 Detach the harness from the fuel rails and disconnect the harness connectors and fuel rail pressure sensor connector from the rear of the engine.
11 Remove the fuel rail mounting bolts.
12 Carefully pull the fuel rail with the fuel injectors out of the intake manifold at the same angle as the fuel injectors. Use a slight wiggle motion to assist in breaking the fuel injectors free from the cylinder head.

CAUTION:

Any injectors that remain stuck in the cylinder head must be removed using a special injector removal tool that uses a slide-hammer type procedure.

→ Note: If the isolator on the bottom of the fuel injector falls off during removal, replace the fuel injector.

13 If the injectors are to be serviced, disconnect the fuel injector connectors, remove and discard the injector clips and pull the injectors from the fuel rail. See the Fuel injector seals – replacement information in this section.

INSTALLATION

14 Installation is reverse of removal, noting the following points:
   a) *Coat the injector O-rings with clean motor oil prior to installation into the fuel rail.*
   b) *Install the fuel injectors to the fuel rail using new injector clips.*
   c) *DO NOT lubricate the injector teflon seals before installing the fuel rail and injectors to the engine.*
   d) *Press firmly to seat the injectors when installing the fuel rail to the engine.*
   e) *Use NEW high-pressure fuel feed pipes. Refill the cooling system (see Chapter 1). Turn the ignition on and check for leaks.*

FUEL RAIL PRESSURE/TEMPERATURE SENSOR - REPLACEMENT

CAUTION:

If the fuel rail pressure/temperature sensor is removed, it cannot be reused and must be replaced with a new one.

15 The fuel rail pressure sensor is located on the front of the driver's side (left) fuel rail. Remove the intake manifold to access the sensor (see Chapter 2B, Section 5), then disconnect the electrical connector and unscrew the sensor from the fuel rail.
16 Coat the O-ring with clean engine oil prior to installation.
17 Tighten the sensor securely.

FUEL INJECTOR SEALS – REPLACEMENT

▶ Refer to illustration 19.18

→ Note: Even if you only removed the fuel rail assembly to replace a single injector or a leaking O-ring, it's a good idea to remove all of the injectors from the fuel rail and replace all of the O-rings at the same time.

CAUTION:

Use care not to gouge or damage the injector surface when replacing the O-rings and seals.

18 Inspect the Teflon ring at the bottom of the injector. If replacement is required, pull the ring away from the injector body as far as possible (to prevent damage to the injector body) and cut the old Teflon ring

FUEL AND EXHAUST SYSTEMS 4-25

from the injector. A special tool is used to install the new Teflon ring (see illustration).

19 Inspect the inlet O-ring and support ring. If replacement is required, note the installed position of the support ring as one side is different and has a bevel (the bevel faces the injector body).

20 Inspect the fuel inlet caps. Remove any debris clogging the cap. If the cap is deteriorated or comes off, it is OK to discard as it is not needed for proper operation of the fuel injector. If the cap is missing, check that it is not in the fuel rail and remove as necessary.

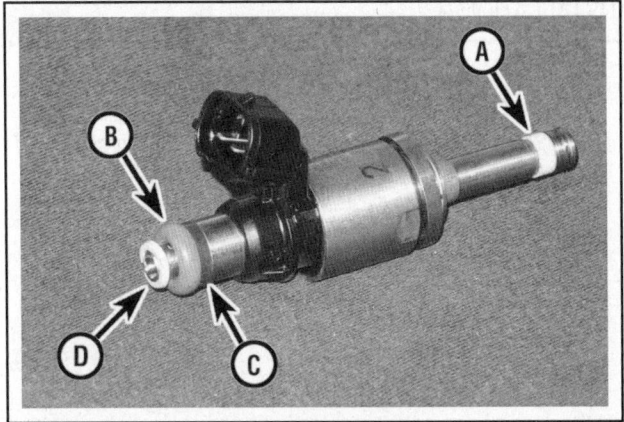

19.18 Identifying DI injector components

A	Teflon ring	C	Support ring
B	Inlet O-ring	D	Fuel inlet cap

20 Turbocharger(s) (3.5L V6 engine) - removal and installation

⁂ WARNING:

Wait until the engine is completely cool before beginning this procedure.

➡ **Note:** The turbochargers must be handled with care and extreme cleanliness when they are removed, as they are precision high-speed components. When parts are disconnected from the turbochargers, the openings on the turbocharger must be protected from entry of dirt or chemicals. Use high-strength tape to cover them as components are detached.

➡ **Note:** The procedure is virtually the same for either the left or right turbocharger.

➡ **Note:** Some turbochargers are equipped with heat shields, some are not. Some replacement turbochargers do not have provisions for heat shields as the manufacturer stopped installing them during production.

REMOVAL

1 The engine must be cold before removing the turbocharger(s). Before disconnecting any hoses, pipes or connections from the turbocharger, remove the heat insulation material.

2 Remove the engine cover

3 Loosen the wheel lug nuts, then raise and support the front of the vehicle on jackstands. Remove the wheel.

4 Drain the cooling system (see Chapter 1).

5 Remove the inner fender splash shield (see Chapter 11).

6 Remove the under-vehicle splash shield and transmission cover and/or skid plate (as equipped).

7 Loosen the clamps and remove the tube between the air intake tube to the air filter housing and the turbocharger inlet for the turbo to be serviced.

8 Loosen the clamps and remove the tube between the intercooler and turbocharger outlet for the turbo to be serviced.

9 Tape off the openings on the turbocharger assembly.

10 If servicing the passenger (right) turbocharger, remove the drivebelt (see Chapter 1). Remove the three air conditioning compressor mounting bolts and position the AC compressor aside (see Chapter 3).

⁂ WARNING:

Do not disconnect the refrigerant lines!

11 Remove the nuts and position the intercooler hose bracket aside to allow access to the AC compressor bolts.

12 Remove the nuts attaching the exhaust flange to the turbocharger. The studs should not be reused.

13 Disconnect the quick-connect fittings at the turbo and disconnect the coolant return line from the turbocharger.

14 Disconnect the vacuum line at the wastegate and secure out of the way.

15 Remove the bolt for the coolant and oil supply lines bracket on the side of the block.

16 Remove the bolts and disconnect the oil lines from the bottom of the turbocharger.

17 Disconnect the quick-connect fitting at the engine block for the coolant supply line.

18 Remove the bolts attaching the turbocharger mounting flange to the exhaust manifold and remove the turbocharger assembly. The coolant supply line will come out with the turbocharger so use care not to damage the line.

19 Disconnect the quick-connect fitting at the turbo to remove the coolant supply line

4-26 FUEL AND EXHAUST SYSTEMS

INSTALLATION

20 Installation is the reverse of removal, noting the following:
   a) *Use new gaskets at the turbo-to-exhaust junction on each side, and new gaskets and O-rings where the oil and coolant supply and return lines connect to the turbocharger housing and engine block*
   b) *Replace the filter on the engine block end of the oil supply line.*
   c) *Lubricate the oil line O-rings with clean engine oil. Lubricate the coolant line O-rings with coolant.*
   d) *It is recommended to use new studs, nuts and bolts for all exhaust-side components.*
   e) *Install the coolant supply line to the turbo before installing the turbo to the cylinder head.*
   f) *When installing the turbocharger and coolant line, install the turbo mounting flange nuts and bolt loosely and connect the coolant supply line to the block, then tighten the flange nuts and bolt.*
   g) *Use anti-seize compound on the new turbocharger-to-exhaust nuts/bolts.*
   h) *Tighten all fasteners to the torque listed in this Chapter's Specifications.*
   i) *Refill and bleed the cooling system and install a new oil filter and fresh engine oil.*

21 Wastegate control actuator (3.5L V6 engine) - replacement and adjustment

REPLACEMENT

1 Loosen the front wheel lug nuts, then raise the front of the vehicle and support it securely on jackstands. Remove the wheel.
2 Remove the inner fender splash shield (see Chapter 11).
3 Loosen the clamps and remove the tube between the air intake tube to the air filter housing and the turbocharger inlet for the wastegate to be serviced.
4 Loosen the clamps and remove the tube between the intercooler and turbocharger outlet for the wastegate to be serviced.
5 For the passenger's (right) side, remove the intercooler tube bracket nuts and position the hoses and bracket aside.
6 Tape off any openings on the turbocharger assembly.
7 Remove the wastegate actuator lower jam nut and disconnect the linkage.
8 Disconnect the wastegate actuator vacuum line.
9 Remove the fasteners and remove the wastegate actuator from the turbocharger.
10 Installation is reverse of removal. Install the lower jam nut loosely and do not tighten until instructed during the adjustment procedure.

ADJUSTMENT

11 With the wastegate installed and the lower jam nut installed finger-tight, push the linkage to the closed position (towards the actuator).
12 While holding the linkage closed, finger-tighten the lower jam nut.
13 While still holding the linkage closed, using a wrench, tighten the upper jam nut.
14 Release the linkage.
15 Now, pull the linkage to the open position (away from the actuator). Now, pull the linkage to the open position (away from the actuator).
16 While holding the linkage open, loosen the upper jam nut 4.5 turns exactly.
17 While holding the linkage open, hold the upper jam nut with a wrench and tighten the lower jam nut.
18 A scan tool will be necessary to clear PCM DTCs, reset the Keep Alive Memory (KAM) and monitor the PCM PID WGATE_A_V (right side) and WGATE_B_V (left side).
19 While monitoring the correct PID, press the accelerator pedal to the floor for 3 seconds and slowly release. Note the minimum voltage of the PID after the pedal is fully released. It should be 1.1V to 1.3V.
20 If it is not within range, the actuator linkage require adjustment again.

22 Intercooler (3.5L V6 engine) - removal and installation

1 Raise and support the front of the vehicle on jackstands.
2 Remove the under vehicle cover and/or skid plate (as equipped).
3 Loosen the clamps and disconnect the two intercooler inlet pipes from the passenger's (right) side of the intercooler.
4 Release the clip and disconnect the intercooler outlet pipe from the driver's (left) side of the intercooler
5 Disconnect the turbocharger bypass valve electrical connector and quick-connect fitting from the intercooler.
6 Remove the two bolts attaching the intercooler bracket to the frame and remove the intercooler from the vehicle.
7 Separate the intercooler from the bracket and remove the turbocharger bypass valve if necessary (see Chapter 6).
8 Installation is reverse of removal.

FUEL AND EXHAUST SYSTEMS 4-27

Specifications

Fuel pressure

Fuel system pressure	
Models with Return fuel system - 1997 through 2004 (at idle)*	
Vacuum hose attached	28 to 45 psi
Vacuum hose detached	40 to 50 psi
Models with Returnless fuel system (at idle)	
2001 through 2004 models	35 to 55 psi
2005 through 2009 models	35 to 70 psi
2010 through 2014 models	55 to 60 psi
2015 and later models	51 to 67 psi
Fuel system hold pressure (after 5 minutes)	30 to 40 psi
Fuel pump pressure (maximum)	75 psi

***Warning:** NEVER attempt to check fuel pressure on the high-pressure side of the 3.5L V6 engine. F-Series trucks are 28 to 50 psi.

Injector resistance

All except 3.5L V6 engine	13.5 to 19 ohms
3.5L V6 engine Unspecified**	

**Although unspecified, testing can be done for comparison purposes and to check for an open or shorted coil.

Torque specifications   Ft-lbs (unless otherwise indicated)

➥ **Note:** One foot-pound (ft-lb) of torque is equivalent to 12 inch-pounds (in-lbs) of torque. Torque values below approximately 15 ft-lbs are expressed in inch-pounds, since most foot-pound torque wrenches are not accurate at these smaller values.

Air intake plenum mounting bolts	
V6 engine	
2000 and earlier models	59 in-lbs
2001 through 2003 models	
Step 1	53 in-lbs
Step 2	89 in-lbs
V8 engines	
Step 1	18 in-lbs
Step 2	100 in-lbs
Intake manifold spacer	
Step 1	53 in-lbs
Step 2	89 in-lbs
Throttle body mounting bolts/nuts	
V6 models	
4.2L V6 (2010 through 2003)	89 in-lbs
3.5L V6 (2015 and later	
Step 1	89 in-lbs
Step 2	Tighten an additional 90 degrees
V8 engines	
2000 and earlier models	80 in-lbs
2001 and later models	
Step 1	80 in-lbs
Step 2	Tighten an additional 90-degrees (1/4 turn)
EGR valve-to-air intake plenum/manifold	12 to 18

4-28 FUEL AND EXHAUST SYSTEMS

Torque specifications (continued) — Ft-lbs (unless otherwise indicated)

➡ **Note:** One foot-pound (ft-lb) of torque is equivalent to 12 inch-pounds (in-lbs) of torque. Torque values below approximately 15 ft-lbs are expressed in inch-pounds, since most foot-pound torque wrenches are not accurate at these smaller values.

Fuel rail mounting bolts	
All except 3.5L V6 engine	89 in-lbs
3.5L V6 engine	
Step 1	89 in-lbs
Step 2	Tighten an additional 45 degrees
Fuel rail pressure sensor	24
Exhaust pipe-to-exhaust manifold bolts	25 to 35
High-pressure fuel pump nuts	
Step 1	89 in-lbs
Step 2	Tighten an additional 45 degrees
High-pressure fuel pump mounting plate bolt	89 in-lbs
High-pressure fuel feed lines	
Fittings	
Step 1	62 in-lbs
Step 2	89 in-lbs
Step 3	Tighten an additional 25 to 30 degrees
Mounting bolts	
Step 1	89 in-lbs
Step 2	Tighten an additional 45 degrees
Turbocharger mounting flange bolts/nuts	24
Turbocharger exhaust flange nuts	30

5
ENGINE ELECTRICAL SYSTEMS

Section

1 General information
2 Battery - removal and installation
3 Battery - emergency jump starting
4 Battery cables - check and replacement
5 Ignition system - general information
6 Ignition system - check
7 Ignition coils - check and replacement
8 Ignition timing - check
9 Charging system - general information and precautions
10 Charging system - check
11 Alternator - removal and installation
12 Alternator components - replacement
13 Starting system - general information and precautions
14 Starter motor and circuit - in-vehicle check
15 Starter motor - removal and installation
16 Starter solenoid - replacement

Reference to other Chapters

Battery check and maintenance - See Chapter 1
CHECK ENGINE light - See Chapter 6
Crankshaft timing sensor - See Chapter 6
Drivebelt check, adjustment and replacement - See Chapter 1
Drivebelt deflection - See Chapter 1
Spark plug replacement - See Chapter 1
Spark plug wire, distributor cap and rotor - check and replacement - See Chapter 1

5-2 ENGINE ELECTRICAL SYSTEMS

1 General information

The engine electrical systems include all ignition, charging and starting components. Because of their engine-related functions, these components are considered separately from chassis electrical devices like the lights, instruments, etc.

Be very careful when working on the engine electrical components. They are easily damaged if checked, connected or handled improperly. The alternator is driven by an engine drivebelt which could cause serious injury if your hands, hair or clothes become entangled in it with the engine running. Both the starter and alternator are connected directly to the battery and could arc or even cause a fire if mishandled, overloaded or shorted out.

Never leave the ignition switch on for long periods of time with the engine off. Don't disconnect the battery cables while the engine is running. Correct polarity must be maintained when connecting battery cables from another source, such as another vehicle, during jump starting. Always disconnect the negative cable first and hook it up last or the battery may be shorted by the tool being used to loosen the cable clamps.

Additional safety related information on the engine electrical systems can be found in *Safety first* near the front of this manual. It should be referred to before beginning any operation included in this Chapter.

2 Battery - removal and installation

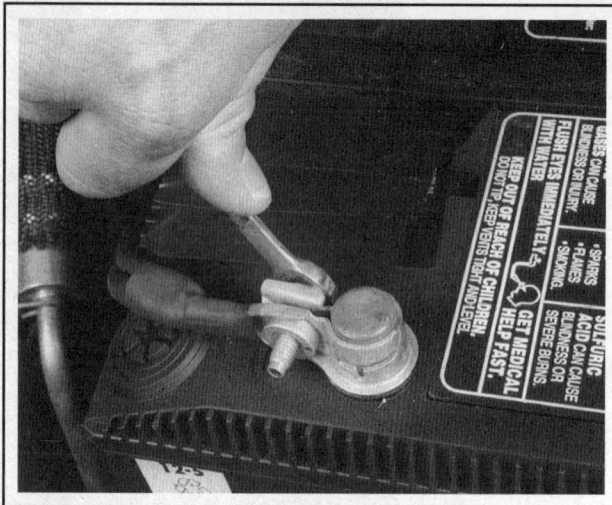

2.1 Removing the cable from a battery post with a wrench - sometimes special battery pliers are required for this procedure if corrosion has caused deterioration of the nut hex (always remove the ground cable first and hook it up last!)

▶ Refer to illustrations 2.1, 2.2 and 2.4

1 Disconnect both cables from the battery terminals (see illustration).

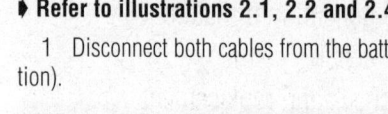

Always disconnect the negative cable first and hook it up last or the battery may be shorted by the tool being used to loosen the cable clamps.

→Note: When the battery is disconnected and reconnected, the vehicle may experience abnormal driving symptoms while the computer (PCM) relearns its adaptive strategy. The vehicle may need to be driven 10 miles or more to regain smooth operation.

2 Remove the bolt and wedge from the battery tray (see illustration).

3 Lift out the battery. Be careful - it's heavy. Special lifting straps that attach to the battery posts are available at auto parts stores - lifting and moving the battery is much easier if you use one.

4 If necessary for access to other components, remove the bolts that secure the battery tray (see illustration).

5 Installation is the reverse of removal.

2.2 Remove the bolt (arrow) and the wedge that holds the base of the battery to the battery tray

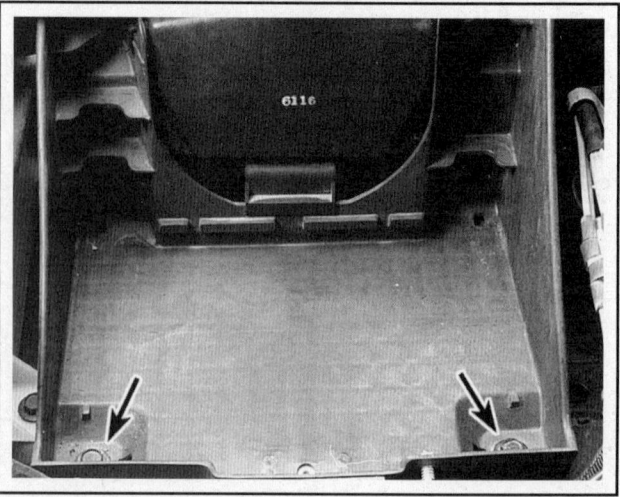

2.4 Remove the mounting bolts for the battery tray (arrows)

ENGINE ELECTRICAL SYSTEMS 5-3

3 Battery - emergency jump starting

Refer to the *Booster battery (jump) starting* procedure at the front of this manual.

4 Battery cables - check and replacement

1 Periodically inspect the entire length of each battery cable for damage, cracked or burned insulation and corrosion. Poor battery cable connections can cause starting problems and decreased engine performance.

2 Check the cable-to-terminal connections at the ends of the cables for cracks, loose wire strands and corrosion. The presence of white, fluffy deposits under the insulation at the cable terminal connection is a sign that the cable is corroded and should be replaced. Check the terminals for distortion, missing mounting bolts and corrosion.

3 When replacing the cables, always disconnect the negative cable first and hook it up last or the battery may be shorted by the tool used to loosen the cable clamps. Even if only the positive cable is being replaced, be sure to disconnect the negative cable from the battery first.

4 Disconnect and remove the cable. Make sure the replacement cable is the same length and diameter.

5 Clean the threads of the relay or ground connection with a wire brush to remove rust and corrosion. Apply a light coat of petroleum jelly to the threads to prevent future corrosion.

6 Attach the cable to the relay or ground connection and tighten the mounting nut/bolt securely.

7 Before connecting the new cable to the battery, make sure that it reaches the battery post without having to be stretched. Clean the battery posts thoroughly and apply a light coat of petroleum jelly to prevent corrosion (see Chapter 1).

8 Connect the positive cable first, followed by the negative cable.

5 Ignition system - general information

ELECTRONIC INTEGRATED (EI) IGNITION SYSTEM

1 The Electronic Integrated (EI) Ignition system is a complete electronically controlled ignition system that does not incorporate a distributor or rotor and cap. The EI system consists of a crankshaft timing sensor (variable reluctance sensor), camshaft sensor, ignition coil packs, an EEC-V module (PCM), the spark plug wires and the spark plugs. This engine is equipped with an ignition coil for each pair of spark plugs. The EI system features a waste-spark method of spark distribution. Each cylinder is paired with its companion cylinder in the firing order (1-5, 2-6, 3-4 [V6 engine]) or (1-6, 5-3, 4-7, 2-8 [V8 engine]) so one cylinder under compression fires simultaneously with its opposing cylinder, where the piston is on the exhaust stroke. Since the cylinder on the exhaust stroke requires very little of the available voltage to fire its plug, most of the voltage is used to fire the plug under compression.

➡**Note 1: The EI system is not equipped with an ignition module. Here the PCM functions as the overall controller of the ignition system by receiving engine speed, camshaft and crankshaft position signals and determining the correct ignition timing and injector ON-TIME (rich/lean) but also functions as the controller of the ignition coil(s) primary circuit which was basically the job of the ignition module in earlier distributorless ignition systems.**

➡**Note 2: The camshaft sensor on 4.2L engines is a Hall-Effect switching device mounted in a distributor-like housing on the intake manifold. The camshaft sensor on 4.6L and 5.4L engines is a variable reluctance device mounted on the front cover near the camshaft sprocket. Refer to Chapter 6 for additional information.**

2 This ignition system does not have any moving parts (no distributor) and all engine timing and spark distribution is handled electronically. This system has fewer parts that require replacement and provides more accurate spark timing. During engine operation, the EI ignition module (PCM) calculates spark angle and determines the turn-on and firing time of the ignition coil.

3 The crankshaft timing sensor is a variable reluctance sensor mounted above the front pulley timing gear. This electromagnetic device senses movement of the teeth on the pulley timing gear and generates an A/C voltage signal which increases with engine rpm. This sensor provides engine speed and crankshaft position signals to the PCM. The main function of the EI module is to synchronize the ignition coils so they are turned ON an OFF in the proper sequence for accurate spark control. Refer to Chapter 6 for additional information and testing procedures on the crankshaft sensor.

EI IGNITION COIL OVER PLUG (COP) SYSTEM

4 This system works basically the same as the EI system on the other models except each cylinder is equipped with its own coil and there are no ignition wires to the spark plugs. Each cylinder is fired sequentially on its compression stroke, thus eliminating the waste spark method. COP ignition systems operate in three different modes; *engine crank, engine running* and *CMP Failure Mode Effects Management (FMEM)*. Although the system operates sequentially in engine running mode, the PCM fires two cylinders simultaneously (companion cylinders) like the waste spark systems previously described in *engine crank* and *CMP FMEM Modes*. This is only to enhance driveability during warm-up or limp home modes of operation.

5 The COP ignition system uses the camshaft position sensor to identify the TDC of the compression stroke to fire the individual coils.

5-4 ENGINE ELECTRICAL SYSTEMS

6 Ignition system - check

WARNING:

Because of the very high secondary (spark plug) voltage generated by the ignition system, extreme care should be taken when this check is done.

→Note: Beginning in 1994, the manufacturer introduced a second generation self diagnosis system specified by EPA regulations called On Board Diagnosis (OBD) II. This system incorporates a series of diagnostic monitors that detect and identify emissions systems faults and store the information in the computer memory. This updated system also tests sensors and output actuators, diagnoses drive cycles, freezes data and clears codes. This powerful diagnostic computer must be accessed using the new OBD II SCAN tool and 16 pin Data Link Connector (DTC) located under the driver's dash area. All engines and powertrain combinations described in this manual are equipped with the On Board Diagnosis II (OBD-II) system. Refer to Chapter 6 for additional information on the OBD II system and its diagnostic capabilities.

CALIBRATED IGNITION TESTER METHOD (ALL SYSTEMS)

▶ Refer to illustration 6.2

1 If the engine turns over but won't start, disconnect the spark plug lead from any spark plug and attach it to a calibrated ignition tester (available at most auto parts stores). Make sure the tester is designed for Ford ignition systems if a universal tester isn't available.

→Note: On COP ignition systems, remove the coil assembly from the valve cover and attach the spark tester to the individual coil pack to check for spark.

2 Connect the clip on the tester to a bolt or metal bracket on the engine (see illustration), crank the engine and watch the end of the tester to see if bright blue, well-defined sparks occur.

3 If sparks occur, sufficient voltage is reaching the plug to fire it (repeat the check at the remaining plug wires to verify that the distributor cap and rotor are OK). However, the plugs themselves may be fouled, so remove and check them as described in Chapter 1 or install new ones.

ELECTRONIC INTEGRATED (EI) IGNITION SYSTEM

General checks

▶ Refer to illustration 6.5 and 6.8

4 If no sparks or intermittent sparks occur, check for a bad spark plug wire by swapping wires.

5 If the problem isn't caused by the spark plug wire, check for battery voltage to the ignition coil with the ignition key ON (engine not running). Attach a 12 volt test light to the battery negative (-) terminal. Disconnect the coil electrical connector and check for power at the positive (+) terminal (see illustration). Battery voltage should be available. If there is no battery voltage, check the 20 amp fuse that protects the ignition circuit (see Chapter 12 for additional information on the fuses and the wiring schematics).

6 Be sure to check the primary and secondary resistances of the ignition coils (see Section 7).

7 Check the ignition coil electrical connectors for dirt, corrosion and damage.

8 If battery voltage is available to the ignition coils, attach an LED test light to the battery positive (+) terminal and each negative (-) terminal to the coil (on the vehicle harness side) (see illustration), then crank the engine (be sure to check each negative terminal. Confirm that the test light flashes. This test checks for the trigger signal (ground) from the computer.

CAUTION:

Use only an LED test light to avoid damaging the PCM.

6.2 To use a calibrated ignition tester, simply disconnect a spark plug wire, clip the tester to a convenient ground (like a valve cover bolt) and operate the starter - if there is enough power to fire the plug, sparks will be visible between the electrode tip and the tester body

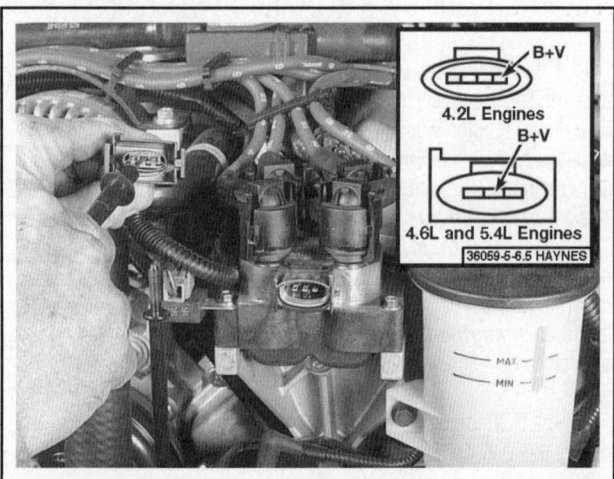

6.5 Disconnect the electrical connector from the ignition coil and check for battery voltage to the coil with the ignition key on

ENGINE ELECTRICAL SYSTEMS 5-5

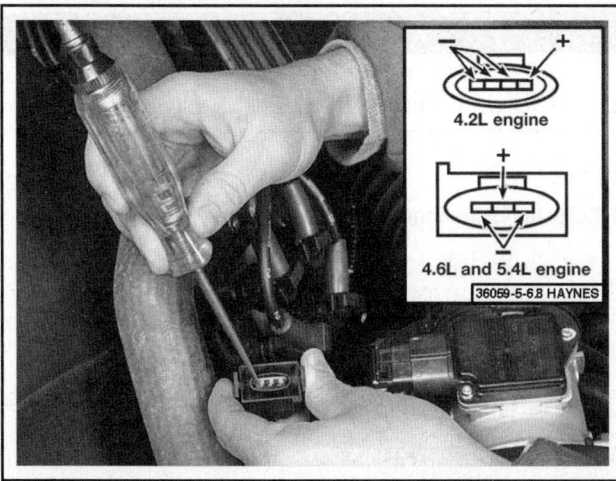

6.8 Connect an LED test light to the positive battery terminal and the coil negative (-) terminals on the ignition coil harness connectors and watch for a blinking light when the engine is cranked

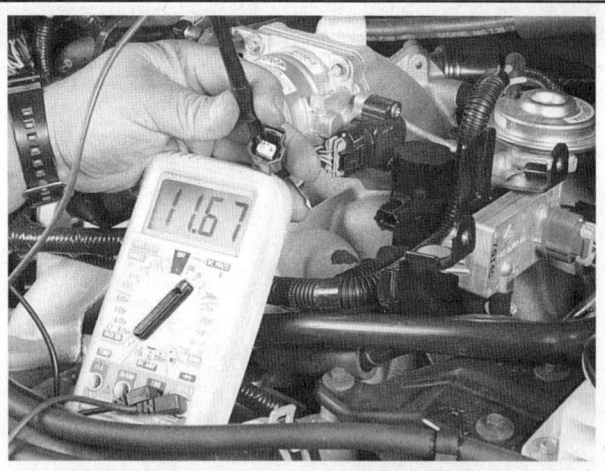

6.12 Disconnect the electrical connectors from each individual coil assembly and check for battery voltage to the coil with the ignition switch ON (engine not running)

9 If the test light does not flash, check the crankshaft position sensor (see Chapter 6). If the crankshaft sensor checks out OK, have the PCM checked by a dealer service department or other qualified automotive repair facility.

Sensor checks

10 These models are equipped with a camshaft sensor as well as a crankshaft sensor. The camshaft sensor signals the PCM to begin sequential pulsation of the fuel injectors. This camshaft sensor on 4.2L engines is a Hall Effect switching device activated by a single vane. This camshaft sensor is mounted on the top of the engine in the normal location of the distributor. The camshaft sensor in the 4.6L and 5.4L engines is a variable reluctance device which is triggered by the high point mark on the camshaft sprocket. This sensor is mounted in the timing cover on the left cylinder head near the camshaft sprocket. This type of camshaft sensor can be checked on the bench using an A/C voltmeter.

11 The crankshaft sensor is located near the crankshaft front pulley mounted in a bracket. These sensors are difficult to reach for testing purposes but it is of major importance that they be checked when dealing with ignition system diagnostics. In the event the crankshaft sensor or camshaft sensor is defective (or both), replace them with new parts and continue checking the ignition system to verify the working condition of all ignition system components. Refer to Chapter 6 for all the locations, checking and replacement procedures on the crankshaft and camshafts sensors.

EI IGNITION WITH COIL OVER PLUG (COP) SYSTEM

▶ **Refer to illustration 6.12 and 6.15**

12 If no sparks or intermittent sparks occur, check for battery voltage to the ignition coil (see illustration) with the ignition key ON (engine not running). Attach a 12 volt test light to the battery negative (-) terminal. Disconnect the coil harness connector and check for power to the positive (+) terminal. Battery voltage should be available with the ignition key ON (engine not running). If there is no battery voltage, check

6.15 Connect an LED test light to the battery positive (+) terminal and check for a trigger signal on each of the COP harness connector negative terminals (-) while an assistant cranks the engine

the 20 amp fuse that governs the ignition circuit (see Chapter 12 for additional information on the fuses and the wiring diagrams).

13 Check the individual ignition coils (see Section 7).

14 Unplug the ignition coil wiring harness connectors and inspect them for dirt, corrosion and damage.

15 If battery voltage is available to the ignition coils, attach an LED test light to the battery positive (+) terminal and to each coil negative (-) terminal, one at a time (see illustration) and crank the engine.

➡**Note: It will be necessary to disconnect all of the COP electrical connectors while testing to prevent the engine from starting.**

This test checks for the trigger signal (ground) from the computer.

16 The test light should flash as the engine is cranked over. If the test light does not flash, check the crankshaft position sensor (see Chapter 6). If the crankshaft sensor checks out OK, have the PCM checked by a dealer service department or other qualified automotive repair facility.

5-6 ENGINE ELECTRICAL SYSTEMS

7 Ignition coils - check and replacement

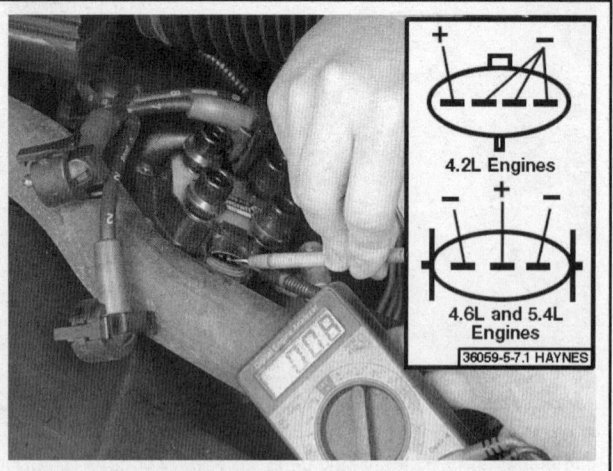

7.1 To check the primary resistance of the EI coil, connect the probes to the positive (+) terminal and each negative (-) terminal of the coil. The resistance should be the same for each check

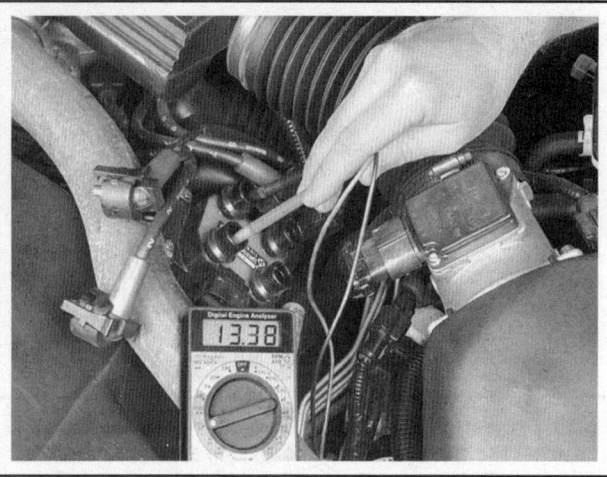

7.2 Check the coil secondary resistance by probing the paired companion cylinders

ELECTRONIC INTEGRATED (EI) IGNITION SYSTEM

Check

▶ Refer to illustrations 7.1 and 7.2

1 With the ignition off, disconnect the electrical connector(s) from the coil. Connect an ohmmeter across the coil positive (+) terminal and each negative (-) terminal (see illustration). The resistance should be as listed in this Chapter's Specifications. If not, replace the coil.

2 Connect an ohmmeter between the secondary terminals (see illustration) (the one that the spark plug wires connect to) of each coil pack. The resistance should be as listed in this Chapter's Specifications. If not, replace the coil.

➡ Note: Each coil pack is paired according to the companion cylinders. Be sure to check resistance with these designated terminals only:

4.2L V6 engines 1/5, 2/6, 3/4
4.6L engines .... 7/4, 8/2, 1/6 and 3/5

Replacement

▶ Refer to illustration 7.6

3 Disconnect the negative cable from the battery.

4 Disconnect the ignition coil electrical connector(s) from each individual coil pack.

5 Disconnect the spark plug wires by squeezing the locking tabs and twisting while pulling. DO NOT just pull on the wires to disconnect them. Disconnect all of the spark plug wires.

6 Remove the bolts securing the ignition coil to the mounting bracket on the engine (see illustration).

7 Installation is the reverse of the removal procedure with the following additions:

   a) *Prior to installing the spark plug wire into the ignition coil, coat the entire interior of the rubber boot with silicone dielectric compound.*
   b) *Insert each spark plug wire into the proper terminal of the ignition coil. Push the wire into the terminal and make sure the boots are fully seated and both locking tabs are engaged properly.*

7.6 Remove the coil pack mounting screws (arrows) and lift it from the engine

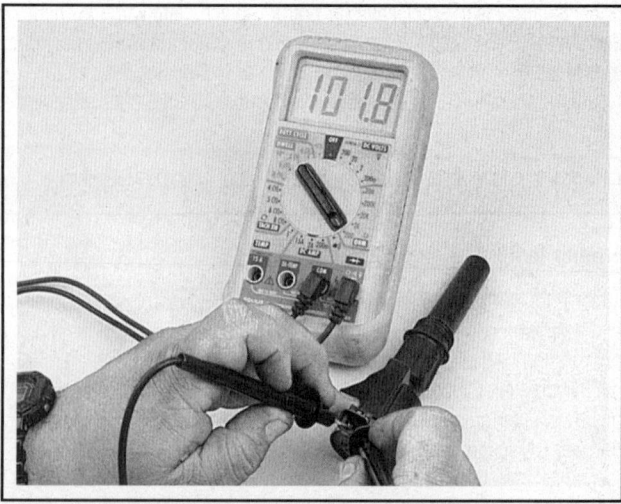

7.8 Check the COP assembly primary resistance

ENGINE ELECTRICAL SYSTEMS 5-7

EI IGNITION WITH COIL OVER PLUG (COP) SYSTEM

Check

♦ Refer to illustrations 7.8 and 7.9

➡ Note: On 3.5L V6 engines, lift the fuel pump insulator up and off of the driver's side valve cover to access the ignition coils.

8 With the ignition off, disconnect the electrical connector(s) from each coil assembly. Connect an ohmmeter across the coil primary terminal (+) and the negative terminal (-) (see illustration). The resistance should be as listed in this Chapter's Specifications. If not, replace the coil.

9 Remove the COP assembly from the cylinder head. Connect an ohmmeter between the secondary terminals (see illustration) (the one that fits over the spark plug). The resistance should be as listed in this Chapter's Specifications.

Replacement

♦ Refer to illustration 7.12

10 Disconnect the negative cable from the battery.

11 Disconnect the ignition coil electrical connector(s) from each individual coil. Mark each electrical connector with tape to prevent mix-ups during reassembly.

12 Remove the bolt securing the ignition coil (see illustration), then pull the coil from the cylinder head.

➡ Note: On later model engines it may be necessary to first disconnect the electrical connectors for the fuel injectors to gain access to each ignition coil mounting bolt.

13 Installation is the reverse of the removal procedure with the following additions:

a) *Prior to installing the coil over plug (COP) assembly into the cylinder head, coat the entire interior of the assembly with silicone dielectric compound.*
b) *Connect each COP electrical connector to its correct coil and make sure they are tight and secure.*

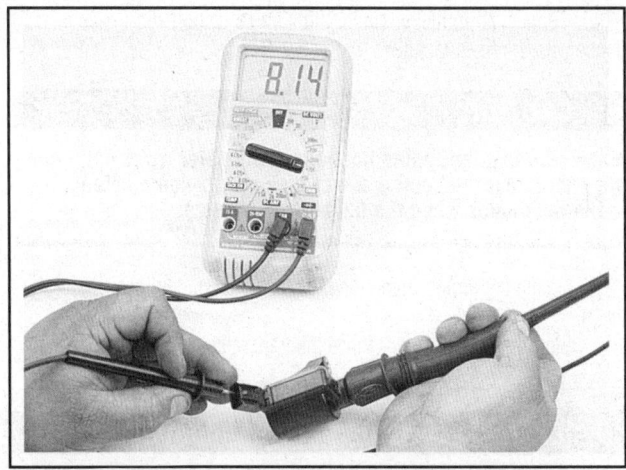

7.9 Check the COP assembly secondary resistance

7.12 Remove the COP mounting bolt and lift the assembly from the cylinder head

8 Ignition timing - check

♦ Refer to illustration 8.3

➡ Note 1: This procedure applies only to 2004 and earlier models.

➡ Note 2: This ignition timing procedure only checks the base timing setting specified by the factory. Timing cannot be adjusted, therefore the purpose of this check is to verify that the computer is controlling the ignition timing and that the base setting is correct. In most cases, the ignition system can be checked (see Section 6) but if the base setting remains incorrect, the PCM (computer) is defective. Take the vehicle to the dealer service department to verify and repair the ignition system problem(s).

➡ Note 3: These systems are equipped with a shorting bar inserted into the SPOUT (spark output) connector. This harness disconnect is used to remove the computer from the ignition timing control functions. Removal of the bar from the connector will retard the timing 2 to 3 degrees. The SPOUT connector is located in the right rear corner of the engine compartment. Do not remove the Shorting Bar except for checking ignition base timing.

1 Apply the parking brake and block the wheels. Turn off all accessories (heater, air conditioner, etc.).

2 Start the engine and warm it up. Once it has reached operating temperature, turn it off.

3 Unplug the shorting bar from the SPOUT connector located in the wiring harness in the right rear corner of the engine compartment (see illustration).

8.3 Remove the shorting bar (arrow) from the SPOUT connector to check base timing

5-8 ENGINE ELECTRICAL SYSTEMS

4 Connect an inductive timing light and a tachometer in accordance with the manufacturer's instructions.

> ※※ **CAUTION:**
>
> **Make sure that the timing light and tachometer wires don't hang anywhere near the cooling fan or they may become entangled in the fan blades when the fan begins to rotate.**

5 Locate the timing marks on the crankshaft pulley (see Chapter 2).
6 Start the engine again. Place the transmission in DRIVE (parking brake applied).
7 Point the timing light at the pulley timing marks and note whether the specified timing mark is aligned with the timing pointer on the front of the timing chain cover. Refer to the Specifications listed in this Chapter.
8 If the proper mark isn't aligned with the stationary pointer, have the PCM checked by a dealer service department or other qualified repair shop.
9 Turn off the engine.
10 Insert the shorting bar into the SPOUT connector.
11 Remove the timing light and tachometer from the engine compartment.

9 Charging system - general information and precautions

The charging system includes the alternator, a voltage regulator a charge indicator or warning light, the battery, a large fuse (called a mega fuse) and the wiring between all the components. The charging system supplies electrical power for the ignition system, the lights, the radio, etc. The alternator is driven by a drivebelt at the front of the engine.

The purpose of the voltage regulator is to limit the alternator's voltage to a preset value. This prevents power surges, circuit overloads, etc., during peak voltage output. On integral voltage regulator systems, a solid state regulator is housed inside a plastic module mounted on the alternator itself.

These models are equipped with either a Motorcraft 95 amp or a 130 amp output rated alternator. On 2009 and earlier models, the voltage regulator can be removed from the backside of the alternator but the alternator must be removed from the engine first (see Section 11). On 2010 and later models, the voltage regulator is inside the alternator, and cannot be replaced separately.

The charging system is protected by a series of large fuses (MEGA fuses) located in a box mounted on the firewall. In the event of charging system problems, check these fuses for damage or broken contacts.

The charging system doesn't ordinarily require periodic maintenance. However, the drivebelt, battery and wires and connections should be inspected at the intervals outlined in Chapter 1.

Be very careful when making electrical circuit connections to a vehicle equipped with an alternator and note the following:

a) *When reconnecting wires to the alternator from the battery, be sure to note the polarity.*
b) *Before using electric welding equipment to repair any part of the vehicle, disconnect the wires from the alternator and the battery terminals.*
c) *Never start the engine with a battery charger connected.*
d) *Always disconnect both battery cables before using a battery charger (negative cable first, positive cable last).*

10 Charging system - check

GENERAL CHECKS

♦ **Refer to illustration 10.2**

1 If a malfunction occurs in the charging circuit, the charging system warning light on the instrument panel may be illuminated, but do not immediately assume that the alternator is causing the problem. First, check the following items:

a) *The battery cables where they connect to the battery. Make sure the connections are clean and tight.*
b) *The battery electrolyte specific gravity. If it is low, charge the battery.*
c) *Check the external alternator wiring and connections.*
d) *Check the drivebelt condition and tension (see Chapter 1).*
e) *Check the alternator mounting bolts for tightness.*
f) *Run the engine and check the alternator for abnormal noise.*
g) *Use a scan tool to check for trouble codes in the charging system*

2 Using a voltmeter, check the battery voltage with the engine off. It should be approximately 12-volts (see illustration).
3 Start the engine and check the battery voltage again. It should now be approximately 14 to 15-volts.
4 If the indicated voltage reading is less or more than the specified charging voltage, replace the voltage regulator (see Section 13).

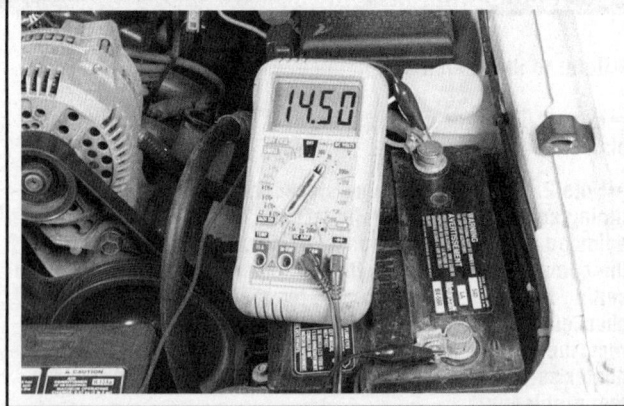

10.2 To measure charging voltage, attach the voltmeter leads to the battery terminals, start the engine and record the voltage reading

If replacing the regulator fails to restore the voltage to the specified range, the problem may be within the alternator.

➥**Note: The following checks are intended to direct the home mechanic to circuit problems that may be interfering with the charging system's ability to function properly. Many times a**

ENGINE ELECTRICAL SYSTEMS 5-9

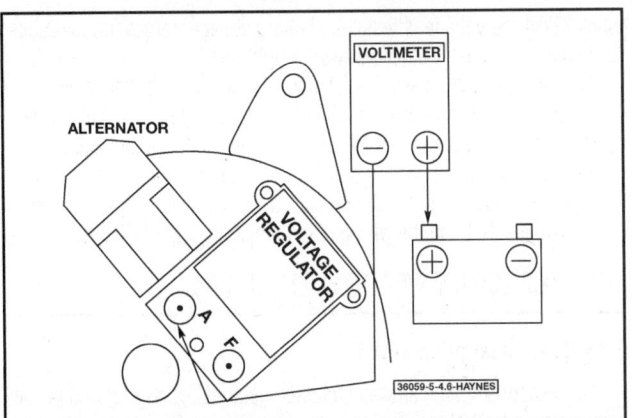

10.5 Check for a voltage drop between terminal A on the voltage regulator and the battery positive terminal (+). The voltage drop (from battery voltage) should be 0.25 volts or less

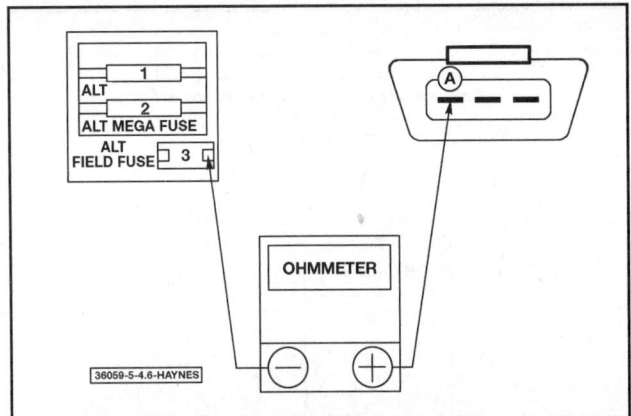

10.7 Measure the resistance between the voltage regulator harness connector A and the ALT FIELD fuse pin number 2 (it should be less than 10,000 ohms)

charging system problem results from corroded, damaged or broken terminals or harness connectors that operate within the charging system. Due to the special equipment necessary to test or service the alternator, it is recommended that if a fault is suspected the vehicle be taken to a dealer or a shop with the proper equipment.

OVERCHARGING CONDITION

▶ **Refer to illustration 10.5 and 10.7**

5 Most models are equipped with a voltmeter on the instrument panel that indicates battery voltage with the ignition key ON (engine not running), and alternator output when the engine is running. Observe the voltmeter at idle and high rpm. If the gauge reads high (15 volts and over), check for a overcharging condition. Measure the voltage drop between the voltage regulator test point A and the battery positive post (see illustration). The voltage drop should be 0.25 volts or less. If the voltage drop exceeds this value, check and repair the circuit to the MEGA fuse junction panel (orange/light blue wire) and the circuit to the secondary MEGA fuse junction panel (yellow/white wire).

6 Next, check the field circuit drain for a possible internal problem that could cause overcharging. With the ignition key OFF, measure the voltage between the voltage regulator test point F and the chassis ground (see illustration 10.5). If battery voltage is available, replace the voltage regulator (internal short). If there is no voltage, have the alternator tested by a dealer service department or automotive electrical repair facility.

7 Another possible cause of overcharging is excessive resistance in the junction panel or circuit. Disconnect the negative and positive battery cables and measure the resistance between the voltage regulator harness connector A and the ALT FIELD FUSE (20A) pin number 2 (see illustration). The resistance should be less than 10,000 ohms. If the resistance is higher, repair the circuit (orange/light blue wire).

8 Also, check the junction panel resistance between the ALT MEGA fuse (175 amp) and the positive battery cable (cable removed from the battery). The resistance should be 10,000 ohms or less. If the resistance is higher, repair the circuit (yellow/white wire).

CHARGE LIGHT REMAINS ON

9 The charge light on the instrument panel illuminates with the key ON and engine not running, and should go out when the engine runs. If the light remains ON, check the alternator FIELD fuse (see illustra-tion 10.7). Refer to Chapter 12 for the location or check your owner's manual.

10 Also, with the ignition key OFF, measure the voltage bestrewn the voltage regulator test point A and chassis ground (see illustration 10.5). There should be battery voltage. If there is no voltage available repair the circuit (refer to the wiring diagrams at the end of Chapter 5).

11 Check for a defective one-way circuit. Remove the alternator one-way terminal (light green/red), install a jumper wire from the battery positive terminal (+) and check that the light illuminates. Because the light is canceled (OFF) when the harness is disconnected from the battery and the light illuminates when battery voltage is applied, the circuit from the alternator to the light and the light bulb in the dash is correct. If the charge light remains ON when the engine is running, check the voltage regulator and electrical connections (see Section 12).

12 Also, check the ALT fuse and the ALT FIELD fuse in the junction box mounted on the engine compartment bulkhead. They should be intact and making positive connections (see illustration 10.7). Replace any broken fuses.

CHARGE LIGHT FLICKERS INTERMITTENTLY

▶ **Refer to illustration 10.13**

13 If the charge light on the dash flickers intermittently, check the voltage regulator connections, the alternator one-way connection, the alternator B+ eyelet connection, the power distribution box eyelets (see illustration) and the battery cables. They should be clean and tight.

10.13 Remove the covers (arrows) and check the power distribution eyelets for corrosion

5-10   ENGINE ELECTRICAL SYSTEMS

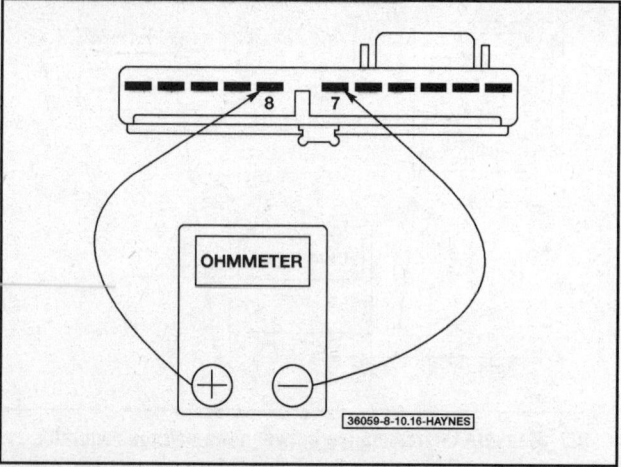

10.16 Disconnect the instrument cluster electrical connector and check the resistance between terminals 7 and 8. It should be between 445 and 495 ohms

14 With the ignition key OFF, measure the voltage between the voltage regulator test point F and the chassis ground (see Step 6). If battery voltage is available to test port F, check for a loose fuse (see Step 12). If the voltage is less than battery voltage, remove the brushes and check the brush holder screws for tightness.

15 With the ignition key ON (engine not running), connect a wire to the alternator test point A and the positive battery post. If the charge light flickers, the voltage regulator is defective. Have the alternator checked by a dealer service department or other qualified automobile electric repair facility. If the charge light does not flicker, repair the alternator harness. Refer to the wiring diagrams at the end of Chapter 12.

VOLTAGE GAUGE READS HIGH OR LOW

▶ **Refer to illustration 10.16**

16 Disconnect the instrument cluster electrical connector and probe terminals 7 and 8 of the cluster with an ohmmeter. Resistance should be between 445 and 495 ohms (see illustration). If the resistance is incorrect, replace the instrument cluster (see Chapter 12).

BATTERY DOES NOT HOLD A CHARGE

17 Measure the voltage between the B+ terminal on the alternator and chassis ground (-). There should be battery voltage present with the ignition key OFF.

18 Check the condition of the battery. Replace it if necessary.

11 Alternator - removal and installation

▶ **Refer to illustrations 11.6a, 11.6b and 11.6c**

1 Detach the cable from the negative terminal of the battery.
2 Unplug the electrical connectors from the alternator.
3 Remove the drivebelt (see Chapter 1).
4 On 4.6L engines, remove the ignition wire assembly from the intake manifold area.
5 On 3.5L V6 engines, remove the turbocharger inlet and outlet tubes (see Chapter 4) then the turbocharger bracket nuts and bolt.
6 Remove the bolts and separate the alternator from the engine (see illustrations). On 2005 and later 5.4L engines, remove the four bolts and the mounting bracket above the alternator, then remove the two lower alternator mounting bolts.
7 Installation is the reverse of removal.
8 After the alternator is installed, install the drivebelt and reconnect the cable to the negative terminal of the battery.

11.6a Remove the lower mounting bolts (arrows) (4.6L engine shown)

11.6b Remove the three mounting bolts (arrows) and separate the alternator bracket from the air intake plenum (4.6L engine shown)

11.6c Alternator mounting bolts - 4.2L V6 engine

12 Alternator components - replacement

♦ Refer to illustrations 12.3, 12.4, 12.5 and 12.9

➡ **Note:** *This Section only applies to 2009 and earlier models.*

➡ **Note:** *Before disassembling the alternator, make sure that replacement parts are available.*

1. Remove the alternator (see Section 11).
2. Set the alternator on a clean workbench.
3. Remove the four voltage regulator mounting screws (see illustration).
4. Detach the voltage regulator (see illustration).
5. Detach the rubber plugs and remove the brush lead retaining screws and nuts to separate the brush leads from the holder (see illustration). Note that the screws have Torx heads and require a special screwdriver.
6. After noting the relationship of the brushes to the brush holder assembly, remove both brushes. Don't lose the springs.
7. If you're installing a new voltage regulator, insert the old brushes into the brush holder of the new regulator. If you're installing new brushes, insert them into the brush holder of the old regulator. Make sure the springs are properly compressed and the brushes are properly inserted into the recesses in the brush holder.
8. Install the brush lead retaining screws and nuts.
9. Insert a short section of wire, like a paper clip, through the hole in the voltage regulator (see illustration) to hold the brushes in the retracted position during regulator installation.
10. Carefully install the regulator. Make sure the brushes don't hang up on the rotor.
11. Install the voltage regulator screws and tighten them securely.
12. Remove the wire or paper clip.
13. Install the alternator (see Section 11).

12.3 To detach the voltage regulator/brush holder assembly, remove the four screws

12.4 Lift the assembly from the alternator

12.5 To remove the brushes from the voltage regulator/brush holder assembly, detach the rubber plugs from the two brush lead screws and remove both screws (arrows)

12.9 Before installing the voltage regulator/brush holder assembly, insert a paper clip as shown to hold the brushes in place during installation - after installation, simply pull the paper clip out

5-12 ENGINE ELECTRICAL SYSTEMS

13 Starting system - general information and precautions

1 The function of the starting system is to crank the engine fast enough to start it. The system is composed of the starter motor, starter relay, battery, switch and connecting wires.

2 Turning the ignition key to the Start position actuates the starter relay through the starter control circuit. The starter relay then connects the battery to the starter solenoid.

3 These models are equipped with a starter/solenoid assembly that is mounted to the transmission bellhousing.

4 Vehicles equipped with an automatic transmission are equipped with a Transmission Range sensor in the starter control circuit, which prevents operation of the starter unless the shift lever is in Neutral or Park. Manual transmission vehicles are equipped with a starter clutch pedal position switch. The starter will not crank with the foot off the clutch pedal.

5 The starter circuit is equipped with a starter relay. This relay is located on the firewall in the right rear corner of the engine compartment. On 2003 and later models, the starter relay is located in the underhood fuse/relay box.

6 Never operate the starter motor for more than 15 seconds at a time without pausing to allow it to cool for at least two minutes. Excessive cranking can cause overheating, which can seriously damage the starter.

14 Starter motor and circuit - in-vehicle check

▶ Refer to illustration 14.6

→ **Note:** *Before diagnosing starter problems, make sure the battery is fully charged.*

1 If the starter motor doesn't turn at all when the switch is operated, make sure the shift lever is in Neutral or Park.

2 Make sure the battery is charged and that all cables at the battery and starter solenoid terminals are secure.

3 If the starter motor spins but the engine doesn't turn over, then the drive assembly in the starter motor is slipping and the starter motor must be replaced (see Section 15).

4 If, when the switch is actuated, the starter motor doesn't operate at all but the starter solenoid operates (clicks), then the problem lies with either the battery, the starter solenoid contacts or the starter motor connections.

5 If the starter solenoid doesn't click when the ignition switch is actuated, either the starter solenoid is defective, the starter relay is bad, the ignition switch is faulty, the Transmission Range sensor (automatic) or clutch switch (manual) is bad, or there is a problem in the wiring between the components. Check the starter circuit.

6 To check the starter circuit, remove the push-on connector from the relay - this is the signal wire from the ignition switch (see illustration). Make sure that the connection is clean and secure. If the connections are good, check the operation of the relay with a jumper wire. To do this, place the transmission in Park (automatic transmission) or Neutral (manual transmission). Remove the push-on connector from the relay. Connect a jumper wire between the battery positive terminal and the exposed terminal on the relay. If the starter motor now operates, the starter relay is okay. The problem is in the ignition switch, Transmission Range sensor Clutch Pedal Position switch or in the wiring between these components (look for open or loose connections).

7 If the starter motor still doesn't operate, bridge the two large terminals on the relay with a screwdriver. If the starter now works, replace the relay. If it doesn't operate, check for voltage to the relay (it should be available on one of the large terminals). If voltage isn't present, check the Mega Fuses, the two main battery fuses (adjacent to the relay, under the plastic covers) and the cable to the relay.

8 If voltage is present, check for voltage to the starter motor while the ignition key is turned to Start. If voltage is present, replace the starter assembly. If voltage is not present and the relay checked out OK, trace the wiring between the relay and the starter for an open circuit condition.

9 If the starter motor cranks the engine at an abnormally slow speed, first make sure the battery is fully charged and all terminal connections are clean and tight. Also check the connections at the starter solenoid and battery ground. Eyelet terminals should not be easily rotated by hand. If the engine is partially seized, or has the wrong viscosity oil in it, it will crank slowly.

14.6 The starter relay is mounted on the firewall, in the right rear corner of the engine compartment on 2002 and earlier models

- A *Signal wire from ignition switch*
- B *Relay output (to starter)*
- C *Relay input (from battery)*

ENGINE ELECTRICAL SYSTEMS 5-13

15 Starter motor - removal and installation

▶ Refer to illustration 15.4

1 Detach the cable from the negative terminal of the battery.
2 Raise the vehicle and support it securely on jackstands.

※※ WARNING:

Some models covered by this manual are equipped with self leveling suspension systems. Always disconnect electrical power to the suspension system before lifting or towing the vehicle (see Chapter 10). Failure to perform this procedure may result in unexpected shifting or movement of the vehicle which could cause personal injury.

3 Disconnect the large cable from the terminal on the starter motor and the solenoid terminal connections.
4 Remove the starter motor mounting bolt and nut (see illustration) and detach the starter from the engine. V6 models have two bolts and V8 models have three bolts.
5 Installation is the reverse of removal.

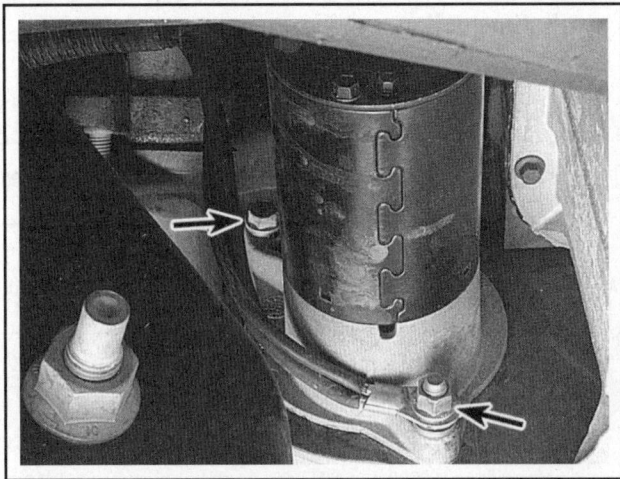

15.4 Remove the starter mounting bolt and nut (arrows) and separate the assembly from the transmission bellhousing

16 Starter solenoid - replacement

▶ Refer to illustrations 16.3a and 16.3b

1 Remove the starter assembly from the engine compartment (see Section 15).
2 Remove the electrical connector from the solenoid M terminal.
3 Remove the solenoid mounting bolts and separate the solenoid from the starter body (see illustrations).
4 Installation is the reverse of removal.

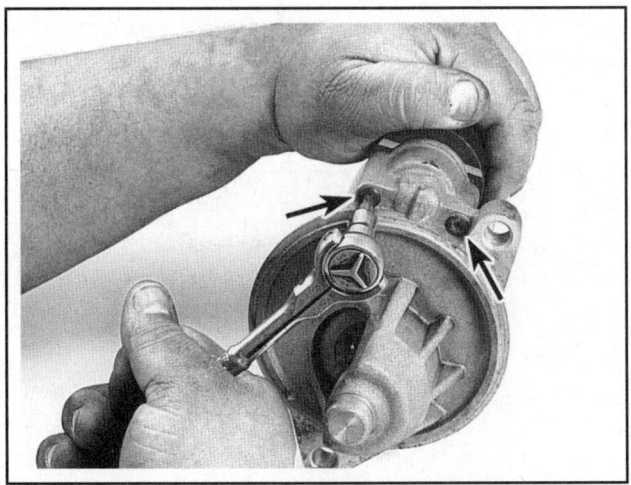

16.3a Remove the solenoid mounting bolts (arrows)

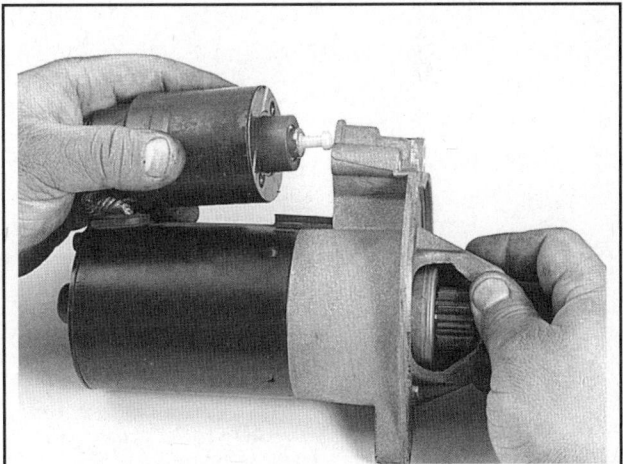

16.3b Separate the solenoid from the starter

5-14 ENGINE ELECTRICAL SYSTEMS

Specifications

Battery voltage
    Engine off   12-volts
    Engine running   14-to-15 volts

Firing order   See Chapter 2

Ignition coil-to-distributor cap wire resistance   5,000 ohms per foot

Ignition coil resistance
    Electronic Integrated Ignition (EI) Systems
        Primary resistance   0.3 to 1.0 ohms
        Secondary resistance   6.5 to 11.5 K-ohms
    EI systems with Coil Over Plug (COP)
        Primary resistance   0.55 ohms
        Secondary resistance   5,500 ohms

Ignition timing (base setting)
    All engines   10-degrees BTDC with SPOUT disconnected

Alternator brush length
    New   1/2 inch
    Minimum   1/4 inch

Section

1 General information
2 On Board Diagnosis (OBD) system and trouble codes
3 Powertrain Control Module (PCM) - replacement
4 Information sensors
5 Exhaust Gas Recirculation (EGR) system
6 Evaporative Emissions Control System (EVAP)
7 Positive Crankcase Ventilation (PCV) system
8 Catalytic converter
9 Oil pressure control solenoid (3.5L V6 engine) - replacement
10 Turbocharger bypass valve (3.5L V6 engine) - replacement
11 Turbocharger wastegate vacuum sensor - replacement
12 Accelerator Pedal Position (APP) sensor (2005 and later models)
 - replacement and adjustment

6
EMISSIONS AND ENGINE CONTROL SYSTEMS

1 General information

1.1a Typical emission and engine control system component locations - V6 engine

1 SPOUT connector	5 IAC valve	9 Power distribution box
2 Starter relay (under cover)	6 IMRC actuators	10 EDIS coil pack
3 DPFE sensor and EVR solenoid	7 MAF sensor	11 EGR valve
4 TPS	8 Vapor management valve	

▶ **Refer to illustrations 1.1a, 1.1b and 1.7**

To prevent pollution of the atmosphere from incompletely burned and evaporating gases, and to maintain good driveability and fuel economy, a number of emission control systems are incorporated (see illustrations). They include:

The Electronic Engine Control system (EEC-V) OBD-II
The Evaporative Emission Control system (EVAP)
Positive Crankcase Ventilation (PCV) system
Exhaust Gas Recirculation (EGR) system
Catalytic converter

All of these systems are linked, directly or indirectly, to the emission control system.

The Sections in this Chapter include general descriptions, checking procedures within the scope of the home mechanic and component replacement procedures (when possible) for each of the systems listed above.

Before assuming that an emissions control system is malfunctioning, check the fuel and ignition systems carefully. The diagnosis of some emission control devices requires specialized tools, equipment and training. If checking and servicing become too difficult or if a procedure is beyond your ability, consult a dealer service department. Remember, the most frequent cause of emissions problems is simply a loose or broken vacuum hose or wire, so always check the hose and wiring connections first.

This doesn't mean, however, that emission control systems are particularly difficult to maintain and repair. You can quickly and easily perform many checks and do most of the regular maintenance at home with common tune-up and hand tools.

EMISSIONS AND ENGINE CONTROL SYSTEMS

1.1b Typical emission and engine control system component locations - V8 engine

1. Shorting bar
2. Starter relay
3. IAC valve
4. DPFE sensor and EVR solenoid
5. EGR valve
6. Vapor management valve
7. Power distribution box
8. MAF sensor
9. EDIS coil pack (4.6L)
10. Fuel pressure regulator
11. TPS
12. EDIS coil pack (4.6L)

→ **Note:** Because of a Federally mandated extended warranty which covers the emission control system components, check with your dealer about warranty coverage before working on any emissions-related systems. Once the warranty has expired, you may wish to perform some of the component checks and/or replacement procedures in this Chapter to save money.

Pay close attention to any special precautions outlined in this Chapter. It should be noted that the illustrations of the various systems may not exactly match the system installed on the vehicle you're working on because of changes made by the manufacturer during production or from year-to-year.

A Vehicle Emissions Control Information (VECI) label is located in the engine compartment (see illustration). This label contains important emissions specifications and adjustment information, as well as a vacuum hose schematic with emissions components identified. When servicing the engine or emissions systems, the VECI label in your particular vehicle should always be checked for up-to-date information.

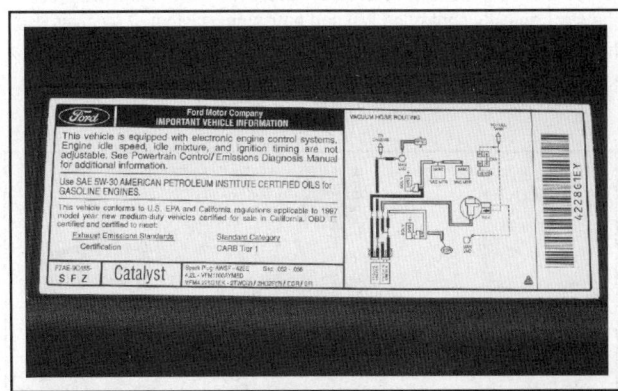

1.7 The Vehicle Emission Control Information (VECI) label is located in the engine compartment on the radiator support and contains information on the emission devices on your vehicle, vacuum line routing, etc. (V6 engine shown)

6-4 EMISSIONS AND ENGINE CONTROL SYSTEMS

2 On Board Diagnosis (OBD) system and trouble codes

SCAN TOOL INFORMATION

▶ Refer to illustration 2.2

1 Hand-held scanners are the most powerful and versatile tools for analyzing engine management systems used on later model vehicles.

➥ **Note: An aftermarket generic scanner should work with any model covered by this manual. Before purchasing a generic scan tool, verify that it will work properly with the OBD-II system you want to scan. If necessary, of course, you can always have the codes extracted by a dealer service department or an independent repair shop with a professional scan tool.**

2 With the arrival of the Federally mandated emission control system (OBD-II), specially designed aftermarket scanners have been developed. Several tool manufacturers have released OBD-II scan tools for the home mechanic (see illustration).

OBD SYSTEM GENERAL DESCRIPTION

▶ Refer to illustration 2.4

3 All models are equipped with the second-generation on-board diagnostic (OBD-II) system. This system consists of an on-board computer known as the Powertrain Control Module (PCM), and information sensors, which monitor various functions of the engine and send data to the PCM. This system incorporates a series of diagnostic monitors that detect and identify fuel injection and emissions control systems faults and store the information in the computer memory. This updated system also tests sensors and output actuators, diagnoses drive cycles, freezes data and clears codes.

4 This powerful diagnostic computer must be accessed using an OBD-II scan tool (see illustration) and the 16 pin Data Link Connector (DLC) located under the driver's dash area. The PCM is the brain of the electronically controlled fuel and emissions system. It receives data from a number of sensors and other electronic components (switches, relays, etc.). Based on the information it receives, the PCM generates output signals to control various relays, solenoids (fuel injectors) and other actuators. The PCM is specifically calibrated to optimize the emissions, fuel economy and drivability of the vehicle.

5 It isn't a good idea to attempt diagnosis or replacement of the PCM or emission control components at home while the vehicle is under warranty. Because of a Federally mandated warranty which covers the emissions system components and because any owner-induced damage to the PCM, the sensors and/or the control devices may void this warranty, take the vehicle to a dealer service department if the PCM or a system component malfunctions.

INFORMATION SENSORS

6 When battery voltage is applied to the air conditioning compressor solenoid, a signal is sent to the PCM, which interprets the signal as an added load created by the compressor and increases engine idle speed accordingly to compensate.

7 The **Intake Air Temperature sensor (IAT)**, positioned in the air intake duct (see Section 4), provides the PCM with fuel/air mixture temperature information. The PCM uses this information to control fuel flow, ignition timing and EGR system operation.

8 The **Engine Coolant Temperature (ECT)** sensor, which is threaded into a coolant passage in the intake manifold, monitors engine coolant temperature. The ECT sends the PCM a voltage signal which influences PCM control of the fuel mixture, ignition timing and EGR operation.

9 The **Heated Exhaust Gas Oxygen (HEGO)** sensors, which are threaded into the exhaust manifolds before and after the catalytic converter, constantly monitor the oxygen content of the exhaust gases. A voltage signal which varies in accordance with the difference between the oxygen content of the exhaust gases and the surrounding atmosphere is sent to the PCM. The PCM converts this exhaust gas oxygen content signal to the fuel/air ratio, compares it to the ideal ratio for current engine operating conditions and alters the signal to the injectors accordingly.

10 The **Throttle Position Sensor (TPS)**, which is mounted on the side of the throttle body (see Section 4) and connected directly to

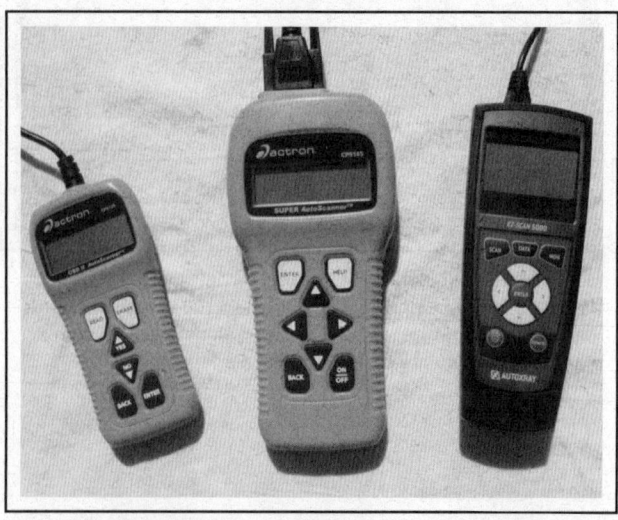

2.2 Scanners like these from Actron and AutoXray are powerful diagnostic aids - they can tell you just about anything you want to know about your engine management system

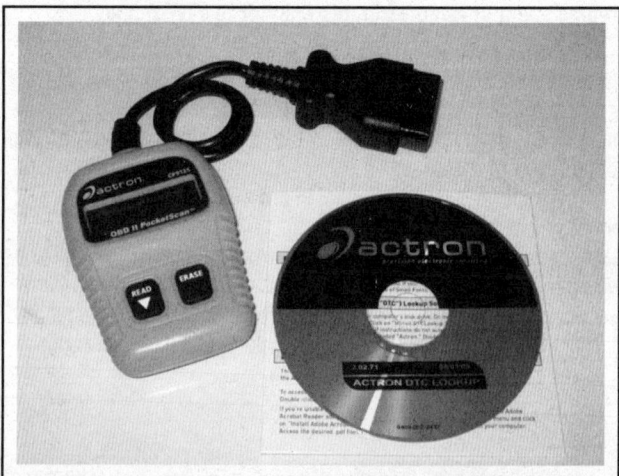

2.4 OBD-II trouble code readers are an economical choice for the home mechanic - they can read stored generic trouble codes (some can also read various manufacturer-specific codes) and turn off the CHECK ENGINE light after repairs have been made

EMISSIONS AND ENGINE CONTROL SYSTEMS   6-5

the throttle shaft, senses throttle movement and position, then transmits an electrical signal to the PCM. This signal enables the PCM to determine when the throttle is closed, in its normal cruise condition or wide open.

11 The **Mass Air Flow (MAF)** sensor, which is mounted in the air cleaner intake passage, measures the mass of the air entering the engine (see Section 4). Because air mass varies with air temperature (cold air is denser than warm air), measuring air mass provides the PCM with a very accurate way of determining the correct amount of fuel to obtain the ideal fuel/air mixture.

OUTPUT ACTUATORS

12 The **PCM power relay**, which is activated by the ignition switch, supplies battery voltage to the EEC-V system components when the switch is in the Start or Run position.

➡ **Note: The fuel pump relay and the PCM power relay are located in the Power Distribution Box in the engine compartment. Refer to Chapter 12 or your owner's manual for additional information for relay location.**

13 The **Vapor Management Valve (VMV)** switches manifold vacuum to operate the VMV when a signal is received from the PCM. The EEC-V computer (PCM) activates the VMV, allowing fuel vapor to flow from the canister to the intake manifold to be burned in the combustion process.

14 The solenoid-operated **fuel injectors** are located above the intake ports (see Chapter 4). The PCM controls the length of time the injector is open. The "open" time of the injector determines the amount of fuel delivered. For information regarding injector replacement, refer to Chapter 4.

15 The **fuel pump relay** is activated by the PCM with the ignition switch in the On position. When the ignition switch is turned to the On position, the relay is activated to supply initial line pressure to the system. For information regarding fuel pump check and replacement, refer to Chapter 4.

16 The **Electronic Integrated (EI) Ignition module** (see Chapter 5) is incorporated into the PCM. The PCM uses a signal from the camshaft sensor to determine piston position. Ignition timing is determined by the PCM, which then signals the module (PCM) to fire the coil. For further information regarding the ignition system, refer to the appropriate Section in Chapter 5.

2.18 Typical Data Link Connector (DLC) on an OBD-II vehicle

OBTAINING AND CLEARING OBD-II SYSTEM CODES

♦ **Refer to illustration 2.18**

17 On OBD-II systems, the PCM will illuminate the Malfunction Indicator Light on the dash if it recognizes a component fault for two consecutive drive cycles. It will continue to set the light until the PCM does not detect any malfunction for three or more consecutive drive cycles. Because the OBD-II system requires a SCAN tool to reset the light, if the tool is not available for diagnostics, have the system checked by a dealer service department or other qualified repair facility.

18 The diagnostic codes for the EEC-V (OBD-II) systems can be extracted from the PCM using a special SCAN tool that is programmed to interface with this new system by plugging into the DLC (see illustration). If the tool is not available, have the vehicle checked at a dealer service department.

19 To clear the codes from the PCM memory, install the OBD-II SCAN tool, scroll the menu for the function that describes "CLEARING CODES' and follow the prescribed method for that particular SCAN tool. If necessary, have the codes cleared by a dealer service department or other qualified repair facility.

✲✲ CAUTION:

Do not disconnect the battery from the vehicle to clear the codes. This will erase stored operating parameters from the memory and cause the engine to run rough for a period of time while the computer relearns the information.

OBD-II TROUBLE CODES

Code	Probable cause
P0010	Intake camshaft position actuator circuit open (bank 1)
P0011	"A" Camshaft position - timing - over-advanced (bank 1)
P0012	"A" Camshaft position - timing over-retarded (bank 1)
P0014	B Camshaft Position Timing - Over-Advanced (Bank 1)
P0015	B Camshaft Position Timing - Over-Retarded (Bank 1)
P0016	Crankshaft Position/Camshaft Position Correlation (Bank 1 Sensor A)
P0019	Crankshaft Position/Camshaft Position Correlation (Bank 2 Sensor B)
P0020	Intake camshaft position actuator circuit open (bank 2)

Code	Probable cause
P0021	Intake camshaft position timing - over-advanced (bank 2)
P0022	Intake camshaft position timing - over-retarded (bank 2)
P0023	B Camshaft Position Actuator A Control Circuit/Open Bank 2
P0024	B Camshaft Position Timing - Over-Advanced (Bank 2)
P0025	B Camshaft Position Timing - Over-Retarded (Bank 2)
P0030	Upstream oxygen sensor, bank 1, heater circuit
P0034	Turbocharger/Supercharger Bypass Valve A Control Circuit Low
P0035	Turbocharger/Supercharger Bypass Valve A Control Circuit High
P0036	HO2S Heater Control Circuit (Bank 1 Sensor 2)

OBD-II TROUBLE CODES (CONTINUED)

Code	Probable cause
P0037	HO2S Heater Control Circuit Low (Bank 1 Sensor 2)
P0038	HO2S Heater Control Circuit High (Bank 1 Sensor 2)
P0040	Upstream oxygen sensors swapped from bank to bank (HO2S - bank 1, sensor 1/bank 2, sensor 1)
P0041	Downstream oxygen sensors swapped from bank to bank (HO2S - bank 1, sensor 2/bank 2, sensor 2)
P0050	B Camshaft Position Actuator A Control Circuit/Open Bank 2
P0053	Oxygen sensor heater control circuit resistance, Bank 1, sensor 1
P0054	Oxygen sensor heater control circuit resistance, Bank 1, sensor 2
P0056	Oxygen sensor heater control circuit resistance, Bank 2, sensor 2
P0059	Oxygen sensor heater control circuit resistance, Bank 2, sensor 1
P0060	Downstream oxygen sensor heater circuit open or short, Bank 2, sensor 2
P0068	Throttle position (TP) sensor inconsistent with mass air flow (MAF) sensor
P0071	B Camshaft Position Actuator A Control Circuit/Open Bank 2
P0072	Ambient Air Temperature Sensor Circuit A Low
P0073	Ambient Air Temperature Sensor Circuit A High
P0074	Ambient Air Temperature Sensor Circuit A Intermittent/Erratic
P007B	Charge Air Cooler Temperature Sensor Circuit Range/Performance (Bank 1)
P007C	Charge Air Cooler Temperature Sensor Circuit Low (Bank 1)
P007D	Charge Air Cooler Temperature Sensor Circuit High (Bank 1)
P060A	Internal Powertrain Control Module (PCM) error
P060B	Internal Powertrain Control Module (PCM) analog to digital processing error
P060C	Internal Powertrain Control Module (PCM) main processor error
P061B	Internal Powertrain Control Module (PCM) torque calculation error
P061C	Internal Powertrain Control Module (PCM) engine rpm calculation error
P061D	Internal Powertrain Control Module (PCM) air mass error
P061F	Internal Powertrain Control Module (PCM) Throttle Actuator Controller (TAC) error
P0087	Fuel Rail/System Pressure - Too Low (Bank 1)
P0088	Fuel Rail/System Pressure - Too High (Bank 1)
P008A	Low Pressure Fuel System Pressure - Too Low
P008B	Low Pressure Fuel System Pressure - Too High
P0093	Fuel System Leak Detected - Large Leak
P0094	Fuel System Leak Detected - Small Leak

Code	Probable cause
P0096	Intake Air Temperature Sensor 2 Circuit Range/Performance (Bank 1)
P0097	Intake Air Temperature Sensor 2 Circuit Low (Bank 1)
P0098	Intake Air Temperature Sensor 2 Circuit High (Bank 1)
P0100	Mass Or Volume Air Flow Sensor A Circuit
P0102	Mass Airflow (MAF) sensor circuit low input
P0103	Mass Airflow (MAF) sensor circuit high input
P0106	Manifold absolute pressure or barometric pressure circuit, range or performance problem
P0107	Manifold absolute pressure or barometric pressure circuit, low input
P0108	Manifold absolute pressure or barometric pressure circuit, high input
P0109	Manifold absolute pressure or barometric pressure circuit, Intermittent
P0112	Intake Air Temperature (IAT) sensor circuit low input
P0113	Intake Air Temperature (IAT) sensor circuit high input
P0116	Engine coolant temperature circuit range/performance problem
P0117	Engine Coolant Temperature (ECT) sensor circuit low input
P0118	Engine Coolant Temperature (ECT) sensor circuit high input
P0121	In range Throttle Position Sensor (TPS) fault
P0122	Throttle Position Sensor (TPS) circuit low input
P0123	Throttle Position Sensor (TPS) circuit high input
P0124	Throttle position or pedal position sensor/switch circuit, intermittent
P0125	Insufficient coolant temperature for closed loop fuel control
P0128	Coolant thermostat (coolant temperature below thermostat regulating temperature)
P0131	Upstream heated O2 sensor circuit low voltage (Bank 1)
P0132	O2 sensor circuit, high voltage (bank 1, sensor 1)
P0133	Upstream heated O2 sensor circuit slow response (Bank 1)
P0135	Upstream heated O2 sensor heater circuit fault (Bank 1)
P0136	Downstream heated O2 sensor fault (Bank 1)
P0138	O2 sensor circuit, high voltage (bank 1, sensor 2)
P0141	O2 sensor heater circuit malfunction (bank 1, sensor 2)
P0148	Fuel delivery error
P0151	O2 sensor circuit, low voltage (bank 2, sensor 1)
P0152	O2 sensor circuit, high voltage (bank 2, sensor 1)
P0153	O2 sensor circuit, slow response (bank 2, sensor 1)
P0155	O2 sensor heater circuit malfunction (bank 2, sensor 1)

EMISSIONS AND ENGINE CONTROL SYSTEMS 6-7

Code	Probable cause
P0156	O2 sensor circuit malfunction (bank 2, sensor 2)
P0158	O2 sensor circuit, high voltage (bank 2, sensor 2)
P0161	O2 sensor heater circuit malfunction (bank 2, sensor 2)
P0171	System Adaptive fuel too lean (Bank 1)
P0172	System Adaptive fuel too rich (Bank 1)
P0174	System Adaptive fuel too lean (Bank 2)
P0175	System Adaptive fuel too rich (Bank 2)
P0180	Fuel temperature sensor A circuit malfunction
P0181	Fuel temperature sensor A circuit, range or performance problem
P0183	Fuel temperature sensor A circuit, high input
P0190	Fuel rail pressure sensor circuit malfunction
P0191	Injector Pressure sensor system performance
P0192	Injector Pressure sensor circuit low input
P0193	Injector Pressure sensor circuit high input
P0196	Fuel rail pressure sensor circuit, range or performance problem
P0197	Fuel rail pressure sensor circuit, low input
P0198	Fuel rail pressure sensor circuit, high input
P0201	Injector circuit malfunction - cylinder no. 1
P0202	Injector circuit malfunction - cylinder no. 2
P0203	Injector circuit malfunction - cylinder no. 3
P0204	Injector circuit malfunction - cylinder no. 4
P0205	Injector circuit malfunction - cylinder no. 5
P0206	Injector circuit malfunction - cylinder no. 6
P0207	Injector circuit malfunction - cylinder no. 7
P0208	Injector circuit malfunction - cylinder no. 8
P0217	Engine coolant overtemperature condition
P0218	Transmission overheating condition
P0219	Engine overspeed condition
P0221	Throttle position or pedal position sensor/switch B, range or performance problem
P0222	Throttle position or pedal position sensor/switch B circuit, low input
P0223	Throttle position or pedal position sensor/switch B circuit, high input
P0224	Throttle position or pedal position sensor/switch B circuit, intermittent
P0230	Fuel pump primary circuit malfunction
P0231	Fuel pump secondary circuit, low
P0232	Fuel pump secondary circuit, high

Code	Probable cause
P0234	Turbocharger/Supercharger A Overboost Condition
P0236	Turbocharger/Supercharger Boost Sensor A Circuit Range/Performance
P0237	Turbocharger/Supercharger Boost Sensor A Circuit Low
P0238	Turbocharger/Supercharger Boost Sensor A Circuit High
P023A	Charge Air Cooler Coolant Pump Control Circuit/Open
P023B	Charge Air Cooler Coolant Pump Control Circuit Low
P023C	Charge Air Cooler Coolant Pump Control Circuit High
P0243	Turbocharger/Supercharger Wastegate Actuator A
P0244	Turbocharger/Supercharger Wastegate Actuator A Range/Performance
P0245	Turbocharger/Supercharger Wastegate Actuator A Low
P0246	Turbocharger/Supercharger Wastegate Actuator A High
P0247	Turbocharger/Supercharger Wastegate Actuator B
P0248	Turbocharger/Supercharger Wastegate Actuator B Range/Performance
P0249	Turbocharger/Supercharger Wastegate Actuator B Low
P0250	Turbocharger/Supercharger Wastegate Actuator B High
P025A	Fuel Pump Module A Control Circuit/Open
P025B	Fuel Pump Module A Control Circuit Range/Performance
P025C	Fuel Pump Module A Control Circuit Low
P025D	Fuel Pump Module A Control Circuit High
P025E	Turbocharger/Supercharger Boost Sensor A Intermittent/Erratic
P026A	Charge Air Cooler Efficiency Below Threshold
P027B	Fuel Pump Module B Control Circuit Range/Performance
P028D	Charge Air Cooler Cooling Fan Control Circuit Low
P028E	Charge Air Cooler Cooling Fan Control Circuit High
P0297	Vehicle Overspeed Condition
P0298	Engine Oil Over Temperature Condition
P0299	Turbocharger/Supercharger A Underboost Condition
P02EE	Cylinder 1 Injector Circuit Range/Performance
P02EF	Cylinder 2 Injector Circuit Range/Performance
P02F0	Cylinder 3 Injector Circuit Range/Performance
P02F1	Cylinder 4 Injector Circuit Range/Performance
P02F2	Cylinder 5 Injector Circuit Range/Performance
P02F3	Cylinder 6 Injector Circuit Range/Performance
P02FC	Cold Start Fuel Injection Control Circuit Low
P02FD	Cold Start Fuel Injection Control Circuit High
P0297	Vehicle over-speed condition
P0298	Engine oil over temperature

6-8 EMISSIONS AND ENGINE CONTROL SYSTEMS

OBD-II TROUBLE CODES (CONTINUED)

Code	Probable cause
P0300	Random/multiple cylinder misfire detected
P0301	Cylinder no. 1 misfire detected
P0302	Cylinder no. 2 misfire detected
P0303	Cylinder no. 3 misfire detected
P0304	Cylinder no. 4 misfire detected
P0305	Cylinder no. 5 misfire detected
P0306	Cylinder no. 6 misfire detected
P0307	Cylinder no. 7 misfire detected
P0308	Cylinder no. 8 misfire detected
P0315	Crankshaft position system - variation not learned
P0316	Misfire detected during startup - first 1000 revolutions
P0325	Knock sensor circuit fault
P0326	Knock sensor circuit performance
P0330	Knock sensor no. 2 circuit malfunction (bank 2)
P0331	Knock sensor no. 2 circuit, range or performance problem (bank 2)
P0340	Camshaft position sensor "A", circuit malfunction (bank 1)
P0345	Camshaft position sensor "A", circuit malfunction (bank 2)
P0350	Ignition coil primary or secondary circuit malfunction
P0351	Ignition coil no. 1 primary circuit fault
P0352	Ignition coil no. 2 primary circuit fault
P0353	Ignition coil no. 3 primary circuit fault
P0354	Ignition coil no. 4 primary circuit fault
P0355	Ignition coil no. 5 primary circuit fault
P0356	Ignition coil no. 6 primary circuit fault
P0357	Ignition coil no. 7 primary circuit fault
P0358	Ignition coil no. 8 primary circuit fault
P0400	EGR flow fault
P0401	EGR insufficient flow detected
P0402	EGR excessive flow detected
P0403	Exhaust gas recirculation circuit malfunction
P0405	Exhaust gas recirculation valve position sensor A circuit low
P0406	Exhaust gas recirculation valve position sensor A circuit high
P0411	Secondary air injection system, incorrect flow detected
P0412	Secondary air injection system switching valve A, circuit malfunction
P0420	Catalyst system efficiency below threshold (Bank 1)
P0421	Catalyst system efficiency below threshold (Bank 1)

Code	Probable cause
P0430	Catalyst system efficiency below threshold (Bank 2)
P0431	Catalyst system efficiency below threshold (Bank 2)
P0442	EVAP small leak detected
P0443	EVAP VMV circuit fault
P0446	Evaporative emission control system, vent control circuit malfunction
P0451	Evaporative emission control system, pressure sensor range or performance problem
P0452	EVAP fuel tank pressure sensor low input
P0453	EVAP fuel tank pressure sensor high input
P0454	Evaporative emission system pressure sensor/switch intermittent
P0455	Evaporative emission (EVAP) control system leak detected (no purge flow or large leak)
P0456	Evaporative emission (EVAP) control system leak detected (very small leak)
P0457	Evaporative emission control system leak detected (fuel cap loose/off)
P0460	Fuel level sensor circuit malfunction
P0461	Fuel level sensor circuit, range or performance problem
P0462	Fuel level sensor circuit, low input
P0463	Fuel level sensor circuit, high input
P0480	Cooling fan no. 1, control circuit malfunction
P0481	Cooling fan no. 2, control circuit malfunction
P0482	Cooling fan no. 3, control circuit malfunction
P0500	VSS fault
P0501	Vehicle speed sensor, range or performance problem
P0503	Vehicle speed sensor circuit, Intermittent, erratic or high input
P0505	IAC valve system fault
P0506	Idle control system, rpm lower than expected
P0507	Idle control system, rpm higher than expected
P0511	Idle air control (IAC) system - circuit malfunction
P0512	Starter request circuit
P0528	Fan speed sensor - no signal
P0532	A/C Refrigerant Pressure Sensor A - circuit low
P0533	A/C Refrigerant Pressure Sensor A - circuit high
P0534	A/C refrigerant charge loss
P0537	A/C evaporator temperature sensor - circuit low
P0538	A/C evaporator temperature sensor - circuit high
P0552	Power steering pressure sensor circuit, low input

EMISSIONS AND ENGINE CONTROL SYSTEMS 6-9

Code	Probable cause
P0553	Power steering pressure sensor circuit, high input
P0562	System voltage low
P0563	System voltage high
P0579	Cruise control multi-function input A - circuit range/performance problem
P0581	Cruise control multi-function input A - circuit high
P0602	Control module, programming error
P0603	PCM Keep Alive Memory test error
P0605	PCM Read Only Memory test error
P0606	PCM processor fault
P0620	Alternator control circuit malfunction
P0622	Alternator lamp F - control circuit malfunction
P0625	Alternator field terminal - circuit low
P0626	Alternator field terminal - circuit high
P0645	A/C clutch relay control circuit
P0657	Actuator supply voltage A circuit/open
P065B	Alternator control - circuit range/performance problem
P0660	Intake manifold tuning valve control circuit (bank 1)
P0701	Transmission control system range/performance problem
P0702	Transmission control system electrical problem
P0703	Torque converter/brake switch B, circuit malfunction
P0704	Clutch switch input circuit malfunction
P0705	Transmission range sensor, circuit malfunction (PRNDL input)
P0706	Transmission range sensor A circuit range/performance problem
P0707	Transmission range sensor circuit, low input
P0708	Transmission range sensor circuit, high input
P0709	Transmission range sensor A circuit intermittent problem
P0710	Transmission fluid temperature sensor A circuit problem
P0711	Transmission fluid temperature sensor A circuit range/performance problem
P0712	Transmission fluid temperature sensor A circuit low
P0713	Transmission fluid temperature sensor A circuit high
P0715	Turbine/input shaft speed sensor A circuit problem
P0717	Turbine/input shaft speed sensor A circuit no signal
P0718	Turbine/input shaft speed sensor A circuit intermittent problem
P0720	Output shaft speed sensor circuit problem
P0721	Output shaft speed sensor circuit range/performance problem
P0722	Output shaft speed sensor circuit no signal
P0729	6th gear incorrect ratio incorrect or problem

Code	Probable cause
P0731	1st gear incorrect ratio incorrect or problem
P0732	2nd gear incorrect ratio incorrect or problem
P0733	3rd gear incorrect ratio incorrect or problem
P0734	4th gear incorrect ratio incorrect or problem
P0735	5th gear incorrect ratio incorrect or problem
P0740	Torque converter clutch solenoid circuit/open
P0741	Torque converter clutch solenoid circuit performance/stucck OFF
P0742	Torque converter clutch solenoid circuit stuck ON
P0743	Torque converter clutch solenoid circuit electrical problem
P0744	Torque converter clutch solenoid circuit intermittent problem
P0748	Pressure control solenoid A electrical problem
P0750	Shift solenoid A problem
P0751	Shift solenoid A performance/stuck OFF
P0752	Shift solenoid A stuck ON
P0753	Shift solenoid A electrical problem
P0754	Shift solenoid A intermittent problem
P0755	Shift solenoid B problem
P0756	Shift solenoid B performance/stuck OFF
P0757	Shift solenoid B stuck ON
P0758	Shift solenoid B electrical problem
P0759	Shift solenoid B intermittent problem
P0760	Shift solenoid C problem
P0761	Shift solenoid C performance/stuck OFF
P0762	Shift solenoid C stuck ON
P0763	Shift solenoid C electrical problem
P0764	Shift solenoid C intermittent problem
P0765	Shift solenoid D problem
P0766	Shift solenoid D performance/stuck OFF
P0767	Shift solenoid D stuck ON
P0768	Shift solenoid D electrical problem
P0769	Shift solenoid D intermittent problem
P0770	Shift solenoid E problem
P0771	Shift solenoid E performance/stuck OFF
P0772	Shift solenoid E stuck ON
P0773	Shift solenoid E electrical problem
P0774	Shift solenoid E intermittent problem
P07A8	Transmission friction element D performance/stuck OFF
P07A9	Transmission friction element D stuck ON

6-10 EMISSIONS AND ENGINE CONTROL SYSTEMS

OBD-II TROUBLE CODES (CONTINUED)

Code	Probable cause
P07AA	Transmission friction element E performance/stuck OFF
P0815	Upshift switch circuit problem
P0816	Downshift switch circuit problem
P0826	Up and down switch circuit problem
P0882	TCM power input signal low
P0883	TCM power input signal high
P0960	Pressure control solenoid A control circuit/open
P0961	Pressure control solenoid A control circuit range/performance problem
P0962	Pressure control solenoid A control circuit low
P0963	Pressure control solenoid A control circuit high
P0973	Shift solenoid A control circuit low
P0974	Shift solenoid A control circuit high
P0976	Shift solenoid B control circuit low
P0977	Shift solenoid B control circuit high
P0979	Shift solenoid C control circuit low
P0980	Shift solenoid C control circuit high
P0982	Shift solenoid D control circuit low
P0983	Shift solenoid D control circuit high
P0984	Shift solenoid E control circuit range/performance problem
P1260	Theft detected, vehicle immobilized
P1285	Cylinder head overtemperature condition
P1299	Cylinder head overtemperature protection active
P1464	A/C demand out of self-test range
P1572	Brake Pedal Position (BPP) switch circuit
P1602	Immobilizer/ECM communication error
P1622	Immobilizer ID does not match
P1639	Vehicle ID block corrupted - not programmed
P1703	Brake switch out of self-test range

3 Powertrain Control Module (PCM) - replacement

▶ Refer to illustration 3.7

※ WARNING:

The models covered by this manual are equipped with airbags. Always disconnect the negative battery cable, then the positive battery cable and wait 2 minutes before working in the vicinity of the impact sensors, steering column or instrument panel to avoid the possibility of accidental deployment of the airbag, which could cause personal injury (see Chapter 12).

→ **Note:** Because of a Federally mandated extended warranty which covers the emission control system components, check with your dealer about warranty coverage before working on any emissions-related systems. Once the warranty has expired, you may wish to perform some of the component checks and/or replacement procedures in this Chapter to save money.

→ **Note:** On 2006 and later models, you can remove the PCM, but if it must be replaced, it will have to be done by a dealer service department or other properly equipped repair facility because module configuration must be programmed into the new PCM or the vehicle will experience a Passive Anti-Theft System (PATS) no-start condition. This will occur even if the vehicle is not equipped with Passive Anti-Theft System.

1 The Powertrain Control Module (PCM) is located inside the passenger compartment under the passenger side dashboard, tucked into the corner. The electrical connector retaining bolt must be loosened from the engine compartment and the module must be removed from inside the passenger compartment. The PCM is easily distinguished by the aluminum casing surrounding the module. On 2005 and later Expedition/Navigator 5.4L engines, the PCM is located on the engine side of the firewall, at the right side.

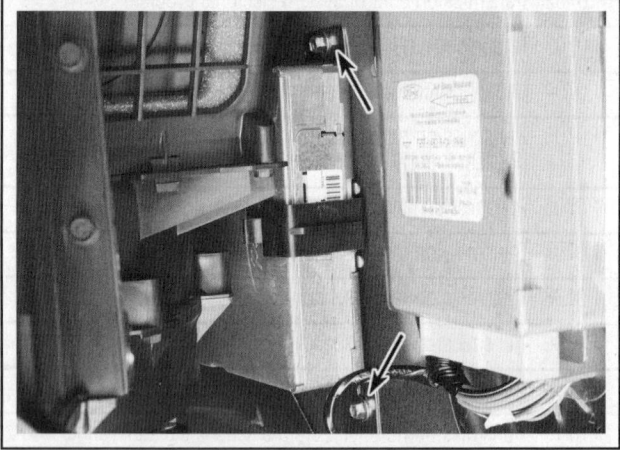

3.7 Remove the PCM bracket mounting bolts (arrows)

2 Disconnect the negative battery cable from the battery, then the positive cable and wait at least two minutes before proceeding.
3 Working in the engine compartment, remove the bolt that retains the electrical connector to the PCM.
4 Remove the scuff plate from the passenger's side door (see Chapter 11).
5 Remove the kick panel from the passenger's side foot area.
6 Remove the PCM bracket clips.
7 Remove the PCM bracket retaining bolts (see illustration).
8 Carefully slide the PCM out.

→ **Note:** Avoid any static electricity damage to the computer by using gloves and a special anti-static pad to store the PCM on once it is removed.

EMISSIONS AND ENGINE CONTROL SYSTEMS

4 Information sensors

➡ **Note:** The OBD-II system can detect a variety of different sensor problems and set codes to indicate the specific trouble area. If an OBD-II SCAN tool is not available, have the codes extracted from the PCM by a dealer service department or other qualified automotive repair facility.

➡ **Note:** If any of the following checks indicate that a sensor is good (and not the cause of the driveability problem or trouble code), check the wiring harness and electrical connectors between the sensor and the PCM for an open or short-circuit condition. If no problems are found, have the vehicle checked by a dealer service department or other qualified repair shop.

ENGINE COOLANT TEMPERATURE (ECT) SENSOR

♦ Refer to illustrations 4.2, 4.3 and 4.4

General description

1   The coolant sensor is a thermistor (a resistor which varies the value of its voltage output in accordance with temperature changes). The change in the resistance values will directly affect the voltage signal from the coolant sensor. As the sensor temperature DECREASES, the resistance values will INCREASE. As the sensor temperature INCREASES, the resistance values will DECREASE.

Check

2   Check the resistance value of the coolant temperature sensor while it is completely cold (50 to 65-degrees F = 58,750 to 40,500 ohms). Next, start the engine and warm it up until it reaches operating temperature (see illustration). The resistance should be lower (180 to 220-degrees F = 3,800 to 1,840 ohms).

➡ **Note:** Access to the coolant temperature sensor makes it difficult to position electrical probes on the terminals. If necessary, remove the sensor and perform the tests in a pan of heated water to simulate the conditions.

3   If the resistance values on the sensor are correct, check the refer

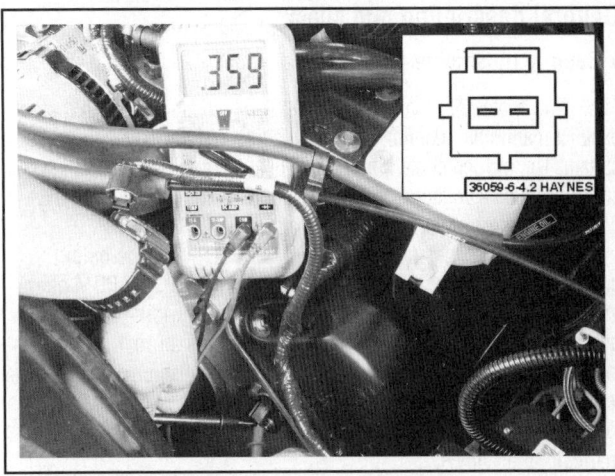

4.2 Check the resistance of the coolant temperature sensor with the engine completely cold and then with the engine at operating temperature. Resistance should decrease as temperature increases

ence voltage from the PCM to the sensor (see illustration). The reference voltage should be approximately 5.0 volts.

Replacement

4   Before installing the new sensor, wrap the threads with Teflon sealing tape to prevent leakage and thread corrosion (see illustration).

5   To remove the sensor, unplug the electrical connector, then carefully unscrew it.

※※ **CAUTION:**

Handle the coolant sensor with care. Damage to this sensor will affect the operation of the entire fuel injection system. Install the sensor and tighten it securely.

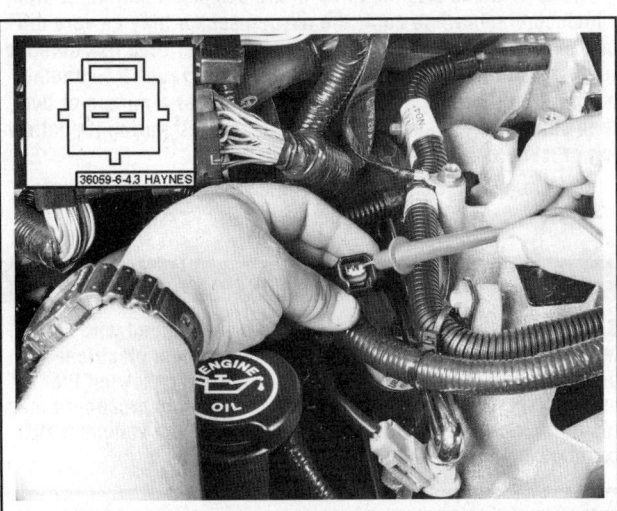

4.3 Working on the harness side, check the reference voltage from the PCM to the coolant temperature sensor with the ignition key ON and the engine not running. It should be approximately 5.0 volts

4.4 To prevent leakage, wrap the threads of the coolant temperature sensor with Teflon tape before installing it

OXYGEN SENSOR

General description and check

◆ Refer to illustration 4.9

6 The heated oxygen sensors (HO2S), which are located in the exhaust manifolds, monitor the oxygen content of the exhaust gas stream. The oxygen content in the exhaust reacts with the oxygen sensor to produce a voltage output which varies from 0.1-volt (high oxygen, lean mixture) to 0.9-volts (low oxygen, rich mixture).

7 The PCM constantly monitors this variable voltage output to determine the ratio of oxygen to fuel in the mixture. The PCM alters the air/fuel mixture ratio by controlling the pulse width (open time) of the fuel injectors. A mixture ratio of 14.7 parts air to 1 part fuel is the ideal mixture ratio for minimizing exhaust emissions, thus allowing the catalytic converter to operate at maximum efficiency. It is this ratio of 14.7 to 1 which the PCM and the oxygen sensor attempt to maintain at all times.

8 The oxygen sensor produces no voltage when it is below its normal operating temperature of about 600-degrees F. During this initial period before warm-up, the PCM operates in open loop mode.

9 Allow the engine to reach normal operating temperature and check that the oxygen sensor is producing a varying signal voltage between 0.1 and 0.9-volts (see illustration).

➡ **Note: Post-catalytic oxygen sensors will produce a much slower fluctuating voltage value, reflecting the results of the catalyzed exhaust mixture. Keep this in mind when testing a post-catalytic converter oxygen sensor.**

10 Also check to make sure the oxygen sensor heaters are supplied with battery voltage (see illustration 4.9). Backprobe the HTR GND and the B+ Volt. terminals with the ignition key ON. Because battery voltage is supplied to the oxygen sensors through a relay, voltage will only be delivered for a very short time (3 seconds) when the ignition key is cycled. Have an assistant turn the ignition key to ON while observing the voltmeter. Refer to Chapter 12 for additional information on the wiring schematics and relays.

11 When any of the above codes occur, the PCM operates in the open loop mode - that is, it controls fuel delivery in accordance with a programmed default value instead of feedback information from the oxygen sensor.

12 The proper operation of the oxygen sensor depends on four conditions:

  a) **Electrical** - *The low voltages generated by the sensor depend upon good, clean connections which should be checked whenever a malfunction of the sensor is suspected or indicated.*
  b) **Outside air supply** - *The sensor is designed to allow air circulation to the internal portion of the sensor. Whenever the sensor is removed and installed or replaced, make sure the air passages are not restricted.*
  c) **Proper operating temperature** - *The PCM will not react to the sensor signal until the sensor reaches approximately 600-degrees F. This factor must be taken into consideration when evaluating the performance of the sensor.*
  d) **Unleaded fuel** - *The use of unleaded fuel is essential for proper operation of the sensor. Make sure the fuel you are using is of this type.*

13 In addition to observing the above conditions, special care must be taken whenever the sensor is serviced.

  a) *The oxygen sensor has a permanently attached pigtail and elec-*

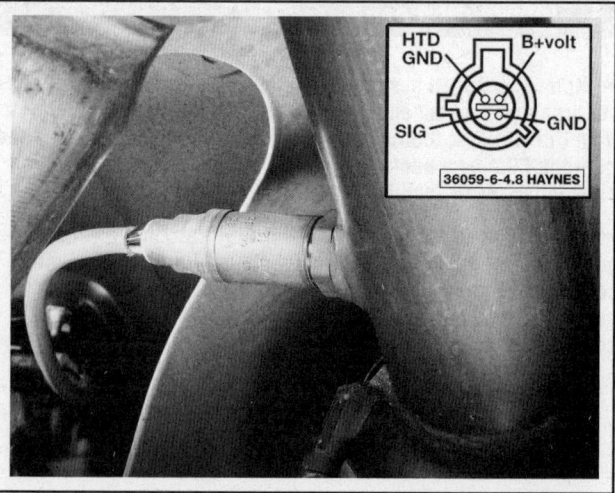

4.9 Using pins or paper clips, backprobe SIG (+) and GND (-) and check for a varying millivolt signal as the system adjusts the air/fuel ratio. Voltage should remain steady as the engine warms-up (open loop) and then vary from 0.10 volts (100 millivolts) to 0.9 volts (900 millivolts) if the system is operating properly (closed loop)

  trical connector which should not be removed from the sensor. Damage or removal of the pigtail or electrical connector can adversely affect operation of the sensor.
  b) *Grease, dirt and other contaminants should be kept away from the electrical connector and the louvered end of the sensor.*
  c) *Do not use cleaning solvents of any kind on the oxygen sensor.*
  d) *Do not drop or roughly handle the sensor.*
  e) *The silicone boot must be installed in the correct position to prevent the boot from being melted and to allow the sensor to operate properly.*

Replacement

◆ Refer to illustration 4.17

➡ **Note: Because it is installed in the exhaust manifold or pipe, which contracts when cool, the oxygen sensor may be very difficult to loosen when the engine is cold. Rather than risk damage to the sensor (assuming you are planning to reuse it in another manifold or pipe), start and run the engine for a minute or two, then shut it off. Be careful not to burn yourself during the following procedure.**

14 Disconnect the cable from the negative terminal of the battery.
15 Raise the vehicle and place it securely on jackstands.

※※ WARNING:

Some models covered by this manual are equipped with self leveling suspension systems. Always disconnect electrical power to the suspension system before lifting or towing the vehicle (see Chapter 10). Failure to perform this procedure may result in unexpected shifting or movement of the vehicle which could cause personal injury.

16 Disconnect the electrical connector from the sensor.
17 Carefully unscrew the sensor from the exhaust manifold (see illustration).

EMISSIONS AND ENGINE CONTROL SYSTEMS   6-13

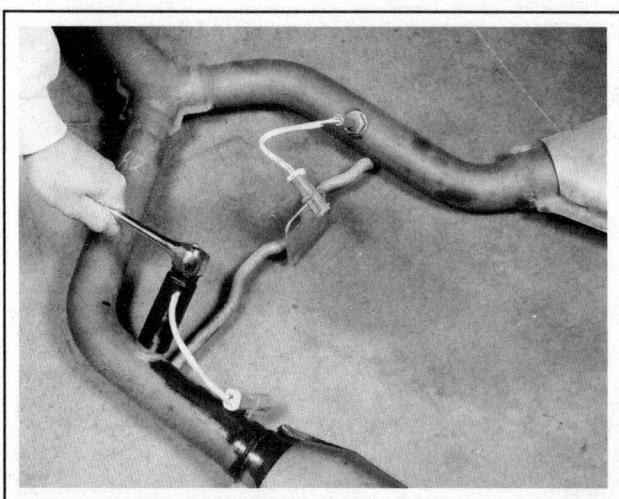

4.17 Removing an oxygen sensor using a special socket (exhaust system removed for clarity)

✳ CAUTION:

Excessive force may damage the threads.

18 Anti-seize compound must be used on the threads of the sensor to facilitate future removal. The threads of new sensors will already be coated with this compound, but if an old sensor is removed and reinstalled, recoat the threads.
19 Install the sensor and tighten it securely.
20 Reconnect the electrical connector of the pigtail lead to the main engine wiring harness.
21 Lower the vehicle and reconnect the cable to the negative terminal of the battery.

THROTTLE POSITION SENSOR (TPS)

General description

22 The Throttle Position Sensor (TPS) is located on the end of the throttle shaft on the throttle body. By monitoring the output voltage from the TPS, the PCM can determine fuel delivery based on throttle valve angle (driver demand). A broken or loose TPS can cause intermittent bursts of fuel from the injector and an unstable idle because the PCM thinks the throttle is moving.

Check

▶ **Refer to illustrations 4.23 and 4.25**

23 To check for the proper signal voltage from the TPS, install the probes of the voltmeter into the ground wire (GND) (-) and signal wire (SIG) (+) on the backside of the electrical connector (see illustration).

➡ **Note: Be careful when backprobing the electrical connector. Do not damage the wiring harness or pull on any connectors to make clean contact.**

24 Turn the ignition switch to ON (engine not running), the sensor should read 0.50 to 1.0-volt at closed throttle. Have an assistant depress the accelerator pedal to simulate full throttle and the sensor should increase voltage to 4.0 to 5.0-volts. If the TPS voltage readings are incorrect, replace it with a new unit.

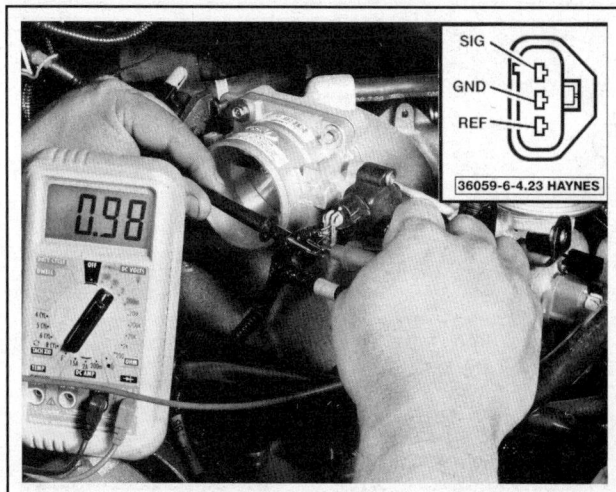

4.23 Backprobe the TPS using straight pins or other suitable probes on SIG (+) and GND (-). Check the signal voltage, it should be 0.5 to 1.0 volt at closed throttle. Rotate the accelerator completely to wide open throttle and confirm the voltage increases steadily to 4.5 to 5.0 volts

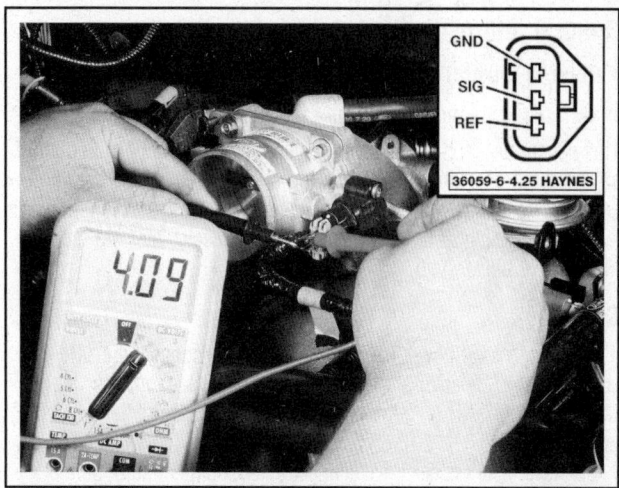

4.25 Check the reference voltage to the TPS with a voltmeter. Backprobe terminal REF with the positive (+) probe and GND (-) with the negative probe of the voltmeter and make sure the reference voltage is approximately 5.0 volts

25 Also, check the TPS reference voltage. Install the positive probe of the voltmeter (see illustration) onto the voltage reference wire (REF) and the negative probe of the voltmeter onto the ground wire (GND). There should be approximately 5.0 volts sent from the PCM to the TPS.

Replacement

26 The TPS is a non-adjustable unit. Remove the two retaining screws and separate the TPS from the throttle body. On 2005 and later Expedition/Navigator models, the TPS screws are installed with thread-locking compound during manufacture. To remove these screws, use a screwdriver to loosen them two full turns, then use a power screwdriver to remove them fully.
27 Installation is the reverse of removal. On 2005 and later Expedition/Navigator models, install the TPS with new screws, using a power screwdriver.

6-14 EMISSIONS AND ENGINE CONTROL SYSTEMS

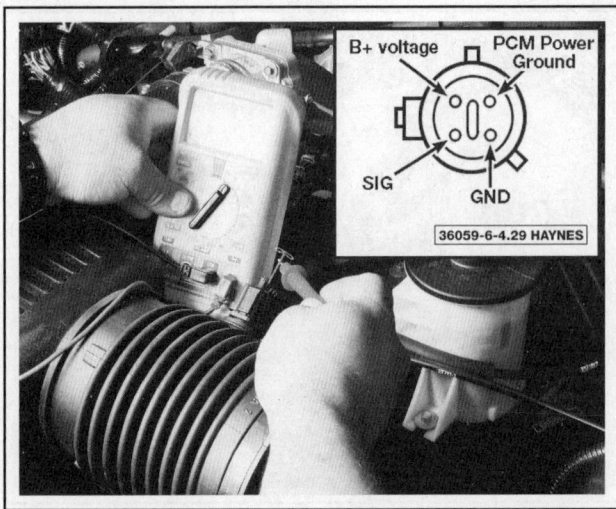

4.29 Check for battery voltage to the B+ terminal on the MAF sensor (key ON engine not running)

4.36 Disconnect the air inlet duct, remove the nuts (arrows) and separate the MAF sensor from the air cleaner housing

MASS AIRFLOW SENSOR (MAF)

General Information

▶ Refer to illustrations 4.29 and 4.36

28 The Mass Airflow (MAF) sensor is located on the air intake duct. This sensor uses a hot wire sensing element to measure the amount of air entering the engine. The air passing over the hot wire causes it to cool. Consequently, this change in temperature can be converted into an analog voltage signal to the PCM which in turn calculates the required fuel injector pulse width.

Check

29 To check for power to the MAF sensor, disconnect the MAF sensor electrical connector and connect the probes of a voltmeter to the B+ (+) and the GND (-) terminals on the harness side of the connector (see illustration).

30 Reconnect the electrical connector and, using straight pins or other suitable probes, backprobe the MAF SIG (+) and GND (-) terminals with the voltmeter. Start the engine and check the voltage. The voltage should be 0.2 to 1.5 volts at idle.

31 Raise the engine rpm. The signal voltage from the MAF sensor should increase to about 2.0 volts. It is impossible to simulate driving conditions in the driveway but it is necessary to observe the voltmeter for a fluctuation in signal voltage as the engine speed is raised. The vehicle will not be under load conditions but it should manage to vary slightly. This parameter is easily monitored on an OBD-II SCAN tool. Look for a steady increase in the voltage signal.

32 If you suspect a defective MAF sensor, stop the engine, disconnect the MAF harness connector. Using an ohmmeter, probe the MAF terminals SIG (+) and GND (-). If the hot wire element inside the sensor has been damaged it will be indicated by an open circuit (infinite resistance).

33 If the voltage readings are correct, check the wiring harness for open circuits or a damaged harness (see Chapter 12).

Replacement

34 Disconnect the electrical connector from the MAF sensor.

35 On 2009 and earlier models, remove the upper section of the air filter assembly (see Chapter 4).

36 Remove the nuts (or bolts) (see illustration) and remove the MAF sensor.

37 Installation is the reverse of removal.

TRANSMISSION RANGE (TR) SENSOR (AUTOMATIC TRANSMISSION)

General description

▶ Refer to illustration 4.38

38 The Transmission Range (TR) sensor, located on the transmission (see illustration), indicates to the PCM when the transmission is in Park, Neutral, Drive or Reverse. This information is used for starting, Transmission Converter Clutch (TCC), Exhaust Gas Recirculation (EGR) and Idle Air Control (IAC) valve operation. For example, if the signal wire(s) become grounded, it may be difficult to start the engine in Park or Neutral.

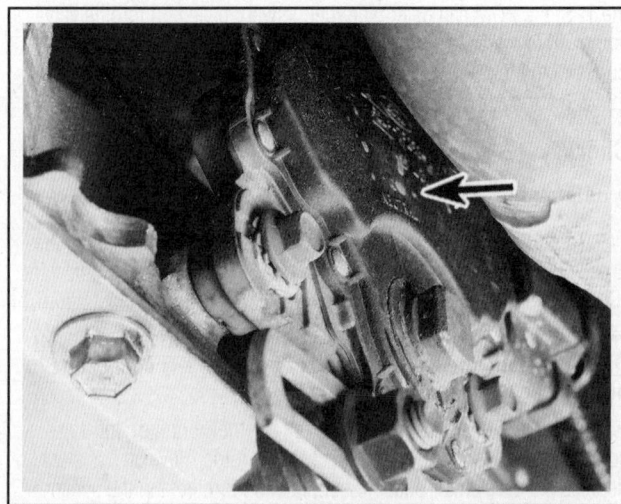

4.38 Location of the Transmission Range sensor (TR)

EMISSIONS AND ENGINE CONTROL SYSTEMS 6-15

Check

◆ Refer to illustration 4.40

→ Note: *The following checks do not specifically test for or indicate a Transmission Range sensor problem. These checks test for power and cranking voltage to the TR sensor during startup and cranking modes. This will verify that the circuit to the TR sensor (PCM, starter relay, wiring harness, etc.) is working properly and providing power to the TR sensor. Have the diagnostic codes extracted from the PCM before finalizing a defective TR sensor. If necessary, have the vehicle diagnosed by a dealer service department.*

39 In the event there is a problem with the Transmission Range (TR) sensor, first check the terminal connectors for proper attachment.

40 Working on the TR sensor harness on the computer side, use a voltmeter and with the ignition key ON (engine not running), check for power to the switch on terminal number 2 (vehicles produced before 6/24/96) or terminal number 9 (vehicles produced after 6/24/96) (see illustration). There should be voltage present.

41 Disable the ignition system (see Chap-ter 5) and have an assistant turn the ignition key to Start and check for cranking voltage to the TR sensor on terminal number 1 (vehicles produced before 6/24/96) or terminal number 10 (vehicles produced after 6/24/96). Cranking voltage should be more than 9 volts but less than battery voltage.

42 The remaining TR sensor checks must be performed with a specialized SCAN tool. Have the system checked by a dealer service department or other qualified automotive repair facility.

43 Check the adjustment of the switch. If the switch is out of adjustment, perform the procedure and clear the codes. Check the system for any other problems.

Adjustment

44 Follow the transmission shift control cable adjustment in Chapter 7 and check for a distinct "click" when the shift lever selects each gear (Park, Reverse, Neutral, Drive etc.).

45 The shift button should release smoothly and there should not be any cable binding preventing smooth transition be-tween gears.

AIR CONDITIONING CLUTCH CONTROL

46 During air conditioning operation, the PCM monitors the application of the air conditioning compressor clutch. The PCM commands the IAC valve to adjust the idle speed of the engine to compensate for the additional load.

47 First, check for battery voltage to the air conditioning clutch with the engine running, the air conditioning system activated and the air conditioning clutch harness connector disconnected. Battery voltage should be available. If not, check the air conditioning clutch solenoid.

48 Use a jumper wire from the battery positive terminal (+) and apply voltage to the air conditioning clutch. There should be a definite "click" when the clutch is activated.

49 Check for battery voltage to the air conditioning clutch solenoid. Battery voltage should be present with the air conditioning system properly charged and the air conditioning selected.

→ Note: *If no power exists, then check the air conditioning high pressure cut-out fan switch, the air conditioning clutch cycling pressure switch, the air conditioning/heater control assembly (see Chapter 3) and the 15 amp fuse that protects the circuit (see Chapter 12).*

50 In most cases, if the air conditioning does not function, the problem is probably related to the air conditioning system relays and

4.40 Check for battery voltage to the TR sensor on terminal number 2 (vehicles produced before 6/24/96) or terminal number 9 (vehicles produced after 6/24/96)

switches and not the PCM. Refer to Chapter 3 for additional information on the air conditioning system and diagnostics.

VEHICLE SPEED SENSOR (VSS)

General description

51 The Vehicle Speed Sensor (VSS) is located near the rear section of the transmission. This sensor is a variable reluctance sensor that produces a pulsing voltage proportional to the speed of the vehicle. These pulses are translated by the PCM and provided for other systems for fuel and transmission shift control. The VSS is part of the Transmission Converter Clutch (TCC) system.

Check

◆ Refer to illustrations 4.52 and 4.53

52 To check the vehicle speed sensor, remove the electrical connector in the wiring harness near the sensor. Using a voltmeter, check for reference voltage to the sensor (see illustration). With the ignition key ON (engine not running), the reference wire should have 5 volts avail-

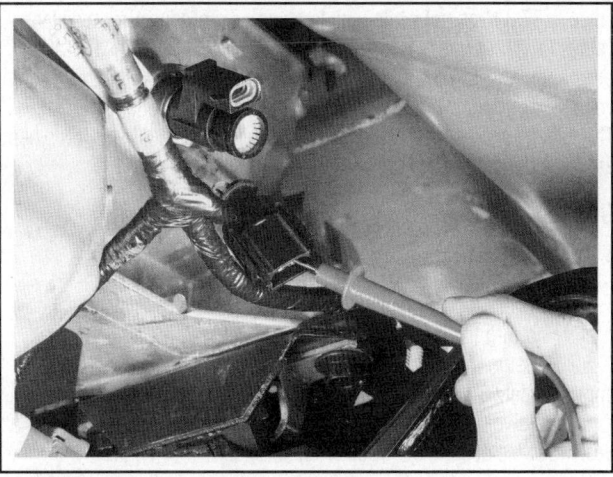

4.52 Disconnect the VSS harness connector and check for voltage at the connector

6-16 EMISSIONS AND ENGINE CONTROL SYSTEMS

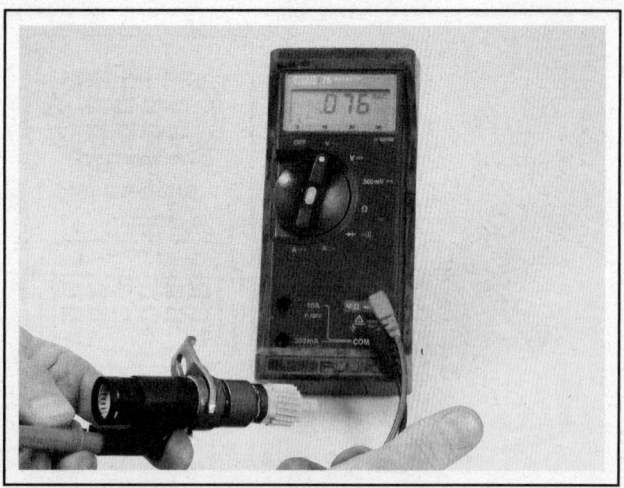

4.53 Remove the VSS and check for a pulsing AC voltage signal as the VSS gear is slowly turned

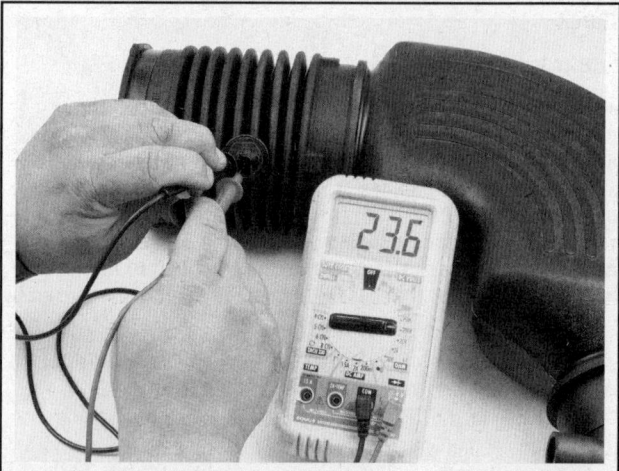

4.61 Remove the electrical connector from the IAT sensor (located in air cleaner housing), then check the resistance of the IAT sensor cold and warm. It may be necessary to use simulated conditions to obtain accurate results

able. If there is no voltage available, have the PCM diagnosed by a dealer service department or other qualified repair shop.

53 Place the VSS on a bench and check for a pulsing voltage signal. Backprobe the harness connector using two pins and while slowly spinning the VSS gear, observe that the voltage signal pulses from 0 to 0.5 volts. Use the AC scale on the voltmeter (see illustration).

Replacement

54 To replace the VSS, disconnect the electrical connector from the VSS.

55 Remove the retaining bolt and withdraw the VSS from the transmission.

56 Installation is the reverse of removal.

INTAKE AIR TEMPERATURE (IAT) SENSOR

General description

57 The Intake Air Temperature (IAT) sensor is located inside the air intake duct. This sensor acts as a resistor which changes value according to the temperature of the air entering the engine. Low temperatures produce a high resistance value (for example, at 68-degrees F the resistance is 37.3 K-ohms) while high temperatures produce low resistance values (at 212-degrees F the resistance is 2.0 K-ohms). The PCM supplies approximately 5-volts (reference voltage) to the IAT sensor. The voltage will change according to the temperature of the incoming air. The voltage will be high when the air temperature is cold and low when the air temperature is warm.

Check

▶ Refer to illustrations 4.61

58 To check the IAT sensor, disconnect the two prong electrical connector and turn the ignition key ON but do not start the engine.

59 Measure the reference voltage. Reference voltage should be approximately 5-volts.

60 If the voltage signal is not correct, have the PCM diagnosed by a dealer service department or other repair shop.

61 Measure the resistance across the sensor terminals (see illustration). The resistance should be HIGH when the air temperature is LOW. Next, start the engine and let it idle (cold). Wait until engine reaches operating temperature. Turn the ignition OFF, disconnect the IAT sensor and measure the resistance across the terminals. The resistance should be LOW when the air temperature is HIGH. If the sensor does not exhibit this change in resistance, replace it with a new part.

POWER STEERING PRESSURE SWITCH

62 Turning the steering wheel increases power steering fluid pressure and engine load. The pressure switch will close before the load can cause an idle problem.

63 A pressure switch that will not open or an open circuit from the PCM will cause timing to retard at idle and this will affect idle quality.

64 A pressure switch that will not close or an open circuit may cause the engine to die when the power steering system is used heavily.

65 Any problems with the power steering pressure switch or circuit should be repaired by a dealer service department or other qualified repair shop.

CRANKSHAFT POSITION (CKP) SENSOR

General information

▶ Refer to illustrations 4.66

66 Some models are equipped with a crankshaft position sensor. The crankshaft position sensor (see illustration) is mounted adjacent to a pulse wheel located on the crankshaft. The crankshaft sensor monitors the pulse wheel as the teeth pass under the magnetic field created by the sensor. The pulse wheel has 35 teeth and a spot where one tooth is missing. By monitoring the lost tooth, the crankshaft sensor determines the piston travel, crankshaft position and speed information and sends the information to the PCM.

EMISSIONS AND ENGINE CONTROL SYSTEMS   6-17

4.66 Location of the crankshaft sensor (arrow) on the V6 engine

4.68 With the voltmeter set on the AC scale, check for a pulsing voltage signal between 0 and 0.05 volts as the engine is slowly rotated

Check

▸ Refer to illustration 4.68

67 Disconnect the CKP sensor electrical connector and with the ignition key ON (engine not running), check for battery voltage to the CKP sensor.

68 Connect a voltmeter to the crankshaft sensor and using the AC scale, check the voltage pulses as the gear is slowly rotated (see illustration). Use a large socket and breaker bar to rotate the crankshaft pulley.

69 If no pulsing voltage signal is produced, replace the crankshaft sensor.

Replacement

70 Remove the electrical connector and the retaining bolt.

➡ Note: On some models it may be necessary to remove the A/C compressor to gain access to the sensor (see Chapter 3).

71 Installation is the reverse of removal.

CAMSHAFT POSITION (CMP) SENSOR

General information

▸ Refer to illustration 4.72

72 These models are equipped with a camshaft sensor as well as a crankshaft sensor. The camshaft sensor signals the PCM to begin sequential pulsation of the fuel injectors. On V6 models, it is a Hall Effect switching device activated by a single vane. This camshaft sensor is mounted on the top of the engine in the normal location of the distributor (see illustration). On V8 models, it is a variable reluctance device which is triggered by the high point mark on the camshaft. The sensor is mounted on the front of the cylinder head near the camshaft sprocket.

4.72 Location of the camshaft position sensor on the V6 engine

Check

V6 models

73 With the ignition key ON (engine not running), check the signal voltage. Backprobe the signal wire (+) (dark green) from the computer and the ground wire (-) (black/white) and while slowly rotating the engine, observe that the voltage pulses from 0 to 5.0 volts.

➡ Note: Do not crank the engine over using the starter but instead install a breaker bar and socket onto the front crankshaft pulley bolt and rotate the engine slowly by hand.

74 If there is no pulsing signal voltage but instead a steady 5.0 volt signal, then most likely the camshaft sensor is defective. If there is no voltage available, then the PCM is not supplying voltage to the camshaft sensor. In the former case, replace the camshaft sensor. In the latter case, have the PCM checked by a dealer service department or other qualified automotive repair facility.

6-18 EMISSIONS AND ENGINE CONTROL SYSTEMS

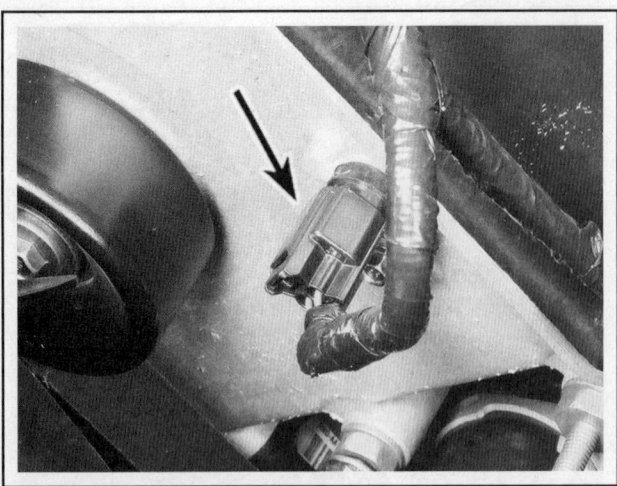

4.76 Location of the camshaft position sensor on a V8 engine

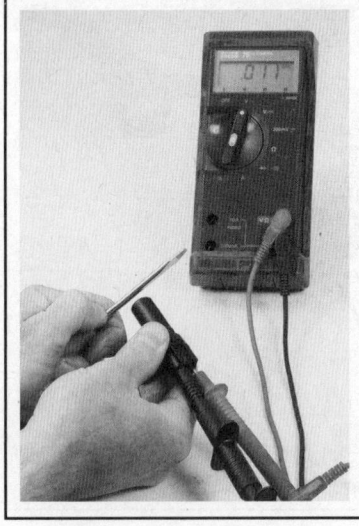

4.77 Working on the bench, carefully pass a metal object close to the tip of the camshaft sensor and see if the AC voltage fluctuates - if the camshaft sensor does not exhibit any reaction as the magnetic field is broken, the sensor must be replaced

V8 models

▸ Refer to illustrations 4.76 and 4.77

75 Check for battery voltage to the camshaft position sensor with the ignition key ON (engine not running).

76 Remove the camshaft sensor from the engine and place it on a clean workbench (see illustration).

77 Check the AC voltage output. Install the probes of a voltmeter set on the AC scale onto the camshaft sensor and observe that it produces a voltage pulse as a metal object is passed over the tip of the sensor (see illustration).

78 If no pulsing voltage signal is produced, replace the camshaft sensor.

Replacement

V6 models

▸ Refer to illustrations 4.85 and 4.89

→ Note: V6 models equipped with a camshaft position sensor are mounted on a synchronizer assembly, which is essentially a drive unit for the sensor. If you are simply replacing the cam position sensor, it is not necessary to remove the synchronizer assembly from the engine - just remove the screws from the sensor, detach it from the synchronizer and install the new sensor. However, many engine repair procedures require removal of the synchronizer assembly, in which case it will be necessary to perform the following procedure to time the synchronizer. This procedure requires the use of a special tool, available at most auto parts stores, to properly align the sensor; read through the entire procedure and obtain the necessary tool before beginning.

79 Position the number 1 piston at TDC. Refer to Chapter 2B for the procedure.

80 Disconnect the cable from the negative battery terminal.

81 Mark the relative position of the camshaft position sensor electrical connector so the assembly can be oriented properly upon installation (this is only necessary if the synchronizer assembly will be removed). Disconnect the electrical connector from the camshaft position sensor. Remove the screws and detach the sensor from the synchronizer assembly.

82 Partially drain the cooling system (see Chapter 1).

83 Remove the EGR valve and EGR tube (see Section 6).

84 Disconnect the heater hose outlet line (see Chapter 3).

85 Remove the sensor mounting screws (see illustration) and lift the sensor from the housing.

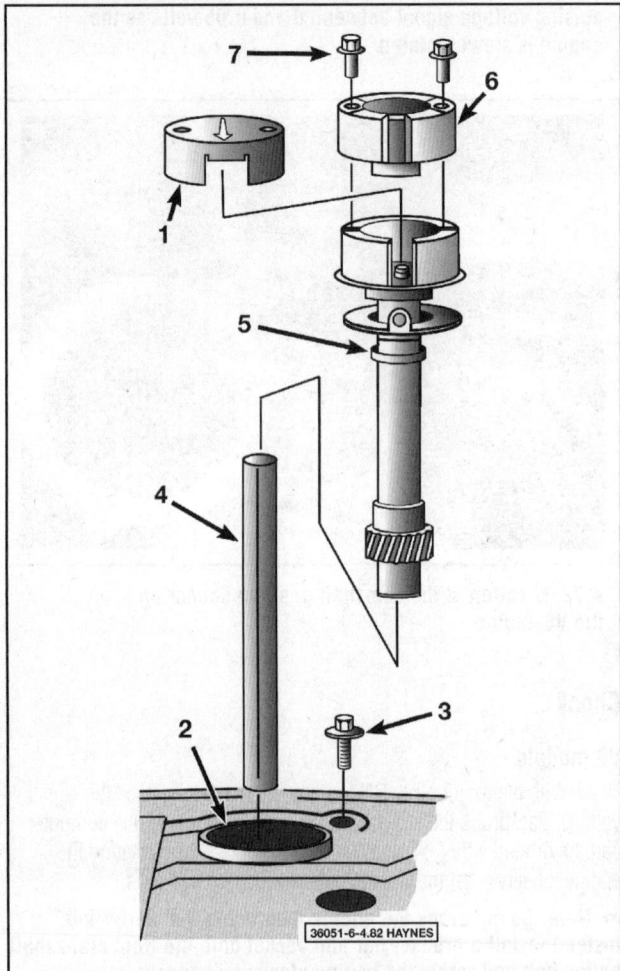

4.85 Exploded view of the camshaft position sensor and synchronizer assembly - V6 engine

1 Synchro positioning tool
2 Timing chain cover
3 Clamp
4 Oil pump intermediate shaft
5 Synchronizer
6 Camshaft Position sensor
7 Bolt

EMISSIONS AND ENGINE CONTROL SYSTEMS   6-19

86 If you will be removing the synchronizer assembly, remove the bolt and withdraw the synchronizer from the engine.

➡ **Note: Remove the oil pump intermediate shaft along with the camshaft position sensor synchronizer assembly.**

87 Place the special alignment tool onto the synchronizer assembly. Align the vane of the synchronizer with the radial slot in the special tool.

88 Turn the tool on the synchronizer until the boss on the tool is engaged with the notch on the synchronizer.

89 Transfer the oil pump intermediate shaft onto the synchronizer assembly. Lubricate the gear, thrust washer and lower bearing of the synchronizer assembly with clean engine oil. Insert the assembly into the engine, with the arrow tool pointing 54-degrees clockwise from the engine's centerline (see illustration).

90 Turn the tool clockwise a little so the synchronizer engages with the oil pump intermediate shaft. Push down on the synchronizer, turning the tool gently until the gear on the synchronizer engages with the gear on the camshaft.

91 Install the hold-down bolt and tighten it securely. Remove the positioning tool.

92 Install the EGR assembly.

93 Reconnect the heater hose outlet to the engine block.

94 Install the camshaft position sensor and tighten the screws securely.

※※ **CAUTION:**

Check the position of the electrical connector on the sensor to make sure it is aligned with the mark you made in Step 81. If it isn't oriented correctly, DO NOT rotate the synchronizer to reposition it - doing so will result in the fuel system being out of time with the engine, possibly damaging the engine (and at the very least cause driveability problems). If the connector is not oriented properly, repeat the synchronizer installation procedure.

95 Plug in the electrical connector to the sensor and reconnect the cable to the negative terminal of the battery.

V8 models

96 On 2009 and earlier models, remove the power steering fluid reservoir from the left cylinder head (see Chapter 10). On 2010 and later models, remove the air filter outlet tube for access to the left CMP sensor (see Chapter 4).

97 Remove the retaining fastener and separate the camshaft sensor from the cylinder head (see illustration 4.76).

98 Installation is the reverse of removal.

BRAKE ON/OFF (BOO) SWITCH

General Information

99 The brake On/Off switch informs the PCM when the brakes are being applied. The switch closes when brakes are applied and opens when the brakes are released. The switch is located on the brake pedal assembly.

100 The brake light circuit and bulbs are wired into the switch circuit so it is important in diagnosing any driveability problems to make sure all the brake light bulbs are working properly (not burned out) or the driver may feel poor idle quality (see Chapter 7).

Check

101 Disconnect the electrical connector from the Brake On/Off switch and using a 12 volt test light, check for battery voltage to the switch.

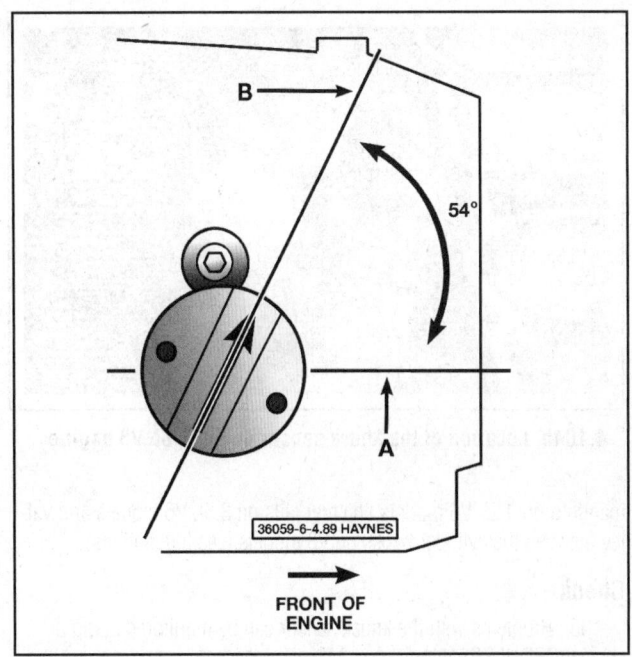

4.89 With the cam synchronizer installed and fully seated in the timing cover, the arrow on the tool (B) should point 54 degrees from the centerline of the engine (A)

102 Also, check continuity from the switch to the brake light bulbs. Change any burned out bulbs or damaged wire looms.

Replacement

103 Refer to Chapter 9 for the replacement procedure.

KNOCK SENSOR

General description

▸ **Refer to illustrations 4.104a and 4.104b**

104 The knock sensor detects abnormal vibration in the engine. The sensor produces an AC output voltage which increases with the severity of the knock. The signal is fed into the PCM and the timing is retarded up to 20 degrees to compensate for severe detonation. The knock sensor is located on the right side of the engine block under the exhaust

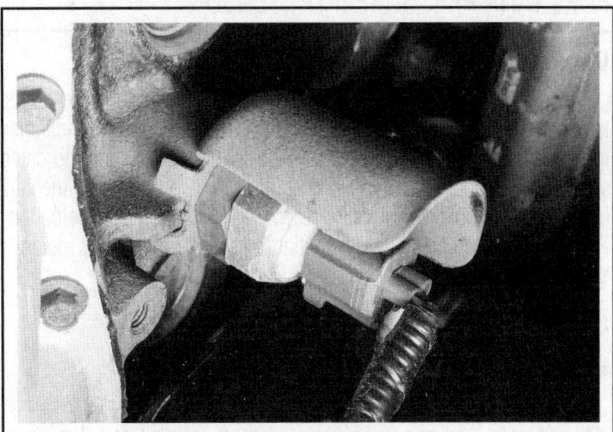

4.104a The knock sensor is located on the right side of the lower engine block on the 4.2L V6 engine

6-20 EMISSIONS AND ENGINE CONTROL SYSTEMS

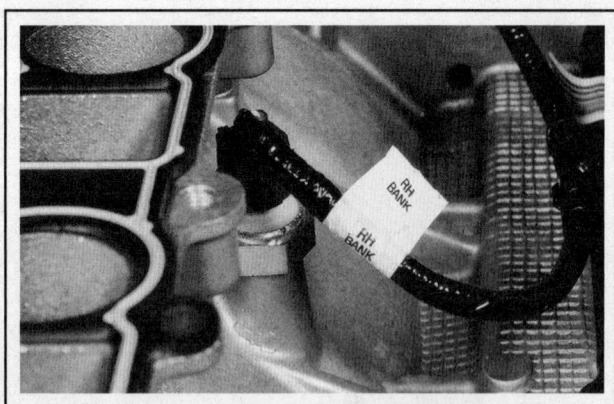

4.104b Location of the knock sensor on the 4.6L V8 engine

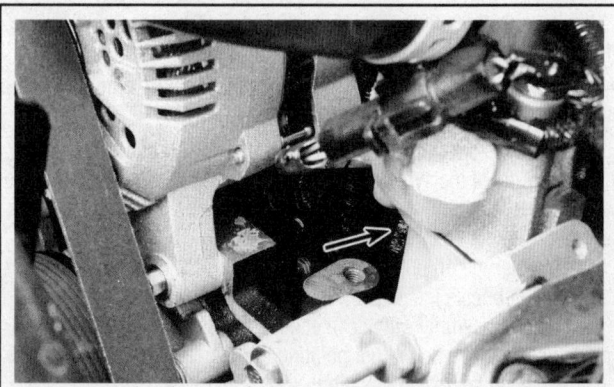

4.106 Location of the cylinder head temperature sensor on the 5.4L V8 engine

manifold on 4.2L V6 models, the fuel rails on 3.5L V6 models and valley between the cylinder banks on V8 models (see illustrations).

Check

105 Problems with the knock sensor can be monitored using a special OBD-II SCAN tool, have a dealer service department check the system for correct operation.

CYLINDER HEAD TEMPERATURE (CHT) SENSOR

Replacement

▶ Refer to illustration 4.106

→ Note: On 2005 and later Expedition/Navigator models, it is usually necessary to remove the intake manifold for access to the CHT sensor (see Chapter 2B).

106 To remove the sensor, unplug the electrical connector, then carefully unscrew it (see illustration).

❋❋ CAUTION:

Handle the CHT sensor with care. Damage to this sensor will affect the operation of the entire fuel injection system. Install the sensor and tighten it securely.

FUEL TANK PRESSURE (FTP) SENSOR

General information

107 The fuel tank pressure (FTP) sensor is used to monitor the fuel tank pressure or vacuum during the OBD-II test portion for emissions integrity. This test scans various sensors and output actuators to detect abnormal amounts of fuel vapors that may not be purging into the canister and/or the intake system for recycling. The FTP sensor helps the PCM monitor this pressure differential (pressure vs. vacuum) inside the fuel tank.

Check

▶ Refer to illustration 4.108

108 With the ignition key ON (engine not running), check for REF voltage to the fuel tank pressure sensor (see illustration). Voltage should be available. It may be difficult to access the harness with the fuel tank

in place. Find a location in the harness near the tank to check for voltage without removing the FTP sensor and fuel tank from the vehicle.

109 If voltage is available, the remaining checks must be performed with a specialized SCAN tool. Have the FTP sensor and EVAP system checked by a dealer service department or other qualified repair facility.

CLUTCH PEDAL POSITION SWITCH (MANUAL TRANSMISSION)

General description

110 The Clutch Pedal Position switch is located on the clutch release arm (see Chapter 7) under the driver's dash. The clutch pedal position switch acts as a start inhibitor, preventing power to reach the starter relay until the clutch pedal is depressed. This switch works similar as the Transmission Range sensor on automatic transmission models.

Check

111 Working on the clutch pedal position switch harness connector (ignition switch side), use a voltmeter and with the ignition key ON (engine not running), check for power to the switch. There should be voltage present.

112 Working on the clutch pedal position switch side of the connector, use an ohmmeter to check the resistance with the pedal up and

4.108 The fuel tank pressure sensor is located on the top of the fuel tank

EMISSIONS AND ENGINE CONTROL SYSTEMS

then check it with the pedal depressed. The switch should register high to infinite resistance with the pedal released, and zero resistance with the pedal depressed (refer to the wiring diagrams at the end of Chapter 12 for the proper terminals to check). If there is no difference, replace the switch.

113 If both these test results are correct, check the starter system (see Chapter 5).

Adjustment

114 Follow the clutch pedal position switch replacement and adjustment procedures in Chapter 8.

VARIABLE CAMSHAFT TIMING (VCT)

115 2005 and later Expedition/Navigator 5.4L engines have a variable camshaft timing system, designed to improve performance and emissions. When signaled by the PCM, the VCT solenoid moves a valve to deliver engine oil pressure to the camshaft phaser/sprocket assemblies (one at the front of each camshaft), altering camshaft timing under certain conditions. The VCT phaser and the camshaft sprocket are a unit (see Chapter 2B or 2C for phaser/sprocket removal).

ENGINE OIL PRESSURE (EOP) SENSOR - REPLACEMENT

➡ **Note: On 3.5L turbocharged and 5.0L models, the EOP sensor is located on the oil cooler/filter adapter housing.**

116 Loosen the left-front wheel lug nuts, then raise and support the front of the vehicle on jackstands.
117 Remove the under vehicle cover and/or skid plate (as equipped).
118 Remove the driver's side inner fender splash shield (see Chapter 11).
119 Locate the EOP sensor installed in the oil cooler adapter housing.
120 Disconnect the electrical connector and unscrew the EOP sensor to remove.
121 Installation is reverse of removal. Tighten the sensor securely.

TURBOCHARGER BOOST PRESSURE (TCBP) AND CHARGE AIR COOLER TEMPERATURE (CACT) SENSOR – REPLACEMENT

➡ **Note: The 3.5L V6 engines are equipped with a combination Turbocharger Boost Pressure (TCBP) and Charge Air Cooler Temperature (CACT) sensor. A problem in the sensor circuit will set a diagnostic trouble code.**

122 Locate the TCBP/CACT sensor in the intake pipe between the intercooler and the throttle body.
123 Disconnect the electrical connector and remove the fasteners.
124 Remove the TCBP/CACT sensor from the air intake pipe.
125 Installation is reverse of removal.

FUEL TANK PRESSURE (FTP) SENSOR - REPLACEMENT

▶ **Refer to illustration 4.128**

❄ **WARNING:**

Gasoline is extremely flammable, so take extra precautions when you work on any part of the fuel system. Don't smoke or allow open flames or bare light bulbs near the work area, and don't work in a garage where a gas-type appliance (such as a water heater or clothes dryer) is present. Since gasoline is carcinogenic, wear fuel-resistant gloves when there's a possibility of being exposed to fuel, and, if you spill any fuel on your skin, rinse it off immediately with soap and water. Mop up any spills immediately and do not store fuel-soaked rags where they could ignite. When you perform any kind of work on the fuel system, wear safety glasses and have a Class B type fire extinguisher on hand. The fuel system is under pressure, so if any lines must be disconnected, the pressure in the system must be relieved first (see Chapter 4 for more information).

126 The fuel tank pressure (FTP) sensor is used to monitor the fuel tank pressure or vacuum during the OBD-II test portion for emissions integrity. This test scans various sensors and output actuators to detect abnormal amounts of fuel vapors that may not be purging into the canister or the intake system for recycling. The FTP sensor helps the PCM monitor this pressure differential (pressure vs. vacuum) inside the fuel tank. A problem in the fuel tank pressure sensor circuit will set a diagnostic trouble code.

➡ **Note: The Fuel Tank Pressure (FTP) sensor is part of the vapor tube assembly and serviced as one unit.**

127 Remove the fuel tank (see Chapter 4).
128 Disconnect the vapor tube from the quick-disconnect fittings and remove the vapor tube and FTP sensor (see illustration).
129 Installation is the reverse of removal.
130 Turn the ignition on and check for leaks.

MANIFOLD ABSOLUTE PRESSURE (MAP) SENSOR (3.5L ENGINES) – REPLACEMENT

131 Remove the two nuts at the front of the engine cover. Lift up at the front and unhook the retainers at the rear of the cover to remove.
132 Locate the MAP sensor on the top of the intake manifold and disconnect the electrical connector.
133 Remove the mounting bolt and remove the MAP sensor.
134 Installation is reverse of removal. Lubricate the MAP sensor O-ring with clean engine oil prior to installation.

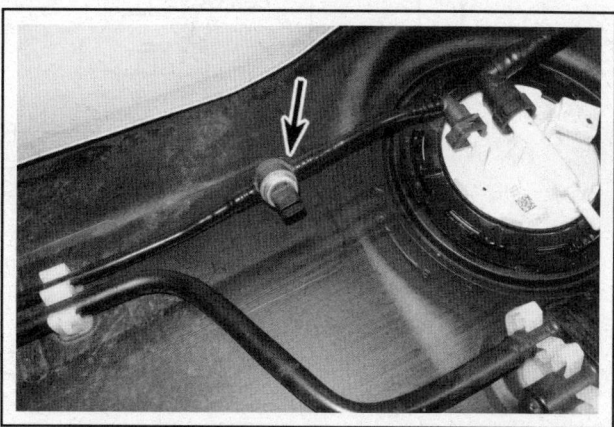

4.128 Identifying the fuel pressure sensor and tube on top of the fuel tank

5 Exhaust Gas Recirculation (EGR) system

GENERAL DESCRIPTION

Refer to illustrations 5.2

1 The EGR system is used to lower oxides of nitrogen (NOx) emission levels caused by high combustion temperatures. The EGR recirculates a small amount of exhaust gases into the intake manifold. The additional mixture lowers the temperature of combustion thereby reducing the formation of NOx compounds.

2 The EGR flow rate is determined by monitoring the pressure across a fixed metering orifice as exhaust gasses pass through it. This system is called the Differential Pressure Feedback (DPFE) system (see illustration). The pressure sensor monitors upstream (before) and downstream (after) exhaust backpressure. This backpressure coefficient is relayed to the PCM and the correct amount of EGR (duty cycle) is applied to the EGR vacuum regulator control (EVR). By calculating the difference between the two pressures, the PCM determines exactly the EGR flow rate at all driving conditions. The DPFE is more accurate than early systems in that the computer does not have to guess at the upstream pressure coefficient to determine EGR flow rate as the engine drives through various road conditions such as hard acceleration, downshifting, engine misfire, poor fuel combustion, etc. All these conditions will cause the exhaust backpressure to vary and requires more strict and responsive EGR control to limit NOx emission levels.

3 Here is a list of the various components on the DPFE EGR system: the EGR valve, the EGR vacuum regulator, differential feedback pressure sensor, the PCM, the EGR pipe and the various vacuum and pressure lines for the EGR system.

4 The OBD-II system can detect a variety of different EGR system problems and set codes to indicate the specific trouble area. Trouble codes P1400 through P1410 are designated EGR system trouble codes. If an OBD-II SCAN tool is not available, have the codes extracted from the PCM by a dealer service department or other qualified automotive repair facility.

CHECK

5 Too much EGR flow tends to weaken combustion, causing the engine to run rough or stop. When EGR flow is excessive, the engine can stop after a cold start or at idle after deceleration, the vehicle can surge at cruising speeds or the idle may be rough. If the EGR valve remains constantly open, the engine may not idle at all.

6 Too little or no EGR flow allows combustion temperatures to get too high during acceleration and load conditions. This can cause spark knock (detonation), engine overheating or emission test failure.

7 The following checks will help you pinpoint problems in the EGR system. Where the procedure says to lift up on the EGR valve diaphragm, it's a good idea to wear a heat-resistant glove to prevent burns.

EGR valve

8 The EGR valve is controlled by a normally open EGR vacuum regulator (EVR) which allows vacuum to pass when energized. The PCM energizes the EVR to turn on the EGR. The PCM controls the EGR when three conditions are present: engine coolant is above 113-degrees F, the TPS is at part throttle and the MAF sensor is in its mid-range.

9 Make sure the vacuum hoses are in good condition and hooked up correctly.

10 To perform a leakage test, hook up a vacuum pump to the EGR valve. Apply a vacuum of 5 to 6 in-Hg to the valve. The vacuum pump should hold vacuum.

11 If access is possible, position your finger tip under the vacuum diaphragm and apply vacuum to the EGR valve. You should feel movement of the EGR diaphragm.

✳✳ WARNING:

The EGR valve becomes very hot during engine operation - it's a good idea to wear a glove when performing this check.

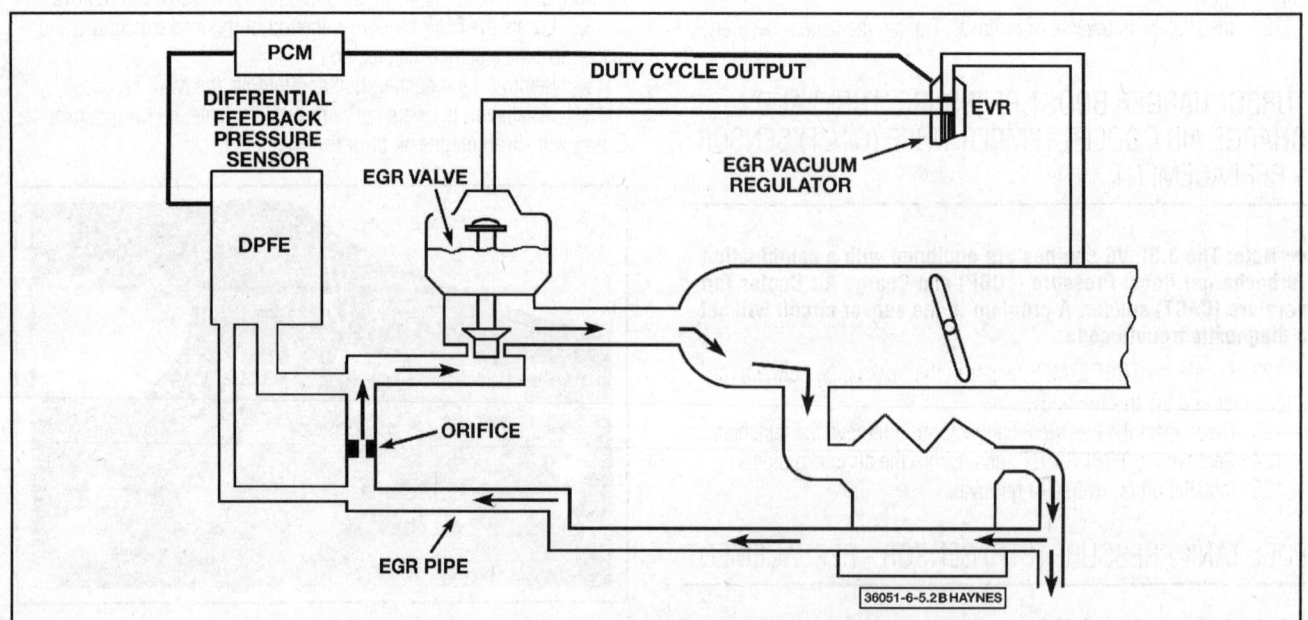

5.2 Typical Differential Pressure Feedback EGR (DPFE) system

EMISSIONS AND ENGINE CONTROL SYSTEMS 6-23

5.16 Working on the harness side of the Electronic Vacuum Regulator (arrow) electrical connector, check for battery voltage

5.17 Check the resistance of the EGR vacuum regulator. It should be 30 to 70 ohms

12 Remove the EGR valve (see Step 20) and clean the inlet and outlet ports with a wire brush or scraper. Do not sandblast the valve or clean it with gasoline or solvents. These liquids will destroy the EGR valve diaphragm.

13 If the specified conditions are not met, replace the EGR valve.

EGR control system

▶ Refer to illustrations 5.16, 5.17 and 5.18

14 If a code is displayed there are several possibilities for EGR failure. Engine coolant temperature (ECT) sensor, TPS, MAF sensor, TCC system and the engine rpm govern the parameters the EGR system use for distinguishing the correct ON time.

15 These systems use an Electronic Vacuum Regulator (EVR) to control the amount of exhaust gas through the EGR valve. The valve is normally open (engine at operating temperature) and the vacuum source is a ported signal. The PCM uses a controlled "pulse width" or electronic signal to turn the EGR ON and OFF (the "duty cycle"). The duty cycle should be zero percent (no EGR) when in Park or Neutral, when the TPS input is below the specified value or when Wide Open Throttle (WOT) is indicated.

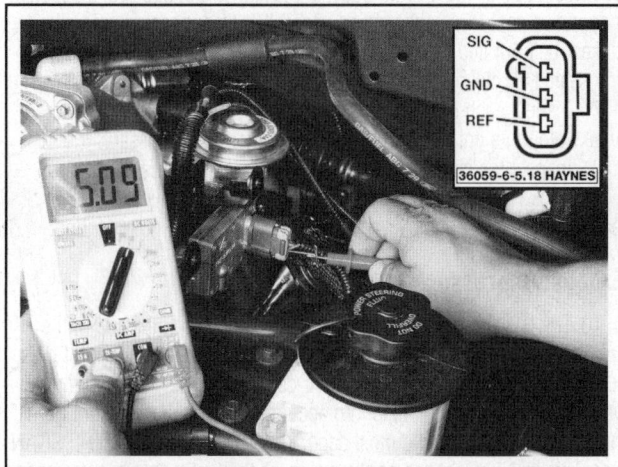

5.18 Check for the correct reference voltage to the DPFE sensor with the ignition key ON (engine not running)

16 To check the EGR vacuum regulator, disconnect the electrical connector to the EGR vacuum regulator, turn the ignition key ON (engine not running) and check for battery voltage to the solenoid (see illustration). Battery voltage should be present.

17 Next, use an ohmmeter and check the resistance of the EGR vacuum regulator. It should be between 30 and 70 ohms (see illustration).

18 Check for reference voltage to the DPFE sensor. With the ignition key on (engine not running), check for voltage on the harness side of the electrical connector (see illustration) on terminal VREF. It should be between 4.0 and 6.0 volts. If the test results are incorrect, replace the DPFE sensor.

19 Check the operation of the Differential Pressure Feedback (DPFE) sensor.

➡ **Note:** *The DPFE sensor on the Differential Pressure Feedback EGR systems have two exhaust lines hooked into the EGR tube.*

Check for signal voltage to the sensor. Backprobe the correct terminals and check for a voltage signal while the engine is running first at cold temperatures and then at warm operating temperatures. With the engine cold there should be NO EGR therefore the voltage should be approximately 0.20 to 0.70 volts. As the engine starts to warm and EGR is signaled by the computer, voltage values should increase to approximately 4.0 to 6.0 volts.

COMPONENT REPLACEMENT

EGR valve

20 When buying a new EGR valve, make sure that you have the right EGR valve. Use the stamped code located on the top of the EGR valve.

21 Detach the cable from the negative terminal of the battery.

22 Remove the air cleaner housing assembly (see Chapter 4).

23 Detach the vacuum line from the EGR valve.

24 Remove the EGR pipe from the exhaust manifold.

➡ **Note:** *On some models, it may be necessary to remove the brake booster bracket and nut assembly to gain access to the EGR pipe.*

25 Remove the bolts securing the EGR valve to the intake manifold/air intake plenum.

6-24 EMISSIONS AND ENGINE CONTROL SYSTEMS

26 Remove the EGR valve and gasket from the manifold. Discard the gasket.

27 With a wire wheel, buff the exhaust deposits from the EGR valve mounting surface on the manifold and, if you plan to use the same valve, the mounting surface of the valve itself. Look for exhaust deposits in the valve outlet. Remove deposit build-up with a screwdriver.

✱✱ CAUTION:

Never wash the valve in solvents or degreaser - both agents will permanently damage the diaphragm. Sand blasting is also not recommended because it will affect the operation of the valve.

28 If the EGR passage contains an excessive build-up of deposits, clean it out with a wire wheel. Make sure that all loose particles are completely removed to prevent them from clogging the EGR valve or from being ingested into the engine.

29 Installation is the reverse of removal.

EGR vacuum regulator

30 Detach the cable from the negative terminal of the battery.
31 Unplug the electrical connector from the solenoid.
32 Clearly label and detach both vacuum hoses.
33 Remove the solenoid mounting screw and remove the solenoid.
34 Installation is the reverse of removal.

DPFE sensor

▶ **Refer to illustration 5.38**

35 Detach the cable from the negative terminal of the battery.
36 Unplug the electrical connector from the sensor.
37 Clearly label and detach both vacuum hoses.
38 Remove the sensor mounting nuts (see illustration) and remove the assembly.
39 Installation is the reverse of removal.

5.38 Remove the DPFE sensor mounting bolts

6 Evaporative Emissions Control System (EVAP)

GENERAL DESCRIPTION

1 This system is designed to trap and store fuel vapors that evaporate from the fuel tank, throttle body and intake manifold during non-operation or idling, store them in the charcoal canister and then route them into the combustion chamber to be burned during engine operation.

2 The Evaporative Emission Control System (EVAP) consists of a charcoal-filled canister and the lines connecting the canister to the fuel tank, fuel vapor management valve (VMV), a fuel tank pressure sensor, fuel filler cap, fuel vapor valve, ported vacuum and intake manifold vacuum.

➡ **Note 1: The evaporative charcoal canister is mounted on the frame near the fuel tank on all models.**

➡ **Note 2: The fuel tank pressure sensor location and checks are covered in Section 4, Information Sensors.**

3 Fuel vapors are transferred from the fuel tank, throttle body and intake manifold to a canister where they are stored when the engine is not operating. When the engine is running, the fuel vapors are purged from the canister by a vapor management valve (VMV) which is PCM controlled, and consumed in the normal combustion process. The fuel tank pressure sensor relays the inside fuel tank pressure to the PCM which in turn regulates the EVAP system purge control system.

4 The OBD-II system can detect a variety of different EVAP system problems and set codes to indicate the specific trouble area. If an OBD-II SCAN tool is not available, have the codes extracted from the PCM by a dealer service department or other qualified automotive repair facility.

CHECK

General system checks

5 Poor idle, stalling and poor driveability can be caused by an inoperative vapor management valve, a damaged canister, split or cracked hoses or hoses connected to the wrong tubes.

6 Evidence of fuel loss or fuel odor can be caused by fuel leaking from fuel lines or the throttle body, a cracked or damaged canister, an inoperative vapor management valve (VMV), disconnected, misrouted, kinked, deteriorated or damaged vapor or control hoses or an improperly seated air cleaner or air cleaner gasket.

7 Inspect each hose attached to the canister for kinks, leaks and breaks along its entire length. Repair or replace as necessary.

8 Inspect the canister. If it is cracked or damaged, replace it.

9 Look for fuel leaking from the bottom of the canister. If fuel is leaking, replace the canister and check the hoses and hose routing.

Excessive pressure in fuel tank

10 The easiest way to check for excess fuel vapor pressure in the fuel tank is simply remove the gas cap and listen for the sound of pressure release similar to a flat tire or air compressor discharge. If the weather is extremely hot, take into account for the extra pressure from the heated molecules. The most accurate test is using the OBD-II SCAN tool. This will run a series of checks using the fuel tank pressure sensor and other output actuators to detect excess pressure. Have the vehicle diagnosed by a dealer service department or other qualified automotive repair facility.

EMISSIONS AND ENGINE CONTROL SYSTEMS 6-25

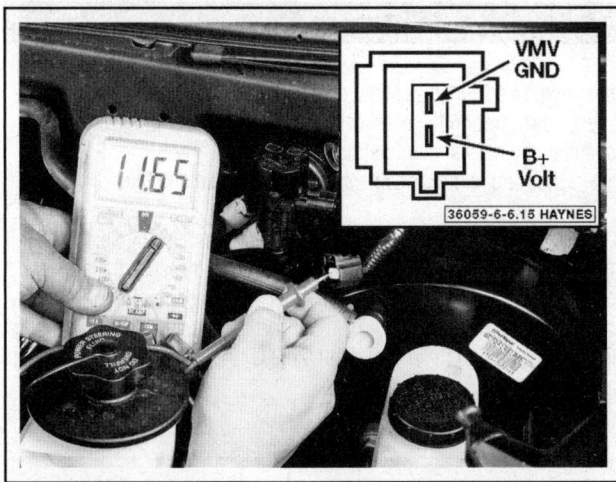

6.15 Check for battery voltage to the Vapor Management valve with the ignition key ON (engine not running)

6.17 Remove the charcoal canister mounting bolt

11 If excess pressure is detected, check the canister fuel vapor hose and inlet port for blockage or collapsed hoses. Also inspect the hoses near the VMV and between the fuel tank and body for kinks and damage.

12 Also, check the evaporative emission valve. Remove the fuel tank (see Chapter 4) and check the evaporative emission valve to make sure the passage through the orifice is open to atmospheric pressure. If it is plugged, replace the valve.

13 Check the charcoal canister for tightness. Remove the close-off line near the canister and install a hand-held pressure pump. Pump the canister to approximately 2.5 psi and confirm that pressure vents through the close-off line. Now install the pump directly at the canister and apply 2.5 psi. The pressure should hold steady.

Fuel vapor odor in engine compartment

▶ **Refer to illustration 6.15**

14 Check the hoses around the VMV for damage or incorrectly routed lines. Correct if necessary.

15 Check the VMV for battery voltage. With the ignition key ON (engine not running), check for battery voltage on the B+ VOLT terminal on the computer side of the VMV harness connector (see illustration). If there is no battery voltage available, have the PCM diagnosed by a dealer service department or other qualified automotive repair facility. If reference voltage is available to the VMV, have the VMV checked using an OBD-II SCAN tool at a dealer service department or other qualified repair facility.

EVAP CANISTER REPLACEMENT

▶ **Refer to illustration 6.17**

16 Clearly label, then detach, all vacuum lines from the canister.
17 Loosen the canister mounting clamp bolt and pull the canister out (see illustration).
18 Installation is the reverse of removal.

CANISTER PURGE VALVE

➡ **Note: On later models, the canister purge valve is located on the left (driver's) side of the intake manifold, just behind the throttle body.**

19 On 3.5L V6 models, remove the engine cover.
20 Disconnect the electrical connector and quick-connect fitting from the valve and intake manifold as necessary
21 On 3.5L V6 models, slide the valve from the retainer on the intake manifold. On later V8 models, remove the two bolts and remove the purge valve from the intake manifold.
22 Installation is the reverse of removal.

6-26 EMISSIONS AND ENGINE CONTROL SYSTEMS

7 Positive Crankcase Ventilation (PCV) system

♦ **Refer to illustration 7.1**

1   The Positive Crankcase Ventilation (PCV) system reduces hydrocarbon emissions by scavenging crankcase vapors. It does this by circulating fresh air from the air cleaner through the crankcase, where it mixes with blow-by gases and is then rerouted through a PCV valve to the intake manifold (see illustration).

2   The main components of the PCV system are the PCV valve, a fresh air filtered inlet and the vacuum hoses connecting these two components with the engine and the EECS system.

3   To maintain idle quality, the PCV valve restricts the flow when the intake manifold vacuum is high. If abnormal operating conditions arise, the system is designed to allow excessive amounts of blow-by gases to flow back through the crankcase vent tube into the air cleaner to be consumed by normal combustion.

4   Checking and replacement of the PCV valve and filter is covered in Chapter 1.

5   On 2005 and later Expedition/Navigator models, an electric heating element is added to the PCV system, to further reduce emissions in cold weather conditions by more complete combustion of the hydrocarbons introduced to the intake manifold by the PCV system. The heating element is bolted to the intake manifold with the PCV hose from the left valve cover attached to it.

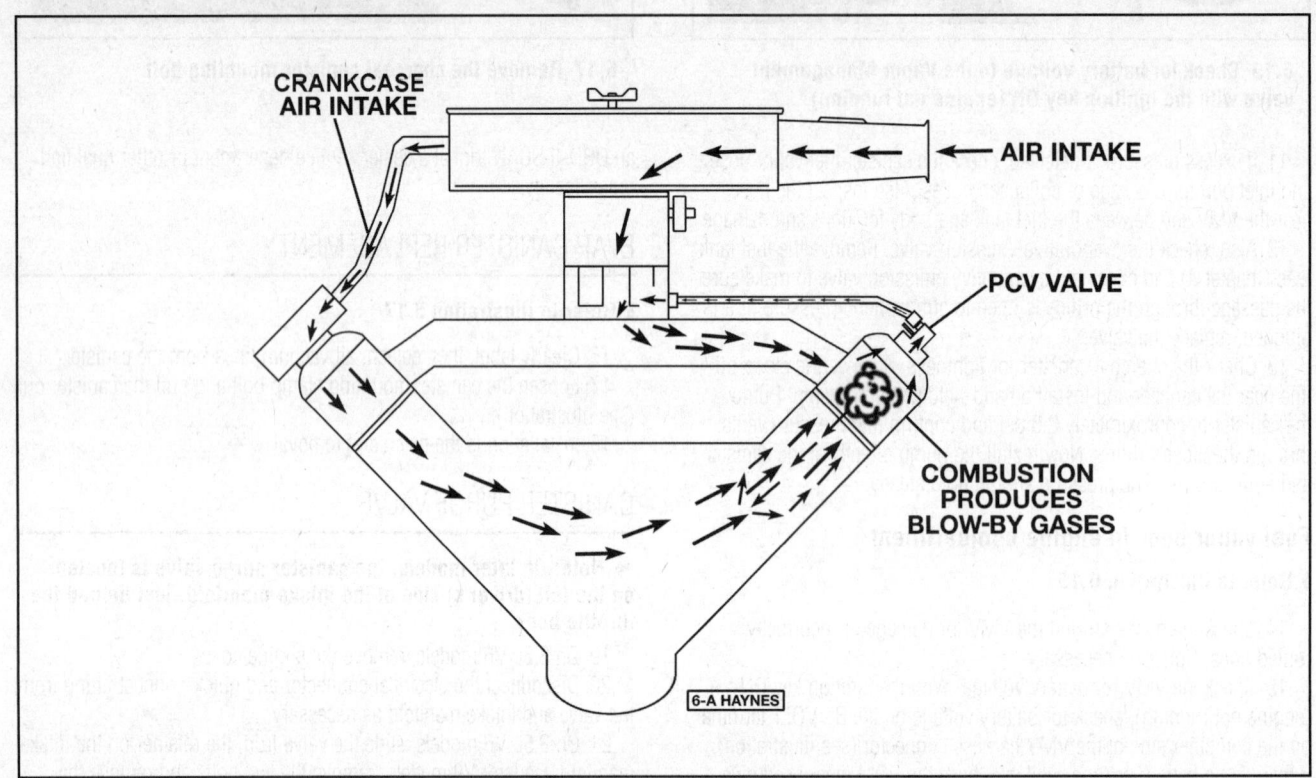

7.1 Gas flow in a typical PCV system

8 Catalytic converter

GENERAL DESCRIPTION

1   The catalytic converter is an emission control device added to the exhaust system to reduce pollutants from the exhaust gas stream. A single-bed converter design is used in combination with a three-way (reduction) catalyst. The catalytic coating on the three-way catalyst contains platinum and rhodium, which lowers the levels of oxides of nitrogen (NOx) as well as hydrocarbons (HC) and carbon monoxide (CO).

CHECK

2   The test equipment for a catalytic converter is expensive and highly sophisticated. If you suspect that the converter on your vehicle is malfunctioning, take it to a dealer or authorized emissions inspection facility for diagnosis and repair.

3   Whenever the vehicle is raised for servicing of underbody components, check the converter for leaks, corrosion and other damage. If damage is discovered, the converter should be replaced.

REPLACEMENT

4   Because the converter is part of the exhaust system, converter replacement requires removal of the exhaust pipe assembly (see Chapter 4). Take the vehicle, or the exhaust system, to a dealer service department or a muffler shop.

EMISSIONS AND ENGINE CONTROL SYSTEMS 6-27

9 Oil pressure control solenoid (3.5L V6 engine) – replacement

1 The oil pressure control solenoid is located inside of the engine and requires the engine front cover to be removed to access and replace the solenoid (see Chapter 2B, Section 9).
2 Disconnect the electrical connector.
3 Remove the retainer and remove the solenoid from the front of the engine.
4 Installation is reverse of removal

10 Turbocharger bypass valve (3.5L V6 engine) – replacement

▶ Refer to illustration 10.6

➡ Note: 3.5L V6 models are equipped with a turbocharger bypass valve. A problem in the sensor circuit will set a diagnostic trouble code.

1 Raise the vehicle and support it securely on jackstands.
2 Remove the under-vehicle splash shield and/or skid plate (as equipped).
3 Locate the turbocharger bypass valve attached to the intercooler, above the outlet pipe.
4 Disconnect the turbocharger bypass valve quick-disconnect fitting.
5 Disconnect the electrical connector.
6 Remove the fasteners and the bypass valve from the intercooler (see illustration).
7 Installation is the reverse of removal.

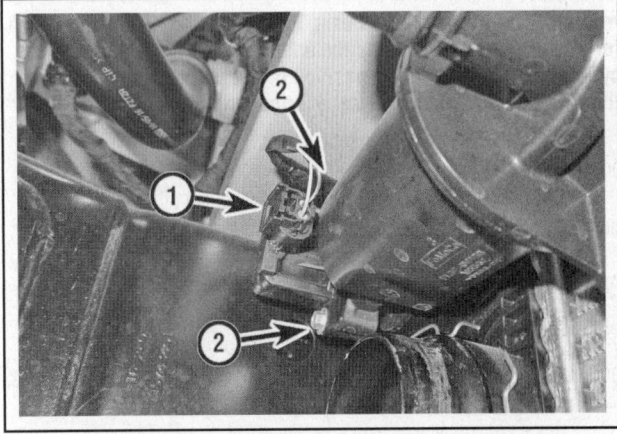

10.6 Disconnect the electrical connector (1) and remove the two bolts (2) attaching the bypass valve to the intercooler

11 Turbocharger wastegate vacuum sensor - replacement

▶ Refer to illustrations 11.2 and 11.4

➡ Note: The 3.5L (turbocharged) models are equipped with a wastegate vacuum sensor. A problem in the sensor circuit will set a diagnostic trouble code.

1 Remove the two nuts at the front of the engine cover. Lift up at the front and unhook the retainers at the rear of the cover to remove.
2 Locate the wastegate vacuum sensor on the top of the intake manifold (see illustration) and disconnect the electrical connector.
3 Remove the mounting bolt and remove the wastegate vacuum sensor.
4 Disconnect the vacuum line that runs to the wastegates (see illustration).
5 Installation is the reverse of removal.

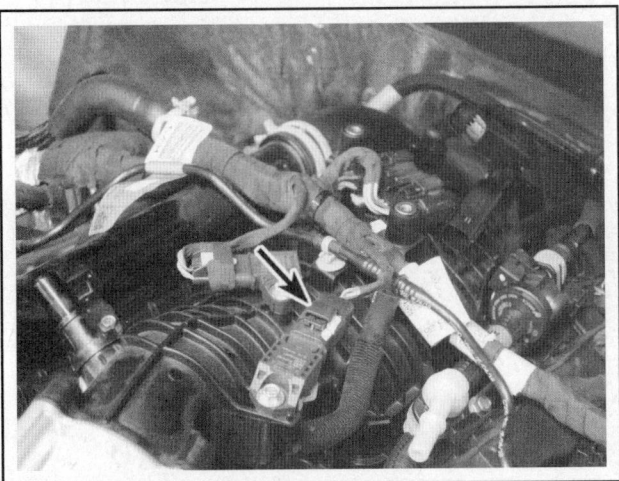

11.2 Locating the wastegate vacuum sensor

11.4 Disconnect the vacuum line after unbolting the sensor

12 Accelerator Pedal Position (APP) sensor (2005 and later models) - replacement and adjustment

▶ Refer to illustration 12.4

➡ **Note:** 2005 and later models are "drive-by-wire" and do not have an accelerator cable. The APP sensor communicates the desired pedal position to the PCM, the PCM then actuates the throttle body to the desired throttle angle.

➡ **Note:** On models with adjustable pedals, after removing and installing or replacing the APP sensor, the cable between the brake pedal and accelerator pedal requires adjustment.

➡ **Note:** The APP sensor is used to monitor the driver's demand for power. The PCM adjusts the amount of fuel injected into the cylinders according to the position of the pedal. A problem in the APP sensor circuit will set a diagnostic trouble code.

1 Disconnect the APP sensor electrical connector. On models with adjustable pedals, disconnect both connectors.

2 On models with adjustable pedals, disconnect the cable running between the brake pedal and the accelerator pedal, at the brake pedal.

3 On all models, remove the mounting fasteners and remove the APP sensor and accelerator pedal (see illustration).

4 Installation is reverse of removal.

5 On models with adjustable pedals, perform the cable adjustment procedure (see Chapter 9, Section 15).

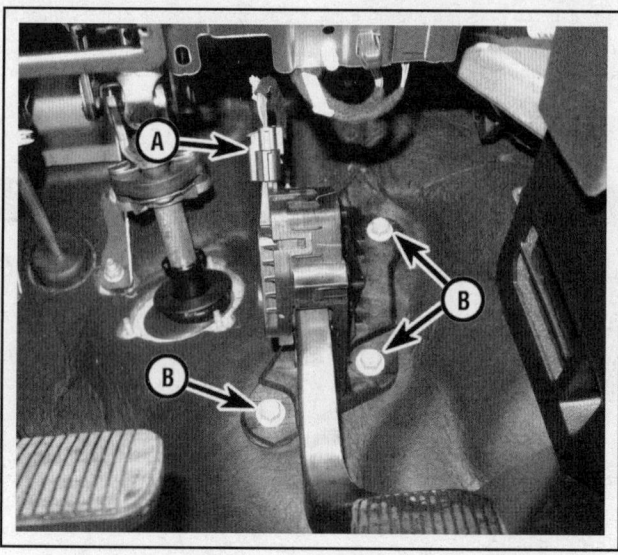

12.4 Slide the connector lock up then depress the tab (A) and disconnect the electrical connector, then remove the APP sensor mounting fasteners (B)

7A
MANUAL TRANSMISSION

Section

1 General information
2 Shift lever - removal and installation
3 Oil seal - replacement
4 Transmission mount - check and replacement
5 Manual transmission - removal and installation
6 Manual transmission overhaul - general information

Reference to other Chapters

Manual transmission lubricant - change - See Chapter 1
Manual transmission lubricant level - check - See Chapter 1

7A-2 MANUAL TRANSMISSION

1 General information

Vehicles covered by this manual are equipped with either a five-speed manual or a four-speed automatic transmission. Information on the manual transmission is included in this Part of Chapter 7. Information on the automatic transmission can be found in Part B of this Chapter. You'll also find certain procedures common to both transmissions - such as oil seal replacement - in Part A. Information on the transfer case used on 4WD models can be found in Part C.

The M50D is a fully-synchronized, five-speed manual transmission with an overdrive fifth gear. If you're planning to replace the transmission, look for the service identification tag on the right (driver's) side of the transmission. If the tag is missing or indecipherable, go to the Vehicle Safety Compliance Label on the left (driver's) side door pillar; it has the correct transmission identification codes.

Depending on the expense involved in having a transmission overhauled, it might be a better idea to consider replacing it with either a new or rebuilt unit. Your local dealer or transmission shop should be able to supply information concerning cost, availability and exchange policy. Regardless of how you decide to remedy a transmission problem, you can still save a lot of money by removing and installing the unit yourself.

2 Shift lever - removal and installation

▶ Refer to illustrations 2.1, 2.2a and 2.2b

1 Remove the shift lever boot retainer screws (see illustration) and pull up the boot.

2 Remove the shift lever nut and bolt (see illustrations) and pull off the shift lever.

3 If the inner shift lever boot (the black rubber square-shaped boot below the shift lever), remove the four boot retaining screws (see illustration 2.2b) and replace it.

4 Installation is the reverse of removal. Be sure to tighten the shift lever bolt to the torque listed in this Chapter's Specifications.

2.1 To remove the shift lever boot, remove the screws retaining the bezel

2.2a To remove the shift lever, remove this nut . . .

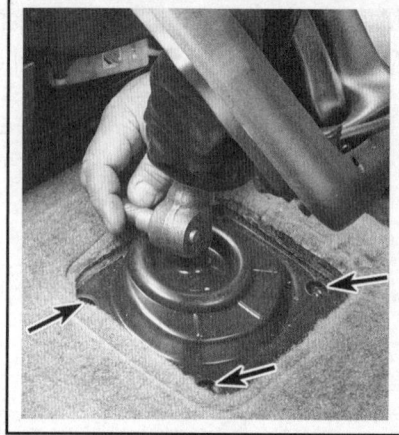

2.2b . . . pull out the bolt and lift off the lever; to remove the inner shift lever boot, remove these screws (arrows)

3 Oil seal - replacement

EXTENSION HOUSING SEAL

▶ Refer to illustrations 3.4 and 3.5

➡ Note: This procedure applies to both manual and automatic transmissions.

1 Oil leaks frequently occur due to wear of the extension housing oil seal or the vehicle speed sensor seal. Replacement of these seals is relatively easy, since the repairs can usually be performed without removing the transmission from the vehicle.

2 If you suspect a leak at the extension housing seal, raise the vehicle and support it securely on jackstands. If the vehicle is equipped with air suspension, turn off the air suspension system. The switch is located in the area of the right kick panel (see Chapter 10).

✸✸ WARNING:

On models with air suspension, electrical power to the air suspension system must be turned off before raising the vehicle. Failure to do so can result in a sudden inflation or deflation of the air springs, causing instability of the vehicle while it's off the ground. The extension housing seal is located at the rear end of the transmission, where the driveshaft is attached.

If the extension housing seal is leaking, transmission lubricant will be evident on the front of the driveshaft and may be dripping from the rear of the transmission.

3 Remove the driveshaft (see Chapter 8).

MANUAL TRANSMISSION   7A-3

3.4 Remove the extension housing seal with a seal removal tool

3.5 Install the new extension housing seal with a large deep socket or section of pipe

4   Using a soft-face hammer, carefully tap off the dust shield, if equipped. Be careful not to distort it. Using a screwdriver, pry bar or seal removal tool, carefully pry out the extension housing seal (see illustration). Do not damage the splines on the transmission output shaft.

5   Using a large section of pipe or a very large deep socket as a drift, install the new extension housing seal (see illustration). Drive it into the bore squarely and make sure it's completely seated. Install a new bushing using the same method. Install the dust shield (if equipped) by carefully tapping it into place.

6   Lubricate the splines of the transmission output shaft and the outside of the driveshaft sleeve yoke with light-weight grease, then install the driveshaft (see Chapter 8). Be careful not to damage the lip of the new seal. Lower the vehicle. Turn on the switch for the air suspension system, if applicable.

VEHICLE SPEED SENSOR O-RING

▶ Refer to illustrations 3.8, 3.9 and 3.11

7   The vehicle speed sensor is located on the left side of the extension housing. If you suspect a leak at the vehicle speed sensor, raise the vehicle and support it securely on jackstands. If the vehicle is equipped with air suspension, turn off the air suspension system. The switch is located in the area of the right kick panel (see Chapter 10).

✳✳ WARNING:

On models with air suspension, electrical power to the air suspension system must be turned off before raising the vehicle. Failure to do so can result in a sudden inflation or deflation of the air springs, causing instability of the vehicle while it's off the ground.

If the sensor seal is leaking, transmission lubricant will be dripping from the left side of the extension housing.

8   Unplug the electrical connector from the vehicle speed sensor (see illustration).
9   Remove the sensor retaining bolt (see illustration).
10 Remove the sensor by pulling it straight out.
11 Remove the O-ring (see illustration).
12 Install a new O-ring. Be sure to lubricate it with a light coat of clean oil to protect it from damage when the sensor is installed.
13 Installation is the reverse of removal. Be sure to lubricate the seal and the sensor with a light coat of clean engine oil before installation. Lower the vehicle. Turn on the switch for air suspension system, if applicable.

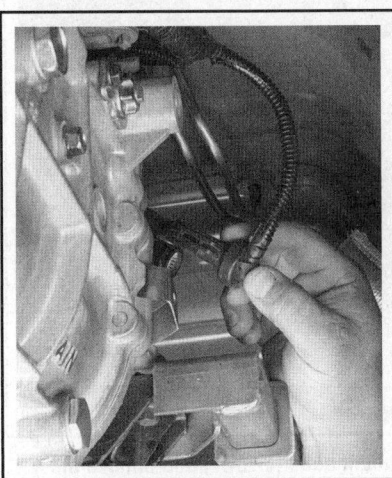

3.8 Unplug the electrical connector from the vehicle speed sensor

3.9 Remove the vehicle speed sensor retaining bolt

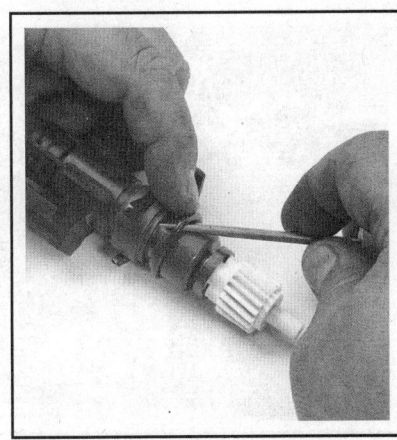

3.11 Remove the vehicle speed sensor O-ring and install a new one; be sure to lubricate the new O-ring to protect it during installation of the sensor

7A-4 MANUAL TRANSMISSION

4 Transmission mount - check and replacement

4.2 To check the transmission mount, insert a prybar or large screwdriver and try to lever the transmission up and down; if the transmission is easily moved, look closely at the rubber insulator part of the mount - it's probably torn or cracked, and must be replaced

CHECK

▶ Refer to illustration 4.2

1 If the vehicle is equipped with air suspension, turn off the air suspension system. The switch is located in the area of the right kick panel (see Chapter 10).

※ WARNING:

On models with air suspension, electrical power to the air suspension system must be turned off before raising the vehicle. Failure to do so can result in a sudden inflation or deflation of the air springs, causing instability of the vehicle while it's off the ground.

2 Insert a large screwdriver or pry bar into the space between the transmission and the crossmember and try to pry the transmission up slightly (see illustration). The transmission should not move away from the insulator much. If there is any separation of the rubber, the mount is worn out.

REPLACEMENT

▶ Refer to illustrations 4.3, 4.4a and 4.4b

3 Remove the nuts attaching the mount to the crossmember (see illustration).
4 Raise the transmission slightly with a jack, remove the bolts attaching the mount to the transmission (see illustrations). Remove the mount.
5 Installation is the reverse of the removal procedure. Be sure to tighten the nuts/bolts securely.
6 Remove the jackstands and lower the vehicle. Turn on the switch for the air suspension system, if applicable.

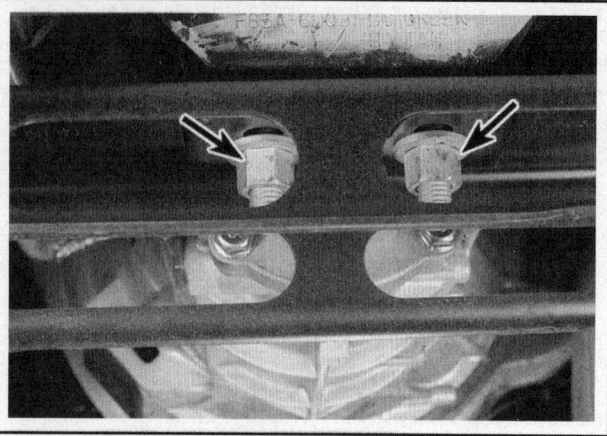

4.3 Remove the transmission mount-to-crossmember nuts (arrows)

4.4a Place a transmission jack or floor jack under the transmission and raise the transmission just enough to allow you to slip the old mount out from between the transmission and the crossmember . . .

4.4b . . . remove the mount bolts (arrows) and remove the mount (crossmember removed for clarity, but crossmember removal is not necessary)

MANUAL TRANSMISSION 7A-5

5 Manual transmission - removal and installation

REMOVAL

▶ Refer to illustrations 5.10, 5.11, 5.13, 5.15a, 5.15b, 5.17, 5.19, 5.20a, 5.20b, 5.20c, 5.21a, 5.21b and 5.23

1 Disconnect the negative battery cable from the battery.
2 Place the transmission in Neutral.
3 Remove the shift lever and inner shift lever boot (see Section 2).
4 If the vehicle is equipped with air suspension, turn off the air suspension system. The switch is located in the area of the right kick panel (see Chapter 10).

✺ WARNING:

On models with air suspension, electrical power to the air suspension system must be turned off before raising the vehicle. Failure to do so can result in a sudden inflation or deflation of the air springs, causing instability of the vehicle while it's off the ground.

5 Raise the vehicle and support it securely on jackstands.
6 Remove the driveshaft(s) (see Chapter 8).
7 On 4WD models, remove the skid plate and transfer case (see Chapter 7C).
8 Disconnect the clutch hydraulic line (see Chapter 8).
9 Remove the starter motor (see Chapter 5).
10 Unplug the electrical connectors from the two heated oxygen sensors (see illustration).
11 Place a floor jack under the engine (see illustration). Put a block of wood between the jack head and the engine oil pan.
12 Remove the two transmission mount-to-crossmember nuts (see illustration 4.3).
13 Detach the left and right heat shields from the crossmember; there's one bolt for each shield (see illustration).
14 Place a transmission jack or a floor jack under the transmission and secure the transmission to the jack with safety chains.
15 Raise the transmission slightly to take the weight off the crossmember, then remove the rear crossmember (see illustrations).

5.10 Unplug the electrical connectors (upper arrows) for the oxygen sensors; detach the fuel line bracket (lower arrow)

5.11 Support the engine with a floor jack; be sure to put a block of wood between the oil pan and the jack head to protect the pan from damage

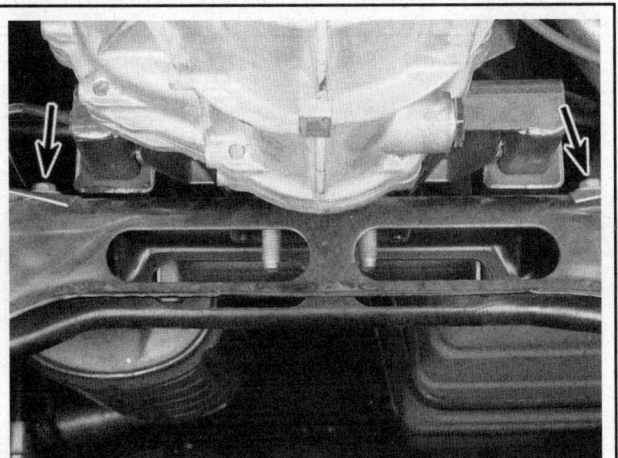

5.13 Remove the heat shield bolts (one per shield) which attach the heat shields to the crossmember, and detach the heat shields (right heat shield shown, left similar)

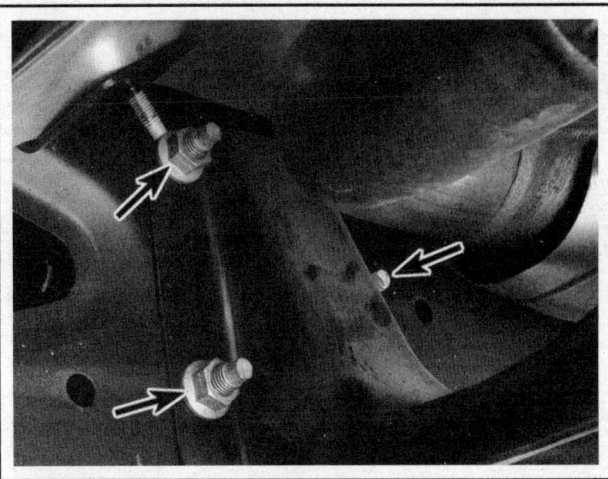

5.15a To detach the rear crossmember from the frame, remove these nuts (arrows) from the left end . . .

7A-6 MANUAL TRANSMISSION

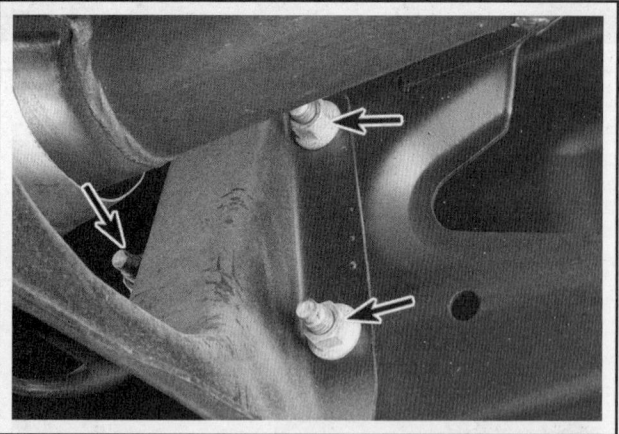

5.15b ... and these nuts (arrows) from the right end

5.17 Remove these exhaust pipe bracket nuts (arrows) and detach the bracket

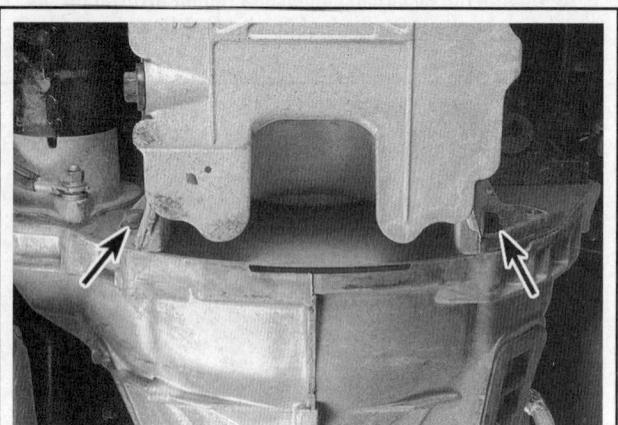

5.19 On V6 engines, remove the oil pan-to-engine bolts (arrows)

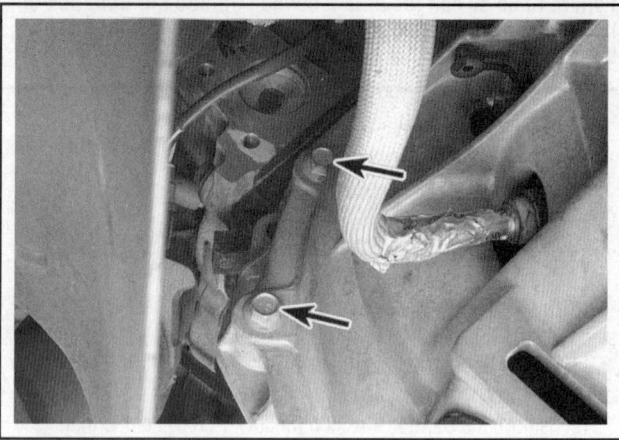

5.20a On V6 engines, remove the six engine-to-transmission bolts: there are two on the left (arrows) ...

16 Remove the transmission mount (see Section 4).
17 Remove the exhaust pipe bracket nuts (see illustration) and detach the pipe from the brackets.
18 Remove the fuel line bracket nut (see illustration 5.10).

19 On vehicles with a V6 engine, remove the oil pan-to-transmission bolts (see illustration).
20 On vehicles with a V6 engine, remove the six engine-to-transmission bolts (see illustrations).

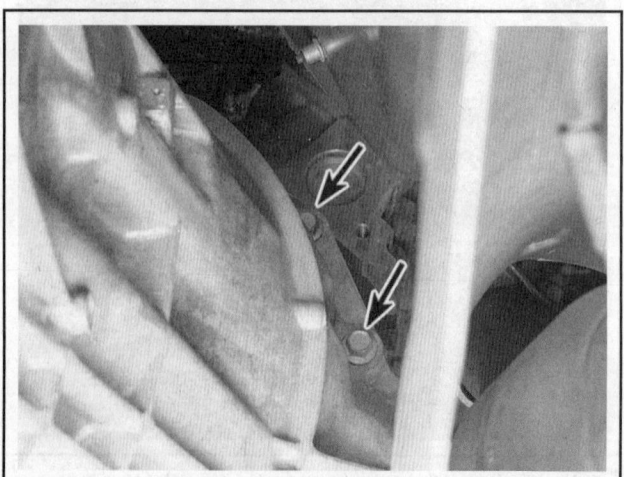

5.20b ... two more on the right (arrows) ...

5.20c ... and the two up top which can be removed with a long extension

MANUAL TRANSMISSION 7A-7

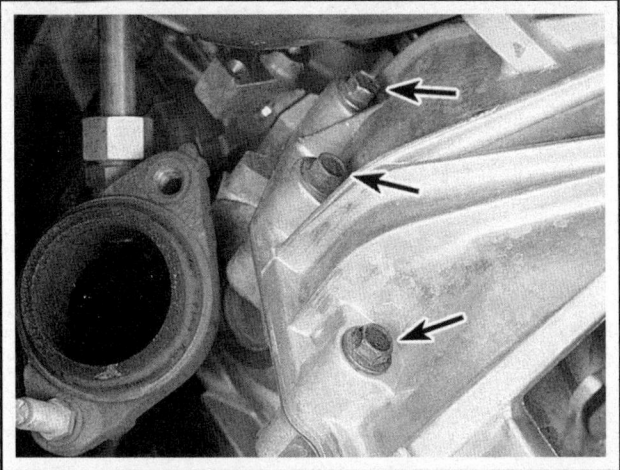

5.21a On V8 engines, remove the transmission-to-engine bolts (arrows) from the left side . . .

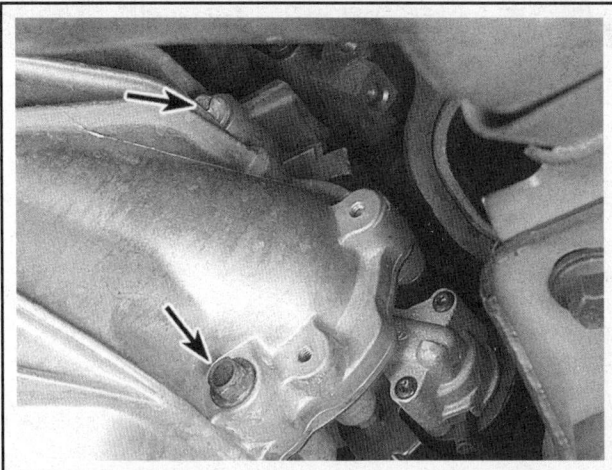

5.21b . . . and transmission-to-engine bolts (arrows) from the right side

21 On vehicles with a V8 engine, remove the six transmission-to-engine bolts (see illustrations).

22 Remove the exhaust pipe-to-exhaust manifold nuts from both exhaust pipe flanges (see Chapter 4).

23 Make a final check that all wires have been disconnected from the transmission, then move the transmission and jack toward the rear of the vehicle until the transmission input shaft is clear of the clutch or clutch housing. If the transmission input shaft is difficult to disengage from the clutch hub, use a prybar to separate the transmission from the engine (see illustration). Keep the transmission level as you pull it to the rear.

24 Once the input shaft is clear, lower the transmission and remove it from under the vehicle.

✳✳ CAUTION:

Do not depress the clutch pedal while the transmission is out of the vehicle.

5.23 If the transmission input shaft hangs up on the clutch hub, the transmission may be difficult to separate from the engine; if this happens, carefully separate the two with a large prybar as shown

25 Inspect the clutch components. Generally speaking, new clutch components should always be installed anytime the transmission is removed (see Chapter 8).

INSTALLATION

26 Install the clutch components, if they were removed (see Chapter 8).

27 With the transmission secured to the jack, raise it into position behind the engine and carefully slide it forward, engaging the input shaft with the clutch plate hub. Do not use excessive force to install the transmission - if the input shaft won't slide into place, readjust the angle of the transmission or turn the input shaft so the splines engage properly with the clutch.

28 Once the transmission is flush with the clutch housing or engine, install the transmission-to-engine or transmission-to-clutch housing bolts. Tighten the bolts to the torque listed in this Chapter's Specifications.

29 Install the transmission mount and crossmember. Tighten all nuts and bolts securely.

30 Remove the jacks supporting the transmission and the engine.

31 Install the various components removed previously. Be sure to tighten all transmission-to-engine and (on V6 engines) oil pan-to-transmission bolts to the torque listed in this Chapter's Specifications. Refer to Chapter 7, Part C, for installation of the transfer case (if equipped), Chapter 8 for the installation of the driveshaft(s) and Chapter 4 for information regarding the exhaust system components. To connect the clutch hydraulic line, refer to Chapter 8.

32 Make a final check to verify all wires and hoses have been reconnected and the transmission has been filled with lubricant to the proper level (see Chapter 1). Lower the vehicle. Turn on the switch for the air suspension system, if applicable.

33 Install the inner shift lever boot, the shift lever and the outer boot (see Section 2).

34 Connect the negative battery cable. Road test the vehicle and check for leaks.

7A-8 MANUAL TRANSMISSION

6 Manual transmission overhaul - general information

Overhauling a manual transmission is a difficult job for the do-it-yourselfer. It involves the disassembly and reassembly of many small parts. Numerous clearances must be precisely measured and, if necessary, changed with select fit spacers and snap-rings. As a result, if transmission problems arise, it can be removed and installed by a competent do-it-yourselfer, but overhaul should be left to a transmission repair shop. Rebuilt transmissions may be available - check with your dealer parts department and auto parts stores. At any rate, the time and money involved in an overhaul is almost sure to exceed the cost of a rebuilt unit.

Nevertheless, it's not impossible for an inexperienced mechanic to rebuild a transmission if the special tools are available and the job is done in a deliberate step-by-step manner so nothing is overlooked.

The tools necessary for an overhaul include internal and external snap-ring pliers, a bearing puller, a slide hammer, a set of pin punches, a dial indicator and possibly a hydraulic press. In addition, a large, sturdy workbench and a vise or transmission stand will be required.

During disassembly of the transmission, make careful notes of how each piece comes off, where it fits in relation to other pieces and what holds it in place.

Before taking the transmission apart for repair, it will help if you have some idea what area of the transmission is malfunctioning. Certain problems can be closely tied to specific areas in the transmission, which can make component examination and replacement easier. Refer to the *Troubleshooting* Section at the front of this manual for information regarding possible sources of trouble.

Torque specifications — Ft-lbs (unless otherwise indicated)

Note: One foot-pound (ft-lb) of torque is equivalent to 12 inch-pounds (in-lbs) of torque. Torque values below approximately 15 ft-lbs are expressed in inch-pounds, since most foot-pound torque wrenches are not accurate at these smaller values.

Shift lever retaining bolt	89 to 123 in-lbs
Transmission-to-engine bolts	
V6 engine	
Transmission-to-engine block bolts	30 to 41
Transmission-to-oil pan bolts	28 to 38
V8 engines	30 to 41

Section

1   General information
2   Diagnosis - general
3   Shift lever - removal and installation
4   Shift indicator cable adjustment
5   Transmission control switch - description, check and replacement
6   Shift cable - removal, installation and adjustment
7   Transmission Range (TR) sensor - description, adjustment and replacement
8   Shift interlock system - description, check and actuator replacement
9   Automatic transmission - removal and installation

Reference to other Chapters

Automatic transmission fluid and filter change - See Chapter 1
Automatic transmission fluid level check - See Chapter 1
Oil seal - replacement - See Chapter 7A
Transmission mount - check and replacement - See Chapter 7A

7B
AUTOMATIC TRANSMISSION

7B-2 AUTOMATIC TRANSMISSION

1 General information

> **⁕⁕ CAUTION:**
> If a vehicle with an automatic transmission is disabled, do NOT tow it at speeds greater than 30 mph or distances over 50 miles.

Early models equipped with an automatic transmission use either a 4R70W, an E4OD, or a 4R100. Though different in design, they're similar in operation: electronic-shift four-speed with a lock-up torque converter, known as a torque converter clutch (TCC). The TCC provides a direct connection between the engine and the drive wheels for improved efficiency and fuel economy. The transmission kickdown function is controlled electronically. All three of these transmissions use a shift cable for manual shifting.

2005 and later Expedition/Navigator models have six-speed 6R75 or 6R80 automatic transmissions. Transmission functions are controlled by a Transmission Control Module (TCM), which is housed inside the transmission. The assembly includes the sensors, six pressure regulators and the transmission range sensor.

Because of the complexity of the clutches and the electronic and hydraulic control systems, and because of the special tools and expertise needed to overhaul an automatic transmission, diagnosis and repair of this transmission must be handled by a dealer service department or a transmission repair shop. The procedures in this Chapter are limited to general diagnosis, routine maintenance and adjustment: replacing the shift lever, replacing and adjusting the shift cable, and similar jobs. Serious repair work, however, must be done by a transmission specialist. But if the transmission must be rebuilt or replaced, you can save money by removing and installing it yourself, so instructions for that procedure are included as well.

2 Diagnosis - general

➥**Note:** Automatic transmission malfunctions may be caused by five general conditions: poor engine performance, improper adjustments, hydraulic malfunctions, mechanical malfunctions or malfunctions in the computer or its signal network. Diagnosis of these problems should always begin with a check of the easily repaired items: fluid level and condition (see Chapter 1), and shift cable adjustment. Next, perform a road test to determine if the problem has been corrected or if more diagnosis is necessary. If the problem persists after the preliminary tests and corrections are completed, additional diagnosis should be done by a dealer service department or transmission repair shop. Refer to the *Troubleshooting* Section at the front of this manual for information on symptoms of transmission problems.

PRELIMINARY CHECKS

1 Drive the vehicle to warm the transmission to normal operating temperature.

2 Check the fluid level as described in Chapter 1:
   a) *If the fluid level is unusually low, add enough fluid to bring the level within the designated area of the dipstick, then check for external leaks (see below).*
   b) *If the fluid level is abnormally high, drain off the excess, then check the drained fluid for contamination by coolant. The presence of engine coolant in the automatic transmission fluid indicates that a failure has occurred in the internal radiator walls that separate the coolant from the transmission fluid (see Chapter 3).*
   c) *If the fluid is foaming, drain it and refill the transmission, then check for coolant in the fluid or a high fluid level.*

3 Check the engine idle speed.

➥**Note:** If the engine is malfunctioning, do not proceed with the preliminary checks until it has been repaired and runs normally.

4 Inspect the shift cable (see Section 6). Make sure it's properly adjusted and operates smoothly.

FLUID LEAK DIAGNOSIS

5 Most fluid leaks are easy to locate visually. Repair usually consists of replacing a seal or gasket. If a leak is difficult to find, the following procedure may help.

6 Identify the fluid. Make sure it's transmission fluid and not engine oil or brake fluid (automatic transmission fluid is a deep red color).

7 Try to pinpoint the source of the leak. Drive the vehicle several miles, then park it over a large sheet of cardboard. After a minute or two, you should be able to locate the leak by determining the source of the fluid dripping onto the cardboard.

8 Make a careful visual inspection of the suspected component and the area immediately around it. Pay particular attention to gasket mating surfaces. A mirror is often helpful for finding leaks in areas that are hard to see.

9 If the leak still cannot be found, clean the suspected area thoroughly with a degreaser or solvent, then dry it.

10 Drive the vehicle for several miles at normal operating temperature and varying speeds. After driving the vehicle, visually inspect the suspected component again.

11 Once the leak has been located, the cause must be determined before it can be properly repaired. If a gasket is replaced but the sealing flange is bent, the new gasket will not stop the leak. The bent flange must be straightened.

12 Before attempting to repair a leak, check to make sure the following conditions are corrected or they may cause another leak.

➥**Note:** Some of the following conditions cannot be fixed without highly specialized tools and expertise. Such problems must be referred to a transmission repair shop or a dealer service department.

Gasket leaks

13 Check the pan periodically. Make sure the bolts are tight, no bolts are missing, the gasket is in good condition and the pan is flat (dents in the pan may indicate damage to the valve body inside).

14 If the pan gasket is leaking, the fluid level or the fluid pressure may be too high, the vent may be plugged, the pan bolts may be too tight, the pan sealing flange may be warped, the sealing surface of the transmission housing may be damaged, the gasket may be damaged or the transmission casting may be cracked or porous. If sealant instead of gasket material has been used to form a seal between the pan and the transmission housing, it may be the wrong sealant.

AUTOMATIC TRANSMISSION 7B-3

Seal leaks

15 If a transmission seal is leaking, the fluid level or pressure may be too high, the vent may be plugged, the seal bore may be damaged, the seal itself may be damaged or improperly installed, the surface of the shaft protruding through the seal may be damaged or a loose bearing may be causing excessive shaft movement.

16 Make sure the dipstick tube seal is in good condition and the tube is properly seated. Periodically check the area around the speedometer gear or sensor for leakage. If transmission fluid is evident, check the O-ring for damage.

Case leaks

17 If the case itself appears to be leaking, the casting is porous and will have to be repaired or replaced.

18 Make sure the oil cooler hose fittings are tight and in good condition.

Fluid comes out vent pipe or fill tube

19 If this condition occurs, the transmission is overfilled, there is coolant in the fluid, the case is porous, the dipstick is incorrect, the vent is plugged or the drain back holes are plugged.

3 Shift indicator cable adjustment

1 Remove the upper steering column cover (see Chapter 11).
2 Put the shift lever in the Drive position.
3 Move the shift lever clockwise until it stops (all the way to the 1st gear position), then move it back two detent positions to the Drive position.
4 Hang an eight-pound weight on the shift lever to hold it in place.
5 Center the indicator needle (the pointer) in the middle of the Drive position.
6 Rotate the thumbwheel on the bottom of the steering column to remove all slack from the indicator cable.
7 Install the upper steering column cover.

4 Shift lever - removal and installation

1 Remove the ignition key lock cylinder (see Chapter 12).
2 Remove the upper steering column shroud.
3 Unplug the electrical connector for the transmission control switch (TCS).
4 Remove the shift lever pin and discard it.
5 Remove the shift lever assembly.
6 Installation is the reverse of removal. Be sure to replace the shift lever pin; do NOT use the old pin.

5 Transmission control switch - description, check and component replacement

DESCRIPTION

1 Normally, the powertrain control module (PCM) allows automatic shifts from first through fourth gear. When the transmission control switch (TCS) is pressed, overdrive is overridden, and the PCM allows shifts from first through third only. (The PCM also turns on the transmission control indicator lamp (TCIL), an LED which indicates that "overdrive cancel mode" has been activated. If the TCIL flashes instead, there's either a sensor failure or a short in the electronic pressure control circuit (EPC); in either event, take the vehicle to a dealer to have the system serviced.) When the switch is pressed again, normal operation is resumed.

CHECK

2 The TCS *circuit* can be fully tested only at the dealer. However, there are some simple tests you can do to determine whether the switch itself is bad:

 a) *Check fuse 29 (see Chapter 12). Fuse 29 is a 10A fuse on 1997 pick-ups and a 5A fuse on 1998 and later pick-ups, Expeditions and Navigators.*
 b) *Remove the TCS (see below) and check the resistance of the switch. When the TCS button is pressed and held down (it's a momentary-contact switch, so you have to hold it down to measure the resistance), the resistance should be less than 5 ohms; when the button is released, resistance should be more than 10 K-ohms. If the TCS doesn't perform as described, replace it. If the indicated resistance is within the specified range, go to the next test.*
 c) *Apply battery voltage to the TCS and verify that the TCIL comes on. If it doesn't, replace the TCS. If it does come on, the switch is okay. Take the vehicle to a dealer to have the remainder of the system checked out.*

COMPONENT REPLACEMENT

3 Remove the TCS cover.
4 Remove the TCS.
5 Installation is the reverse of removal.

7B-4 AUTOMATIC TRANSMISSION

6 Shift cable - removal, installation and adjustment

REMOVAL AND INSTALLATION

1 Detach the shift cable from the steering column shift tube lever.
2 Detach the shift cable from the steering column bracket.
3 Push the rubber grommet and shift cable through the firewall.
4 If the vehicle is equipped with air suspension, turn off the air suspension system. The switch is located in the area of the right kick panel.

✳✳ WARNING:

On models with air suspension, electrical power to the air suspension system must be turned off before raising the vehicle (see Chapter 10). Failure to do so can result in a sudden inflation or deflation of the air springs, causing instability of the vehicle while it's off the ground.

5 Raise the vehicle and place it securely on jackstands.
6 Detach the shift cable from the manual lever. On 2005 and later Expedition/Navigator models, release the cover by moving the locking tab for access to the ball-stud on the lever.
7 Detach the shift cable from the cable bracket.
8 Installation is the reverse of removal. Be sure to adjust the cable before reattaching it to the manual lever (see below). If the vehicle is equipped with air suspension, reactivate the air suspension system by turning on the switch.

ADJUSTMENT

9 Put the shift lever in the Drive position and put an eight-pound weight on the lever to hold it there.
10 If the vehicle is equipped with air suspension, deactivate the air suspension system by turning off the air suspension switch, which is located on the right kick panel.

✳✳ WARNING:

On models with air suspension, electrical power to the air suspension system must be turned off before raising the vehicle. Failure to do so can result in a sudden inflation or deflation of the air springs, causing instability of the vehicle while it's off the ground.

11 Raise the vehicle and support it securely on jackstands.
12 Detach the shift cable from the manual lever.
13 Unlock the lock tab on the shift cable.
14 Move the manual lever to the First gear position, then move it back two detent positions to the Drive position.
15 Reattach the shift cable to the manual lever.
16 Lock the shift cable locking tab.
17 If the vehicle is equipped with air suspension, reactivate the air suspension system by turning on the switch.
18 Remove the jackstands and lower the vehicle.
19 Remove the weight from the shift lever.
20 Move the shift lever through all gear positions and verify that the indicated positions correspond with the actual gear positions at the manual lever. Also verify that the engine will start only in Park and Neutral, and that the back-up lights come on when the shift lever is placed in Reverse. If necessary, readjust the cable until these conditions are met. It may also be necessary to adjust the transmission range sensor (see Section 7).

7 Transmission Range (TR) sensor - description, adjustment and replacement

DESCRIPTION

➟ **Note:** *On 2005 and later Expedition/Navigator models, the Transmission Range sensor is inside the transmission as part of the TCM. It is not adjustable and can only be replaced as an assembly with the TCM.*

1 The Transmission Range (TR) sensor, which is located at the manual lever on the transmission, is an information sensor for the powertrain control module (PCM). Among its functions are those normally handled by a conventional Park/Neutral switch: it prevents the engine from starting in any gear other than Park or Neutral, and closes the circuit for the back-up lights when the shift lever is moved to Reverse. The TR sensor also contains the sensing circuit for Neutral in 4X4 Low range on 4WD models. For information on the TR sensor's other functions, refer to Chapter 6.

ADJUSTMENT

◆ **Refer to illustration 7.5**

2 If the engine starts in any position other Park or Neutral, the TR sensor is either out of adjustment or defective. First, perform a quick functional check to verify that the sensor is operating properly.

3 If the vehicle is equipped with air suspension, deactivate the air suspension system by turning off the air suspension switch, which is located on the right kick panel.

✳✳ WARNING:

On models with air suspension, electrical power to the air suspension system must be turned off before raising the vehicle (see Chapter 10). Failure to do so can result in a sudden inflation or deflation of the air springs, causing instability of the vehicle while it's off the ground.

4 Raise the vehicle and support it securely on jackstands.
5 The factory recommends a special transmission range (TR) sensor alignment tool, available at most auto parts stores, but there's a quick and easy method to verify whether the sensor is adjusted, and to adjust it if it isn't:
  a) Turn the ignition switch to On, put the shift lever in Reverse and verify that the back-up lights come on;
  b) If they do, but the engine can't be started in Park or Neutral, or it can be started in any gear other than Park or Neutral, then the sensor is probably defective. The complete sensor check procedure is in Chapter 6.

AUTOMATIC TRANSMISSION

c) If they don't, detach the shift cable from the manual lever (see Section 6), loosen the sensor retaining bolts (see illustration) and move the sensor slightly until the back-up lights come on. Tighten the sensor retaining bolts to the torque listed in this Chapter's Specifications and reattach the shift cable.

d) If you can't get the back-up lights to come on by moving the sensor slightly, verify that the back-up lights and the back-up light circuit are okay (see Chapter 12 and the Wiring Diagrams at the end of Chapter 12). If the back-up lights and circuit are okay, the sensor is probably bad. Refer to Chapter 6.

6 Remove the jackstands and lower the vehicle. If the vehicle is equipped with air suspension, reactivate the air suspension system by turning on the switch.

REPLACEMENT

7 If the vehicle is equipped with air suspension, deactivate the air suspension system by turning off the air suspension switch, which is located on the right kick panel.

※※ WARNING:

On models with air suspension, electrical power to the air suspension system must be turned off before raising the vehicle (see Chapter 10). Failure to do so can result in a sudden inflation or deflation of the air springs, causing instability of the vehicle while it's off the ground.

8 Raise the vehicle and place it securely on jackstands.
9 Unplug the electrical connector from the TR sensor.

7.5 To adjust the TR sensor without the special tools: detach the shift cable, loosen the sensor retaining bolts, move the sensor slightly until the back-up lights come on, then tighten the sensor retaining bolts to the torque listed in this Chapter's Specifications and reattach the shift cable

10 Detach the shift cable from the manual lever (see Section 6).
11 On vehicles equipped with a 4R70W transmission, remove the manual lever; on models with an E4OD or 4R100, the manual lever is between the TR sensor and the transmission, so it doesn't need to be removed.
12 Remove the TR sensor retaining bolts (see illustration 7.5).
13 Remove the TR sensor.
14 Installation is the reverse of removal. Be sure to tighten the sensor retaining screws to the torque listed in this Chapter's Specifications and adjust the sensor (see Step 5). Remove the jackstands and lower the vehicle. If the vehicle is equipped with air suspension, reactivate the air suspension system by turning on the switch.

8 Shift interlock system - description, check and actuator replacement

DESCRIPTION

1 The shift interlock system prevents the shift lever from being moved out of the Park position unless the brake pedal is depressed. The system consists of a shift lock actuator mounted on the steering column. When the ignition key is turned to the Run position, the actuator is energized unless the brake pedal is depressed. If the shift lever cannot be moved out of the Park position when the brake pedal is applied, the following series of simple checks will help you quickly pinpoint the problem:

CHECK

2 The shift lock actuator receives voltage when the ignition key is in the Run position. This circuit energizes the actuator and it prevents you from moving the shift lever out of the Park position. The actuator also receives voltage from another circuit, through the brake light switch, that is closed only when the brake pedal is depressed. It's this second circuit that de-energizes the solid state actuator when the brake pedal is depressed. So first, try to verify that the actuator is working.

3 Get inside the vehicle, close the doors and windows, start the engine, let it settle down to a fully warmed-up idle, put your head under the dash so that your ear is close to the actuator (it's mounted on the steering column), then depress the brake pedal and listen carefully for the sound of the actuator clicking.

4 If you don't hear the actuator click when you depress the brake pedal, check the 5A fuse for the actuator and the 15A fuse for the brake light switch (see Chapter 12). Replace either fuse if it's bad and recheck the actuator.

5 If the actuator and brake light switch fuses are good but the actuator still doesn't click when the brake pedal is depressed, verify that the actuator is getting battery voltage through both circuits (one is hot in the Run position, one is hot only when the brake light switch is closed).

6 If the actuator isn't getting voltage through the first "hot-in-Run-only" circuit, repair that circuit and retest.

7 If the actuator isn't getting voltage through the brake light switch circuit, apply the brake pedal and verify that the brake lights come on.

  a) If the brake lights don't come on, troubleshoot the brake light circuit and determine whether the circuit itself or the brake light switch is defective (see Chapter 9), make the necessary repairs or component replacement, then retest the actuator.
  b) If the brake lights come on, the brake light switch and circuit are okay. Repair the circuit between the brake light switch and the actuator and retest.
  c) If the actuator still doesn't work, replace it (see below).

ACTUATOR REPLACEMENT

➡ **Note:** *On 2005 and later Expedition/Navigator models with floor-shift, the actuator is an integral part of the shifter assembly, which must be replaced as a unit.*

8 Remove the steering column covers (see Chapter 11).

7B-6 AUTOMATIC TRANSMISSION

9 Remove the three shift lock actuator bolts.
10 Remove the insert plate and shift lock actuator.
11 Remove and discard the shift lock actuator clip. The shift lock actuator clip is an assembly aid and doesn't need to be replaced. Separate the insert plate from the shift lock actuator.
12 Installation is the reverse of removal.

9 Automatic transmission - removal and installation

→Note: *On 2009 and later Expedition/Navigator models, if the transmission is going to be replaced, the module configuration must be retrieved from the PCM. This will have to be done by a dealer service department or other properly equipped repair facility because the Programmable Module Installation (PMI) configuration must be followed using the Vehicle Communication Module (VCM) and Integrated Diagnostic System (IDS) software after the new transmission is installed. Without the correct module configuration, the new transmission will not work correctly.*

REMOVAL

▶ Refer to illustrations 9.7a, 9.7b, 9.8, 9.11, 9.15 and 9.16

1 Disconnect the negative battery cable from the battery.
2 Put the shift lever in the Neutral position.
3 Remove the transmission harness electrical connector retaining bolt and detach and unplug the connector.
4 If the vehicle is equipped with air suspension, deactivate the air suspension system by turning off the air suspension switch, which is located on the right kick panel.

※※ WARNING:

On models with air suspension, electrical power to the air suspension system must be turned off before raising the vehicle (see Chapter 10). Failure to do so can result in a sudden inflation or deflation of the air springs, causing instability of the vehicle while it's off the ground.

Raise the vehicle and support it securely on jackstands.
5 Remove the driveshaft(s) (see Chapter 8).
6 On 4WD models, remove the skid plate, if equipped, and transfer case (see Chapter 7C).
7 On vehicles equipped with a V6 engine and a 4R70W or 4R100 transmission, remove the transmission inspection cover (see illustration); on vehicles with a V8 engine and a 4R70W transmission, remove both the metal inspection cover and remove the rubber plug from the access hole in the left rear part of the block (see illustration). Models with an E4OD transmission are also equipped with a metal access cover and the access hole in the block; remove the rubber plug and the metal cover on these models, as well.
8 Mark the relationship of the torque converter to the driveplate, then remove the four torque converter retaining nuts (see illustration). Rotate the crankshaft to bring each nut within reach through the inspection cover hole.
9 Remove the starter motor (see Chapter 5).
10 Disconnect the shift cable from the manual lever and from the cable bracket (see Section 6).
11 Disconnect the transmission oil cooler lines (see illustration).
12 Unplug the electrical connectors from the two heated oxygen sensors (see Chapter 6), the vehicle speed sensor, the output shaft sensor, the transmission range sensor (all on the left side, and the solenoid body assembly (on the right side).

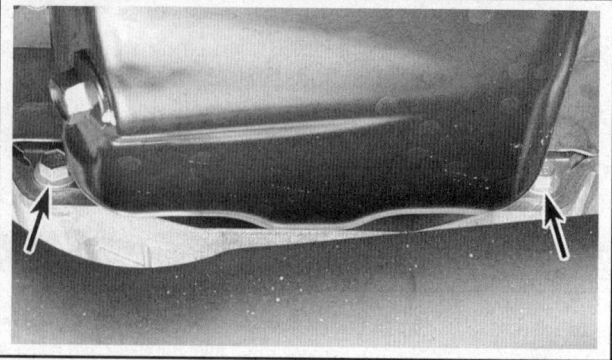

9.7a On vehicles equipped with a V6 engine and a 4R70W transmission, remove the transmission inspection cover bolts (arrows) and remove the cover

9.7b On vehicles with a V8 engine and a 4R70W or a 4R100 transmission, remove both the metal inspection cover and remove this plastic plug from the access hole in the left rear part of the block; models with an E4OD transmission are also equipped with a metal access cover and an access hole in the block

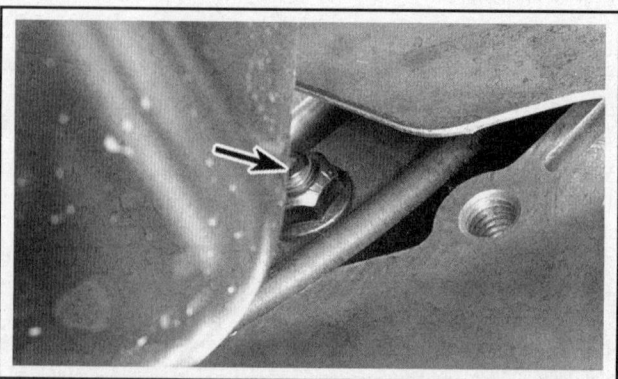

9.8 Remove the four torque converter retaining nuts (arrow); rotate the crankshaft to bring each nut within reach through the inspection cover hole

AUTOMATIC TRANSMISSION 7B-7

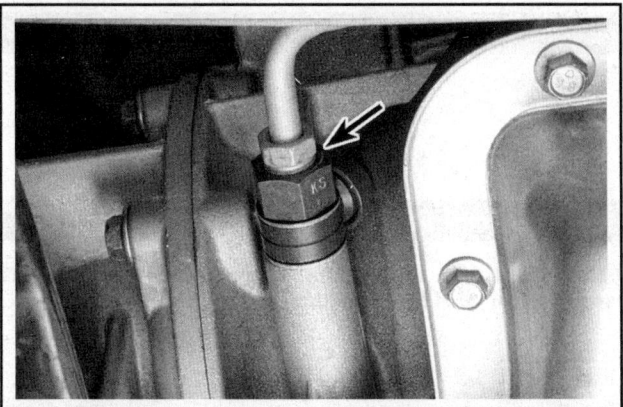

9.11 Disconnect the transmission oil cooler lines (arrow) (other line, not visible in this photo, is located above and behind this one)

9.15 Remove the two nuts (arrows) that attach the transmission mount to the crossmember, then raise the transmission slightly, unbolt the crossmember and remove it

13 Place a floor jack under the engine. Place a wood block between the jack head and the engine oil pan.

14 Place a transmission jack or a floor jack under the transmission and secure the transmission to the jack with safety chains.

15 Remove the two nuts (see illustration) that attach the transmission mount to the crossmember.

16 Remove the two heat shield bolts (see illustration).

17 Raise the transmission slightly to take the weight off the crossmember, then remove the crossmember bolts and remove the crossmember.

18 Remove the nuts or bolts that attach the transmission mount to the transmission, then remove the mount.

19 Remove the dipstick tube. On some models, the tube is a two-piece design: the upper half has a welded-on bracket that's bolted to the head; once unbolted from the head, it can be removed separately from the lower part of the tube.

20 Remove the exhaust pipe bracket nuts and the bracket (see Chapter 4).

21 Remove the fuel line bracket (see Chapter 4).

22 Remove the transmission-to-engine bolts.

23 Detach the exhaust pipes from the exhaust manifolds (see Chapter 4).

24 Make a final check that all wires have been disconnected from the transmission, then move the transmission and jack toward the rear of the vehicle until the torque converter is separated from the driveplate. Secure the torque converter to the transmission so it won't fall out during removal.

INSTALLATION

25 Prior to installation, make sure the torque converter hub is securely engaged with the pump gear.

26 With the transmission secured to the jack, raise it into position. Be sure to keep it level so the torque converter doesn't fall out and disengage itself from the pump gear.

27 Turn the torque converter until the marks on the converter and driveplate are aligned.

28 Move the transmission forward carefully until the dowel pins engage with the holes in the bellhousing.

29 Attach the exhaust pipes to the exhaust manifolds (see Chapter 4).

30 Install the transmission-to-engine bolts and tighten them to the torque listed in this Chapter's Specifications.

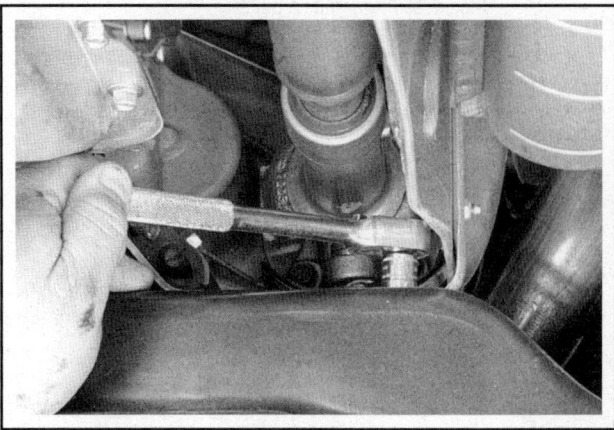

9.16 Remove the two heat shield bolts

31 Attach the fuel line bracket and the exhaust pipe bracket. Tighten all nuts securely.

32 Install the transmission mount and crossmember and tighten all nuts and bolts securely.

33 Install the heat shield bolts and tighten them securely.

34 Install the two transmission mount nuts and tighten them securely.

35 Remove the jacks supporting the transmission and the engine.

36 Plug in the two heated oxygen sensor connectors.

37 Attach the two oil cooler lines.

38 Attach the shift cable to the manual lever (see Section 6).

39 Install the four torque converter nuts and tighten them to the torque listed in this Chapter's Specifications.

40 Install the transmission inspection cover and tighten the bolts securely. Install the rubber access plug (if equipped).

41 Install the starter motor (see Chapter 5).

42 Install the dipstick tube.

43 If the vehicle is equipped with 4WD, install the transfer case (see Chapter 7C).

44 Install the driveshaft(s) (see Chapter 8).

45 Remove jacks and jackstands and lower the vehicle.

46 Plug in the electrical connector for the transmission harness. Install the connector retaining bolt and tighten it securely.

47 Attach the negative battery cable.

48 Fill the transmission with the specified fluid (Chapter 1), run the engine and check for fluid leaks.

7B-8 AUTOMATIC TRANSMISSION

Specifications

General

Transmission fluid type	See Chapter 1

Torque specifications   Ft-lbs (unless otherwise indicated)

➙**Note:** One foot-pound (ft-lb) of torque is equivalent to 12 inch-pounds (in-lbs) of torque. Torque values below approximately 15 ft-lbs are expressed in inch-pounds, since most foot-pound torque wrenches are not accurate at these smaller values.

Manual lever nut	
4R70W	22 to 26
6R80	159 in-lbs
Transmission range sensor retaining screws	
4R70W	62 to 89 in-lbs
E4OD and 4R100	71 to 88 in-lbs
Transmission-to-engine bolts	30 to 40
Torque converter nuts	22 to 30

Section

1 General information
2 Shift lever (manual-shift models) - removal and installation
3 4WD indicator switch (manual-shift models) - replacement
4 Shift range selector switch (electric-shift models) - replacement
5 Electric shift motor (electric-shift models) - replacement
6 Oil seal - replacement
7 Transfer case - removal and installation

7C
TRANSFER CASE

7C-2 TRANSFER CASE

1 General information

Four-wheel drive (4WD) models are equipped with a Borg-Warner 44-06 manual-shift (44-06-000-10) or electric-shift (44-06-000-11) transfer case mounted on the rear of the transmission. Drive is transmitted from the engine, through the transmission and the transfer case, to the front and rear axles by driveshafts.

We don't recommend trying to rebuild any of these transfer cases at home. They're difficult to overhaul without special tools, and rebuilt units are available for less than it would cost to rebuild your own. However, there are a number of components that you can check, adjust and/or replace - and those are the items covered in this Chapter.

MECHANICAL SHIFT-ON-THE-FLY (MSOF) SYSTEM (PICK-UP ONLY)

The mechanical shift-on-the-fly (MSOF) system on trucks allows the driver to manually select one of three ranges, 2WD High, 4WD Low or 4WD High. The driver can switch between 2WD High and 4WD High at speeds up to 55 mph (up to 45 mph at temperatures below 32 degrees F). 4WD Low can only be engaged or disengaged with the brake pedal depressed, the transmission in Neutral, and the speed under three mph (this is not a synchronized shift).

When shifting from 2WD High to 4WD High while the vehicle is moving, an electromagnetic clutch inside the transfer case brings the front driveline up to speed as follows: When the manual shift lever is moved from 2WD High to 4WD High, a 4WD indicator light is illuminated, and the 4WD electric clutch relay and electromagnetic clutch are energized. When the transfer is shifted into 4WD Low, an indicator is also illuminated. Both indicator lights are turned on by a 4WD indicator switch on the transfer case.

ELECTRONIC SHIFT-ON-THE-FLY (ESOF) SYSTEM (PICK-UP ONLY)

The electronic shift-on-the-fly (ESOF) system on trucks allows the driver the same three ranges (2WD High, 4WD High, 4WD Low) as the MSOF system, and the same rules apply for shifting from 2WD High to 4WD High, and for shifting to 4WD Low. However, the means by which a shift is initiated are electronic: When 4WD High is selected at a switch on the dash, the generic electronic module (GEM) receives a voltage signal commanding it to energize the electromagnetic clutch inside the transfer case and the relays which energize the transfer case shift motor. When the shift motor reaches the correct position (determined by the position of contact plates which send inputs to the GEM), power to the shift relays and motor is cut. When the transfer case front and rear output shafts are turning at the same speed, a spring-loaded lock-up collar mechanically engages the mainshaft hub to the drive sprocket, the front axle collar is engaged and the electromagnetic clutch is de-energized.

AUTOMATIC FOUR-WHEEL-DRIVE (A4WD) SYSTEM (EXPEDITION AND NAVIGATOR SPORT-UTILITY VEHICLES)

The automatic four-wheel-drive (A4WD) system on sport-utility vehicles allows the driver the same ranges at the ESOF system for pick-ups described above. In addition, it offers a third 4WD option, automatic four-wheel-drive (A4WD). When A4WD is selected, the generic electronic module (GEM) allocates torque to the front and rear driveshafts by controlling a transfer case clutch. When negotiating a corner on dry pavement, the transfer case clutch allows for the slight differences that occur in rotating speed between the front and rear driveshafts. But when the rear wheels are slipping, the GEM detects this condition and the duty cycle to the transfer case clutch is increased until the speed difference between driveshafts is reduced.

The GEM uses a variety of sensors, solenoids and relays to monitor and control the A4WD system. A vehicle speed sensor (VSS) monitors vehicle speed, and a pair of Hall effect sensors (one for each driveshaft) monitor the speed of the front and rear driveshafts (all three sensors are located on the transfer case). An electric shift motor, mounted on the rear of the transfer case, drives a rotary cam which moves the mode fork and range fork inside the transfer case to select between 2WD High/A4WD, 4WD High and 4WD Low ranges. A shift motor sensing plate on the shift motor monitors the range in which the transfer case is operating. A transmission range (TR) sensor (see Chapter 7B) monitors the gear position of the transmission (which must be in Neutral to shift into Low). A pair of solenoids (one for 2WD, one for 4WD), mounted on the upper right firewall, route vacuum to the vacuum motor which engages and disengages the center axle disconnect collar in the front axle assembly. A 4X4 shift motor relay module contains two GEM-controlled relays which shift the transfer case shift motor between 2WD and 4WD modes. A solid-state clutch relay activates the A4WD clutch inside the transfer case.

On 2005 and later Expedition/Navigator models with 4WD, vacuum-operated hubs are used in an Integrated Wheel End (IWE) system. When the vehicle is in 2WD mode, the system applies vacuum to the hubs so that they are disengaged from the driveaxles, thus reducing wear and improving fuel economy.

2 Shift lever (manual-shift models) - removal and installation

1 Unscrew the shift lever knob.
2 Remove the four shift lever boot retaining screws and remove the boot.
3 Remove the four shift lever cover bolts and remove the cover.
4 Remove the shift lever retaining bolt and remove the shift lever.
5 Installation is the reverse of removal.

3 4WD indicator switch (manual-shift models) - replacement

1 If the vehicle is equipped with air suspension, turn off the air suspension system. The switch is located in the area of the right kick panel (see Chapter 10).

※※ WARNING:

On models with air suspension, electrical power to the air suspension system must be turned off before raising the vehicle

TRANSFER CASE   7C-3

(see Chapter 10). Failure to do so can result in a sudden inflation or deflation of the air springs, causing instability of the vehicle while it's off the ground.

2 Raise the vehicle and place it securely on jackstands.
3 Unplug the electrical connector from the 4WD indicator switch.
4 Unscrew the switch.
5 Installation is the reverse of removal.

4 Shift range selector switch (electric-shift models) - replacement

1 Remove the center trim panel (see Chapter 11).
2 Remove the shift range selector switch knob.
3 Unplug the electrical connector from the shift range selector switch.
4 Remove the two shift range selector switch retaining screws.
5 Remove the shift range selector switch.
6 Installation is the reverse of removal.

5 Electric shift motor - replacement

▶ Refer to illustration 5.6

1 If the vehicle is equipped with air suspension, turn off the air suspension system. The switch is located in the area of the right kick panel (see Chapter 10).

※ WARNING:

On models with air suspension, electrical power to the air suspension system must be turned off before raising the vehicle (see Chapter 10). Failure to do so can result in a sudden inflation or deflation of the air springs, causing instability of the vehicle while it's off the ground.

2 Raise the vehicle and place it securely on jackstands.
3 Unplug the shift motor electrical connector.
4 Remove the wire spacer from the electrical connector for the electric shift motor.
5 Remove the coil wire and pin from the connector.
6 Remove the four electric shift motor bolts (see illustration).
7 Remove the electric shift motor.
8 Installation is the reverse of removal.

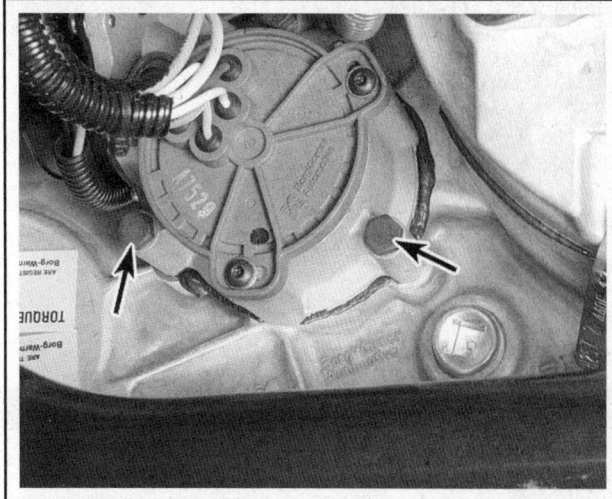

5.6 Remove the four electric shift motor retaining bolts (two are shown by the arrows)

6 Oil seal - replacement

▶ Refer to illustrations 6.4 and 6.6

→Note: This procedure applies to the rear seal only; front seal replacement requires the removal and disassembly of the transfer case.

1 If the vehicle is equipped with air suspension, turn off the air suspension system. The switch is located in the area of the right kick panel (see Chapter 10).

※ WARNING:

On models with air suspension, electrical power to the air suspension system must be turned off before raising the vehicle (see Chapter 10). Failure to do so can result in a sudden inflation or deflation of the air springs, causing instability of the vehicle while it's off the ground.

2 Raise the vehicle and support it securely on jackstands.
3 Remove the rear driveshaft (see Chapter 8).
4 Pry out the seal with a screwdriver or a seal removal tool (see illustration). Don't damage the seal bore.

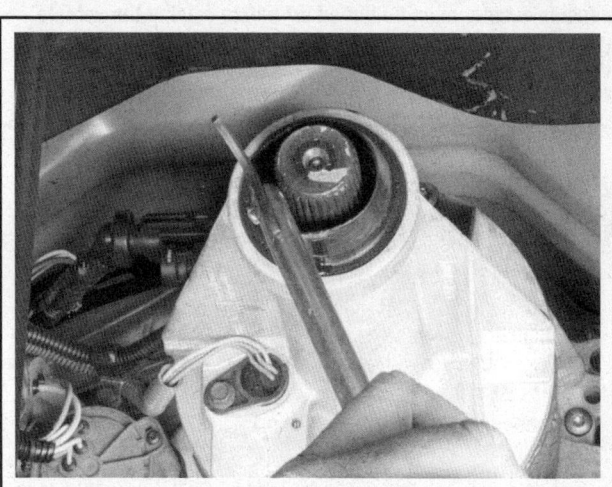

6.4 Pry out the rear transfer case seal with a seal removal tool or large screwdriver

7C-4 TRANSFER CASE

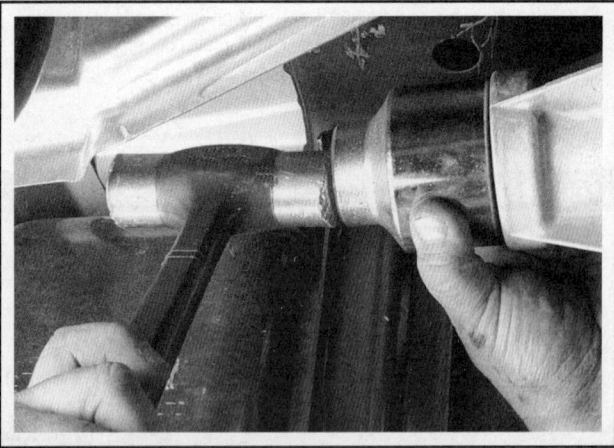

6.6 Drive the new rear transfer case seal into place with a large socket or section of tubing

5 Lubricate the new seal lips with petroleum jelly.
6 Drive the seal into place with a large socket or section of tubing (see illustration). The outside diameter of the socket should be slightly smaller than the outside diameter of the seal.
7 The remainder of installation is the reverse of removal.

7 Transfer case - removal and installation

REMOVAL

1 If the vehicle is equipped with air suspension, turn off the air suspension system. The switch is located in the area of the right kick panel (see Chapter 10).

❄ WARNING:

On models with air suspension, electrical power to the air suspension system must be turned off before raising the vehicle (see Chapter 10). Failure to do so can result in a sudden inflation or deflation of the air springs, causing instability of the vehicle while it's off the ground.

2009 and earlier models

▸ Refer to illustration 7.13

2 Raise the vehicle and support it securely on jackstands.
3 Remove the transfer skid plate bolts, then remove the skid plate.
4 Drain the transfer case lubricant (see Chapter 1).
5 Remove the front driveshaft shield, then remove the front and rear driveshafts (see Chapter 8).
6 Remove the torsion bars and the rear torsion bar support (see Chapter 10).
7 On manual shift vehicles, disconnect the shift rod.
8 Unplug the electrical connector from the vehicle speed sensor.
9 On manual-shift vehicles, unplug the electrical connector from the 4WD indicator switch.

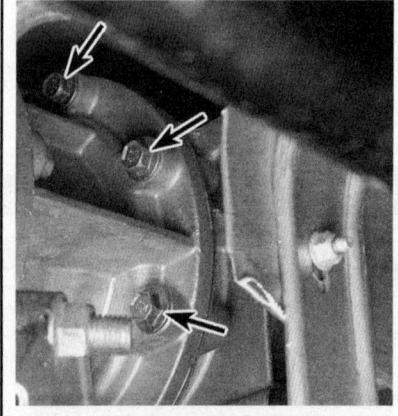

7.13 Remove the transfer case-to-transmission bolts (arrows)

10 On manual-shift vehicles, unplug the electrical connector for the transfer case coil.
11 On electric-shift vehicles, unplug the electric shift motor connector.
12 Support the transfer case with a jack - preferably a special jack made for this purpose. Safety chains or tie-downs will help steady the transfer case on the jack.
13 Remove the transfer case-to-transmission bolts (see illustration).
14 Make a final check that all wires and hoses have been disconnected from the transfer case, then move the transfer case and jack toward the rear of the vehicle until the transfer case is clear of the transmission. Keep the transfer case level as this is done. Once the input shaft is clear, lower the transfer case and remove it from under.

2010 and later models

15 Disconnect the cable from the negative terminal of the battery (see Chapter 5).
16 Raise the vehicle and support it securely on jackstands.
17 Remove the skid plate bolts, then remove the skid plate.
18 Place a transmission jack under the transmission pan.
19 Drain the transfer case lubricant (see Chapter 1).
20 Remove the front and rear driveshafts (see Chapter 8).
21 Disconnect the transfer case wiring harness and the electrical connector for the electric shift motor (see Section 5).
22 Remove the transfer case vent tube.
23 Remove the right and left exhaust shields and remove the EVAP canister bolt (see Chapter 6).
24 Remove the transmission crossmember and transmission mounting bolts (see Chapter 7A or 7B).
25 Support the transfer case with a jack - preferably a special jack made for this purpose. Safety chains or tie-downs will help steady the transfer case on the jack.
26 Remove and discard the transfer case-to-transmission bolts (see illustration 7.13).
27 Make a final check that all wires and hoses have been disconnected from the transfer case, then move the transfer case and jack toward the rear of the vehicle until the transfer case is clear of the transmission. Keep the transfer case level as this is done. Once the input shaft is clear, lower the transfer case and remove it from under the vehicle.
28 Remove all traces of old gasket material from the transfer case-to-transmission mounting surfaces.

INSTALLATION

➡**Note: On models with aluminum transfer case-to-transmission bolts, do not install a gasket.**

29 Installation is the reverse of removal. Tighten the NEW transfer case-to-transmission bolts to the torque listed in this Chapter's Specifications. Refill the transfer case with the recommended type and quantity of fluid (see Chapter 1).

Torque specifications — Ft-lbs (unless otherwise indicated)

Note: One foot-pound (ft-lb) of torque is equivalent to 12 inch-pounds (in-lbs) of torque. Torque values below approximately 15 ft-lbs are expressed in inch-pounds, since most foot-pound torque wrenches are not accurate at these smaller values.

Transfer case-to-transmission bolts
- 2007 and earlier models — 30 to 40
- 2008 and 2009 models — 15
- 2010 and later models — 150 in-lbs

7C-6 TRANSFER CASE

Notes

Section
1 General information
2 Clutch - description and check
3 Clutch master cylinder - removal, inspection and installation
4 Clutch release bearing - removal, inspection and installation
5 Clutch release cylinder - removal, inspection and installation
6 Clutch components - removal, inspection and installation
7 Pilot bearing - inspection and replacement
8 Clutch hydraulic system - bleeding
9 Clutch pedal position switch - check and replacement
10 Driveshafts and universal joints - general information
11 Driveline inspection
12 Driveshaft - removal and installation
13 Universal joints - replacement
14 Driveshaft slip yoke boot and center bearing (two-piece driveshafts) - inspection and replacement
15 Axles - description and check
16 Axleshaft (rear) - removal and installation
17 Axleshaft oil seal (rear) - replacement
18 Axleshaft bearing (rear) - replacement
19 Differential pinion seal - replacement
20 Axle/differential (rear) - removal and installation
21 Driveaxles - removal and installation
22 Driveaxle boot replacement (4WD models)
23 Vacuum shift system (4WD models) - description, check and component replacement
24 Axle (front) (4WD models) - removal and installation

Reference to other Chapters
Differential lubricant change - See Chapter 1
Differential lubricant level check - See Chapter 1
Driveaxle CV joint boot check (4WD models) - See Chapter 1

8
CLUTCH AND DRIVELINE

8-2 CLUTCH AND DRIVELINE

1 General information

The Sections in this Chapter deal with the components from the rear of the engine to the rear wheels (except for the transmission and transfer case, which are dealt with in Chapter 7), and to the front wheels on four-wheel drive (4WD) models. In this Chapter, the components are grouped into three categories: clutch, driveshaft(s) and axle(s). Separate Sections within this Chapter cover checks and repair procedures for components in each of these three groups.

Since nearly all these procedures involve working under the vehicle, make sure it's safely supported on sturdy jackstands or a hoist where the vehicle can be safely raised and lowered.

2 Clutch - description and check

1 All vehicles with a manual transmission have a single dry plate, diaphragm spring-type clutch. The clutch disc has a splined hub which allows it to slide along the splines of the transmission input shaft. The clutch and pressure plate are held in contact by spring pressure exerted by the diaphragm in the pressure plate.

2 The clutch release system is operated by hydraulic pressure. The hydraulic release system consists of the clutch pedal, a master cylinder and fluid reservoir, a release (or slave) cylinder and the hydraulic line connecting the two components. The release cylinder and release bearing are integral parts of a single assembly which is installed concentric to the input shaft and is bolted to the front of the transmission.

3 When the clutch pedal is depressed, a pushrod pushes against brake fluid inside the master cylinder, applying hydraulic pressure against the release bearing, which pushes against the diaphragm fingers of the clutch pressure plate.

4 Terminology can be a problem when discussing the clutch components because common names are in some cases different from those used by the manufacturer. For example, the driven plate is also called the clutch plate or disc, the clutch release bearing is sometimes called a throwout bearing, the release cylinder is sometimes called the slave cylinder.

5 Unless you're replacing components with obvious damage, do these preliminary checks to diagnose clutch problems:
  a) The first check should be of the fluid level in the clutch master cylinder. If the fluid level is low, add fluid as necessary and inspect the hydraulic system for leaks. If the master cylinder reservoir is dry, bleed the system as described in Section 8 and recheck the clutch operation.
  b) To check "clutch spin-down time," run the engine at normal idle speed with the transmission in Neutral (clutch pedal up - engaged). Disengage the clutch (pedal down), wait several seconds and shift the transmission into Reverse. No grinding noise should be heard. A grinding noise would most likely indicate a bad pressure plate or clutch disc.
  c) To check for complete clutch release, run the engine (with the parking brake applied to prevent vehicle movement) and hold the clutch pedal approximately 1/2-inch from the floor. Shift the transmission between 1st gear and Reverse several times. If the shift is rough, component failure is indicated.
  d) Visually inspect the pivot bushing at the top of the clutch pedal to make sure there's no binding or excessive play.

3 Clutch master cylinder - removal, inspection and installation

REMOVAL

▶ Refer to illustrations 3.1, 3.3, 3.6 and 3.8

1 Working under the dash, disconnect the clutch master cylinder pushrod from the clutch pedal (see illustration).

2 Disconnect the clutch pedal position switch (see Section 9).

3 Remove the two clutch master cylinder reservoir push pins (see illustration).

4 If the vehicle is equipped with air suspension, turn off the air

3.1 To disconnect the clutch master cylinder pushrod from the pin on the clutch pedal lever, pry it off with a screwdriver

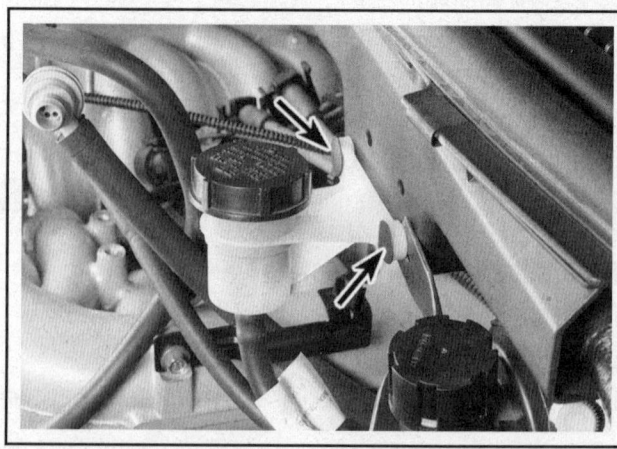

3.3 To detach the clutch master cylinder reservoir from the firewall, remove these two "push pins" (arrows)

CLUTCH AND DRIVELINE 8-3

suspension system. The switch is located in the area of the right kick panel (see Chapter 10).

※ WARNING:

On models with air suspension, electrical power to the air suspension system must be turned off before raising the vehicle. Failure to do so can result in a sudden inflation or deflation of the air springs, causing instability of the vehicle while it's off the ground.

5 Raise the vehicle and place it securely on jackstands.
6 Disconnect the clutch fluid hydraulic line from the transmission (see illustration). Have rags handy, as some fluid will be lost as the line is removed.

※ CAUTION:

Don't allow brake fluid to come into contact with the paint - it will damage the finish.

Also have a plug ready and immediately plug the line to prevent leakage.
7 Lower the vehicle.
8 To detach the master cylinder from the firewall, rotate the master cylinder clockwise 45-degrees (see illustration).

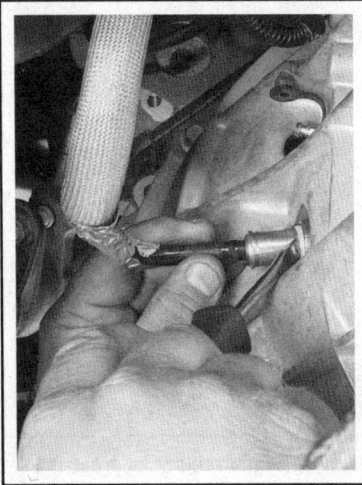

3.6 If you don't have a clutch coupling tool to disconnect the clutch fluid hydraulic line from the transmission, use a screwdriver instead - carefully push the spring-loaded transmission side of the coupling toward the transmission and simultaneously pull straight out on the hydraulic line side of the fitting

INSPECTION

▸ Refer to illustrations 3.9a, 3.9b, 3.9c, 3.9d and 3.9e

9 Disassemble the master cylinder, wash the parts in clean brake fluid and inspect them for damage and wear (see illustrations). If the

3.8 To detach the master cylinder from the firewall, rotate it clockwise 45-degrees and pull it out

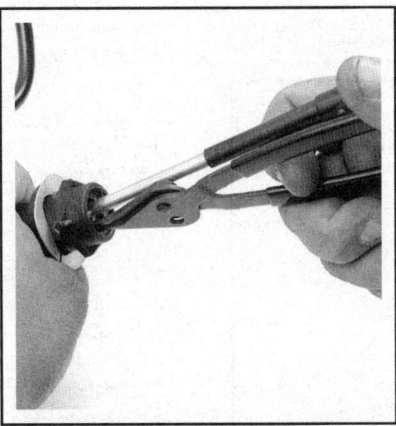

3.9a Depress the master cylinder pushrod and remove the snap-ring with a pair of snap-ring pliers

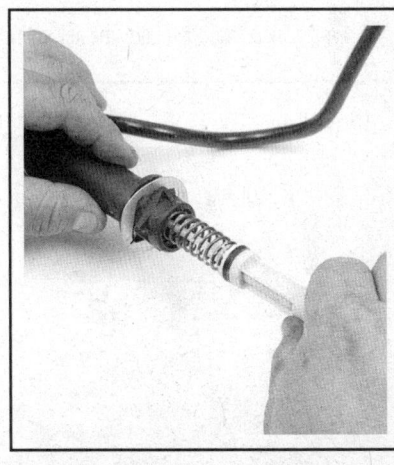

3.9b Pull out the piston and spring assembly

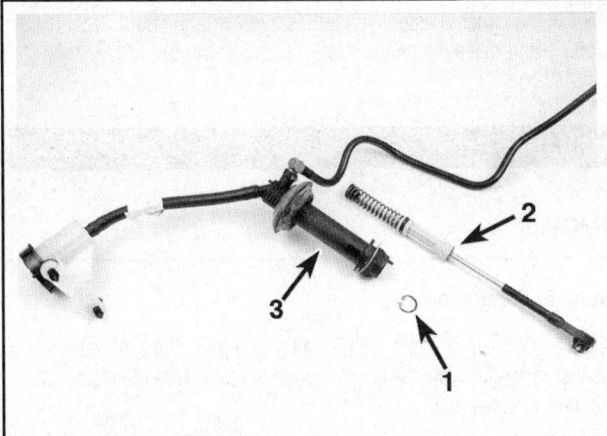

3.9c After disassembling the clutch master cylinder, wash all the parts in clean brake fluid and lay them out for inspection

1 Snap-ring
2 Pushrod/piston/spring assembly
3 Reservoir/master cylinder/hydraulic line assembly

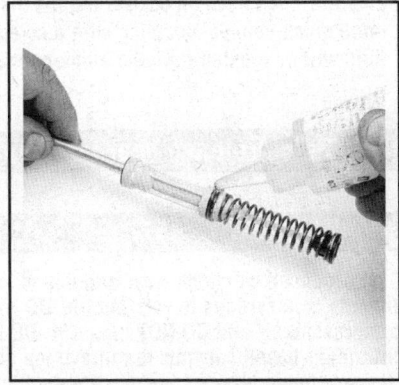

3.9d Before reassembling the master cylinder, lubricate the piston and cups with clean brake fluid or brake assembly lube

8-4 CLUTCH AND DRIVELINE

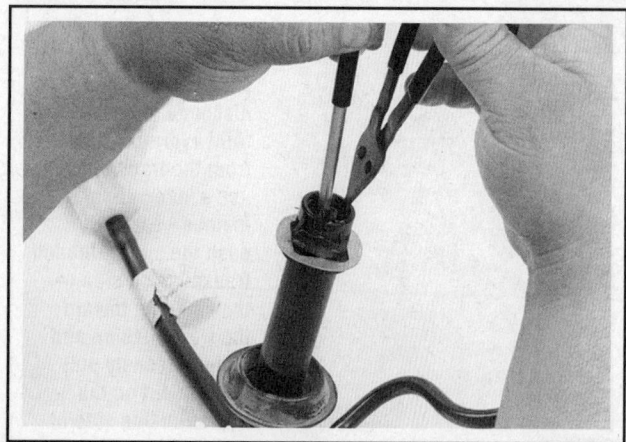

3.9e Install the pushrod/piston/spring assembly, depress the pushrod and install the snap-ring with a pair of snap-ring pliers

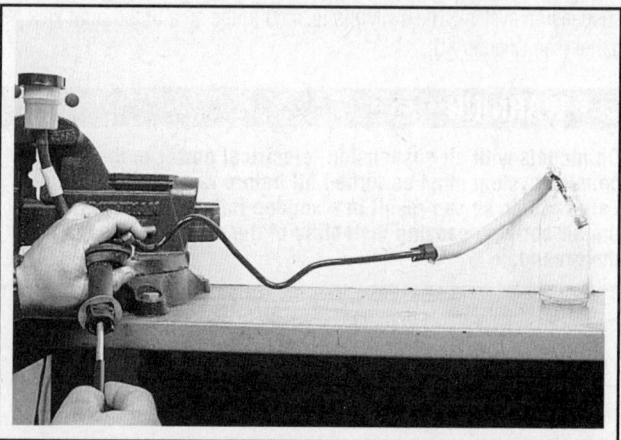

3.10a Secure the master cylinder reservoir in a bench vise as shown; put the release cylinder end of the hydraulic line in a container, fill the clutch master cylinder to the "Full" line . . .

spring is weak, the cups are damaged or worn, or the piston is scored, replace the master cylinder.

INSTALLATION

◆ Refer to illustrations 3.10a and 3.10b

10 Before installing the clutch master cylinder, bleed it as follows:

a) Secure the master cylinder reservoir in a bench vise (see illustration).
b) Fill the clutch master cylinder to the "Full" line.
c) Put the release cylinder end of the hydraulic line in a container of clean brake fluid.
d) Depress and hold the clutch master cylinder pushrod, then have an assistant open the internal fitting of the male quick-connect coupling, allowing fluid to flow out (see illustration).
e) When the stream of fluid subsides, release the internal fitting of the male quick-connect fitting, then release the master cylinder pushrod.
f) Repeat this procedure until a solid stream of brake fluid, with no air bubbles, is discharged from the quick-connect fitting. Check and refill the fluid reservoir, as necessary, during this procedure.

11 Install the master cylinder in the firewall and lock it into place by rotating it 45-degrees counterclockwise (don't rotate it any further or you'll damage it).

12 Install the fluid reservoir, using new push pins.

13 Working inside the vehicle, connect the clutch pedal position switch and connect the clutch master cylinder pushrod to the clutch pedal.

14 Connect the hydraulic line to the transmission with the coupling tool.

15 Lower the vehicle and reactivate the air suspension system, if equipped.

16 Fill the clutch master cylinder reservoir with the fluid specified in Chapter 1 and bleed the clutch system (see Section 8).

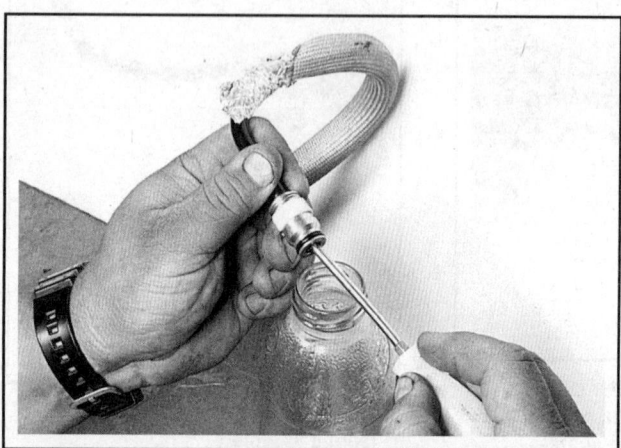

3.10b . . . depress and hold the clutch master cylinder pushrod, then have an assistant open the internal fitting of the male quick-connect coupling with a screwdriver - a stream of fluid will be ejected from the quick-connect coupling

4 Clutch release bearing - removal, inspection and installation

✶✶ WARNING:

Dust produced by clutch wear and deposited on clutch components is hazardous to your health. DO NOT blow it out with compressed air and DO NOT inhale it. DO NOT use gasoline or petroleum-based solvents to remove the dust. Brake system cleaner should be used to flush the dust into a drain pan. After the clutch components are wiped clean with a rag, dispose of the contaminated rags and cleaner in a covered, marked container.

REMOVAL

◆ Refer to illustration 4.4

1 If the vehicle is equipped with air suspension, turn off the air suspension system. The switch is located in the area of the right kick panel (see Chapter 10).

CLUTCH AND DRIVELINE

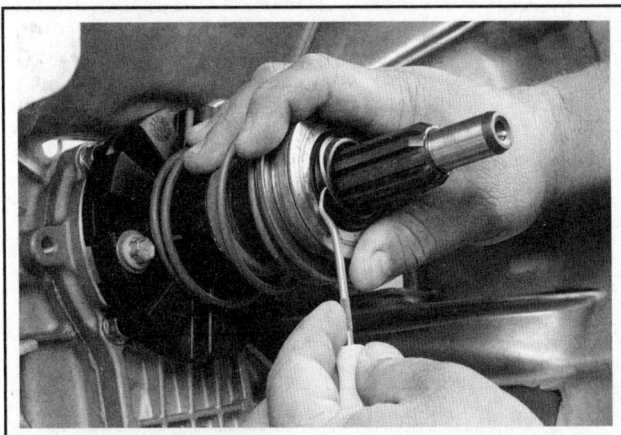

4.4 To disengage the clutch release bearing from the release cylinder, push the release bearing against spring pressure and remove the retaining ring

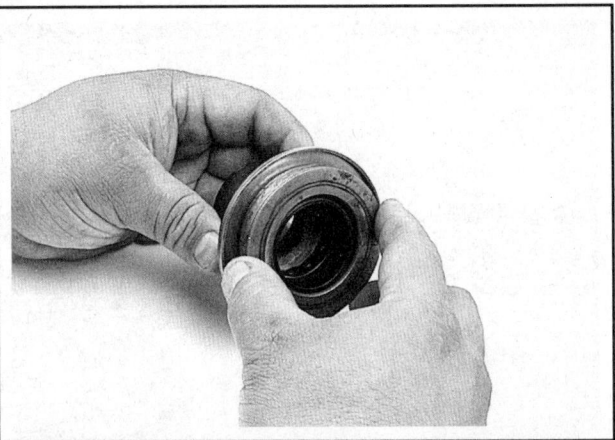

4.6 To check the clutch release bearing, hold the hub (the center) of the bearing and rotate the outer portion while applying pressure; if the bearing doesn't turn smoothly or if it's noisy or rough, replace it

✳✳ WARNING:

On models with air suspension, electrical power to the air suspension system must be turned off before raising the vehicle. Failure to do so can result in a sudden inflation or deflation of the air springs, causing instability of the vehicle while it's off the ground.

2 Raise the vehicle and support it securely on jackstands.
3 Remove the transmission (see Chapter 7A).
4 Push the clutch release hub and bearing against the spring and remove the retaining ring (see illustration).

INSPECTION

▶ **Refer to illustration 4.6**

5 Wipe off the bearing with a clean rag and inspect it for damage, wear and cracks. Don't immerse the bearing in solvent - it's sealed for life and immersion in solvent will ruin it. Look for surface scoring or burrs that might impede the sliding motion of the release bearing. Small imperfections can be removed with fine-grade emery cloth. If the damage is more serious, replace the bearing.

6 Hold the center of the bearing and rotate the outer portion while applying pressure (see illustration). If the bearing doesn't turn smoothly or if it's noisy or rough, replace it.

➡ **Note: Considering the difficulty involved with replacing the release bearing, we recommend replacing the release bearing whenever the clutch components are replaced.**

7 While the release bearing is removed, inspect the release cylinder as well (see Section 5).

INSTALLATION

8 Lightly lubricate the friction surfaces of the release bearing with high-temperature grease.
9 Position the spring and the release bearing on the release cylinder, push the release bearing against the spring and install the retainer ring.
10 Install the transmission (see Chapter 7A).
11 Remove the jackstands and lower the vehicle.
12 Reactivate the air suspension system, if equipped.
13 Bleed the clutch hydraulic release system (see Section 8).

5 Clutch release cylinder - removal, inspection and installation

▶ **Refer to illustration 5.6**

1 If the vehicle is equipped with air suspension, turn off the air suspension system. The switch is located in the area of the right kick panel (see Chapter 10).

✳✳ WARNING:

On models with air suspension, electrical power to the air suspension system must be turned off before raising the vehicle. Failure to do so can result in a sudden inflation or deflation of the air springs, causing instability of the vehicle while it's off the ground.

2 Raise the vehicle and support it securely on jackstands.
3 Remove the transmission (see Chapter 7A).
4 If you're planning to replace only the release cylinder itself, remove the clutch release bearing (see Section 4). If you're planning to replace the release bearing and the release cylinder, don't bother to remove the release bearing.
5 Inspect the clutch release cylinder for leaks. If you see brake fluid, replace the release cylinder.

8-6 CLUTCH AND DRIVELINE

5.6 Remove the clutch release cylinder bolts (arrows) and remove the release cylinder; note the bleeder screw location (arrow)

6 Remove the clutch release cylinder bolts (see illustration) and remove the release cylinder.

INSPECTION

▸ **Refer to illustration 5.7**

7 Clean off the release cylinder and inspect the parts (see illustration). If the spring is weak, the piston is worn or damaged, or the boot is cracked, worn or leaking, replace the release cylinder with a new or rebuilt unit.

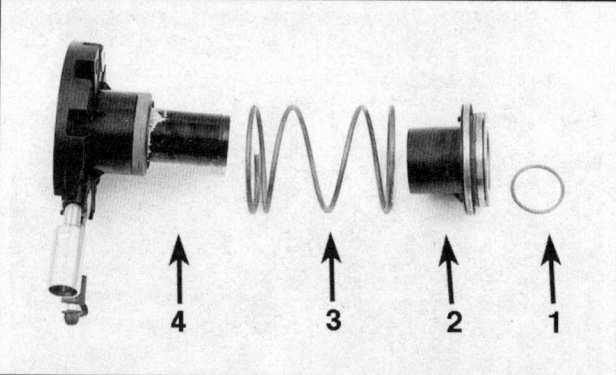

5.7 Clean off the release cylinder, disassemble it and lay out the parts for inspection:

1 Retainer
2 Release bearing
3 Spring
4 Release cylinder assembly

INSTALLATION

8 Position the clutch release cylinder on the input shaft, install the release cylinder retaining bolts and tighten them to the torque listed in this Chapter's Specifications.
9 Install the clutch release bearing (see Section 4).
10 Install the transmission (see Chapter 7A).
11 Remove the jackstands and lower the vehicle.
12 Reactivate the air suspension system, if equipped.
13 Fill the clutch fluid reservoir with the recommended fluid (see Chapter 1).
14 Bleed the clutch hydraulic system (see Section 8).

6   Clutch components - removal, inspection and installation

✳✳ WARNING:

Dust produced by clutch wear and deposited on clutch components is hazardous to your health. DO NOT blow it out with compressed air and DO NOT inhale it. DO NOT use gasoline or petroleum-based solvents to remove the dust. Brake system cleaner should be used to flush the dust into a drain pan. After the clutch components are wiped clean with a rag, dispose of the contaminated rags and cleaner in a covered, marked container.

REMOVAL

▸ **Refer to illustrations 6.4, 6.5 and 6.6**

➡ **Note 1:** The clutch components are normally accessed by removing the transmission. However, anytime the engine is removed, check the clutch for wear and replace worn components as necessary. The relatively low cost of the clutch components compared to the time and trouble spent gaining access to them warrants their replacement anytime the engine or transmission is removed, unless they are new or in near-perfect condition. The following procedures are based on the assumption the engine will stay in place.

➡ **Note 2:** Some pressure plates require adjustment. Check with your clutch pressure plate manufacturer.

1 If the vehicle is equipped with air suspension, turn off the air suspension system. The switch is located in the area of the right kick panel (see Chapter 10).

✳✳ WARNING:

On models with air suspension, electrical power to the air suspension system must be turned off before raising the vehicle. Failure to do so can result in a sudden inflation or deflation of the air springs, causing instability of the vehicle while it's off the ground.

2 Raise the vehicle and support it securely on jackstands.
3 Remove the transmission (see Chapter 7A). Support the engine while the transmission is out. An engine hoist should be used to support it from above. If you use a jack underneath the engine instead, make sure a piece of wood is positioned between the jack and oil pan to spread the load.

✳✳ CAUTION:

The pick-up for the oil pump is very close to the bottom of the oil pan. If the pan is bent or distorted in any way, engine oil starvation could occur.

CLUTCH AND DRIVELINE 8-7

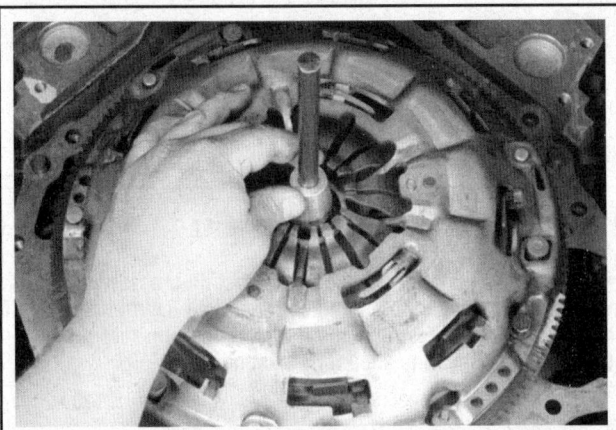

6.4 Use a clutch alignment tool to hold the clutch plate prior to loosening or to center it before tightening the pressure plate bolts

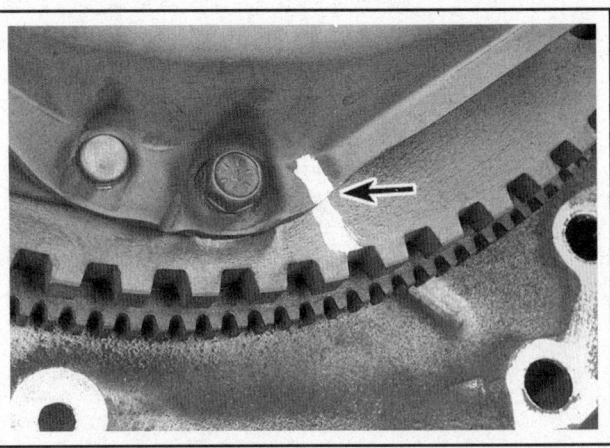

6.5 If you're going to re-use the same pressure plate, mark its relationship to the flywheel (arrow)

4 To support the clutch disc during removal, install an alignment tool through the clutch disc hub (see illustration).

5 Carefully inspect the flywheel and pressure plate for indexing marks. The marks are usually an X, an O or a white letter. If they cannot be found, paint a mark so the pressure plate and the flywheel will be in the same alignment during installation (see illustration).

6 Loosen the six pressure plate-to-flywheel bolts (see illustration) in 1/4-turn increments until they can be removed by hand. Work in a criss-cross pattern until all spring pressure is relieved, then hold the pressure plate securely and completely remove the bolts, followed by the pressure plate and clutch disc.

INSPECTION

▶ Refer to illustrations 6.8, 6.10 and 6.12

7 Ordinarily, when a problem occurs in the clutch, it can be attrib-

uted to wear of the clutch driven plate assembly (clutch disc). However, all components should be inspected at this time.

➥**Note: If the clutch components are contaminated with oil, there will be shiny, black glazed spots on the clutch disc lining, which will cause the clutch to slip. Replacing clutch components won't completely solve the problem - be sure to check the crankshaft rear oil seal and the transmission input shaft seal for leaks. If it looks like a seal is leaking, be sure to install a new one to avoid the same problem with the new clutch.**

8 Check the flywheel for cracks, heat checking, grooves and other obvious defects (see illustration). If the imperfections are slight, a machine shop can machine the surface flat and smooth, which is highly recommended regardless of the surface appearance. Refer to Chapter 2, Part A, for the flywheel removal and installation procedure.

9 Inspect the pilot bearing (see Section 7).

10 Check the lining on the clutch disc. There should be at least 1/16-inch of lining above the rivet heads. Check for loose rivets,

6.6 Loosen the six pressure plate-to-flywheel bolts (arrows) in 1/4-turn increments until they can be removed by hand. Work in a criss-cross pattern until all spring pressure is relieved, then hold the pressure plate securely and completely remove the bolts, followed by the pressure plate and clutch disc

6.8 Inspect the surface of the flywheel for cracks, dark-colored areas (signs of overheating) and other obvious defects; resurfacing will correct minor defects (the surface of this flywheel is in fairly good condition, but resurfacing is always a good idea)

8-8 CLUTCH AND DRIVELINE

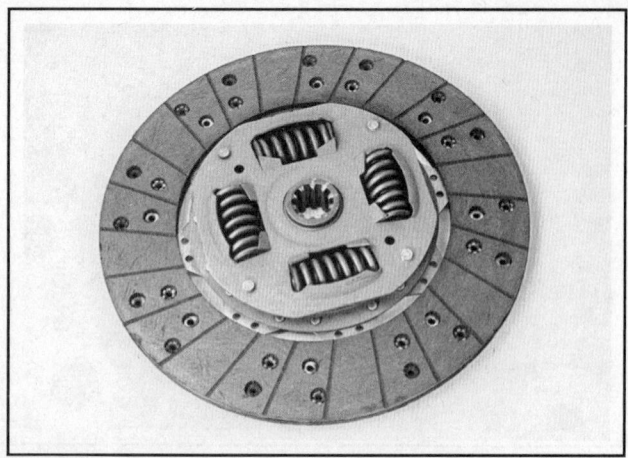

6.10 Inspect the clutch plate lining, springs and splines for wear

distortion, cracks, broken springs and other obvious damage (see illustration). As mentioned above, ordinarily the clutch disc is routinely replaced, so if you're in doubt about its condition, replace it.

11 The release bearing should also be replaced along with the clutch disc (see Section 4).

12 Check the machined surfaces and the diaphragm spring fingers of the pressure plate (see illustration). If the surface is grooved or otherwise damaged, replace the pressure plate. Also check for obvious damage, distortion, cracks, etc. Light glazing can be removed with medium-grit emery cloth. If a new pressure plate is required, new and factory-rebuilt units are available.

INSTALLATION

▶ Refer to illustrations 6.14, 6.15a and 6.15b

➡ Note: Before installation, the self-adjusting pressure plate must be pre-adjusted, and to do so requires a hydraulic press. If you don't have a press, take the flywheel and pressure plate to an automotive machine shop for adjustment.

13 Remove the flywheel (see Chapter 2), if you haven't already done so.

14 Position the pressure plate on the flywheel, place both of them in a hydraulic press, then depress the clutch diaphragm fingers until the adjusting ring moves freely (see illustration).

15 Using a screwdriver, move the adjusting ring counterclockwise until the tension springs are compressed to the dimension listed in this Chapter's Specifications (see illustrations).

16 Holding the adjusting ring in place, release pressure on the diaphragm fingers.

17 Remove the flywheel and pressure plate from the press and clean their machined surfaces with lacquer thinner or acetone. It's important that no oil or grease is on these surfaces or the lining of the clutch

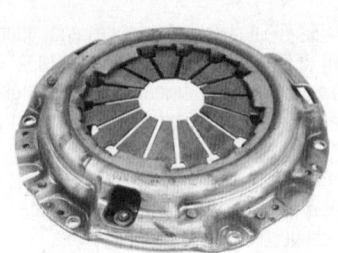
NORMAL FINGER WEAR

EXCESSIVE FINGER WEAR

BROKEN OR BENT FINGERS

6.12 Replace the pressure plate if excessive wear or damage is noted

6.14 Position the pressure plate on the flywheel and place both components in a hydraulic press as shown, then, using a suitable adapter, depress the clutch diaphragm fingers until the adjusting ring (arrow) moves freely

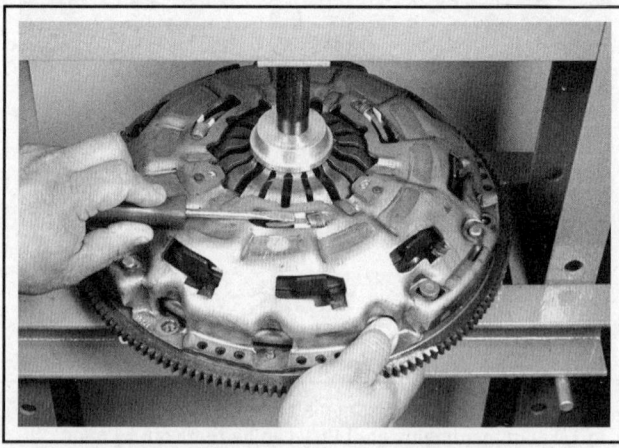

6.15a Using a screwdriver, move the adjusting ring counterclockwise . . .

disc. Handle the parts only with clean hands.

18 Position the clutch disc and pressure plate against the flywheel with the clutch held in place with an alignment tool (see illustration 6.4). Make sure it's installed properly (most replacement clutch plates will be marked "flywheel side" or something similar - if it's not marked, install the clutch disc with the damper springs toward the transmission).

19 Tighten the pressure plate-to-flywheel bolts only finger-tight, working around the pressure plate.

20 Center the clutch disc by ensuring the alignment tool extends through the splined hub and into the pilot bearing in the crankshaft. Wiggle the tool up, down or from side-to-side as needed to bottom the tool in the pilot bearing. Tighten the pressure plate-to-flywheel bolts a little at a time, working in a criss-cross pattern, to prevent distorting the cover. After all the bolts are snug, tighten them to the torque listed in this Chapter's Specifications. Remove the alignment tool.

21 Using high-temperature grease, lubricate the inner groove of the release bearing (see Section 6). Also place grease on the release lever contact areas and the transmission input shaft bearing retainer.

22 Install the clutch release cylinder (see Section 5) and the clutch release bearing, if removed (see Section 4).

23 Install the transmission and all components removed previously.

6.15b . . . until the tension springs are compressed to the specified dimension

24 Remove the jackstands and lower the vehicle.
25 Reactivate the air suspension system, if equipped (see Chapter 10).

7 Pilot bearing - inspection and replacement

▶ **Refer to illustrations 7.5, 7.9 and 7.10**

1 The clutch pilot bearing is a needle roller type bearing which is pressed into the rear of the crankshaft. It's greased at the factory and does not require additional lubrication. Its primary purpose is to support the front of the transmission input shaft. The pilot bearing should be inspected whenever the clutch components are removed from the engine. Because of its inaccessibility, replace it with a new one if you have any doubt about its condition.

➡ **Note: If the engine has been removed from the vehicle, disregard the following Steps which don't apply.**

2 Remove the transmission (see Chapter 7A).
3 Remove the clutch components (see Section 6).
4 Using a flashlight, inspect the bearing for excessive wear, scoring, dryness, roughness and any other obvious damage. If any of these conditions are noted, replace the bearing.

5 Removal can be accomplished with a special puller available at most auto parts stores (see illustration), but an alternative method also works very well.

6 Find a solid steel bar which is slightly smaller in diameter than the bearing. Alternatives to a solid bar would be a wood dowel or a socket with a bolt fixed in place to make it solid.

7 Check the bar for fit - it should just slip into the bearing with very little clearance.

8 Pack the bearing and the area behind it (in the crankshaft recess) with heavy grease. Pack it tightly to eliminate as much air as possible.

9 Insert the bar into the bearing bore and strike the bar sharply with a hammer, which will force the grease to the back side of the bearing and push it out (see illustration). Remove the bearing and clean all grease from the crankshaft recess.

7.5 One way to remove the pilot bearing is with a special puller designed for the job

7.9 You can also remove the pilot bearing by packing the recess behind the bearing with heavy grease and forcing it out hydraulically with a steel rod slightly smaller than the bore in the bearing - when the hammer strikes the rod, the bearing will pop out of the crankshaft

8-10 CLUTCH AND DRIVELINE

7.10 Tap the bearing into place with a bushing driver or a socket slightly smaller than the outside diameter of the bearing

10 To install the new bearing, lightly lubricate the outside surface with grease, then drive it into the recess with a soft-face hammer (see illustration). The seal must face out (toward the transmission).

11 Install the clutch components, transmission and all other components removed previously. Tighten all fasteners to the recommended torque.

8 Clutch hydraulic system - bleeding

▶ Refer to illustrations 8.6a, 8.6b and 8.7

1 The hydraulic system should be bled to remove all air anytime some part of the system has been removed or the fluid level has been allowed to fall so low that air has been drawn into the master cylinder.

2 Fill the clutch master cylinder reservoir with new brake fluid conforming to DOT 3 specifications.

✲✲ CAUTION:

Do not re-use any of the fluid coming from the system during the bleeding operation or use fluid which has been inside an open container for an extended period of time.

3 Quickly depress the clutch pedal five to ten times. Wait one to three minutes, then repeat this procedure again. Wait another one to three minutes and repeat the procedure again. Wait, then do it one more time.

4 If the vehicle is equipped with air suspension, turn off the air suspension system. The switch is located in the area of the right kick panel (see Chapter 10).

✲✲ WARNING:

On models with air suspension, electrical power to the air suspension system must be turned off before raising the vehicle. Failure to do so can result in a sudden inflation or deflation of the air springs, causing instability of the vehicle while it's off the ground.

5 Raise the vehicle and support it securely on jackstands.

6 Attach a hose to the bleeder screw (see illustrations) and place the other end of the hose into a container partially filled with clean brake fluid (make sure the end of the hose is submerged). Loosen the bleeder screw then have an assistant depress the clutch pedal to the

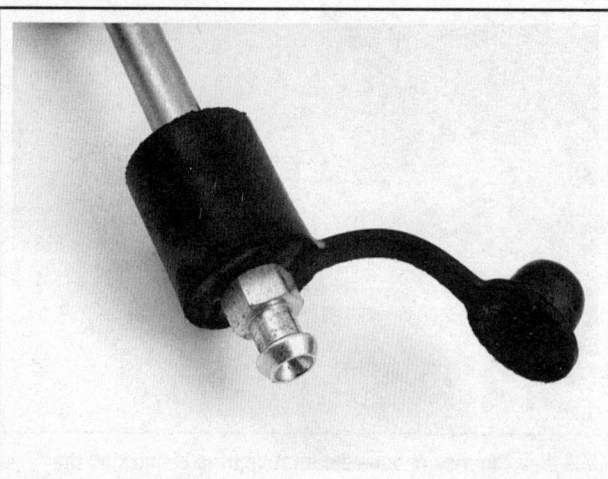

8.6a The bleeder screw for the clutch release cylinder looks just like the bleeder on a brake caliper or wheel cylinder, and is loosened the same way

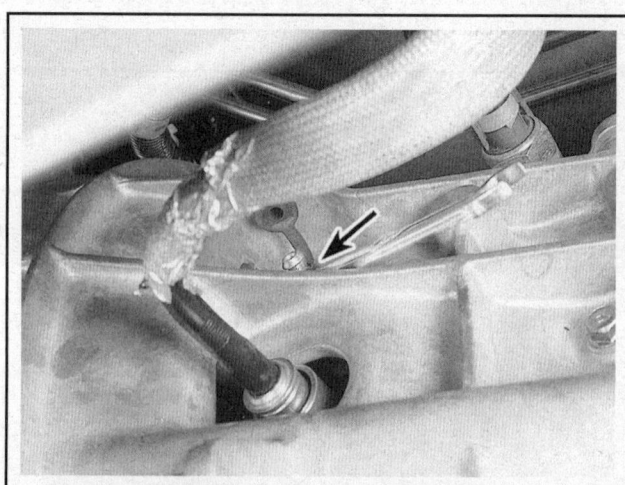

8.6b Locate the bleeder screw (arrow) right above the release cylinder hydraulic line fitting on the left side of the transmission

CLUTCH AND DRIVELINE

floor and hold it down. Tighten the bleeder screw, then have your assistant release the clutch pedal.

7 Depress the clutch pedal, measure pedal travel (see illustration) and compare your measurement to the clutch pedal travel listed in this Chapter's Specifications. If pedal travel is within the specified range, the clutch hydraulic system is properly bled. If pedal travel is not within the specified range, repeat the bleeding procedure.

8.7 To verify that the clutch hydraulic system has been properly bled, depress the clutch pedal, measure pedal travel and compare your measurement to the clutch pedal travel listed in this Chapter's Specifications; if pedal travel is within the specified range, the clutch hydraulic system is fully bled, but if pedal travel isn't within range, the system isn't yet fully bled

9 Clutch pedal position switch - check and replacement

CHECK

1 The clutch pedal position switch, which is mounted on a bracket at the right end of the clutch pedal shaft, directly above the accelerator pedal, prevents the engine from being started unless the clutch pedal is depressed. It's also an information sensor for the powertrain control module; for more information on its other functions, refer to Chapter 6.

2 If the engine starts without depressing the clutch pedal, check the switch (see "Information sensors" in Chapter 6).

REPLACEMENT

3 Disconnect the clutch master cylinder pushrod from the clutch pedal (see illustration 3.1).

4 Unplug the electrical connector from the clutch pedal position switch.

5 Pull down the retaining clip underneath the switch, push the locking tabs together on top of the switch and slide off the plastic retainer.

6 Remove the switch.

7 Installation is the reverse of removal.

10 Driveshafts and universal joints - general information

1 A driveshaft is a tube, or a pair of tubes, that transmits power between the transmission (or transfer case on 4WD models) and the differential in the rear axle (2WD models) (and, on 4WD models, the differential in the front axle).

UNIVERSAL JOINTS

2 Universal joints are used at both ends of all driveshafts, and right behind the center bearing on two-piece rear driveshafts. A universal joint, or U-joint, is a double-pivoted connection for transmitting power from a driving to a driven shaft through an angle.

3 Most U-joints are the single-cardan type. A single-cardan design is a type of U-joint in which the shaft ends are connected by two Y-shaped yokes at right angles to each other. Torque is transmitted from one yoke to the other via a cross-shaped bearing known as a spider, or cross. Bearings pressed into each of the yoke arms ride on each of the four arms of the spider.

4 A double-cardan U-joint is used on 4WD sport utility models without air suspension. A double-cardan design is simply two single-cardan spiders in series, one after the other.

REAR DRIVESHAFTS

5 All rear driveshafts employ a "slip yoke" (a combination splined slip joint and U-joint) at the front, which slips into the extension housing of the transmission (2WD models) or transfer case (4WD models). This arrangement allows the driveshaft to slide back-and-forth within the transmission during vehicle operation. An oil seal prevents leakage of fluid at this point and keeps dirt from entering the transmission. If leakage is evident at the front of the driveshaft, replace the oil seal (see Chapter 7A or 7C).

6 A center bearing supports the driveline on vehicles with a two-piece rear driveshaft. The center bearing is a ball-type bearing mounted in a rubber cushion attached to the frame. The center bearing is pre-lubricated and sealed at the factory. A slip yoke at the front of the rear driveshaft section allows the rear driveshaft to slide back-and-forth dur-

8-12 CLUTCH AND DRIVELINE

ing vehicle operation. A dust boot over the slip joint prevents dirt and dust from entering the slip joint. If the slip yoke boot is torn or damaged, clean, inspect and lubricate the slip joint, then replace the boot.

7 The rear end of all rear driveshafts is connected to the rear axle differential pinion flange by a flange yoke with four bolts. The pinion shaft uses a seal to prevent lubricant from leaking out of the differential. If you see a leak here, replace the pinion seal.

FRONT DRIVESHAFTS (4WD MODELS)

8 Front driveshafts on 4WD models employ a slip yoke at the *front* end of the driveshaft on trucks, but at the *rear* end on sport utility vehicles. On trucks, the *rear* end of the front driveshaft is equipped with a *double*-cardan (two spiders) U-joint; on sport utility vehicles, the *front* end of the driveshaft is equipped with a *single*-cardan U-joint. All front driveshafts are bolted to the transfer case output shaft companion flange and to the front axle differential pinion flange.

ALL DRIVESHAFTS

9 The driveshaft assembly requires very little service. The universal joints are lubricated for life and must be replaced if problems develop. The driveshaft must be removed from the vehicle for this procedure.

10 Since the driveshaft is a balanced unit, it's important that no undercoating, mud, etc. be allowed to stay on it. When the vehicle is raised for service it's a good idea to clean the driveshaft and inspect it for any obvious damage. Also, make sure the small weights used to originally balance the driveshaft are in place and securely attached. Whenever the driveshaft is removed it must be reinstalled in the same relative position to preserve the balance.

11 Problems with the driveshaft are usually indicated by a noise or vibration while driving the vehicle. A road test should verify if the problem is the driveshaft or another vehicle component. Refer to the *Troubleshooting* Section at the front of this manual. If you suspect trouble, inspect the driveline (see Section 11).

11 Driveline inspection

1 Raise the rear of the vehicle and support it securely on jackstands. Block the front wheels to keep the vehicle from rolling off the stands.

2 Crawl under the vehicle and visually inspect the driveshaft. Look for any dents or cracks in the tubing. If any are found, the driveshaft must be replaced.

3 Check for oil leakage at the front and rear of the driveshaft. Leakage where the driveshaft enters the transmission or transfer case indicates a defective transmission/transfer case seal (see Chapter 7). Leakage where the driveshaft joins the differential indicates a defective pinion seal (see Section 19).

4 While under the vehicle, have an assistant rotate a rear wheel so the driveshaft will rotate. As it does, make sure the universal joints are operating properly without binding, noise or looseness. The universal joint can also be checked with the driveshaft motionless, by gripping your hands on either side of the joint and attempting to twist the joint. Any movement at all in the joint is a sign of considerable wear. Lifting up on the shaft will also indicate movement in the universal joints.

5 Check the driveshaft-to-pinion flange bolts to make sure they're tight.

6 Listen for any noise from the center bearing, if equipped. If the center bearing is worn or damaged, replace it (see Section 14). Inspect the rubber portion of the center bearing support for cracking or separation, and replace it too if necessary. If the slip yoke on the rear driveshaft is leaking, replace the slip yoke boot (see Section 14).

7 On 4WD models, check the front driveshaft as well. Also look for grease leakage around the slip yoke. A leak at the slip yoke on sport utility vehicles indicates failure of the yoke seal. Replace the driveshaft with a new or rebuilt unit. A leaking slip yoke on trucks is easily fixed by replacing the dust boot protecting the splined portion of the driveshaft that slides in and out of the slip yoke.

8 On 4WD models, inspect for leaks where the driveshafts connect to the transfer case and where the front driveshaft connects to the front differential. Leakage indicates worn oil seals. You can easily replace the rear transfer case seal (see Chapter 7C) but not the front transfer case seal, which requires removal and disassembly of the transfer case. To replace the front differential pinion seal, refer to Section 19.

9 On 4WD models, check for looseness in the front driveaxle CV joints. And look for grease in the area in front of, above and behind the driveaxles. If you find grease in this area, it's been flung out of a torn CV joint rubber boot. Oil leakage at the left inner CV joint can indicate a defective side gear oil seal; oil leakage at the right inner CV joint can indicate a leaking axleshaft housing tube seal. Replacement of these seals requires special tools; remove the front axle assembly (see Section 24) and have the seal(s) replaced by a dealer or by a 4WD specialist shop.

12 Driveshaft - removal and installation

1 If the vehicle is equipped with air suspension, turn off the air suspension system. The switch is located in the area of the right kick panel (see Chapter 10).

WARNING:
On models with air suspension, electrical power to the air suspension system must be turned off before raising the vehicle. Failure to do so can result in a sudden inflation or deflation of the air springs, causing instability of the vehicle while it's off the ground.

2 Raise the vehicle and support it securely on jackstands. Place the transmission in Neutral with the parking brake off. Block the front wheels to prevent the vehicle from rolling.

REAR DRIVESHAFT

Removal

▶ Refer to illustrations 12.3 and 12.4

3 Using a scribe, a hammer and punch, or paint, mark the relationship of the driveshaft to the differential pinion flange (see illustration). These marks ensure that the driveshaft's dynamic balance will be restored when the driveshaft is installed again.

CLUTCH AND DRIVELINE 8-13

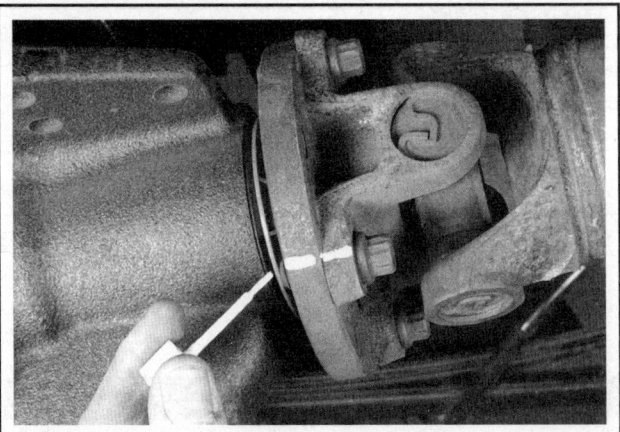

12.3 Mark the relationship of the rear driveshaft U-joint to the differential pinion flange

12.4 Insert a screwdriver through the U-joint to prevent the driveshaft from turning as you break loose the four U-joint-to-pinion flange bolts

4 Remove the rear universal joint bolts (see illustration). Turn the driveshaft (or wheels) as necessary to bring the bolts into the most accessible position.

5 On vehicles with a two-piece driveshaft, remove the two center bearing support bracket bolts.

6 Lower the rear of the driveshaft, slide the slip yoke at the front end of the driveshaft out of the transmission or transfer case, and remove the driveshaft assembly.

7 Wrap a plastic bag over the transmission or transfer case extension housing and hold it in place with a rubber band. This will prevent loss of fluid and protect against contamination while the driveshaft is out.

8 While the driveshaft is removed, check and, if necessary, replace the extension housing seal (see Chapter 7A or 7C) and/or the differential pinion seal (see Section 19). If the driveshaft is a two-piece type, check the center bearing for any damage or wear (see Section 14).

Installation

9 If you removed the center bearing and support bracket from a two-piece driveshaft, install them now and reattach the front and rear parts of the driveshaft (see Section 14). If you removed the old extension housing seal, install a new one now (see Chapter 7).

10 Remove the plastic bag from the transmission or transfer case extension housing and wipe the area clean. Slide the slip yoke at the front of the driveshaft onto the transmission or transfer case output shaft; make sure you don't damage the lip of the seal.

11 On vehicles with a two-piece driveshaft, raise the center bearing assembly into position, install the support bracket bolts and tighten them to the torque listed in this Chapter's Specifications.

12 Raise the rear of the driveshaft into position, checking to be sure the marks are in alignment. If not, turn the rear wheels to match the pinion flange and the driveshaft. Tighten all bolts to the torque listed in this Chapter's Specifications.

FRONT DRIVESHAFT (4WD MODELS)

Removal

▶ Refer to illustrations 12.14 and 12.15

13 Remove the three front driveshaft shield nuts and remove the front driveshaft shield.

14 Mark the relationship of the front driveshaft to the front axle differential pinion flange and to the transfer case output shaft companion

12.14 Mark the yoke at each end of the driveshaft to its companion flange

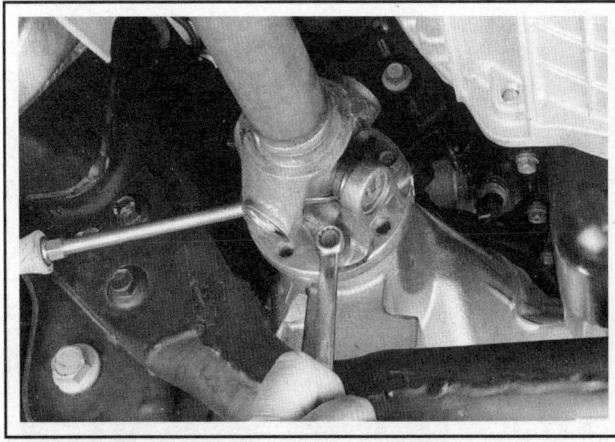

12.15 Remove the bolts that attach the front driveshaft to the front axle differential pinion flange (shown), and the bolts that attach the front driveshaft to the transfer case output shaft companion flange

flange (see illustration).

15 Remove the bolts that attach the front driveshaft to the front axle pinion flange and to the transfer case output shaft companion flange (see illustration).

8-14 CLUTCH AND DRIVELINE

16 Carefully lower and remove the front driveshaft.

17 While the front driveshaft is removed, check and, if necessary, replace the differential pinion seal (see Section 19). Also, check the transfer case output shaft seal. If this seal is leaking, remove the transfer case (see Chapter 7C) and take it to a dealer or a 4WD specialist. The transfer case must be disassembled before this seal can be replaced.

Installation

18 Bolt the front end of the driveshaft to the front axle differential pinion flange and tighten the bolts to the torque listed in this Chapter's Specifications.

19 Extend or compress the driveshaft as necessary, attach the rear end to the transfer case output shaft companion flange and tighten the and bolts to the torque listed in this Chapter's Specifications.

20 Install the front driveshaft shield and tighten the bolts securely.

ALL MODELS

21 Remove the jackstands and lower the vehicle.

22 Reactivate the air suspension system, if equipped (see Chapter 10).

13 Universal joints - replacement

➡ Note: Purchase a universal joint service kit for your model vehicle before beginning this procedure. Also, read through the entire procedure before beginning work.

SINGLE-CARDAN U-JOINTS

▶ Refer to illustrations 13.2, 13.4 and 13.9

➡ Note: A press or large vise will be required for this procedure. It may be advisable to take the driveshaft to a local dealer service department, service station or machine shop where the universal joints can be replaced for you, normally at a reasonable charge.

1 Remove the driveshaft as outlined in the previous Section.

2 Using a small pair of pliers, remove the snap-rings from the spider (see illustration).

3 Supporting the driveshaft, place it in position on a workbench equipped with a vise.

4 Place a piece of pipe or a large socket with the same inside diameter over one of the bearing caps. Position a socket which is of slightly smaller diameter than the cap on the opposite bearing cap (see illustration) and use the vise or press to force the cap out (inside the pipe or large socket), stopping just before it comes completely out of the yoke. Use the vise or large pliers to work the cap the rest of the way out.

5 Transfer the sockets to the other side and press the opposite bearing cap out in the same manner.

6 Pack the new universal joint bearings with grease. Instructions for lubrication may be included with the universal joint servicing kit and should be followed carefully.

7 Position the spider in the yoke and partially install one bearing cap in the yoke.

8 Start the spider into the bearing cap and then partially install the other cap. Align the spider and press the bearing caps into position, being careful not to damage the dust seals.

9 Install the snap-rings. If difficulty is encountered in seating the snap-rings, strike the driveshaft yoke sharply with a hammer. This will

13.2 A pair of needle-nose pliers can be used to remove the universal joint snap-rings

13.4 To press the universal joint out of the driveshaft yoke, set it up in a vise with the small socket pushing the joint and bearing cap into the large socket

13.9 If the snap-ring will not seat in the groove, strike the yoke with a brass hammer - this will relieve the tension that has set up in the yoke and slightly spring the yoke ears (this should also be done if the joint feels tight when assembled)

CLUTCH AND DRIVELINE   8-15

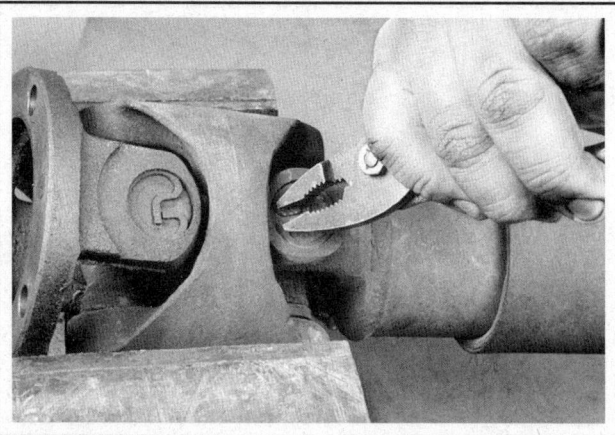

13.12 Remove all the old snap-rings

spring the yoke ears slightly and allow the snap-rings to seat in the groove (see illustration).

10 Install the grease fitting and fill the joint with grease. Be careful not to overfill the joint, as this could blow out the grease seals.

11 Install the driveshaft, tightening the companion flange bolts to the torque listed in this Chapter's Specifications.

DOUBLE-CARDAN U-JOINTS

▶ Refer to illustrations 13.12, 13.13, 13.14a, 13.14b, 13.15a, 13.15b, 13.16a, 13.16b, 13.17, 13.18, 13.19, 13.20a, 13.20b, 13.22a, 13.22b, 13.22c, 13.22d and 13.22e

12 Remove the eight snap-rings that retain the bearing caps (see illustration).

13 Mark the relationship of the companion flange, the center yoke and the driveshaft yoke (see illustration) or, if you're overhauling the front U-joint, the slip yoke, the centering socket yoke, the center yoke and the front driveshaft yoke.

❋❋ CAUTION:

The U-joints must be assembled with these pieces in their original positions so the driveshaft does not become imbalanced.

13.14a Once you have pushed out a cap about 3/8-inch, twist it out with a pair of water pump pliers . . .

13.13 Marking the relationship of the parts to one another is essential, if you want to preserve dynamic balance and fit; if you're rebuilding the rear U-joint (the one depicted in these photos), mark the companion flange, the center yoke and the driveshaft yoke; if you're overhauling the front U-joint, mark the slip yoke, the centering socket yoke, the center yoke and the front driveshaft yoke

14 Starting with the two bearing caps in the companion flange end of the center yoke, position the U-joint in a bench vise, with a smaller socket on one side, to push against the bearing cap and spider, and a larger socket on the other side, with an inside diameter large enough to allow the opposite bearing cap to protrude into it without interference (see illustration 4.13). Tighten the jaws of the vise until the bearing protrudes about 3/8-inch out of the yoke. Loosen the vise jaws, rotate the U-joint 90-degrees so that you can get at the protruding bearing cap, work the cap out with a pair of water pump pliers (see illustration) and remove the cap (see illustration). Then rotate the U-joint another 90-degrees, install the large and small socket again and press out and remove the opposite bearing cap (the one you just pushed in to drive out the first cap). Then do the other two bearing caps for that spider. It's easier to start with the outermost spider, regardless of which U-joint you're overhauling.

13.14b . . . and remove the cap

13.15a Once you have removed the center yoke, you can remove the two bearing caps from the driveshaft yoke in the same fashion: Place a large socket on one side and a smaller socket on the opposite side, then push the cap on the right into the larger socket until it protrudes about 3/8-inch as described above, twist the cap off with a pair of water pump pliers . . .

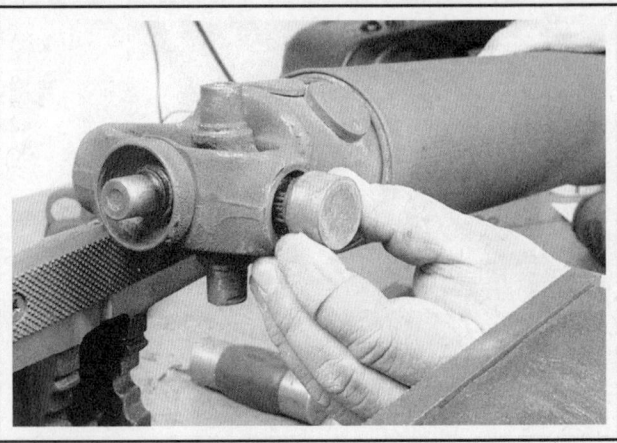

13.15b . . . and remove the cap; then flip the U-joint around and remove the other bearing cap the same way

13.16a Place the companion flange in a bench vise as shown and remove the dust cover and centering ball: Since you're not going to reuse it, you can knock the old dust cover off with a hammer and chisel

13.16b Once you've got the dust cover off, rotate the centering ball as shown, grab it with the bench vise jaws and carefully tap off the companion flange

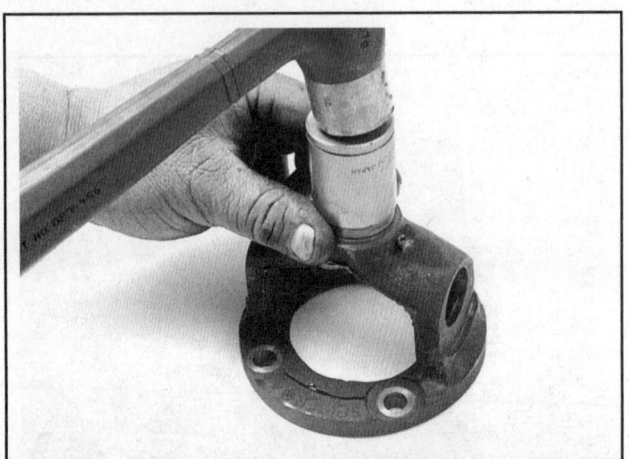

13.17 Carefully tap a new centering ball and dust cover onto the companion flange with a large socket; just make sure you don't damage the bearing surface of the centering ball (Dana units simply use a circular, flat seal, not a dust cover, to protect the centering ball)

15 After you've got the center yoke off, remove the two bearing caps from the driveshaft yoke in the same fashion as described above (see illustrations).

16 Place the companion flange in a bench vise as shown and remove the dust cover and centering ball (see illustrations).

17 Install a new centering ball and dust cover on the companion flange (see illustration).

18 Install the new spider bearing in the driveshaft yoke and secure it with a couple of new bearing caps (see illustration). Make sure the grease fitting on the spider faces away from the center yoke; if the spider is installed with the grease fitting facing toward the center yoke, it could create a clearance problem, and you won't be able to grease the spider. Don't forget to install a new seal with each new bearing cap (see illustration 4.12a).

➡ **Note:** The seals are integral with the caps on some rebuild kits.

19 Place the driveshaft yoke in the bench vise and press the two caps into the yoke until they're flush with the driveshaft yoke (see illustration).

CLUTCH AND DRIVELINE 8-17

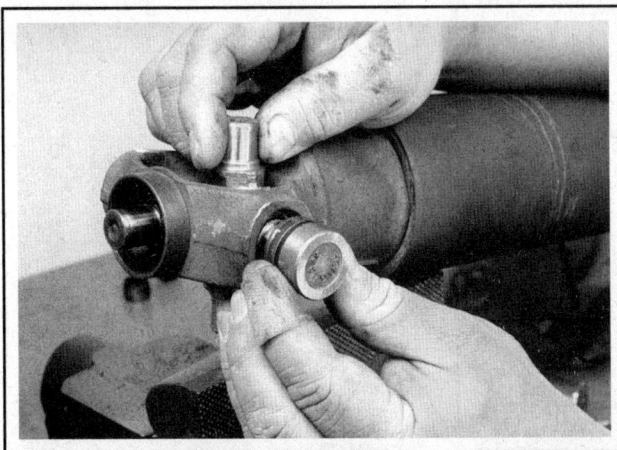

13.18 Install a new spider bearing into the driveshaft yoke and secure it with a couple of new bearing caps (be sure to install a new seal with each new bearing cap unless, of course, the seals are integral with the caps)

13.19 Place the driveshaft yoke in the bench vise and press two opposite caps into the yoke until they're flush with the driveshaft yoke, then install a couple of sockets and press the caps into the yoke until they're fully seated (tops of caps flush with snap-ring grooves)

13.20a Install the center yoke on the other two legs of the new spider (make sure the marks you made on the center yoke and the driveshaft yoke are lined up) . . .

13.20b . . . install the new bearing caps and seals, then press the two caps into the center yoke

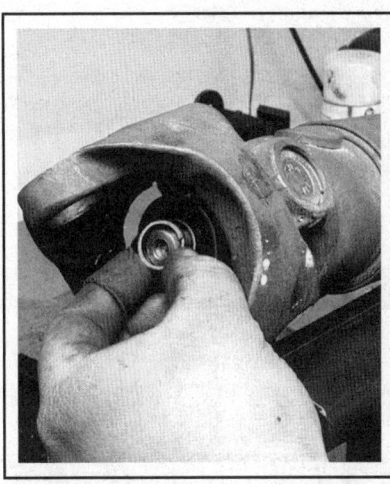

13.22a Install the centering spring and, if applicable, a new dust seal for the centering ball

20 Install the center yoke on the other two legs of the new spider and hold it in place with a couple more new bearing caps and seals (see illustrations). Make sure the marks you made on the center yoke and the driveshaft yoke are lined up. Place the driveshaft in the vise and press the two caps into the center yoke until they're flush with the yoke.

21 Install a pair of sockets between each opposing pair of bearing caps and press in the caps until they're fully seated (the tops of the bearing caps must be flush with the lower edges of the snap-ring grooves in the driveshaft yoke, center yoke and companion flange (or the driveshaft yoke and center yoke, if you're rebuilding a front U-joint).

22 Install the centering spring (see illustration) and a new center seal on the centering stud and guide the centering stud into the centering ball. Connect the companion flange to the other end of the center yoke with a new spider (see illustration). Make sure the grease fitting on the spider faces away from the center yoke (see illustration). If the spider is installed with the grease fitting facing toward the center yoke, it could create a clearance problem, and you won't be able to grease the spider. Tap the new bearing caps into the center yoke to hold everything together (see illustration), press in the caps so they're flush with the

13.22b Attach the companion flange to the center yoke with the other new spider

8-18 CLUTCH AND DRIVELINE

13.22c Make sure the grease fitting on the spider faces away from the center yoke (toward the companion flange) to avoid clearance problems, and to provide access to the fitting for lubrication

13.22d Tap the new bearing caps into place to hold the spider, then press in the caps the same way you did the caps for the other spider, with a pair of sockets

13.22e After the new bearing caps are fully seated (flush with the snap-ring grooves in the yokes), install new snap-rings

center yoke, press the other two caps into the companion flange until they're flush with the companion flange, then press in all four caps with a pair of sockets until they're fully seated. Install new snap-rings (see illustration).

→Note: If difficulty is encountered in seating the snap-rings, strike the driveshaft yoke(s) sharply with a hammer (see illustration 13.9). This will spring the yoke ears slightly and allow the snap-rings to seat in the groove.

14 Driveshaft slip yoke boot and center bearing (two-piece driveshafts) - replacement

1 If the vehicle is equipped with air suspension, turn off the air suspension system. The switch is located in the area of the right kick panel (see Chapter 10).

⁂ WARNING:
On models with air suspension, electrical power to the air suspension system must be turned off before raising the vehicle. Failure to do so can result in a sudden inflation or deflation of the air springs, causing instability of the vehicle while it's off the ground.

2 Raise the vehicle and place it securely on jackstands.
3 Remove the driveshaft (see Section 12).

SLIP YOKE BOOT

4 Put the driveshaft assembly on the bench and cut off the slip yoke boot clamps.
5 Mark the relationship of the slip yoke to the driveshaft. Separate the slip yoke from the splined stub shaft on the driveshaft.
6 Remove the driveshaft slip yoke boot. Inspect the boot for cracks and tears. If necessary, discard it and get a new one.
7 Inspect the slip yoke grease for signs of dirt or water contamination. If the grease is contaminated, pull off the slip yoke, clean off the old grease, and inspect the stub shaft and slip yoke splines for signs of excessive wear or corrosion. If the splines are damaged, replace the driveshaft with a new or rebuilt unit.
8 Install the center bearing, if removed (see below).
9 Slide the end of the new boot with the smaller opening onto the splined stub shaft. Push it on as far as it will go.
10 Attach the boot to the stub shaft with a new boot clamp. You'll need a special crimp clamp pliers, available at most auto parts stores, to crimp the new clamp.
11 Lubricate the stub shaft splines with high temperature grease and slide the slip yoke onto the splines, making sure the marks you made in Step 5 are in alignment.
12 Fill the slip yoke boot with about 10 grams of the same lubricant.
13 Slip the larger boot clamp onto the slip yoke.
14 Slip the boot onto the slip yoke, then adjust the distance from the weld bead on the driveshaft to the centerline of the U-joint to the dimension listed in this Chapter's Specifications.
15 Pry the lip of the boot open with a screwdriver to bleed air from the slip yoke boot, then install the large clamp and crimp it with suitable pliers (see Step 10).
16 Install the driveshaft (see Section 12).
17 Remove the jackstands, lower the vehicle and reactivate the air suspension system, if equipped.

CENTER BEARING

18 The center bearing is pressed onto the driveshaft and cannot be removed without a press and special adapter. Take the driveshaft to an automotive machine shop and have the old bearing pressed off and a new bearing installed. If the rubber support is damaged, have it replaced at this time as well.
19 Install the driveshaft (see Section 12).
20 Remove the jackstands, lower the vehicle and reactivate the air suspension system, if equipped.

CLUTCH AND DRIVELINE   8-19

15 Axles - description and check

DESCRIPTION

1   The rear axle assembly is a hypoid (the centerline of the pinion gear is below the centerline of the ring gear), semi-floating type. When the vehicle goes around a corner, the differential allows the outer rear tire to turn more quickly than the inner tire. The axleshafts are splined to the differential side gears, so when the vehicle goes around a corner, the inner tire, which turns more slowly than the outer tire, turns its side gear more slowly than the outer tire turns its side gear. The differential pinion gears roll around the slower side gear, driving the outer side gear - and tire - more quickly. The differential is housed within a casting (known as a "carrier") with a pressed steel cover. The steel axle tubes are pressed into and welded to the carrier.

2   On 4WD models, a fully independent front axle assembly is used. This consists of a differential and a pair of driveaxles. Each driveaxle has an inner and outer constant velocity (CV) joint. The outer CV joints are splined into the wheel hub and bearing assemblies; the inner CV joints are bolted to the axleshaft flanges. Because the differential is offset to the left (to clear the engine and to align it with the transfer case), the distance between the differential and the right front wheel is greater than the distance from the differential to the left wheel. In order to use two equal-length driveaxles, a longer axleshaft, housed inside a tube, is employed on the right side to make up the difference.

3   An optional locking limited-slip rear axle is also available. This differential allows for normal operation until one wheel loses traction. A limited-slip unit is similar in design to a conventional differential, except for the addition - at either side of the differential side gears - of a series of alternating clutch friction discs and plates (not unlike a modern clutch pack on a motorcycle) which slow the rotation of the differential case when one wheel is on a firm surface and the other on a slippery one. The difference in wheel rotational speed produced by this condition applies additional force to the pinion gears and through the clutch friction discs, which are splined to the axleshafts, equalizes the rotation speed of the axleshaft driving the wheel with traction.

CHECK

4   Often, a suspected "axle" problem lies elsewhere. Do a thorough check of other possible causes before assuming the axle is the problem.

5   The following noises are those commonly associated with axle diagnosis procedures:

   a) Road noise is often mistaken for mechanical faults. Driving the vehicle on different surfaces will determine if the road surface is the cause of the noise. Road noise will remain the same if the vehicle is under power or coasting.
   b) Tire noise is sometimes mistaken for mechanical problems. Tires which are worn or low on pressure are particularly susceptible to emitting vibrations and noises. Tire noise will remain about the same during varying driving situations, where axle noise will change during coasting, acceleration, etc.
   c) Engine and transmission noise can be deceiving because it will travel along the driveline. To isolate engine and transmission noises, make a note of the engine speed at which the noise is most pronounced. Stop the vehicle and place the transmission in Neutral and run the engine to the same speed. If the noise is the same, the axle is not at fault.

6   Because of the special tools needed, overhauling the differential isn't cost effective for a do-it-yourselfer. The procedures included in this Chapter describe axleshaft removal and installation, axleshaft oil seal replacement, axleshaft bearing replacement and removal of the entire unit for repair or replacement. Any further work should be left to a dealer service department or other qualified repair shop.

IDENTIFICATION

7   If the axle must be replaced, refer to the axle identification tag bolted to the differential cover (rear axle) or bolted to the inner end of the right axle tube (front axle). This tag provides the plant code, which denotes the design and specific ratio of the axle, and indicates whether the axle is a conventional design or a locking differential (known as "Traction Loc"). The tag also indicates the axle ratio, ring gear diameter, build year, build month and build day, all of which are essential to obtaining the right axle.

16 Axleshaft (rear) - removal and installation

▶ Refer to illustrations 16.4a, 16.4b and 16.5

1   If the vehicle is equipped with air suspension, turn off the air suspension system. The switch is located behind the right kick panel (see Chapter 10).

※ WARNING:
On models with air suspension, electrical power to the air suspension system must be turned off before raising the vehicle.

Failure to do so can result in a sudden inflation or deflation of the air springs, causing instability of the vehicle while it's off the ground.

2   Raise the rear of the vehicle, support it securely on jackstands and block the front wheels. Remove the wheel and brake drum or disc (see Chapter 9).

3   Remove the cover from the differential carrier and allow the lubricant to drain into a container.

8-20 CLUTCH AND DRIVELINE

16.4a Position a large screwdriver between the rear axle case and a ring gear bolt to keep the differential case from turning when removing the pinion shaft lock bolt

16.4b Rotate the differential case 180-degrees and slide the pinion shaft out of the case until the stepped part of the shaft contacts the ring gear

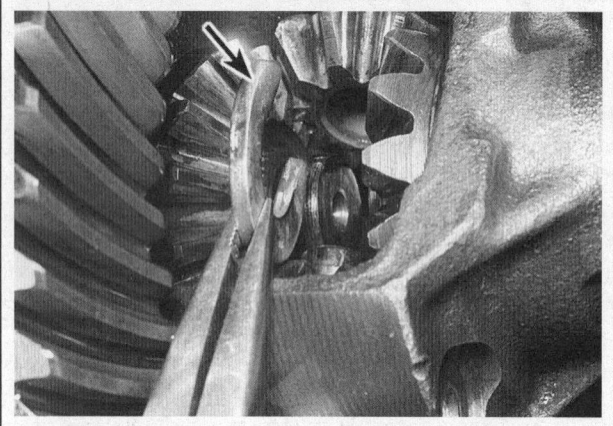

16.5 Push in on the axle flange and remove the C-lock (arrow) from the inner end of the axle shaft

4 Remove the lock bolt from the differential pinion shaft. Slide the notched end of the pinion shaft out of the differential case as far as it will go (see illustrations).

5 Push the outer (flanged) end of the axleshaft in and remove the C-lock from the inner end of the shaft (see illustration).

6 Withdraw the axleshaft, taking care not to damage the oil seal in the end of the axle housing as the splined end of the axleshaft passes through it.

7 Installation is the reverse of removal. The manufacturer recommends that a new pinion shaft lock bolt be used, but if one is not available, coat the threads with a non-hardening thread locking compound. Install the pinion shaft lock bolt and tighten it to the torque listed in this Chapter's Specifications.

8 Install the differential cover (see Chapter 1). Install the brake drum or disc/caliper (see Chapter 9).

9 Refill the axle with the correct quantity and grade of lubricant (see Chapter 1). Tighten the wheel lug nuts to the torque listed in the Chapter 1 Specifications.

17 Axleshaft oil seal (rear) - replacement

▶ **Refer to illustrations 17.2 and 17.3**

1 Remove the axleshaft or driveaxles on Expedition/Navigator models (see Section 16).

2 Pry the oil seal out of the end of the axle housing with a large screwdriver or the inner end of the axleshaft (see illustration).

17.2 You can pry out the old seal with the axleshaft

3 Apply multi-purpose grease to the oil seal recess and tap the new seal evenly into place with a hammer and seal installation tool, large socket or piece of pipe so the lips are facing in and the metal face is visible from the end of the axle housing (see illustration). When correctly installed, the face of the oil seal should be flush with the end of the axle housing.

4 Install the axleshaft (see Section 16).

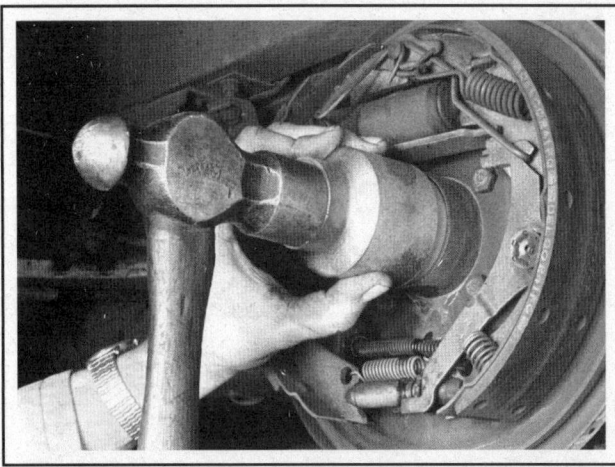

17.3 Install the new axleshaft oil seal with a seal driver or large socket

18 Axleshaft bearing (rear) - replacement

1 Remove the axleshaft (see Section 16) and the oil seal (see Section 17).

2 A bearing puller which grips the bearing from behind will be required for this job. These tools, adapted to screw onto slide hammers, are available from tool stores and many auto parts stores.

3 Attach a slide hammer to the puller and extract the bearing from the axle housing.

4 Clean out the bearing recess and drive in the new bearing with a bearing installer or a piece of pipe positioned against the outer bearing race. Make sure the bearing is tapped in to the full depth of the recess and the numbers on the bearing are visible from the outer end of the axle housing.

5 Install a new oil seal (see Section 16), then install the axleshaft.

19 Differential pinion seal (rear) - replacement

▶ Refer to illustrations 19.4, 19.5, 19.7, 19.9, 19.10 and 19.11

1 If the vehicle is equipped with air suspension, turn off the air suspension system. The switch is located in the area of the right kick panel (see Chapter 10).

❋❋ WARNING:
On models with air suspension, electrical power to the air suspension system must be turned off before raising the vehicle. Failure to do so can result in a sudden inflation or deflation of the air springs, causing instability of the vehicle while it's off the ground.

2 Loosen the rear wheel lug nuts, raise the rear of the vehicle and support it securely on jackstands. Block the front wheels to keep the vehicle from rolling off the stands. Remove the wheels (this will allow you to obtain a more accurate pinion shaft preload reading).

3 Disconnect the driveshaft and fasten it out of the way.

4 Use an inch-pound torque wrench to check the torque required to rotate the pinion. Record it for use later (see illustration).

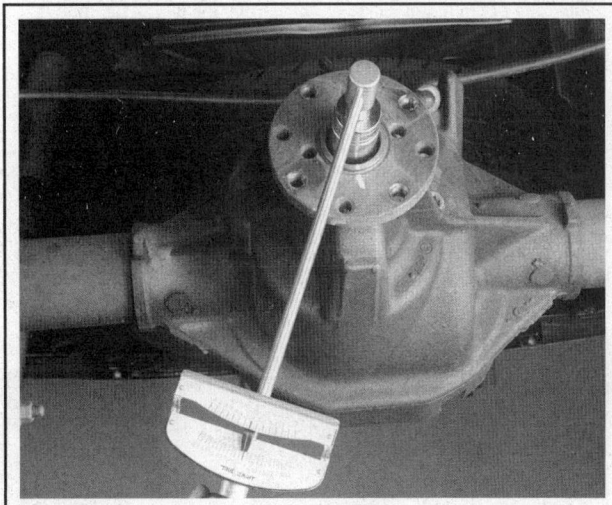

19.4 Use an inch-pound torque wrench to check the torque necessary to rotate the pinion shaft

8-22 CLUTCH AND DRIVELINE

19.5 Mark the relative positions of the pinion, nut and flange (arrows) before removing the nut

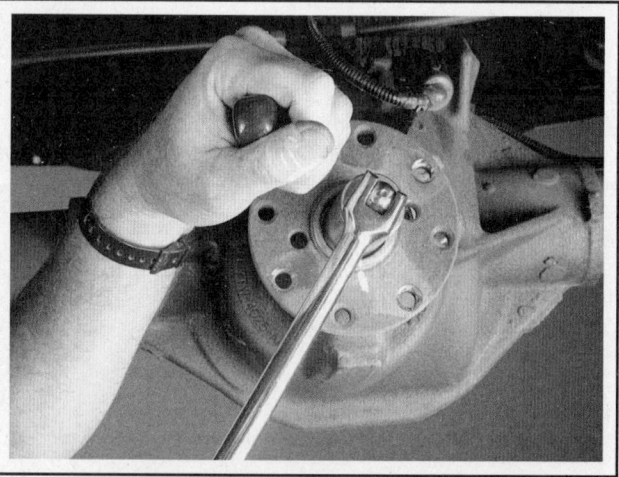

19.7 If you don't have the factory tool to hold the pinion flange while loosening and backing off the pinion flange locknut, use a screwdriver inserted through a hole in the flange and jammed against the top of the reinforcing rib on the differential carrier

5 Put alignment marks on the pinion shaft, nut and flange (see illustration).

6 Count the number of threads visible between the end of the nut and the end of the pinion shaft and record it for use later.

7 A special tool can be used to keep the companion flange from moving while the self-locking pinion nut is loosened. If the special tool isn't available, insert a screwdriver through one of the bolt holes (see illustration).

8 Remove the pinion nut. If you're working on a 2010 or later model, discard the nut; it must be replaced with a new one.

➥**Note:** *The drive pinion nut is color coded - the same color pinion nut must be used during installation. The color is visible on the back side of the nut.*

9 Remove the companion flange. It may be necessary to use a two or three-jaw puller to pull off the flange (see illustration). Do NOT attempt to pry behind the flange or hammer on the end of the pinion shaft.

10 Pry out the old seal (see illustration) and discard it.

11 Lubricate the lips of the new seal with high-temperature grease and tap it evenly into position with a seal installation tool or a large socket. Make sure it enters the housing squarely and is tapped in to its full depth (see illustration).

12 Align the mating marks made before disassembly and install the companion flange. If necessary, tighten the pinion nut to draw the flange into place. Do not try to hammer the flange into position.

13 Apply non-hardening sealant to the ends of the splines visible in the center of the flange so oil will be sealed in.

14 Install the washer (if equipped) and pinion nut. Tighten the nut carefully until the original number of threads are exposed.

15 Measure the torque required to rotate the pinion and tighten the nut in small increments until it matches the figure recorded in Step 5. In order to compensate for the drag of the new oil seal, the nut should be tightened more until the rotational torque of the pinion slightly exceeds what was recorded earlier, but not by more than 5 in-lbs.

16 Connect the driveshaft, install the wheels and lower the vehicle. Tighten the lug nuts to the torque listed in the Chapter 1 Specifications.

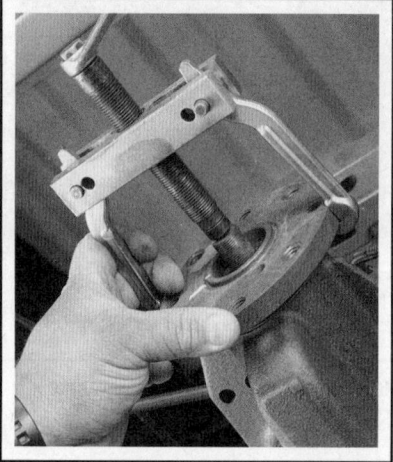

19.9 If the pinion flange is difficult to remove, pull it off with a small puller

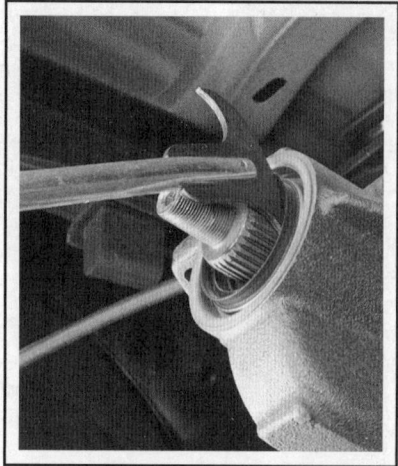

19.10 Pry out the old pinion seal with a seal removal tool

19.11 Lubricate the lips of the new pinion seal and seat it squarely in the bore, then drive it into the carrier with a seal driver or a large socket

CLUTCH AND DRIVELINE 8-23

20 Axle/differential (rear) - removal and installation

REAR AXLE ASSEMBLY REMOVAL (ALL EXCEPT 2003 AND LATER EXPEDITION/NAVIGATOR)

All models

▶ Refer to illustrations 20.7 and 20.8

1 If the vehicle is equipped with air suspension, turn off the air suspension system. The switch is located in the area of the right kick panel (see Chapter 10).

※ WARNING:

On models with air suspension, electrical power to the air suspension system must be turned off before raising the vehicle. Failure to do so can result in a sudden inflation or deflation of the air springs, causing instability of the vehicle while it's off the ground.

2 Loosen the rear wheel lug nuts, raise the rear of the vehicle and support it securely on jackstands placed under the frame (not under the axle). Block the front wheels to keep the vehicle from rolling off the stands. Remove the rear wheels.
3 Disconnect the driveshaft from the rear axle differential pinion flange (see Section 12).
4 Fasten the driveshaft out of the way with a piece of wire from the underbody.
5 Disconnect the parking brake cables from the rear drum brakes or disc brakes (see Chapter 9).
6 Remove the brake assemblies and backing plates (see Chapter 9).
7 Unplug the electrical connector from the rear ABS sensor (see illustration).
8 Disconnect the two rear axle brake lines, the axle vent tube and the rear brake hose from the junction block on top of the axle (see illustration).

Truck models

9 Position a floor jack under the differential case and raise it slightly.

10 Disconnect the lower ends of both shock absorbers from their brackets on the rear axle (see Chapter 10), then compress them to get them out of the way.
11 At each end of the axle, remove the four spring U-bolt nuts, the two U-bolts and the spring plate that attach the axle to each spring (see Section 14, Chapter 10). On models with air suspension, you'll also need to unbolt the anti-wind bar from the axle.
12 Check to be sure nothing is still attached between the axle assembly and the vehicle.
13 Slowly lower the axle to the ground with the floor jack. Since the axle is very long, it can be unstable. It's a good idea to have an assistant steady the axle as it is being lowered. On 4WD models, remove the rear spring spacers as the axle is being lowered.

Expedition and Navigator models

14 Remove the two bolts from each of the two brackets that hold the stabilizer bar to the axle, then remove the two brackets (see Chapter 10).
15 Loosen the stabilizer bar links and swing the stabilizer bar down, out of the way of the axle (see Chapter 10).
16 Disconnect the air suspension system height sensor from the axle housing.
17 Position a floor jack under the rear axle differential case and raise the axle assembly slightly, taking the weight off the suspension components.
18 Disconnect the rear axle track bar from the axle and tie the track bar up, out of the way with a piece of heavy wire (see Chapter 10).
19 Remove the rubber jounce bumper from the top of the axle housing assembly.
20 Disconnect the suspension upper arms from the axle housing (see Chapter 10).
21 Disconnect the suspension lower arms and position them out of the way (see Chapter 10).
22 Disconnect the lower shock absorbers from the axle housing and compress the shock absorbers up, out of the way (see Chapter 10).
23 Check to be sure there is nothing else attached between the axle housing and vehicle, then carefully lower the axle assembly to the floor. Since the axle is very long, it can be unstable. It's a good idea to have an assistant steady the axle as it is being lowered.

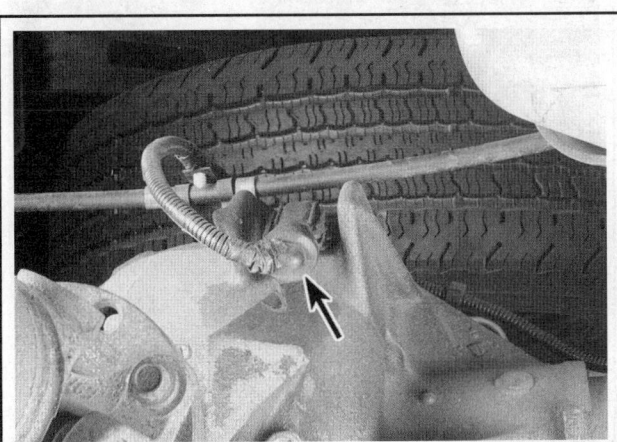

20.7 Unplug the electrical connector from the rear ABS sensor

20.8 Disconnect the two rear brake lines, the axle vent tube and the rear brake hose from the junction block on top of the axle

8-24 CLUTCH AND DRIVELINE

REAR DIFFERENTIAL REMOVAL (2003 AND LATER EXPEDITION/NAVIGATOR)

24 Loosen the wheel lug nuts. Raise the rear of the vehicle and support it securely on jackstands. Block the front wheels to prevent the vehicle from rolling.

25 Remove the rear driveshaft (see Section 12).

26 Remove the driveaxles (see Section 21).

27 Remove the spare tire.

28 Remove the differential vent hose.

29 At the top of the differential, disconnect the electrical connector from the vehicle speed sensor.

30 Support the differential safely with a floor jack. Remove the two nuts securing the differential to the mounting brackets, then remove the front torque arm bolt. Move the differential forward enough for the studs to clear, then carefully lower the differential.

31 Installation is the reverse of removal, noting the following points:
   a) *Follow the information in Steps 32 through 34.*
   b) *Be sure to install the upper and the lower axle housing bolts.*
   c) *Tighten all fasteners to the torque listed in this Chapter's Specifications.*

INSTALLATION (ALL MODELS)

32 Installation is the reverse of removal. Lower the vehicle weight onto the wheels before tightening the suspension arm/stabilizer bar fasteners completely.

33 Reactivate the air suspension system, if equipped.

34 Bleed the brakes (see Chapter 9).

21 Driveaxles - removal and installation

FRONT DRIVEAXLES (4WD MODELS)

2004 and earlier models

▶ Refer to illustrations 21.3 and 21.4

1 If the vehicle is equipped with air suspension, turn off the air suspension system. The switch is located in the area of the right kick panel (see Chapter 10).

❋❋ WARNING:

On models with air suspension, electrical power to the air suspension system must be turned off before raising the vehicle. Failure to do so can result in a sudden inflation or deflation of the air springs, causing instability of the vehicle while it's off the ground.

2 Loosen the wheel lug nuts, raise the vehicle and support it securely on jackstands. Remove the wheel.

3 Remove the cotter pin and nut lock. Place a prybar or large screwdriver between the wheel studs to hold the driveaxle and break the driveaxle hub nut loose with a large breaker bar (see illustration). Remove the nut and washer.

4 Using the same setup, hold the driveaxle and remove the driveaxle flange bolts (see illustration). Be sure to mark the relationship of the inner CV joint to the flange before removing the bolts.

5 Remove the steering knuckle (see Chapter 10).

6 Remove the driveaxle assembly.

7 Installation is the reverse of removal. Be sure to tighten the driveaxle hub nut and the CV joint flange bolts to the torque listed in this Chapter's Specifications. Use a new cotter pin, too.

8 Reactivate the air suspension system, if equipped.

21.3 To remove the driveaxle hub nut, hold the driveaxle by inserting a large screwdriver or prybar between the wheel studs as shown and use a large breaker bar to loosen the nut

21.4 Mark the relationship of the inner CV joint to the axleshaft flange, then, with the prybar again between the wheel studs, loosen the flange bolts

CLUTCH AND DRIVELINE 8-25

2005 and later models

9 Remove the hub dust cap and the driveaxle nut.

10 Loosen the wheel lug nuts, raise the front of the vehicle and support it securely on jackstands under the frame. Block the rear wheels to keep the vehicle from rolling off the stands.

11 Disconnect the vacuum tube(s) at the rear of the IWE hub.

12 Remove the three IWE retainer-to-knuckle bolts.

13 Disconnect the tie-rod end from the steering knuckle (see Chapter 10).

14 Disconnect the upper balljoint from the steering knuckle (see Chapter 10).

15 While pushing inward on the driveaxle, pull the steering knuckle outward.

16 A special tool that clamps to the inner driveaxle CV joint is required to pull the joint away from the differential.

17 Installation is the reverse of the removal procedure. Use a new circlip and seal at the inboard ends of the driveaxle, and use new nuts on both the tie-rod ends and the upper balljoint.

REAR DRIVEAXLES (2003 AND LATER EXPEDITION/NAVIGATOR)

Removal

18 Block the front wheels to prevent the vehicle from rolling. Loosen the wheel lug nuts, raise the rear of the vehicle and support it securely on jackstands. Remove the wheel.

19 Have an assistant depress the brake pedal while you unscrew the driveaxle/hub nut.

20 Remove the brake disc (see Chapter 9). Be sure to use mechanic's wire to suspend the brake caliper and avoid straining the hose.

21 Remove the bolt that retains the parking brake cable to the control arm.

22 Cover the stabilizer bar link stud threads with a piece of rubber hose to prevent the CV boots from tearing.

23 Detach the lower control arm and the toe link from the rear knuckle (see Chapter 10).

※※ CAUTION:

Don't let the driveaxle hang by the inner CV joint.

24 Use a puller tool to push the driveaxle out of the hub.

25 Raise the lower end of the knuckle outward to gain clearance, then drive the inner end of the driveaxle out of the differential and remove it from the vehicle.

Installation

26 Pry the old spring clip from the inner end of the driveaxle and install a new one.

27 Apply a light film of grease to the area on the inner CV joint stub shaft where the seal rides, then insert the splined end of the inner CV joint into the differential. Make sure the spring clip locks in its groove.

28 Apply a light film of grease to the outer CV joint splines, pull the knuckle assembly outward and insert the outer end of the driveaxle into the hub.

29 Connect the lower control arm and toe link to the knuckle (see Chapter 10). Tighten the fasteners to the torque listed in the Chapter 10 Specifications.

30 Install a new driveaxle/hub nut. Tighten the hub nut securely, but don't try to tighten it to the actual torque specification yet.

31 Install the parking brake cable to the lower control arm.

32 Install the brake disc and caliper (see Chapter 9).

33 Have an assistant push on the brake pedal, then tighten the driveaxle/hub nut to the torque listed in this Chapter's Specifications. Install the wheel cover or hub cap.

34 Install the wheel and lug nuts, then lower the vehicle. Tighten the lug nuts to the torque listed in the Chapter 1 Specifications.

22 Driveaxle boot replacement (4WD models)

➡ **Note:** If the CV joints exhibit signs of wear indicating need for an overhaul (usually due to torn boots), explore all options before beginning the job. Complete rebuilt driveaxles are available on an exchange basis, which eliminates much time and work. Whichever route you choose to take, check on the cost and availability of parts before disassembling the vehicle.

INNER CV JOINT

Disassembly

♦ **Refer to illustrations 22.3, 22.4, 22.5, 22.7, 22.9, 22.10 and 22.11**

1 Remove the driveaxle from the vehicle (see Section 21).

2 Mount the driveaxle in a vise. The jaws of the vise should be lined with wood or rags to prevent damage to the axleshaft.

3 Cut the boot clamps from the boot and discard them (see illustration).

22.3 Cut off the boot clamps and discard them

8-26 CLUTCH AND DRIVELINE

22.4 Pry the wire retainer ring from the CV joint housing with a small screwdriver

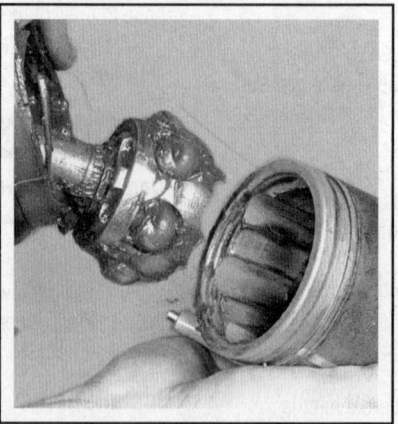

22.5 With the retainer removed, the outer race can be pulled off the bearing assembly

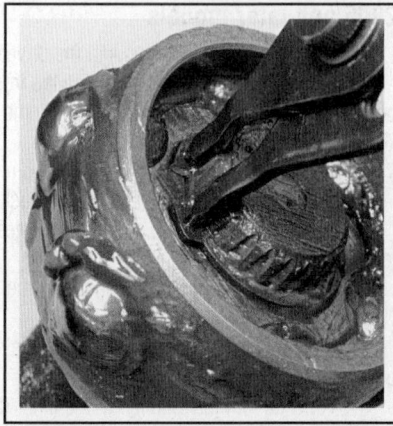

22.7 Remove the snap-ring from the end of the axleshaft

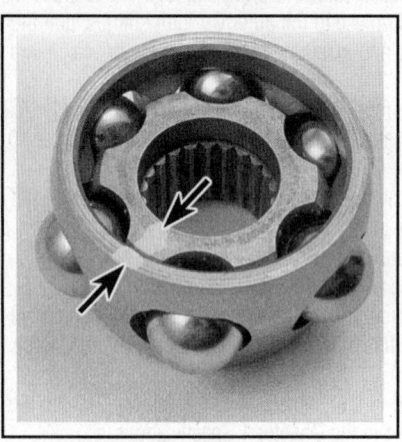

22.9 Make index marks on the inner race and cage so they'll both be facing the same direction when reassembled

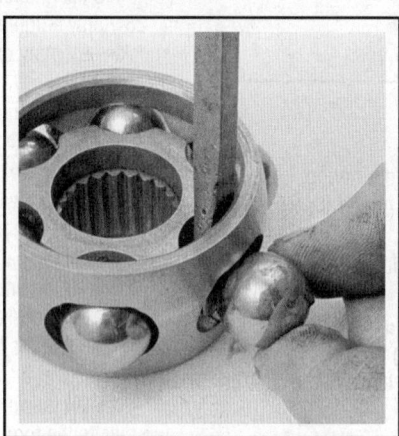

22.10 Pry the balls from the cage with a screwdriver (be careful not to nick or scratch them)

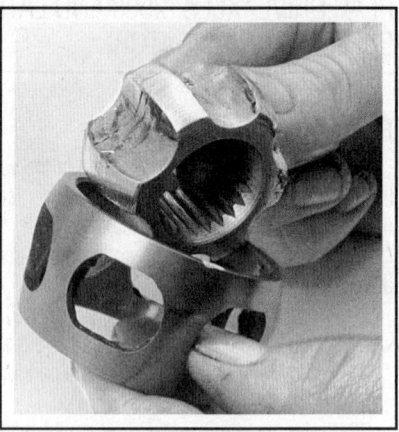

22.11 Tilt the inner race 90-degrees and rotate it out of the cage

4 Slide the boot back on the axleshaft and pry the wire ring ball retainer from the outer race (see illustration).

5 Pull the outer race off the inner bearing assembly (see illustration).

6 Wipe as much grease off the inner bearing as possible.

7 Remove the snap-ring from the end of the axleshaft (see illustration).

8 Slide the inner bearing assembly off the axleshaft.

9 Mark the inner race and cage to ensure that they are reassembled with the correct sides facing out (see illustration).

10 Using a screwdriver or piece of wood, pry the balls from the cage (see illustration). Be careful not to scratch the inner race, the balls or the cage.

11 Rotate the inner race 90-degrees, align the inner race lands with the cage windows and rotate the race out of the cage (see illustration).

Inspection

▶ Refer to illustrations 22.12a and 22.12b

12 Clean the components with solvent to remove all traces of grease. Inspect the cage and races for pitting, score marks, cracks and other signs of wear and damage. Shiny, polished spots are normal and will not adversely affect CV joint performance (see illustrations).

Reassembly

▶ Refer to illustrations 22.14, 22.16, 22.17, 22.20, 22.22, 22.23 and 22.24

13 Insert the inner race into the cage. Verify that the matchmarks are on the same side. However, it's not necessary for them to be in direct alignment with each other.

14 Press the balls into the cage windows with your thumbs (see illustration).

15 Wrap the axleshaft splines with tape to avoid damaging the boot.

16 Slide the small boot clamp and boot onto the axleshaft, then remove the tape (see illustration).

17 Install the inner race and cage assembly on the axleshaft with the larger diameter side or "bulge" of the cage facing the axleshaft end (see illustration).

18 Install the snap-ring (see illustration 22.7).

19 Fill the outer race with CV joint grease (normally included with the new boot kit).

20 Pack the inner race and cage assembly with grease, by hand, until grease is worked completely into the assembly (see illustration).

CLUTCH AND DRIVELINE 8-27

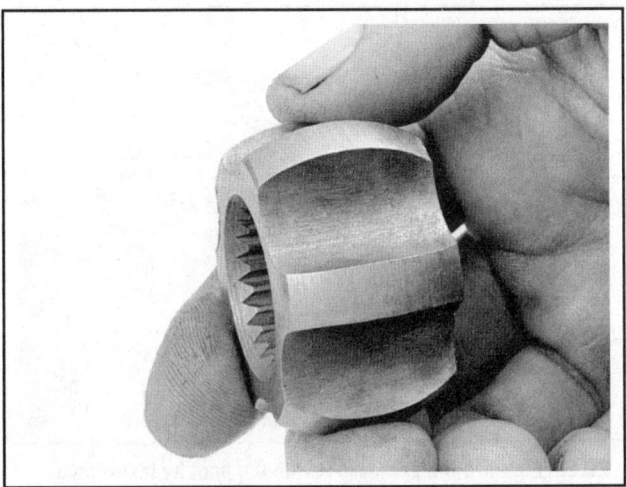

22.12a Inspect the inner race lands and grooves for pitting and score marks

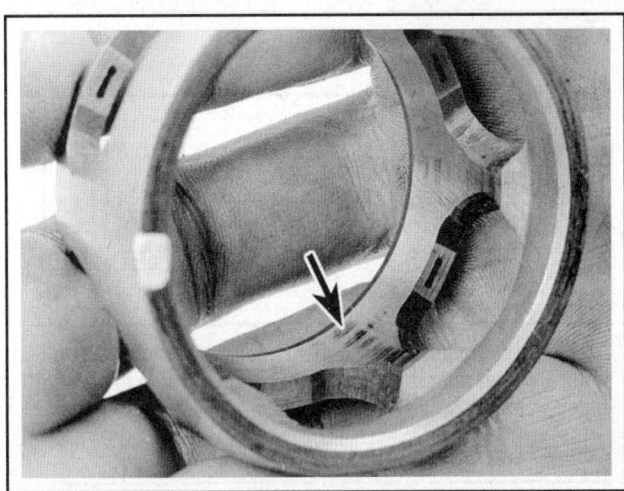

22.12b Inspect the cage for cracks, pitting and score marks (shiny spots are normal and don't affect operation)

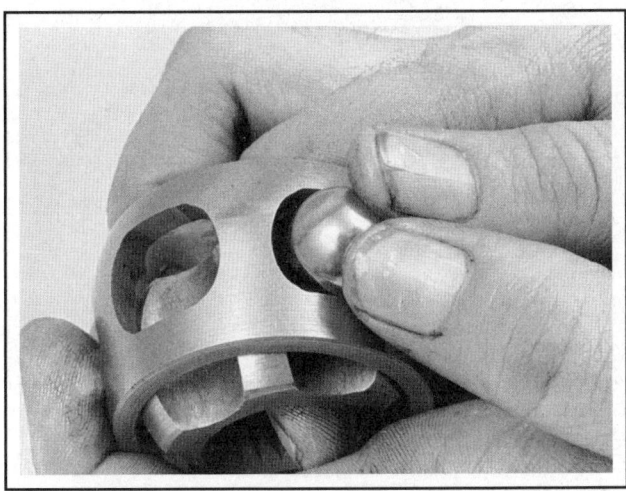

22.14 Press the balls into the cage through the windows

22.16 Wrap the splined area of the axle with tape to prevent damage to the boot, then install the small clamp and the boot

22.17 Install the inner race and cage assembly with the "bulge" facing the end of the axleshaft

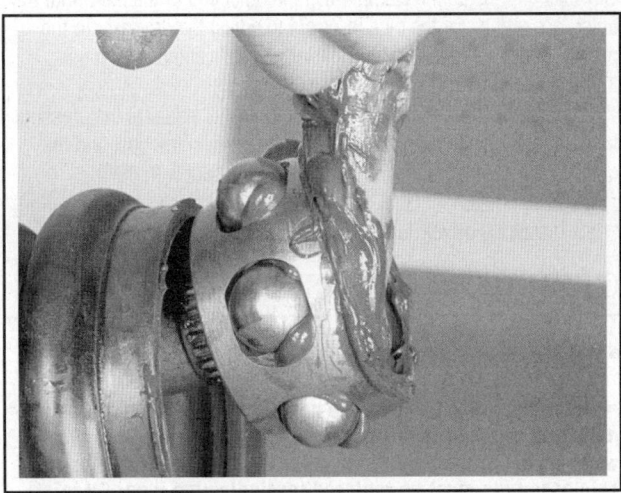

22.20 Pack grease into the bearing until it's completely full

8-28 CLUTCH AND DRIVELINE

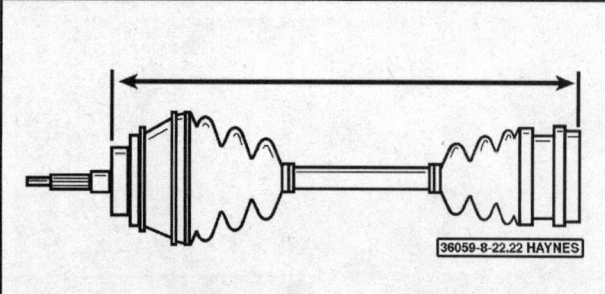

22.22 Measure between the points indicated and move the inner CV joint in or out until the driveaxle is set to the length listed in this Chapter's Specifications

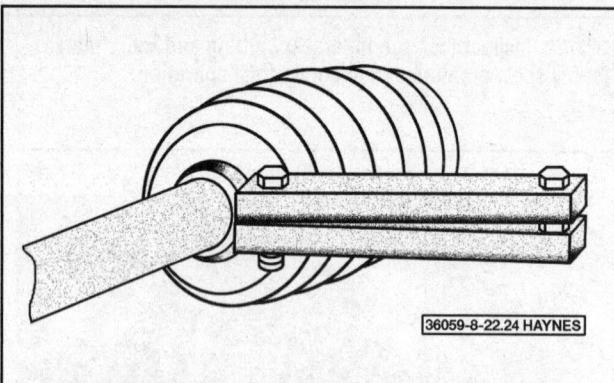

22.24 A special tool is required to crimp the boot clamps in place; tighten the bolt nearest the clamp until the ends of the tool touch - this tool is available at most auto parts stores (it generates much more clamping force than plier-type crimping tools)

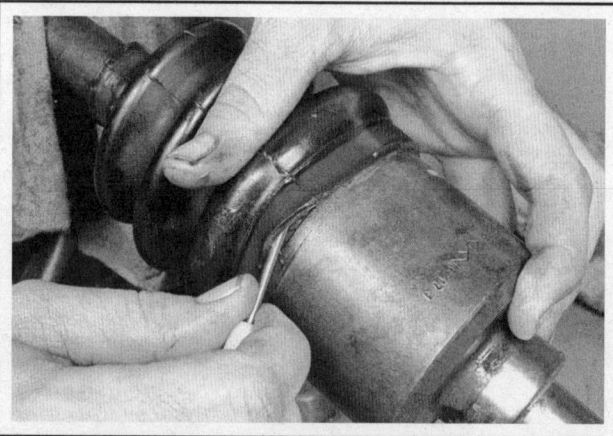

22.23 Equalize the pressure inside the boot by inserting a small screwdriver between the boot and the outer race

22.30 After the old grease has been rinsed away and the cleaning solvent has been blown out with compressed air, rotate the outer joint housing through its full range of motion and inspect the bearing surfaces for wear or damage - if any of the balls, the race or cage look damaged, replace the driveaxle and outer joint

21 Slide the outer race down onto the inner race and install the wire ring retainer.

22 Wipe any excess grease from the axle boot groove on the outer race. Seat the small diameter of the boot in the recessed area on the axleshaft and install the clamp. Push the other end of the boot onto the outer race and move the race in-or-out to adjust the driveaxle to the proper length, as listed in this Chapter's Specifications (see illustration).

23 With the driveaxle set to the proper length, equalize the pressure in the boot by inserting a dull screwdriver between the boot and the outer race (see illustration). Don't damage the boot with the tool.

24 Install the boot clamps and crimp them in place (see illustration).

25 Install the driveaxle (see Section 21).

OUTER CV JOINT AND BOOT

♦ Refer to illustration 22.30

➡ **Note:** *The outer CV joint is a non-serviceable item and is permanently retained to the driveaxle. If any damage or excessive wear occurs to the axle or the outer CV joint, the entire driveaxle assembly must be replaced (excluding the inner CV joint).*

26 Remove the driveaxle from the vehicle (see Section 21).

27 Mount the driveaxle in a vise. The jaws of the vise should be lined with wood or rags to prevent damage to the axleshaft.

28 Cut the boot clamps from both inner and outer boots and discard them (see illustration 22.3).

29 Remove the inner CV joint and boot (see Steps 4 through 11).

30 Remove the outer CV joint boot. Wash the outer CV joint assembly in solvent and inspect it as described in Step 12 (see illustration). Replace the axle assembly if any CV joint components are excessively worn. Install the new, outer boot and clamps onto the axleshaft (see illustration 22.16).

31 Repack the outer CV joint with CV joint grease and spread grease inside the new boot as well.

32 Position the outer boot on the CV joint and install new boot clamps (see illustration 22.24).

33 Reassemble the inner CV joint and boot (see Steps 13 through 24).

34 Install the driveaxle (see Section 21).

CLUTCH AND DRIVELINE

23 Vacuum shift system (4WD models) - description, check and component replacement

DESCRIPTION

1 When a 4WD range is selected by the switch on the dash, a vacuum shift motor on the front axle engages the differential with the right axleshaft via a shift lever which moves a shift fork which slides a coupler cluster gear, meshing it with a cluster gear shaft.

CHECK

♦ Refer to illustration 23.5a and 23.5b

2 If the vehicle is equipped with air suspension, turn off the air suspension system. The switch is located in the area of the right kick panel (see Chapter 10).

WARNING:

On models with air suspension, electrical power to the air suspension system must be turned off before raising the vehicle. Failure to do so can result in a sudden inflation or deflation of the air springs, causing instability of the vehicle while it's off the ground.

3 Raise the vehicle and support it securely on jackstands.
4 Remove the shift motor cover.
5 There are two vacuum lines to the shift motor diaphragm. Disconnect the line from the inner port (see illustration), hook up a hand-operated vacuum pump in its place, apply 10 in-Hg vacuum and watch the shift lever (see illustration). If it moves freely, the vacuum shift motor is working properly; if it doesn't, check to see if vacuum is showing on the vacuum pump gauge.
6 If vacuum is holding (the vacuum gauge reading remains steady), the problem is something binding in the shift fork and mechanism or a clogged vacuum port. Check for binding by trying to move the shift mechanism by hand.
7 If vacuum does not hold (the gauge moves back to zero), the vacuum motor is faulty - replace it.
8 Check for vacuum to the motor by connecting the vacuum hose at the shift motor to a vacuum gauge and, with the engine running in Park (parking brake set), press the 4WD button on the dash (during this test, be sure to use a long vacuum hose for safety so you don't have to be under the vehicle). If there's a strong vacuum reading on the gauge, the system is working properly and the problem lies with the axle shift mechanism - these problems are best remedied by a dealer service department or other qualified shop. If there is no vacuum reading on the gauge, the problem is with the vacuum hoses between the shift motor and the switch or in the switch itself.

VACUUM MOTOR REPLACEMENT

9 Detach the vacuum hoses from the vacuum shift motor diaphragm (see illustration 23.5a).
10 Peel back the dust boot and remove the E-ring retainer.
11 Disengage the shift rod from the shift lever and pull out the vacuum shift motor.
12 Installation is the reverse of removal.
13 Lower the vehicle and reactivate the air suspension system, if equipped.

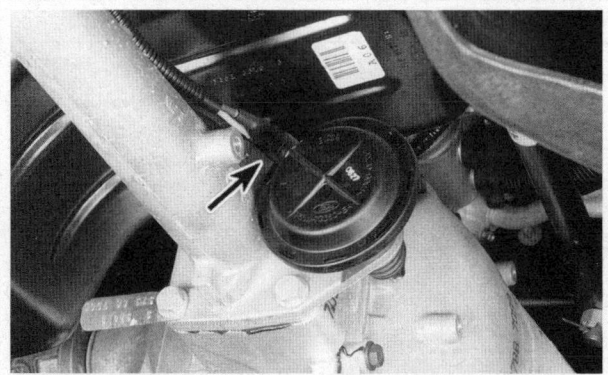

23.5a To check the vacuum shift motor, locate the two vacuum line connections (arrow), detach the line from the inboard port, hook up a hand-operated vacuum pump, apply 10 in-Hg vacuum . . .

23.5b . . . and verify that the vacuum motor shift rod moves the shift lever (upper arrow); to remove the vacuum shift motor, peel back the dust boot (lower arrow) and remove the E-ring retainer

24 Axle (front) removal and installation

♦ Refer to illustrations 24.6, 24.7, 24.8 and 24.11

1 If the vehicle is equipped with air suspension, turn off the air suspension system. The switch is located in the area of the right kick panel (see Chapter 10).

WARNING:

On models with air suspension, electrical power to the air suspension system must be turned off before raising the vehicle. Failure to do so can result in a sudden inflation or deflation of the air springs, causing instability of the vehicle while it's off the ground.

8-30 CLUTCH AND DRIVELINE

2 Loosen the wheel lug nuts, raise the front of the vehicle and support it securely on jackstands placed under the frame. Block the rear wheels to keep the vehicle from rolling off the stands. Remove the front wheels.

3 Remove the front driveshaft (see Section 12).

4 Disconnect the front driveaxles from the axleshaft flanges (see illustration 21.4).

5 Disconnect the vacuum lines from the vacuum shift motor (see Section 23).

6 Disconnect the vent hose from the front axle tube (see illustration). On 2005 and later Expedition/Navigator models, remove the four bolts and remove the front crossmember.

7 Position a jack under the front axle differential case (see illustration).

8 Remove the front carrier bushing bolts (see illustration).

9 Remove the four front differential support bolts.

10 Remove the front differential support.

11 Remove the right axle tube bushing bolt (see illustration).

12 Remove the differential cover bushing bolt.

13 Installation is the reverse of removal. Be sure to tighten all fasteners securely.

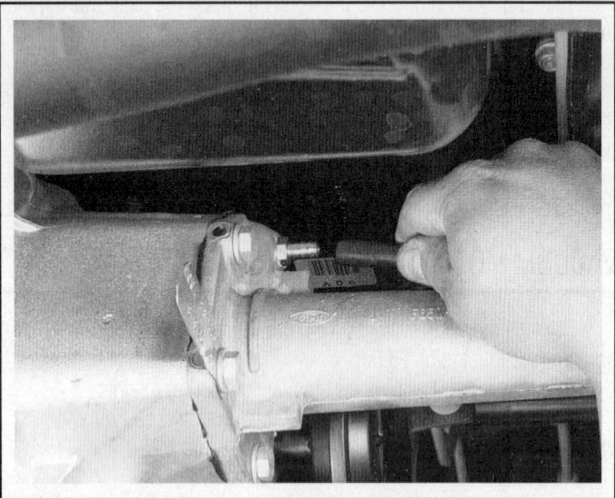

24.6 Disconnect the vent hose from the front axle tube

24.7 Position a jack under the front axle differential case

24.8 Remove the front carrier bushing bolts (arrows)

24.11 Remove the right axle tube bushing bolt (arrow)

CLUTCH AND DRIVELINE 8-31

Specifications

General

Clutch fluid type	See Chapter 1
Clutch disc lining thickness (minimum)	1/16-inch
Clutch pedal travel	6 to 7 inches
Pressure plate tension spring dimension	0.55 (35/64) inch
Driveshaft slip yoke adjustment (from weld bead on driveshaft to U-joint centerline)	10 inches
Front driveaxle length (4WD models)	
2004 and earlier models	16.43 inches
2005 and 2006 models	16.93 inches
2007 and later models	
Left	16.00 inches
Right	16.21 inches
Rear driveaxle length (2003 and later Expedition/Navigator)	
2003 through 2006 models	
Left	39.29 inches
Right	37.65 inches
2007 through 2009 models	
Left	37.49 inches
Right	35.86 inches
2010 and later models	
Left	37.24 inches
Right	35.63 inches

Torque specifications — Ft-lbs (unless otherwise indicated)

➡ **Note:** One foot-pound (ft-lb) of torque is equivalent to 12 inch-pounds (in-lbs) of torque. Torque values below approximately 15 ft-lbs are expressed in inch-pounds, since most foot-pound torque wrenches are not accurate at these smaller values.

Clutch

Pressure plate-to-flywheel bolts	35 to 46
Release cylinder bolts	14 to 19

Rear driveshaft

Flange yoke-to-rear axle pinion flange bolts	70 to 95
Center bearing support bracket bolts	
2004 and earlier models	39 to 53
2005 and later models	35

Front driveshaft

Flange yoke-to-front axle pinion flange bolts	
2004 and earlier models	65 to 88
2005 and later models	41
Flange yoke-to-transfer case flange bolts	
2004 and earlier models	65 to 88
2005 and later models	41

Rear axle

Differential pinion shaft lock bolt	15 to 30
Rear differential mounting bolts (Expedition/Navigator)	100

8-32 CLUTCH AND DRIVELINE

Torque specifications (continued) — Ft-lbs (unless otherwise indicated)

➥ Note: One foot-pound (ft-lb) of torque is equivalent to 12 inch-pounds (in-lbs) of torque. Torque values below approximately 15 ft-lbs are expressed in inch-pounds, because most foot-pound torque wrenches are not accurate at these smaller values.

Driveaxles
- Driveaxle hub nut (new)
  - Front driveaxle
    - 2004 and earlier models — 221
    - 2005 and later models (with IWE hubs) — 20
  - Rear driveaxle (2003 and later Expedition/Navigator)
    - 2003 through 2009 — 221
    - 2010 and later — 184
- Front driveaxle-to-axleshaft flange bolts
  - (4WD models) — 51 to 67
- Front IWE retainer-to-knuckle bolts
  - (2005 and later 4WD models) — 108 in-lbs

Section

1 General information
2 Anti-lock brake system - general information
3 Disc brake pads - replacement
4 Disc brake caliper - removal, overhaul and installation
5 Brake disc - inspection, removal and installation
6 Drum brake shoes - replacement
7 Wheel cylinder - removal, overhaul and installation
8 Master cylinder - removal, overhaul and installation
9 Brake hoses and lines - check and replacement
10 Brake hydraulic system - bleeding
11 Brake light switch - check and replacement
12 Power brake booster - check, removal and installation
13 Parking brake cables - replacement
14 Parking brake shoes (models with rear disc brakes) - replacement
15 Adjustable brake pedal and bracket - removal, installation and indexing

Reference to other Chapters

Brake check - See Chapter 1
Brake fluid level check - See Chapter 1
Front wheel bearing check, repack and adjustment (2WD models only) -
    See Chapter 1

9

BRAKES

9-2 BRAKES

1 General information

GENERAL

All models covered by this manual are equipped with hydraulically operated, power-assisted brake systems. All front brakes are discs. Trucks use either drum or disc rear brakes; sport utility models use rear discs.

All brakes are self-adjusting. The front disc brakes automatically compensate for pad wear, while the rear drum brakes incorporate an adjustment mechanism which is activated as the brakes are applied.

The hydraulic system has separate circuits for the front and rear brakes. If one circuit fails, the other circuit will remain functional and a warning indicator will light up on the dashboard when a substantial amount of brake fluid is lost, showing that a failure has occurred.

MASTER CYLINDER

The master cylinder is mounted on the front of the power brake booster, and can be identified by the large fluid reservoir on top. The master cylinder has separate primary and secondary piston assemblies for the front and rear circuits.

POWER BRAKE BOOSTER

The power brake booster uses engine manifold vacuum to provide assistance to the brakes. It is mounted on the firewall in the engine compartment, directly behind the master cylinder.

ANTI-LOCK BRAKE SYSTEM (ABS)

A rear wheel anti-lock brake system (RABS) is used on these vehicles to improve directional stability and control during hard braking. Some models are equipped with a 4-wheel anti-lock brake system (4WABS) that prevents wheel skid at all four wheels.

PARKING BRAKE

The parking brake mechanically operates the rear brakes only. The parking brake cables pull on a lever attached to the brake shoe assembly, causing the shoes to expand against the drum.

PRECAUTIONS

There are some general precautions and warnings related to the brake system:

a) *Use only brake fluid conforming to DOT 3 specifications.*
b) *The brake pads and linings contain fibers which are hazardous to your health if inhaled. Whenever you work on brake system components, clean all parts with brake system cleaner or denatured alcohol. Do not allow the fine dust to become airborne.*
c) *Safety should be paramount whenever any servicing of the brake components is performed. Do not use parts or fasteners which are not in perfect condition, and be sure all clearances and torque specifications are adhered to. If you are at all unsure about a certain procedure, seek professional advice. Upon completion of any brake system work, test the brakes carefully in a controlled area before driving the vehicle in traffic. If a problem is suspected in the brake system, don't drive the vehicle until it's fixed.*

2 Anti-lock brake system - general information

DESCRIPTION

The Anti-lock brake system is designed to maintain vehicle maneuverability, directional stability and optimum deceleration under severe braking conditions on most road surfaces. It does so by monitoring the rotational speed of the wheels and controlling the brake line pressure during braking. This prevents the wheels from locking up prematurely.

Two types of systems are used: rear wheel anti-lock brake system (RABS) and 4-wheel anti-lock (4WABS). RABS only controls lockup on the rear wheels, whereas 4WABS prevents lockup on all four wheels.

Actuator assembly

The actuator assembly includes the master cylinder and a control valve which consists of a dump valve and an isolation valve. The valve operates by changing the brake fluid pressure in response to signals from the control unit.

Control unit

The control unit for the anti-lock brakes on RABS-equipped models is called the anti-lock electronic control module; it's located inside the center of the dash. The control unit on 4WABS-equipped models is called the electronic hydraulic control unit (EHCU); it's located inside the engine compartment. Either control unit is the "brain" for the system. The function of the control unit is to accept analog voltage inputs from the speed sensors, process that data, and control hydraulic line pressure to avoid wheel lockup.

The control units for both systems constantly monitor the system, even under normal driving conditions, to detect malfunctions. If a problem develops within the system, the control unit illuminates a yellow ABS warning light on the instrument cluster, and may even shut down the anti-lock system if it's a serious malfunction. A diagnostic trouble code will also be stored, which, when retrieved by a service technician, will indicate the problem area or component.

Speed sensor

A speed sensor produces an "analog" (continuously variable) voltage output, which is transmitted to the control unit, where it's converted to digital information, compared to the control unit's program, and interpreted as wheel rotation speed. On both systems, a single rear wheel speed sensor is located in the top of the differential carrier. 4WABS systems also use a front wheel speed sensor in each front steering knuckle.

Brake light switch

The brake light switch, known as the brake on-off (or BOO) switch on these models, signals the control unit when the driver steps on the brake pedal. Without this signal the anti-lock system won't activate. The RABS or RWABS system is de-activated when the brake pedal is released.

BRAKES 9-3

Diagnosis and repair

If the yellow ABS warning light on the instrument cluster comes on and stays on, make sure the parking brake is released and there's no problem with the brake hydraulic system. If neither of these is the cause, the anti-lock system is probably malfunctioning. Although special test procedures are necessary to properly diagnose the system, the home mechanic can perform a few preliminary checks before taking the vehicle to a dealer service department.

a) Make sure the brakes, calipers and wheel cylinders are in good condition.
b) Check the electrical connectors at the control unit.
c) Check the fuses.
d) Follow the wiring harness to the speed sensors and brake light on-off (BOO) switch and make sure all connections are secure and the wiring isn't damaged.

If the above preliminary checks don't rectify the problem, the vehicle should be diagnosed by a dealer service department.

3 Disc brake pads - replacement

▸ Refer to illustration 3.3

※ WARNING:

Disc brake pads must be replaced on both wheels at the same time - never replace the pads on only one wheel. Also, brake system dust is hazardous to your health. DO NOT blow it out with compressed air and DO NOT inhale it. An approved filtering mask should be worn when working on the brakes. DO NOT use gasoline or solvents to remove the dust. Use brake system cleaner only!

➡ Note: Some models have brake designs with a long retaining spring on the outside of each caliper.

1 If the vehicle is equipped with air suspension, turn off the air suspension system. The switch is located in the area of the right kick panel (see Chapter 10).

※ WARNING:

On models with air suspension, electrical power to the air suspension system must be turned off before raising the vehicle. Failure to do so can result in a sudden inflation or deflation of the air springs, causing instability of the vehicle while it's off the ground.

2 Loosen the wheel lug nuts, raise the end of the vehicle to be worked on and support it securely on jackstands. Apply the parking brake. Remove the wheels.

3 Remove about two-thirds of the fluid from the master cylinder reservoir and discard it; as the pistons are pushed in for clearance to allow the pads to be removed, the fluid will be forced back into the reservoir. Position a drain pan under the brake assembly and clean the caliper and surrounding area with brake system cleaner. Push the piston(s) back into the bore(s) with a C-clamp to provide room for the new brake pads (see illustration). As the piston is depressed to the bottom of the caliper bore, the fluid in the master cylinder will rise. Make sure it doesn't overflow. If necessary, siphon off some of the fluid.

F-150/F-250 AND 2002 AND EARLIER EXPEDITION/NAVIGATOR MODELS

▸ Refer to illustrations 3.4a through 3.4p (front) and 3.4q through 3.4w (rear)

4 To replace the front brake pads, follow the accompanying photos, beginning with illustration 3.4a; to replace the rear brake pads, start at illustration 3.4q. Be sure to stay in order and read the caption under each illustration. Work on one brake assembly at a time so that you'll have something to refer to if necessary.

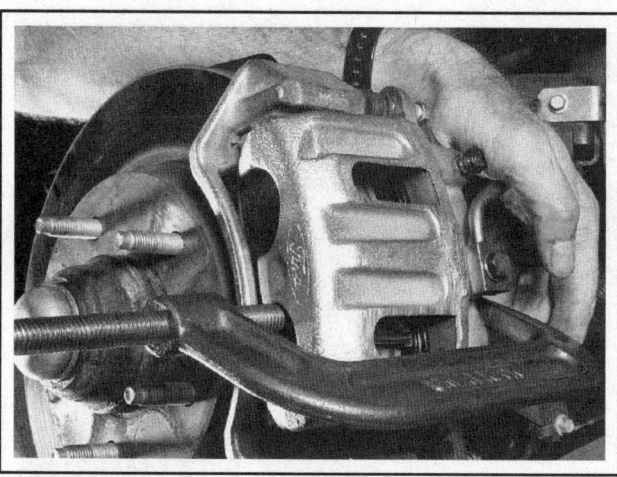

3.3 To make room for the new pads, use a C-clamp to depress the piston(s)

3.4a Wash down the disc and brake pads with brake cleaner to remove brake dust; DO NOT blow off brake dust with compressed air

9-4 BRAKES

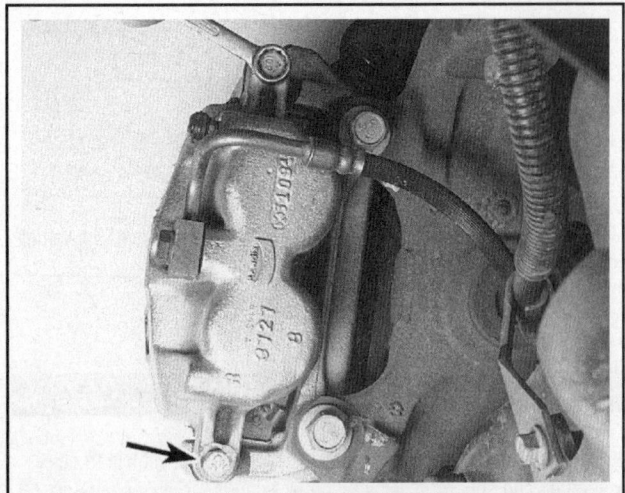

3.4b Remove the two caliper mounting bolts, check them for thread damage and replace as necessary

3.4c Pull off the caliper . . .

3.4d . . . and hang it from the upper control arm with a piece of wire; DON'T allow the caliper to hang by the brake hose!

3.4e Remove the outer brake pad

3.4f Remove the inner brake pad

3.4g Pry off the upper anti-rattle clip with a small screwdriver (check it for cracks and make sure it fits tightly)

BRAKES 9-5

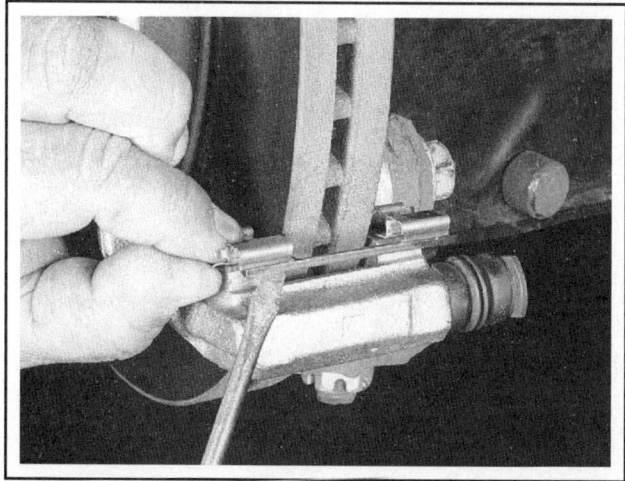

3.4h Pry off the lower anti-rattle clip with a small screwdriver

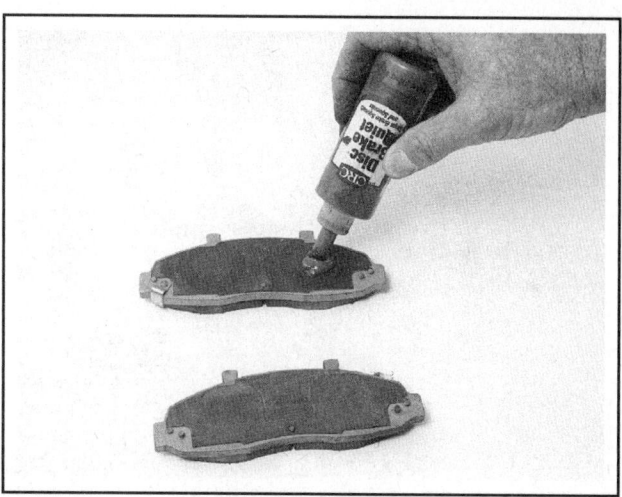

3.4i Apply some anti-squeal compound to the backing plates of the new pads

3.4j Install the upper anti-rattle clip; make sure it's fully seated on the anchor plate

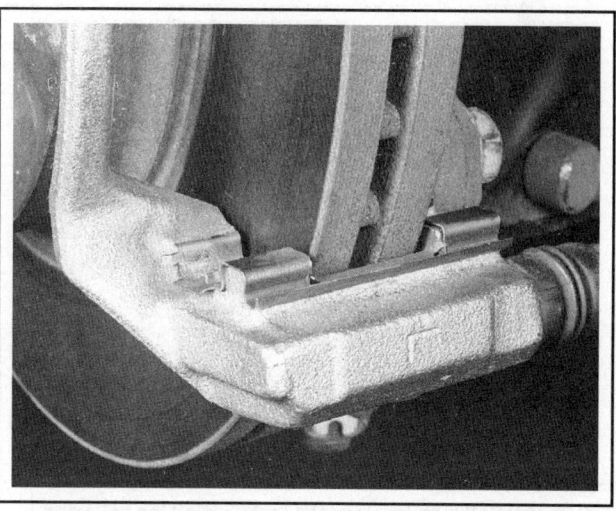

3.4k Install the lower anti-rattle clip; make sure it's fully seated on the anchor plate

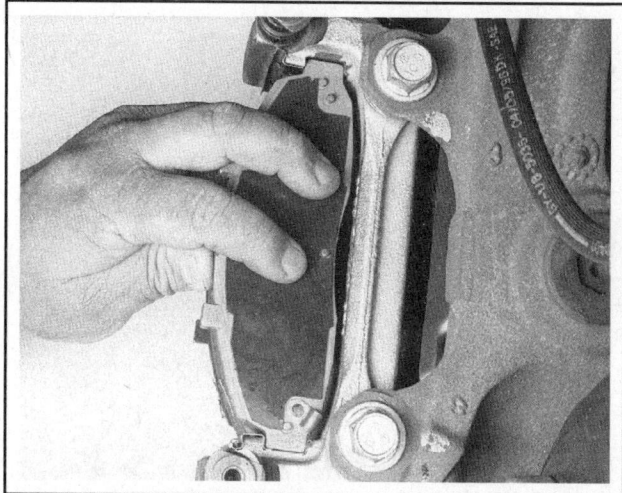

3.4l Install the new inner brake pad; make sure both ends of the pad are properly seated in the anchor plate and anti-rattle clips

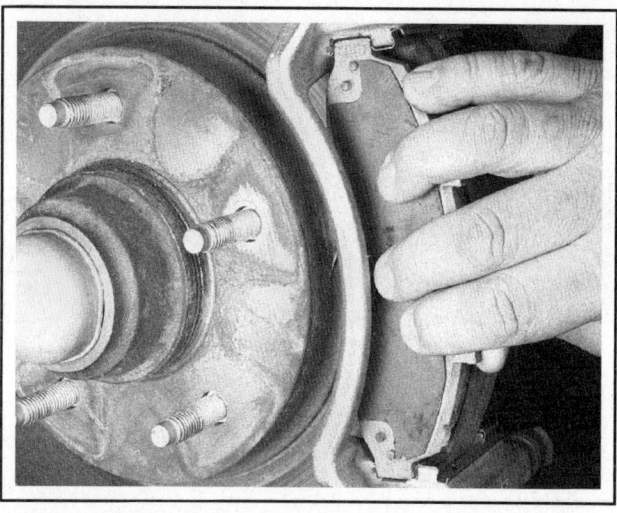

3.4m Install the new outer brake pad; make sure both ends of the pad are properly seated in the anchor plate and anti-rattle clips

9-6 BRAKES

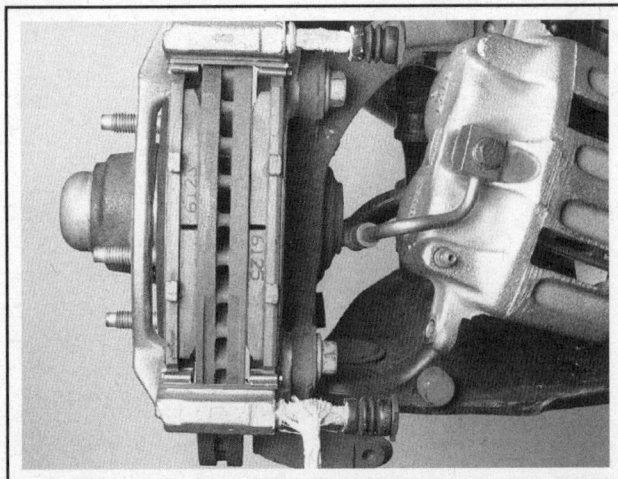

3.4n Apply high temperature grease to the caliper slider pins

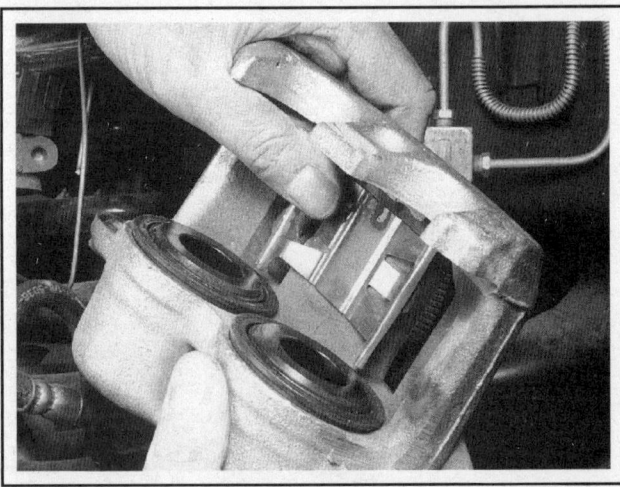

3.4o Make sure that the big anti-rattle clip inside the caliper is in good condition and fully seated

3.4p Install the caliper over the pads, install the bolts and tighten them to the torque listed in this Chapter's Specifications

3.4q Wash down the disc and brake pads with brake cleaner to remove brake dust; DO NOT blow off brake dust with compressed air

3.4r Remove the two caliper mounting bolts (arrow indicates upper bolt), check them for thread damage and replace as necessary

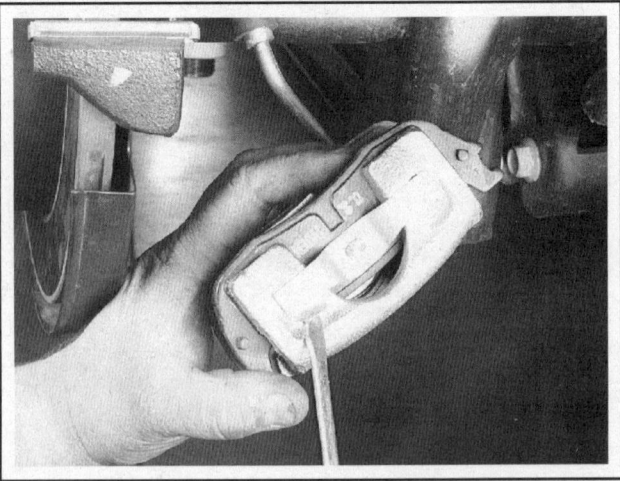

3.4s Remove the rear caliper and pry off the old outer brake pad

BRAKES 9-7

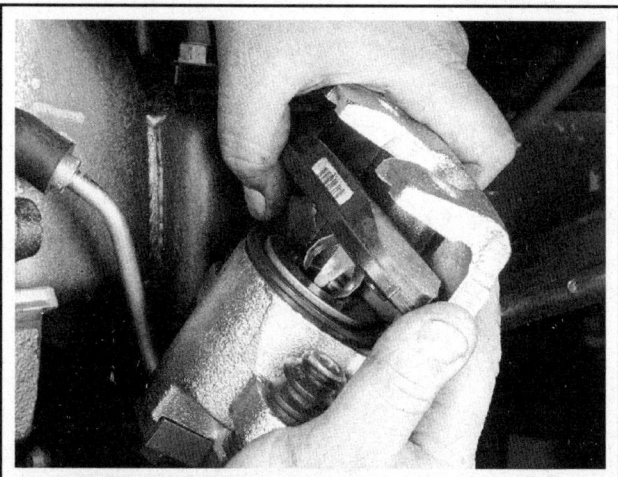

3.4t Pop the inner brake pad retainer out of the caliper piston face and remove the inner pad

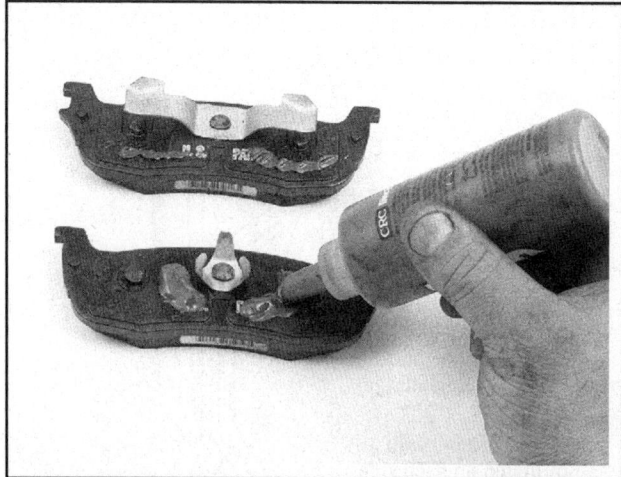

3.4u Apply some anti-squeal compound to the backing plates of the new pads

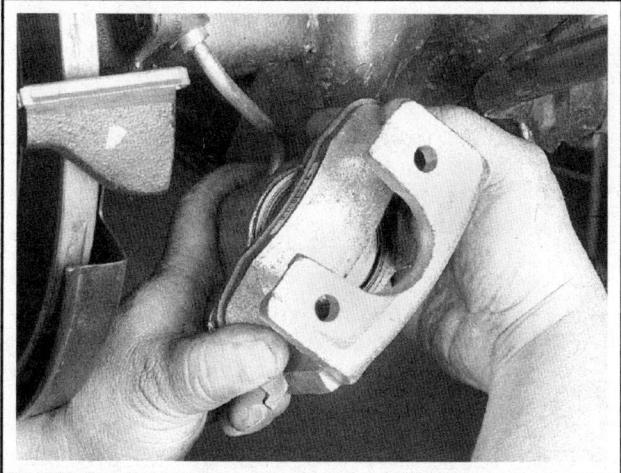

3.4v Install the new inner brake pad; make sure the retainer is seated all the way into the hole in the piston face

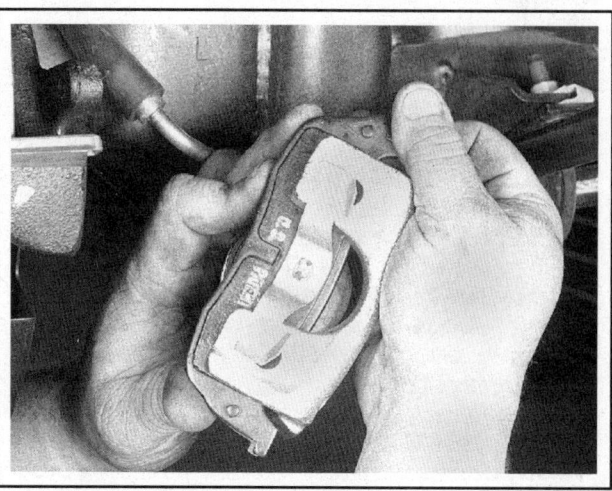

3.4w Install the new outer brake pad; make sure the retainer is fully seated on the caliper bridge; install the caliper and tighten the caliper bolts to the torque listed in this Chapter's Specifications

5 While the pads are removed, inspect the caliper for brake fluid leaks and ruptures in the piston boot. Overhaul or replace the caliper as necessary (see Section 4). Also inspect the brake disc carefully (see Section 5). If machining is necessary, follow the information in that Section to remove the disc.

6 Before installing the caliper mounting bolts, clean them and check them for corrosion and damage. If they're significantly corroded or damaged, replace them. Be sure to tighten the caliper mounting bolts to the torque listed in this Chapter's Specifications.

7 Install the brake pads on the opposite wheel, then install the wheels and lower the vehicle. Tighten the lug nuts to the torque listed in the Chapter 1 Specifications.

8 Add brake fluid to the reservoir until it's full (see Chapter 1). Pump the brakes several times to seat the pads against the disc, then check the fluid level again.

9 Check the operation of the brakes before driving the vehicle in traffic. Try to avoid heavy brake applications until the brakes have been applied lightly several times to seat the pads.

2003 AND LATER EXPEDITION/NAVIGATOR MODELS

▶ **Refer to illustrations 3.10a, 3.10b, 3.10c and 3.10d**

10 Follow the photo sequence (see illustrations 3.4a through 3.4w), with the following differences:

  a) *2003 through 2006 models use a long spring (anchor housing spring) attached on the outside of each caliper.*

❋❋ WARNING:

These anchor housing springs have different tabs on the top and bottom and must be reinstalled in the exact same position as they were originally (see illustration).

➡ **Note: Remember to work on one caliper at a time so that the other caliper can be referred to if necessary.**

9-8 BRAKES

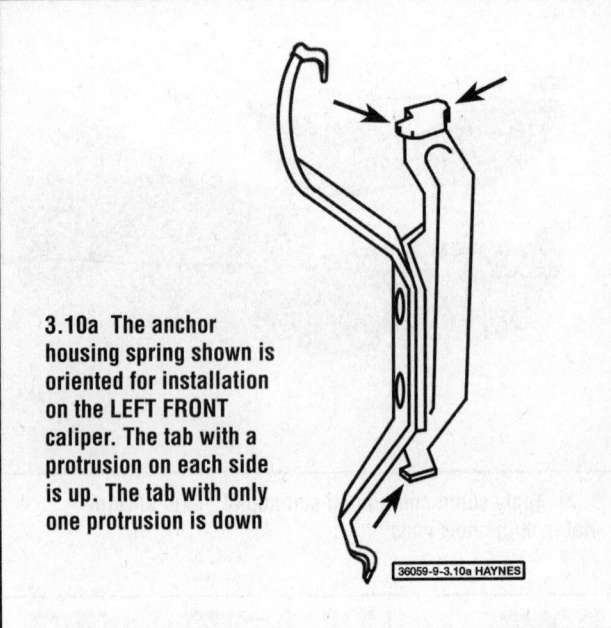

3.10a The anchor housing spring shown is oriented for installation on the LEFT FRONT caliper. The tab with a protrusion on each side is up. The tab with only one protrusion is down

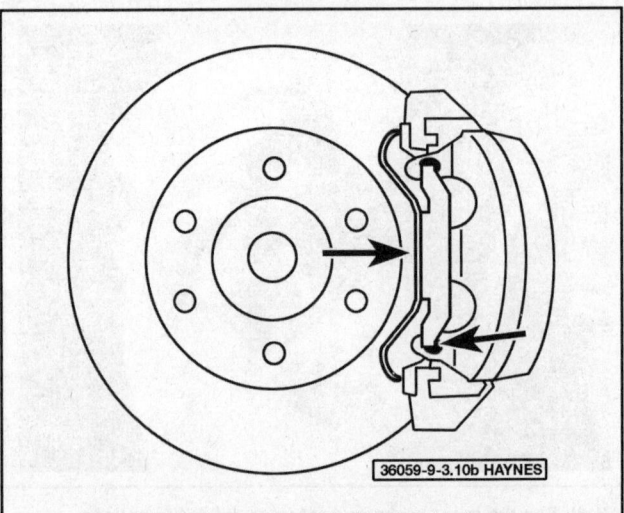

3.10b Push the center of the spring rearward while simultaneously pulling the lower tab from the hole in the caliper

b) To release the spring on the LEFT FRONT (driver's side) caliper, push the center of the spring inward (toward the rear) while pulling the BOTTOM tab out of the hole in the caliper (see illustration). Swivel and rotate the spring so that the TOP tab can be guided out of its hole in the caliper (see illustration).

c) To release the spring on the RIGHT FRONT (passenger's side) caliper, push the center of the spring inward (toward the rear) while pulling the TOP tab out of the hole in the caliper. Swivel and rotate the spring so that the BOTTOM tab can be guided out of its hole in the caliper.

d) Remove the pads from the caliper mounting bracket. On the outer brake pads of 2007 through 2009 models, rotate the clips 90-degrees to release them from their notches in the mounting bracket (see illustration).

e) Remove and discard the caliper bushing caps, the caliper bolts and the caliper bushings. The manufacturer recommends using new hardware for reassembly.

f) Apply a small amount of lubricant to the inside of the new caliper bushings before installing them into the caliper bores. Install the replacement caliper bolts and tighten them to the torque listed in this Chapter's Specifications.

g) After installing the brake pads, install the anchor housing spring to its original position. It is installed in the exact reverse of removal. If the anchor housing spring is damaged, replace it with a new one.

h) Rear disc brake pad replacement is similar, except that the rear caliper does not have the long springs and the pads do not have the rotating clips.

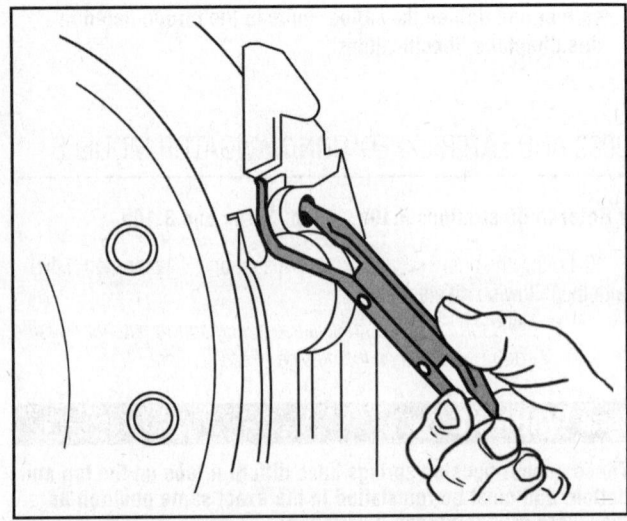

3.10c Swivel and rotate the spring to release the upper tab from the hole in the caliper

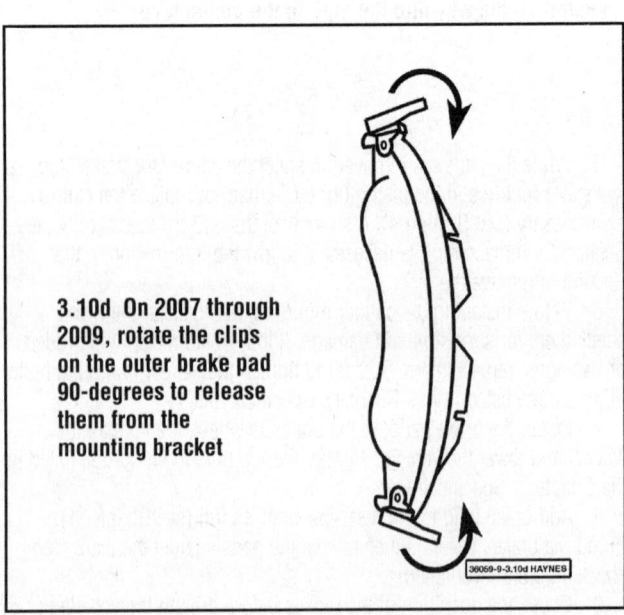

3.10d On 2007 through 2009, rotate the clips on the outer brake pad 90-degrees to release them from the mounting bracket

BRAKES 9-9

4 Disc brake caliper - removal, overhaul and installation

Refer to illustrations 4.3, 4.6, 4.7, 4.8, 4.12 and 4.14

→**Note:** If an overhaul is indicated (usually because of fluid leaks, a stuck piston or broken bleeder screw) explore all options before beginning this procedure. New and factory rebuilt calipers are available on an exchange basis, which makes this job quite easy. If it's decided to rebuild the calipers, make sure rebuild kits are available before proceeding. Always rebuild or replace the calipers in pairs - never rebuild just one of them.

1 If the vehicle is equipped with air suspension, turn off the air suspension system. The switch is located in the area of the right kick panel (see Chapter 10).

WARNING:

On models with air suspension, electrical power to the air suspension system must be turned off before raising the vehicle. Failure to do so can result in a sudden inflation or deflation of the air springs, causing instability of the vehicle while it's off the ground.

2 Loosen the wheel lug nuts, raise the vehicle and support it securely on jackstands. Apply the parking brake. Remove the wheels.
3 Remove the brake hose-to-caliper banjo bolt (see illustration).

→**Note:** If you're removing the caliper simply to access other components, do NOT disconnect the brake hose.

Discard the two copper sealing washers on each side of the banjo fitting; use new ones when you reattach the brake hose to the caliper. Wrap a plastic bag around the end of the hose to prevent fluid loss and contamination.

4 Remove the caliper mounting bolts (see illustration 3.4b or 3.4r) and remove the caliper.
5 Clean the caliper with brake system cleaner. DO NOT use kerosene or petroleum-based solvents.
6 Carefully knock the dust boot(s) from the caliper (see illustration).
7 Place a block of wood between the caliper bridge and the piston(s), then force the piston(s) out of the bore(s) with compressed air applied through the hydraulic fluid inlet (see illustration). Use low air pressure; a small tire pump may be adequate.

WARNING:

Keep your hands away from the pistons when blowing them out!

8 Remove the piston seal(s) (see illustration).

4.3 Remove the brake hose banjo bolt

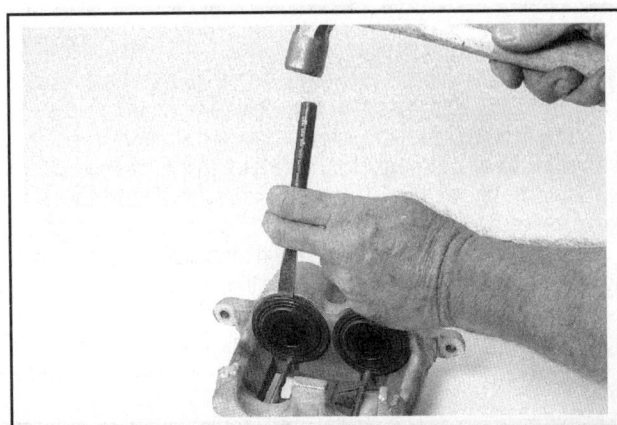

4.6 Gently tap the dust boots loose with a hammer and chisel

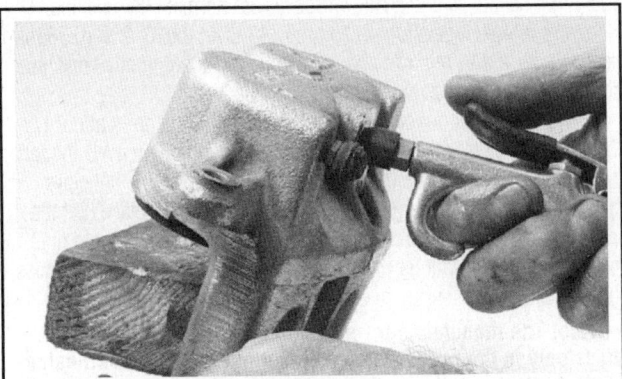

4.7 Using a wooden block as a cushion, ease the piston(s) out of the caliper bore(s) with low-pressure compressed air (on dual-piston calipers, make sure the block is thick enough that neither piston can be popped all the way out; otherwise, you'll have a difficult time removing the other piston!)

4.8 To remove the seal from the caliper bore, use a plastic or wooden tool, such as a pencil

9-10 BRAKES

4.12 Position the pistons square to the bores, then compress them into the caliper bores with a C-clamp and block of wood; make sure you leave each piston sticking out a little so that you can seat the dust boot into the caliper

4.14 Make sure that the big anti-rattle clip is in good condition and is fully seated

9 Clean the piston(s) and cylinder bore(s) with brake system cleaner, clean brake fluid or denatured alcohol only.

※ WARNING:

DO NOT, under any circumstances, clean brake system parts with gasoline or petroleum-based solvents.

10 Inspect the surfaces of the piston for nicks and burrs and loss of plating. If surface defects are present, the pistons must be replaced. Check the caliper bore in a similar way. Light polishing with crocus cloth is permissible to remove light corrosion and stains. If damage isn't removable with a crocus cloth, replace the caliper assembly. Do NOT hone the cylinder bore. If the bores show signs of wear, replace the caliper assembly.

11 Lubricate the new piston seal(s) with clean brake fluid and position the seal(s) in the cylinder groove(s) with your fingers.

12 Install the new dust boot(s) onto the piston(s), making sure the boot seats in the groove at the top of the piston. Lubricate the caliper bore(s) and the piston(s) with clean brake fluid. Insert the piston(s) into the bore(s). Make sure the piston(s) are "square" to the caliper bore(s), then depress the piston(s) most of the way into the bore with a C-clamp and a block of wood (see illustration). Do NOT push the pistons all the way into the bore yet.

13 Make sure the outer circumference of each new boot is fully seated in the caliper bore and the inner lip is properly seated into the groove in the piston, then slowly and carefully push each piston and boot into the bore until it's fully bottomed. If the boot is difficult to seat, use a hammer and a blunt punch to gently tap it into place.

14 Install the large anti-rattle clip, if removed (see illustration).

15 Install the caliper and tighten the mounting bolts to the torque listed in this Chapter's Specifications.

16 Connect the brake hose to the caliper, using new sealing washers. Make sure the brake hose is angled properly and that nothing interferes with the hose, then tighten the brake hose banjo bolt to the torque listed in this Chapter's Specifications.

17 Lower the vehicle and reactivate the air suspension system, if equipped (see Chapter 10).

18 Bleed the hydraulic system (see Section 10).

5 Brake disc - inspection, removal and installation

INSPECTION

▸ Refer to illustrations 5.4, 5.5a and 5.5b

1 If the vehicle is equipped with air suspension, turn off the air suspension system. The switch is located in the area of the right kick panel (see Chapter 10).

※ WARNING:

On models with air suspension, electrical power to the air suspension system must be turned off before raising the vehicle. Failure to do so can result in a sudden inflation or deflation of the air springs, causing instability of the vehicle while it's off the ground.

2 Loosen the wheel lug nuts, raise the vehicle and support it securely on jackstands. Apply the parking brake. Remove the wheels.

3 Visually inspect the disc surface for score marks and other damage. Light scratches and shallow grooves are normal after use and won't affect brake operation. Deep grooves - over 0.015-inch (0.38 mm) deep - require disc removal and refinishing by an automotive machine shop. Be sure to check both sides of the disc.

4 To check disc runout, place a dial indicator at a point about 1/2-inch from the outer edge of the disc (see illustration). On 4WD model front discs and all rear discs, install the lug nuts, with the flat sides facing in, and tighten them securely to hold the disc in place. Set the indicator to zero and turn the disc. The indicator reading should not exceed the runout limit listed in this Chapter's Specifications. If it does, the disc should be refinished by an automotive machine shop.

➡**Note: The manufacturer recommends resurfacing the brake discs only in the event of pedal pulsations or hard (overheated) spots on the disc. If you elect not to have the discs resurfaced, deglaze them with sandpaper or emery cloth.**

5 The disc must not be machined to a thickness less than the specified minimum thickness, which is cast into the disc (see illustration). The disc thickness can be checked with a micrometer (see illustration).

BRAKES

5.4 Measure the brake disc runout with a dial indicator

5.5a The minimum (discard) thickness of the brake disc is cast into the disc

5.5b Measure the brake disc thickness at several points with a micrometer

5.7 The front caliper anchor bracket is retained by two bolts (arrows)

REMOVAL AND INSTALLATION

▶ Refer to illustration 5.7

6 Remove the brake calipers (don't disconnect the brake hoses) and hang them out of the way (see Section 4).

7 If you're removing a front disc, remove the caliper anchor bracket (see illustration).

8 On 2WD model front discs, remove the grease cap, wheel bearing retainer nut, spindle nut, outer bearing retainer washer and outer wheel bearing, then remove the disc/hub (see *Front wheel bearing check, repack and adjustment* in Chapter 1). On 4WD model front discs and all rear discs, simply remove the lug nuts installed in Step 4 and pull the disc off the hub.

9 Installation is the reverse of removal.

10 Lower the vehicle and reactivate the air suspension system, if equipped. Tighten the wheel lug nuts to the torque listed in the Chapter 1 Specifications.

6 Drum brake shoes - replacement

▶ Refer to illustrations 6.4 and 6.5a through 6.5qq

❈❈ WARNING:
Brake shoes must be replaced on both wheels at the same time - never replace the shoes on only one wheel. Also, brake system dust is hazardous to your health. DO NOT blow it out with compressed air and DO NOT inhale it. DO NOT use gasoline or solvent to remove the dust. Use brake system cleaner or denatured alcohol only.

1 If the vehicle is equipped with air suspension, turn off the air suspension system. The switch is located in the area of the right kick panel (see Chapter 10).

❈❈ WARNING:
On models with air suspension, electrical power to the air suspension system must be turned off before raising the vehicle. Failure to do so can result in a sudden inflation or deflation of the air springs, causing instability of the vehicle while it's off the ground.

2 Loosen the wheel lug nuts, raise the rear of the vehicle and support it securely on jackstands. Block the front wheels to keep the vehicle from rolling off the stands. Remove the rear wheels.

3 Remove the brake drum. If the drum is stuck because of corrosion between the axle flange and the wheel studs or brake drum, spray penetrating oil around the flange and studs and allow it to soak in. Tap

9-12 BRAKES

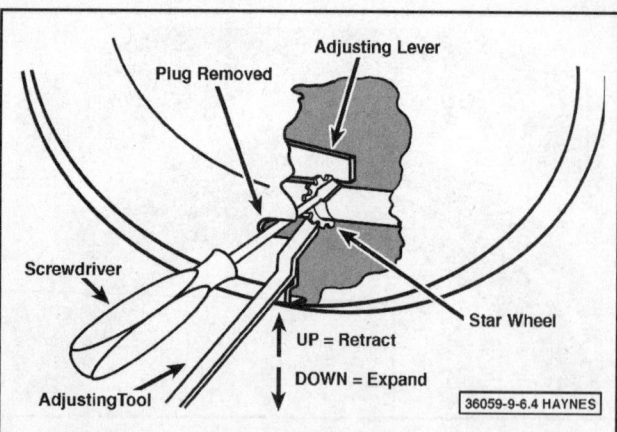

6.4 Use a thin screwdriver to push the lever away, then use an adjusting tool or another screwdriver to back off the star wheel

6.5a Wash down the brake assembly with brake cleaner; DO NOT blow it out with compressed air!

6.5b Remove the secondary brake shoe retracting spring (the spring tool shown here is available at most auto parts stores and makes this job much easier and safer) . . .

around the studs with a hammer and flange to break the drum loose, then tap around the back edge of the drum to remove it.

4 If the drum is locked onto the shoes because of excessive drum wear, knock out the plug in the access hole in the backing plate, insert a screwdriver through the hole, push the actuator lever off the star adjuster (see illustration) and, using another screwdriver or a brake adjuster tool, back off the adjuster wheel to retract the shoes.

➡Note: If the plug is already knocked out, make sure it's not lodged somewhere inside the brake assembly.

Be sure to install a plug (available at most auto parts stores) in the hole when you're done to prevent water from entering the brake assembly.

5 Clean the brake assembly with brake system cleaner before beginning work (see illustration). Follow illustrations 6.5b through 6.5qq for the inspection and replacement of the brake shoes. Be sure to stay in order and read the caption under each illustration.

6 Clean the brake drum and check it for score marks, deep

6.5c . . . then remove the primary shoe retracting spring

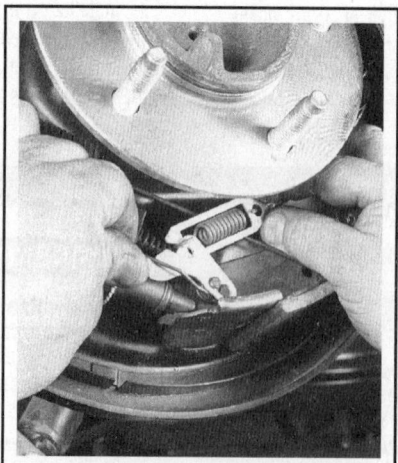

6.5d Disengage the brake shoe adjusting lever cable from the brake shoe adjusting lever

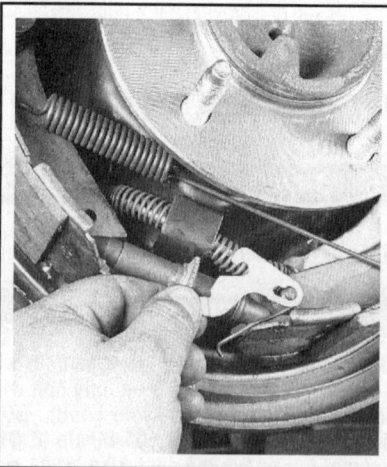

6.5e Remove the brake shoe adjusting lever

Brakes 9-13

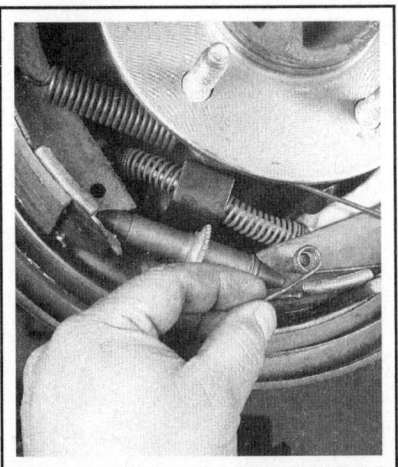

6.5f Remove the brake shoe adjusting lever spring

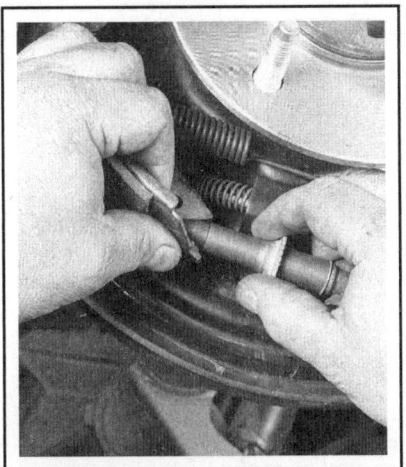

6.5g Remove the brake shoe adjuster assembly

6.5h Remove the upper end of the brake shoe adjusting lever cable from the anchor pin

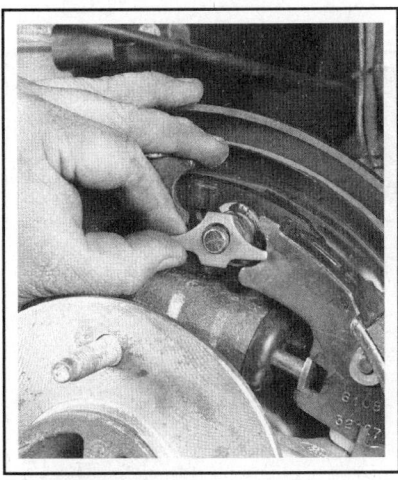

6.5i Remove the anchor pin guide plate from the anchor pin

6.5j Remove the cable guide

6.5k Remove the parking brake link and spring

6.5l Using a hold-down spring tool (available at most auto parts stores), remove the front hold-down spring assembly . . .

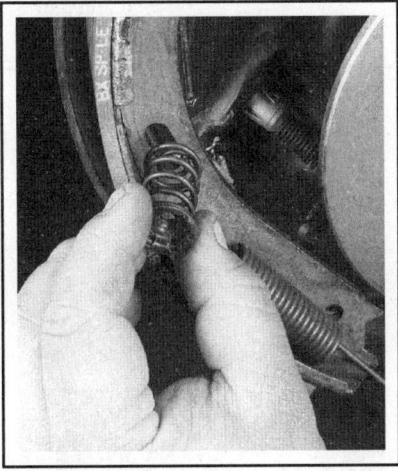

6.5m . . . which includes the upper washer, the spring, the lower washer and the pin

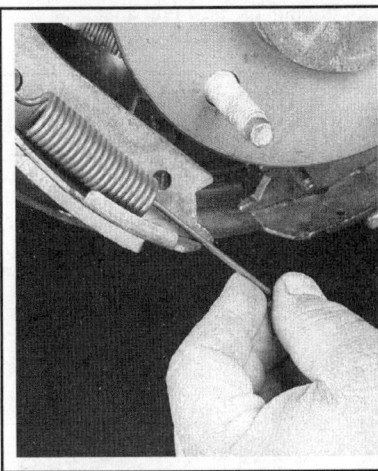

6.5n Disengage the brake shoe adjusting screw spring from the brake shoes

9-14 BRAKES

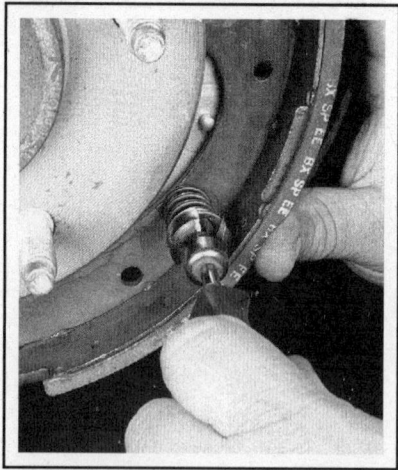
6.5o Remove the rear hold-down spring assembly

6.5p Pry off this E-clip retainer with a small screwdriver, disengage the parking brake lever pin from the rear shoe and remove the parking brake lever (don't lose the washer on the other side, between the lever and the shoe)

6.5q Lubricate the shoe contact points on the backing plate with high-temperature brake grease

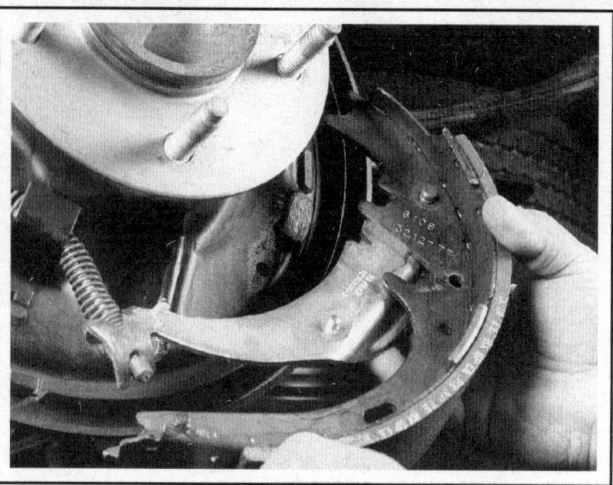
6.5r Put the washer on the parking brake lever pin, insert the pin through the new rear shoe . . .

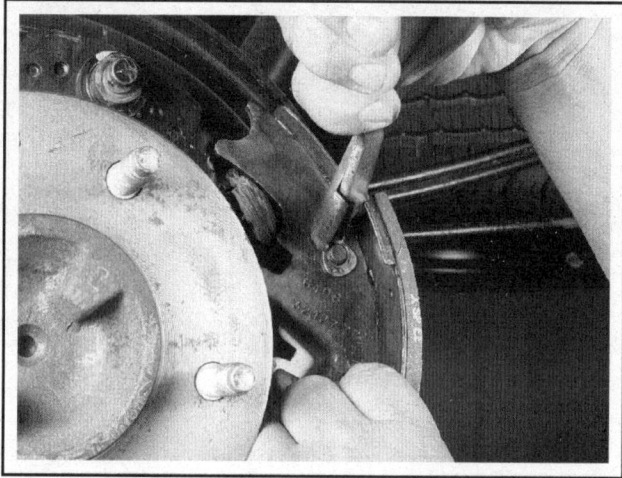
6.5s . . . and install the E-clip retainer, squeezing its end together

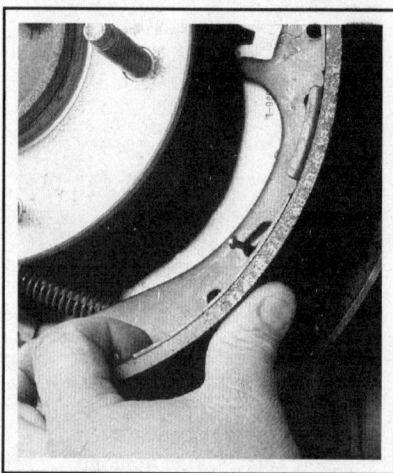
6.5t Install the hold-down pin through the holes in the backing plate and rear shoe . . .

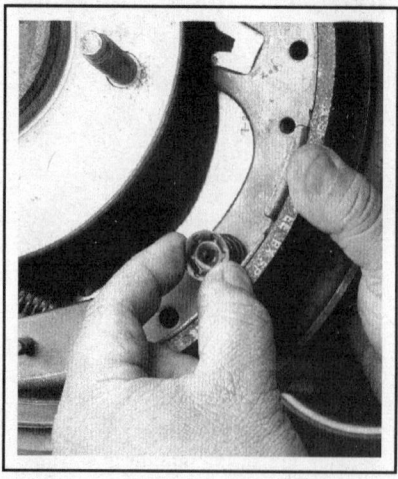
6.5u . . . install the rear hold-down spring assembly . . .

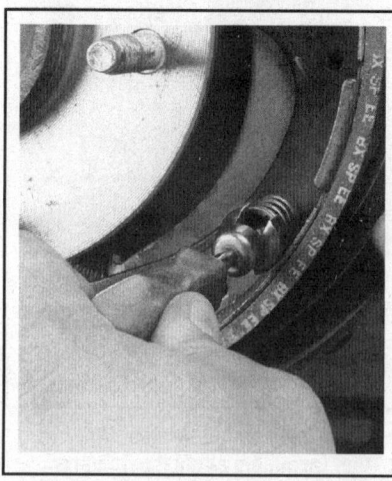
6.5v . . . and give it a 90-degree twist with a hold-down tool to lock it into place

BRAKES 9-15

6.5w Install the guide plate on the anchor pin

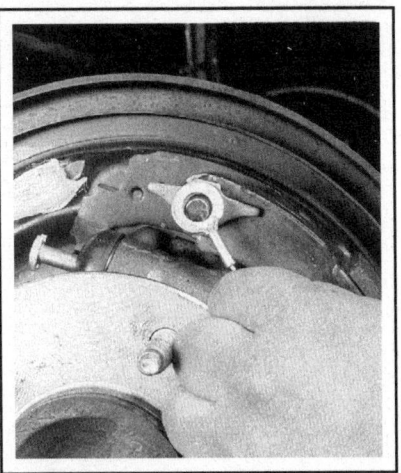

6.5x Install the brake shoe adjusting lever cable on the anchor pin

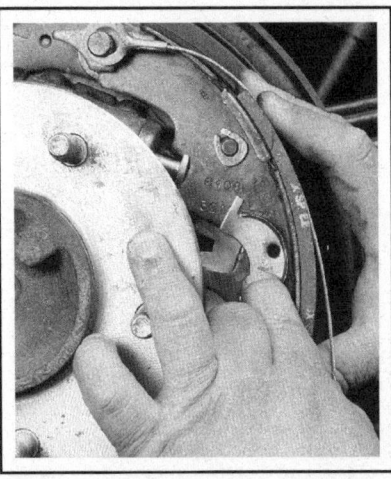

6.5y Install the brake shoe adjusting lever cable guide on the rear shoe

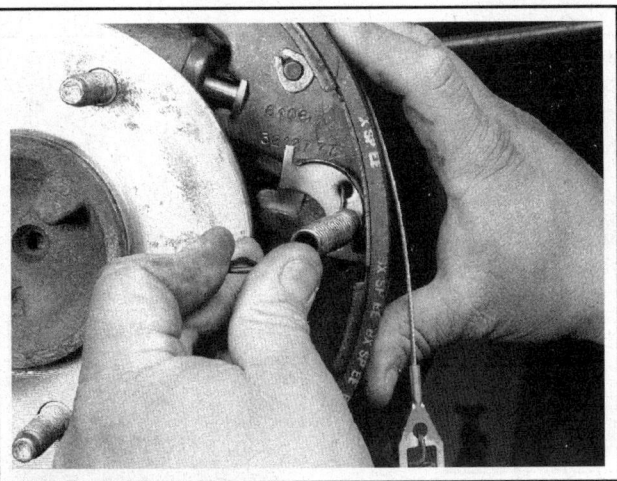

6.5z Hook the short end of the secondary (long) shoe retracting spring into the hole in the cable guide and into the rear shoe as shown . . .

6.5aa . . . and hook the long end of the spring over the anchor pin. Warning: *Wear safety goggles during this step to protect your eyes if the spring slips off the spring tool*

6.5bb Thread the brake shoe adjusting lever cable over the guide

6.5cc Place the parking brake link and spring in position, making sure that the link is correctly engaged with the parking brake lever and secondary shoe

6.5dd Place the primary (front) brake shoe assembly in position, making sure that it's properly engaged with the parking brake lever link

9-16 BRAKES

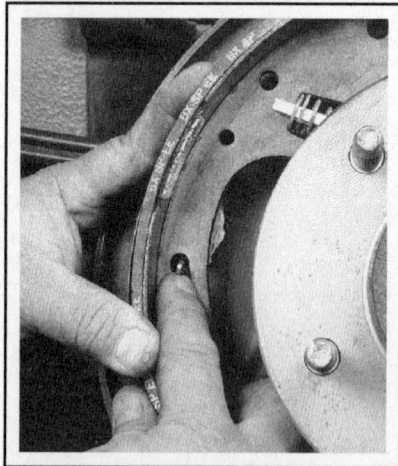

6.5ee Insert the hold-down pin through the holes in the backing plate and the front brake shoe . . .

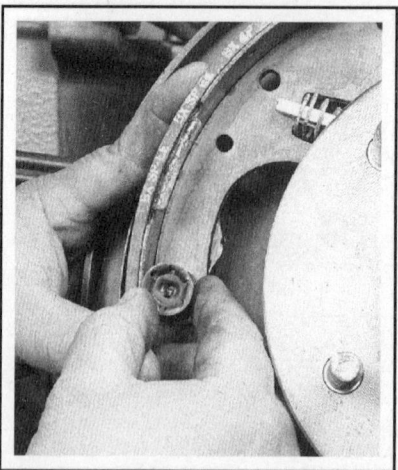

6.5ff . . . install the hold-down spring assembly . . .

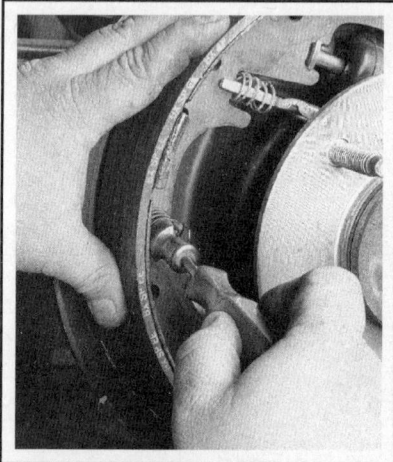

6.5gg . . . and give the hold-down spring assembly a 90-degree twist with a hold-down tool to lock it into place

6.5hh Hook the short end of the primary (shorter) brake shoe retracing spring into this hole in the front shoe . . .

6.5ii . . . and hook the other end of the spring over the anchor pin. Warning: *Wear safety goggles during this step to protect your eyes in the event that the spring slips off the spring tool*

6.5jj Install the brake shoe adjusting lever spring with the straight end of the spring pointing toward the rear brake shoe and the hooked end facing forward

6.5kk Install the brake shoe adjuster lever

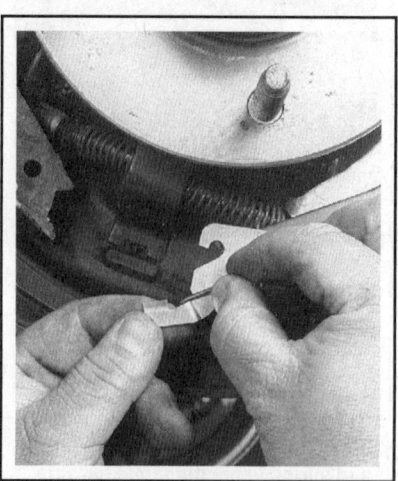

6.5ll Hook the end of the adjusting lever spring over the adjusting lever

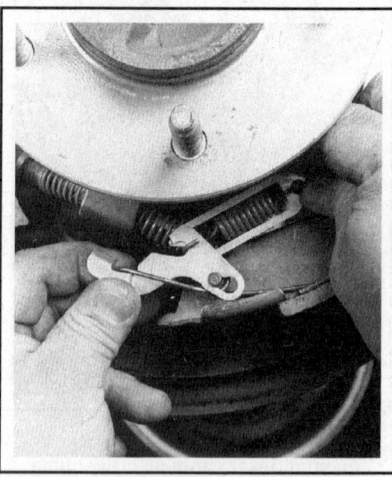

6.5mm Hook the lower end of the brake shoe adjusting lever cable over the adjusting lever

BRAKES 9-17

6.5nn Hook the short end of the lower shoe retracting spring into the front shoe . . .

6.5oo . . . and hook the long end of the spring into the rear shoe

6.5pp Engage the rear end of the adjuster assembly with its notch in the rear shoe . . .

grooves, hard spots (which will appear as small discolored areas) and cracks. If the drum is worn, scored or out-of-round, it can be resurfaced by an automotive machine shop.

➡ **Note:** Professionals recommend resurfacing the drums whenever a brake job is done. Resurfacing will eliminate the possibility of out-of-round drums. If the drums are worn so much they can't be resurfaced without exceeding the maximum allowable diameter (stamped into the drum), new ones will be required. At the very least, if you elect not to have the drums resurfaced, remove the glazing from the surface with medium-grit emery cloth using a swirling motion.

7 Repeat this procedure for the other rear brake assembly.
8 Install the brake drums. Pump the brake several times, then turn the adjuster star wheels using a screwdriver inserted through the hole in the backing plate until the shoes slightly drag on the drums as the drums are turned. Now, back off the adjuster until the shoes don't drag on the drums.
9 Install the rear wheels, install the lug nuts, lower the vehicle and tighten the wheel lug nuts to the torque listed in the Chapter 1 Specifications.
10 Check the brake pedal position. If the brake pedal goes too close to the floor, further adjustment of the brakes is required. Back the vehi-

6.5qq . . . pull the front shoe forward slightly and engage the front end of the adjuster assembly with its notch in the front shoe

cle up, making repeated stops, to actuate the self-adjusters, which work only when the vehicle is in reverse. Test the brakes for proper operation before driving in traffic.

7 Wheel cylinder - removal, overhaul and installation

➡ **Note:** If an overhaul is indicated (usually because of fluid leakage or sticky operation) explore all options before beginning the job. New wheel cylinders are available, which makes this job quite easy. If you decide to rebuild the wheel cylinder, make sure a rebuild kit is available before proceeding. Never overhaul only one wheel cylinder. Always rebuild both of them at the same time.

REMOVAL

▸ **Refer to illustration 7.2**

1 Remove the brake drum and brake shoes (see Section 6).
2 Unscrew the brake line fitting from the rear of the wheel cylinder (see illustration). If available, use a flare-nut wrench to avoid rounding off the corners on the fitting. Don't pull the metal line out of the wheel cylinder - it could bend, making installation difficult.

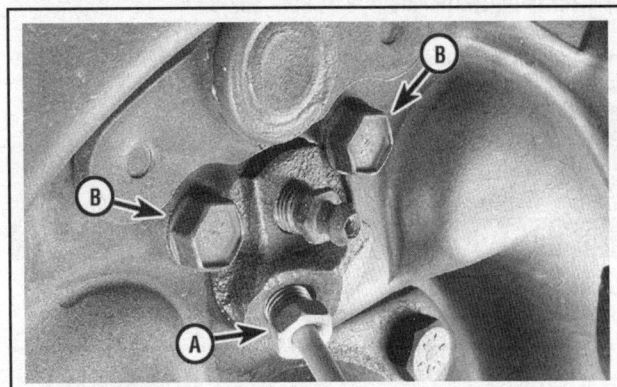

7.2 To remove the wheel cylinder, loosen the brake line threaded fitting (A) with a flare-nut wrench, then remove the wheel cylinder mounting bolts (B)

9-18 BRAKES

3 Remove the two bolts securing the wheel cylinder to the brake backing plate.
4 Remove the wheel cylinder.
5 Plug the end of the brake line to prevent the loss of brake fluid and the entry of dirt.

OVERHAUL

♦ Refer to illustration 7.6

6 To disassemble the wheel cylinder, remove the pushrod and rubber boot from each end of the cylinder, push out the two pistons and cups, and remove the expander spring (see illustration). Discard the rubber parts and use new ones from the rebuild kit when reassembling the wheel cylinder.
7 Inspect the pistons for scoring and scuff marks. If present, the pistons should be replaced with new ones.
8 Examine the inside of the cylinder bore for score marks and corrosion. If these conditions exist, the cylinder can be honed slightly to restore it, but replacement is recommended.
9 If the cylinder is in good condition, clean it with brake system cleaner or denatured alcohol.

※※ WARNING:

DO NOT, under any circumstances, use gasoline or petroleum-based solvents to clean brake parts!

10 Remove the bleeder screw and make sure the hole is clean.
11 Lubricate the cylinder bore, cups and pistons with clean brake fluid, then insert one of the new cups into the bore. Make sure the lip on the cup faces in.
12 Place the expander spring in the opposite end of the bore and push it in until it contacts the cup.
13 Install the other cup in the cylinder bore.
14 Attach the rubber boots to the pistons, then install the pistons, boots and pushrods.
15 The wheel cylinder is now ready for installation.

INSTALLATION

16 Installation is the reverse of removal. Attach the brake line to the wheel cylinder before installing the mounting bolts and tighten the line fitting after the wheel cylinder mountings bolts have been tightened. If available, use a flare-nut wrench to tighten the line fitting.
17 Bleed the brakes (see Section 10). Don't drive the vehicle in traffic until the operation of the brakes has been thoroughly tested.

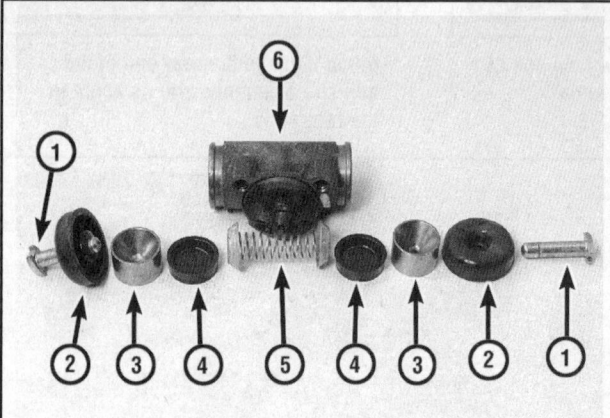

7.6 An exploded view of a typical wheel cylinder

1 Pushrods
2 Dust boots
3 Pistons
4 Cups
5 Expander spring
6 Wheel cylinder

8 Master cylinder - removal, overhaul and installation

REMOVAL

♦ Refer to illustrations 8.2, 8.3, 8.4a and 8.4b

→Note: Before deciding to overhaul the master cylinder, check on the availability and cost of a new or factory rebuilt unit and also the availability of a rebuild kit.

1 Place rags under the brake line fittings and prepare caps or plastic bags to cover the ends of the lines once they're disconnected.

※※ CAUTION:

Brake fluid will damage paint. Cover all painted surfaces and avoid spilling fluid during this procedure.

2 If the vehicle is equipped with cruise control, unplug the electrical connector from the brake pressure switch (see illustration). On 2005 through 2009 models, remove the air filter housing (see Chapter 4) and unbolt and set aside the degas bottle (see Chapter 3). On 2010

8.2 If the vehicle is equipped with cruise control, unplug the electrical connector from the brake pressure switch

and later models, remove the air filter housing/coolant expansion tank assembly (see Chapter 4). On models so equipped, disconnect any hose locating mounts from the master cylinder.

Brakes

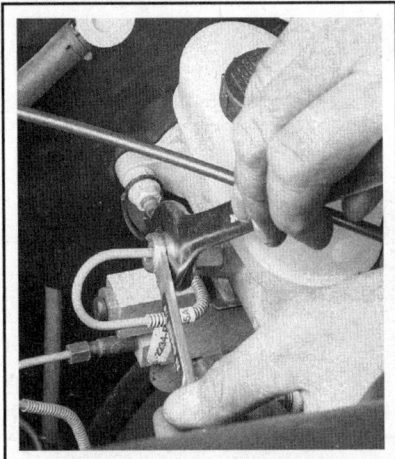 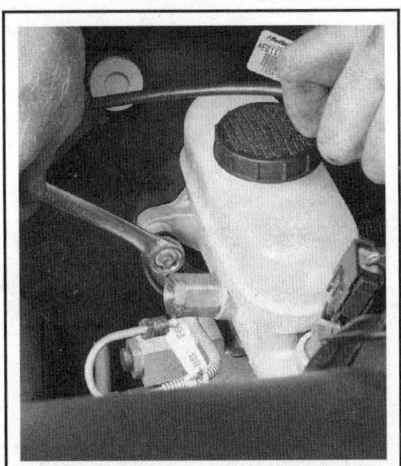 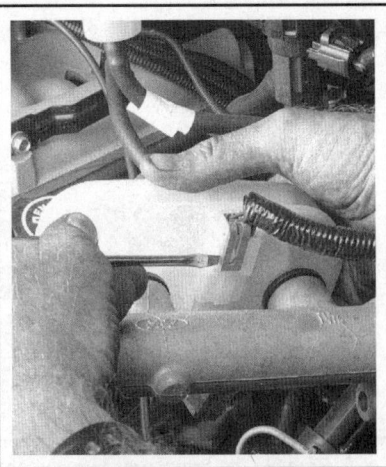

8.3 Unscrew the brake line threaded fittings with a flare-nut wrench; use a back-up wrench to hold the fluid control valve while loosening the rear line fitting

8.4a To detach the master cylinder assembly from the power brake booster, remove the two master cylinder mounting nuts (left nut not visible in this photo)

8.4b Pull the master cylinder off its mounting studs, turn it on its side and unplug the electrical connector from the low fluid level warning switch

3 Loosen the tube nuts at the ends of the brake lines where they enter the master cylinder. To prevent rounding off of the flats on these nuts, a flare-nut wrench, which wraps around the nut, should be used (see illustration). Pull the brake lines away from the master cylinder slightly and plug the ends to prevent contamination.

4 Remove the two master cylinder mounting nuts (see illustration), pull the master cylinder off its mounting studs, turn it on its side and unplug the electrical connector for the low fluid level warning switch (see illustration). Remove the master cylinder from the vehicle.

5 Remove the reservoir cap, then discard any fluid remaining in the reservoir.

OVERHAUL AND INSTALLATION

▶ **Refer to illustrations 8.6, 8.7, 8.8, 8.9, 8.10, 8.11, 8.16, 8.17 and 8.24**

6 Pry the reservoir off the master cylinder (see illustration). Pry out the old reservoir grommets and inspect them for wear. If they're cracked or torn, discard them and get new ones.

7 Depress the primary piston with a screwdriver and remove the snap-ring with a pair of snap-ring pliers (see illustration).

8 Remove the fluid control valve (see illustration).

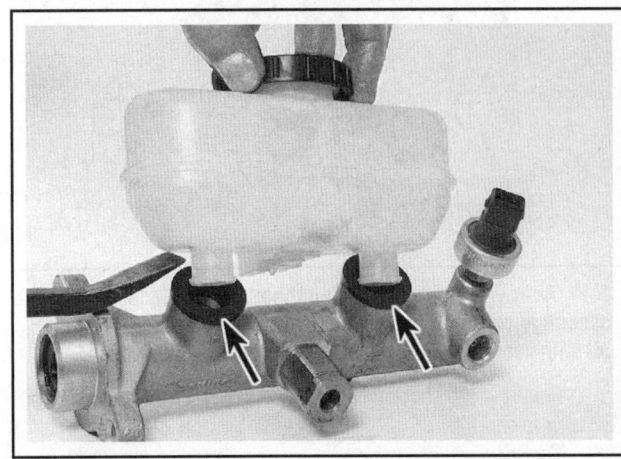

8.6 Pry the reservoir off the master cylinder with a small prybar or a big screwdriver, then pry out the reservoir rubber grommets (arrows)

8.7 Push the primary piston into the master cylinder with a Phillips screwdriver and remove the snap-ring with a pair of snap-ring pliers

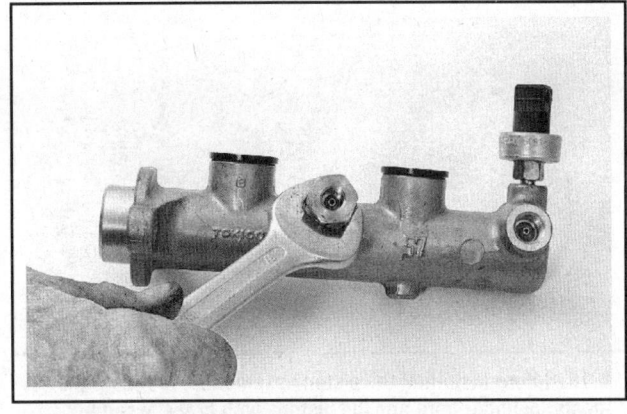

8.8 Remove the fluid control valve from the master cylinder

9-20 BRAKES

8.9 Remove the brake pressure switch, if equipped, from the master cylinder

8.10 Remove the primary and secondary pistons from the master cylinder; if the secondary piston sticks, tap the cylinder against a block of wood to eject it

9 Remove the brake pressure switch, if equipped (see illustration).
10 Remove the primary piston and secondary piston assemblies from the cylinder bore. It may be necessary to tap the master cylinder against a block of wood to expel the secondary piston (see illustration).
11 Wash all the parts with clean brake fluid, blow them dry and lay them out for inspection (see illustration).
12 Inspect the cylinder bore for damage or excessive wear. If any damage is found, replace the master cylinder body; abrasives cannot be used on the bore.
13 Inspect the primary and secondary piston assemblies for signs of wear, corrosion or damage. If any of these parts look corroded or damaged, rebuild the master cylinder or replace it with a new or rebuilt unit.
14 Remove the old seals from the pistons. Install the new secondary piston seals with the lips facing away from each other. The lip on the primary seal must face toward the secondary piston.
15 Attach the springs to the piston assemblies.
16 Lubricate the cylinder bore and the piston assemblies (see illustration) with brake assembly lube or clean brake fluid and install both piston assemblies.
17 Depress the primary piston with a screwdriver and install the snap ring with a pair of snap-ring pliers (see illustration).

8.11 After you've disassembled the master cylinder assembly, wash all the parts in clean brake fluid, blow them off with compressed air, and lay them out for inspection:

1 Reservoir
2 Snap ring
3 Primary piston/spring assembly
4 Secondary piston/spring assembly
5 Reservoir grommets
6 Master cylinder

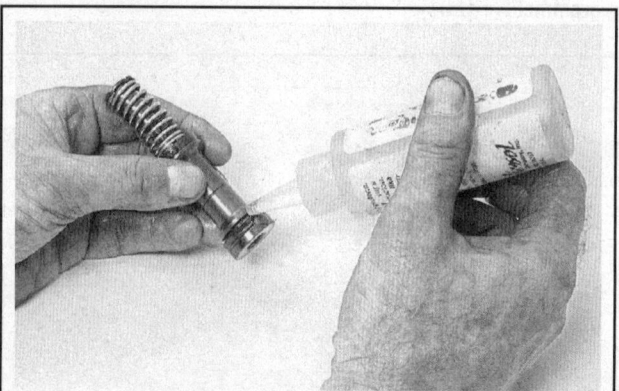

8.16 Before installing the piston assemblies, lubricate the master cylinder bore and the pistons with brake assembly lube or clean brake fluid

8.17 Depress the primary piston with a Phillips screwdriver and install the snap-ring with a pair of snap-ring pliers

BRAKES 9-21

18 Lubricate the new reservoir grommets with brake fluid and press them into the master cylinder body. Make sure they're properly seated.

19 Lay the reservoir on a hard surface and press the master cylinder body onto the reservoir, using a rocking motion.

20 Inspect the reservoir cover and diaphragm for cracks and deformation. Replace any damaged parts with new ones and attach the diaphragm to the cover.

21 Whenever the master cylinder is removed, the entire hydraulic system must be bled. The time required to bleed the system can be reduced if the master cylinder is filled with fluid and bench bled (refer to Steps 22 through 24) before the master cylinder is installed on the vehicle.

22 Fill the reservoirs with brake fluid. The master cylinder should be supported so the brake fluid won't spill during the bench bleeding procedure.

23 Hold your fingers tightly over the holes where the brake lines normally connect to the master cylinder to prevent air from being drawn back into the master cylinder.

24 Stroke the piston several times to ensure all air has been expelled. A large Phillips screwdriver can be used to push on the piston assembly (see illustration). Wait several seconds each time for brake fluid to be drawn from the reservoir into the piston bore, then depress the piston again, removing your finger as brake fluid is expelled. Be sure to put your fingers back over the holes each time before releasing the piston. When the bleeding procedure is complete, temporarily install plugs in the holes.

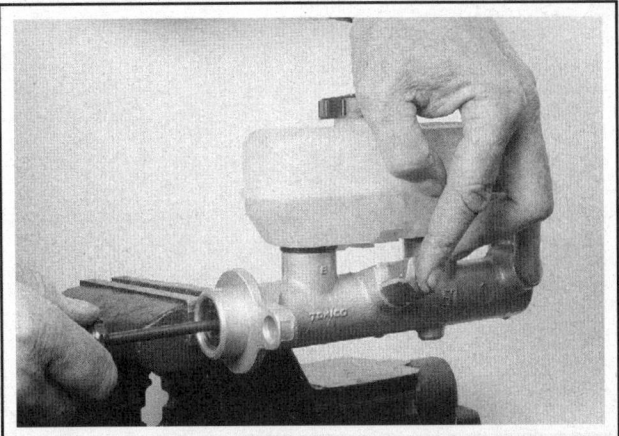

8.24 With your fingers plugging both outlets, stroke the piston several times to ensure all air has been expelled: wait several seconds each time for brake fluid to be drawn from the reservoir into the piston bore, then depress the piston again, slightly loosening finger pressure at each outlet as brake fluid is expelled, then reapplying pressure each time before releasing the piston

25 Carefully install the master cylinder by reversing the removal steps.

26 Bleed the brake system (see Section 10).

9 Brake hoses and lines - check and replacement

1 If the vehicle is equipped with air suspension, turn off the air suspension system. The switch is located in the area of the right kick panel (see Chapter 10).

※※ WARNING:

On models with air suspension, electrical power to the air suspension system must be turned off before raising the vehicle. Failure to do so can result in a sudden inflation or deflation of the air springs, causing instability of the vehicle while it's off the ground.

2 About every six months, with the vehicle raised and placed securely on jackstands, the flexible hoses which connect the steel brake lines with the front and rear brake assemblies should be inspected for cracks, chafing of the outer cover, leaks, blisters and other damage. These are important and vulnerable parts of the brake system and inspection should be complete. A light and mirror will be needed for a thorough check. If a hose exhibits any of the above defects, replace it with a new one.

FLEXIBLE HOSES

▶ Refer to illustrations 9.3, 9.4a and 9.4b

3 Clean all dirt away from the ends of the hose. To disconnect a

9.3 To disconnect a front brake hose junction block, unscrew the tube nuts (the threaded fittings) with a flare-nut wrench (so you don't round off the corners of the nuts) then remove the junction bracket bolt (arrow)

front brake hose from the brake lines, unscrew the brake line fittings from the hose junction block (see illustration). Be careful not to bend the junction bracket or kink the lines. If necessary, soak the connections with penetrating oil.

9-22 BRAKES

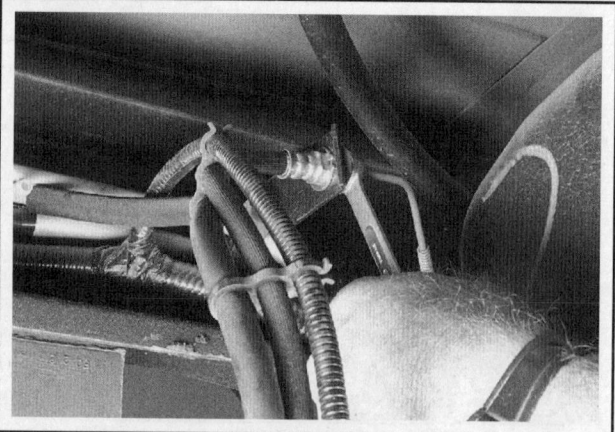

9.4a To disconnect a rear brake hose from the metal line, unscrew the tube nut with a flare-nut wrench . . .

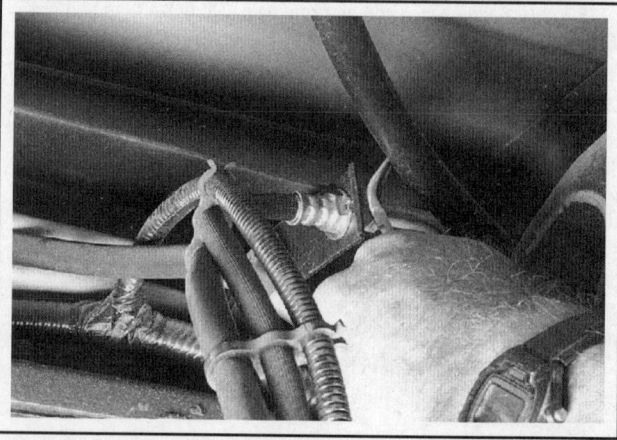

9.4b . . . then remove the retainer clip with a pair of needle-nose pliers

4 To disconnect a rear brake hose from the brake line, unscrew the metal tube nut with a flare nut wrench (see illustration), then remove the U-clip from the female fitting at the bracket (see illustration) and remove the hose from the bracket.

5 Disconnect the hose from the front caliper, or from the rear wheel cylinder or caliper (see Section 4), discarding the copper washers on either side of the fitting.

6 Using new copper washers, attach the new brake hose to the caliper or wheel cylinder.

7 To reattach a front brake hose to the metal lines, bolt the hose junction bracket to the frame, tighten the bolt securely, and screw in the tube nuts.

8 To reattach a rear brake hose to the metal line, insert the end of the hose through the frame bracket, make sure the hose isn't twisted, then attach the metal line by tightening the tube nut fitting securely. Install the U-clip at the frame bracket.

➥Note: *The weight of the vehicle must be on the suspension, so the vehicle should not be raised while positioning the hose.*

9 Carefully check to make sure the suspension or steering components don't make contact with the hose. Have an assistant push down on the vehicle and also turn the steering wheel lock-to-lock during inspection.

10 Bleed the brake system (see Section 10).

METAL BRAKE LINES

11 When replacing brake lines, be sure to use the correct parts. Don't use copper tubing for any brake system components. Purchase steel brake lines from a dealer parts department or auto parts store.

12 Prefabricated brake line, with the tube ends already flared and fittings installed, is available at auto parts stores and dealer parts departments. These lines are also bent to the proper shapes.

13 When installing the new line make sure it's well supported in the brackets and has plenty of clearance between moving or hot components.

14 After installation, check the master cylinder fluid level and add fluid as necessary. Bleed the brake system as outlined in the next Section and test the brakes carefully before placing the vehicle into normal operation.

10 Brake hydraulic system - bleeding

▸ Refer to illustration 10.8

✱✱ WARNING:

Wear eye protection when bleeding the brake system. If the fluid comes in contact with your eyes, immediately rinse them with water and seek medical attention.

➥Note: *Bleeding the brake system is necessary to remove any air that's trapped in the system when it's opened during removal and installation of a hose, line, caliper, wheel cylinder or master cylinder.*

1 It will probably be necessary to bleed the system at all four brakes if air has entered the system due to low fluid level, or if the brake lines have been disconnected at the master cylinder.

2 If a brake line was disconnected only at a wheel, then only that caliper or wheel cylinder must be bled.

3 If a brake line is disconnected at a fitting located between the master cylinder and any of the brakes, that part of the system served by the disconnected line must be bled.

4 Remove any residual vacuum (or hydraulic pressure) from the brake power booster by applying the brake several times with the engine off.

5 Remove the master cylinder reservoir cover and fill the reservoir with brake fluid. Reinstall the cover.

➥Note: *Check the fluid level often during the bleeding operation and add fluid as necessary to prevent the fluid level from falling low enough to allow air bubbles into the master cylinder.*

6 Have an assistant on hand, as well as a supply of new brake fluid, an empty clear plastic container, a length of plastic, rubber or vinyl tubing to fit over the bleeder valve and a wrench to open and close the bleeder valve.

7 Beginning at the right rear wheel, loosen the bleeder screw slightly, then tighten it to a point where it's snug but can still be loosened quickly and easily.

BRAKES 9-23

10.8 When bleeding the brakes, a hose is connected to the bleed screw at the caliper or wheel cylinder and then submerged in brake fluid - air will be seen as bubbles in the tube and container (all air must be expelled before moving to the next wheel)

8 Place one end of the tubing over the bleeder screw fitting and submerge the other end in brake fluid in the container (see illustration).

9 Have the assistant pump the brakes a few times to get pressure in the system, then hold the pedal firmly depressed.

10 While the pedal is held depressed, open the bleeder screw just enough to allow a flow of fluid to leave the valve. Watch for air bubbles to exit the submerged end of the tube. When the fluid flow slows after a couple of seconds, tighten the screw and have your assistant release the pedal.

11 Repeat Steps 9 and 10 until no more air is seen leaving the tube, then tighten the bleeder screw and proceed to the left rear wheel, the right front wheel and the left front wheel, in that order, and perform the same procedure. Be sure to check the fluid in the master cylinder reservoir frequently.

12 Never use old brake fluid. It contains moisture which will deteriorate the brake system components.

13 Refill the master cylinder with fluid at the end of the operation.

14 Check the operation of the brakes. The pedal should feel solid when depressed, with no sponginess. If necessary, repeat the entire process.

✸✸ WARNING:

Do not operate the vehicle if you are in doubt about the effectiveness of the brake system. It is possible for air to become trapped in the anti-lock brake system valve assembly, so, if the pedal continues to feel spongy after repeated bleedings or the BRAKE or ANTI-LOCK light stays on, have the vehicle towed to a dealer service department or other qualified shop to be bled with the aid of a scan tool.

11 Brake light switch - check and replacement

CHECK

▶ **Refer to illustration 11.1**

1 The brake light switch (see illustration) illuminates the rear brake lights when the brake pedal is applied. The manufacturer refers to the brake light switch as the brake on-off (or BOO) switch because it also functions as a digital information switch for the powertrain control module. The switch is located at the upper end of the brake pedal, at the connection between the pedal pin and the power brake booster pushrod. You'll need to remove the trim panel beneath the steering column to get to the switch and connector.

2 With the brake pedal in the fully released position, the switch opens the brake light circuit. When the brake pedal is depressed, the switch closes the circuit and sends current to the brake lights.

3 If the brake lights are inoperative, check the fuse and the bulbs (see Chapter 12).

4 If the fuse and bulbs are okay, check the switch itself. To check the switch, refer to the *Information sensors* Section in Chapter 6.

REPLACEMENT

Early models

▶ **Refer to illustrations 11.7 and 11.8**

5 Remove the trim panel below the steering column.
6 Unplug the electrical connector from the switch.
7 Working under the dash, remove the cotter pin (see illustration).

11.1 The brake light switch, or brake on-off (BOO) switch, is located near the upper end of the brake pedal, where it's attached to the pedal pin and to the power brake booster pushrod (electrical connector unplugged from switch in this photo)

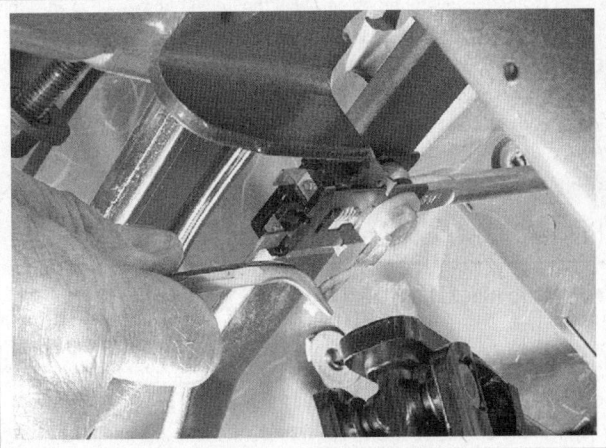

11.7 To remove the brake light or BOO switch from the brake pedal pin, remove this cotter pin and pull the switch and power brake booster pushrod off the pin

9-24 BRAKES

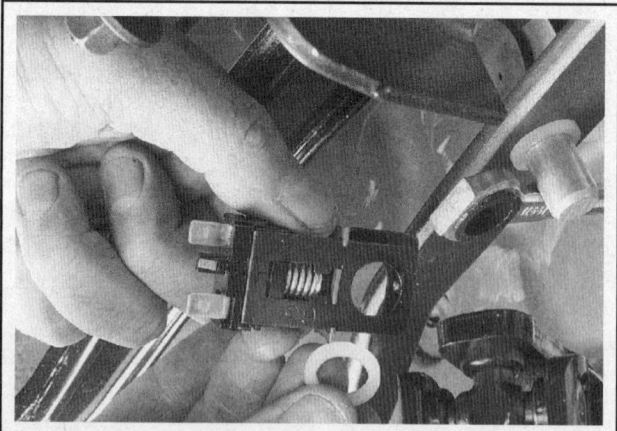

11.8 To remove the brake light or BOO switch from the power brake booster pushrod, simply slide it off; don't lose the small nylon washers, and make sure you install them when installing the switch

8 Disengage the brake light switch from the master cylinder pushrod and from the pin on the brake pedal (see illustration).
9 Installation is the reverse of removal.

LATER MODELS

♦ Refer to illustrations 11.10

11.10 Brake light switch details
1 Electrical Conector   2 Brake light switch

✷✷ WARNING:

Do not press or pull on the brake pedal while removing or installing the brake switch; otherwise it can be set out of adjustment or damage the switch.

10 Disconnect the electrical connector from the switch (see illustration).
11 Rotate the switch a quarter-turn clockwise and remove the switch.
12 Installation is the reverse of removal.

12 Power brake booster - check, removal and installation

OPERATING CHECK

1 Depress the brake pedal several times with the engine off and make sure that there is no change in the pedal reserve distance.
2 Depress the pedal and start the engine. If the pedal goes down slightly, operation is normal.

AIRTIGHTNESS CHECK

3 Start the engine and turn it off after one or two minutes. Depress the brake pedal several times slowly. If the pedal goes down farther the first time but gradually rises after the second or third depression, the booster is airtight.
4 Depress the brake pedal while the engine is running, then stop the engine with the pedal depressed. If there is no change in the pedal reserve travel after holding the pedal for 30 seconds, the booster is airtight.

REMOVAL AND INSTALLATION

♦ Refer to illustrations 12.7 and 12.9

5 Disassembly of the power unit requires special tools and is not ordinarily performed by the home mechanic. If a problem develops, it's recommended that a new or factory rebuilt unit be installed.
6 In the engine compartment, remove the nuts attaching the master cylinder to the booster and carefully pull the master cylinder, along with the RWAL control module and isolation/dump valve, forward until it clears the mounting studs. Be careful not to bend or kink the brake

12.7 Before removing the power brake booster, detach the intake manifold vacuum hose from the booster housing

lines. On 2005 and later Expedition/Navigator models, remove the air filter housing (see Chapter 4) and unbolt and set aside the degas bottle (see Chapter 3).
7 Disconnect the vacuum hose where it attaches to the power brake booster (see illustration).
8 In the passenger compartment, remove the cotter pin and the brake light switch and disconnect the power brake pushrod from the top of the brake pedal (see Section 11).

BRAKES   9-25

9 Remove the nuts attaching the booster to the firewall (see illustration).

10 Carefully lift the booster unit away from the firewall and out of the engine compartment.

11 To install the booster, place it into position and tighten the retaining nuts. Connect the pushrod to the brake pedal and install the brake light switch (see Section 11).

12 Install the master cylinder and vacuum hose.

13 Carefully test the operation of the brakes before placing the vehicle in normal operation.

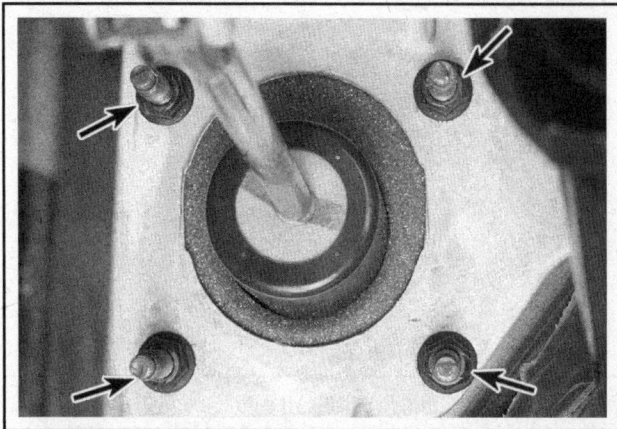

12.9 To detach the power brake booster from the firewall, remove these four nuts (arrows)

13 Parking brake cables - replacement

1 If the vehicle is equipped with air suspension, turn off the air suspension system. The switch is located in the area of the right kick panel (see Chapter 10).

✲✲ WARNING:

On models with air suspension, electrical power to the air suspension system must be turned off before raising the vehicle. Failure to do so can result in a sudden inflation or deflation of the air springs, causing instability of the vehicle while it's off the ground.

2 Release the parking brake. If you're removing a rear cable, loosen the rear wheel lug nuts for that side. Raise the vehicle and place it securely on jackstands. Remove the wheel.

FRONT CABLE

◆ Refer to illustrations 13.3a, 13.3b, 13.4a, 13.4b, 13.4c, 13.4d, 13.6, 13.7, 13.8 and 13.9

3 The forward end of the front cable is attached to the parking brake pedal assembly under the left end of the dash. To take tension off the parking brake cable assembly, have an assistant under the vehicle pull down on the cable (see illustration), then insert a drill bit through

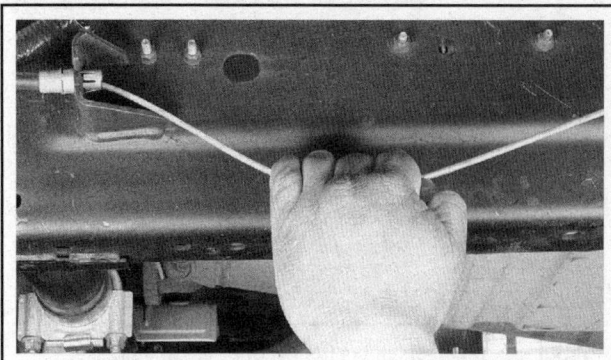

13.3a To take tension off the parking brake cable assembly, have an assistant under the vehicle pull down on the front cable . . .

the hole in the parking brake pedal assembly (see illustration). Don't remove this pin until the new cable has been installed. The hole for the pin on 2005 and later Expedition/Navigator models is at the 9 'o-clock position when looking straight at the assembly from the right side of the vehicle.

4 Working under the vehicle, disengage the rear end of the front

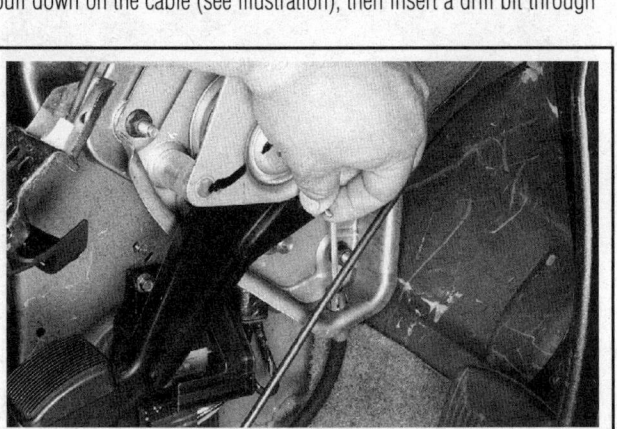

13.3b . . . while you insert a drill bit through the hole in the parking brake pedal assembly; don't remove this pin until the new cable has been installed (F-150 shown)

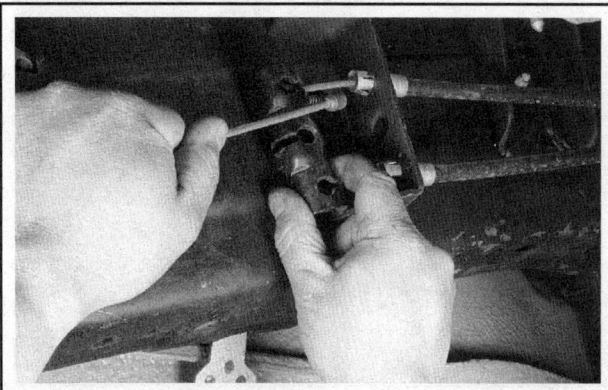

13.4a Working under the vehicle, disengage the rear end of the cable from the parking brake cable equalizer (F-150/F-250 models)

9-26 BRAKES

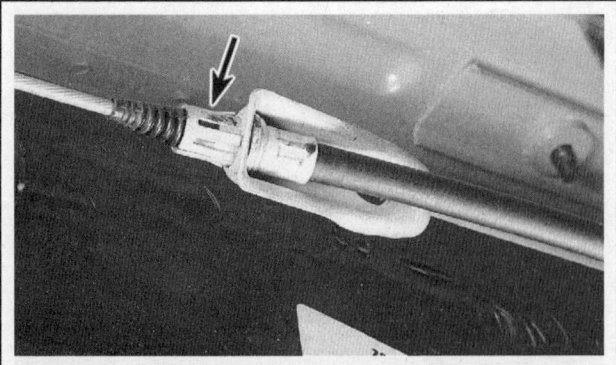

13.4b On sport utility vehicles, locate the equalizer under the left side of the vehicle, disengage the left rear cable from the equalizer by squeezing the tangs of this cable ferrule (arrow) . . .

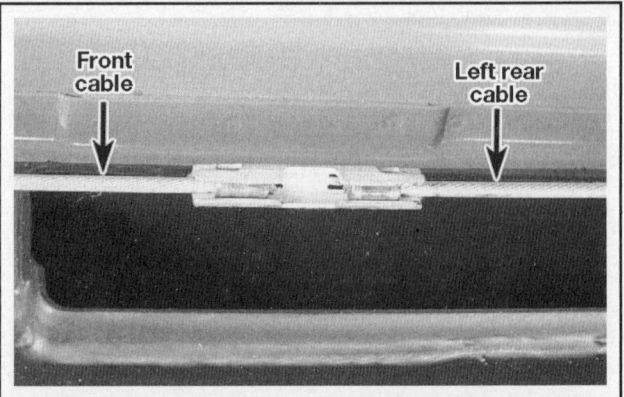

13.4c . . . trace the left rear cable forward to this connection right in front of the equalizer, remove both cables from the connection . . .

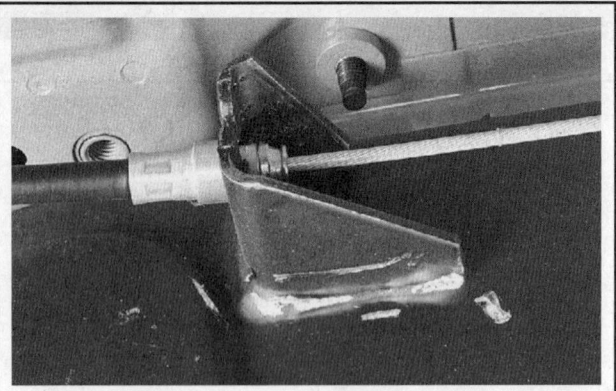

13.4d . . . then detach the front cable from this bracket (Expedition shown)

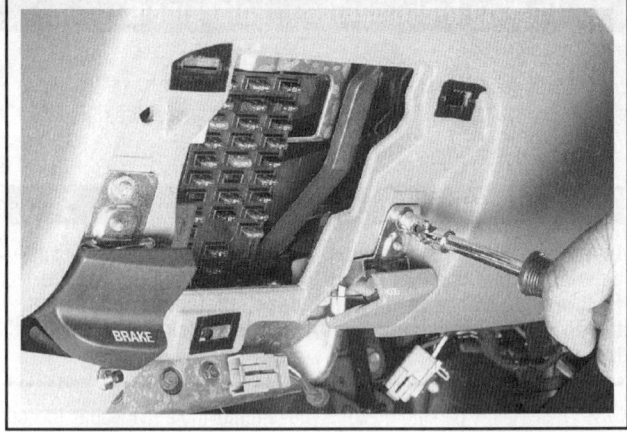

13.6 Remove the retaining screw from the parking brake release handle assembly, then snake the release handle through the opening in the dash (F-150 shown)

cable from the parking brake cable equalizer (see illustrations).

5   Remove the front door scuff plate, the cowl side trim panel and the lower instrument panel steering column cover (see Chapter 11).

6   Remove the retaining screw from the parking brake release handle assembly (see illustration) and route the release handle through the opening in the dash.

7   At the top of the parking brake pedal assembly, unplug the elec-

trical connector for the parking brake switch (see illustration).

8   Remove the three parking brake pedal assembly retaining nuts (see illustration) and remove the parking brake pedal assembly from its mounting bracket.

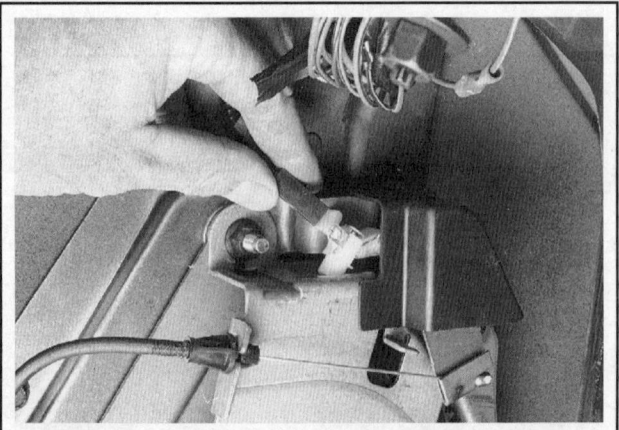

13.7 At the top of the parking brake pedal assembly, unplug the electrical connector for the parking brake switch (F-150 shown)

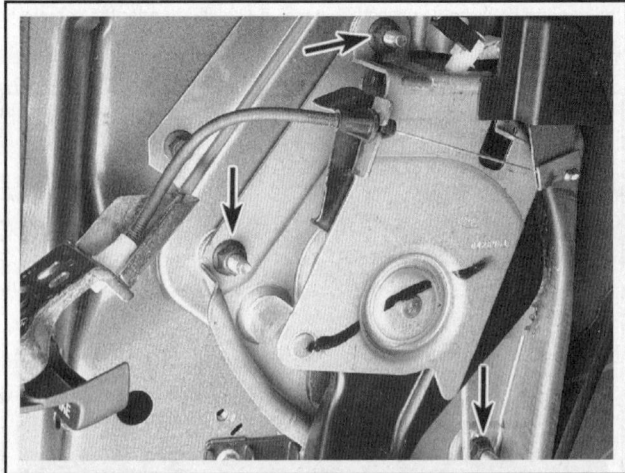

13.8 Remove the three parking brake pedal assembly retaining nuts (arrows) and remove the parking brake pedal assembly from its mounting bracket (F-150 shown)

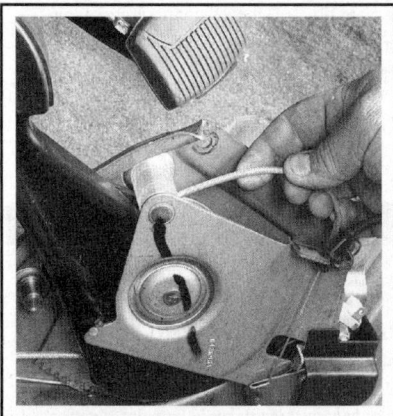

13.9 Disengage the plug on the end of the front parking brake cable from the parking brake pedal (F-150 shown)

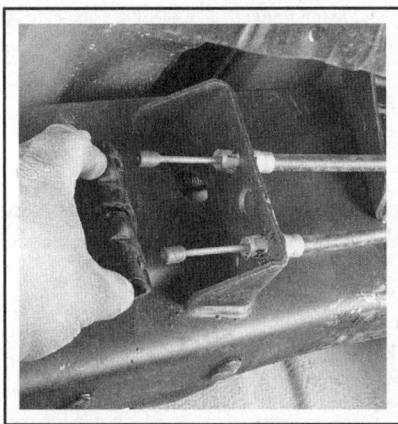

13.14 Have an assistant take tension off the parking brake cable assembly while you simultaneously disengage the rear cables from the equalizer (F-150 shown)

13.15 Squeeze the tangs on the parking brake cable ferrule and slip the rear cable you're replacing through its bracket (F-150 shown)

9 Disengage the plug on the end of the front parking brake cable from the parking brake pedal (see illustration).

10 Squeeze the cable ferrule and slide the cable through the parking brake pedal bracket.

11 Pry the rubber seal from the front floorpan.

12 Working underneath the vehicle again, squeeze the cable ferrule at the bracket located forward of the equalizer, and remove the front parking brake cable.

13 Installation is the reverse of removal.

REAR CABLES

▶ **Refer to illustrations 13.14, 13.15, 13.16a, 13.16b, 13.17a and 13.17b**

14 On trucks, take tension off the parking brake cable assembly (see illustration 13.3a) and simultaneously disengage the equalizer from the rear cables (see illustration). On sport utility vehicles, disengage the left rear cable from the equalizer and from the front cable/rear cable connector (see illustrations 13.4b and 13.4c), then disengage the right rear cable from the equalizer.

15 Squeeze the tangs on the parking brake cable ferrule and slip the

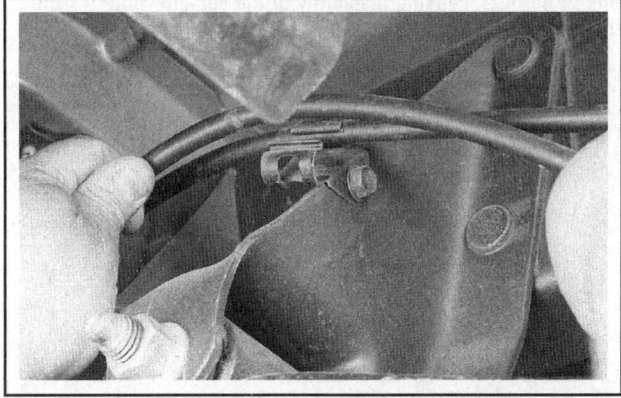

13.16a To detach either rear cable from the frame, disengage it from this clip above the front end of the left leaf spring-to-frame bracket bolt (F-150 only)

rear cable you're replacing through its bracket (see illustration).

16 Unclip the rear cable from the frame (see illustrations).

17 On models with rear drum brakes, remove the brake drum and brake shoe assembly (see Section 6). Disconnect the parking brake

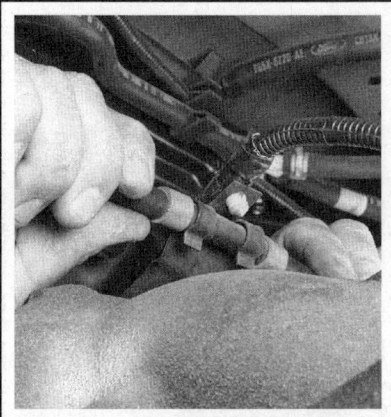

13.16b To detach the right rear cable from the axle, disengage it from these two clips above the differential (F-150 only)

13.17a On models with rear drum brakes, disconnect the parking brake cable from the parking brake lever . . .

13.17b . . . squeeze the fingers on the cable ferrule in the backing plate, and slide the cable through the backing plate (F-150 shown)

9-28 BRAKES

cable from the parking brake lever (see illustration), squeeze the fingers on the cable ferrule in the backing plate (see illustration) and slide the cable through the backing plate.

18 On models with rear disc brakes, squeeze the tangs on the cable ferrule at the caliper anchor plate bracket, then disengage the cable from the parking brake lever.

19 Installation is the reverse of removal.

14 Parking brake shoes (models with rear disc brakes) - replacement

▶ Refer to illustrations 14.4a through 14.4l

1 If the vehicle is equipped with air suspension, turn off the air suspension system. The switch is located in the area of the right kick panel (see Chapter 10).

※ WARNING:

On models with air suspension, electrical power to the air suspension system must be turned off before raising the vehicle. Failure to do so can result in a sudden inflation or deflation of the air springs, causing instability of the vehicle while it's off the ground.

2 Release the parking brake. Loosen the rear wheel lug nuts. Raise the vehicle and place it securely on jackstands. Remove the wheels.

3 Remove the brake calipers (see Section 4) and the brake discs (see Section 5).

4 Clean the parking brake assembly with brake system cleaner before beginning work. Follow illustrations 14.4a through 14.4l for the inspection and replacement of the parking brake shoes. Be sure to stay in order and read the caption under each illustration.

5 Clean the brake disc/parking brake drum and check it for score marks, deep grooves, hard spots (which will appear as small discolored areas) and cracks. If the disc/drum is worn, scored or out-of-round, it can be resurfaced by an automotive machine shop.

14.4a Wash down the brake assembly with brake cleaner; DO NOT blow it out with compressed air!

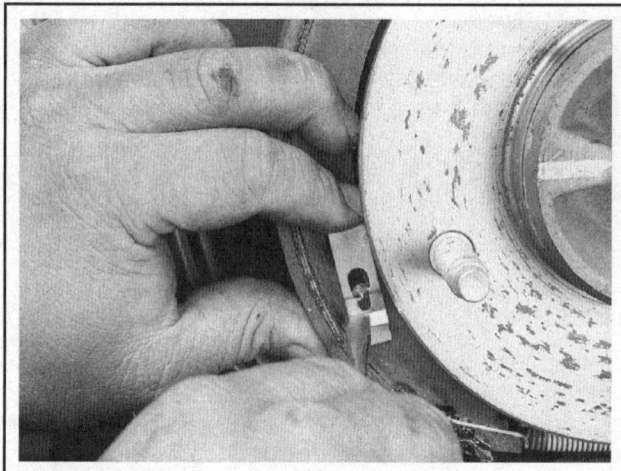

14.4b Remove the front hold-down clip . . .

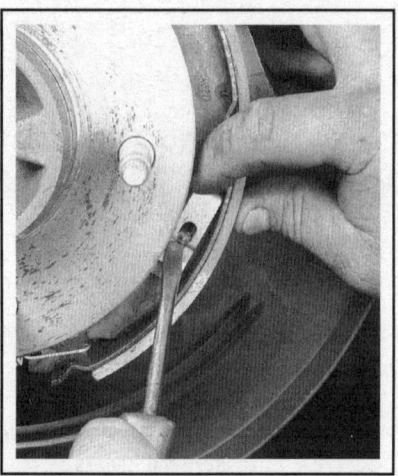

14.4c . . . and the rear hold-down clip

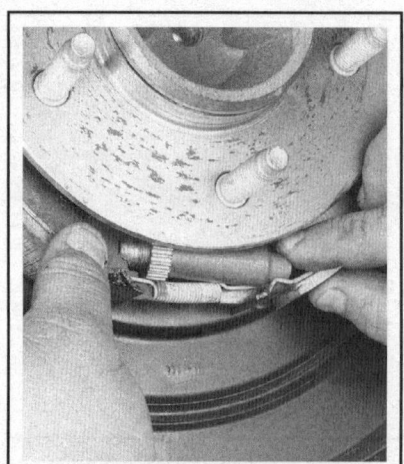

14.4d Remove the parking brake shoe adjuster . . .

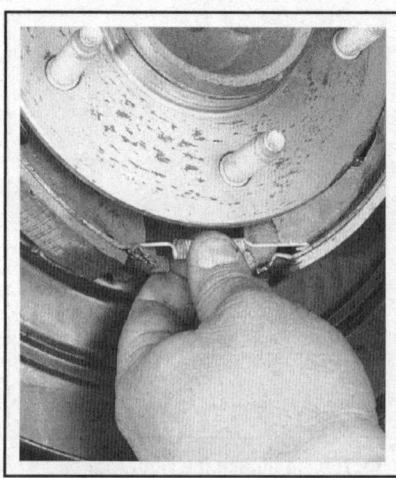

14.4e . . . remove the lower return spring

BRAKES 9-29

14.4f Spread the parking brake shoes apart as shown and lift off the parking brake shoe assembly; if you're replacing the old shoes, remove the upper return spring (arrow) and transfer it to the new shoes

6   Repeat this procedure for the other rear brake assembly.
7   Install the brake discs (see Section 5) and the brake calipers (see

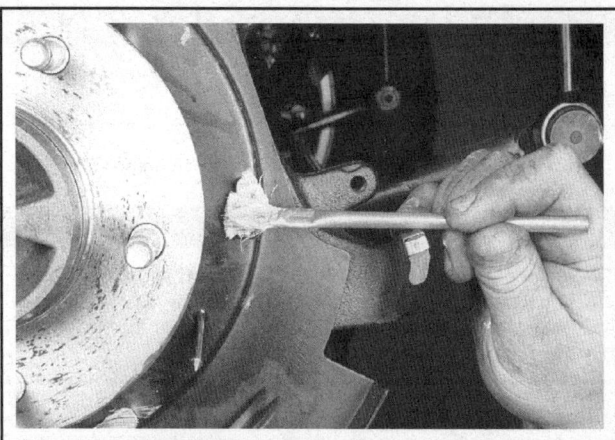

14.4g Lubricate the shoe contact points on the backing plate with high-temperature brake grease

Section 4).
8   Install the rear wheels, install the lug nuts, lower the vehicle and tighten the wheel lug nuts to the torque listed in the Chapter 1 Specifications.
9   Reactivate the air suspension, if equipped.

14.4h With the upper return spring installed into the shoes as shown, spread the lower ends of the shoes apart and install the parking brake shoe assembly

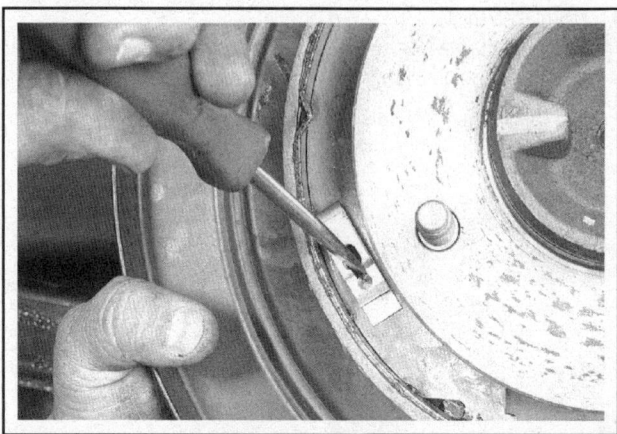

14.4i Install the front shoe hold-down clip . . .

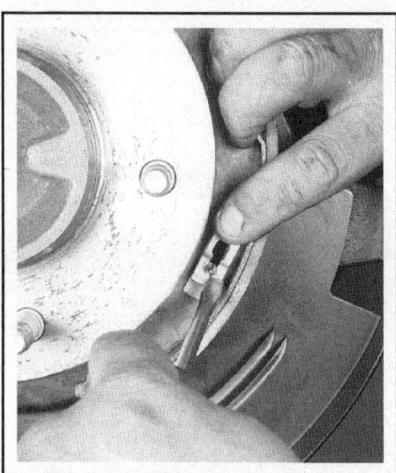

14.4j . . . and the rear shoe hold-down clip

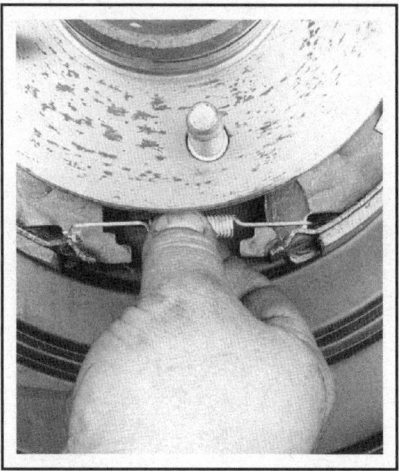

14.4k Install the lower return spring . . .

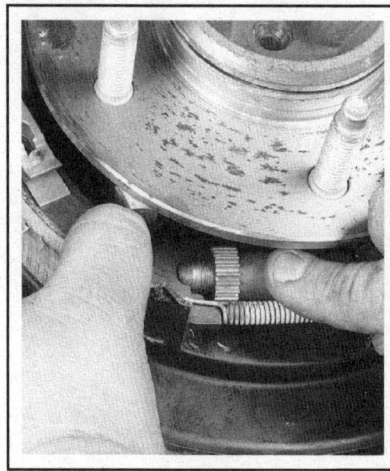

14.4l . . . and the adjuster

9-30 BRAKES

15 Adjustable brake pedal and bracket - removal, installation and indexing

REMOVAL AND INSTALLATION

1 Move the pedals to the most forward position (towards the firewall).

※ WARNING:

The brake pedal and accelerator pedal should be in the same position whether you are installing a new cable or pedal assembly. They should either be all the way forward or all the way rearward to avoid damaging the components.

2 Remove the driver's side knee bolster trim.
3 Remove the four bolts securing the steering column reinforcement panel and remove it.
4 Disconnect the drive cable from the adjustable cable drive. Push in on the sides of the connector and pull the cable off.
5 Remove the brake light switch (see Section 11).
6 Remove the pin and brake booster rod from the brake pedal.
7 If equipped, remove the trailer lighting module from the brake booster bracket.
8 Remove the four booster to pedal assembly bracket nuts.
9 Push the brake booster towards the engine and lift the pedal assembly out of the vehicle.
10 Installation is the reverse of removal, except the pedal assembly will need to be indexed to allow for full forward and rearward adjustment.

11 Torque all the fasteners to the torque listed in this Chapter's Specifications.

INDEXING

➡**Note: The pedal assembly must be indexed whenever it is removed.**

12 With the adjustment cable disconnected from the brake pedal drive actuator, operate the accelerator pedal to the furthest rear (towards the driver) position.
13 Connect the drive cable to the brake pedal drive actuator.
14 Operate the pedals until the brake pedal pin that holds the pedal to the booster rod is aligned with the witness mark (small scribed line) on the side of the brake pedal bracket.
15 Disconnect the drive cable from the brake pedal once again.
16 Operate the accelerator pedal alone in the forward (towards the engine) until there is a 4 mm gap between the screw head attaching the assembly to the drive motor and the moving part of the accelerator pedal.
17 Reconnect the cable to the brake pedal drive actuator.
18 Check the movement of the pedal assembly. There should be full movement forward and rearward on both pedals, and they should operate evenly with each other. If not, repeat the procedure.

BRAKES 9-31

Specifications

General
Brake fluid type	See Chapter 1

Disc brakes
Brake pad minimum thickness	See Chapter 1
Disc lateral runout limit	0.003 inch
Disc minimum (discard) thickness	Cast into disc

Drum brakes
Minimum brake lining thickness	See Chapter 1
Maximum drum diameter	Cast into drum

Torque specifications — Ft-lbs (unless otherwise indicated)

→ **Note:** One foot-pound (ft-lb) of torque is equivalent to 12 inch-pounds (in-lbs) of torque. Torque values below approximately 15 ft-lbs are expressed in inch-pounds, since most foot-pound torque wrenches are not accurate at these smaller values.

Brake hose-to-caliper banjo bolt (front or rear)	
2009 and earlier	23 to 29
2010 and later	30
Caliper mounting bolts	
Front	
F-150 and F-250	21 to 26
Expedition/Navigator	
2002 and earlier models	21 to 26
2003 through 2009 models	41
2010 and later models	27
Rear	
F-150 and F-250	20
Expedition/Navigator	
2002 and earlier models	20
2003 through 2007 models	26
2008 and later models	28
Caliper mounting bracket fasteners	
Front	
F-150 and F-250	136
Expedition/Navigator	
2002 and earlier models	136
2003 through 2009 models	148
2010 and later models	184
Rear	
F-150 and F-250	80
Expedition/Navigator	
1997 models	70
1998 through 2002 models	40
2003 and later models	140
Master cylinder mounting nuts	15 to 21
Power brake booster mounting nuts	16 to 21
Wheel cylinder mounting bolts	11 to 19
Wheel lug nuts	See Chapter 1
Power brake booster-to-adjustable pedal bracket nuts	18

Notes

Chapter 10
Suspension and steering systems

Section
1. General information
2. Shock absorber (1997 through 2002 SUV models/2003 pick-up models) or shock absorber/coil spring assembly (2003 and later SUV models) (front) - removal and installation
3. Stabilizer bar (front) - removal and installation
4. Balljoints - check and replacement
5. Upper control arm - removal and installation
6. Lower control arm - removal and installation
7. Coil spring (2WD models) - removal and installation
8. Steering knuckle (2WD models) - removal and installation
9. Hub and bearing assembly (2003 and later 2WD and all 4WD models) – replacement
10. Steering knuckle (4WD models) - removal and installation
11. Torsion bar (4WD models) - removal and installation
12. Suspension Load Leveling Control/Vehicle Dynamic Suspension systems - general information
13. Shock absorber (rear) - removal and installation
14. Leaf spring/air spring - removal and installation
15. Track bar - removal and installation
16. Rear stabilizer bar (2002 and earlier Expedition/Navigator) - removal and installation
17. Rear suspension arms (2002 and earlier Expedition/Navigator) - removal and installation
18. Rear coil spring/air spring (2002 and earlier Expedition/Navigator) - removal and installation
19. Steering wheel and clockspring - removal and installation
20. Steering linkage (2002 and earlier Pick-ups and Expedition/Navigator) - inspection, removal and installation
21. Steering gear - removal and installation
22. Power steering pump - removal and installation
23. Power steering system - bleeding
24. Independent rear suspension (2003 and later Expedition/Navigator) - component replacement
25. Wheels and tires - general information
26. Front end alignment - general information

Reference to other Chapters
Front wheel bearing check, repack and adjustment (2WD models) - See Chapter 1
Power steering fluid level check - See Chapter 1
Suspension and steering check - See Chapter 1
Tire and tire pressure checks - See Chapter 1
Tire rotation - See Chapter 1

10-2 SUSPENSION AND STEERING SYSTEMS

1 General information

Refer to illustrations 1.1a, 1.1b, 1.1c, 1.1d, 1.3 and 1.4

The front suspension (see illustrations) is fully independent. Each wheel is connected to the frame by a steering knuckle, upper and lower balljoints and upper and lower control arms. Coil springs and shock absorbers are used on 2WD models; 4WD models use shocks and torsion bars. On 1997 through 2002 SUV models/2003 pick-up models, the coil springs are mounted between the spring pockets on the frame and the lower control arms. On 2003 and later SUV models, the shock absorbers are mounted inside an assembly called the shock absorber/coil spring assembly. The shocks or shock absorber/coil spring assemblies are attached to the lower control arms by bolts and nuts, the upper end of each shock or shock absorber/coil spring assembly is attached to a bracket on the frame. A stabilizer bar, connected to the frame and to the two lower control arms, reduces vehicle roll during cornering.

The steering linkage consists of a Pitman arm (the manufacturer calls this part a "sector shaft arm"), idler arm, center link (the manufacturer calls it a "steering sector shaft arm drag link"), and two adjustable tie-rod assemblies (each consisting of an inner tie-rod, adjuster tube and outer tie-rod). When the steering wheel is turned, the gear rotates the Pitman arm which forces the center link to one side. The tie-rods, which are connected to the center link by ball studs, transfer steering

1.1a Front suspension and steering components (2WD models)

1 Stabilizer bar	6 Inner tie-rod ends	10 Lower balljoint ballstud/nut
2 Stabilizer bar bushings	7 Tie-rod adjuster tubes	11 Lower control arm
3 Pitman arm	8 Outer tie-rod ends	12 Shock absorber
4 Center link	9 Steering knuckles	13 Lower control arm pivot bolts/bushings
5 Idler arm		

SUSPENSION AND STEERING SYSTEMS

force to the steering knuckles. The tie-rods are adjustable and are used for toe-in adjustments. The center link is positioned by the Pitman arm and the idler arm. The idler arm pivots on a support attached to the right frame rail. The steering knuckles on both 2WD and 4WD models are similar, but the front wheel bearing setups are different: 2002 and earlier 2WD models use inner and outer tapered roller bearings riding on a spindle that's part of the steering knuckle; 2003 and earlier 2WD models, and 4WD models use a sealed hub and bearing assembly that's bolted to the steering knuckle. The procedure for servicing the 2WD front bearing assembly is in Chapter 1; the procedure for replacing the hub and bearing assembly on a 2003 and earlier models, and 4WD models is in this Chapter.

The rear suspension on trucks (see illustration) consists of a pair of multi-leaf springs and two shock absorbers. The rear axle assembly is attached to the leaf springs by U-bolts. The front ends of the springs are attached to the frame at the front hangers, through rubber bushings. The rear ends of the springs are attached to the frame by shackles which allow the springs to alter their length when the vehicle is in operation. Trucks with rear air suspension are also equipped with an "anti-wind" bar, which is bolted between the front leaf spring hanger and the left end of the axle. The anti-wind spring prevents the axle and the rear spring from twisting (or winding up) during braking and acceleration.

The rear suspension on sport utility (see illustration) vehicles consists of a pair of coil springs and a pair of shock absorbers. The rear axle assembly is positioned by a pair of upper and lower suspension arms. These models also use a stabilizer bar, which is connected to the frame via a pair of links and is bolted to brackets on the rear axle, to reduce vehicle roll during cornering. These models are also equipped

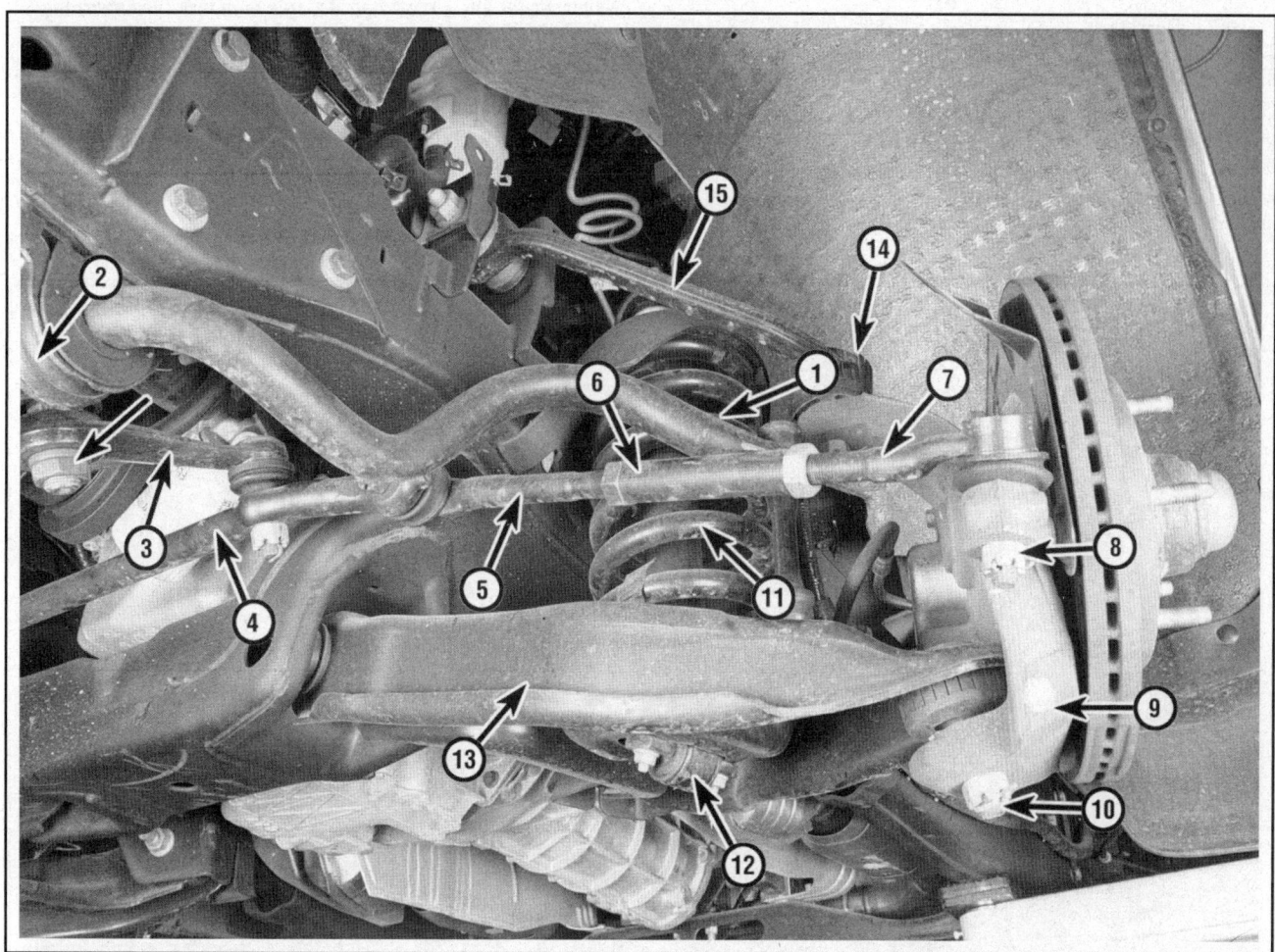

1.1b A closer look at the front suspension and steering components (2WD models)

1. Stabilizer bar
2. Stabilizer bar bushing and mounting clamp
3. Pitman arm
4. Center link
5. Inner tie-rod end
6. Tie-rod adjuster tube
7. Outer tie-rod end
8. Tie-rod balljoint ballstud/nut
9. Steering knuckle
10. Lower balljoint ballstud/nut
11. Coil spring
12. Shock absorber
13. Lower control arm
14. Upper control arm balljoint
15. Upper control arm

10-4 SUSPENSION AND STEERING SYSTEMS

with a track bar bolted to the frame at its left end, and to the rear axle at its opposite end.

Some models are equipped with either a rear-only or a four-wheel air suspension system. Both systems are referred to by the manufacturer as Suspension Load Leveling Control (1997 and 1998 models) or Vehicle Dynamic Suspension (1999 and later). Models with the rear air suspension system use air springs instead of coil springs. Models with the four-wheel system also use air springs at the rear, and air shocks up front. For more information on Suspension Load Leveling Control/Vehicle Dynamic Suspension, refer to Section 12.

Frequently, when working on the suspension or steering system components, you may come across fasteners which seem impossible to loosen. These fasteners on the underside of the vehicle are continually subjected to water, road grime, mud, etc., and can become rusted or "frozen," making them extremely difficult to remove. In order to unscrew these stubborn fasteners without damaging them (or other components), be sure to use lots of penetrating oil and allow it to soak in for a while. Using a wire brush to clean exposed threads will also ease removal of the nut or bolt and prevent damage to the threads. Sometimes a sharp blow with a hammer and punch is effective in breaking the bond between a nut and bolt threads, but care must be taken to prevent the punch from slipping off the fastener and ruining the threads. Heating the stuck fastener and surrounding area with a torch sometimes helps too, but isn't recommended because of the obvious dangers associated with fire. Long breaker bars and extension, or "cheater," pipes will increase leverage, but never use an extension pipe on a ratchet - the ratcheting mechanism could be damaged. Sometimes, turning the nut or bolt in the tightening (clockwise) direction first will help to break it loose. Fasteners that require drastic measures to unscrew should always be replaced with new ones.

Since most of the procedures that are dealt with in this Chapter involve jacking up the vehicle and working underneath it, a good pair of jackstands will be needed. A hydraulic floor jack is the preferred type of jack to lift the vehicle, and it can also be used to support certain components during various operations.

※※ WARNING:

Never, under any circumstances, rely on a jack to support the vehicle while working on it. Also, whenever any of the suspension or steering fasteners are loosened or removed they must be inspected and, if necessary, replaced with new ones of the same part number or of original equipment quality and design. Torque specifications must be followed for proper reassembly and component retention. Never attempt to heat or straighten suspension or steering components. Instead, replace bent or damaged parts with new ones.

1.1c Front suspension and steering components (4WD models)

1 Stabilizer bar
2 Center link
3 Tie-rod adjuster tubes
4 Outer tie-rod ends
5 Steering knuckle
6 Lower balljoint ballstud nuts
7 Lower control arms
8 Torsion bars

SUSPENSION AND STEERING SYSTEMS

1.1d A closer look at the front suspension and steering components (4WD models)

1. Stabilizer bar
2. Stabilizer bar link
3. Center link
4. Inner tie-rod end
5. Tie-rod adjuster tube
6. Outer tie-rod end
7. Balljoint ballstud nut
8. Steering knuckle
9. Upper control arm pivot bolt/nut
10. Upper control arm
11. Upper control arm balljoint
12. Stabilizer bar bushing/bracket
13. Lower control arm
14. Shock absorber
15. Height sensor
16. Steering gear mounting bolts

1.3 Rear suspension components (trucks)

1. Shock absorbers
2. Lower shock absorber mounts
3. Leaf springs
4. Leaf spring-to-anchor plate U-bolts

10-6 SUSPENSION AND STEERING SYSTEMS

1.4 Rear suspension (2002 and earlier Expedition/Navigator)

1. Stabilizer bar
2. Stabilizer bar bushing brackets
3. Stabilizer bar links
4. Track bar
5. Track bar-to-frame bracket bolt/nut
6. Track bar-to-axle bracket bolt
7. Air springs
8. Air spring solenoid
9. Lower suspension arms
10. Lower suspension arm-to-frame bracket bolts
11. Lower suspension arm-to-axle bracket bolts/nuts

2 Shock absorber (1997 through 2002 SUV models/2003 pick-up models) or shock absorber/coil spring assembly (2003 and later SUV models) - removal and installation

1997 THROUGH 2002 SUV MODELS AND 2003 PICK-UP MODELS

▶ Refer to illustrations 2.3, 2.4a and 2.4b

1 If the vehicle is equipped with air suspension, turn OFF the air suspension system. The switch is located in the area of the right kick panel (see Section 12).

✸✸ WARNING:

On models with air suspension, electrical power to the air suspension system must be turned off before raising the vehicle. Failure to do so can result in a sudden inflation or deflation of the air springs, causing instability of the vehicle while it's off the ground.

If the vehicle is equipped with four-wheel air suspension, disconnect the upper end of the front height sensor from the upper frame bracket behind and above the left shock, then firmly push in the red ring on the shock body, hold it there, and pull out the air line.

2 Loosen the front wheel lug nuts, raise the front of the vehicle and support it securely on jackstands. Apply the parking brake. Remove the wheels.

3 Using an open-end wrench to hold the shock from turning, remove the upper shock mounting nut (see illustration). Remove the retainer and bushing.

2.3 To detach the upper end of the shock absorber from the frame bracket, remove this nut (upper arrow), the retainer and the bushing; you'll need a back-up wrench on the lower nut (lower arrow) to prevent the shock from turning when you loosen the nut

SUSPENSION AND STEERING SYSTEMS

2.4a To detach the lower end of the shock absorber from the lower control arm on a 2WD model, remove these two nuts

2.4b To detach the lower end of the shock absorber from the lower control arm on a 4WD model, remove this bolt and nut (the nut is on the front side of the lower arm)

4 Working from underneath the vehicle, remove the two nuts (2WD models) or bolt and nut (4WD models) which attach the lower end of the shock absorber to the lower control arm (see illustrations) and pull the shock out from below.

5 Installation is the reverse of removal. Be sure to tighten the upper mounting nut and the lower mounting bolts to the torque listed in this Chapter's Specifications. On vehicles with four-wheel air suspension, make sure that at least 1/8-inch of air line is inserted into the red fitting. Tighten the upper nut and lower nut and bolt to the torque listed in this Chapter's Specifications.

6 Lower the vehicle and reactivate the air suspension system, if equipped.

2003 AND LATER SUV MODELS

▸ Refer to illustrations 2.8 and 2.11

✳✳ WARNING:

The manufacturer states to discard removed suspension component fasteners (nuts and bolts) and replace them with new ones.

➡ Note: It is possible to replace the shocks or springs individually but the unit will have to be disassembled by a qualified repair shop with the proper equipment, and this will add considerable cost to the project. You can compare the cost of replacing the complete assemblies yourself to the cost of replacing individual components (with the help of a shop).

7 Loosen the front wheel lug nuts. Raise the vehicle and support it securely on jackstands. Remove the front wheels.

8 Remove the nuts that attach the upper end of the shock to the frame (see illustration).

9 Separate the tie-rod end from the steering knuckle and secure it aside (see Section 13).

10 Separate the upper balljoint from the control arm (see Section 5).

11 Remove the fasteners attaching the lower end of the shock absorber to the lower control arm (see illustration).

12 Remove the shock absorber/coil spring assembly.

13 Inspect the shock absorber for leaking fluid, dents, cracks and other damage. Inspect the coil spring for chips and cracks which could cause premature failure. Inspect the spring seats for hardness and general deterioration. If any of the components of the assembly are worn or damaged, have the unit serviced by a qualified repair shop or replace it.

14 Installation is the reverse of removal. Be sure to tighten the fas-

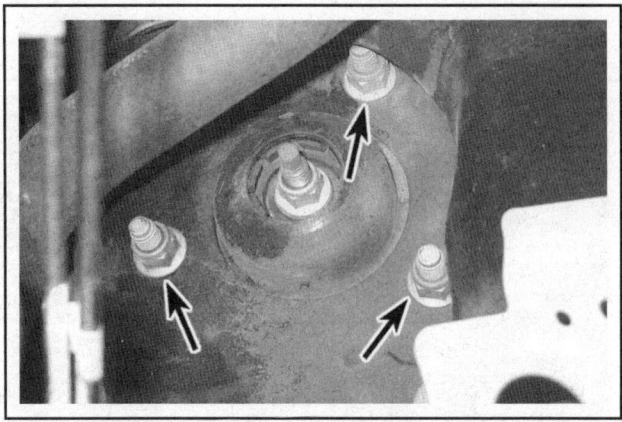

2.8 Shock absorber/coil spring upper mounting fasteners - DO NOT remove the damper rod nut in the middle

teners to the torque listed in this Chapter's Specifications. Tighten the wheel lug nuts to the torque listed in the Chapter 1 Specifications.

➡ Note: The shock absorber lower mounting fasteners should be tightened with the vehicle at normal ride height. This can be done after the vehicle has been lowered to the ground (on vehicles with adequate clearance), or can be simulated by raising the lower control arm with a floor jack.

2.11 Shock absorber/coil spring lower mounting bolt - the nut is on the other side

10-8 SUSPENSION AND STEERING SYSTEMS

3 Stabilizer bar (front) - removal and installation

▶ Refer to illustrations 3.4a, 3.4b and 3.5

1 If the vehicle is equipped with air suspension, turn off the air suspension system. The switch is located in the area of the right kick panel (see Section 12).

※ WARNING:
On models with air suspension, electrical power to the air suspension system must be turned off before raising the vehicle. Failure to do so can result in a sudden inflation or deflation of the air springs, causing instability of the vehicle while it's off the ground.

2 Raise the vehicle and support it securely on jackstands.
3 On 4WD models, remove the skid plate, if equipped.
4 Remove the nuts from the link bolts (see illustrations) and remove the link bolts. Keep any metal washers and rubber bushings in order, and be sure to keep the parts for the left and right sides separate.
5 Remove the stabilizer bar bushing bracket bolts (see illustration) and remove the bushing brackets.
6 Remove the stabilizer bar. Remove the rubber bushings from the stabilizer bar.
7 Inspect all rubber bushings for wear and damage. If any of the rubber parts are cracked, torn or generally deteriorated, replace them.
8 When you install the rubber bushings on the stabilizer bar, position them so that the slits face toward the front of the vehicle.
9 Installation is otherwise the reverse of removal. Be sure to tighten all fasteners to the torque listed in this Chapter's Specifications.

※ CAUTION:
New nuts should be used upon reassembly. On 2008 and later Expedition/Navigator models, it is suggested that the bolts (attached to plates) should also be replaced with new plates/bolts. These components are assembled with high torque levels.

 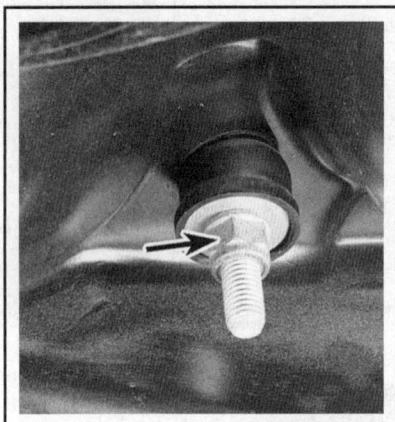 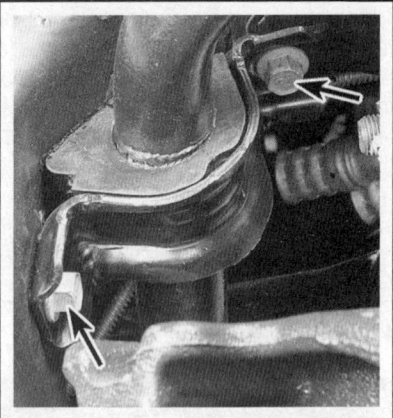

3.4a To disconnect the stabilizer bar link bolt (arrow) from the lower control arm . . .

3.4b . . . remove this nut (arrow); be sure to keep all the retainers (washers), bushings and spacers in order when you pull out the link bolt

3.5 To separate the stabilizer bar from the frame, remove the bushing bracket bolts (arrows) from both frame rails (left bracket shown, right bracket identical)

4 Balljoints - check and replacement

1 Inspect the control arm balljoints for looseness anytime either of them is separated from the steering knuckle. See if you can turn the ballstud in its socket with your fingers. If the balljoint is loose, or if the ballstud can be turned, replace the balljoint. You can also check the balljoints with the suspension assembled as follows.

2 If the vehicle is equipped with air suspension, turn off the air suspension system. The switch is located in the area of the right kick panel (see Section 12).

※ WARNING:
On models with air suspension, electrical power to the air suspension system must be turned off before raising the vehicle. Failure to do so can result in a sudden inflation or deflation of the air springs, causing instability of the vehicle while it's off the ground.

3 Loosen the wheel lug nuts, raise the front of the vehicle and support it securely on jackstands. Remove the wheels. Place a floor jack under the lower control arm as close to each balljoint as possible and raise it slightly.

UPPER BALLJOINTS

4 Position a dial indicator against the wheel rim, grasp the top and bottom of the tire and "rock" the tire, alternately pushing the top and pulling the bottom, and vice versa. The dial indicator should indicate no more than 0.125-inch deflection. If the indicated reading exceeds this figure, the balljoint is faulty and should be replaced.

LOWER BALLJOINTS

5 Position a dial indicator against the wheel rim and insert a prybar

SUSPENSION AND STEERING SYSTEMS   10-9

between the lower control arm and the steering knuckle. As you lever the prybar, the needle should not deflect more than 0.125-inch. If it does, the balljoint is faulty and should be replaced.

REPLACEMENT

6 Balljoint replacement, on these models, has evolved from replacing the balljoints and control arms (as an assembly) to replacing the balljoint individually; this is due to the availability of balljoints from aftermarket suppliers. As a result, it may be possible for you to replace an individual balljoint on your specific year and model vehicle although the manufacturer has not addressed the procedure for your specific vehicle. Check with your local auto parts store for availability on replacement balljoints for your vehicle. A special balljoint removal/installer tool is necessary and can usually be obtained from an auto parts store or tool rental business.

7 Detach the steering knuckle from the control arm (see Section 8 or 10).

8 Remove the snap ring for the balljoint and discard it.

9 Place the special balljoint removal tool on the balljoint according to the tool manufacturer's instructions, then remove the balljoint.

10 Using the special balljoint installer tool, install the balljoint.

✱✱ CAUTION:

Be careful not to damage the rubber balljoint boot during installation.

11 Install the balljoint snap ring.
12 Attach the steering knuckle to the control arm.

5 Upper control arm - removal and installation

REMOVAL

▶ Refer to illustrations 5.3, 5.5 and 5.6

➡ Note: This procedure applies to both 2WD and 4WD models.

1 If the vehicle is equipped with air suspension, turn off the air suspension system. The switch is located in the area of the right kick panel (see Section 12).

✱✱ WARNING:

On models with air suspension, electrical power to the air suspension system must be turned off before raising the vehicle. Failure to do so can result in a sudden inflation or deflation of the air springs, causing instability of the vehicle while it's off the ground.

2 Measure the height of the center of the hub to the ground before proceeding. Loosen the wheel lug nuts, raise the front of the vehicle and support it securely on jackstands. Remove the wheel. Position a floor jack, with a wood block on the jack head (to act as a cushion), under the lower control arm in the area between the spring seat and the balljoint. Raise the jack slightly to take the spring pressure off the upper control arm.

✱✱ WARNING:

The jack must remain in this position throughout the entire procedure.

3 Mark the position of the upper control arm camber adjustment cam (see illustration).

4 If the vehicle is equipped with a 4-wheel anti-lock brake system (4WABS), remove the bracket bolt for the wheel speed sensor wire and place the sensor wire out of harm's way.

5 To disconnect the upper control arm from the steering knuckle, remove the cotter pin from the balljoint castle nut, loosen the nut a few turns (don't remove it), install a small puller (see illustration) and break the balljoint loose from the knuckle. Now remove the nut.

6 Remove the upper control arm pivot bolts and nuts (see illustration). Remove the control arm.

7 Inspect the pivot bolt bushings for wear. If the bushings are worn, you'll have to replace the control arm. Inspect the upper balljoint seal. If

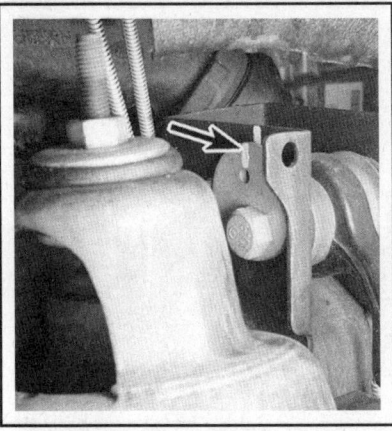

5.3 Mark the relationship of the eccentric camber adjustment cams to the frame brackets before removing the front and rear pivot nuts and bolts

5.5 Using a suitable puller, separate the upper balljoint from the steering knuckle; note that the loosened ballstud nut has been left on the ballstud to prevent the parts from separating violently

5.6 To detach the upper control arm from the frame brackets, remove the pivot bolts and nuts

10-10 SUSPENSION AND STEERING SYSTEMS

it's leaking, replace the control arm. The balljoint is not serviceable.

INSTALLATION

8 Position the arm in the frame brackets and install new bolts and nuts. They must be installed with their heads facing toward each other. Make sure the marks you made prior to disassembly are aligned, then tighten, but don't torque, the pivot bolt nuts.

9 Attach the balljoint to the steering knuckle and tighten the ballstud nut to the torque listed in this Chapter's Specifications.
10 Reattach the 4WABS front brake anti-lock sensor wire bracket and tighten the bolt securely.
11 Install the wheel and lug nuts, then lower the vehicle. Tighten the lug nuts to the torque listed in the Chapter 1 Specifications.
12 Tighten the upper arm pivot bolt nuts, front nut first, to the torque listed in this Chapter's Specifications.
13 Reactivate the air suspension system, if equipped.

6 Lower control arm - removal and installation

2WD MODELS

Removal

▶ Refer to illustrations 6.6, 6.8, 6.9, 6.10, 6.11a and 6.11b

1 If the vehicle is equipped with air suspension, turn off the air suspension system. The switch is located in the area of the right kick panel (see Section 12).

※ WARNING:

On models with air suspension, electrical power to the air suspension system must be turned off before raising the vehicle. Failure to do so can result in a sudden inflation or deflation of the air springs, causing instability of the vehicle while it's off the ground.

2 Loosen the wheel lug nuts, raise the vehicle and support it securely on jackstands. Remove the wheel.
3 Remove the disc brake caliper, brake pads, caliper anchor bracket and disc (see Chapter 9).
4 Remove the disc brake splash shield.
5 Remove the shock absorber (see Section 2).
6 Detach the brake hose bracket from the lower control arm (see illustration).
7 Remove the front stabilizer bar link nut (see Section 3).
8 Using a suitable coil spring compressor (see illustration), install the compressor in accordance with the manufacturer's instructions and compress the coil spring.

6.6 Remove this brake hose bracket bolt (arrow) and detach the brake hose from the lower control arm (2WD models)

9 Remove the cotter pin, loosen - don't remove - the castellated ballstud nut, install a suitable puller (see illustration) and separate the lower control arm from the steering knuckle.
10 If you're removing the left control arm, be sure to mark the relationship of the left rear adjustment cam (see illustration).
11 Remove the lower control arm pivot bolts (see illustrations) and pull the lower arm from its frame brackets. Remove the spring.

Installation

12 Installation is the reverse of removal. Don't tighten the pivot bolt nuts or bolts until the vehicle is back on the ground, at normal ride height, then tighten all fasteners to the torque listed in this Chapter's Specifications.

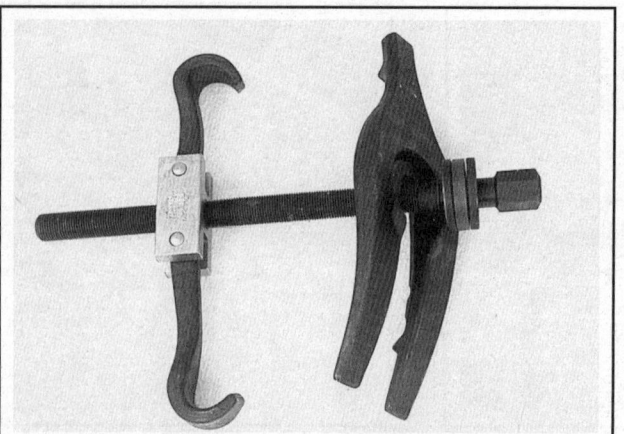

6.8 A typical aftermarket internal spring compressor tool: the hooked arms grip the upper coils of the spring, the plate is inserted below the lower coil, and when the nut on the threaded rod is turned, the spring is compressed

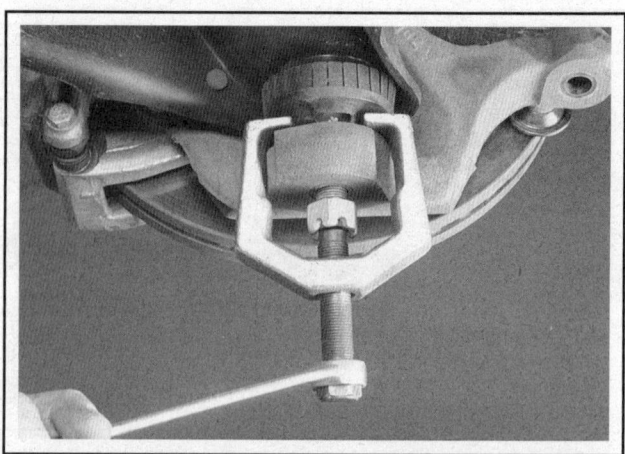

6.9 Using a suitable puller, separate the lower balljoint from the steering knuckle; note that the loosened ballstud nut has been left on the ballstud to prevent the parts from separating violently (2WD models)

SUSPENSION AND STEERING SYSTEMS

6.10 If you're removing the left lower control arm, be sure to mark the relationship of the left rear adjustment cam to the frame bracket (2WD models)

➡ **Note:** You can also raise the lower control arm with a floor jack to simulate normal ride height, then tighten the nuts/bolts.

Note that the tightening torque for the left rear cam bolt nut is considerably higher than any of the other bolts/nuts. The manufacturer recommends that all suspension bolts/nuts removed in this procedure be replaced with new ones upon reassembly.

13 Reactivate the air suspension system, if equipped.

4WD MODELS

Removal

♦ **Refer to illustrations 6.22a and 6.22b**

14 If the vehicle is equipped with air suspension, turn off the air suspension system. The switch is located in the area of the right kick panel (see Section 12).

❊❊ WARNING:

On models with air suspension, electrical power to the air suspension system must be turned off before raising the vehicle. Failure to do so can result in a sudden inflation or deflation of the air springs, causing instability of the vehicle while it's off the ground.

15 Loosen the wheel lug nuts, raise the vehicle and support it securely on jackstands. Remove the wheel.
16 Remove the skid plate, if equipped.
17 If the vehicle is equipped with a 4-wheel antilock brake system (4WABS), detach the front brake anti-lock sensor wire bracket from the control arm.

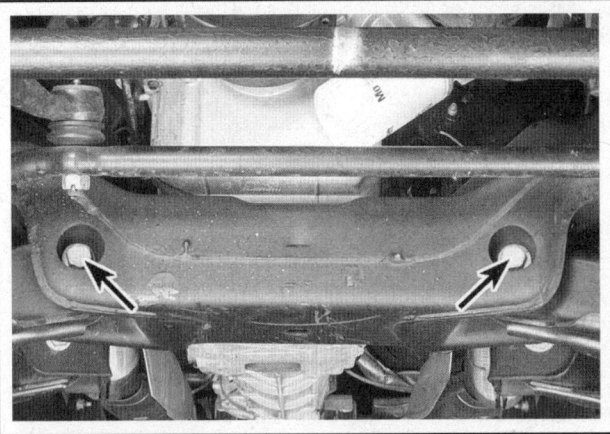

6.11a The lower control arm front pivot bolts (arrows) are hidden inside these holes in the lower crossmember (2WD models)

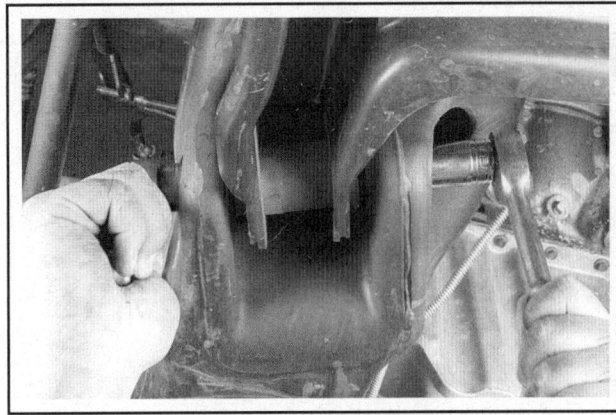

6.11b Remove the lower control arm pivot bolts and nuts (2WD models)

18 Remove the torsion bar (see Section 11).
19 Disconnect the shock absorber from the lower control arm (see Section 2).
20 Disconnect the stabilizer bar from the lower control arm (see Section 3).
21 Detach the lower balljoint from the steering knuckle (see illustration 6.9).
22 Remove the lower control arm pivot bolts, nuts and washers (see illustrations). Remove the lower control arm.
23 Check the bushings for damage or wear. Some models have

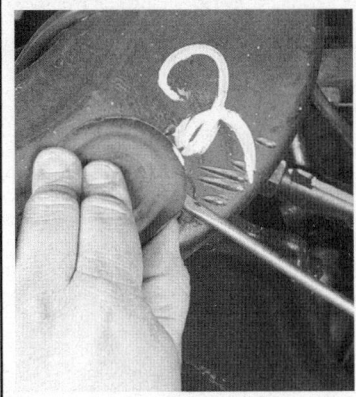

6.22a The front pivot bolts (arrows) for the lower control arm are hidden inside these holes in the crossmember (4WD models)

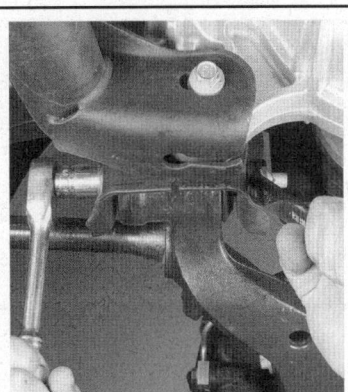

6.22b Remove the lower control arm pivot bolts and nuts (4WD models)

10-12 SUSPENSION AND STEERING SYSTEMS

welded-in bushings that can't be replaced. If this is the case, replace the control arm. Other models have replaceable bushings, but a press and special adapters are required to remove and install them. Take the control arm to a dealer service department or other repair shop to have the bushings replaced for you.

Installation

24 Raise the control arm into position and install the control arm pivot bolts, washers and nuts, but don't tighten them completely at this time. Make sure the threaded portions of both bolts are facing forward, i.e. the nuts should be at the front.

25 Carefully raise the lower control arm with the floor jack until the balljoint stud can be inserted into the hole in the steering knuckle. Install the balljoint stud nut, tighten it to the torque listed in this Chapter's Specifications, then install a new cotter pin.

26 Connect the stabilizer bar to the lower control arm (see Section 3).

27 Install the shock absorber (see Section 2).

28 Install the torsion bar (see Section 11).

29 If the vehicle is equipped with a 4-wheel antilock brake system (4WABS), reattach the front brake anti-lock sensor wire bracket to the control arm.

30 Using a floor jack, raise the lower control arm to simulate normal ride height, then tighten the lower control arm pivot bolts to the torque listed in this Chapter's Specifications.

31 Install the wheel and lug nuts. Lower the vehicle and tighten the lug nuts to the torque listed in the Chapter 1 Specifications.

32 Install the skid plate, if equipped.

33 Measure the vehicle's ride height on each side, from equal points on the frame to the ground. If the side that has been worked on is higher or lower than the other side, turn the torsion bar adjusting screw accordingly until the vehicle sits level. This may take a few tries, and it's important to roll the vehicle back and forth and jounce the front end between adjustments, to settle the suspension and get an accurate reading.

7 Coil spring (2WD models) - removal and installation

REMOVAL

1 Refer to Section 6 and remove the coil spring along with the lower control arm.

2 Check the coil spring upper insulator for cracks and other signs of deterioration and replace it if necessary.

INSTALLATION

3 Place the insulator on top of the coil spring (the upper end of the spring is the flat end).

4 Install the top of the spring into the spring pocket and the bottom in the lower control arm. There are two drain holes in the spring seat in the lower control arm. When it's seated correctly, the lower end of the spring covers the first hole and covers about half of the second hole.

5 Refer to Section 6 and install the coil spring and lower control arm.

8 Steering knuckle (2WD models) - removal and installation

1 If the vehicle is equipped with air suspension, turn off the air suspension system. The switch is located in the area of the right kick panel (see Section 12).

WARNING:

On models with air suspension, electrical power to the air suspension system must be turned off before raising the vehicle. Failure to do so can result in a sudden inflation or deflation of the air springs, causing instability of the vehicle while it's off the ground.

2 Raise the front of the vehicle and support it securely on jackstands. Apply the parking brake.

3 Support the lower control arm with a floor jack. Raise the jack slightly.

WARNING:

The jack must remain in this position throughout the entire procedure.

4 Remove the wheel.

5 Remove the brake caliper (see Chapter 9) and the brake disc/hub assembly (see *Front wheel bearing check, repack and adjustment* in Chapter 1).

6 Remove the disc splash shield from the steering knuckle.

7 Disconnect the tie-rod end from the knuckle (see Section 20).

8 Disconnect the balljoints from the steering knuckle (see Sections 5 and 6).

9 Remove the steering knuckle.

10 Installation is the reverse of removal. Be sure to tighten all fasteners to the torque listed in this Chapter's Specifications and reactivate the air suspension system, if equipped, after lowering the vehicle.

11 Adjust the front wheel bearings (see Chapter 1) and have the front wheel alignment checked by a dealer service department or alignment shop.

SUSPENSION AND STEERING SYSTEMS

9 Hub and bearing assembly (2003 and later 2WD and all 4WD models) – replacement

▶ Refer to illustrations 9.5 and 9.6

➡ **Note 1:** The hub and bearing assembly is a sealed unit and isn't serviceable. If it's defective, it must be replaced.

➡ **Note 2:** On Expedition/Navigator models with independent rear suspension, the rear hub/bearing assembly is replaced with the same procedure as for front hub/bearing assemblies.

1 Put the vehicle in gear, apply the parking brake and break loose the driveaxle/hub nut with a socket and large breaker bar (see Chapter 8).

2 If the vehicle is equipped with air suspension, turn off the air suspension system. The switch is located in the area of the right kick panel (see Section 12).

✳✳ WARNING:

On models with air suspension, electrical power to the air suspension system must be turned off before raising the vehicle. Failure to do so can result in a sudden inflation or deflation of the air springs, causing instability of the vehicle while it's off the ground.

9.5 To detach the disc splash shield from the steering knuckle on 4WD models, remove these three bolts (arrows)

3 Loosen the wheel lug nuts, raise the vehicle and support it securely on jackstands. Remove the wheel. Remove the driveaxle/hub nut.

4 Unbolt the brake caliper and hang it out of the way with a piece of wire, then remove the caliper anchor bracket and the brake disc (see Chapter 9).

5 On models with 4-wheel ABS (4WABS), remove the disc splash shield from the steering knuckle (see illustration), then remove the speed sensor retaining bolt, remove the sensor from the knuckle and set the sensor and wire harness safely aside. On Expedition/Navigator models with independent rear suspension, separate the outboard end of the driveaxles from the hub, using a puller (see Chapter 8).

6 Remove the hub assembly-to-steering knuckle bolts (see illustration).

7 Tap the hub assembly from side-to-side to break it loose from the steering knuckle. Pull the hub assembly off the end of the driveaxle. Wrap the end of the driveaxle with a rag to prevent damaging it. If the hub is stuck on the splines on the end of the driveaxle, use a puller to free it.

8 Installation is the reverse of the removal procedure. Be sure to lubricate the driveaxle splines with multi-purpose grease if equipped,

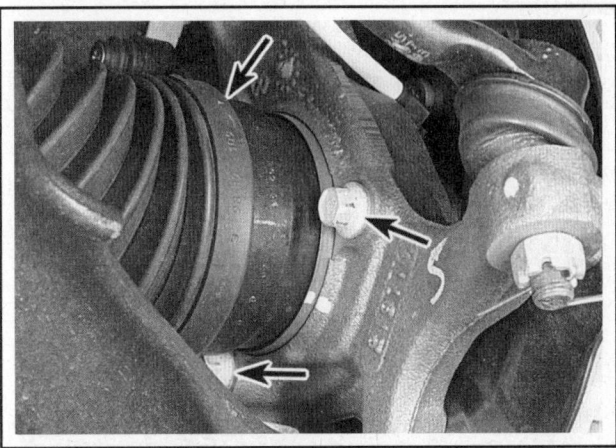

9.6 To detach thev hub and bearing assembly from the steering knuckle, remove the three bolts (4WD model shown, 2WD similar)

and tighten all of the fasteners to the torque listed in this Chapter's Specifications. If the hub-to-knuckle interface was equipped with an O-ring, install a new O-ring upon reassembly.

10 Steering knuckle (4WD models) - removal and installation

REMOVAL

1 Put the vehicle in gear, apply the parking brake and break loose the driveaxle/hub nut with a socket and large breaker bar.

2 If the vehicle is equipped with air suspension, turn off the air suspension system. The switch is located in the area of the right kick panel (see Section 12).

✳✳ WARNING:

On models with air suspension, electrical power to the air suspension system must be turned off before raising the vehicle. Failure to do so can result in a sudden inflation or deflation of the air springs, causing instability of the vehicle while it's off the ground.

10-14 SUSPENSION AND STEERING SYSTEMS

3 Loosen the wheel lug nuts, raise the front of the vehicle and support it securely on jackstands. Apply the parking brake. Remove the wheel. Remove the driveaxle/hub nut.

4 Unbolt the brake caliper, hang it out of the way with a piece of wire, then remove the caliper anchor bracket and brake disc (see Chapter 9). Remove the disc splash shield (see illustration 9.5) and, if equipped, the 4WABS speed sensor.

5 Disconnect the tie-rod end from the steering knuckle (see Section 20).

6 Support the lower control arm with a floor jack and detach the steering knuckle from the lower balljoint (see Section 6).

✶✶ WARNING:

The jack must remain in this position throughout the entire procedure.

7 Support the steering knuckle and separate it from the upper balljoint (see Section 5).

8 Using a puller, push the driveaxle out of the hub while withdrawing the steering knuckle and hub assembly. On 2005 and later Expedition/Navigator models, remove the bolts securing the IWE hub to the knuckle (see Chapter 8).

9 If you're planning to replace the hub and bearing assembly or the knuckle, remove the hub and bearing assembly from the knuckle (see Section 9).

10 Inspect the seal. If it's damaged or shows signs of deterioration, pry it out with a large screwdriver or prybar. Install a new seal by driving it in with a socket that has an outside diameter slightly smaller than the seal.

INSTALLATION

11 Installation is the reverse of the removal procedure. Be sure to lubricate the driveaxle splines with multi-purpose grease and tighten all of the fasteners to the torque listed in this Chapter's Specifications.

11 Torsion bar (4WD models) - removal and installation

▶ Refer to illustrations 11.3a, 11.3b, 11.4, 11.5, 11.6, 11.7a and 11.7b

1 If the vehicle is equipped with air suspension, turn off the air suspension system. The switch is located in the area of the right kick panel (see Section 12).

✶✶ WARNING:

On models with air suspension, electrical power to the air suspension system must be turned off before raising the vehicle. Failure to do so can result in a sudden inflation or deflation of the air springs, causing instability of the vehicle while it's off the ground.

2 Loosen the front wheel lugs nuts, raise the vehicle and place it securely on jackstands. Remove the wheel.

3 Mark the relationship of the torsion bar and the torsion bar crossmember support (see illustration). Count the number of threads showing on the torsion bar adjuster bolt and mark the relationship of the bolt to the torsion bar adjuster nut (see illustration).

4 In the torsion bar adjuster arm, there's a small dimple. Install a small puller with its bolt centered on this dimple (see illustration).

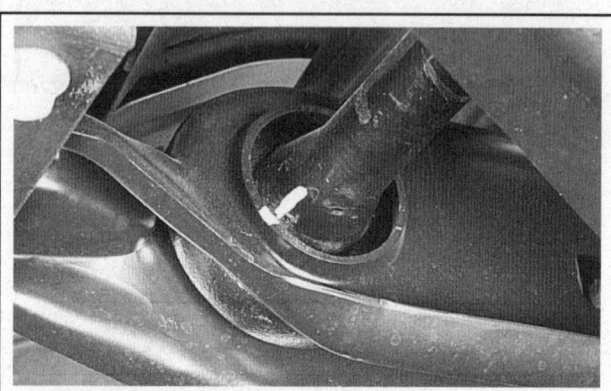

11.3a To ensure proper adjustment of the torsion bar upon reassembly, mark the relationship of the torsion bar to the crossmember . . .

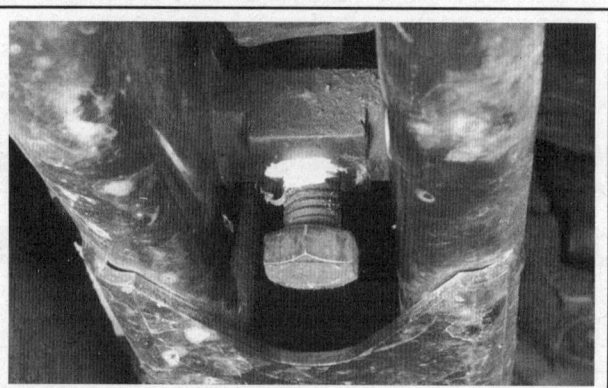

11.3b . . . and count the number of threads showing on the torsion bar adjuster bolt and mark the relationship of the bolt to the torsion bar adjuster nut as insurance

11.4 Install a puller, with the fingers hooked around the flange running along each side of the crossmember; make sure the puller bolt is centered on the dimple in the torsion bar adjuster arm; tighten the puller bolt until all tension is removed from the adjuster nut

SUSPENSION AND STEERING SYSTEMS

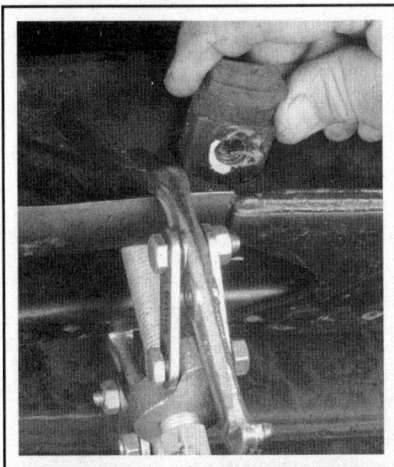

11.5 With tension removed from the adjuster nut, remove the adjuster nut

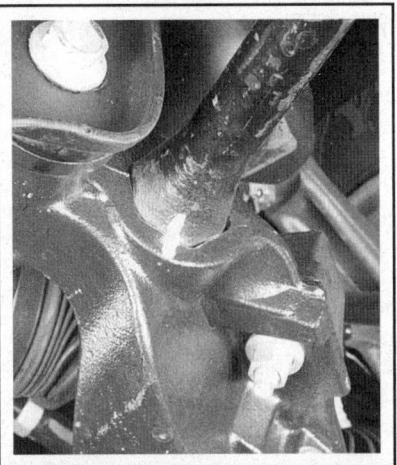

11.6 Mark the relationship of the torsion bar to the lower control arm as shown

11.7a Slide the torsion bar forward through the lower control arm far enough to pull the rear end of the bar out of the crossmember . . .

5 Turn the puller bolt until all tension is removed from the torsion bar adjuster arm, then remove the torsion bar adjuster nut (see illustration). Remove the puller.

6 Mark the relationship of the forward end of the torsion bar to the lower control arm (see illustration).

7 Push the torsion bar forward, through the lower control arm, until the rear end of the bar clears the crossmember (see illustration) and remove the torsion bar adjuster arm (see illustration).

8 Pull the torsion bar down and to the rear as far as it will go. If the front end of the bar hangs up in the lower control arm, drive it out of the control arm with a brass drift.

9 Installation is the reverse of removal. Be sure to clean out the hexagonal hole in the lower control arm and lube it with multi-purpose grease before inserting the torsion bar into the arm. Also apply some grease to the hex ends of the torsion bar, to the top of the adjuster arm and to the adjuster bolt. Make sure that the marks you made on the rear end of the torsion bar and the crossmember, and on the front end of the torsion bar and the control arm, line up. And make sure that the torsion bar adjuster bolt is tightened until the same number of threads are showing and the marks you made on the adjuster bolt and nut are lined up.

10 Install the wheel, remove the jackstands and lower the vehicle.

11 Tighten the wheel lug nuts to the torque listed in the Chapter 1 Specifications.

12 Measure the vehicle's ride height on each side, from equal points

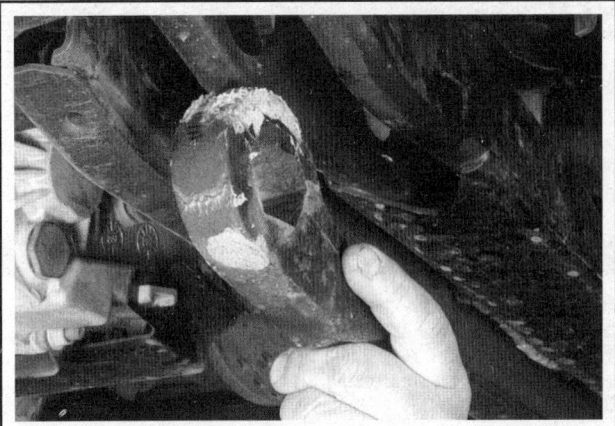

11.7b . . . and remove the torsion bar adjuster arm. Hold your hand under the arm as you slide out the torsion bar to prevent the arm from falling

on the frame to the ground. If the side that has been worked on is higher or lower than the other side, turn the torsion bar adjusting bolt accordingly until the vehicle sits level. This may take a few tries, and it's important to roll the vehicle back and forth and jounce the front end between adjustments, to settle the suspension and get an accurate reading.

12 Suspension Load Leveling Control/Vehicle Dynamic Suspension systems - general information

REAR-ONLY SYSTEM

The rear-only suspension load leveling control system offers a softer ride, and improves handling by maintaining a constant vehicle attitude. It does this by automatically adjusting the vehicle's height above the road and its front-to-rear attitude as the vehicle is loaded or unloaded. The rear-only system consists of a pair of rear air springs, an air compressor, air lines, air spring solenoids, a height sensor, and a control module. On F-250 trucks, the air springs are used in addition to the leaf springs; on sport utility vehicles, they're used instead of the standard coil springs.

4-WHEEL SYSTEM

The 4-wheel system offers three vehicle heights: "kneel," "trim" and "off-road." In kneel height mode, the vehicle is lowered one inch lower than its trim height to facilitate entering and exiting the vehicle. Trim height is the normal vehicle ride height. Off-road height is one inch higher than trim height. The 4-wheel system uses components similar to (but not necessarily interchangeable with) the rear-only system, as well as a pair of air shocks up front. It also uses two height sensors, a steering sensor and several other sensors on the transfer case and elsewhere on the vehicle to monitor driver and road inputs. The control

10-16 SUSPENSION AND STEERING SYSTEMS

module used in this system is also used on the rear-only system.

TURNING OFF THE SYSTEM

Models with a switch

▶ Refer to illustration 12.3

The air suspension switch (see illustration) is located behind the right kick panel. To disable the system, remove the kick panel (see Chapter 11) and turn off the switch. This cuts the power to the air suspension control module, which deactivates the system. The system should be turned off anytime the vehicle is going to be raised off the ground so that it won't react to the higher vehicle height.

Models with a message center

To disable and enable the system, turn the ignition key to ON, and sit in the vehicle with the doors closed. Wait until all warning chimes have finished.

12.3 To deactivate the air suspension system, remove the right kick panel and turn off the switch

Press the SETUP button until "AIR SUSPENSION <ON> OFF" appears in the display. The brackets indicate whether the system is on or off.

Press the RESET button to switch the air suspension to either on or off.

13 Shock absorber (rear) - removal and installation

▶ Refer to illustrations 13.4a, 13.4b and 13.4c

1 If the vehicle is equipped with air suspension, turn off the air suspension system. The switch is located in the area of the right kick panel (see Section 12).

※ WARNING:

On models with air suspension, electrical power to the air suspension system must be turned off before raising the vehicle. Failure to do so can result in a sudden inflation or deflation of the air springs, causing instability of the vehicle while it's off the ground.

2 Loosen the wheel lug nuts, raise the rear of the vehicle and support it securely on jackstands. Block the front wheels so the vehicle doesn't roll off the stands. Remove the rear wheels.
3 Support the rear axle with a floor jack placed under the differential.
4 Remove the shock absorber upper and lower mounting fasteners (see illustrations).
5 Remove the shock absorber.
6 Installation is the reverse of removal. Make sure you install the nuts and bolts facing in the proper direction. Tighten all fasteners to the torque listed in this Chapter's Specifications. Reactivate the air suspension system, if equipped, after lowering the vehicle.

13.4b To detach the lower end of the shock absorber from the axle bracket on a truck, remove this nut and bolt (arrows)

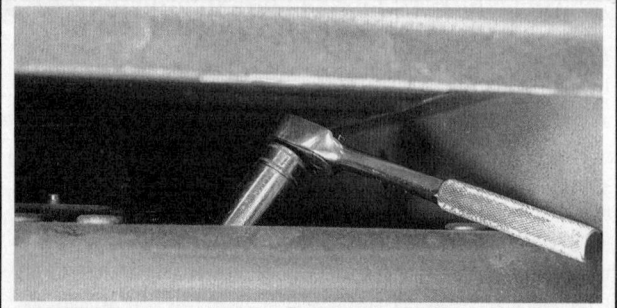

13.4a To detach the upper end of the shock absorber from the frame on a truck, unscrew the nut with a ratchet placed between the frame rail and bed

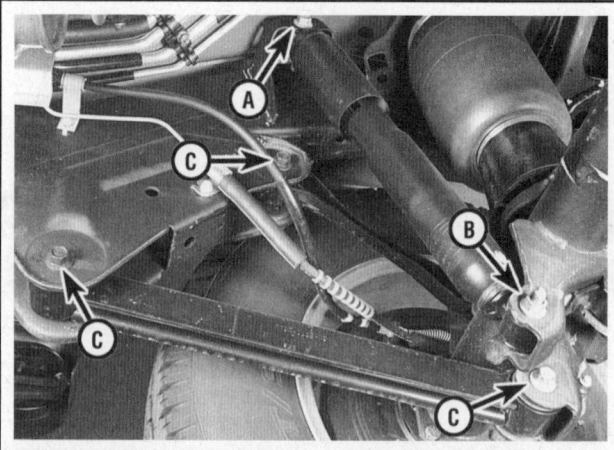

13.4c To detach the upper end of the shock absorber from the frame on a sport utility vehicle, remove this nut and bolt (A); to detach the lower end, remove this nut and bolt (B); if you're removing an upper or lower suspension arm, remove the indicated bolts (C) (upper arm-to-frame bracket bolt not visible)

SUSPENSION AND STEERING SYSTEMS

14 Leaf spring/air spring - removal and installation

◆ Refer to illustrations 14.4, 14.5 and 14.6

➥ Note: This procedure applies only to trucks.

LEAF SPRING

1 If the vehicle is equipped with air suspension, turn off the air suspension system. The switch is located in the area of the right kick panel (see Section 12).

⁕⁕ WARNING:

On models with air suspension, electrical power to the air suspension system must be turned off before raising the vehicle. Failure to do so can result in a sudden inflation or deflation of the air springs, causing instability of the vehicle while it's off the ground.

2 Loosen the rear wheel lug nuts, raise the rear of the vehicle and support it securely on jackstands. Block the front wheels to keep the vehicle from rolling off the stands. Remove the rear wheels. Support the axle with a floor jack and raise it slightly to relieve the tension on the leaf springs.
3 On vehicles with air suspension, unbolt the anti-wind bar from the axle.
4 Remove the four U-bolt nuts (see illustration), the spring plate, the rear spring spacer (4WD models only), and the two U-bolts.
5 Lower the floor jack to allow the spring to unload. At the front end of the spring, remove the nut and bolt from the spring-to-front bracket (see illustration).
6 At the rear end of the spring, remove the upper spring-to-shackle nut and bolt (see illustration).
7 Remove the spring assembly.
8 If the bushings at the ends of the spring are worn or deteriorated, an automotive machine shop or dealer service department can press the old ones out and press new ones in.
9 Installation is the reverse of removal. Gradually tighten the U-bolt nuts in a criss-cross pattern. Then tighten all the fasteners to the torque listed in this Chapter's Specifications.
10 Lower the vehicle. Reactivate the air suspension system, if equipped.

14.4 Remove the four U-bolt nuts (arrows), then remove the spring plate, the rear spring spacer (4WD models only), and the two U-bolts

AIR SPRING

11 Turn off the air suspension system. The switch is located in the area of the right kick panel (see Section 12).

⁕⁕ WARNING:

On models with air suspension, electrical power to the air suspension system must be turned off before raising the vehicle. Failure to do so can result in a sudden inflation or deflation of the air springs, causing instability of the vehicle while it's off the ground.

12 Loosen the rear wheel lug nuts, raise the rear of the vehicle and support it securely on jackstands. Block the front wheels to keep the vehicle from rolling off the stands. Remove the rear wheels.
13 Support the axle with a floor jack and raise it slightly to relieve the tension on the leaf springs.
14 Depress the red plastic retaining ring and disconnect the air line from the solenoid valve.

14.5 At the front end of the leaf spring, remove the nut and bolt (arrows) that attach the spring to the front frame bracket

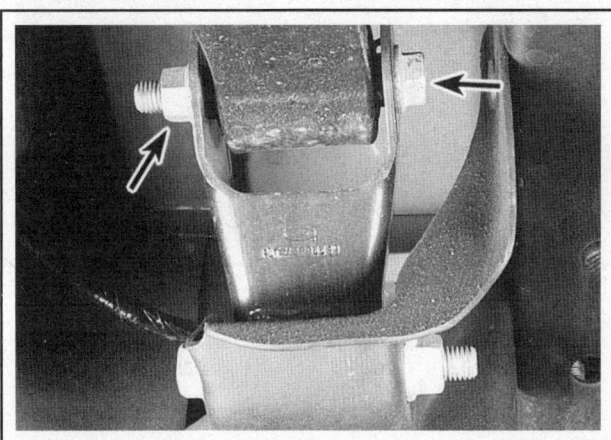

14.6 At the rear end of the spring, remove the upper spring-to-shackle nut and bolt (arrows)

10-18 SUSPENSION AND STEERING SYSTEMS

15 Unplug the electrical connector from the solenoid valve.
16 Remove the retaining clip from the top of the air spring.
17 Compress the air spring and remove it.
18 Installation is the reverse of removal.

19 Lower the vehicle and reactivate the air suspension system.
20 Start the engine and allow the compressor to fill the air suspension system. Do NOT drive the vehicle until the air suspension system has been restored.

15 Track bar - removal and installation

⦁ Refer to illustrations 15.3, 15.4 and 15.5

➡ Note: This procedure applies only to sport utility vehicles.

1 If the vehicle is equipped with air suspension, turn off the air suspension system. The switch is located in the area of the right kick panel (see Section 12).

✳✳ WARNING:

On models with air suspension, electrical power to the air suspension system must be turned off before raising the vehicle. Failure to do so can result in a sudden inflation or deflation of the air springs, causing instability of the vehicle while it's off the ground.

2 Loosen the rear wheel lug nuts, raise the rear of the vehicle and support it securely on jackstands. Block the front wheels to keep the vehicle from rolling off the stands. Remove the rear wheels.
3 If the vehicle is equipped with air suspension, unplug the electrical connector from the height sensor, then unclip the plastic retainer and set the wiring harness aside. Squeeze the retaining tabs and disengage the height sensor (see illustration) from its mounting studs.
4 Unbolt the right end of the track bar from the axle bracket (see illustration).
5 Unbolt the left end of the track bar from the frame bracket (see illustration).
6 Remove the track bar.
7 Installation is the reverse of removal. Be sure to tighten the track bar fasteners to the torque listed in this Chapter's Specifications. Activate the air suspension system, if equipped.

15.3 If the vehicle is equipped with air suspension, unplug the electrical connector from the height sensor, then squeeze these retainers (arrows) to detach the sensor from its ball stud mounts

15.4 To detach the right end of the track bar from the axle bracket, remove this bolt (arrow)

15.5 To detach the left end of the track bar from the frame bracket, remove this nut and bolt (arrows)

SUSPENSION AND STEERING SYSTEMS

16 Rear stabilizer bar (2002 and earlier Expedition/Navigator) - removal and installation

♦ Refer to illustrations 16.3 and 16.5

1 If the vehicle is equipped with air suspension, turn off the air suspension system. The switch is located in the area of the right kick panel (see Section 12).

WARNING:

On models with air suspension, electrical power to the air suspension system must be turned off before raising the vehicle. Failure to do so can result in a sudden inflation or deflation of the air springs, causing instability of the vehicle while it's off the ground.

2 Loosen the rear wheel lug nuts, raise the rear of the vehicle and support it securely on jackstands. Block the front wheels to keep the vehicle from rolling off the stands. Remove the rear wheels.
3 Remove the stabilizer bar-to-link nuts (see illustration).
4 Remove the bolts from the upper ends of the links and remove the links.
5 Remove the stabilizer bar bushing bracket bolts (see illustration) and remove the stabilizer bar assembly.
6 Inspect the stabilizer bar bushings and link bushings for cracks, tears and other signs of deterioration. Replace as necessary.
7 Installation is the reverse of removal. Be sure to tighten all fasteners to the torque listed in this Chapter's Specifications.
8 Reactivate the air suspension system, if equipped.

16.3 To remove the stabilizer link, remove the lower nut and upper bolt (arrows)

16.5 To detach the stabilizer bar from the rear axle, remove these bolts (arrows) from both bushing brackets

17 Rear suspension arms (2002 and earlier Expedition/Navigator) - removal and installation

➡Note: 2003 and later Expedition and Navigator models are equipped with independent rear suspension. Removal of rear suspension components on these models is covered in Section 24.

1 If the vehicle is equipped with air suspension, turn off the air suspension system. The switch is located in the area of the right kick panel (see Section 12).

WARNING:

On models with air suspension, electrical power to the air suspension system must be turned off before raising the vehicle. Failure to do so can result in a sudden inflation or deflation of the air springs, causing instability of the vehicle while it's off the ground.

2 Loosen the rear wheel lug nuts, raise the rear of the vehicle and support it securely on jackstands. Block the front wheels to keep the vehicle from rolling off the stands. Remove the rear wheels.
3 Support the rear axle by placing a floor jack under the differential.
4 To remove a suspension arm, unbolt it from the frame and from the axle (see illustration 13.4c).
5 Installation is the reverse of removal.
6 Reactivate the air suspension system, if equipped.

10-20 SUSPENSION AND STEERING SYSTEMS

18 Rear coil spring/air spring (2002 and earlier Expedition/Navigator) - removal and installation

→ Note: The following procedure applies only to sport utility vehicles.

COIL SPRING

♦ Refer to illustration 18.3

1 Loosen the rear wheel lug nuts, raise the rear of the vehicle and support it securely on jackstands. Block the front wheels to keep the vehicle from rolling off the stands. Remove the rear wheels.
2 Remove the driveshaft (see Chapter 8).
3 Unplug the electrical connector for the rear ABS speed sensor (see illustration).
4 Make sure the parking brake system is fully released, then release cable tension and disconnect the rear parking brake cables from the parking brake levers on the rear brake calipers (see Chapter 9). Remove the calipers and hang them safely out of the way (do not disconnect the brake lines from the calipers).
5 Detach the axle vent tube.
6 Detach the rear stabilizer bar from the axle (see Section 16).
7 Support the rear axle by placing a floor jack under the differential.
8 Disconnect the lower ends of the shock absorbers from the axle (see Section 13).
9 Disconnect the track bar from the axle (see Section 15).
10 Disconnect the lower and upper suspension arms from the axle (see Section 17).
11 Carefully lower the axle and remove the coil springs. Installation is the reverse of removal. Don't tighten any suspension fasteners to the torque listed in this Chapter's Specifications until the vehicle is back on the ground.

AIR SPRING

♦ Refer to illustrations 18.14 and 18.16

12 Turn off the air suspension system. The switch is located in the area of the right kick panel (see Section 12).

✱✱ WARNING:

On models with air suspension, electrical power to the air suspension system must be turned off before raising the vehicle. Failure to do so can result in a sudden inflation or deflation of the air springs, causing instability of the vehicle while it's off the ground.

→ Note: On 2005 and later Expedition/Navigator models, replacement of the rear air spring/shock should be performed by a dealer service department or other qualified shop.

13 Loosen the rear wheel lug nuts, raise the rear of the vehicle and support it securely on jackstands. Block the front wheels to keep the vehicle from rolling off the stands. Remove the rear wheels.
14 Remove the air spring retainer (see illustration).
15 Lift the bottom of the air spring off the axle.
16 Unplug the electrical connector from the air spring solenoid (see illustration).
17 Carefully and slowly disconnect the air line from the air spring and remove the air spring.
18 Installation is the reverse of removal. Make sure that at least 1/8-inch of white air line is inserted into the solenoid.
19 Be sure to reactivate the air suspension system, then start the engine and allow the compressor to pump up the system to normal trim height.

✱✱ WARNING:

Do NOT drive the vehicle until the system has restored it to its normal trim height.

18.3 Unplug the electrical connector for the rear ABS speed sensor

18.14 Remove the air spring retainer clip

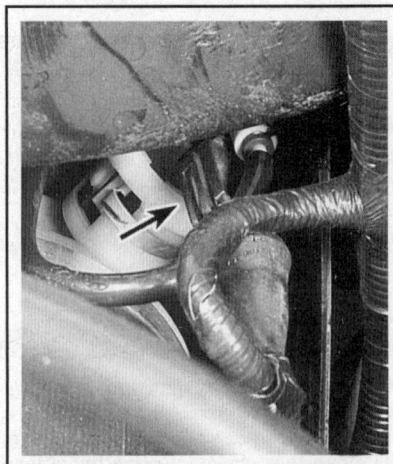

18.16 Unplug the electrical connector (arrow) from the solenoid

SUSPENSION AND STEERING SYSTEMS

19 Steering wheel and clockspring - removal and installation

STEERING WHEEL

♦ Refer to illustrations 19.3a, 19.3b, 19.4, 19.5, 19.6, 19.7 and 19.8

WARNING:

The models covered by this manual are equipped with Supplemental Restraint systems (SRS), more commonly known as airbags. Always disconnect the negative battery cable, then the positive battery cable and wait two minutes before working in the vicinity of the impact sensors, steering column or instrument panel to avoid the possibility of accidental deployment of the airbag, which could cause personal injury (see Chapter 12). Do not use electrical test equipment on any of the airbag system wiring or tamper with them in any way.

1 Park the vehicle with the front wheels pointing directly forward and the steering wheel centered.

WARNING:

Disconnect the negative battery cable, then the positive battery cable from the battery and wait at least two minutes for the backup power supply to be depleted.

2 Disconnect the battery-to-starter relay cable (see *Airbag system - general information* in Chapter 12).
3 Remove the two airbag module retaining screws (see illustrations) and lift off the airbag module.

WARNING:

Carry the airbag module with the trim cover facing away from you, and set the airbag module in a safe location with the trim cover facing up.

4 Unplug the electrical connectors for the airbag and horn (see illustration).
5 Remove the steering wheel bolt (see illustration).
6 Mark the relationship of the steering wheel to the steering shaft and unplug the electrical connector from the steering wheel (see illustration).
7 Using a two-jaw puller, remove the steering wheel (see illustration).

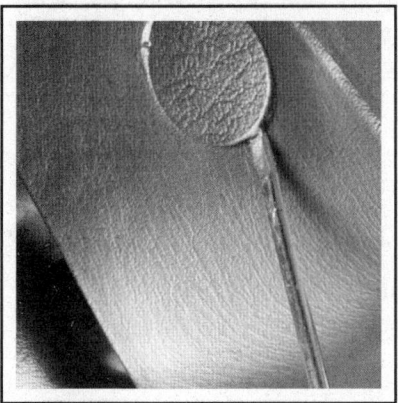

19.3a To detach the airbag module from the steering wheel, pry off the screw covers (left cover shown, right cover on other side of steering wheel) . . .

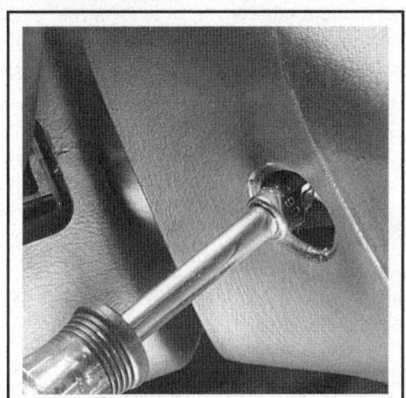

19.3b . . . and remove the airbag module retaining screws

19.4 Lift the airbag off the steering wheel and unplug the electrical connectors for the horn (upper arrow) and for the airbag (lower arrow)

19.5 Remove the steering wheel retaining bolt

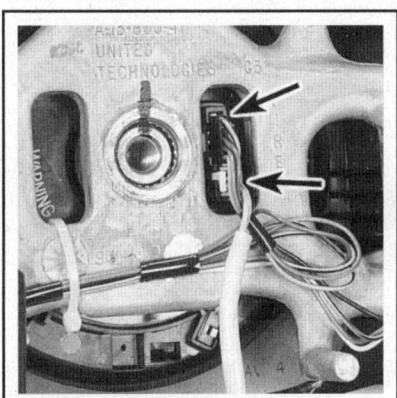

19.6 Mark the relationship of the steering wheel to the steering shaft before removing the wheel, and unplug these two electrical connectors (arrows)

19.7 Remove the steering wheel with a two-jaw puller - do not try to hammer off the steering wheel or you will damage the column bearings

10-22 SUSPENSION AND STEERING SYSTEMS

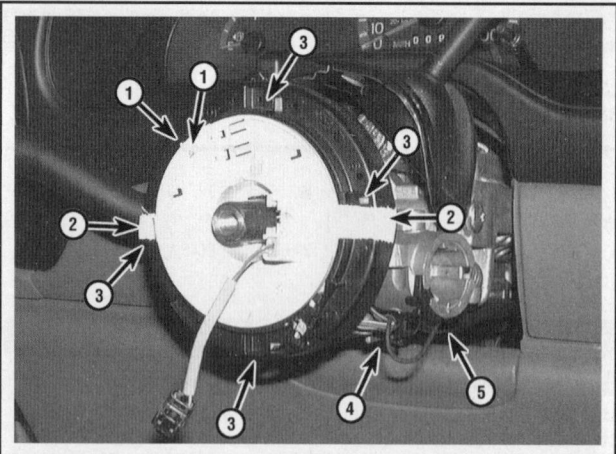

19.8 Clockspring and related details:

1 Index marks
2 Tape
3 Release tab locations
4 PATS transmitter location
5 KEY-IN warning indicator

19.18 Releasing a clockspring tab using a small screwdriver

20.12 To separate the outer tie-rod end from the steering knuckle, remove the cotter pin from the castellated nut, back off - but don't remove - the nut, install a small puller and force the ballstud out of the knuckle

✳ CAUTION:

Any attempt to remove the steering wheel without using a puller to do so can damage the column.

8 **Do not rotate the clockspring rotor!** Apply two strips of tape to the clockspring to keep the rotor from turning. Note the alignment index marks as equipped (see illustration).

✳ CAUTION:

If either the clockspring rotor or the steering shaft is turned while the steering wheel is removed, the clockspring will become un-centered and could be damaged when the steering wheel is reinstalled and used. This will disable the airbag system. To avoid damage, re-center the clockspring (see Step 9).

CLOCKSPRING

centering

9 Verify that the front wheels are pointing straight ahead. Turn the clockspring rotor lightly by hand until slight resistance is felt (do not exert pressure on the rotor when turning it or you will break the clockspring). Now, turn the clockspring in the opposite direction as specified below. The clockspring will be centered when the index marks are aligned (see illustration 19.8).

✳ CAUTION:

The clockspring must be in the centered state before installing the steering wheel.

a) On 1997 and 1998 models, rotate the clockspring counter-clockwise until it reaches its stop. Rotate the assembly from this position 2 turns clockwise.
b) On 1999 and 2000 models, the manufacturer does not provide a procedure for centering the clockspring. However, you may be able to center the clockspring by rotating the rotor in the clockspring housing end to end while counting the number of turns it

takes. Do this a couple of times. Then, divide the number of turns in half while noting any index marks. Align the index marks.
c) On 2001 and 2002 Expedition/Navigator and 2001 and later F-150 models, rotate the clockspring clockwise until it reaches its stop. Rotate the assembly from this position 3 turns counterclockwise.
d) On 2003 and 2004 Expedition/Navigator models, rotate the clockspring counterclockwise until it reaches its stop. Rotate the assembly from this position 2-3/4 turns clockwise.
e) On 2005 and later Expedition/Navigator models, rotate the clockspring counterclockwise until it reaches its stop. Rotate the assembly from this position 2-1/4 to 2-1/2 turns clockwise.

Removal and installation

1997 through 2002 Expedition/Navigator or through 2003 F-150 models

▶ Refer to illustration 19.18

10 On tilt steering column systems, remove the tilt handle by turning it counterclockwise.
11 Remove the center trim panel (see Chapter 11).
12 Remove the knee bolster (see Chapter 11).
13 Remove the steering column covers (see Chapter 11).
14 Remove the ignition lock cylinder (see Chapter 12).
15 Remove the Passive Anti-Theft System (PATS) transmitter from below the steering column, if equipped (see illustration 19.8).
16 Remove the ignition KEY-IN warning indicator switch from the lock cylinder housing (see illustration 19.8).
17 Disconnect the clockspring electrical connectors.
18 Release the retaining tabs holding the clockspring to the steering column, then remove it (see illustration 19.8 and the accompanying illustration).
19 Installation is the reverse of removal.

2003 and later Expedition/Navigator models

20 Disconnect the clockspring electrical connector.
21 Remove the multifunction switch (see Chapter 12) if the clockspring harness connector cannot be removed from the steering column.
22 Remove the clockspring mounting screws.
23 Slide the clockspring off the steering column.
24 Installation is the reverse of removal.

SUSPENSION AND STEERING SYSTEMS

20 Steering linkage (2002 and earlier Pick-ups and Expedition/Navigator) - inspection, removal and installation

INSPECTION

1 The steering linkage (see illustrations 1.1a through 1.1d) connects the steering gear to the front wheels and keeps the wheels in proper relation to each other. The linkage consists of the Pitman arm, the idler arm, the center link, two adjustable tie-rods and a steering damper. The Pitman arm, which is fastened to the steering gear shaft, moves the center link back-and-forth. The center link is supported on the other end by a frame-mounted idler arm. The back-and-forth motion of the center link is transmitted to the steering knuckles through a pair of tie-rod assemblies. Each tie-rod is made up of an inner and outer tie-rod end, a threaded adjuster tube and two clamps.

2 Set the wheels in the straight-ahead position and lock the steering wheel.

3 If the vehicle is equipped with air suspension, turn off the air suspension system. The switch is located in the area of the right kick panel (see Section 12).

※ WARNING:

On models with air suspension, electrical power to the air suspension system must be turned off before raising the vehicle. Failure to do so can result in a sudden inflation or deflation of the air springs, causing instability of the vehicle while it's off the ground.

4 Raise one side of the vehicle until the tire is approximately 1-inch off the ground.

5 Mount a dial indicator with the needle resting on the outside edge of the wheel. Grasp the front and rear of the tire and, using light pressure, wiggle the wheel back-and-forth and note the dial indicator reading. The gauge reading should be less than 0.108-inch. If the play in the steering system is more than specified, inspect each steering linkage pivot point and ball stud for looseness and replace parts, if necessary.

6 Raise the vehicle and support it on jackstands. Push up, then pull down on the center link end of the idler arm, exerting a force of approximately 25 pounds each way. Measure the total distance the end of the arm travels. If the play is greater than 1/4-inch, replace the idler arm.

7 Check for torn ball stud boots, frozen joints and bent or damaged linkage components.

8 Lower the vehicle. Reactivate the air suspension system, if equipped.

REMOVAL AND INSTALLATION

9 If the vehicle is equipped with air suspension, turn off the air suspension system. The switch is located in the area of the right kick panel (see Section 12).

※ WARNING:

On models with air suspension, electrical power to the air suspension system must be turned off before raising the vehicle. Failure to do so can result in a sudden inflation or deflation of the air springs, causing instability of the vehicle while it's off the ground.

10 Loosen the wheel lug nuts, raise the vehicle and support it securely on jackstands. Apply the parking brake. Remove the wheel.

Tie-rod

▸ Refer to illustrations 20.12, 20.13, 20.14a, 20.14b and 20.14c

➞Note: This procedure covers replacing the tie-rod ends as well as the entire tie-rod. If you'll only be replacing a tie-rod end, ignore the Steps that don't apply.

11 Remove the cotter pin and loosen, but do not remove, the castellated nut from the ballstud. If only the outer tie-rod end is being replaced, loosen only the ballstud nut at the steering knuckle; if only the inner tie-rod end is being replaced, loosen only the ballstud nut at the center link. If the entire tie-rod assembly (inner and outer tie-rod ends and adjuster tube) is being replaced, loosen both ballstud nuts.

12 If the outer tie-rod end or the entire tie-rod assembly is being replaced, use a puller to separate the tie-rod end from the steering knuckle (see illustration). Remove the castellated nut and pull the tie-rod end ballstud from the knuckle.

13 If the inner tie-rod end or the entire tie-rod is being replaced, separate the inner tie-rod end from the center link, using the same procedure as Steps 11 and 12 (see illustration).

14 If the inner or outer tie-rod end is being replaced, measure the distance from the end of the adjuster tube to the center of the ball stud (see illustration) and record it. If you're going to install the same inner

20.13 To separate the inner tie-rod end from the center link, remove the cotter pin from the castellated nut, back off - but don't remove - the nut, install a suitable small puller and force the ballstud out of the center link

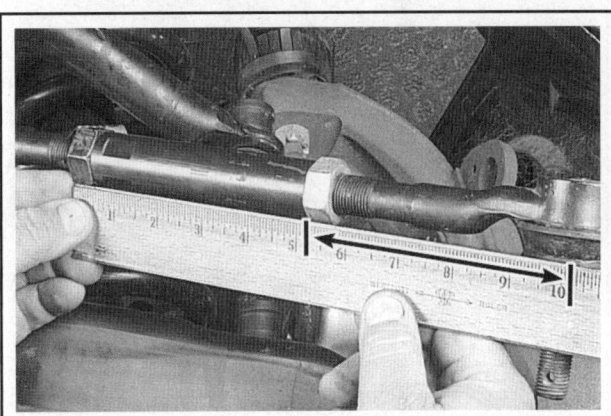

20.14a If you're planning to install a new inner or outer tie-rod end, measure the distance from the end of the adjuster tube to the centerline of the ballstud and record your measurement *before* unscrewing the tie-rod end from the adjuster tube

10-24 SUSPENSION AND STEERING SYSTEMS

20.14b If you're going to install the same inner or outer tie-rod, back off the jam nut . . .

20.14c . . . and paint an alignment mark on the threads adjacent to the end of the tube; when installing the tie-rod end, simply screw it back in until the mark is next to the end of the tube again

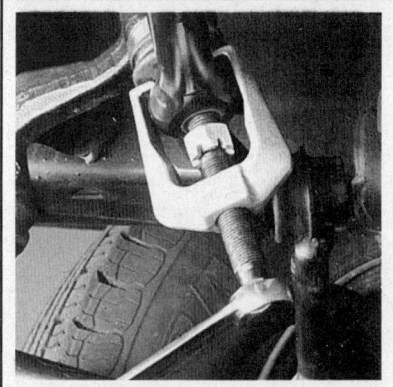

20.21 To separate the idler arm from the center link, remove the cotter pin from the castellated nut, back off - but don't remove - the nut, install a puller as shown, and force the ballstud out of the center link

or outer tie-rod end, simply back off the jam nut and mark the threads (see illustrations) and unscrew the tie-rod end.

15 Lubricate the threaded portion of the tie-rod end with chassis grease. Screw the new tie-rod end into the adjuster tube and adjust the distance from the tube to the ball stud to the previously measured dimension. The number of threads showing on the inner and outer tie-rod ends should be equal within three threads. Don't tighten the jam nuts yet.

16 Connect the disconnected ball stud nuts. Tighten the nuts to the torque listed in this Chapter's Specifications and install a new cotter pin. If the ball stud spins when attempting to tighten the nut, force it into the tapered hole with a large pair of pliers. If necessary, tighten the castellated nut slightly to align a slot in the nut with the cotter hole in the ball stud.

17 Insert the inner tie-rod end ball stud into the center link until it's seated. Install the nut and tighten it to the torque listed in this Chapter's Specifications.

18 Tighten the jam nut(s) securely.

19 Install the wheel and lug nuts, lower the vehicle and tighten the lug nuts to the torque listed in the Chapter 1 Specifications. Reactivate the air suspension system, if equipped. Drive the vehicle to an alignment shop to have the front end alignment checked and, if necessary, adjusted.

Idler arm

▶ Refer to illustrations 20.21 and 20.22

20 Loosen but do not remove the idler arm-to-center link nut.

21 Separate the idler arm from the center link with a small puller (see illustration). Remove the nut.

22 Remove the idler arm-to-frame bolts (see illustration).

23 To install the idler arm, position it on the frame and install the bolts, tightening them to the torque listed in this Chapter's Specifications.

24 Insert the idler arm ball stud into the center link and install the nut. Tighten the nut to the torque listed in this Chapter's Specifications. If the ball stud spins when attempting to tighten the nut, force it into the tapered hole with a large pair of pliers.

25 Install the wheel and lug nuts, lower the vehicle and tighten the lug nuts to the torque listed in the Chapter 1 Specifications. Reactivate the air suspension system, if equipped.

20.22 To detach the idler arm bracket from the frame, remove these bolts (arrows)

20.35 Before removing the Pitman arm, be sure to mark the relationship of the arm to the steering gear sector shaft, then use a Pitman arm puller to separate the arm from the sector shaft

SUSPENSION AND STEERING SYSTEMS

26 Drive the vehicle to an alignment shop to have the front end alignment checked and, if necessary, adjusted.

Center link

27 Separate the two inner tie-rod ends from the center link (see illustration 20.13).
28 Separate the center link from the idler arm (see illustration 20.21).
29 Separate the center link from the Pitman arm.
30 Installation is the reverse of the removal procedure. If the ball studs spin when attempting to tighten the nuts, force them into the tapered holes with a large pair of pliers. Be sure to tighten all of the nuts to the torque listed in this Chapter's Specifications.
31 Install the wheel and lug nuts, lower the vehicle and tighten the lug nuts to the torque listed in the Chapter 1 Specifications. Reactivate the air suspension system, if equipped.
32 Drive the vehicle to an alignment shop to have the front end alignment checked and, if necessary, adjusted.

Pitman arm

◆ Refer to illustration 20.35

33 Using a puller, separate the center link from the Pitman arm.
34 Loosen the Pitman arm nut and washer.
35 Mark the relationship of the Pitman arm to the steering gear sector shaft (see illustration).
36 Remove the Pitman arm with a Pitman arm puller.
37 Inspect the ball stud threads for damage. Inspect the ball stud seals for excessive wear. Clean the threads on the ball stud.
38 Installation is the reverse of removal. Make sure the marks you made on the Pitman arm and steering gear sector shaft are aligned.
39 Install the wheel and lug nuts, lower the vehicle and tighten the lug nuts to the torque listed in the Chapter 1 Specifications. Reactivate the air suspension system, if equipped.
40 Drive the vehicle to an alignment shop to have the front end alignment checked and, if necessary, adjusted.

21 Steering gear - removal and installation

REMOVAL

◆ Refer to illustrations 21.5, 21.7 and 21.8

WARNING:

DO NOT allow the steering column shaft to rotate with the steering gear removed or damage to the airbag system could occur. As a method of preventing the shaft from turning, wrap the seat belt around the rim of the steering wheel and buckle the belt in place.

1 If the vehicle is equipped with air suspension, turn off the air suspension system. The switch is located in the area of the right kick panel (see Section 12).

WARNING:

On models with air suspension, electrical power to the air suspension system must be turned off before raising the vehicle. Failure to do so can result in a sudden inflation or deflation of the air springs, causing instability of the vehicle while it's off the ground.

2 Point the front wheels directly forward. Raise the front of the vehicle and support it securely on jackstands. Apply the parking brake.
3 Remove the skid plate bolts and remove the skid plate, if equipped.
4 Slide back the plastic cover that protects the U-joint connecting the intermediate shaft to the steering gear input shaft.
5 Mark the relationship of the intermediate shaft U-joint to the steering gear input shaft and remove the U-joint pinch bolt (see illustration).

Recirculating ball type

6 On all 2002 and earlier Pick-ups and Expedition/Navigator models, separate the center link from the Pitman arm and remove the Pitman arm from the steering gear selector shaft (see Section 20). On 2003 and later Expedition/Navigator models, disconnect the power steering hoses from the rack-and-pinion steering gear.
7 Disconnect the hydraulic fluid lines from the steering gear (see illustration).

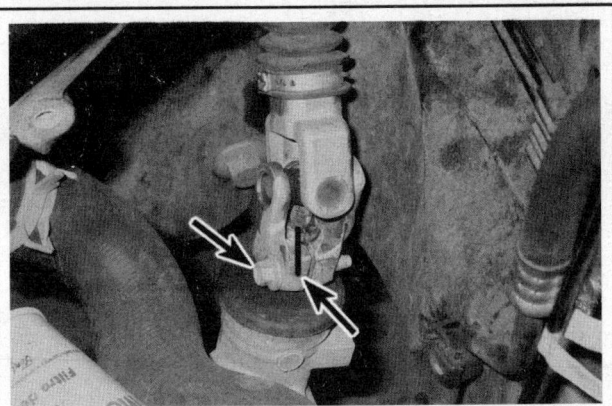

21.5 Mark the relationship of the intermediate shaft to the input shaft and remove the U-joint pinch bolt

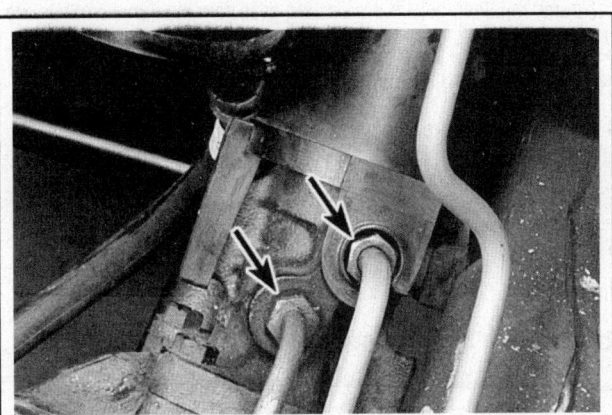

21.7 Disconnect the pressure and return lines from the steering gear

10-26 SUSPENSION AND STEERING SYSTEMS

21.8 To detach the steering gear from the frame, remove these three bolts (arrows)

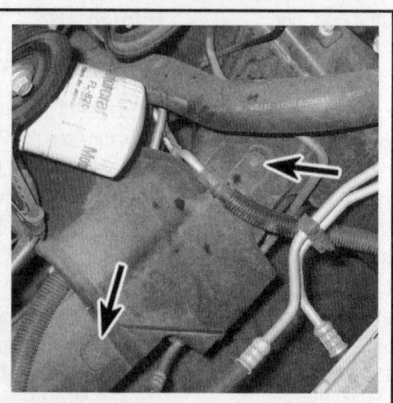

21.9 Oil drip tray mounting fasteners (as equipped)

21.10 The pressure and return line fittings are held by single fastener at the steering gear (rack and pinion)

21.14 Remove the steering gear bracket-to-crossmember fasteners (A) and then remove the fasteners for the steering gear brackets (B) from the steering gear (2006 models shown - other models similar)

8 Remove the steering gear retaining bolts (see illustration) from the frame rail, then detach the steering gear from the frame and remove it.

Rack and pinion type

▶ Refer to illustrations 21.9 and 21.10

9 Remove the oil drip tray (if equipped) (see illustration).
10 Remove the bolt to disconnect the pressure and return lines from the steering gear and discard any O-ring seals (see illustration).
11 Disconnect the tie-rod ends from the steering knuckles (see Section 20).

2003 through 2006 Navigator models

12 Rotate the actuator on the steering gear so the that electrical connector is pointing towards the steering gear input shaft, then disconnect the electrical connector from the actuator.
13 Detach the left (driver's side) lower control arm from the frame bracket (see Section 6). Place a small block of wood (2 X 4) between the frame bracket and the control arm to keep the control arm pushed out.

➡ Note: Be sure to mark its relationship to the frame bracket before removing it.

All models of rack and pinion type

▶ Refer to illustration 21.14

14 On 2003 through 2006 models, remove the steering gear bracket-to-crossmember fasteners and then remove the mounting brackets from the steering gear (see illustration). On 2007 and later models, remove the mounting bolts for the steering gear.

✳✳ CAUTION:

Note the position of the steering gear hydraulic lines and be careful not to damage them during removal.

15 Remove the steering gear assembly through the opening at the left wheel well.

INSTALLATION (ALL TYPES)

16 Installation is the reverse of removal. Be sure to align all matchmarks and tighten all suspension and steering gear fasteners to the torque values listed in this Chapter's Specifications. Also, use new O-rings when assembling the pressure and return lines to the steering gear as equipped.
17 Lower the vehicle and re-activate the air suspension system, if equipped. Tighten the wheel lug nuts to the torque listed in the Chapter 1 Specifications.
18 Refill the power steering fluid reservoir with the proper fluid (see Chapter 1), then bleed the system (see Section 23).

SUSPENSION AND STEERING SYSTEMS 10-27

22 Power steering pump - removal and installation

▶ Refer to illustrations 22.2, 22.3, 22.6, 22.11 and 22.17

1 Remove the fan and shroud (see Chapter 3) and the serpentine drivebelt (see Chapter 1).

2 If you're planning to replace the pump with a new or rebuilt unit, remove the pulley now with a pulley removal tool (see illustration). Power steering pump pulley puller/installer tools are available at most auto parts stores. If you're planning to install the same pump, disregard this step.

3 Remove the three power steering reservoir bolts (see illustration) and set the reservoir aside.

4 Position a drain pan under the power steering pump reservoir hoses. Disconnect the power steering reservoir-to-pump hose and drain the power steering system. Plug the hose to prevent contaminants from entering.

5 Remove the two belt deflector nuts and remove the belt deflector, if equipped.

6 Remove the two upper pump retaining bolts (see illustration). A third bolt underneath is easier to reach from underneath the vehicle.)

7 If the vehicle is equipped with air suspension, turn off the air suspension system. The switch is located in the area of the right kick panel (see Section 12).

✴✴ WARNING:

On models with air suspension, electrical power to the air suspension system must be turned off before raising the vehicle. Failure to do so can result in a sudden inflation or deflation of the air springs, causing instability of the vehicle while it's off the ground.

8 Raise the front of the vehicle and support it securely on jackstands. Apply the parking brake.

9 Remove the skid plate bolts and skid plate, if equipped.

10 Unplug the electrical connector from the electronic variable orifice (EVO) power steering control valve actuator.

11 Remove the third power steering pump bolt from the underside of the pump (see illustration).

12 Lower the vehicle.

13 Remove the power steering pump.

14 Disconnect the reservoir-to-pump hose and the fluid pressure line from the pump (see illustration 22.11).

15 If you removed the power steering pump pulley, inspect the spokes on the pulley for paint marks. If there are already two paint marks, replace the pulley; the two marks indicate that this pulley has already been removed and installed two times and must, therefore, be replaced. If there is one paint mark, or there are no paint marks, the pulley can be installed again. Be sure to put a paint mark on a spoke, near the hub, to indicate that the pulley has been removed and installed.

✴✴ CAUTION:

Do NOT install a pulley that has already been removed and installed two times.

16 Installation is the reverse of removal. Be sure to tighten the three pump mounting bolts to the torque listed in this Chapter's Specifications. Where the power steering pressure line attaches to the pump, install a new Teflon® seal.

22.2 Remove the power steering pump pulley with a pulley removal tool (available at most auto parts stores)

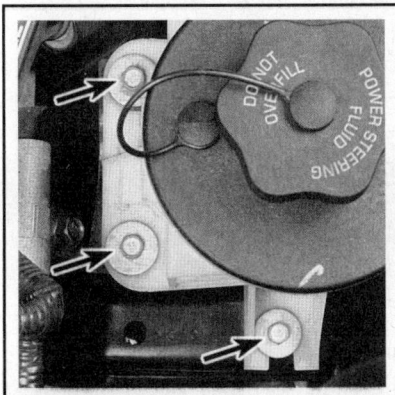

22.3 Remove the three power steering reservoir bolts (arrows) and set the reservoir aside

22.6 Remove the two upper pump retaining bolts (arrows); the lower bolts, on the underside of the pump, are easier to reach from underneath the vehicle

22.11 Remove the other two power steering pump bolts (arrows) and disconnect the fluid hose and line (arrows)

10-28 SUSPENSION AND STEERING SYSTEMS

22.17 Install the pulley on the power steering pump with a pulley installer tool

17 If you removed the power steering pump pulley, install the pulley before installing the serpentine belt (see illustration).
18 Fill the power steering reservoir with the recommended fluid (see Chapter 1) and bleed the system following the procedure described in the next Section.
19 Reactivate the air suspension system, if equipped.

23 Power steering system - bleeding

1 Following any operation in which the power steering fluid lines have been disconnected, the power steering system must be bled to remove all air and obtain proper steering performance.
2 With the front wheels in the straight ahead position, check the power steering fluid level and, if low, add fluid until it reaches the Cold mark on the dipstick.
3 Start the engine and allow it to run at fast idle. Recheck the fluid level and add more if necessary to reach the Cold mark on the dipstick.
4 Bleed the system by turning the wheels from side-to-side, without hitting the stops. This will work the air out of the system. Keep the reservoir full of fluid as this is done.
5 When the air is worked out of the system, return the wheels to the straight ahead position and leave the vehicle running for several more minutes before shutting it off.
6 Road test the vehicle to be sure the steering system is functioning normally and noise free.
7 Recheck the fluid level to be sure it's up to the Hot mark on the dipstick while the engine is at normal operating temperature. Add fluid if necessary (see Chapter 1).

24 Independent rear suspension (2003 and later Expedition/Navigator) - component replacement

✱✱ WARNING 1:

On models with air suspension, electrical power to the air suspension must be turned off before raising the vehicle. Failure to do so can result in a sudden inflation or deflation of the air springs, causing instability of the vehicle while it is off the ground.

✱✱ WARNING 2:

Very high torque is required for removal/installation of many of the critical rear suspension fasteners. You must have a torque wrench capable of measuring torque up to 400 ft-lbs. Because of these high torque values, the manufacturer recommends new OEM bolts be used whenever a suspension component is replaced.

→Note: The switch to turn off the air suspension on 2003 through 2006 is located in the rear storage compartment. On 2007 and later models, the air suspension is turned off through a selection on the instrument panel message center (see your owner's manual for message center operation).

HUB/BEARING ASSEMBLY

1 The rear hub/bearing assemblies are replaced using the procedure outlined in Section 9.

DRIVEAXLES

2 Rear driveaxle replacement is covered in Chapter 8, Section 21.

SHOCK/SPRING ASSEMBLY

3 Loosen the rear wheel lugnuts, then raise and safely support the rear of the vehicle. Remove the rear wheels.
4 Position a jack under the steering knuckle. Make reference marks on the body and the shock upper plate, then remove the three nuts securing the top of the shock/spring assembly to the body and lower the jack.
5 Remove the lower control arm-to-spring bolt and nut and remove the shock/spring assembly.
6 If only the shock is being replaced, use a coil spring compressor (available at tool rental yards) to compress the spring, then remove the shock rod nut and washer at the top of the shock.

SUSPENSION AND STEERING SYSTEMS

⁜⁜ WARNING:

Coil springs can be dangerous when compressed if the tool is not firmly fastened. Proper spring compressors have safety hooks to keep the tool in place on the spring. Keep the ends of the spring pointed in a safe direction - never toward your body.

7 Installation is the reverse of the removal. Use new fasteners and tighten to the torque listed in this Chapter's Specifications.

➡ **Note:** Use the jack to raise the suspension to simulate ride height before final tightening of the lower shock mounting bolt/nut.

LOWER CONTROL ARM

8 See the **Note** and **Warnings** above. Loosen the rear wheel lugnuts, then raise and support the rear of the vehicle. Remove the rear wheels.

9 Disconnect the lower link nut from the stabilizer bar. Use a hex wrench to hold the top of the link while removing the nut from the lower control arm.

10 Remove the nut and bolt securing the shock/spring assembly to the lower control arm.

11 The lower control arm is mounted to the chassis at the front and rear. Remove the front and rear bushing nuts and bolts.

12 Support the steering knuckle so that the driveaxle is not overextended. Remove the nut and bolt securing the lower control arm to the bottom of the wheel knuckle.

13 Installation is the reverse of the removal. Use new fasteners and tighten to the torque listed in this Chapter's Specifications.

⁜⁜ WARNING:

Because of the high torque that must be applied, be extremely careful during final tightening and make sure the vehicle and components are securely supported.

➡ **Note:** Use the jack to raise the suspension to simulate ride height before final tightening of the lower shock mounting bolt/nut and the lower control arm bushing bolts/nuts.

UPPER CONTROL ARM

14 See the **Note** and **Warnings** above. Loosen the rear wheel lugnuts, then raise and support the rear of the vehicle. Remove the rear wheels.

15 Remove the bolt securing the parking brake cable bracket to the upper control arm.

16 Remove the bolt securing the upper arm to the wheel knuckle.

17 Remove the nuts and bolts securing the upper arm bushings to the frame. Remove the control arm.

18 Installation is the reverse of removal. Using new fasteners, tighten the bolts and nuts securely, but not entirely. When the vehicle is back on the ground, tighten fasteners to the torque listed in this Chapter's Specifications.

⁜⁜ WARNING:

Because of the high torque that must be applied, be extremely careful during final tightening and make sure the vehicle is secure and cannot roll.

TOE LINK ROD

19 All models with independent rear suspension have a toe rod, which can be used by alignment personnel to establish the toe-in/toe-out to specifications. When removing the toe link rod, do not loosen the adjuster nut. If the rod is being replaced, adjust the new rod to the same length as the old one, but have the rear wheels aligned at a qualified shop.

20 See the **Note** and **Warnings** above. Loosen the rear wheel lugnuts, then raise and support the rear of the vehicle. Remove the rear wheels.

21 Remove the nut securing the inner end of the toe link to the chassis, then remove the nut at the wheel knuckle. Use a hex wrench to hold the top of the tie-rod end while loosening the nut. Remove the tie rod end from the knuckle and remove the toe link.

22 Installation is the reverse of the removal.

TRAILING ARMS (2007 AND LATER MODELS)

23 See the **Note** and **Warnings** above. Loosen the rear wheel lugnuts, then raise and support the rear of the vehicle. Remove the rear wheels.

24 Later models have two trailing arms that connect the wheel knuckle to the frame.

25 To remove an upper trailing arm, remove the upper trailing arm-to-knuckle nut and bolt, then remove the flag nut and bolt from the frame end of the arm.

26 To remove the lower trailing arm, remove the two bolts securing the arm to the wheel knuckle, then remove the bolt and flag nut securing the arm to the frame.

27 Installation is the reverse of the removal. Install new trailing arm bolts and nuts loosely, tightening them to final torque Specifications in this Chapter only when the vehicle is back on the ground.

⁜⁜ WARNING:

Because of the high torque that must be applied, be extremely careful during final tightening and make sure the vehicle is securely supported and cannot roll.

ST8ABILIZER BAR

28 See the **Note** and **Warnings** above. Loosen the rear wheel lugnuts, then raise and support the rear of the vehicle. Remove the rear wheels.

29 To remove the stabilizer bar, disconnect the end links, using a hex wrench to hold the link while removing the nuts.

30 On models so equipped, remove the rear suspension stone shields.

31 Remove the four stabilizer bar bracket-to-frame bolts, leaving the brackets and rubber bushings on the bar. Remove the stabilizer bar.

32 Installation is the reverse of the removal. When installing the stabilizer bar, the curved potion in the center should face toward the ground when installed. New nuts should be used on the links and on the bracket bolts on the frame.

10-30 SUSPENSION AND STEERING SYSTEMS

25 Wheels and tires - general information

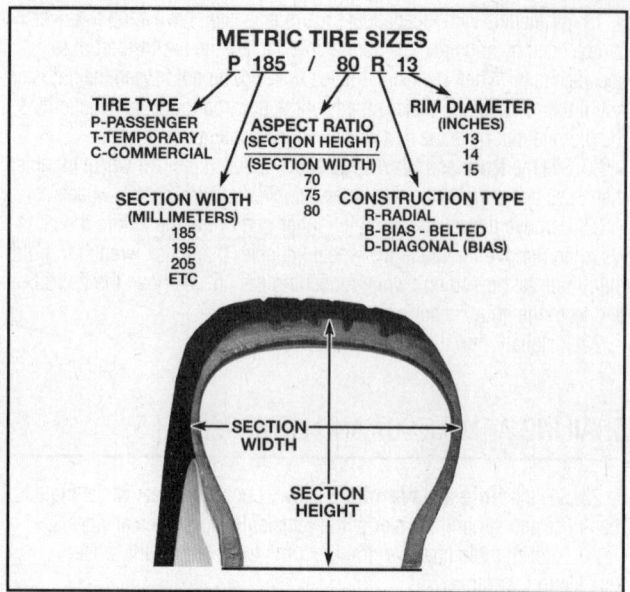

25.1 Metric tire size code

▶ **Refer to illustration 25.1**

All vehicles covered by this manual are equipped with metric-size fiberglass or steel belted radial tires (see illustration). Use of other size or type of tires may affect the ride and handling of the vehicle. Don't mix different types of tires, such as radials and bias belted, on the same vehicle as handling may be seriously affected. It's recommended that tires be replaced in pairs on the same axle, but if only one tire is being replaced, be sure it's the same size, structure and tread design as the other.

Because tire pressure has a substantial effect on handling and wear, the pressure on all tires should be checked at least once a month or before any extended trips (see Chapter 1).

Wheels must be replaced if they're bent, dented, leak air, have elongated bolt holes, are heavily rusted, out of vertical symmetry or if the lug nuts won't stay tight. Wheel repairs that use welding or peening are not recommended.

Tire and wheel balance is important to the overall handling, braking and performance of the vehicle. Unbalanced wheels can adversely affect handling and ride characteristics as well as tire life. Whenever a tire is installed on a wheel, the tire and wheel should be balanced by a shop with the proper equipment.

26 Front end alignment - general information

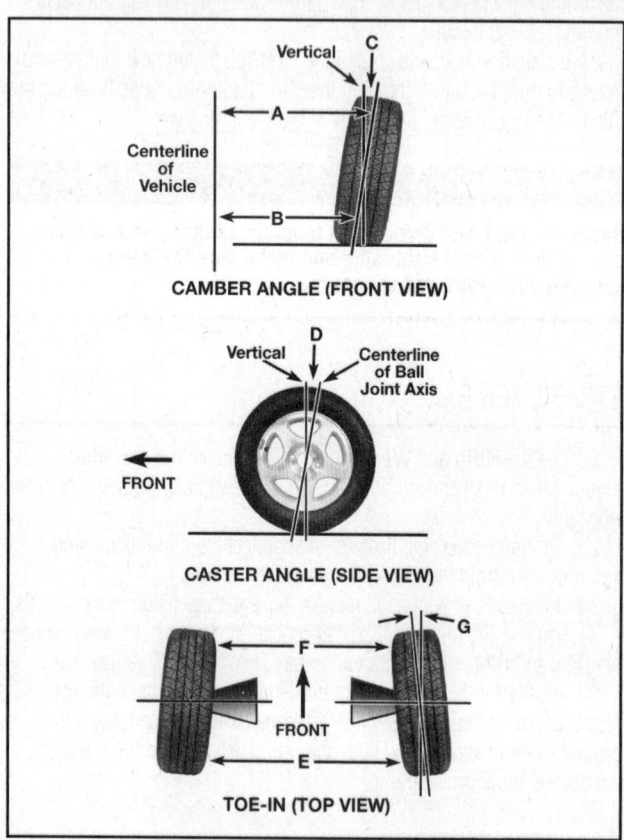

26.1 Front end alignment details

A minus B = C (degrees camber)
D = degrees caster
E minus F = toe-in (measured in inches)
G = toe-in (expressed in degrees)

▶ **Refer to illustration 26.1**

A front end alignment (see illustration) refers to the adjustments made to the front wheels so they're in proper angular relationship to the suspension and the ground. Front wheels that are out of proper alignment not only affect steering control, but also increase tire wear.

Getting the proper front wheel alignment is a very exacting process, one in which complicated and expensive machines are necessary to perform the job properly. Because of this, you should have a technician with the proper equipment perform these tasks. We will, however, use this space to give you a basic idea of what is involved with front end alignment so you can better understand the process and deal intelligently with the shop that does the work.

Toe-in is the turning in of the front wheels. The purpose of a toe specification is to ensure parallel rolling of the front wheels. In a vehicle with zero toe-in, the distance between the front edges of the wheels will be the same as the distance between the rear edges of the wheels. The actual amount of toe-in is normally only a fraction of an inch. Toe-in is adjusted by turning the tie-rod adjuster tube to lengthen or shorten the tie-rod. Incorrect toe-in will cause the tires to wear improperly by making them scrub against the road surface. Toe-in can be adjusted by turning the adjusting sleeves on the tie-rods equal amounts in the same direction.

Camber is the tilting of the front wheels from the vertical when viewed from the front of the vehicle. When the wheels tilt out at the top, the camber is said to be positive (+). When the wheels tilt in at the top the camber is negative (-). The amount of tilt is measured in degrees from the vertical and this measurement is called the camber angle. This angle affects the amount of tire tread which contacts the road and compensates for changes in the suspension geometry when the vehicle is cornering or traveling over an undulating surface. Camber is adjusted by rotating cam-shaped adjusters at the front and rear of the upper control arm.

Caster is the tilting of the top of the front steering axis from the vertical. A tilt toward the rear is positive caster and a tilt toward the front is negative caster. Caster is adjusted by rotating cam-shaped adjusters at the front and rear of the upper control arm.

SUSPENSION AND STEERING SYSTEMS

Torque specifications — Ft-lbs (unless otherwise indicated)

→ **Note:** One foot-pound (ft-lb) of torque is equivalent to 12 inch-pounds (in-lbs) of torque. Torque values below approximately 15 ft-lbs are expressed in inch-pounds, since most foot-pound torque wrenches are not accurate at these smaller values..

Front suspension
- Air shock
  - Lower bolt/nut
    - 2003 and earlier models — 56 to 76
    - 2004 and 2005 models — 293
    - 2006 and later models — 350
  - Upper nut — 37 to 44
- Balljoints - 2WD/4WD
  - Lower balljoint-to-steering knuckle nut
    - 1997 through 2002 — 98
    - 2003 through 2006 — 148
    - 2007 and later — 111
  - Upper balljoint-to-steering knuckle nut
    - 1997 through 2002 — 67
    - 2003 through 2006 — 111
    - 2007 and later — 85
- Lower control arm-to-frame pivot bolts/nuts - 2WD/4WD
  - 1997 through 2002 — 135
  - 2003 through 2006 — 199
  - 2007 and later — 258
- Shock absorber/coil spring assembly
  - Lower nuts - 2WD
    - F-150 — 26
    - F-250
      - 1997 through 1999 — 26
      - 2000 through 2003 — 60
    - Expedition/Navigator
      - 1997 through 2002 — 26
      - 2003 through 2005 — 295
      - 2006 through 2008 — 350
      - 2009 and later — 406
  - Lower bolt/nut - 4WD
    - F-150/F-250 — 76
    - Expedition/Navigator
      - 1997 through 2002 — 76
      - 2003 through 2005 — 295
      - 2006 — 350
      - 2007 and later — 406
  - Upper nuts - 2WD
    - F-150 — 40
    - F-250
      - 1997 through 1999 — 40
      - 2000 through 2003 — 30
    - Expedition/Navigator
      - 1997 through 2002 — 40
      - 2003 through 2005 — 26
      - 2006 and later — 30

10-32 SUSPENSION AND STEERING SYSTEMS

Torque specifications (continued)   Ft-lbs (unless otherwise indicated)

→Note: One foot-pound (ft-lb) of torque is equivalent to 12 inch-pounds (in-lbs) of torque. Torque values below approximately 15 ft-lbs are expressed in inch-pounds, since most foot-pound torque wrenches are not accurate at these smaller values.

Front suspension (continued)

Shock absorber/coil spring assembly (continued)
- Upper nuts - 4WD
  - F-150 — 26
  - F-250
    - 1997 through 1999 — 26
    - 2000 through 2003 — 76
  - Expedition/Navigator
    - 1997 through 2005 — 26
    - 2006 and later — 30

Stabilizer bar
- Bracket-to-frame bolts/nuts
  - 2009 and earlier — 19 to 25
  - 2010 and later — 41
- Link nut(s)
  - 2009 and earlier
    - 2WD — 15 to 21
    - 4WD — 16 to 21
  - 2010 and later — 59

Upper control arm-to-frame pivot bolt nuts — 83 to 112

Hub and bearing assembly bolts
- 2003 through 2009 2WD models — 148
- 2010 and later 2WD models — 129
- All 4WD models — 110 to 148

Rear suspension (pick-ups)

Leaf spring (trucks)
- Leaf spring-to-front bracket nut and bolt
  - 1999 and earlier models — 72 to 97
  - 2000 and later models — 157 to 212
- Shackle-to-frame nuts — 72 to 97
- Shackle-to-spring nuts — 72 to 97
- U-bolt nuts — 72 to 97

Shock absorber
- Upper nut — 22 to 29
- Lower nut — 44 to 60

Rear suspension (Expedition/Navigator)

Suspension arms, 2002 and earlier models
(upper and lower, to frame and axle) — 94 to 127

Lower control arm-to-frame nuts
- 2003 through 2006 models — 184
- 2007 and 2008 models — 221
- 2009 and later models — 266

Lower control arm-to-knuckle nut
- 2003 through 2006 models — 295
- 2007 and 2008 models — 350
- 2009 and later models — 406

SUSPENSION AND STEERING SYSTEMS 10-33

Torque specifications — Ft-lbs (unless otherwise indicated)

Rear suspension (Expedition/Navigator) (continued)

Upper control arm-to-frame	
2003 through 2006 models	
Forward nut	111
Rearward nut	184
2007 and 2008 models	221
2009 and later models	165
Upper control arm-to-knuckle nut	
2003 through 2006 models	111
2007 and later models	76
Track bar, 2002 and earlier models (to frame and axle)	125 to 170
Trailing arms (2007 and later models only)	
Lower trailing arm-to-frame bolt	203
Lower trailing arm-to-knuckle bolt	76
Upper trailing arm-to-frame bolt	203
Upper trailing arm-to-knuckle	
Bolt	184
Nut	76
Shock absorber mounting bolt/nut (2002 and earlier) (upper or lower)	63 to 84
Shock absorber/spring assembly upper mounting nuts (2003 and later)	
2003 through 2009	24
2010 and later	35
Shock absorber/spring lower mounting bolt	
2003 models	350
2004 and 2005 models	332
2006 models	295
2007 and later models	406
Shock rod nut	
2003 through 2006 models	22
2007 and later models	41
Stabilizer bar link nuts	
2002 and earlier models	63 to 84
2003 through 2005 models	66
2006 models	59
2007 and later models	46
Stabilizer bar bushing bracket nuts	
2002 and earlier models	63 to 84
2003 through 2005 models	30
2006 models	41
2007 and later models	35
Toe link-to-frame nut	
2003 and 2004 models	59
2005 models	66
2006 models	52
2007 and later models	203
Toe link-to-knuckle bolt or nut	
2003 through 2006 models	41
2007 and later models	165
Toe link adjuster lock-nut	85
Wheel bearing/hub assembly mounting bolts	
2003 through 2008 models	136
2009 and later models	129

10-34 SUSPENSION AND STEERING SYSTEMS

Torque specifications (continued)   **Ft-lbs (unless otherwise indicated)**

➡ Note: One foot-pound (ft-lb) of torque is equivalent to 12 inch-pounds (in-lbs) of torque. Torque values below approximately 15 ft-lbs are expressed in inch-pounds, since most foot-pound torque wrenches are not accurate at these smaller values.

Steering

Airbag module retaining screws	89 to 115 in-lbs
Idler arm	
Idler arm bracket-to-frame bolts	125 to 169
Idler arm-to-center link ballstud nut	56 to 75
Intermediate shaft-to-steering gear input shaft pinch bolt 30 to 40	
Pitman arm	
Pitman arm-to-steering gear nut	173 to 233
Pitman arm-to-center link ballstud nut	56 to 75
Power steering pump mounting bolts	15 to 20
Steering gear-to-frame mounting bolts	
All 2002 and earlier Pick-ups and Expedition/Navigator	50 to 67
2003 and later Expedition/Navigator (rack-and-pinion)	
2003 to 2006 models	111
2007 and later models	325
Steering wheel bolt	30
Tie-rod ends (linkage)	
Inner tie-rod end-to-center link ballstud nut	56 to 75
Outer tie-rod end-to-steering knuckle ballstud nut	56 to 75
Tie-rod end ballstud nut (rack and pinion)	
2003 to 2006 models	111
2007 and later models	85

Section

1. General information
2. Body - maintenance
3. Vinyl trim - maintenance
4. Upholstery and carpets - maintenance
5. Body repair - minor damage
6. Body repair - major damage
7. Hinges and locks - maintenance
8. Windshield and fixed glass - replacement
9. Hood and rear liftgate support struts - removal and installation
10. Hood - removal, installation and adjustment
11. Hood latch and release cable - removal and installation
12. Radiator grille - removal and installation
13. Bumpers - removal and installation
14. Front fender - removal and installation
15. Door trim panels - removal and installation
16. Door - removal, installation and adjustment
17. Door latch, lock cylinder and handles - removal and installation
18. Door window glass - removal and installation
19. Door window glass regulator - removal and installation
20. Sideview mirrors - removal and installation
21. Liftgate - removal, installation and adjustment
22. Tailgate - removal, installation and adjustment
23. Tailgate latch, handle and lock cylinder - removal and installation
24. Center console (Expedition and Navigator) - removal and installation
25. Instrument cluster bezel - removal and installation
26. Dashboard trim panels - removal and installation
27. Steering column cover - removal and installation
28. Instrument panel - removal and installation
29. Cowl cover - removal and installation
30. Seats - removal and installation

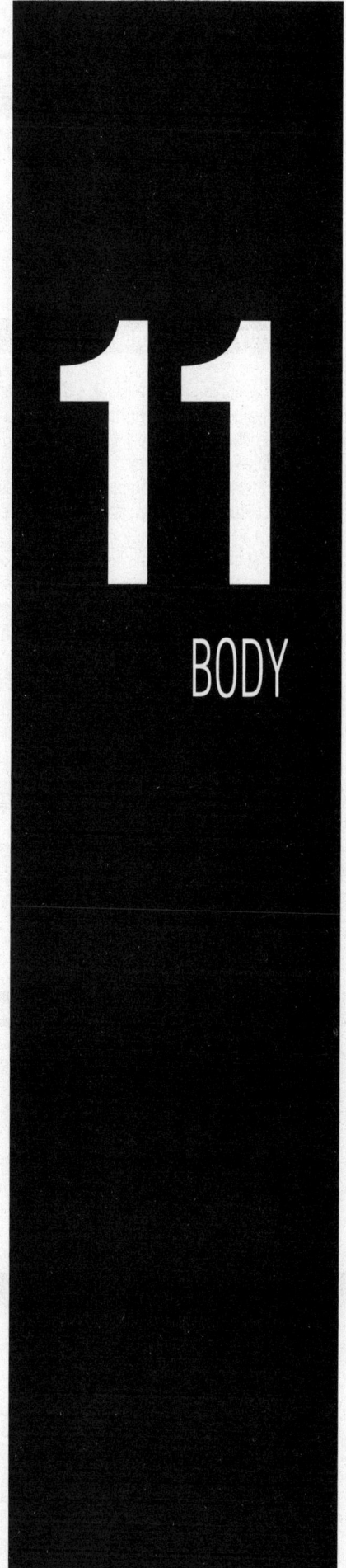

11
BODY

11-2 BODY

1 General information

These models feature a full-frame construction, using frame side rails which support the body components, front and rear suspension systems and other mechanical components. Certain components are particularly vulnerable to accident damage and can be unbolted and repaired or replaced. Among these parts are the body moldings, bumpers, hood, doors, tailgate and all glass.

Only general body maintenance practices and body panel repair procedures within the scope of the do-it-yourselfer are included in this Chapter.

2 Body - maintenance

1 The condition of your vehicle's body is very important, because the resale value depends a great deal on it. It's much more difficult to repair a neglected or damaged body than it is to repair mechanical components. The hidden areas of the body, such as the wheel wells, the frame and the engine compartment, are equally important, although they don't require as frequent attention as the rest of the body.

2 Once a year, or every 12,000 miles, it's a good idea to have the underside of the body steam cleaned. All traces of dirt and oil will be removed and the area can then be inspected carefully for rust, damaged brake lines, frayed electrical wires, damaged cables and other problems. The front suspension components should be greased after completion of this job.

3 At the same time, clean the engine and the engine compartment with a steam cleaner or water soluble degreaser.

4 The wheel wells should be given close attention, since undercoating can peel away and stones and dirt thrown up by the tires can cause the paint to chip and flake, allowing rust to set in. If rust is found, clean down to the bare metal and apply an anti-rust paint.

5 The body should be washed about once a week. Wet the vehicle thoroughly to soften the dirt, then wash it down with a soft sponge and plenty of clean soapy water. If the surplus dirt is not washed off very carefully, it can wear down the paint.

6 Spots of tar or asphalt thrown up from the road should be removed with a cloth soaked in solvent.

7 Once every six months, wax the body and chrome trim. If a chrome cleaner is used to remove rust from any of the vehicle's plated parts, remember that the cleaner also removes part of the chrome, so use it sparingly.

3 Vinyl trim - maintenance

Don't clean vinyl trim with detergents, caustic soap or petroleum-based cleaners. Plain soap and water works just fine, with a soft brush to clean dirt that may be ingrained. Wash the vinyl as frequently as the rest of the vehicle.

After cleaning, application of a high quality rubber and vinyl protectant will help prevent oxidation and cracks. The protectant can also be applied to weatherstripping, vacuum lines and rubber hoses (which often fail as a result of chemical degradation) and to the tires.

4 Upholstery and carpets - maintenance

1 Every three months remove the carpets or mats and clean the interior of the vehicle (more frequently if necessary). Vacuum the upholstery and carpets to remove loose dirt and dust.

2 Leather upholstery requires special care. Stains should be removed with warm water and a very mild soap solution. Use a clean, damp cloth to remove the soap, then wipe again with a dry cloth. Never use alcohol, gasoline, nail polish remover or thinner to clean leather upholstery.

3 After cleaning, regularly treat leather upholstery with a leather wax. Never use car wax on leather upholstery.

4 In areas where the interior of the vehicle is subject to bright sunlight, cover leather seats with a sheet if the vehicle is to be left out for any length of time.

5 Body repair - minor damage

▶ See photo sequence

REPAIR OF MINOR SCRATCHES

1 If the scratch is superficial and does not penetrate to the metal of the body, repair is very simple. Lightly rub the scratched area with a fine rubbing compound to remove loose paint and built-up wax. Rinse the area with clean water.

2 Apply touch-up paint to the scratch, using a small brush. Continue to apply thin layers of paint until the surface of the paint in the scratch is level with the surrounding paint. Allow the new paint at least two weeks to harden, then blend it into the surrounding paint by rubbing with a very fine rubbing compound. Finally, apply a coat of wax to the scratch area.

3 If the scratch has penetrated the paint and exposed the metal of the body, causing the metal to rust, a different repair technique is required. Remove all loose rust from the bottom of the scratch with a pocket knife, then apply rust inhibiting paint to prevent the formation of

BODY 11-3

rust in the future. Using a rubber or nylon applicator, coat the scratched area with glaze-type filler. If required, the filler can be mixed with thinner to provide a very thin paste, which is ideal for filling narrow scratches. Before the glaze filler in the scratch hardens, wrap a piece of smooth cotton cloth around the tip of a finger. Dip the cloth in thinner and then quickly wipe it along the surface of the scratch. This will ensure that the surface of the filler is slightly hollow. The scratch can now be painted over as described earlier in this Section.

REPAIR OF DENTS

4 When repairing dents, the first job is to pull the dent out until the affected area is as close as possible to its original shape. There is no point in trying to restore the original shape completely as the metal in the damaged area will have stretched on impact and cannot be restored to its original contours. It is better to bring the level of the dent up to a point which is about 1/8-inch below the level of the surrounding metal. In cases where the dent is very shallow, it is not worth trying to pull it out at all.

5 If the back side of the dent is accessible, it can be hammered out gently from behind using a soft-face hammer. While doing this, hold a block of wood firmly against the opposite side of the metal to absorb the hammer blows and prevent the metal from being stretched.

6 If the dent is in a section of the body which has double layers, or some other factor makes it inaccessible from behind, a different technique is required. Drill several small holes through the metal inside the damaged area, particularly in the deeper sections. Screw long, self-tapping screws into the holes just enough for them to get a good grip in the metal. Now the dent can be pulled out by pulling on the protruding heads of the screws with locking pliers.

7 The next stage of repair is the removal of paint from the damaged area and from an inch or so of the surrounding metal. This is done with a wire brush or sanding disk in a drill motor, although it can be done just as effectively by hand with sandpaper. To complete the preparation for filling, score the surface of the bare metal with a screwdriver or the tang of a file, or drill small holes in the affected area. This will provide a good grip for the filler material. To complete the repair, see the subsection on filling and painting later in this Section.

REPAIR OF RUST HOLES OR GASHES

8 Remove all paint from the affected area and from an inch or so of the surrounding metal using a sanding disk or wire brush mounted in a drill motor. If these are not available, a few sheets of sandpaper will do the job just as effectively.

9 With the paint removed, you will be able to determine the severity of the corrosion and decide whether to replace the whole panel, if possible, or repair the affected area. New body panels are not as expensive as most people think and it is often quicker to install a new panel than to repair large areas of rust.

10 Remove all trim pieces from the affected area except those which will act as a guide to the original shape of the damaged body, such as headlight shells, etc. Using metal snips or a hacksaw blade, remove all loose metal and any other metal that is badly affected by rust. Hammer the edges of the hole in to create a slight depression for the filler material.

11 Wire brush the affected area to remove the powdery rust from the surface of the metal. If the back of the rusted area is accessible, treat it with rust inhibiting paint.

12 Before filling is done, block the hole in some way. This can be done with sheet metal riveted or screwed into place, or by stuffing the hole with wire mesh.

13 Once the hole is blocked off, the affected area can be filled and painted. See the following subsection on filling and painting.

FILLING AND PAINTING

14 Many types of body fillers are available, but generally speaking, body repair kits which contain filler paste and a tube of resin hardener are best for this type of repair work. A wide, flexible plastic or nylon applicator will be necessary for imparting a smooth and contoured finish to the surface of the filler material. Mix up a small amount of filler on a clean piece of wood or cardboard (use the hardener sparingly). Follow the manufacturer's instructions on the package, otherwise the filler will set incorrectly.

15 Using the applicator, apply the filler paste to the prepared area. Draw the applicator across the surface of the filler to achieve the desired contour and to level the filler surface. As soon as a contour that approximates the original one is achieved, stop working the paste. If you continue, the paste will begin to stick to the applicator. Continue to add thin layers of paste at 20-minute intervals until the level of the filler is just above the surrounding metal.

16 Once the filler has hardened, the excess can be removed with a body file. From then on, progressively finer grades of sandpaper should be used, starting with a 180-grit paper and finishing with 600-grit wet-or-dry paper. Always wrap the sandpaper around a flat rubber or wooden block, otherwise the surface of the filler will not be completely flat. During the sanding of the filler surface, the wet-or-dry paper should be periodically rinsed in water. This will ensure that a very smooth finish is produced in the final stage.

17 At this point, the repair area should be surrounded by a ring of bare metal, which in turn should be encircled by the finely feathered edge of good paint. Rinse the repair area with clean water until all of the dust produced by the sanding operation is gone.

18 Spray the entire area with a light coat of primer. This will reveal any imperfections in the surface of the filler. Repair the imperfections with fresh filler paste or glaze filler and once more smooth the surface with sandpaper. Repeat this spray-and-repair procedure until you are satisfied that the surface of the filler and the feathered edge of the paint are perfect. Rinse the area with clean water and allow it to dry completely.

19 The repair area is now ready for painting. Spray painting must be carried out in a warm, dry, windless and dust free atmosphere. These conditions can be created if you have access to a large indoor work area, but if you are forced to work in the open, you will have to pick the day very carefully. If you are working indoors, dousing the floor in the work area with water will help settle the dust which would otherwise be in the air. If the repair area is confined to one body panel, mask off the surrounding panels. This will help minimize the effects of a slight mismatch in paint color. Trim pieces such as chrome strips, door handles, etc., will also need to be masked off or removed. Use masking tape and several thickness of newspaper for the masking operations.

20 Before spraying, shake the paint can thoroughly, then spray a test area until the spray painting technique is mastered. Cover the repair area with a thick coat of primer. The thickness should be built up using several thin layers of primer rather than one thick one. Using 600-grit wet-or-dry sandpaper, rub down the surface of the primer until it is very smooth. While doing this, the work area should be thoroughly rinsed with water and the wet-or-dry sandpaper periodically rinsed as well. Allow the primer to dry before spraying additional coats.

These photos illustrate a method of repairing simple dents. They are intended to supplement Body repair - minor damage in this Chapter and should not be used as the sole instructions for body repair on these vehicles.

1 If you can't access the backside of the body panel to hammer out the dent, pull it out with a slide-hammer-type dent puller. In the deepest portion of the dent or along the crease line, drill or punch hole(s) at least one inch apart . . .

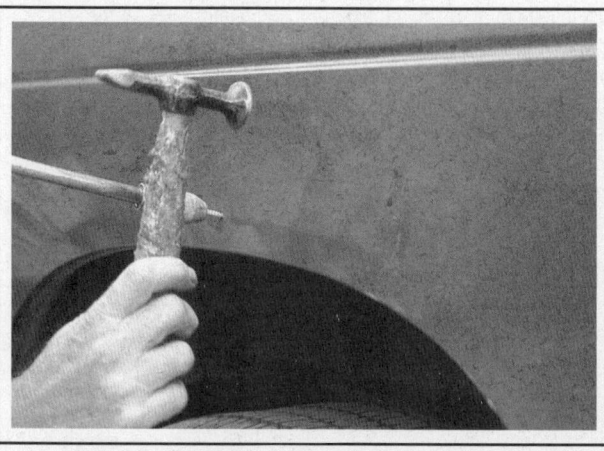

2 . . . then screw the slide-hammer into the hole and operate it. Tap with a hammer near the edge of the dent to help 'pop' the metal back to its original shape. When you're finished, the dent area should be close to its original contour and about 1/8-inch below the surface of the surrounding metal

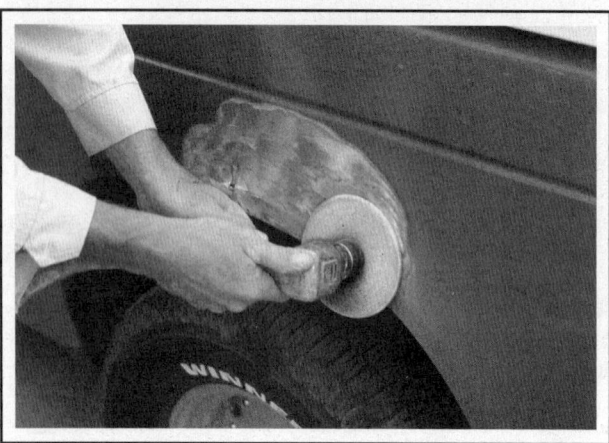

3 Using coarse-grit sandpaper, remove the paint down to the bare metal. Hand sanding works fine, but the disc sander shown here makes the job faster. Use finer (about 320-grit) sandpaper to feather-edge the paint at least one inch around the dent area

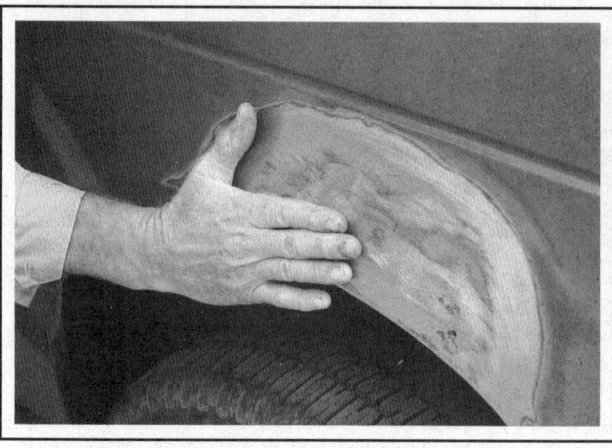

4 When the paint is removed, touch will probably be more helpful than sight for telling if the metal is straight. Hammer down the high spots or raise the low spots as necessary. Clean the repair area with wax/silicone remover

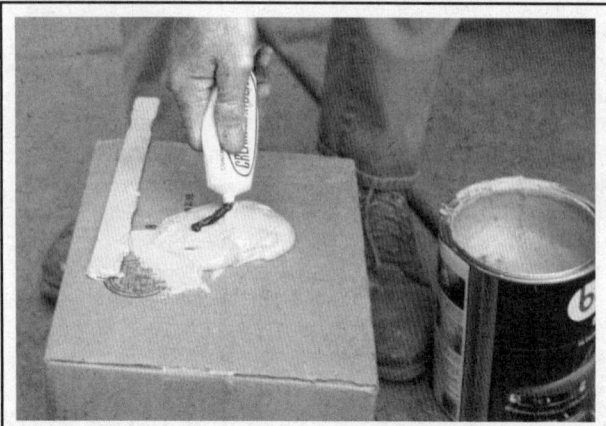

5 Following label instructions, mix up a batch of plastic filler and hardener. The ratio of filler to hardener is critical, and, if you mix it incorrectly, it will either not cure properly or cure too quickly (you won't have time to file and sand it into shape)

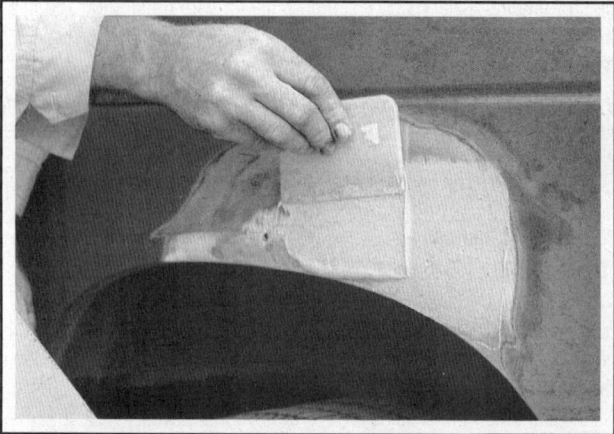

6 Working quickly so the filler doesn't harden, use a plastic applicator to press the body filler firmly into the metal, assuring it bonds completely. Work the filler until it matches the original contour and is slightly above the surrounding metal

7 Let the filler harden until you can just dent it with your fingernail. Use a body file or Surform tool (shown here) to rough-shape the filler

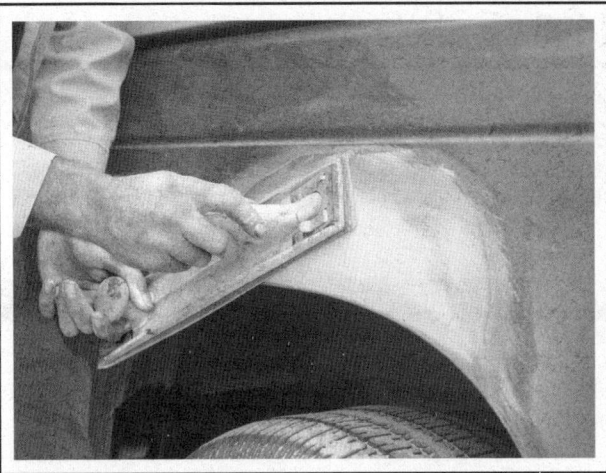

8 Use coarse-grit sandpaper and a sanding board or block to work the filler down until it's smooth and even. Work down to finer grits of sandpaper - always using a board or block - ending up with 360 or 400 grit

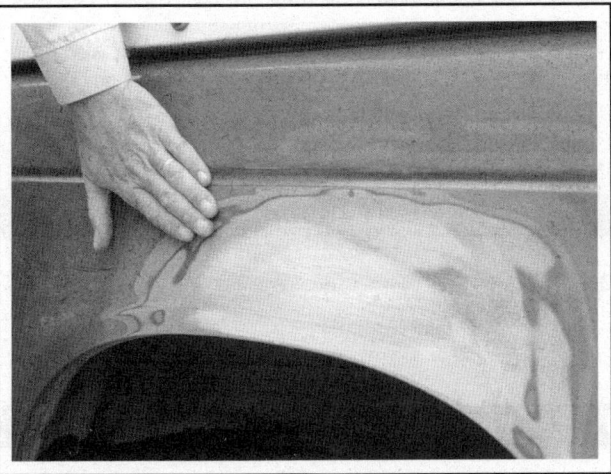

9 You shouldn't be able to feel any ridge at the transition from the filler to the bare metal or from the bare metal to the old paint. As soon as the repair is flat and uniform, remove the dust and mask off the adjacent panels or trim pieces

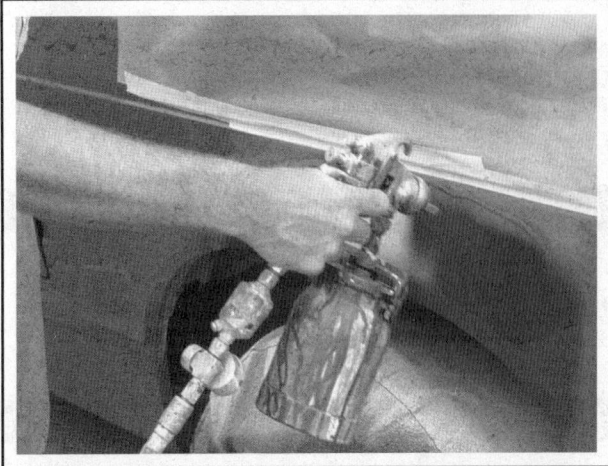

10 Apply several layers of primer to the area. Don't spray the primer on too heavy, so it sags or runs, and make sure each coat is dry before you spray on the next one. A professional-type spray gun is being used here, but aerosol spray primer is available inexpensively from auto parts stores

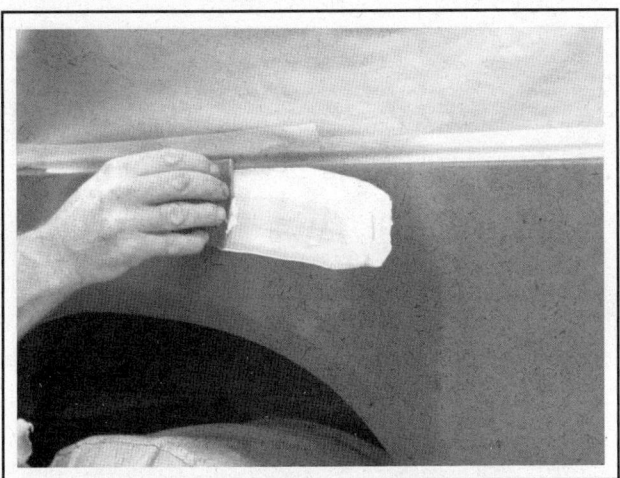

11 The primer will help reveal imperfections or scratches. Fill these with glazing compound. Follow the label instructions and sand it with 360 or 400-grit sandpaper until it's smooth. Repeat the glazing, sanding and respraying until the primer reveals a perfectly smooth surface

12 Finish sand the primer with very fine sandpaper (400 or 600-grit) to remove the primer overspray. Clean the area with water and allow it to dry. Use a tack rag to remove any dust, then apply the finish coat. Don't attempt to rub out or wax the repair area until the paint has dried completely (at least two weeks)

11-6 BODY

21 Spray on the top coat, again building up the thickness by using several thin layers of paint. Begin spraying in the center of the repair area and then, using a circular motion, work out until the whole repair area and about two inches of the surrounding original paint is covered.

Remove all masking material 10 to 15 minutes after spraying on the final coat of paint. Allow the new paint at least two weeks to harden, then use a very fine rubbing compound to blend the edges of the new paint into the existing paint. Finally, apply a coat of wax.

6 Body repair - major damage

1 Major damage must be repaired by an auto body shop specifically equipped to perform these repairs. Most shops have the specialized equipment required to do the job properly.

2 If the damage is extensive, the body must be checked for proper alignment or the vehicle's handling characteristics may be adversely affected and other components may wear at an accelerated rate.

3 Due to the fact that all of the major body components (hood, fenders, etc.) are separate and replaceable units, any seriously damaged components should be replaced rather than repaired. Sometimes the components can be found in a wrecking yard that specializes in used vehicle components, often at considerable savings over the cost of new parts.

7 Hinges and locks - maintenance

Once every 3000 miles, or every three months, the hinges and latch assemblies on the doors, hood and trunk should be given a few drops of light oil or lock lubricant. The door latch strikers should also be lubricated with a thin coat of grease to reduce wear and ensure free movement. Lubricate the door and trunk locks with spray-on graphite lubricant.

8 Windshield and fixed glass - replacement

Replacement of the windshield and fixed glass requires the use of special fast-setting adhesive/caulk materials and some specialized tools. It is recommended that these operations be left to a dealer or a shop specializing in glass work.

9 Hood and rear liftgate support struts - removal and installation

♦ Refer to illustrations 9.2a and 9.2b

→ Note: The hood and rear liftgate are heavy and somewhat awkward to hold - at least two people should perform this procedure.

1 Open the hood or rear liftgate (Expedition/Navigator only) and support it securely.

2 Using a small screwdriver, detach the retaining clips at both ends of the support strut. Then pry or pull sharply to detach it from the vehicle (see illustrations).

3 Installation is the reverse of removal.

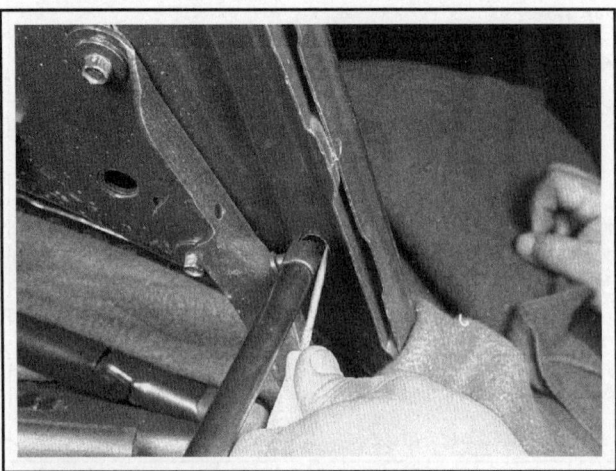

9.2a Use a small screwdriver to pry the clip out of its locking groove, then detach the end of the strut from the locating stud

9.2b Have an assistant support the liftgate before attempting to remove the support strut - pry the clip (arrow) out, then detach the end of the strut from the locating stud

10 Hood - removal, installation and adjustment

→Note: *The hood is heavy and somewhat awkward to remove and install - at least two people should perform this procedure.*

REMOVAL AND INSTALLATION

♦ **Refer to illustrations 10.2 and 10.4**

1 Use blankets or pads to cover the cowl area of the body and fenders. This will protect the body and paint as the hood is lifted off.
2 Make marks or scribe a line around the hood hinge to ensure proper alignment during installation (see illustration).
3 Disconnect any cables or wires that will interfere with removal.
4 Have an assistant support the hood. Detach the support struts (see Section 9). Remove the hinge-to-hood nuts or bolts (see illustration).
5 Lift off the hood.
6 Installation is the reverse of removal.

ADJUSTMENT

♦ **Refer to illustrations 10.10 and 10.11**

7 Fore-and-aft and side-to-side adjustment of the hood is done by moving the hinge plate slot after loosening the bolts or nuts.
8 Scribe a line around the entire hinge plate so you can determine the amount of movement (see illustration 10.2).
9 Loosen the bolts or nuts and move the hood into correct alignment. Move it only a little at a time. Tighten the hinge bolts and carefully lower the hood to check the position.
10 If necessary after installation, the entire hood latch assembly can be adjusted up-and-down as well as from side-to-side on the radiator support so the hood closes securely and flush with the fenders. To make the adjustment, scribe a line or mark around the hood latch mounting bolts to provide a reference point, then loosen them and reposition the latch assembly, as necessary (see illustration). Following adjustment, retighten the mounting bolts.
11 Finally, adjust the hood bumpers on the radiator support so the hood, when closed, is flush with the fenders (see illustration).
12 The hood latch assembly, as well as the hinges, should be periodically lubricated with white, lithium-base grease to prevent binding and wear.

10.2 Before removing the hood, draw a mark around the hinge plate

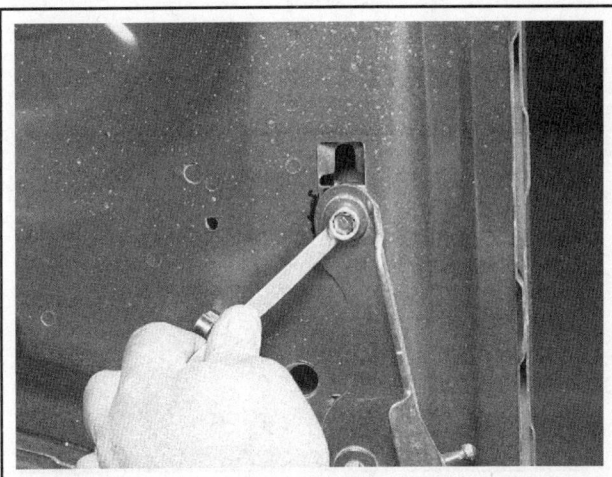

10.4 Remove the hinge-to-hood retaining bolts and lift off the hood with the help of an assistant

10.10 Scribe a line around the latch to use as a reference point. To adjust the hood latch, loosen the retaining bolts (arrows), move the latch and retighten bolts, then close the hood to check the fit

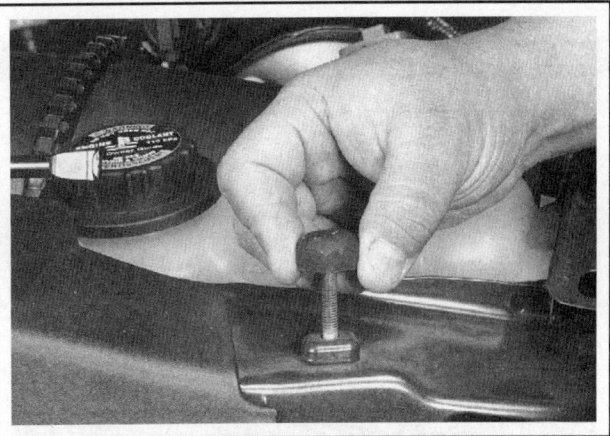

10.11 Adjust the hood closing height by turning the hood bumpers in or out

11-8 BODY

11 Hood latch and release cable - removal and installation

LATCH

▶ Refer to illustration 11.5

1 On 2010 and later models, remove the radiator trim panel (see Section 12).

2 Scribe a line around the latch to aid alignment when reinstalling the latch assembly.

3 On 2009 and earlier Navigator models, remove the handle and cable assembly from the latch assembly.

4 Remove the latch retaining bolts securing the latch to the radiator support (see illustration 10.10) and remove the latch.

5 On all except 2009 and earlier Navigator models, disconnect the hood release cable by disengaging the cable from the back of the latch assembly (see illustration).

6 Installation is the reverse of the removal procedure.

→ **Note:** Adjust the latch so the hood engages securely when closed and the hood bumpers are slightly compressed.

CABLE

▶ Refer to illustration 11.9

7 Disconnect the hood release cable from the latch assembly as described above.

8 Attach a piece of stiff wire to the end of the cable, then follow the cable back to the firewall and detach all the cable retaining clips.

9 Working in the passenger compartment, remove the fuse panel cover (2009 and earlier models) or the lower steering column cover (2010 and later models). Then remove the two release lever mounting bolts and detach the hood release lever (see illustration).

10 Pull the cable and grommet rearward into the passenger compartment until you can see the wire. Ensure that the new cable has a grommet attached, then remove the old cable from the wire and replace it with the new cable.

11 Working from the engine compartment pull the wire back through the firewall.

12 The remainder of installation is the reverse of removal.

→ **Note:** Push on the grommet with your fingers from the passenger compartment to seat the grommet in the firewall correctly.

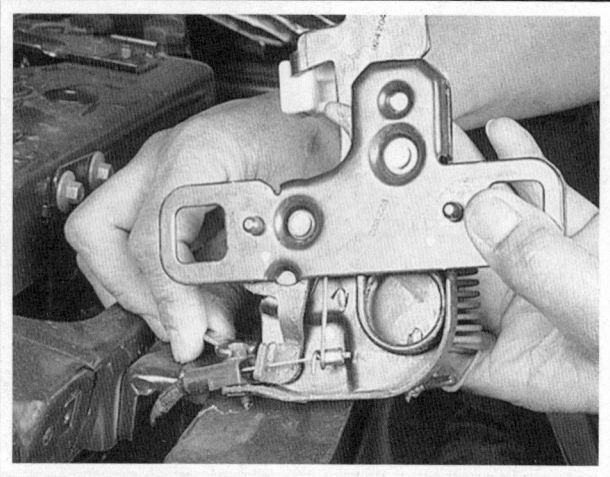

11.5 Disengage the release cable from the backside of the hood latch assembly

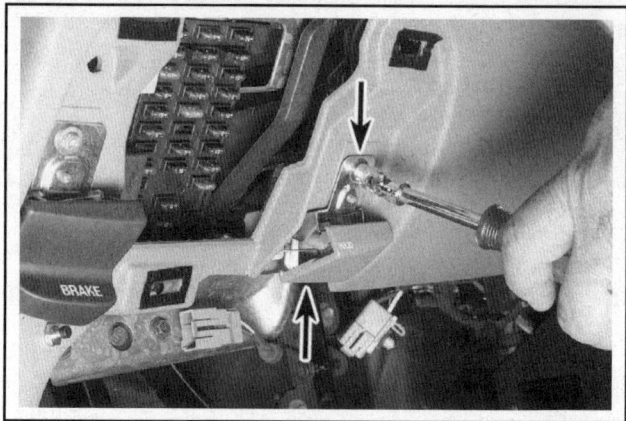

11.9 Remove the hood release lever retaining screws (arrows) and pull the cable rearward into the passenger compartment

12 Radiator grille - removal and installation

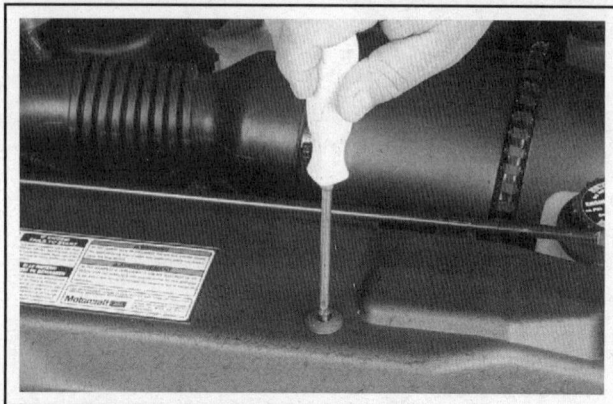

12.1 Remove the plastic trim panel attached to the top of the radiator grille

▶ Refer to illustration 12.1

1 Open the hood and remove the plastic trim panel on top of the grille assembly (see illustration).

2004 AND EARLIER MODELS

Refer to illustrations 12.3, 12.4 and 12.5

2 Remove the headlight housing assemblies (see Chapter 12).

3 Remove the grille retaining screws from the headlight housing openings (see illustration).

4 Remove the clips securing the lower half of the grille (see illustration).

5 Remove the screws securing the upper half of the radiator grille and remove the grille (see illustration).

BODY   11-9

12.3 Remove the retaining screws (arrows) located behind the headlight assemblies

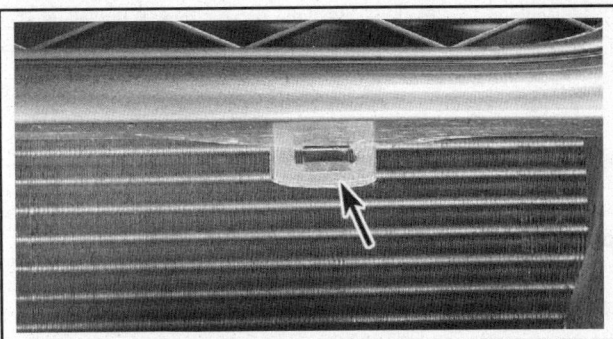

12.4 Detach the clips securing the lower half of the grille

2005 THROUGH 2009 MODELS

6   Remove the front bumper cover (see Section 13).
7   Remove the grille-to-bumper cover screws and clips.
8   Remove the clips securing the lower half of the grille (see illustration 12.4).
9   Remove the screws securing the upper half of the radiator grille and remove the grille (see illustration 12.5).

2010 AND LATER MODELS

10 Remove the front bumper cover (see Section 13).
11 On Expedition models, remove the bumper-to-grille bracket fasteners and remove the bracket.
12 Remove the clips securing the lower half of the grille and remove the grille.

12.5 Detach the retaining screws (arrows) along the top edge and remove the grille from the vehicle

ALL MODELS

13 Installation is the reverse of removal.

13 Bumpers - removal and installation

✱✱ WARNING 1:

The models covered by this manual are equipped with Supplemental Restraint systems (SRS), more commonly known as airbags. Always disconnect the negative battery cable, then the positive battery cable and wait two minutes before working in the vicinity of the impact sensors, steering column or instrument panel to avoid the possibility of accidental deployment of the airbag, which could cause personal injury (see Chapter 12). Do not use electrical test equipment on any of the airbag system wiring or tamper with them in any way.

✱✱ WARNING 2:

Some models covered by this manual are equipped with air suspension systems. Always disconnect electrical power to the suspension system before lifting or towing the vehicle (see Chapter 10). Failure to perform this procedure may result in unexpected shifting or movement of the vehicle which could cause personal injury.

1   Apply the parking brake, raise the vehicle and support it securely on jackstands.
2   Disconnect the negative battery cable, then the positive battery cable and wait two minutes before proceeding any further.

FRONT BUMPER

Pick-up and Expedition

3   Working under the vehicle, disconnect the fog light electrical connections if equipped. On models so equipped, release the pushpins securing the lower air deflector panel from below for access to the fog light connectors.
4   Working from the backside of the bumper, remove the retaining bolts securing the bumper brackets to each frame rail. Then remove the bumper from the vehicle.
5   Installation is the reverse of removal.

Navigator

6   Working under the vehicle, detach the bolts and plastic clips securing the plastic air deflector.
7   Working in the front wheel opening, detach the retaining nuts securing the bumper cover to the fender. On later models, there are four bolts on each side.
8   Remove the retaining bolts at the lower front edge of the wheel opening.
9   Disengage the release tabs located in the grille opening and remove the bumper cover.
10 Disconnect any electrical connections which would interfere with removal.

11-10 BODY

11 Working from the backside of the bumper, remove the retaining bolts securing the bumper brackets to each frame rail. Then remove the bumper from the vehicle.

12 Installation is the reverse of removal.

REAR BUMPER

Pick-up and Expedition

13 Working from the backside of the bumper, remove the license plate light bulbs from the rear bumper. On 2005 and later Expedition/Navigator models, disconnect the backup light connectors.

14 Remove the retaining bolts securing the bumper brackets to each frame rail. Then remove the bumper from the vehicle.

15 Installation is the reverse of removal.

Navigator

16 Working under the vehicle, detach the plastic clips and screws securing the inner splash shield.

17 Remove the nuts and bolts securing the bumper cover the rear quarter panels.

18 Open the liftgate and remove the plastic clips securing the upper edge of the bumper cover. On 2003 through 2006 Expedition/Navigator models, remove the lower rear floor trim panel to access the fasteners.

19 Lift upward and pull outward to remove the bumper cover from the vehicle.

20 Remove the retaining bolts securing the bumper brackets to each frame rail. Then remove the bumper from the vehicle.

21 Installation is the reverse of removal.

14 Front fender - removal and installation

WARNING:

Some models covered by this manual are equipped with air suspension systems. Always disconnect electrical power to the suspension system before lifting or towing the vehicle (see Chap-ter 10). Failure to perform this procedure may result in unexpected shifting or movement of the vehicle which could cause personal injury.

1 Raise the vehicle, support it securely on jackstands and remove the front wheel.

PICK-UP AND EXPEDITION

Refer to illustrations 14.3 and 14.9

2 Remove the radiator grille (see Section 12).

3 Remove the fender retaining bolts located behind the headlight assemblies (see illustration). Access is best with the headlight housings removed (see Chapter 12).

4 Remove the wheel opening molding if equipped.

5 Detach the inner fenderwell bolts and clips, then remove the inner fenderwell.

6 Only early models, if you're removing the passenger side fender, remove the radio antenna also (see Chapter 12).

7 Remove the fender-to-rocker panel bolts.

8 Open the door and remove the fender to door pillar bolts.

9 Remove the remaining fender mounting bolts (see illustration).

Note: On 2007 and later models, remove the cowl cover to access the top fender mounting bolt (see Section 29).

10 Detach the fender. It's a good idea to have an assistant support the fender while it's being moved away from the vehicle to prevent damage to the surrounding body panels.

11 Installation is the reverse of removal.

NAVIGATOR

12 Open the hood and remove the plastic trim panel covering the top of the radiator support and the hood latch assembly.

13 Detach the headlight housing assemblies and remove them from the vehicle (see Chapter 12).

14 Remove the grille opening panels and the grille opening panel reinforcement nut.

15 Remove the front bumper cover assembly (see Section 13).

16 Remove the fender retaining bolts located behind the headlight assemblies (see illustration 14.3).

17 Detach the inner fenderwell bolts and clips, then remove the inner fenderwell and mud shield.

18 On early models, if you're removing the passenger side fender remove the radio antenna (see Chapter 12).

19 Remove the fender-to-rocker panel bolts.

20 Open the door and remove the fender to door pillar bolts.

21 Remove the remaining fender mounting bolts (see illustration 14.9).

22 Detach the fender. It's a good idea to have an assistant support the fender while it's being moved away from the vehicle to prevent damage to the surrounding body panels.

23 Installation is the reverse of removal.

14.3 Detach the retaining bolts (arrows) located behind the headlight assemblies

14.9 Detach the remaining bolts located in the hood opening

BODY 11-11

15 Door trim panels - removal and installation

♦ Refer to illustrations 15.2, 15.3, 15.6a, 15.6b, 15.7a, 15.7b, 15.8a, 15.8b and 15.10

1 Disconnect the negative cable from the battery.
2 On manual window equipped models, remove the window crank, using a hooked tool to remove the retainer clip (see illustration). A special tool is available for this purpose, but it's not essential. With the clip removed, pull off the handle.
3 On power window equipped models, pry out the armrest switch control plate and disconnect the electrical connections (see illustration).
4 On Expedition models, detach the retaining screws located behind the armrest switch control plate and the courtesy lamp housing
5 On Navigator models, pry open the door pull handle covers, detach the screws and remove the door pull handle.
6 On Pick-up and Expedition models, pry off the side view mirror trim cover and the door panel upper retaining clip (see illustrations).
7 On Pick-up models, pry off the door pillar trim cover and the door panel upper retaining clip (see illustrations).

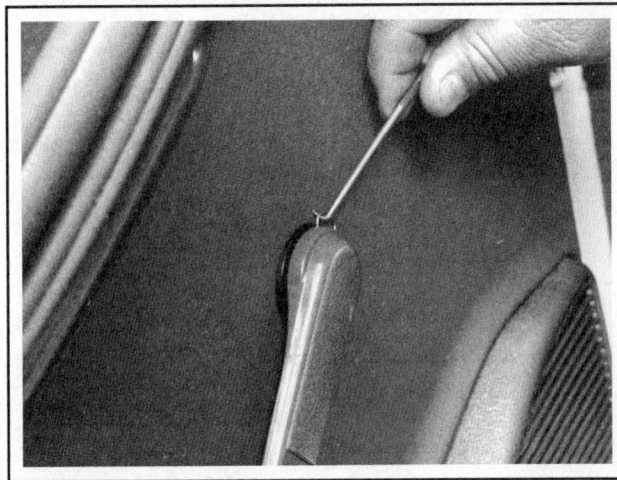

15.2 Use a hooked tool like this to remove the window crank retaining clip

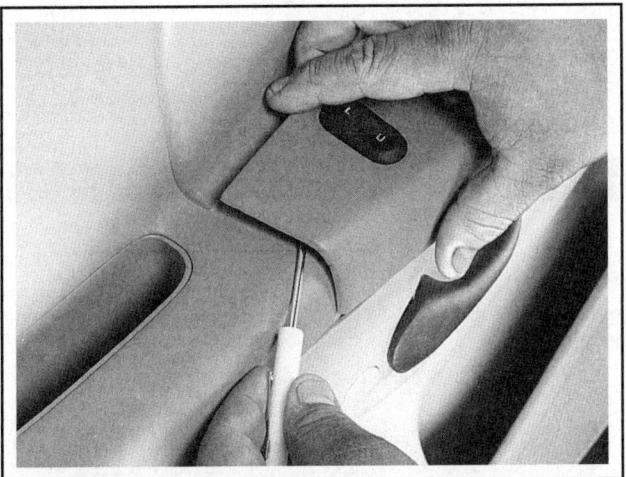

15.3 Using a small screwdriver, pry out the armrest switch control plate

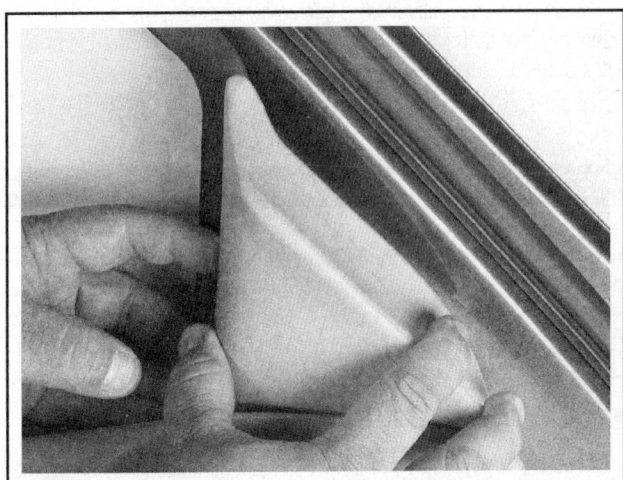

15.6a Pull the top out to remove the side view mirror trim cover

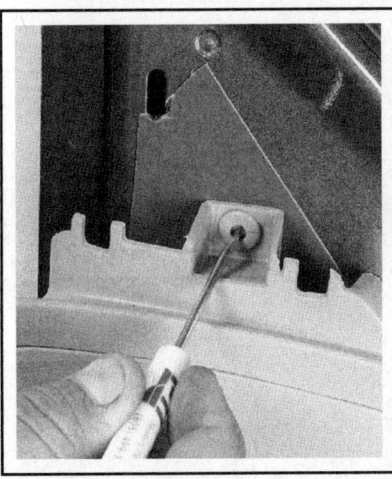

15.6b Detach the clip located behind the side view mirror trim cover

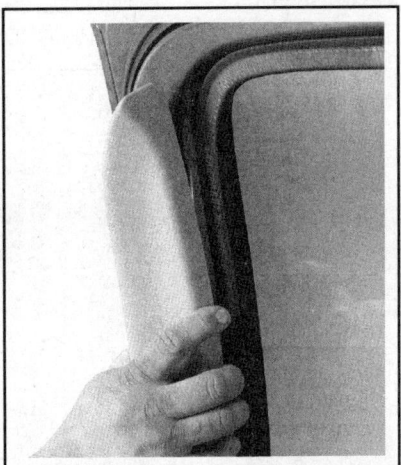

15.7a On Pick-up models, pry off the door pillar trim cover . . .

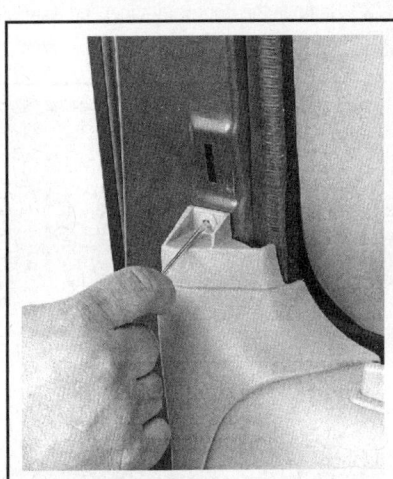

15.7b . . . then detach the clip located behind the trim cover

11-12 BODY

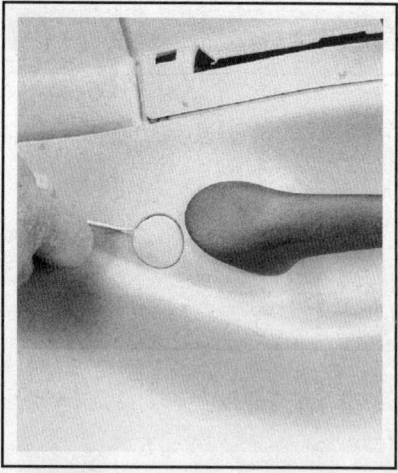

15.8a Pry off the inside handle trim cover (Pick-up shown)

15.8b Remove the inside handle retaining nut

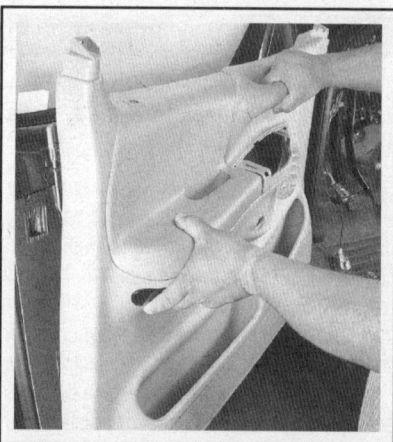

15.10 Lift the door panel up and out to remove it, then disconnect any electrical connections

8 On all models, pry off the inside door handle trim cover, detach the screw and remove the handle (see illustrations).

9 Using a wide putty knife, a thin screwdriver or a special trim panel removal tool, pry out the plastic retaining clip securing the front edge of the door trim panel.

10 Once all of the clips and screws are disengaged, detach the trim panel, disconnect any electrical connectors and remove the trim panel from the vehicle by gently pulling it up and out (see illustration).

11 For access to the inner door, peel back the watershield, taking care not to tear it. To install the trim panel, first press the watershield back into place. If necessary, add more sealant to hold it in place.

12 The remainder of the installation is the reverse of removal.

16 Door - removal, installation and adjustment

→Note: The door is heavy and somewhat awkward to remove and install - at least two people should perform this procedure.

REMOVAL AND INSTALLATION

♦ Refer to illustrations 16.6 and 16.8

1 Raise the window completely in the door and then disconnect the negative cable from the battery

2 Open the door all the way and support it on jacks or blocks covered with rags to prevent damaging the paint.

3 Remove the door trim panel and water deflector as described in (Section 15).

4 Unplug all electrical connections, ground wires and harness retaining clips from the door.

→Note: It is a good idea to label all connections to aid the reassembly process.

5 Working through the door opening, detach the rubber conduit between the body and the door. Then pull wiring harness through the conduit hole and remove from the door.

6 Remove the screws securing the door stop (see illustration).

7 Mark around the door hinges with a pen or a scribe to facilitate realignment during reassembly.

8 Have an assistant hold the door, remove the hinge to door bolts (see illustration) and lift the door off.

9 Installation is the reverse of the removal.

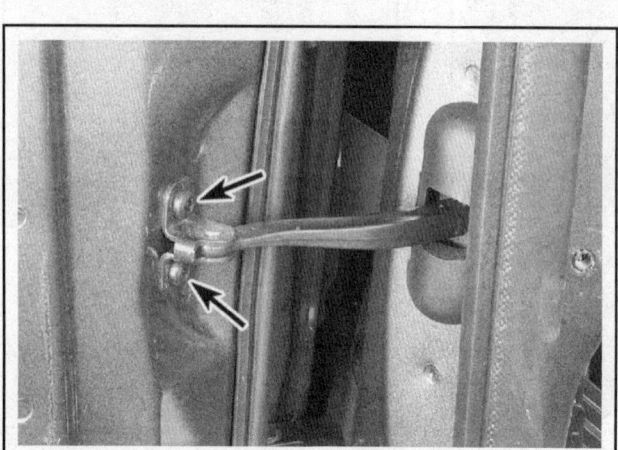

16.6 Remove the door stop retaining screws

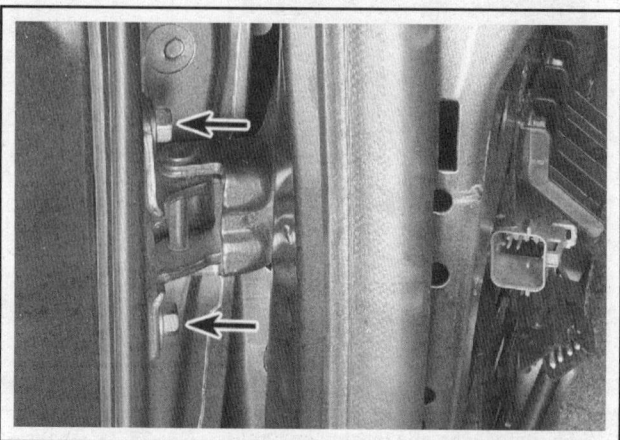

16.8 Before loosening the door retaining bolts (arrows), draw a line around the hinge plate for a reinstallation reference

BODY 11-13

ADJUSTMENT

▶ **Refer to Illustration 16.13**

10 Having proper door to body alignment is a critical part of a well functioning door assembly. First check the door hinge pins for excessive play. Fully open the door and lift up and down on the door without lifting the body. If a door has 1/16-inch or more excessive play, the hinges should be replaced.

11 Door to body alignment adjustments are made by loosening the hinge-to-body or hinge to door bolts and moving the door. Proper body alignment is achieved when the top of the door is aligned with the top of the front fender and rear door or quarter panel and the bottom of the door is aligned with the lower rocker panel. If these goals can't be reached by adjusting the hinge to body or hinge to door bolts, body alignment shims may have to be purchased and inserted behind the hinges to achieve correct alignment.

12 To adjust the door closed position, first check that the door latch is contacting the center of the latch striker. If not remove striker and add or subtract shims to achieve correct alignment.

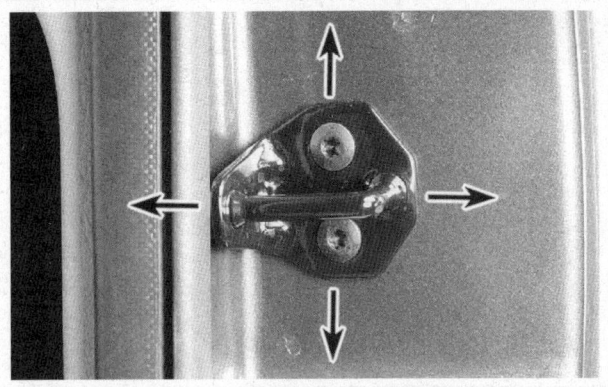

16.13 Adjust the door lock striker by loosening the mounting screws and gently tapping the striker in the desired direction (arrows)

13 Finally adjust latch striker as necessary (up and down or sideways) to provide positive engagement with the latch mechanism (see illustration) and the door panel is flush with rear door or quarter panel.

17 Door latch, lock cylinder and handles - removal and installation

DOOR LATCH

▶ **Refer to illustration 17.2**

1 Raise the window then remove the door trim panel and watershield as described in (see Section 15).
2 Remove the screws securing the latch to the door (see illustration).
3 Working through the large access hole, position the latch as necessary to disengage the outside door handle and outside lock cylinder to latch rods and the inside handle to latch cable.
4 All door locking rods are attached by plastic clips. The plastic clips can be removed by unsnapping the portion engaging the connecting rod and then by pulling the rod out of its locating hole.
5 Position the latch as necessary to disengage the door lock actuator hook. Then remove the latch assembly from the door.
6 Installation is the reverse of removal.

OUTSIDE HANDLE AND DOOR LOCK CYLINDER

▶ **Refer to illustration 17.10**

7 To remove the outside handle and door lock cylinder assembly, raise the window and remove the door trim panel and watershield as described in Section 15.
8 Working through the large access hole, disengage the plastic clips that secure the lock cylinder to latch rod and door handle to latch rod.
9 Disconnect any electrical connectors which would interfere with removal.
10 Remove the outside handle retaining nuts (see illustration) and pull the handle and lock cylinder assembly from the door.

➡ **Note:** Pull the U-wire clip from the outside handle to separate the lock cylinder from the outside handle assembly.

11 Installation is the reverse of removal.

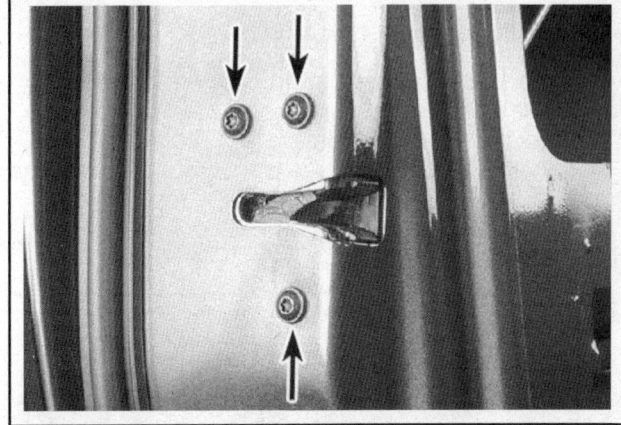

17.2 Remove the latch retaining screws (arrows) from the end of the door, then detach the locking rods and cable and pull the latch assembly through the access hole

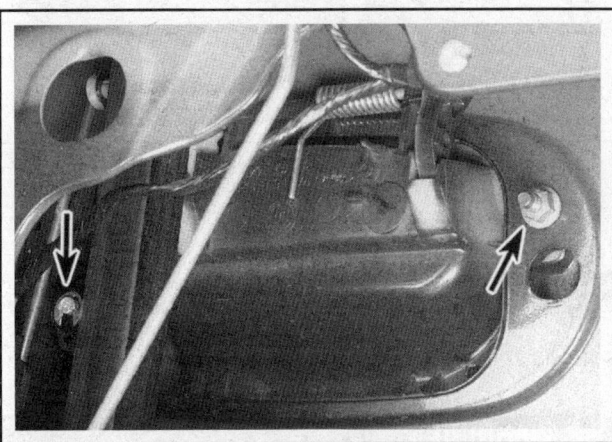

17.10 The outside handle retaining nuts (arrows) can be reached through the access hole in the door frame

11-14 BODY

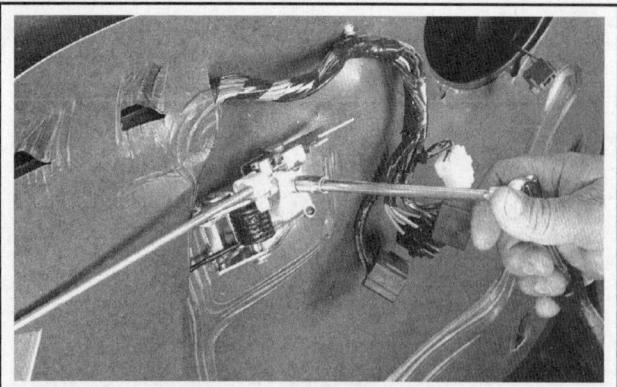

17.13 Remove the handle retaining bolt . . .

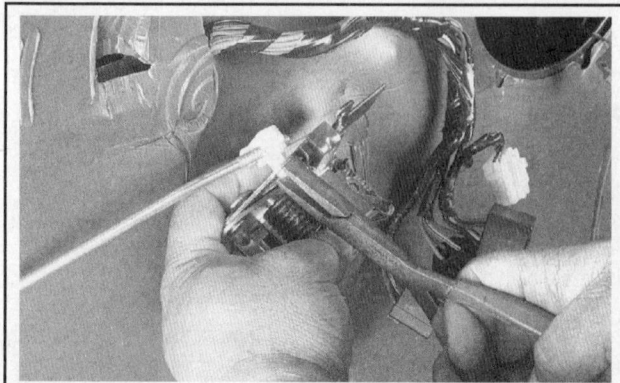

17.14 . . . then rotate the handle out and detach the latch rod retaining clip

INSIDE HANDLE AND CABLE

▶ **Refer to illustrations 17.13 and 17.14**

12 Remove the door trim panel as described in Section 15 and peel away the watershield.

13 Remove the bolt securing the inside handle (see illustration). Pull rearward on the handle to disengage it from the inner door panel.
14 Detach the latch rod from the backside of the handle and remove it from the vehicle (see illustration).
15 Installation is the reverse of removal.

18 Door window glass - removal and installation

▶ **Refer to illustrations 18.4 and 18.5**

1 Remove the door trim panel and the plastic watershield (see Section 15). Remove the front doors speakers (see Chapter 12).
2 Lower the window glass all the way down into the door.
3 Carefully pry the inner and outer weatherstripping out of the door window opening.
4 Remove the rear glass channel bolt and glass channel (see illustration).

5 Raise the window just enough to access the window retaining nuts through the hole in the door frame (see illustration). On later Expedition/Navigator models, loosen - but do not remove - the two glass clamp bolts.
6 Detach the window retaining nuts and remove the glass by pulling it up and out.
7 Installation is the reverse of removal.

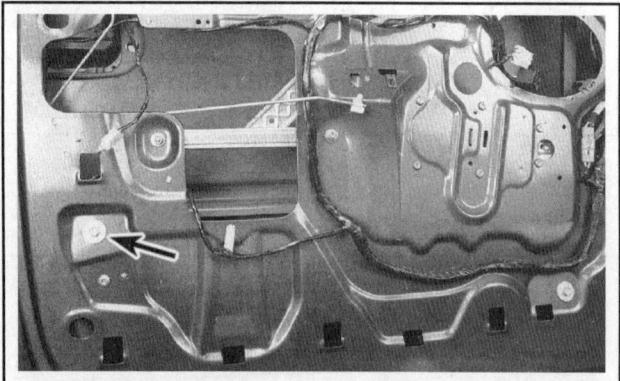

18.4 Detach the rear glass channel bolt (arrow) and glass channel

18.5 Raise the window just enough to access the glass retaining nuts (arrow) through the hole in the door frame - remove the nuts securing the glass to the equalizer arm

19 Door window glass regulator - removal and installation

▶ **Refer to illustrations 19.4a and 19.4b**

✳✳ WARNING 1:

Some models covered by this manual are equipped with side-impact airbags. Always disable the airbag system before working in the vicinity of the impact sensors, steering column, instrument panel, and on later models the doors, headliner or seats. Failure to follow these procedures may cause accidental deployment of an airbag, which could cause personal injury. Do not use electrical test equipment on any component of the airbag system or tamper with the wiring in any way (see Chapter 12).

BODY   11-15

※ WARNING 2:

The regulator arms are under extreme pressure and can cause serious injury if the motor or counter balance spring is removed without locking the sector gear. This can be done by inserting a bolt and nut through the holes in the backing plate and sector gear to lock them together.

1 Remove the door trim panel and the plastic watershield (see Section 15).

2 Remove the window glass assembly (see Section 18).

3 On power operated windows, disconnect the electrical connector from the window regulator motor. On 2007 and later Expedition/Navigator models, disconnect the impact sensor harness.

4 Remove the equalizer arm bracket and the regulator retaining rivets (see illustrations).

5 Pull the equalizer arm and regulator assemblies through the service hole in the door frame to remove it.

6 Installation is the reverse of removal.

7 On 2007 and later models (LF and RF models), the regulator/motor must be initialized if it has been removed from the vehicle.

8 To initialize the motor, turn the ignition key to On and hold the one-touch window switch Down button and hold for two seconds after the glass is fully in the Down position. Wait three seconds and press the one-touch Up button and hold for two seconds after the glass reaches the full Up position.

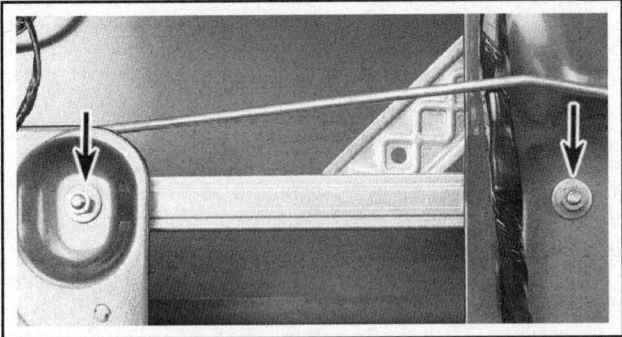

19.4a Detach the window equalizer arm bracket retaining nuts (arrows) . . .

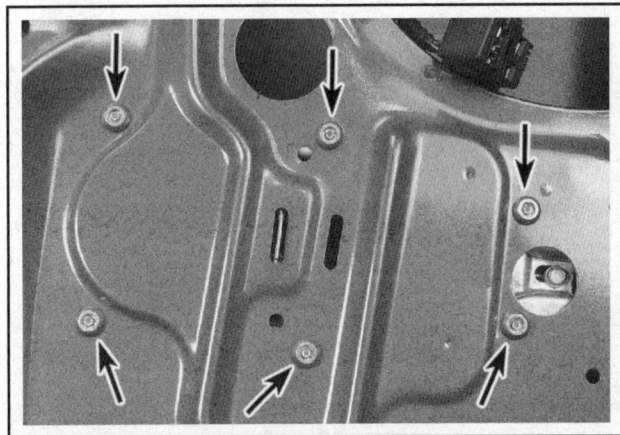

19.4b . . . then drill out the rivets (arrows) securing the window regulator to the door frame

20 Sideview mirrors - removal and installation

▶ Refer to illustrations 20.3a and 20.3b

1 Remove the door trim panel and the plastic watershield (see Section 15).

2 Disconnect the electrical connector from the mirror.

3 Peel back the insulator pad, then remove the three mirror retaining nuts and detach the mirror from the vehicle (see illustrations).

4 Installation is the reverse of removal.

SIDEVIEW MIRROR GLASS REPLACEMENT

※ WARNING:

Place a towel or rag between your hands and the mirror glass for protection just in case the glass breaks while you're working on the mirror.

※ CAUTION:

The mirror glass can be broken or the mirror motor damaged if the glass is not removed properly.

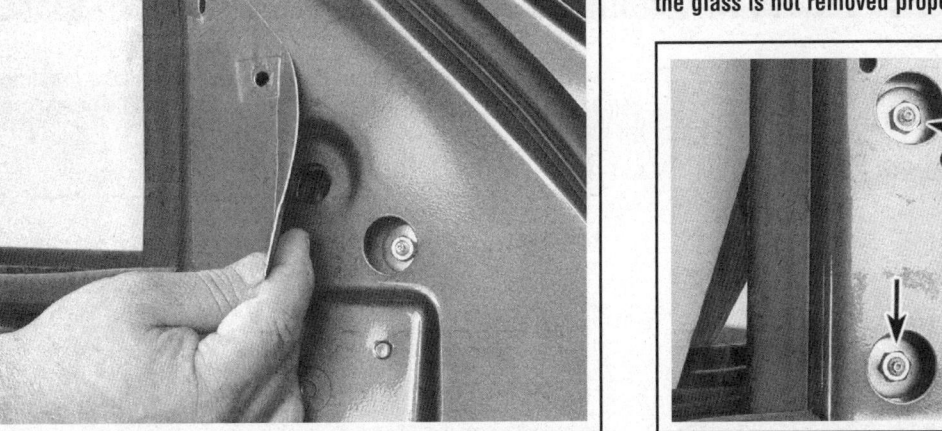

20.3a Peel back the insulator pad . . .

20.3b . . . then detach the mirror retaining nuts (arrows) and remove the mirror from the vehicle

11-16 BODY

5 Reposition the mirror glass to the full upward position, leaving a small gap at the bottom of the mirror.
6 Using a thin flat-blade screwdriver and working from the bottom of the glass, slide the screwdriver behind the glass until it contacts one of the four release tabs.
7 Starting from the left and working to the right, use the screwdriver to push against the release tabs until the mirror glass is released and can be easily removed.
8 Installation is the reverse of removal, making sure all four clips are completely attached.

21 Liftgate - removal, installation and adjustment

21.4 Loosen the hinge bolts (arrows) to remove the liftgate

➙Note: The liftgate is heavy and somewhat awkward to hold - at least two people should perform this procedure.

REMOVAL AND INSTALLATION

▸ Refer to illustration 21.4

1 Open the liftgate and support it securely.
2 Remove the liftgate trim panels and disconnect all wiring harness connectors leading to the liftgate.
3 While an assistant supports the liftgate, detach both ends of the support struts. Then pry or pull sharply to remove them from the vehicle. On Expedition/Navigator models with the power-liftgate option, remove the plastic cover to access the nut securing the power actuator to the liftgate.
4 Detach the hinge to liftgate bolts (see illustration) and remove the liftgate from the vehicle.
5 Installation is the reverse of removal.

ADJUSTMENT

▸ Refer to illustration 21.7

6 Adjustments are made by loosening the hinge-to-liftgate bolts and moving the liftgate. Proper alignment is achieved when the edges of the liftgate are parallel with the rear quarter panel and the top of the tailgate.
7 Finally, adjust the latch striker assembly as necessary (up and down) to provide positive engagement with the latch mechanism (see illustration).

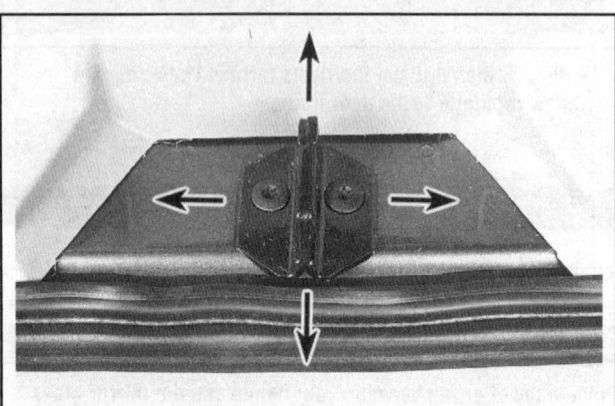

21.7 If the liftgate does not close properly, it will be necessary to loosen the screws to adjust the striker plate in the desired direction

22 Tailgate - removal, installation and adjustment

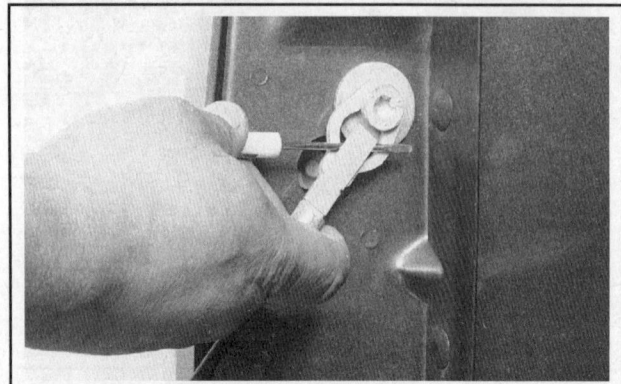

22.4 Remove the tailgate support cables

➙Note: The tailgate is heavy and somewhat awkward to remove and install - at least two people should perform this procedure.

REMOVAL AND INSTALLATION

▸ Refer to illustration 22.4

1 Open the tailgate and remove the trim panels if equipped.
2 Disconnect all wiring harness connectors leading to the tailgate.
3 Cover the lower bumper area around the opening with pads or cloths to protect the painted surfaces when the tailgate is removed.
4 While an assistant supports the tailgate, detach the tailgate support cables (see illustration).
5 Detach the hinge to tailgate bolts and remove the tailgate from

BODY   11-17

the vehicle.

6 Installation is the reverse of removal.

ADJUSTMENT

▶ **Refer to illustration 22.9**

7 Tailgate adjustments are made by loosening the hinge-to-body or hinge-to-tailgate bolts and moving the tailgate. Proper alignment is achieved when the top of the tailgate is aligned with the top of the rear quarter panel. If these goals can't be reached by adjusting the hinge to body or hinge to tailgate bolts, body alignment shims may have to be purchased and inserted behind the hinges to achieve correct alignment.

8 To adjust the tailgate closed position, first check that the latch is contacting the center of the latch striker assembly. If not, remove striker assembly and add or subtract shims to achieve correct alignment.

9 Finally, adjust the latch striker assembly as necessary (up and down or sideways) to provide positive engagement with the latch mechanism (see illustration) and the outside of the tailgate is flush with rear quarter panel.

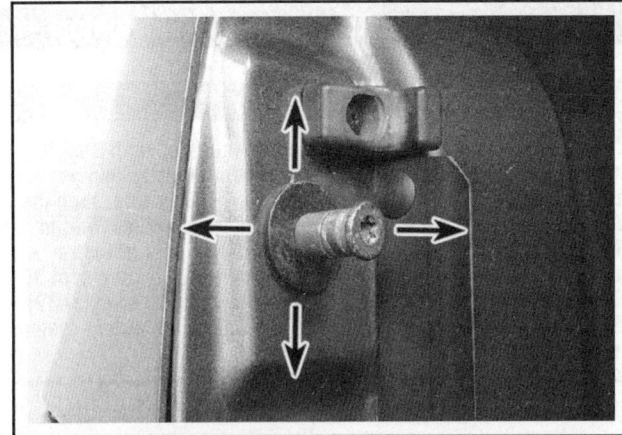

22.9 Loosen the striker and adjust as necessary to provide positive latch engagement

23 Tailgate latch, handle and lock cylinder - removal and installation

▶ **Refer to illustrations 23.1, 23.2 and 23.5**

1 Lower the tailgate and remove the tailgate access cover (see illustration).

LATCH

2 Remove the latch mounting screws (see illustration). It may be necessary to use an impact-driver to loosen them.

3 Disconnect the control rods from the latch and remove the latch from the door.

4 Installation is the reverse of removal.

HANDLE AND LOCK CYLINDER

5 Disconnect the control rods (see illustration).

6 Detach the retaining nuts and remove the handle and lock cylinder assembly from the tailgate.

7 Installation is the reverse of removal.

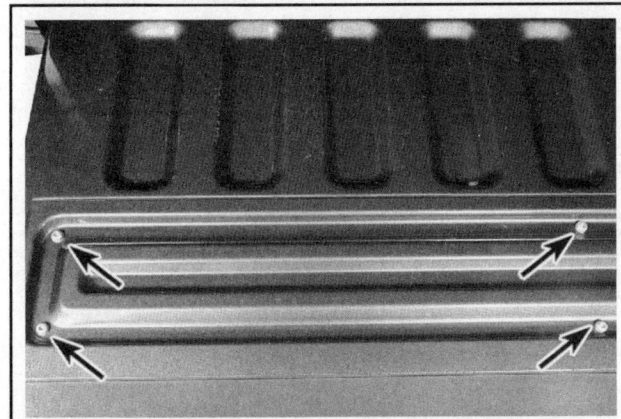

23.1 Remove the cover screws (arrows) to access the inside of the tailgate

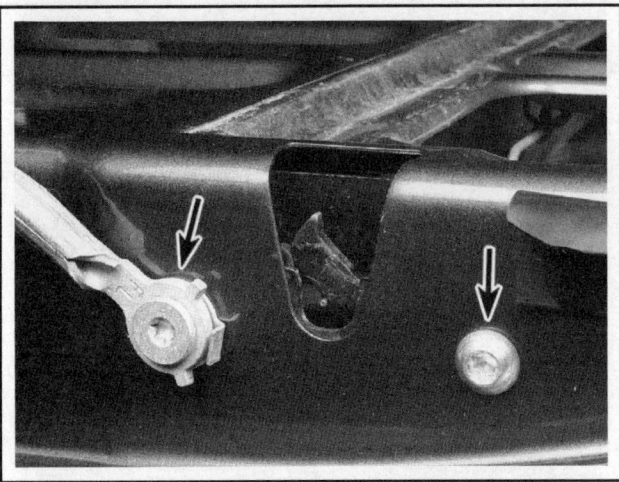

23.2 Remove the tailgate latch retaining screws (arrows)

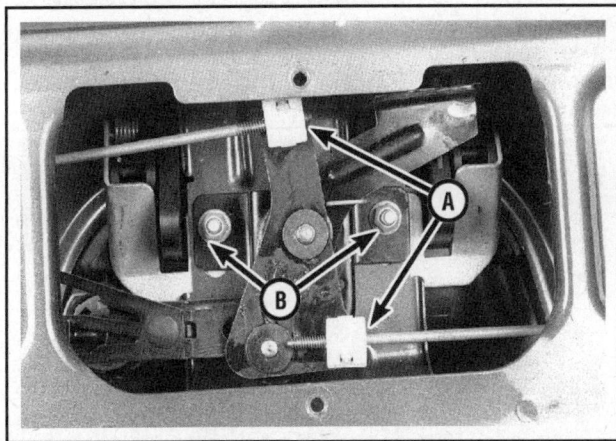

23.5 Disengage the handle-to-latch rods (A) and remove the handle retaining nuts (B)

11-18 BODY

24 Center console (Expedition and Navigator) - removal and installation

WARNING:

The models covered by this manual are equipped with Supplemental Restraint systems (SRS), more commonly known as airbags. Always disconnect the negative battery cable, then the positive battery cable and wait two minutes before working in the vicinity of the impact sensors, steering column or instrument panel to avoid the possibility of accidental deployment of the airbag, which could cause personal injury (see Chapter 12). Do not use electrical test equipment on any of the airbag system wiring or tamper with them in any way.

1 Disconnect the negative battery cable then the positive battery cable and wait two minutes before proceeding any further.

2 Lift up the mat from the floor of the spare change compartment and remove the retaining screw beneath.

3 Open the console glove box and detach the retaining screw, then remove the spare change/drink holder trim bezel.

4 Detach the retaining screws securing the front half of the console. Use a plastic trim tool to lift up the rear edge of the shifter surround panel and lift the panel out.

5 Working along the outside lower edge of the floor console, pry out the trim caps (if equipped) and remove the screws securing the lower edges of the console.

→ Note: On 2007 and later Expedition models, there are two screws on each side of the floor console, one at the rear and one at the middle. On 2007 and later Navigator models, there is only one screw on each side of the floor console located at the rear of the console.

6 On later Navigator models, use a plastic trim tool to pry off the console side panels from the front section of the floor console.

7 Disconnect any electrical connections and remove the console from the vehicle.

8 Installation is the reverse of removal.

25 Instrument cluster bezel - removal and installation

▸ Refer to illustrations 25.4a, 25.4b and 25.4c

WARNING:

The models covered by this manual are equipped with Supplemental Restraint systems (SRS), more commonly known as airbags. Always disconnect the negative battery cable, then the positive battery cable and wait two minutes before working in the vicinity of the impact sensors, steering column or instrument panel to avoid the possibility of accidental deployment of the airbag, which could cause personal injury (see Chapter 12). Do not use electrical test equipment on any of the airbag system wiring or tamper with them in any way.

1 Disconnect the negative battery cable then the positive battery cable and wait two minutes before proceeding any further.

2 Remove the headlight switch (see Chapter 12).

3 On 2002 and earlier models, pry out the center trim panel. On 2003 to 2006 models, remove the screws and the panel below the steering column, then remove the smaller accessory panel at either side of the instrument cluster to access the cluster screws. On 2007 and later Navigator models, remove the panel below the instrument cluster, the steering column multifunction switch (see Chapter 12), then reach up from behind the instrument panel to release the clips on the HVAC duct and pull the duct from the instrument cluster. On 2007 and later Expedition models, remove the panel below the steering column to access the cluster screws.

4 Remove the bezel retaining screws (see illustrations).

5 Tilt the steering wheel down to the lowest position. Pull the top of the instrument cluster bezel outward while disengaging the lower half. Remove the bezel from the vehicle.

6 Installation is the reverse of removal.

25.4a Remove the screws (arrows) securing the upper edge of the instrument cluster bezel . . .

25.4b . . . the screw behind the headlight switch . . .

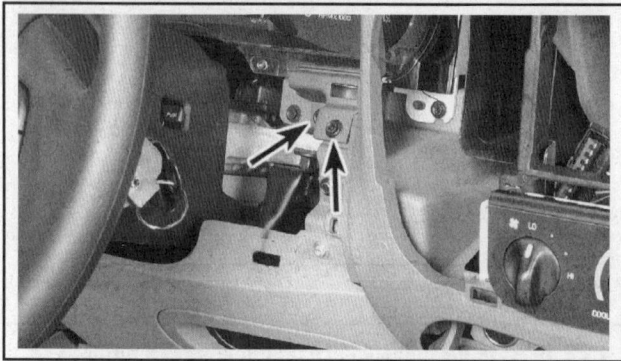

25.4c . . . and the screws along the lower edges

BODY   11-19

26 Dashboard trim panels - removal and installation

> **※ WARNING:**
> The models covered by this manual are equipped with Supplemental Restraint systems (SRS), more commonly known as airbags. Always disconnect the negative battery cable, then the positive battery cable and wait two minutes before working in the vicinity of the impact sensors, steering column or instrument panel to avoid the possibility of accidental deployment of the airbag, which could cause personal injury (see Chapter 12). Do not use electrical test equipment on any of the airbag system wiring or tamper with them in any way.

1 Disconnect the negative battery cable then the positive battery cable and wait two minutes before proceeding any further.

CENTER TRIM PANEL

2003 and earlier models

▶ Refer to illustration 26.2

2 Use a small screwdriver to pry out the clips securing the bottom of the bezel, then unsnap the clips in the top of the trim panel to remove it from the vehicle (see illustration).

2004 and later Expedition models

3 Remove the small panel at the bottom of the instrument panel (just ahead of the console). Remove the center trim panel screws and lift to release the clips at the top, and remove the trim panel.

→ Note: 2010 and later Expeditions do not have screws, the entire trim panel clips into place.

2004 and later Navigator models

4 On 2004 through 2009 Navigator models, remove the shifter boot and the center-top panel of the console.

5 On 2010 and later Navigator models, with the shift lever in Neutral, pry off the lever trim ring, then remove the trim panel from the front of the center console.

6 Remove the screws, remove the lower portion of the two-piece center trim panel and disconnect the electrical connectors. Pry up the defroster grille below the windshield for access to the two screws, then remove the two screws at the lower front end of the panel. Carefully pry up the upper portion of the center trim panel until the electrical connector to the clock can be disconnected and the panel removed.

All models

7 Installation is the reverse of removal.

KNEE BOLSTER

▶ Refer to illustrations 26.10, 26.11 and 26.12

8 On 2002 and earlier models, remove the center trim panel (see Step 2) and detach the passenger compartment fuse panel cover

9 On 2003 and later models, remove the knee bolster screws.

10 Remove the emergency brake handle and the hood release handle (see illustration). Then remove the remaining screws securing the lower edge of the knee bolster.

11 On 2002 and earlier models, pry off the floor heater duct (see illustration).

12 Detach the retaining screws securing the upper edge of the knee bolster and detach it from the vehicle (see illustration). Disconnect all electrical connectors between the bolster and the vehicle.

13 Installation is the reverse of removal.

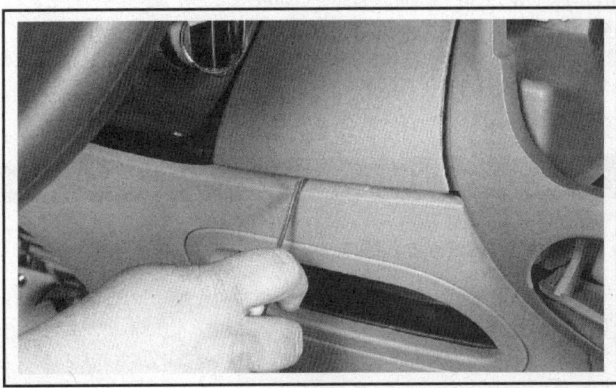

26.2 Carefully pry the center trim panel from the retaining clips

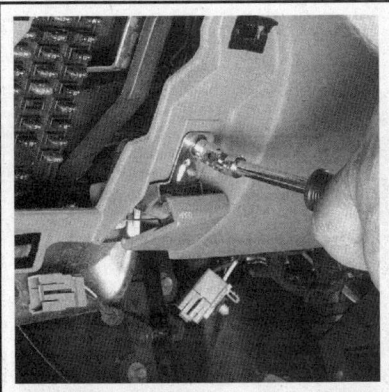

26.10 Detach the fuse panel cover, then remove the emergency brake and the hood release handles - position the handles aside to allow removal of the knee bolster

26.11 Pry off the floor heater duct (2002 and earlier models)

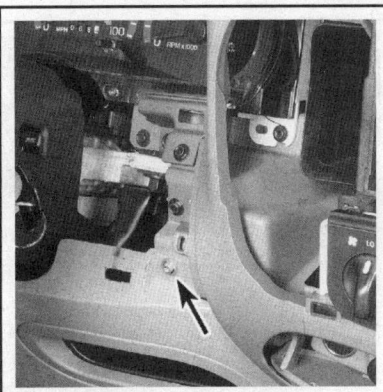

26.12 Remove the remaining screws (arrow) securing the top of the knee bolster

11-20 BODY

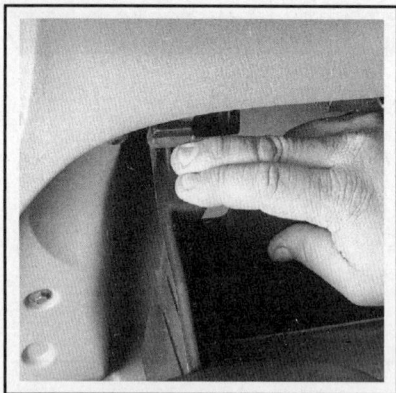

26.14 Press downward on the door stops to release the door stop retaining clips

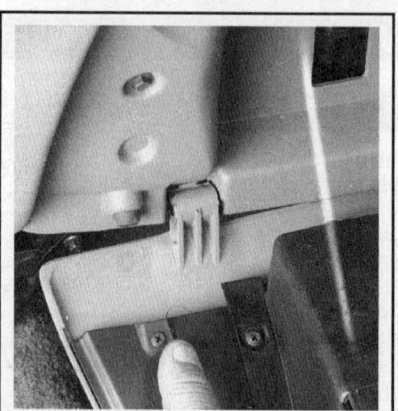

26.15 Tilt the glove box downward, then pull up to remove it

26.17 Pull the radio trim bezel outward to disengage the clips

GLOVE BOX

▸ **Refer to illustrations 26.14 and 26.15**

14 Open the glove box door. Press downward on the door stops to release the upper half of the glove box (see illustration). On models so equipped, disconnect the glove box dampener.

15 To release the lower half simply pull up and out, then remove the glove box from the vehicle (see illustration).

➟**Note: Some models may require removal of the door hinge retaining screws along the lower edge of the glove box.**

16 Installation is the reverse of removal.

RADIO TRIM BEZEL

▸ **Refer to illustration 26.17**

➟**Note: This procedure only applies to early models; on later models, the center trim panel incorporates the audio unit bezel.**

17 To remove the radio trim bezel, simply grasp it with two hands and pull the bezel straight out (see illustration).

18 Installation is the reverse of removal.

27 Steering column cover - removal and installation

▸ **Refer to illustrations 27.3 and 27.5**

1 On 2009 and earlier models, remove the center trim panel (see Section 26) and the key lock cylinder (see Chapter 12).

2 On 2010 and later models, remove the knee bolster trim panel (see Section 26).

3 If equipped, remove the manual tilt wheel lever (see illustration).

4 On 2010 and later models, move the shift lever boot aside (if equipped) and lift the upper column cover from the lower cover.

5 Remove the screws from the lower steering column cover (see illustration). On 2009 and earlier models, separate the upper cover

6 Remove the lower cover from the steering column. On 2010 and later models equipped with power tilt wheels, disconnect the electrical connector.

7 Installation is the reverse of removal.

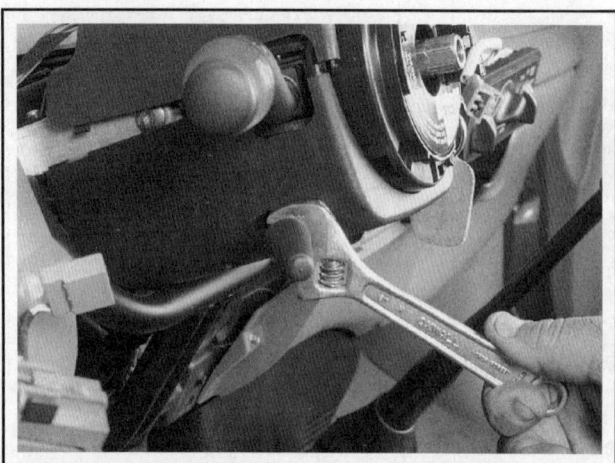

27.3 Use a small wrench to remove the tilt wheel lever (if equipped)

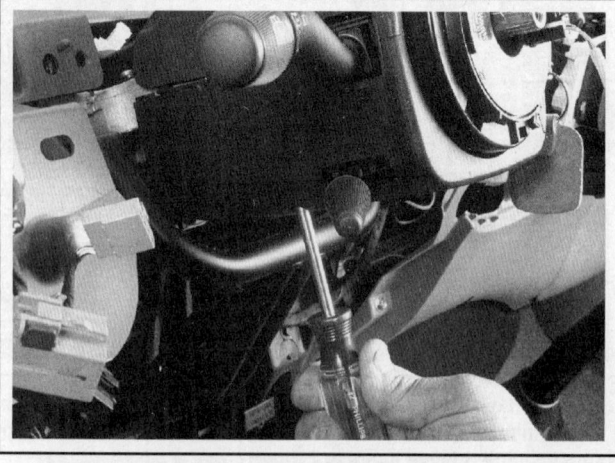

27.5 Detach the screws - then separate the steering column cover halves to remove them from the steering column

BODY 11-21

28 Instrument panel - removal and installation

♦ Refer to illustrations 28.3a, 28.3b, 28.4, 28.7a, 28.7b, 28.8, 28.9, 28.10, 28.11a, 28.11b and 28.12

WARNING:

The models covered by this manual are equipped with Supplemental Restraint systems (SRS), more commonly known as airbags. Always disconnect the negative battery cable, then the positive battery cable and wait two minutes before working in the vicinity of the impact sensors, steering column or instrument panel to avoid the possibility of accidental deployment of the airbag, which could cause personal injury (see Chapter 12). Do not use electrical test equipment on any of the airbag system wiring or tamper with them in any way.

Note: The instrument panel is heavy and somewhat awkward to remove and install - at least two people should perform this procedure.

1 Disconnect the negative battery cable then the positive battery cable and wait two minutes before proceeding any further.

CAUTION:

Always use extreme care when working around airbag modules. Never strike, pry or bump airbag modules to avoid the possibility of accidental deployment.

2 Position the steering wheel in the lock position.

Note: Make sure the steering wheel remains in the lock position during the entire procedure or damage to the airbag sliding contact will occur.

3 Remove the driver and passenger side scuff plates and kick panels (see illustrations).

4 Disconnect the ground cables and electrical connectors located behind the kick panels (see illustration).

5 Remove the dashboard trim panels (see Section 26) and the center floor console (see Section 24).

6 Disconnect the electrical connectors leading to the steering column and the transmission indicator cable (if equipped), then detach the steering column retaining bolts and lower the column to the floor.

7 Pry out the relay panel cover on the passenger's side of the dashboard and disconnect the wiring harness (see illustrations).

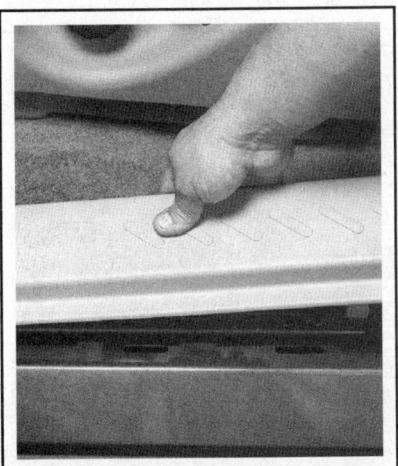

28.3a Pry off the driver and passenger side scuff plates . . .

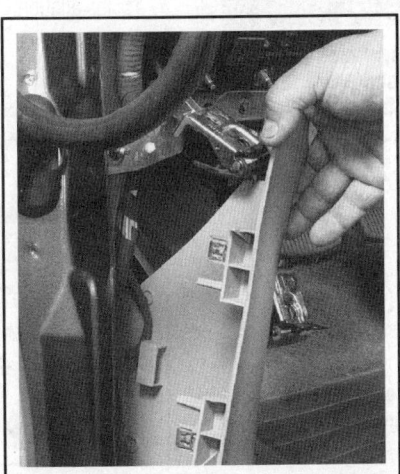

28.3b . . . and kick panels

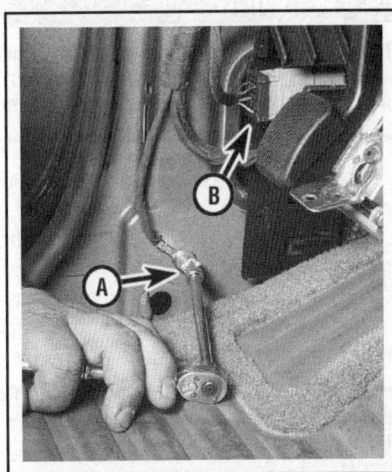

28.4 Disconnect the ground straps (A) and electrical connectors (B) located behind the kick panels

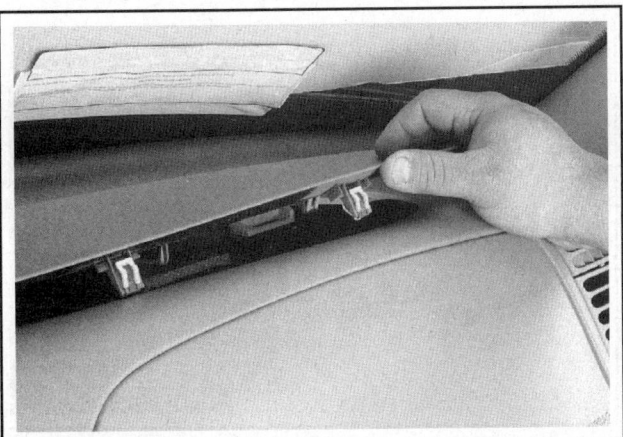

28.7a Pry out the relay cover . . .

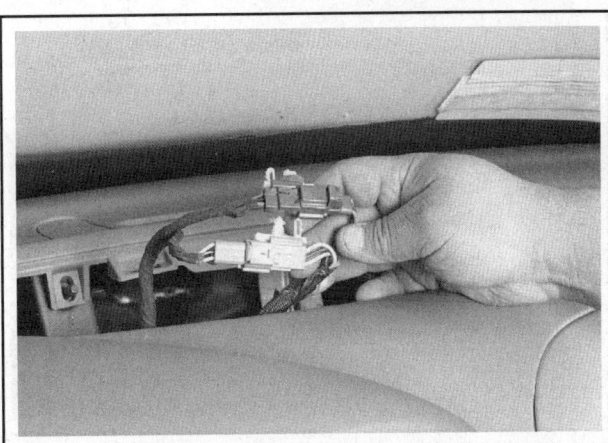

28.7b . . . then disconnect the wiring harness connectors

11-22 BODY

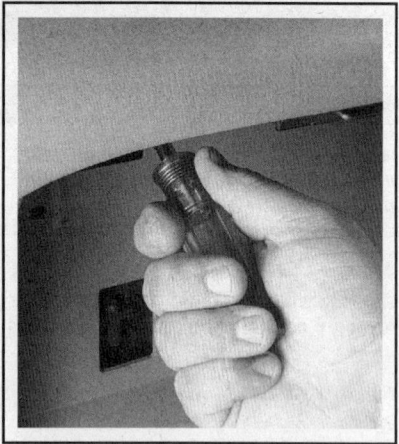

28.8 The passenger side air module retaining screws can be accessed through the relay panel cover and the glove box opening

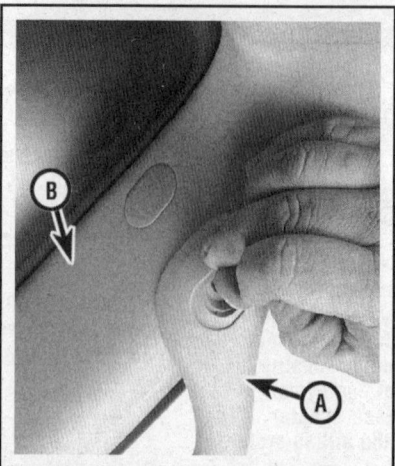

28.9 Remove the passenger assist handle (A) and the door pillar trim panel (B)

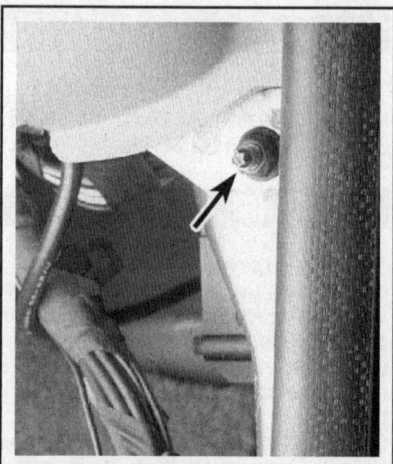

28.10 Remove the nuts and bolts (arrow) securing the lower edge of the instrument panel

8 Remove the passenger side air bag module (see illustration).

✳✳ WARNING:

Handle the airbag module with care! Always carry the module with the trim side facing away from your body and place the module in a secure location with the trim side facing up.

9 Remove the passenger side assist handle and door pillar trim panel (see illustration).
10 Remove the nuts and bolts securing the lower half of the instrument panel (see illustration).
11 Remove the instrument cluster housing cover (see illustrations).
12 Pry out the defroster grille(s) and remove the screws securing the upper edge of the instrument panel (see illustration).

➡ Note: On 2007 and later models, the defrost grille is one long panel covering the entire top edge of the instrument panel below the windshield.

13 On 2007 and later Expedition models, remove the floor console bracket bolts and move the bracket out of the way.

14 Detach any electrical connectors interfering with removal, then pull the instrument panel towards the rear of the vehicle to remove it.
15 Installation is the reverse of removal.

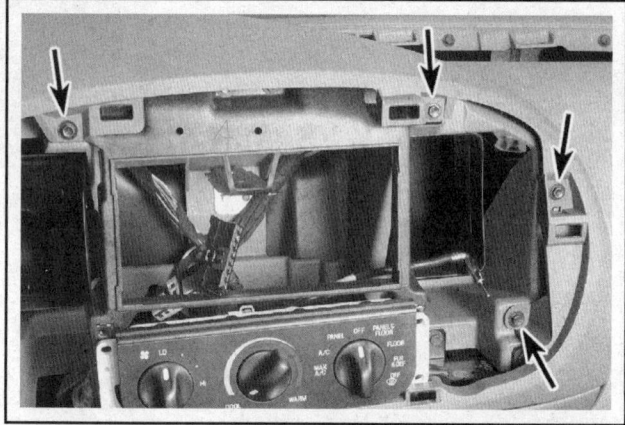

28.11a Remove the instrument cluster housing cover screws from the center opening . . .

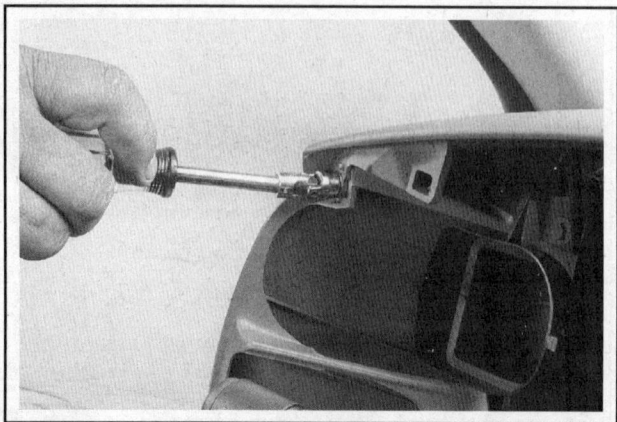

28.11b . . . and at the left corner of the instrument panel

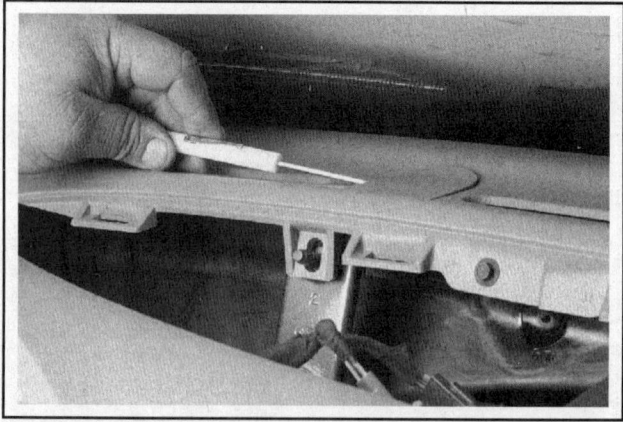

28.12 Pry out the trim covers and remove the bolts retaining the upper half of the instrument panel

BODY 11-23

29 Cowl cover - removal and installation

▶ Refer to illustrations 29.3a, 29.3b and 29.3c

1. Remove the windshield wiper arms (see Chapter 12).
2. Detach the hood seal from the cowl cover (if equipped).
3. Remove the screws and clips securing the left hand and right hand cowl covers (see illustrations).
4. Installation is the reverse of removal.

29.3a Remove the cowl cover retaining screws - the end screws are located under the covers

29.3b Pry off the clips along the front edge

29.3c Remove the retaining clips from each end

30 Seats - removal and installation

FRONT SEAT

▶ Refer to illustration 30.2

1. Position the seat all the way forward or all the way to the rear to access the front seat retaining bolts.
2. Detach any bolt trim covers and remove the retaining bolts (see illustration).
3. Tilt the seat upward to access the underneath, then disconnect any electrical connectors and lift the seat from the vehicle.
4. Installation is the reverse of removal.

2ND ROW SEATS

▶ Refer to illustration 30.5

5. Remove the seat-to-floor mounting bolts, then detach the seat and lift it out of the vehicle (see illustration).

✼✼ CAUTION:

When removing the larger 2nd row seat, the seat base must be in the locked position and the shipping pin (or bolt) inserted into the base before trying to remove the seat. If it is not in the locked position, the base will drop from the seat and the seat linkage may be damaged.

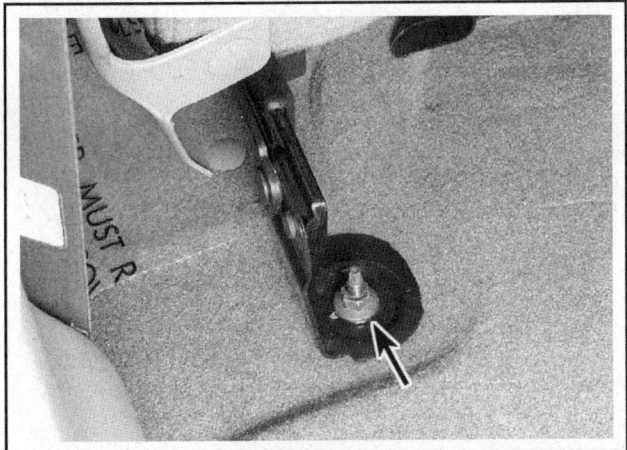

30.2 Detach the trim covers to access the seat retaining bolts

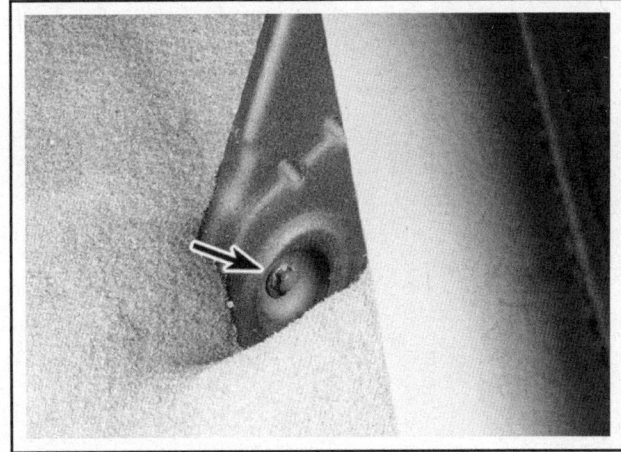

30.5 Remove the rear seat cushion bolts (arrow)

11-24 BODY

6 Installation is the reverse of removal.

THIRD ROW SEAT

7 Open the liftgate and remove the floor storage panel.
8 Remove the mid-floor support panel mounting bolts and retaining pins.

9 Disconnect the electrical connector(s), if equipped from the under the seat(s).
10 Remove the mounting bolts and lift the seat out of the vehicle.

➡ **Note: When removing the 60/40 split seats, the larger side seat is heavy and somewhat awkward to remove and install – it's a good idea to have an assistant help you perform this procedure.**

11 Installation is the reverse of removal.

12 CHASSIS ELECTRICAL SYSTEM

Section

1 General information
2 Electrical troubleshooting - general information
3 Fuses - general information
4 Fusible links - general information
5 Circuit breakers - general information
6 Relays - general information and testing
7 Turn signal/hazard flasher - check and replacement
8 Steering column switches - check and replacement
9 Ignition switch - replacement
10 Ignition lock cylinder - removal and installation
11 Headlight switch - replacement
12 Instrument panel gauges - check
13 Instrument cluster - removal and installation
14 Radio and speakers - removal and installation
15 Antenna - removal and installation
16 Headlights - replacement
17 Headlights - adjustment
18 Bulb replacement
19 Daytime Running Lights (DRL) - general information
20 Wiper motor - check and replacement
21 Horn - check and replacement
22 Rear window defogger - check and repair
23 Cruise control system - description and check
24 Power window system - description and check
25 Power door lock and keyless entry system - description and check
26 Electric side view mirrors - description and check
27 Airbag - general information
28 Wiring diagrams - general information

12-2 CHASSIS ELECTRICAL SYSTEM

1 General information

The electrical system is a 12-volt, negative ground type. Power for the lights and all electrical accessories is supplied by a lead/acid-type battery which is charged by the alternator.

This Chapter covers repair and service procedures for the various electrical components not associated with the engine.

Information on the battery, alternator, distributor and starter motor can be found in Chapter 5.

It should be noted that when portions of the electrical system are serviced, the negative battery cable should be disconnected from the battery to prevent electrical shorts and/or fires.

2 Electrical troubleshooting - general information

A typical electrical circuit consists of an electrical component, any switches, relays, motors, fuses, fusible links or circuit breakers related to that component and the wiring and connectors that link the component to both the battery and the chassis. To help you pinpoint an electrical circuit problem, wiring diagrams are included at the end of this Chapter.

Before tackling any troublesome electrical circuit, first study the appropriate wiring diagrams to get a complete understanding of what makes up that individual circuit. Trouble spots, for instance, can often be narrowed down by noting if other components related to the circuit are operating properly. If several components or circuits fail at one time, chances are the problem is in a fuse or ground connection, because several circuits are often routed through the same fuse and ground connections.

Electrical problems usually stem from simple causes, such as loose or corroded connections, a blown fuse, a melted fusible link or a failed relay. Visually inspect the condition of all fuses, wires and connections in a problem circuit before troubleshooting the circuit.

If test equipment and instruments are going to be utilized, use the diagrams to plan ahead of time where you will make the necessary connections in order to accurately pinpoint the trouble spot.

The basic tools needed for electrical troubleshooting include a circuit tester or voltmeter (a 12-volt bulb with a set of test leads can also be used), a continuity tester, which includes a bulb, battery and set of test leads, and a jumper wire, preferably with a circuit breaker incorporated, which can be used to bypass electrical components. Before attempting to locate a problem with test instruments, use the wiring diagram(s) to decide where to make the connections.

VOLTAGE CHECKS

Voltage checks should be performed if a circuit is not functioning properly. Connect one lead of a circuit tester to either the negative battery terminal or a known good ground. Connect the other lead to a connector in the circuit being tested, preferably nearest to the battery or fuse. If the bulb of the tester lights, voltage is present, which means that the part of the circuit between the connector and the battery is problem free. Continue checking the rest of the circuit in the same fashion. When you reach a point at which no voltage is present, the problem lies between that point and the last test point with voltage. Most of the time the problem can be traced to a loose connection.

➡ **Note: Keep in mind that some circuits receive voltage only when the ignition key is in the Accessory or Run position.**

FINDING A SHORT

One method of finding shorts in a circuit is to remove the fuse and connect a test light or voltmeter in place of the fuse terminals. There should be no voltage present in the circuit. Move the wiring harness from side-to-side while watching the test light. If the bulb goes on, there is a short to ground somewhere in that area, probably where the insulation has rubbed through. The same test can be performed on each component in the circuit, even a switch.

GROUND CHECK

Perform a ground test to check whether a component is properly grounded. Disconnect the battery and connect one lead of a self-powered test light, known as a continuity tester, to a known good ground. Connect the other lead to the wire or ground connection being tested. If the bulb goes on, the ground is good. If the bulb does not go on, the ground is not good.

CONTINUITY CHECK

A continuity check is done to determine if there are any breaks in a circuit - if it is passing electricity properly. With the circuit off (no power in the circuit), a self-powered continuity tester can be used to check the circuit. Connect the test leads to both ends of the circuit (or to the "power" end and a good ground), and if the test light comes on the circuit is passing current properly. If the light doesn't come on, there is a break somewhere in the circuit. The same procedure can be used to test a switch, by connecting the continuity tester to the switch terminals. With the switch turned On, the test light should come on.

FINDING AN OPEN CIRCUIT

When diagnosing for possible open circuits, it is often difficult to locate them by sight because oxidation or terminal misalignment are hidden by the connectors. Merely wiggling a connector on a sensor or in the wiring harness may correct the open circuit condition. Remember this when an open circuit is indicated when troubleshooting a circuit. Intermittent problems may also be caused by oxidized or loose connections.

Electrical troubleshooting is simple if you keep in mind that all electrical circuits are basically electricity running from the battery, through the wires, switches, relays, fuses and fusible links to each electrical component (light bulb, motor, etc.) and to ground, from which it is passed back to the battery. Any electrical problem is an interruption in the flow of electricity to and from the battery.

CHASSIS ELECTRICAL SYSTEM   12-3

3 Fuses - general information

Refer to illustrations 3.1a, 3.1b, 3.1c 3.1d and 3.2

The electrical circuits of the vehicle are protected by a combination of fuses, circuit breakers and cartridge-type fusible links. All models covered in this manual have four fuse blocks: the passenger compartment fuse block is located under the instrument panel on the left or right side of the dashboard, a second fuse block called the power distribution box is located under the hood, on the left inner fenderwell (2009 and earlier) or at the front of the engine compartment (2010 and later); a third fuse block called the engine fuse block is mounted just behind the power distribution box; a fourth fuse block is located next to the battery, and contains the primary battery fuses (see illustrations). Always disconnect the cable from the negative battery terminal before replacing high current fuses.

Miniaturized fuses are employed in the passenger compartment fuse block and the power distribution box. These compact fuses, with blade terminal design, allow fingertip removal and replacement. If an electrical component fails, always check the fuse first. The best way to check the fuses is with a test light. Check for power at the exposed terminal tips of each fuse. If power is present at one side of the fuse but not the other, the fuse is blown. A blown fuse can also be identified by visually inspecting it (see illustration).

Be sure to replace blown fuses with the correct type. Fuses of different ratings are physically interchangeable, but only fuses of the

3.1a The passenger compartment fuse block is located under the drivers side of the instrument panel, behind the fuse panel cover

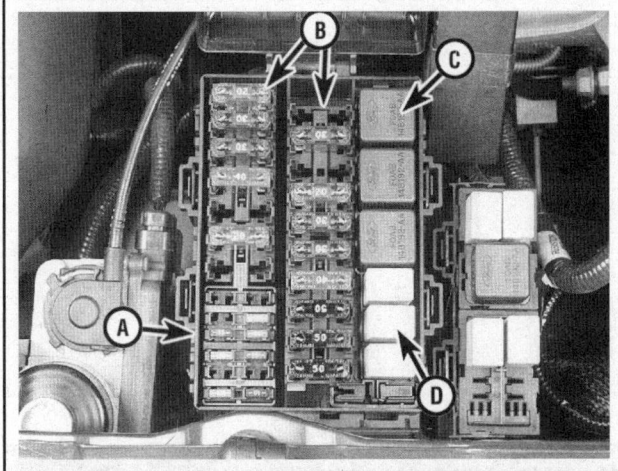

3.1b On 2009 and earlier models, he power distribution box is mounted to the left inner fenderwell in the engine compartment - it contains miniaturized fuses (A), cartridge type fusible links (B) relays (C) and circuit breakers (D)

3.1c The engine fuse block contains miniaturized fuses and is located directly behind the power distribution box

3.1d The primary battery fuses are mounted next to the battery and can be accessed after removing the cover

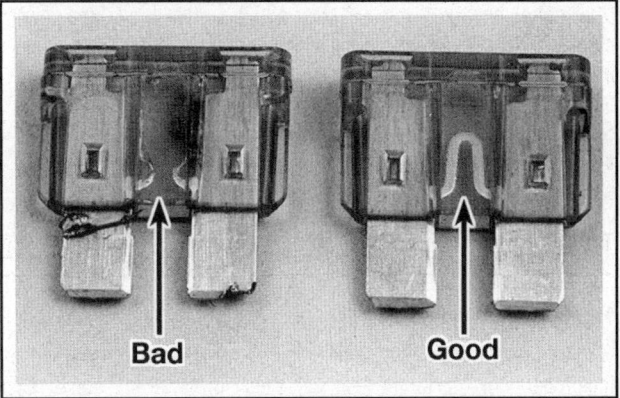

3.2 When a fuse blows, the element between the terminals melts - the fuse on the left is blown, the fuse on the right is good

12-4 CHASSIS ELECTRICAL SYSTEM

proper rating should be used. Replacing a fuse with one of a higher or lower value than specified is not recommended. Each electrical circuit needs a specific amount of protection. The amperage value of each fuse is molded into the fuse body.

If the replacement fuse immediately fails, don't replace it again until the cause of the problem is isolated and corrected. In most cases, the cause will be a short circuit in the wiring caused by a broken or deteriorated wire.

4 Fusible links - general information

Some circuits are protected by fusible links. The links are used in circuits which are not ordinarily fused, such as the ignition circuit.

Cartridge type fusible links are located in the power distribution box and are similar to a large fuses (see illustration 3.1b).

To replace a fusible link, first disconnect the negative cable from the battery. Unplug the burned-out link from the fuse block and replace it with a new unit of the same amperage (available from your dealer or auto parts store). Always determine the cause for the overload which melted the fusible link before installing a new one.

5 Circuit breakers - general information

◆ **Refer to illustration 5.2**

Circuit breakers protect components such as power windows, power door locks and headlights.

On some models the circuit breaker resets itself automatically, so an electrical overload in a circuit breaker protected system will cause the circuit to fail momentarily, then come back on. If the circuit does not come back on, check it immediately (see illustration). Once the condition is corrected, the circuit breaker will resume its normal function.

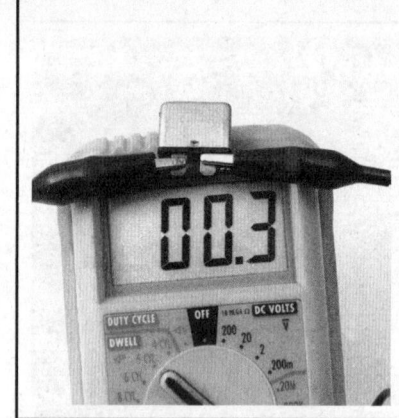

5.2 Perform a continuity test with an ohmmeter to check a circuit breaker - infinite resistance indicates a bad circuit breaker

6 Relays - general information and testing

GENERAL INFORMATION

1 Several electrical accessories in the vehicle, such as the fuel injection system, horns, starter, and fog lamps use relays to transmit the electrical signal to the component. Relays use a low-current circuit (the control circuit) to open and close a high-current circuit (the power circuit). If the relay is defective, that component will not operate properly. The various relays are mounted in engine compartment (see illustration 3.1b) and several locations throughout the vehicle (see Chapter 5). If a faulty relay is suspected, it can be removed and tested using the procedure below or by a dealer service department or other repair shop. Defective relays must be replaced as a unit.

TESTING

◆ **Refer to illustration 6.4**

2 It's best to refer to the wiring diagram for the circuit to determine the proper hook-ups for the relay you're testing. However, if you're not able to determine the correct hook-up from the wiring diagrams, you may be able to determine the test hook-ups from the information that follows.

3 On most relays, two of the terminals are the relay's control circuit (they connect to the relay coil which, when energized, closes the large contacts to complete the circuit). The other terminals are the power

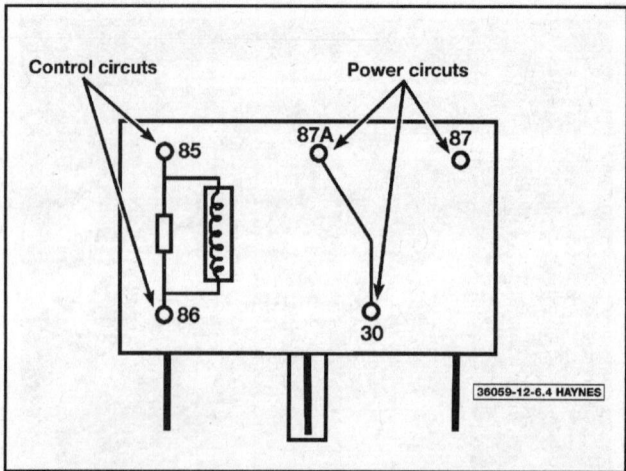

6.4 Most relays are marked on the outside to easily identify the control and power circuits

CHASSIS ELECTRICAL SYSTEM 12-5

circuit (they are connected together within the relay when the control-circuit coil is energized).

4 The relays are marked as an aid to help you determine which terminals are the control circuit and which are the power circuit (see illustration).

5 Remove the relay from the vehicle and check for continuity between the relay power circuit terminals. There should be no continuity between terminals 30 and 87.

6 Connect a fused jumper wire between one of the two control circuit terminals and the positive battery terminal. Connect another jumper wire between the other control circuit terminal and ground. When the connections are made, the relay should click. On some relays, polarity may be critical, so, if the relay doesn't click, try swapping the jumper wires on the control circuit terminals.

7 With the jumper wires connected, check for continuity between the power circuit terminals. Now, there should be continuity between terminals 30 and 87.

8 If the relay fails any of the above tests, replace it.

7 Turn signal/hazard flasher - check and replacement

▶ Refer to illustration 7.1

※※ WARNING:

The models covered by this manual are equipped with Supplemental Restraint systems (SRS), more commonly known as airbags. Always disconnect the negative battery cable, then the positive battery cable and wait two minutes before working in the vicinity of the impact sensors, steering column or instrument panel to avoid the possibility of accidental deployment of the airbag, which could cause personal injury (see Section 27). Do not use electrical test equipment on any of the airbag system wiring or tamper with them in any way.

7.1 The electronic flasher unit is mounted to the right of the steering column under the instrument panel

1 On 2006 and earlier models, the turn signal and hazard flashers are controlled from a single electronic flasher unit which is mounted to the right of the steering column under the instrument panel (see illustration). On 2007 and later models, there is no flasher unit. The flasher function is incorporated into a GEM (Generic Electronic Module) called the Smart Junction Block (SJB), which controls a number exterior lighting functions.

2 When the flasher unit is functioning properly, an audible click can be heard during its operation. If the turn signals fail on one side or the other and the flasher unit does not make its characteristic clicking sound, a faulty turn signal bulb is indicated.

3 If both turn signals fail to blink, the problem may be due to a blown fuse, a faulty flasher unit, a broken switch or a loose or open connection. If a quick check of the fuse box indicates that the turn signal fuse has blown, check the wiring for a short before installing a new fuse.

4 To replace the flasher, simply disconnect the electrical connectors, then remove the screw securing the flasher retaining bracket.

5 Make sure that the replacement unit is identical to the original. Compare the old one to the new one before installing it.

6 Installation is the reverse of removal.

Hazard flasher switch

2007 and later models

7 Remove the center trim panel (see Chapter 11).
8 Disconnect the electrical connector to the switch.
9 Depress the tabs on the switch and push the switch out from the panel.
10 Installation is the reverse of removal.

8 Steering column switches - replacement

※※ WARNING:

The models covered by this manual are equipped with Supplemental Restraint systems (SRS), more commonly known as airbags. Always disconnect the negative battery cable, then the positive battery cable and wait two minutes before working in the vicinity of the impact sensors, steering column or instrument panel to avoid the possibility of accidental deployment of the airbag, which could cause personal injury (see Section 27). Do not use electrical test equipment on any of the airbag system wiring or tamper with them in any way.

MULTI-FUNCTION SWITCH

▶ Refer to illustrations 8.4a and 8.4b

1 The multi-function switch is located on the left side of the steering column. It incorporates the turn signal, headlight dimmer and windshield wiper/washer functions into one switch.

2 Disconnect the negative battery cable, then the positive battery cable and wait two minutes before proceeding any further.

3 Remove the steering column covers (see Chapter 11).

12-6 CHASSIS ELECTRICAL SYSTEM

4 Remove the retaining screws, disconnect the electrical connectors, then detach the switch from the steering column (see illustrations). On some models, there is a tab on the top of the multifunction switch that must be pushed to release the switch.

5 Installation is the reverse of removal.

CRUISE CONTROL SWITCHES

▶ Refer to illustrations 8.6 and 8.9

6 Open the hood and locate the cruise control speed amplifier/servo (see Illustration).

7 Disconnect the negative battery cable, then the positive battery cable and wait two minutes before proceeding any further.

8 Remove the driver's side airbag module from the steering wheel (see Chapter 10). On 2010 and later Navigator models, remove the steering wheel controls trim panel.

9 Detach the switch retaining screw (see illustration). Then disconnect the electrical connections and remove the switch from the steering wheel.

➡ Note: Some later models may not have retaining screws. To remove the cruise control switches, simply pull them from the steering wheel.

10 Installation is the reverse of removal.

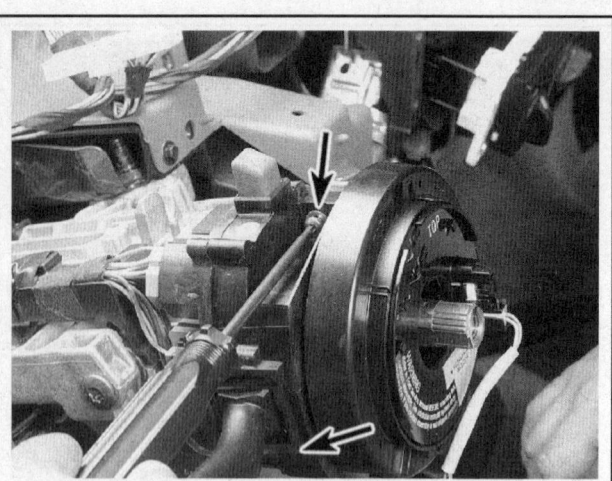

8.4a Remove the retaining screws (arrows) . . .

8.4b . . . unplug the electrical connectors, then remove the multi-function switch

8.6 Location of the cruise control amplifier/servo electrical connector (arrow)

8.9 After the driver's side airbag module has been removed, remove the cruise control actuator switch retaining screw from the rear of the steering wheel (steering wheel removed for clarity)

CHASSIS ELECTRICAL SYSTEM   12-7

9   Ignition switch - replacement

✳✳ WARNING:

The models covered by this manual are equipped with Supplemental Restraint systems (SRS), more commonly known as airbags. Always disconnect the negative battery cable, then the positive battery cable and wait two minutes before working in the vicinity of the impact sensors, steering column or instrument panel to avoid the possibility of accidental deployment of the airbag, which could cause personal injury (see Section 27). Do not use electrical test equipment on any of the airbag system wiring or tamper with them in any way.

▶ Refer to illustration 9.4 and 9.6

1   Disconnect the negative battery cable, then the positive battery cable and wait two minutes before proceeding any further.
2   Turn the ignition key lock cylinder to the Run position.
3   Remove the center trim panel (2002 and earlier models) and the driver side knee bolster (see Chapter 11).
4   Unplug the ignition switch electrical connector and remove the switch retaining screws (see illustration). On some later models, the switch is retained by a tab that fits the switch housing. Depress the tab to remove the switch.
5   Disengage the ignition switch from the actuator pin and remove the switch from the vehicle.
6   Make sure the actuator pin slot in the new ignition switch is in the Run position (see illustration).

➡ Note: A new replacement switch will be set in this position.

7   Place the new switch in position on the actuator pin and install the retaining screws. It may be necessary to move the switch back and forth to line up the screw holes.
8   The remainder of the installation is the reverse of removal. Check for proper operation of the ignition switch in the lock, start and accessory positions.

9.4 Unplug the electrical connector and remove the ignition switch retaining screws (arrows)

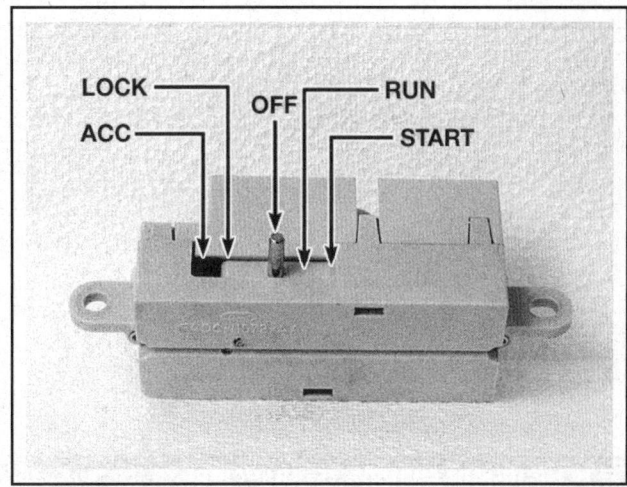

9.6 Ignition switch position details

10   Ignition lock cylinder - removal and installation

▶ Refer to illustration 10.3

✳✳ WARNING:

The models covered by this manual are equipped with Supplemental Restraint systems (SRS), more commonly known as airbags. Always disconnect the negative battery cable, then the positive battery cable and wait two minutes before working in the vicinity of the impact sensors, steering column or instrument panel to avoid the possibility of accidental deployment of the airbag, which could cause personal injury (see Section 27). Do not use electrical test equipment on any of the airbag system wiring or tamper with them in any way.

REMOVAL

1   Disconnect the negative battery cable, then the positive battery cable and wait two minutes before proceeding any further.

➡ Note: On 2003 and later models, the steering column covers must be removed to access the release tab.

2   Turn the ignition key/lock cylinder to the Run position.
3   Insert an 1/8-inch punch into the hole at the bottom of the steering column cover surrounding the lock cylinder. Depress the punch

12-8 CHASSIS ELECTRICAL SYSTEM

while pulling out on the lock cylinder to remove it from the steering column housing (see illustration).

INSTALLATION

4 Depress the retaining pin on the side of the lock cylinder and rotate the ignition key/lock cylinder to the Run position.

5 Install the lock cylinder into the steering column housing, making sure it's fully seated and aligned in the interlocking washer.

6 Rotate the key back to the Off position. This will allow the retaining pin to extend itself back into the locating hole in the steering column housing.

7 Turn the lock to ensure that operation is correct in all positions.

8 The remainder of installation is the reverse of removal.

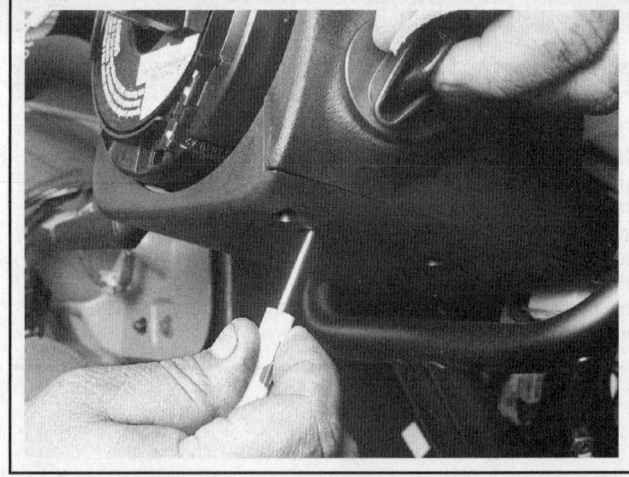

10.3 To remove the ignition lock cylinder, place the key in the "RUN" position, push in on the release tab with a screwdriver or a small punch and pull the cylinder straight out

11 Headlight switch - replacement

✷✷ WARNING:

The models covered by this manual are equipped with Supplemental Restraint systems (SRS), more commonly known as airbags. Always disconnect the negative battery cable, then the positive battery cable and wait two minutes before working in the vicinity of the impact sensors, steering column or instrument panel to avoid the possibility of accidental deployment of the airbag, which could cause personal injury (see Section 27). Do not use electrical test equipment on any of the airbag system wiring or tamper with them in any way.

2002 AND EARLIER MODELS

▸ Refer to illustrations 11.1, 11.2, 11.3a and 11.3b

✷✷ CAUTION:

Failure to follow this procedure will cause damage to the instrument panel and the headlight switch.

1 Disconnect the negative battery cable, then turn the headlight switch to the ON position and pull the knob outward to the fog light ON position. Using a small screwdriver pry the knob off the headlight switch (see illustration).

11.1 Position the headlight switch knob so that the headlights and the fog lights are in the ON position, then pry the knob off where indicated by the arrow

CHASSIS ELECTRICAL SYSTEM   12-9

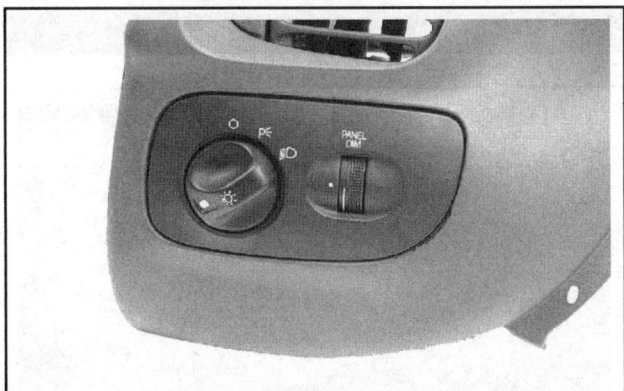

11.2 Reinstall the knob so that it faces 180-degrees from the ON position

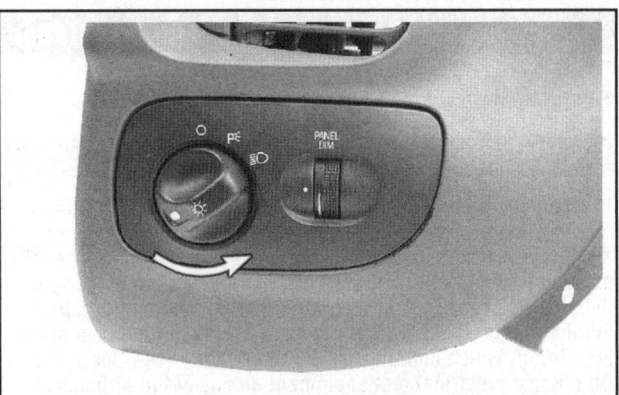

11.3a Turn the knob back to the OFF position

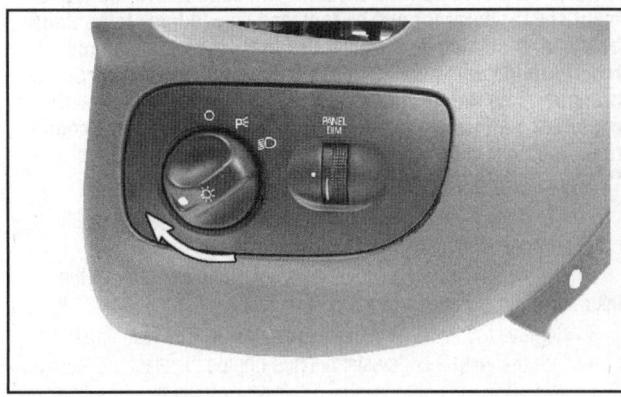

11.3b Rotate the knob clockwise while pulling outward on the headlight switch

2 Turn the knob 180-degrees from the ON position and reinstall it back onto the headlight switch (see illustration).

3 Turn the headlight switch to the OFF position (see illustration). Then rotate the knob clockwise as far as possible and hold it. This will enable the locking tab on the back of the switch to retract and the headlight switch to be removed from the instrument panel (see illustration).

4 Unplug the electrical connector from the back side of the switch, then remove it from the vehicle.

5 Installation is the reverse of removal.

2003 AND LATER MODELS

6 See the **Warning** at the beginning of this Section.

7 Remove the driver's knee bolster (most models) or the lower left under dash panel (2010 and later Expedition models) (see Chapter 11).

8 On models where the switch panel is removable, pry the panel out and disconnect the electrical connectors, then remove the screws securing the switch to the back of the panel.

9 On models where the switch is mounted to the instrument panel, disconnect the electrical connector from behind, then release the tabs securing the switch to the instrument panel.

10 Push out the switch from the back of the panel to remove it.

11 Installation is the reverse of removal.

12 Instrument panel gauges - check

※ WARNING:

Some models covered by this manual are equipped with airbags. Always disable the airbag system (see Section 27) before working in the vicinity of the impact sensors, steering column or instrument panel. Failure to follow these procedures may cause accidental deployment of the airbag, which could cause personal injury. The airbag circuits are easily identified by yellow insulation covering the entire wiring harness. Do not use electrical test equipment on any of these wires or tamper with them in any way.

1 All tests below require the ignition switch to be turned to ON position during testing.

2 If the gauge pointer does not move from the empty, low or cold positions, check the fuse. If the fuse is OK, locate the particular sending unit for the circuit you're working on (see Chapter 4 for fuel sending unit location, Chapter 2 for the oil sending unit location or Chapter 3 for the temperature sending unit location). Connect the sending unit connector to ground momentarily. If the pointer goes to the full, high or hot position replace the sending unit. If the pointer stays in same position, use a jumper wire to ground the sending unit terminal on the back of the gauge. If necessary, refer to the wiring diagrams at the end of this Chapter. If the pointer moves, the problem lies in the wire between the gauge and the sending unit. If the pointer does not move with the sending unit terminal on the back of the gauge grounded, check for voltage at the other terminal of the gauge. If voltage is present, replace the gauge.

12-10 CHASSIS ELECTRICAL SYSTEM

13 Instrument cluster - removal and installation

▸ Refer to illustration 13.3

⁂ WARNING:

The models covered by this manual are equipped with Supplemental Restraint systems (SRS), more commonly known as airbags. Always disconnect the negative battery cable, then the positive battery cable and wait two minutes before working in the vicinity of the impact sensors, steering column or instrument panel to avoid the possibility of accidental deployment of the airbag, which could cause personal injury (see Section 27). Do not use electrical test equipment on any of the airbag system wiring or tamper with them in any way.

→Note: On 2006 and later models, you can remove the instrument cluster but if it must be replaced, it will have to be done by a dealer service department or other properly equipped repair facility, because module configuration must be programmed into the new instrument cluster or the vehicle will experience a Passive Anti-Theft System (PATS) no-start condition. This will occur even if the vehicle is not equipped with Passive Anti-Theft System.

1 Disconnect the negative battery cable, then the positive battery cable and wait two minutes before proceeding any further.
2 Tilt the steering wheel to its lowest position and remove the instrument cluster bezel trim panel (see Chapter 11).
3 Remove the instrument cluster retaining screws (see illustration).
4 Pull the instrument cluster out and unplug the electrical connec-

13.3 The instrument cluster is held in place by screws (arrows) at both sides of the housing

tors from the backside, then remove the cluster from the instrument panel.
5 On models equipped automatic transmissions, detach the transmission range indicator.
6 Installation is the reverse of removal.

14 Radio and speakers - removal and installation

⁂ WARNING:

The models covered by this manual are equipped with Supplemental Restraint systems (SRS), more commonly known as airbags. Always disconnect the negative battery cable, then the positive battery cable and wait two minutes before working in the vicinity of the impact sensors, steering column or instrument panel to avoid the possibility of accidental deployment of the airbag, which could cause personal injury (see Section 27). Do not use electrical test equipment on any of the airbag system wiring or tamper with them in any way.

1 Disconnect the negative battery cable, then the positive battery cable and wait two minutes before proceeding any further.

RADIO

▸ Refer to illustration 14.3

2 Remove the radio trim bezel from the center of the instrument panel (see Chapter 11).

2002 and earlier models

3 On 2002 and earlier models, the radio receiver is retained in the instrument panel by special clips. Releasing these clips requires the use of two sets of removal tool, available at most auto parts stores, or two short lengths of coat hanger wire bent into U-shapes. Insert the tools into the holes at the corners of the radio assembly until you feel the internal clips release. With the clips released, push outward simultaneously on both tools and pull the assembly out of the instrument panel, disconnect the antenna and electrical connectors and remove the unit from the vehicle (see illustration).

4 Install the radio by plugging in the electrical connectors, then sliding the radio along the track and into the instrument panel until the clips can be felt snapping into place.

14.3 Insert the tools until they seat, then push outward simultaneously on both tools to release the clips and withdraw the radio from the dash (2002 and earlier models)

CHASSIS ELECTRICAL SYSTEM   12-11

2003 and later models

➡ **Note:** On 2007 and later models, if the radio (also known as the Audio Control Module [ACM]) must be replaced, it will have to be done by a dealer service department or other properly equipped repair facility because module configuration must be programmed into the new ACM unit. Without the correct module configuration, the new ACM will not work.

5 Remove the screws that secure the radio, then carefully slide the radio assembly out far enough to disengage any electrical connectors before sliding the radio all the way out. Installation is the reverse of removal.

SPEAKERS

♦ Refer to illustration 14.7

6 Remove the door trim panel (see Chapter 11).
7 Remove the mounting screws, withdraw the speaker, unplug the electrical connector and remove the speaker from the vehicle (see illustration).
8 Installation is the reverse of removal.

14.7 After removing the door trim panel, the speaker retaining screws (arrows) are easy to reach

15 Antenna - removal and installation

♦ Refer to illustrations 15.3a, 15.3b and 15.3c

➡ **Note:** 2003 and later Expedition/Navigator models have an antenna grid on the left-rear quarter-glass. There is no mechanical antenna. The grid antennas can be tested and repaired (see Section 22).

AUDIO ANTENNA

1 Remove the right front tire and the right front inner fenderwell as described in the fender removal Section in Chapter 11.
2 Working in the inner fenderwell opening, detach the antenna cable from the antenna base.
3 If you're replacing the antenna, use a small wrench to remove the mast. Then pry off the antenna trim cap and remove the screws securing the antenna to the body (see illustrations).
4 If you're replacing the antenna cable, remove the cable retaining clip securing the cable to the inner fender. Then push the cable and grommet assembly into the passenger compartment.
5 Detach the radio and disconnect the antenna lead from the rear side of the radio (see Section 14).
6 Remove any remaining cable retaining clips, then pull the antenna cable out from under the dash to remove it.
7 Installation is the reverse of removal.

15.3a Use a small wrench to remove the antenna mast

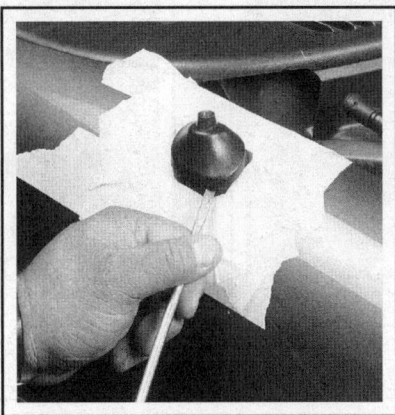

15.3b Cover the fender with a rag or tape before prying off the antenna trim cap to avoid damaging the paint

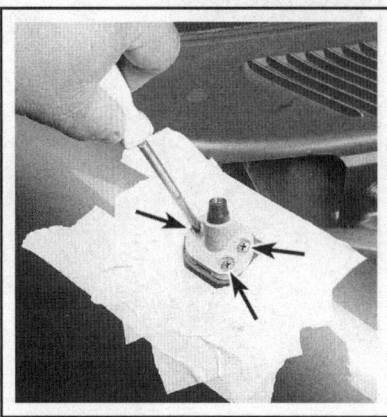

15.3c Remove the retaining screws (arrows) securing the antenna base to the front fender

12-12 CHASSIS ELECTRICAL SYSTEM

SATELLITE RADIO ANTENNA

8 Remove the door-pillar trim panel from the passenger's side of the instrument panel (see Chapter 11).

9 Remove the passenger's side sun visor screws and remove the visor.

10 Disconnect the electrical connector from the satellite radio antenna at the top of the door pillar.

11 Carefully pull down the right corner of the headliner just enough to access the antenna mounting bolt.

※※ CAUTION:

The headliner can be easily damaged if it's pulled down too far.

12 Remove the bolt and detach the antenna from the roof.
13 Installation is the reverse of removal.

16 Headlights - replacement

※※ WARNING:

Halogen gas filled bulbs are under pressure and may shatter if the surface is scratched or the bulb is dropped. Wear eye protection and handle the bulbs carefully, grasping only the base whenever possible. Do not touch the surface of the bulb with your fingers because the oil from your skin could cause it to overheat and fail prematurely. If you do touch the bulb surface, clean it with rubbing alcohol.

HALOGEN HEADLIGHTS

▶ Refer to illustrations 16.2 and 16.8

1 Open the hood.

2 If equipped, push the headlight housing retaining clips rearward, then pull upward on the clips to release the headlight housing (see illustration).

3 On 2003 through 2006 models, loosen, but do not remove, the vertical bolt behind the headlight housing. Remove the two bolts securing the top of the headlight housing to the body.

4 On 2007 and later models, remove the mounting bolts.

5 Pull the headlight housing out and away from the vehicle to access the bulb retaining ring located on the backside.

6 On Navigator models so equipped, remove the headlight bulb cover from the back of the headlight housing.

7 On models so equipped, disconnect the electrical connector from the bulb holder.

8 Rotate the bulb retaining ring counterclockwise as viewed from the rear (see illustration).

9 Withdraw the bulb assembly and retaining ring from the headlight housing.

10 Remove the bulb from the socket assembly by pulling it straight out.

11 Without touching the glass with your bare fingers, insert the new bulb into the socket assembly and then into the headlight housing, install and tighten the retaining ring.

12 Reposition the headlight housing back into place and install the housing retaining clips. Plug in the electrical connector, if equipped. Test headlight operation, then close the hood.

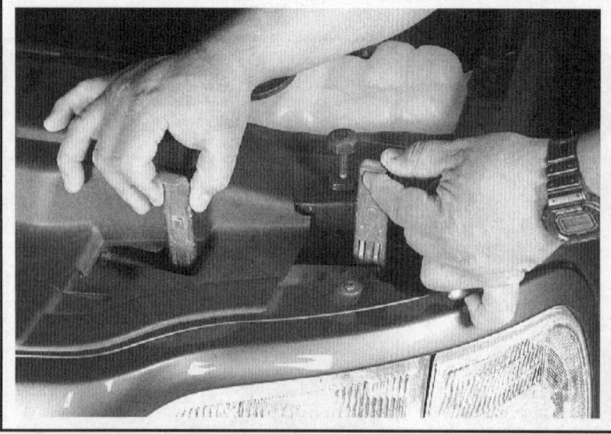

16.2 Disengage the headlight housing retaining clips by pushing the clips rearward - then pull up and out to remove them

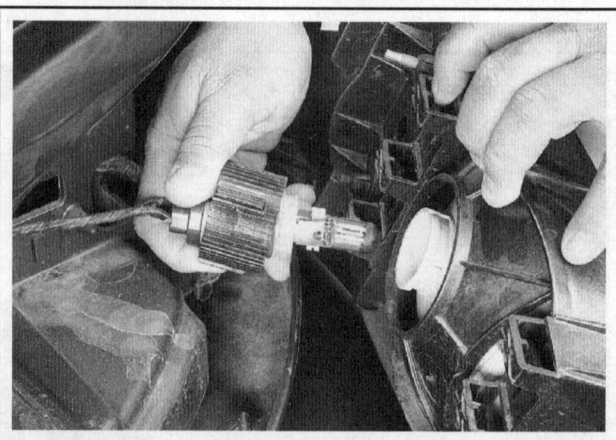

16.8 Rotate the headlight bulb retaining ring counterclockwise and pull the bulb socket assembly out of the housing - when installing the new bulb, don't touch the surface; clean it with rubbing alcohol if you do

CHASSIS ELECTRICAL SYSTEM 12-13

HIGH INTENSITY DISCHARGE (HID) HEADLIGHTS

※ WARNING:

Some models use High Intensity Discharge (HID) bulbs instead of halogen bulbs. These can be identified by the high-voltage warning sticker on the headlight housing. According to the manufacturer, the high voltages produced by this system can be fatal in the event of a shock. Also, the voltage can remain in the circuit even after the headlight switch has been turned to OFF and the ignition key has been removed. Therefore, for your safety, we don't recommend that you try to remove one these headlight housings. Instead, have this service performed by a dealer service department or other qualified repair shop.

17 Headlights - adjustment

▶ Refer to illustrations 17.1 and 17.3

➙Note: The headlights must be aimed correctly. If adjusted incorrectly they could blind the driver of an oncoming vehicle and cause a serious accident or seriously reduce your ability to see the road. The headlights should be checked for proper aim every 12 months and any time a new headlight is installed or front end body work is performed. It should be emphasized that the following procedure is only an interim step which will provide temporary adjustment until the headlights can be adjusted by a properly equipped shop.

1   The headlights have two adjusting screws each, one inboard and one outboard. Both adjusters are accessible from the rear of the headlight assembly and are turned using a small wrench or pliers (see illustration).

➙Note: On Navigator models, turn the horizontal adjustment shafts only (they run parallel with the front edge of the vehicle) to adjust the headlights.

On later Expedition/Navigator models, there is only one adjuster screw, which is for vertical adjustment only.

2   There are several methods of adjusting the headlights. The simplest method requires masking tape, a blank wall and a level floor.

3   Position masking tape vertically on the wall in reference to the vehicle centerline and the centerlines of both headlights (see illustration).

4   Position a horizontal tape line in reference to the centerline of all the headlights.

➙Note: It may be easier to position the tape on the wall with the vehicle parked only a few inches away.

5   Adjustment should be made with the vehicle parked 25 feet from the wall, sitting level, the gas tank half-full and no unusually heavy load in the vehicle.

6   Starting with the low beam adjustment, position the high intensity zone so it is two inches below the horizontal line and two inches to the right of the headlight vertical line. Adjustment is made by turning the top adjusting screw clockwise to raise the beam and counterclockwise to lower the beam. The adjusting screw on the side should be used in the same manner to move the beam left or right.

7   With the high beams on, the high intensity zone should be vertically centered with the exact center just below the horizontal line.

➙Note: It may not be possible to position the headlight aim exactly for both high and low beams. If a compromise must be made, keep in mind that the low beams are the most used and have the greatest effect on safety.

8   Have the headlights adjusted by a dealer service department or service station at the earliest opportunity.

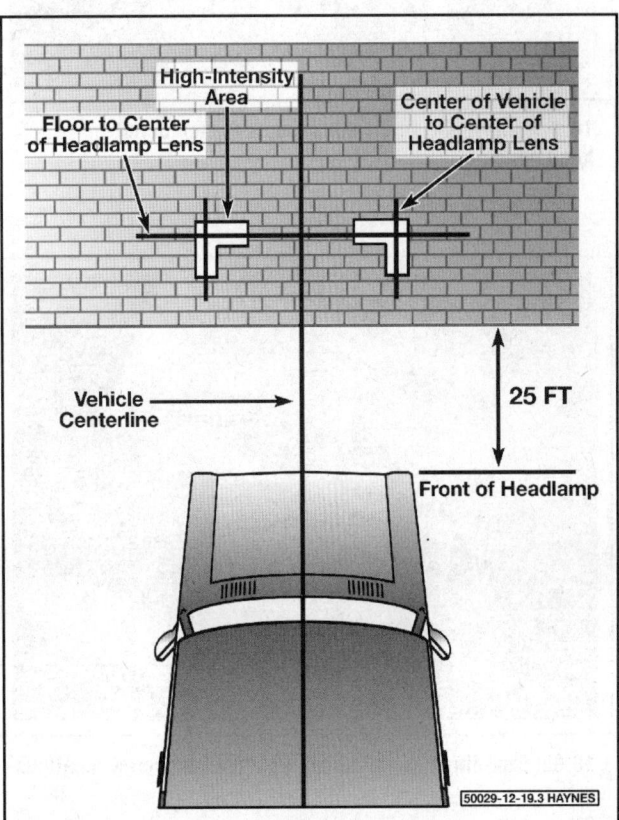

17.3 Headlight adjustment details

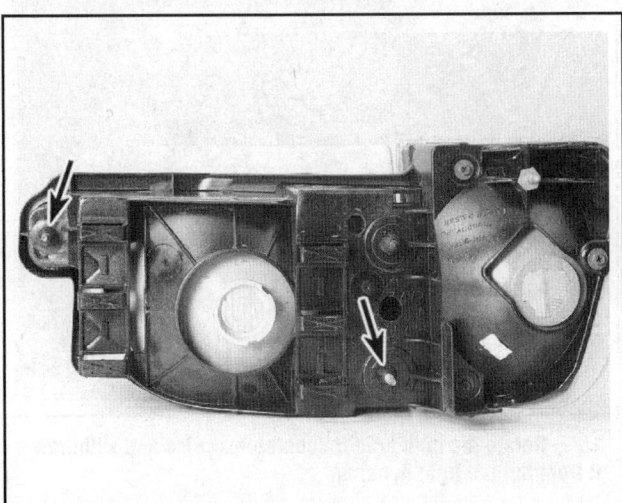

17.1 The headlight adjusting screws (arrows) are located at the rear of the headlight housing

12-14 CHASSIS ELECTRICAL SYSTEM

18 Bulb replacement

FRONT TURN SIGNAL AND SIDE MARKER LIGHTS

▶ Refer to illustrations 18.1

1 Remove the headlight housing (see Section 16), or on some early models, remove the screw at the top of the housing (see illustration).
2 Twist the bulb socket a quarter turn counterclockwise, then remove the bulb assembly from the housing.
3 The defective bulb can then be removed from the socket and replaced.
4 Installation of the headlight housing is the reverse of removal.

REAR TURN SIGNAL, BRAKE, TAIL AND BACK-UP LIGHTS

▶ Refer to illustrations 18.6a, 18.6b and 18.7

5 Open the tailgate or liftgate.
6 Detach the retaining screws securing the rear tail light housing, then pull the tail light assembly outward to access the tail light bulbs (see illustrations).
7 Twist the bulb socket a quarter turn counterclockwise, then remove the bulb assembly from the housing (see illustration).
8 The defective bulb can then be removed from the socket and replaced.
9 Installation of the tail light housing is the reverse of removal.

18.1 Remove the screw that secures the top of the lamp housing

18.6a Taillight housing bolt locations for the F-150/250 (arrows)

18.6b Expedition/Navigator tail light housing screw locations

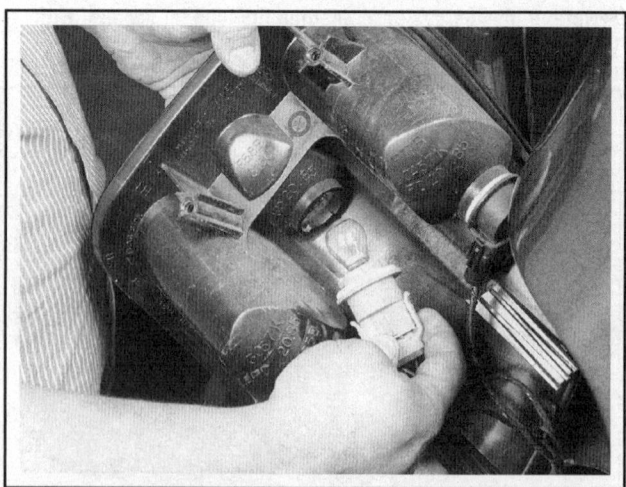

18.7 Rotate the bulb holder counterclockwise and withdraw it from the tail light housing

CHASSIS ELECTRICAL SYSTEM 12-15

LICENSE PLATE LIGHT

▶ **Refer to illustration 18.10**

10 The license plate light bulbs can be accessed from the rear of the bumper (see illustration). Simply twist the bulb socket a quarter turn counterclockwise, then remove the bulb assembly from the housing.

11 The defective bulb can then be pulled straight out of the socket and replaced.

12 Installation is the reverse of removal.

HIGH-MOUNTED BRAKE LIGHT

▶ **Refer to illustrations 18.14a and 18.14b**

13 On F-150/250 models the high mounted brake light bulbs can be accessed from the truck bed. On Expedition/Navigator models the high mounted brake light bulbs can be accessed from the inside of the liftgate.

14 On F-150/250 models, simply remove the lens retaining screws and pull the lamp assembly outward to access the bulbs (see illustrations). On Expedition/Navigator models, open the liftgate. Remove two acorn nuts from the inside of the liftgate, then pull the lamp assembly outward to access the bulbs.

15 On all models, twist the bulb socket a quarter turn counterclockwise, then remove the bulb assembly from the housing. On later models, the bulbs are contained in a housing. Depress the release tabs to remove the housing from the lens, then remove the bulbs.

16 The defective bulb can then be pulled straight out of the socket and replaced.

DOME LIGHT

▶ **Refer to illustration 18.17**

17 Using a small screwdriver, remove the lens and replace the bulb (see illustration).

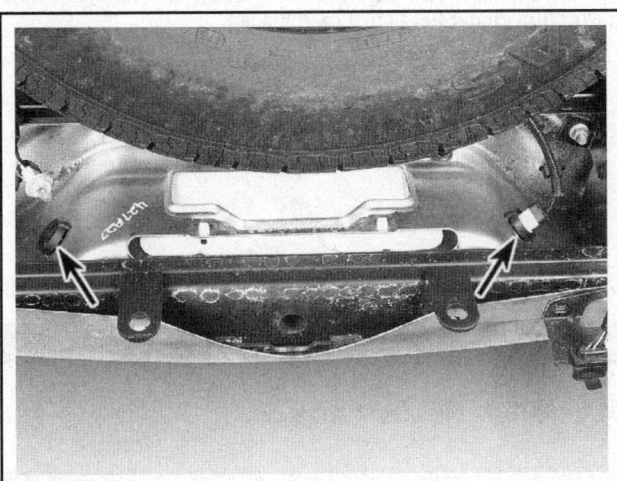

18.10 The license plate light bulbs (arrows) can be accessed from the rear of the bumper assembly

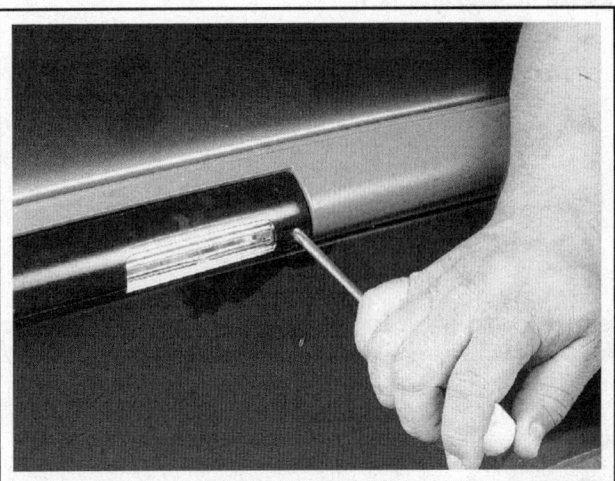

18.14a Remove the screws and detach the lamp assembly to access the high mounted brake light bulbs (F-150/250)

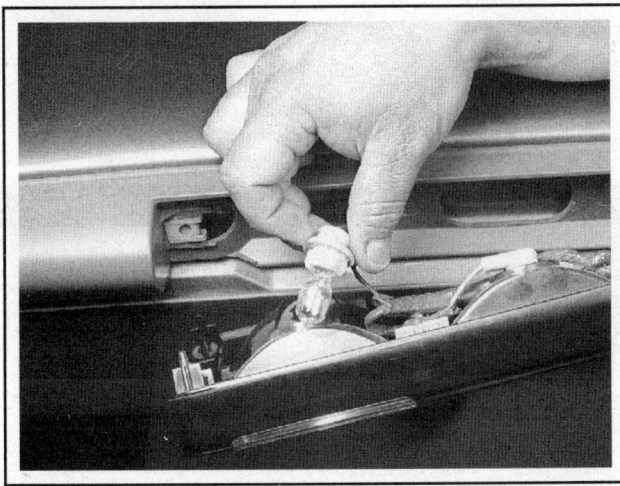

18.14b Twist the bulb holder counterclockwise to remove the high mounted brake light bulbs

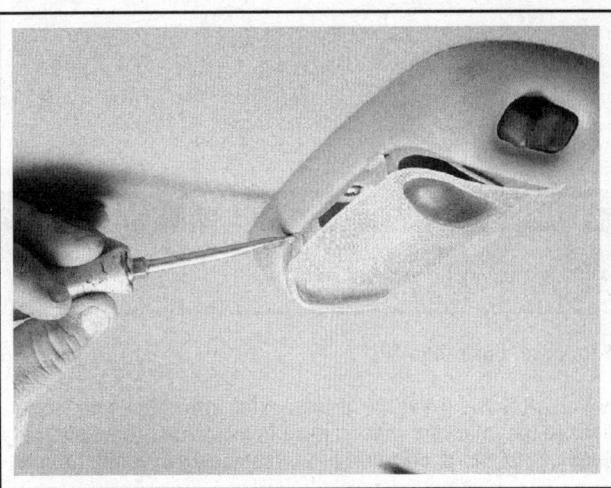

18.17 Use a small screwdriver to pry out the interior domelight lens

12-16 CHASSIS ELECTRICAL SYSTEM

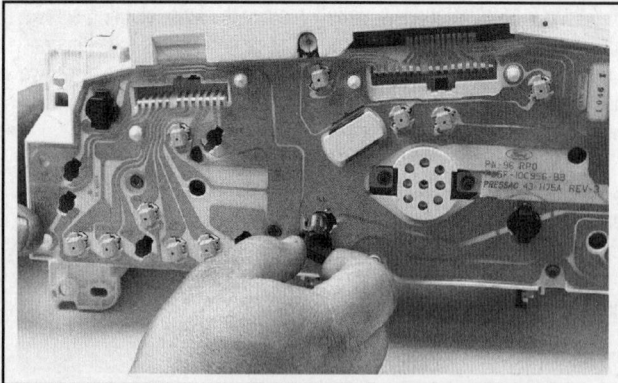

18.18 To remove an instrument cluster bulb, depress the bulb and rotate it counterclockwise

INSTRUMENT CLUSTER ILLUMINATION

▸ **Refer to illustration 18.18**

18 To gain access to the instrument cluster illumination lights, the instrument cluster will have to be removed (see Section 13). The bulbs can then be removed and replaced from the rear of the cluster (see illustration). On 2007 and later Expedition/Navigator models, the cluster illumination is by LED lights, which are not replaceable.

SIDEVIEW MIRROR TURN SIGNAL LIGHT

19 Remove the sideview mirror glass (see Chapter 11).

2006 and earlier models

20 Slide the light assembly outwards while releasing the three retaining tabs on the inside of the mirror and remove the light.
21 Pull the bulb holders out of the light.
22 Installation is the reverse of removal.

2007 and later models

23 Disconnect the electrical connector from the turn signal, then remove the mounting bolt and light assembly from the mirror.

➡**Note: On 2007 and later models, the side mirror turn signal light uses LEDs, which are an integral part of the light housing and are not separately serviceable. If faulty, the entire sideview mirror turn signal light must be replaced.**

24 Installation is the reverse of removal.

19 Daytime Running Lights (DRL) - general information

The Daytime Running Lights (DRL) system illuminates the headlights whenever the engine is running. The only exception is with the engine running and the parking brake engaged. Once the parking brake is released, the lights will remain on as long as the ignition switch is on, even if the parking brake is later applied.

The DRL system supplies reduced power to the headlights so they won't be too bright for daytime use, while prolonging headlight life.

20 Wiper motor - check and replacement

WIPER MOTOR CIRCUIT CHECK

▸ **Refer to illustration 20.2**

➡**Note: Refer to the wiring diagrams for wire colors and locations in the following checks. Keep in mind that power wires are generally larger in diameter and brighter colors, where ground wires are usually smaller in diameter and darker colors. When checking for voltage, probe a grounded 12-volt test light to each terminal at a connector until it lights; this verifies voltage (power) at the terminal.**

1 If the wipers work slowly, make sure the battery is in good condition and has a strong charge (see Chapter 1). If the battery is in good condition, remove the wiper motor (see below) and operate the wiper

CHASSIS ELECTRICAL SYSTEM   12-17

20.2 Windshield wiper motor terminal guide details

1. Park switch terminal
2. Ignition terminal
3. Park switch ground terminal
4. Low speed terminal
5. High speed terminal

20.7 Use a small screwdriver to pry off the cover (arrow) - remove the nut, then pull straight up to remove the wiper arm assembly

arms by hand. Check for binding linkage and pivots. Lubricate or repair the linkage or pivots as necessary. Reinstall the wiper motor. If the wipers still operate slowly, check for loose or corroded connections, especially the ground connection. If all connections look OK, replace the motor.

2 If the wipers fail to operate when activated, check the fuse. If the fuse is OK, connect a jumper wire between the wiper motor and ground, then retest. If the motor works now, repair the ground connection. If the motor still doesn't work, turn the wiper switch to the HI position and check for voltage at the motor (see illustration). If there's voltage at the motor, remove the motor and check it off the vehicle with fused jumper wires from the battery. If the motor now works, check for binding linkage (see Step 1 above). If the motor still doesn't work, replace it. If there's no voltage at the motor, check for voltage at the wiper control module. If there's voltage at the wiper control module and no voltage at the at the wiper motor, check the switch for continuity (see Section 8). If the switch is OK, the wiper control module is probably bad.

3 If the interval (delay) function is inoperative, check the continuity of all the wiring between the switch and wiper control module. If the wiring is OK, check the resistance of the delay control knob of the multi-function switch (see Section 8). If the delay control knob is within the specified resistance, replace the wiper control module.

4 If the wipers stop at the position they're in when the switch is turned off (fail to park), check for voltage at the park feed wire of the wiper motor connector when the wiper switch is OFF but the ignition is ON. If no voltage is present, check for an open circuit between the wiper motor and the fuse panel.

5 If the wipers won't shut off unless the ignition is OFF, disconnect the wiring from the wiper control switch. If the wipers stop, replace the switch. If the wipers keep running, there's a defective limit switch in the motor; replace the motor.

6 If the wipers won't retract below the hood line, check for mechanical obstructions in the wiper linkage or on the vehicle's body which

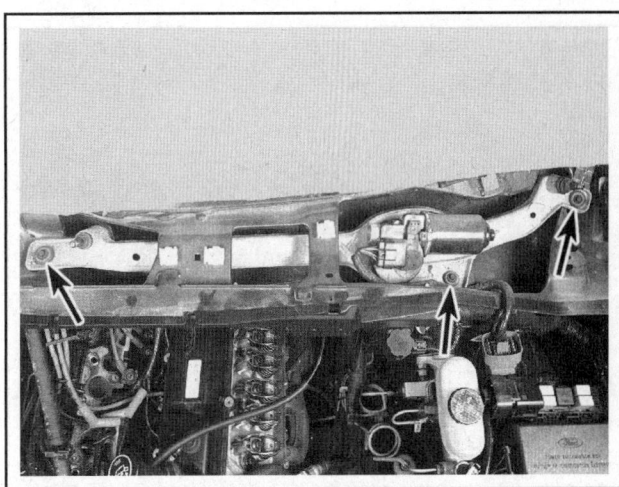

20.10 Detach the wiper mounting bracket retaining bolts (arrows)

would prevent the wipers from parking. If there are no obstructions, check the wiring between the switch and motor for continuity. If the wiring is OK, replace the wiper motor.

WIPER MOTOR REPLACEMENT

▶ **Refer to illustrations 20.7 and 20.10**

7 Remove the windshield wiper arms (see illustration).
8 Remove the cowl cover (see Chapter 11).
9 Disconnect the electrical connector from the wiper motor.
10 Remove the wiper motor mounting bracket retaining bolts (see illustration).
11 Turn the mounting bracket over to access the motor retaining bolts. Remove the bolt securing the wiper arm linkage to the backside of the motor, then remove the motor from the mounting bracket.
12 Installation is the reverse of removal.

12-18 CHASSIS ELECTRICAL SYSTEM

21 Horn - check and replacement

CHECK

▶ **Refer to illustration 21.3**

➡ **Note: Check the fuses before beginning electrical diagnosis.**

1 Disconnect the electrical connector from the horn.

2 To test the horn, connect battery voltage to the two terminals with a pair of jumper wires. If the horn doesn't sound, replace it.

3 If the horn does sound, check for voltage at the terminal when the horn button is depressed (see illustration). If there's voltage at the terminal, check for a bad ground at the horn.

4 If there's no voltage at the horn, check the relay (see Section 6). Note that most horn relays are either the four-terminal or externally grounded three-terminal type.

5 If the relay is OK, check for voltage to the relay power and control circuits. If either of the circuits is not receiving voltage, inspect the wiring between the relay and the fuse panel.

6 If both relay circuits are receiving voltage, depress the horn button and check the circuit from the relay to the horn button for continuity to ground. If there's no continuity, check the circuit for an open. If there's no open circuit, replace the horn button.

7 If there's continuity to ground through the horn button, check for an open or short in the circuit from the relay to the horn.

REPLACEMENT

▶ **Refer to illustration 21.9**

8 The horn(s) are mounted behind the radiator grille. On 2008 and later Expedition/Navigator models, remove the left headlight housing to access the horns.

9 Disconnect the electrical connectors and remove the bracket bolt (see illustration) to remove the horn.

10 Installation is the reverse of removal.

21.3 Connect a voltmeter to the horn wire and ground - test for voltage while the switch is depressed

21.9 Disconnect the electrical connector, remove the bolt (arrow) and detach the horn(s)

22 Rear window defogger - check and repair

➡ **Note: This procedure can also be used to test and repair the grid-type antennas on 2003 and later models.**

1 The rear window defogger consists of a number of horizontal elements baked onto the glass surface.

2 Small breaks in the element can be repaired without removing the rear window.

CHECK

▶ **Refer to illustrations 22.4, 22.5 and 22.7**

3 Turn the ignition switch and defogger system switches to the ON position.

4 When measuring voltage during the next two tests, wrap a piece of aluminum foil around the tip of the voltmeter negative probe and press the foil against the heating element with your finger (see illustration).

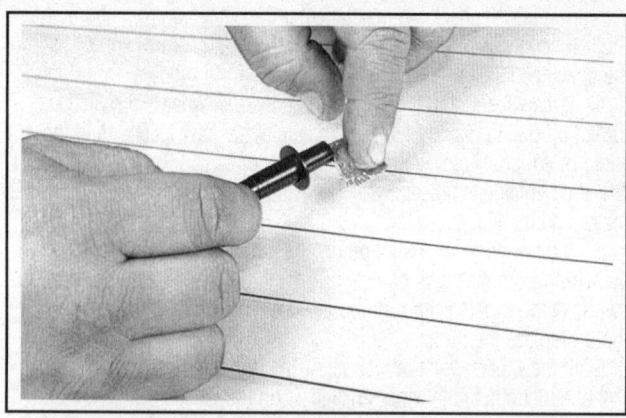

22.4 When measuring the voltage at the rear window defogger grid, wrap a piece of aluminum foil around the negative probe of the voltmeter and press the foil against the wire with your finger

CHASSIS ELECTRICAL SYSTEM   12-19

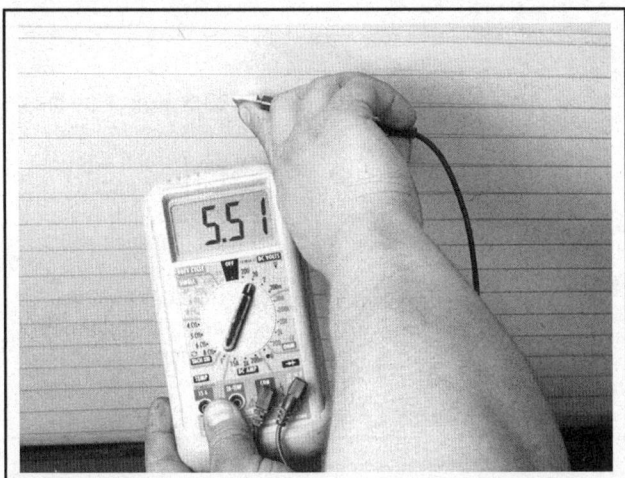

22.5 To determine if a wire has broken, check the voltage at the center of each wire. If the voltage is 6-volts, the wire is unbroken; if the voltage is 12-volts, the wire is broken between the center of the wire and the positive end; if the voltage is 0-volts, the wire is broken between the center of the wire and ground

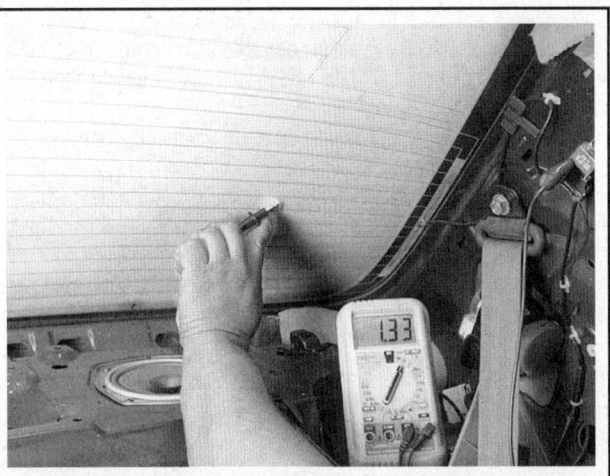

22.7 To find the break, place the voltmeter positive lead against the defogger positive terminal, place the voltmeter negative lead with the foil strip against the heat wire at the positive terminal end and slide it toward the negative terminal end - the point at which the voltmeter deflects from zero to several volts is the point at which the wire is broken

5 Check the voltage at the center of each heating element (see illustration). If the voltage is 6-volts, the element is okay (there is no break). If the voltage is 12-volts, the element is broken between the center of the element and the positive end. If the voltage is 0-volts the element is broken between the center of the element and ground.

6 Connect the negative lead to a good body ground. The reading should stay the same.

7 To find the break, place the voltmeter positive lead against the defogger positive terminal. Place the voltmeter negative lead with the foil strip against the heating element at the positive terminal end and slide it toward the negative terminal end. The point at which the voltmeter deflects from zero to several volts is the point at which the heating element is broken (see illustration).

REPAIR

♦ Refer to illustration 22.13

8 Repair the break in the element using a repair kit specifically recommended for this purpose, such as Dupont paste No. 4817 (or equivalent). Included in this kit is plastic conductive epoxy.

9 Prior to repairing a break, turn off the system and allow it to cool off for a few minutes.

10 Lightly buff the element area with fine steel wool, then clean it thoroughly with rubbing alcohol.

11 Use masking tape to mask off the area being repaired.

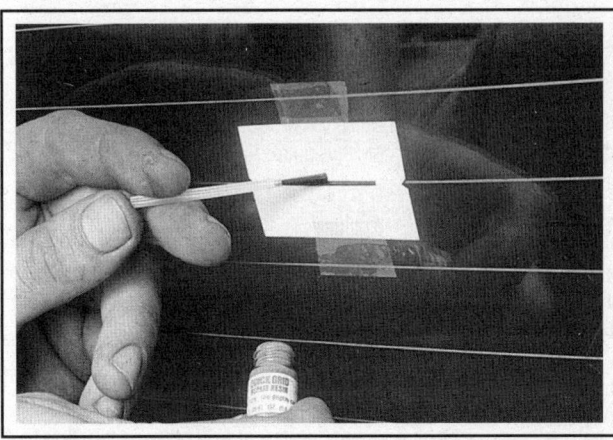

22.13 To use a defogger repair kit, apply masking tape to the inside of the window at the damaged area, then brush on the special conductive coating

12 Thoroughly mix the epoxy, following the instructions provided with the repair kit.

13 Apply the epoxy material to the slit in the masking tape, overlapping the undamaged area about 3/4-inch on either end (see illustration).

14 Allow the repair to cure for 24 hours before removing the tape and using the system.

23 Cruise control system - description and check

♦ Refer to illustration 23.5

1 The cruise control system maintains vehicle speed with a servo motor located in the engine compartment on the driver's side fenderwell, which is connected to the throttle linkage by a cable. The system consists of the servo motor, brake switch, control switches, speed sensors and relays. Some features of the system require special testers and diagnostic procedures which are beyond the scope of this manual. Listed below are some general procedures that may be used to locate common problems.

➡**Note: 2005 and later models are equipped with electronic throttle bodies. The PCM controls the cruise control functions without a cable. Diagnosing the cruise control system requires a scan tool.**

2 Locate and check the fuse (see Section 3).

3 Have an assistant operate the brake lights while you check their operation (voltage from the brake light switch deactivates the cruise control).

4 If the brake lights don't come on or don't shut off, correct the problem and retest the cruise control.

5 Check the control cable between the cruise control servo/amplifier and the throttle linkage and replace as necessary (see illustration).

6 The cruise control system uses a speed sensing device. The speed sensor is located in the transmission. To test the speed sensor, see Chapter 6, Section 4.

7 Test drive the vehicle to determine if the cruise control is now working. If it isn't, take it to a dealer service department or an automotive electrical specialist for further diagnosis.

23.5 Make sure the cruise control and accelerator linkage mounted on the throttle body are not damaged and that they operate smoothly together when the throttle is opened

24 Power window system - description and check

▶ Refer to illustration 24.12

1 The power window system operates electric motors, mounted in the doors, which lower and raise the windows. The system consists of the control switches, the motors, regulators, glass mechanisms and associated wiring.

2 The power windows can be lowered and raised from the master control switch by the driver or by remote switches located at the individual windows. Each window has a separate motor which is reversible. The position of the control switch determines the polarity and therefore the direction of operation.

3 The circuit is protected by a fuse and a circuit breaker. Each motor is also equipped with an internal circuit breaker, this prevents one stuck window from disabling the whole system.

4 The power window system will only operate when the ignition switch is ON. In addition, many models have a window lockout switch at the master control switch which, when activated, disables the switches at the rear windows and, sometimes, the switch at the passenger's window also. Always check these items before troubleshooting a window problem.

5 These procedures are general in nature, so if you can't find the problem using them, take the vehicle to a dealer service department or other properly equipped repair facility.

6 If the power windows won't operate, always check the fuse and circuit breaker first.

7 If only the rear windows are inoperative, or if the windows only operate from the master control switch, check the rear window lockout switch for continuity in the unlocked position. Replace it if it doesn't have continuity.

8 Check the wiring between the switches and fuse panel for continuity. Repair the wiring, if necessary.

9 If only one window is inoperative from the master control switch, try the other control switch at the window.

➡ **Note: This doesn't apply to the driver's door window.**

10 If the same window works from one switch, but not the other, check the switch for continuity.

24.12 If no voltage is found at the motor with the switch depressed - check for voltage at the switch

11 If the switch tests OK, check for a short or open in the circuit between the affected switch and the window motor.

12 If one window is inoperative from both switches, remove the trim panel from the affected door and check for voltage at the switch (see illustration) and at the motor while the switch is operated.

13 If voltage is reaching the motor, disconnect the glass from the regulator (see Chapter 11). Move the window up and down by hand while checking for binding and damage. Also check for binding and damage to the regulator. If the regulator is not damaged and the window moves up and down smoothly, replace the motor. If there's binding or damage, lubricate, repair or replace parts, as necessary.

14 If voltage isn't reaching the motor, check the wiring in the circuit for continuity between the switches and motors. You'll need to consult the wiring diagram for the vehicle. If the circuit is equipped with a relay, check that the relay is grounded properly and receiving voltage.

15 Test the windows after you are done to confirm proper repairs.

CHASSIS ELECTRICAL SYSTEM 12-21

25 Power door lock and keyless entry system - description and check

▶ Refer to illustration 25.9

1 The power door lock system operates the door lock actuators mounted in each door. The system consists of the switches, actuators, and associated wiring. Diagnosis can usually be limited to simple checks of the wiring connections and actuators for minor faults which can be easily repaired.

2 Power door lock systems are operated by bi-directional solenoids located in the doors. The lock switches have two operating positions: Lock and Unlock. On later models with keyless entry the switches activate a module which in turn connects voltage to the door lock solenoids. Depending on which way the switch is activated, it reverses polarity, allowing the two sides of the circuit to be used alternately as the feed (positive) and ground side. On earlier models with out keyless entry the switches directly activate the door lock motors.

3 If you are unable to locate the trouble using the following general steps, consult your a dealer service department.

4 Always check the circuit protection first. On these models the battery voltage passes through the 20 amp circuit breaker located in the passenger compartment fuse block.

5 Operate the door lock switches in both directions (Lock and Unlock) with the engine off. Listen for the faint click of the door lock solenoid (motor) or relay operating.

6 If there's no click, check for voltage at the switches. If no voltage is present, check the wiring between the fuse block and the switches for shorts and opens.

7 If voltage is present but no click is heard, test the switch for continuity. Replace it if there's no continuity in both switch positions.

8 If the switch has continuity but the solenoid doesn't click, check the wiring between the switch and solenoid for continuity. Repair the wiring if there's not continuity.

9 If all but one lock solenoids operate, remove the trim panel from the affected door (see Chapter 11) and check for voltage at the solenoid while the lock switch is operated (see illustration). One of the wires should have voltage in the Lock position; the other should have voltage in the unlock position.

10 If the inoperative solenoid is receiving voltage, replace the solenoid.

25.9 Check for voltage at the lock solenoid while the lock switch is operated

➡ Note: It's common for wires to break in the portion of the harness between the body and door (opening and closing the door fatigues and eventually breaks the wires).

KEYLESS ENTRY SYSTEM

▶ Refer to illustrations 25.13 and 25.14

11 The keyless entry system consists of a remote control transmitter that sends a coded infrared signal to a receiver which then operates the door lock system.

12 Replace the transmitter batteries when the red LED light on the side of the case doesn't light when the button is pushed.

13 Use a small screwdriver to carefully separate the case halves (see illustration).

14 Replace the two 3-volt 2016 lithium batteries (see illustration).

15 Snap the case halves together.

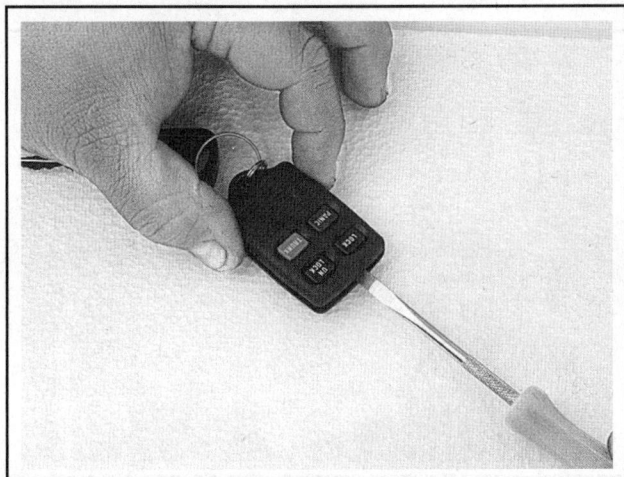

25.13 Use a small screwdriver to separate the transmitter halves

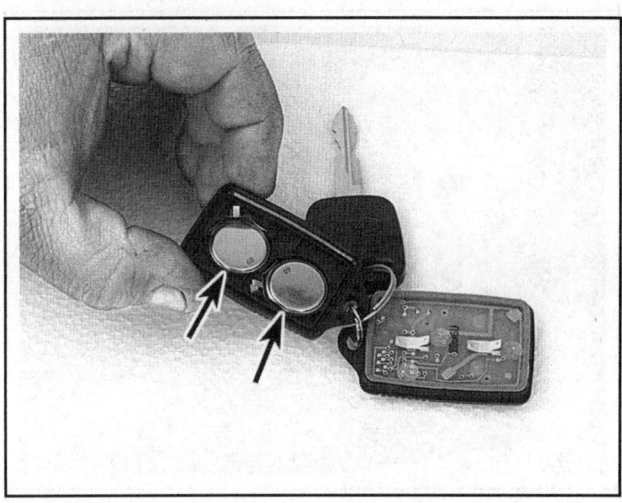

25.14 Replace the lithium batteries (arrows)

26 Electric side view mirrors - description and check

1 Most electric rear view mirrors use two motors to move the glass; one for up and down adjustments and one for left-right adjustments.

2 The control switch has a selector portion which sends voltage to the left or right side mirror. With the ignition ON but the engine OFF, roll down the windows and operate the mirror control switch through all functions (left-right and up-down) for both the left and right side mirrors.

3 Listen carefully for the sound of the electric motors running in the mirrors.

4 If the motors can be heard but the mirror glass doesn't move, there's probably a problem with the drive mechanism inside the mirror. Remove and disassemble the mirror to locate the problem.

5 If the mirrors don't operate and no sound comes from the mirrors, check the fuse (see Section 3).

6 If the fuse is OK, remove the mirror control switch from its mounting without disconnecting the wires attached to it. Turn the ignition ON and check for voltage at the switch. There should be voltage at one terminal. If there's no voltage at the switch, check for an open or short in the wiring between the fuse panel and the switch.

7 If there's voltage at the switch, disconnect it. Check the switch for continuity in all its operating positions. If the switch does not have continuity, replace it.

8 Re-connect the switch. Locate the wire going from the switch to ground. Leaving the switch connected, connect a jumper wire between this wire and ground. If the mirror works normally with this wire in place, repair the faulty ground connection.

9 If the mirror still doesn't work, remove the mirror and check the wires at the mirror for voltage. Check with ignition ON and the mirror selector switch on the appropriate side. Operate the mirror switch in all its positions. There should be voltage at one of the switch-to-mirror wires in each switch position (except the neutral "off" position).

10 If there's not voltage in each switch position, check the wiring between the mirror and control switch for opens and shorts.

11 If there's voltage, remove the mirror and test it off the vehicle with jumper wires. Replace the mirror if it fails this test.

27 Airbag - general information

▶ Refer to illustrations 27.1a and 27.1b

All models are equipped with a Supplemental Restraint System (SRS), more commonly known as an airbag. This system is designed to protect the driver, and the front seat passenger, from serious injury in the event of a head-on or frontal collision. It consists of an airbag module in the center of the steering wheel and the right side of the instrument panel, two crash sensors mounted at the front of the vehicle and a diagnostic module which also contains a safing sensor is located inside the passenger compartment (see illustrations).

AIRBAG MODULE

Steering wheel-mounted

The airbag inflator module contains a housing incorporating the cushion (airbag) and inflator unit, mounted in the center of the steering wheel The inflator assembly is mounted on the back of the housing over a hole through which gas is expelled, inflating the bag almost instantaneously when an electrical signal is sent from the system. A coil assembly on the steering column under the module carries this signal to the module.

27.1a The front crash sensors are mounted behind the front grille

27.1b The airbag diagnostic module is mounted behind the right kick panel

CHASSIS ELECTRICAL SYSTEM 12-23

This coil assembly can transmit an electrical signal regardless of steering wheel position. The igniter in the air bag converts the electrical signal to heat and ignites the sodium azide/copper oxide powder, producing nitrogen gas, which inflates the bag.

Instrument panel-mounted

The airbag is mounted above the glove compartment and designated by the letters SRS (Supplemental Restraint System). It consists of an inflator containing an igniter, a bag assembly, a reaction housing and a trim cover.

The air bag is considerably larger that the steering wheel-mounted unit (8 cu ft vs. 2.3 cu ft) and is supported by the steel reaction housing. The trim cover is textured and painted to match the instrument panel and has a molded seam which splits when the bag inflates. As with the steering-wheel-mounted air bag, the igniter electrical signal converts to heat, converting sodium azide/iron oxide powder to nitrogen gas, inflating the bag.

Side-impact airbags

On later models, side-impact protection for the driver and front seat passenger is provided by side-impact airbags, which are located in the top/outboard side of the front seats. The impact sensors for the side-impact airbags are generally mounted in the front doors. On later Expedition/Navigator models, an optional rear side-curtain airbag system provides side-impact protection for rear seat passengers. The curtain airbags are located behind the headliner, above the doors/windows. They are designed to deploy downward to provide a cushion between the rear passengers and the vehicle sides.

SENSORS

The system has three sensors: two forward crash sensors at the front of the vehicle and a safing sensor mounted inside the airbag diagnostic module which is located behind the right kick panel.

The forward and passenger compartment sensors are basically pressure sensitive switches that complete an electrical circuit during an impact of sufficient G force. The electrical signal from these sensors is sent to the electronic diagnostic monitor which then completes the circuit and inflates the airbag(s).

ELECTRONIC DIAGNOSTIC MONITOR

The electronic diagnostic monitor supplies the current to the airbag system in the event of the collision, even if battery power is cut off. It checks this system every time the vehicle is started, causing the "AIR BAG" light to go on then off, if the system is operating properly. If there is a fault in the system, the light will go on and stay on, flash, or the dash will make a beeping sound. If this happens, the vehicle should be taken to your dealer immediately for service.

DISABLING THE SYSTEM

Whenever working in the vicinity of the steering wheel, steering column or near other components of the airbag system, the system should be disarmed. To do this, perform the following steps:

a) *Turn the ignition switch to Off.*
b) *Detach the cable from the negative battery terminal then detach the positive cable. Wait 2 minutes for the electronic module backup power supply to be depleted.*

ENABLING THE SYSTEM

a) *Turn the ignition switch to the Off position.*
b) *Connect the positive battery cable first, then connect the negative cable.*

12-24 CHASSIS ELECTRICAL SYSTEM

28 Wiring diagrams - general information

Since it isn't possible to include all wiring diagrams for every year covered by this manual, the following diagrams are those that are typical and most commonly needed.

Prior to troubleshooting any circuits, check the fuse and circuit breakers (if equipped) to make sure they're in good condition. Make sure the battery is properly charged and check the cable connections (see Chapter 1).

When checking a circuit, make sure that all connectors are clean, with no broken or loose terminals. When unplugging a connector, do not pull on the wires. Pull only on the connector housings themselves.

CHASSIS ELECTRICAL SYSTEM 12-25

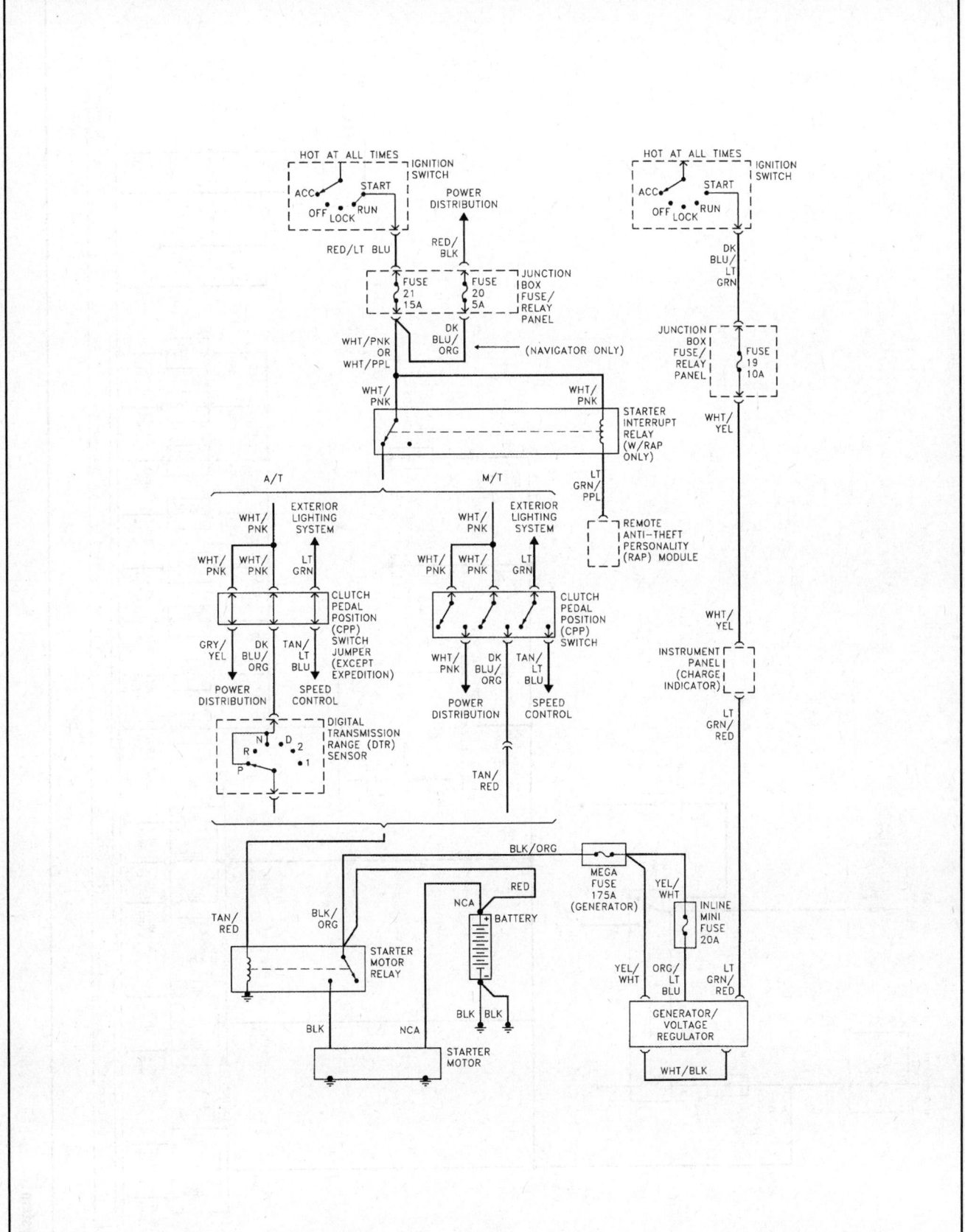

Starting and charging system

12-26 CHASSIS ELECTRICAL SYSTEM

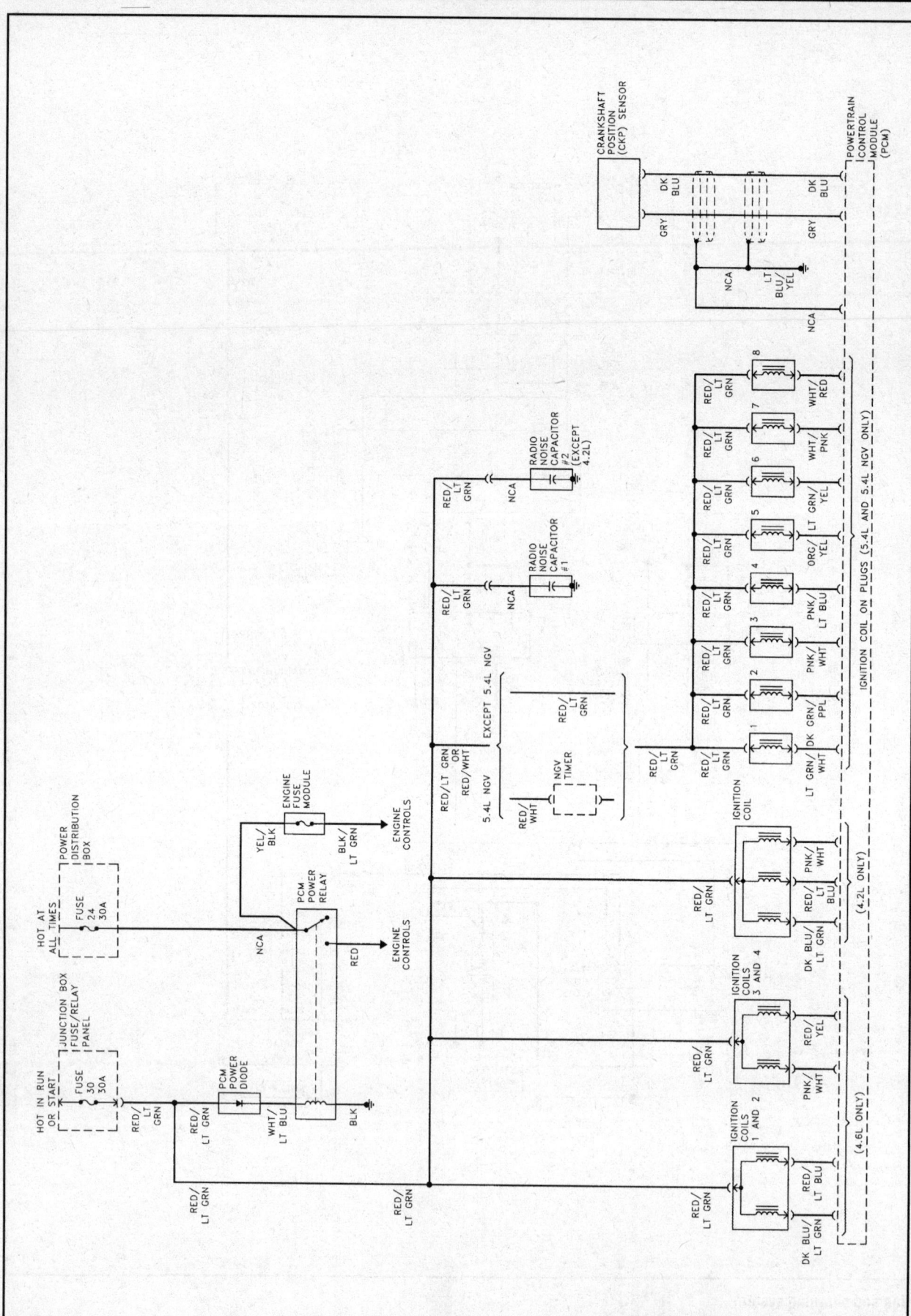

Ignition system

CHASSIS ELECTRICAL SYSTEM 12-27

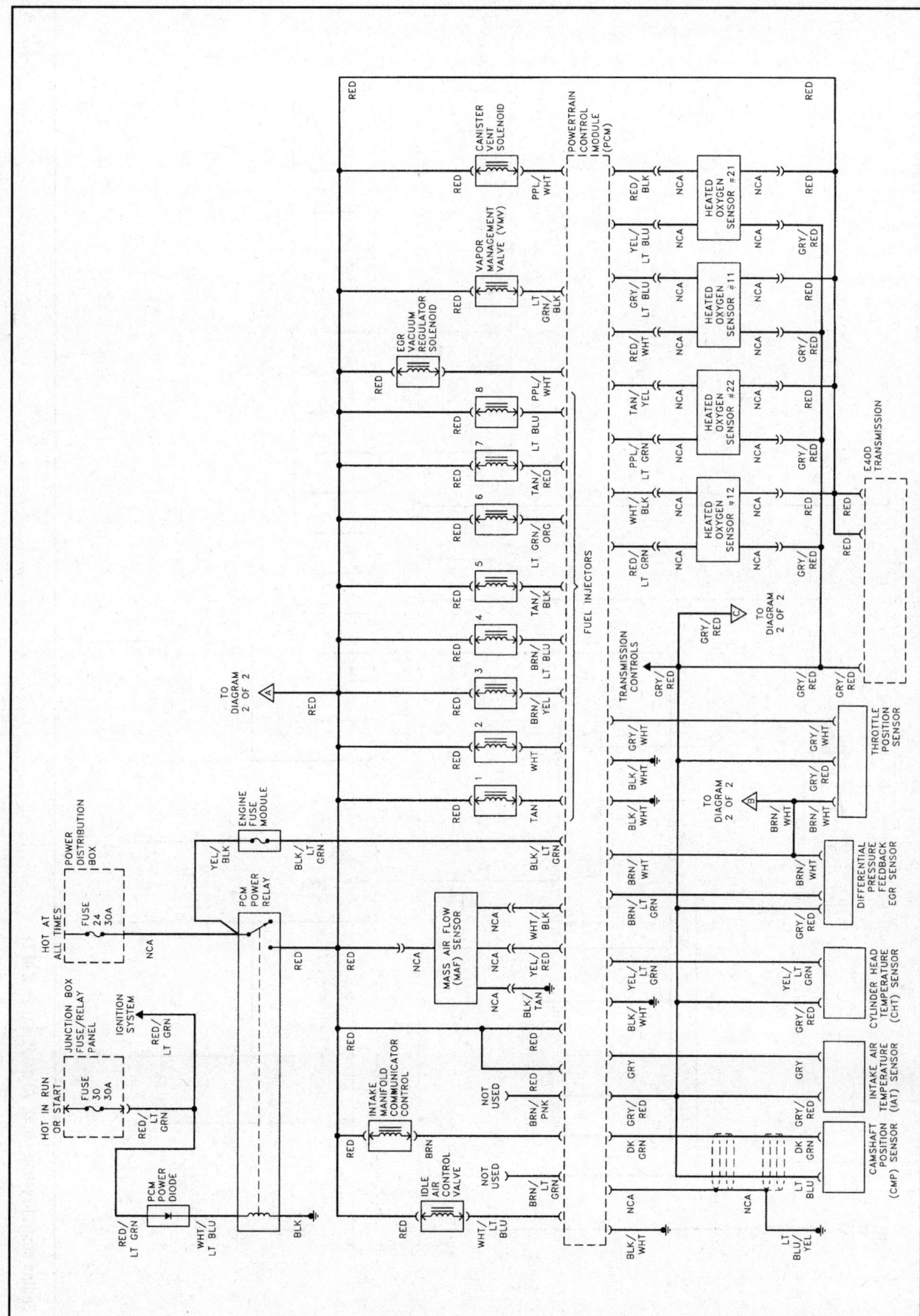

Engine control system - 5.4L V8 models (part 1 of 2)

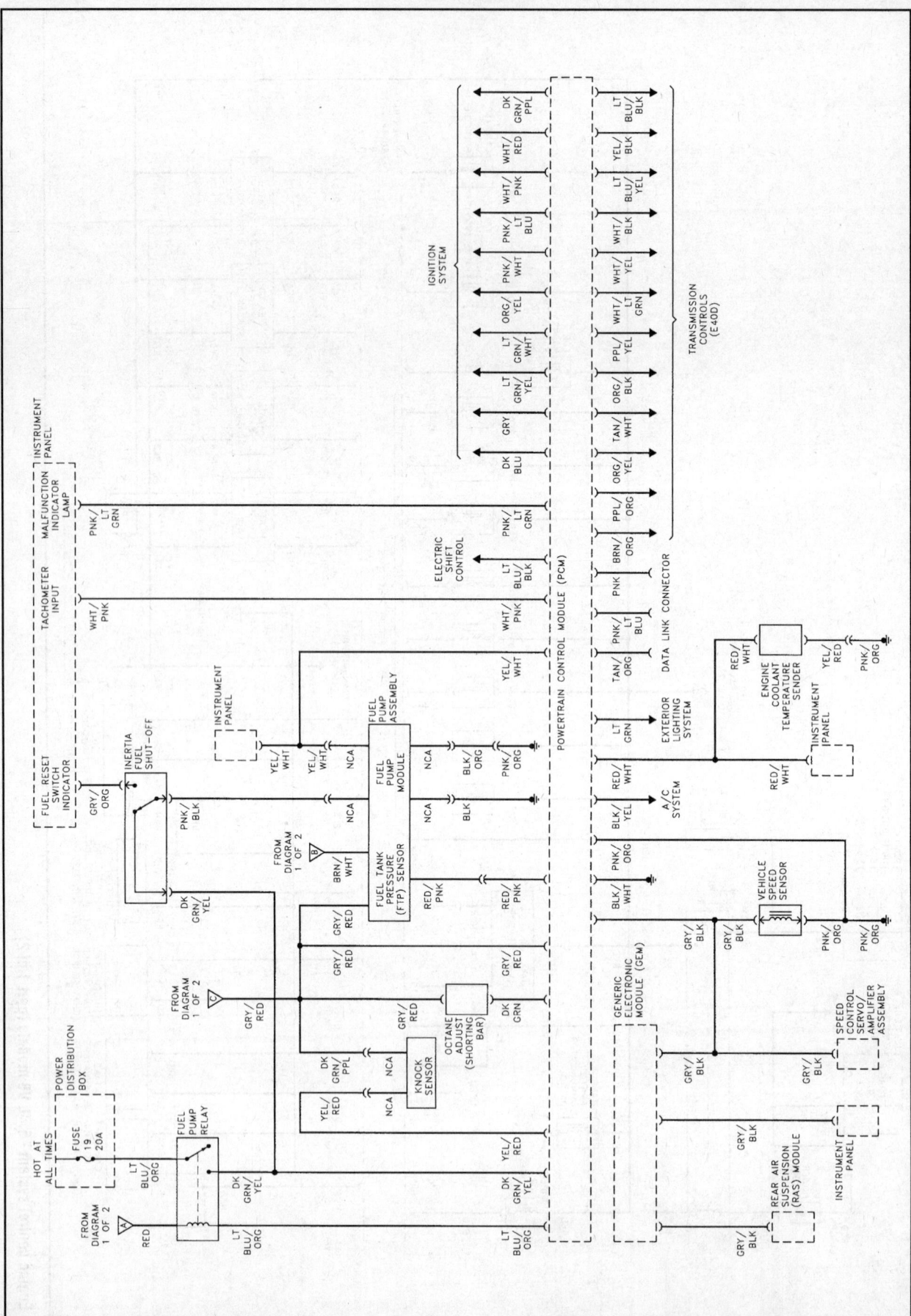

Engine control system - 5.4L V8 models (part 2 of 2)

CHASSIS ELECTRICAL SYSTEM 12-29

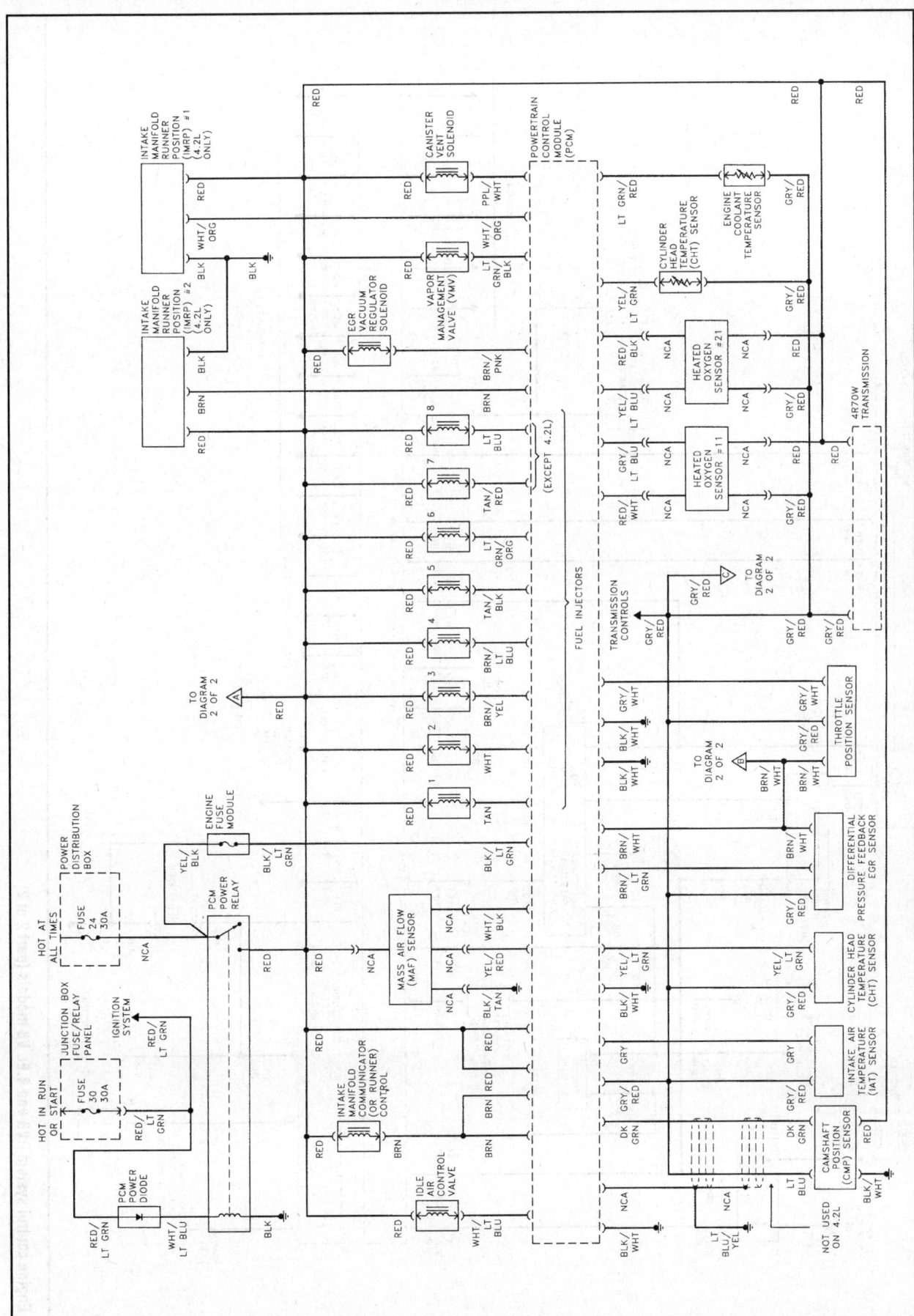

Engine control system - V6 and 4.6L V8 models (part 1 of 2)

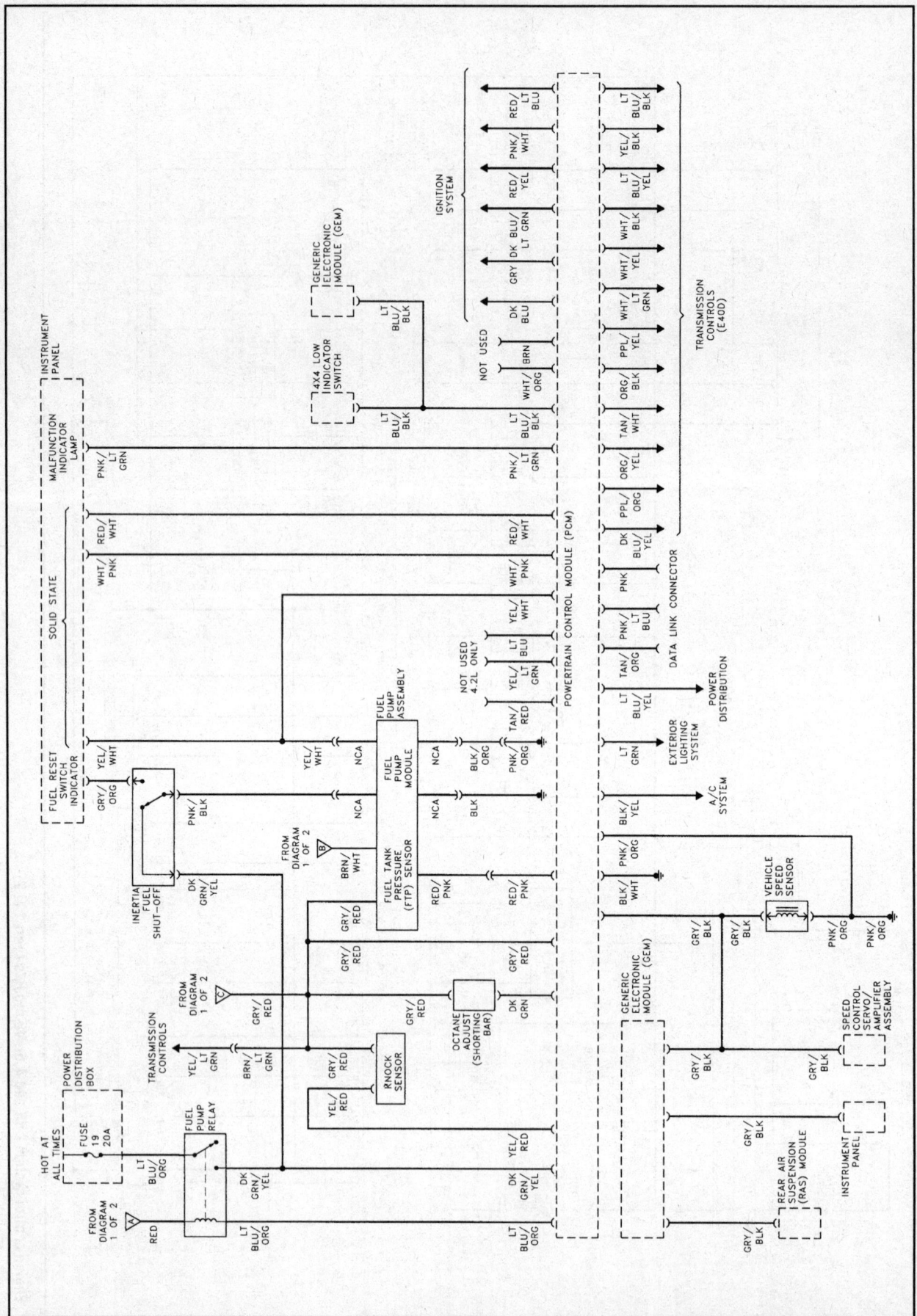

Engine control system - V6 and 4.6L V8 models (part 2 of 2)

CHASSIS ELECTRICAL SYSTEM 12-31

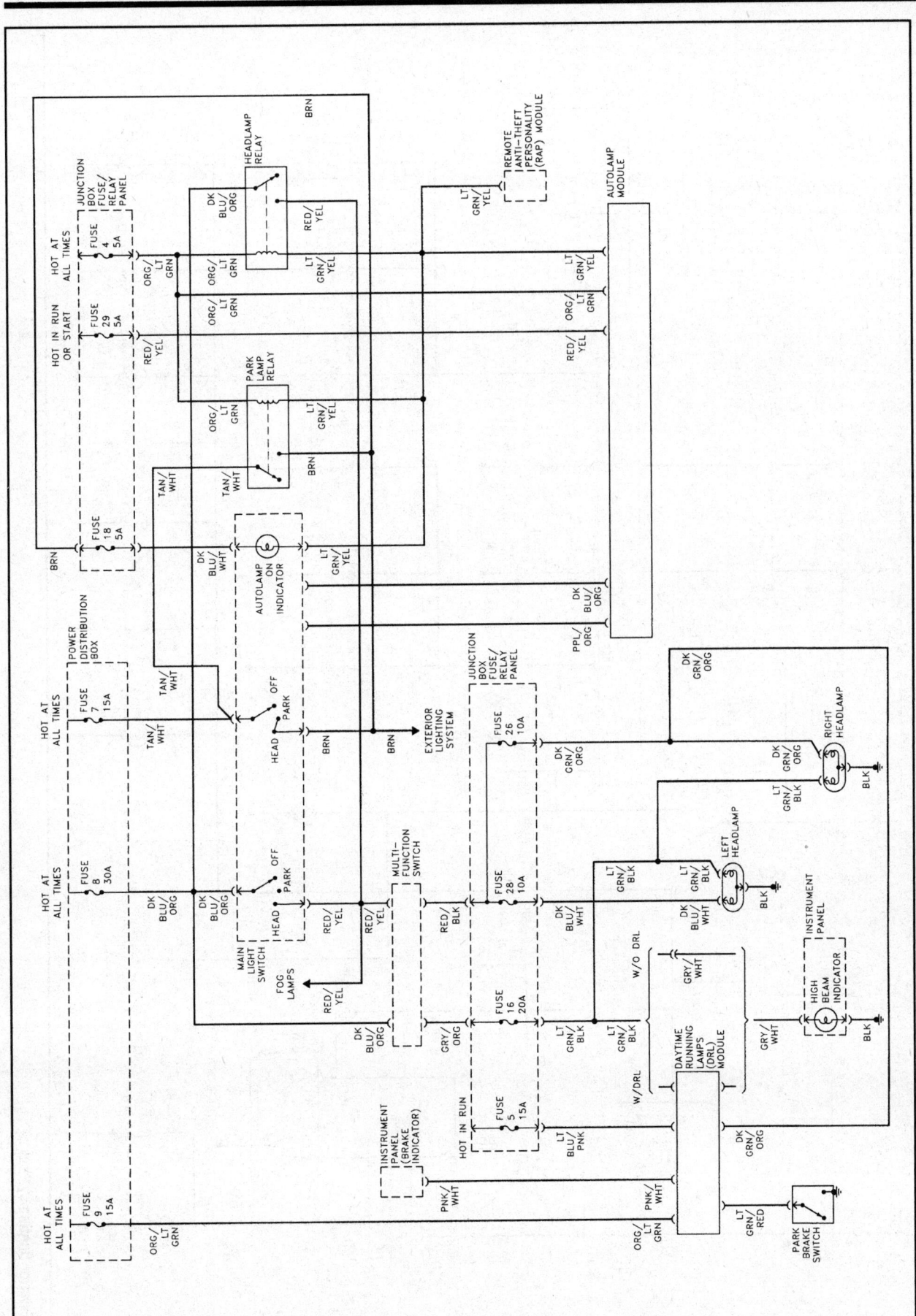

Headlight system

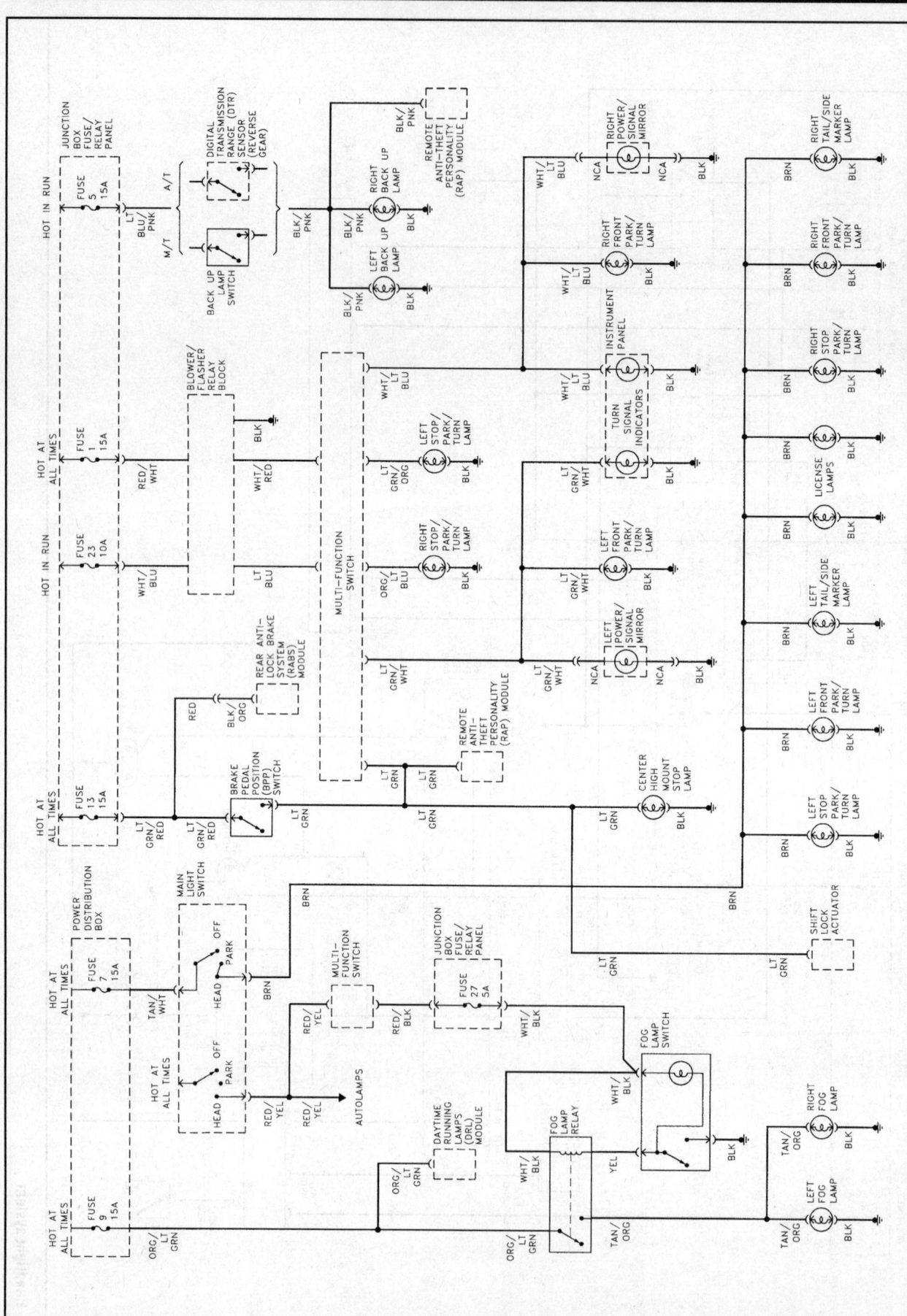

Exterior lighting system

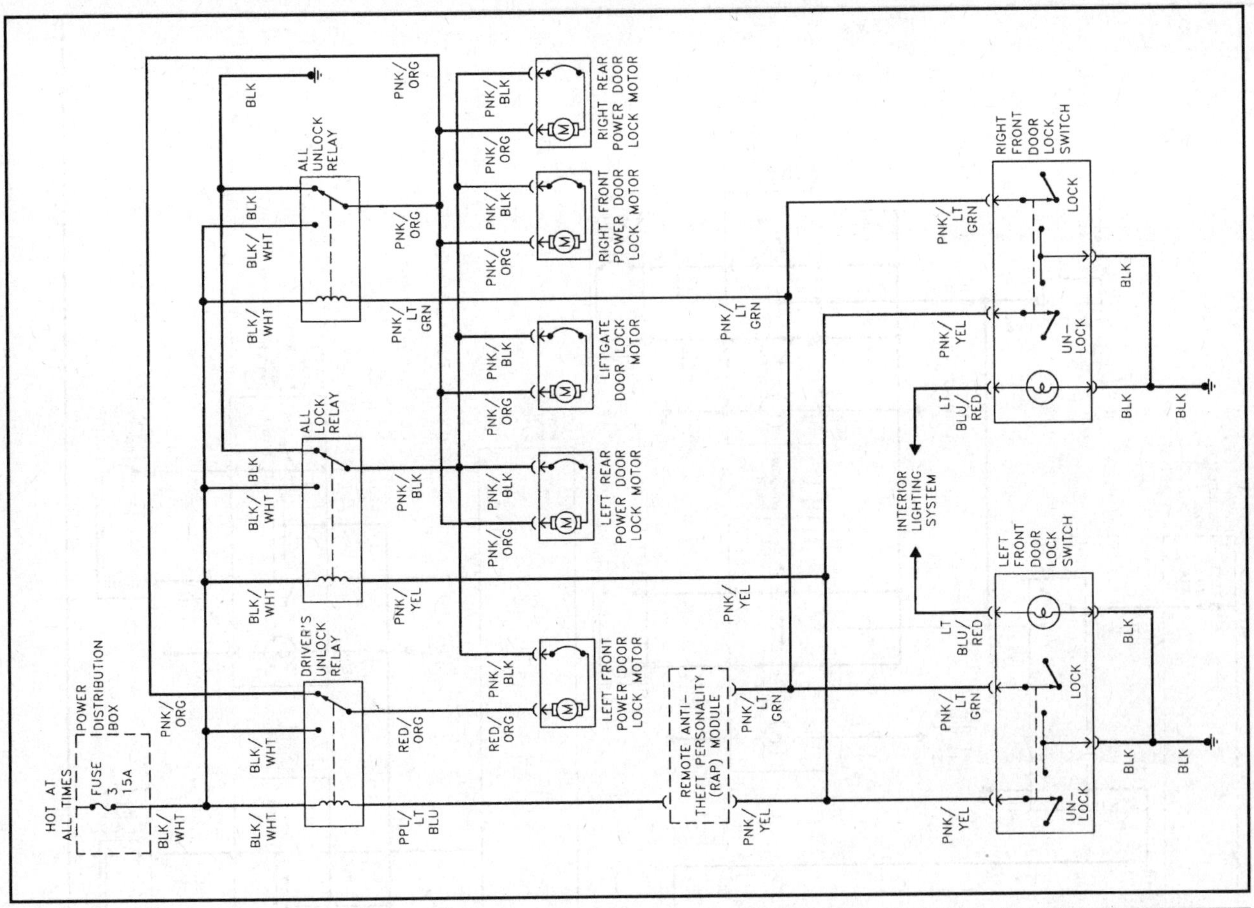

CHASSIS ELECTRICAL SYSTEM 12-33

Power door lock system - with Remote Anti-theft Personality module

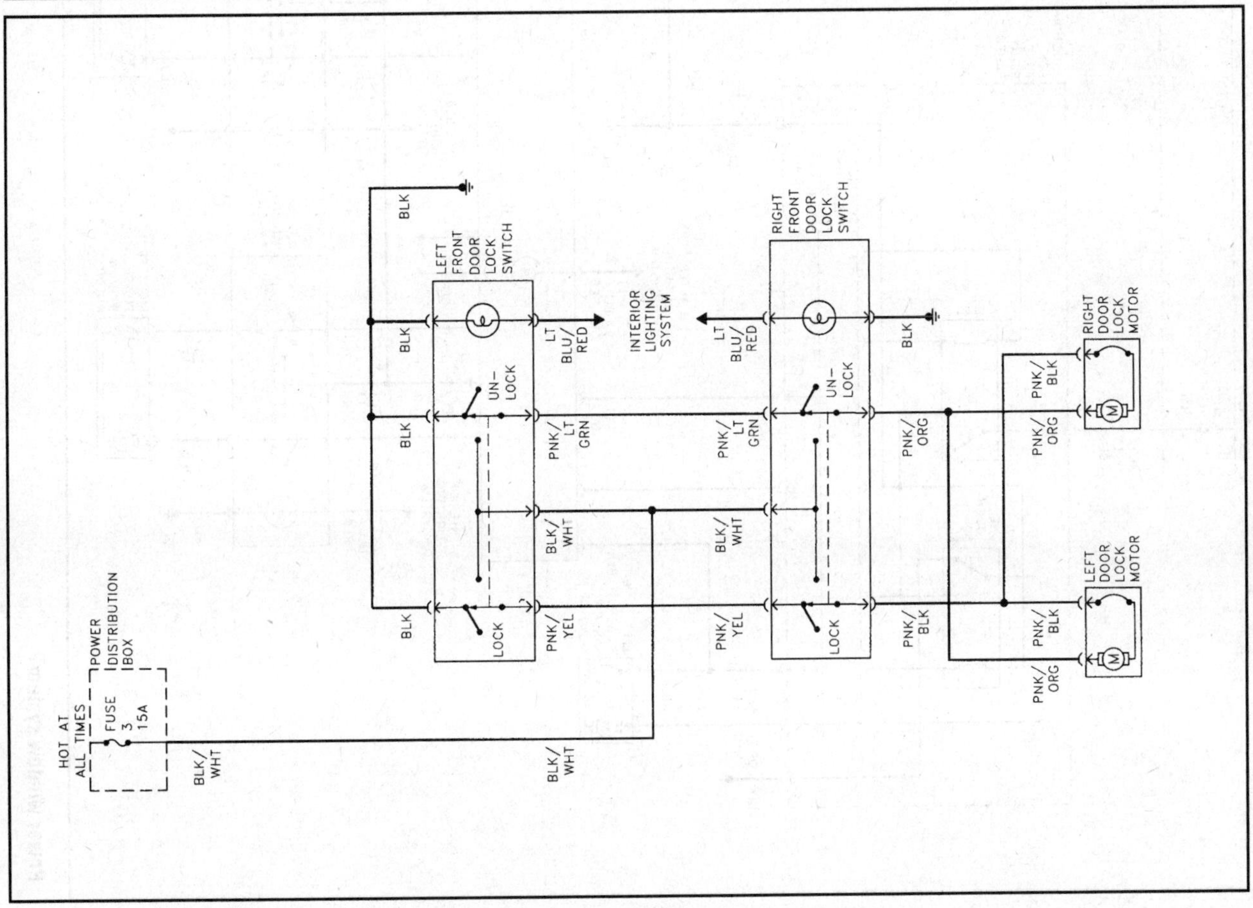

Power door lock system - without Remote Anti-theft Personality module

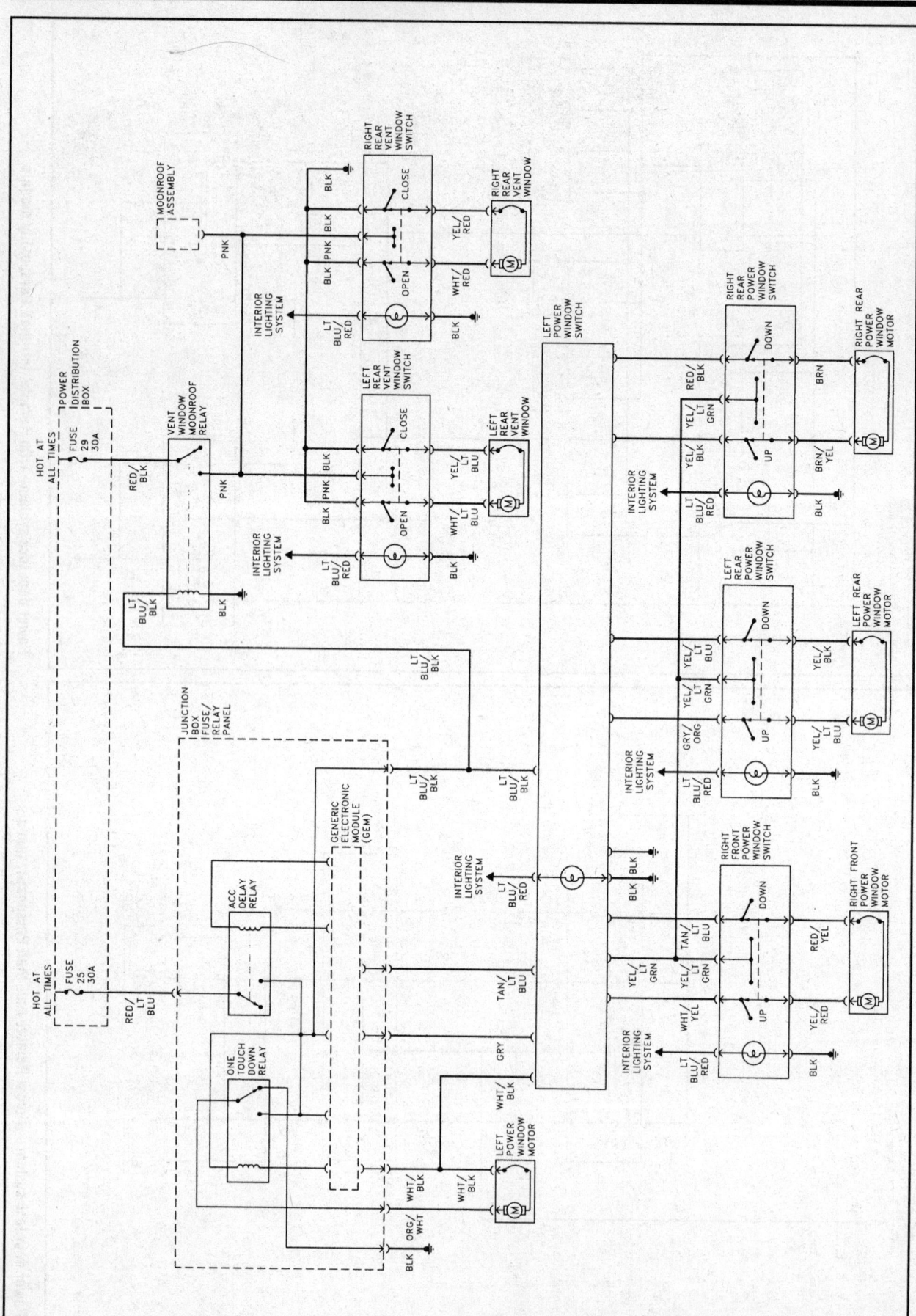

Power window system

CHASSIS ELECTRICAL SYSTEM 12-35

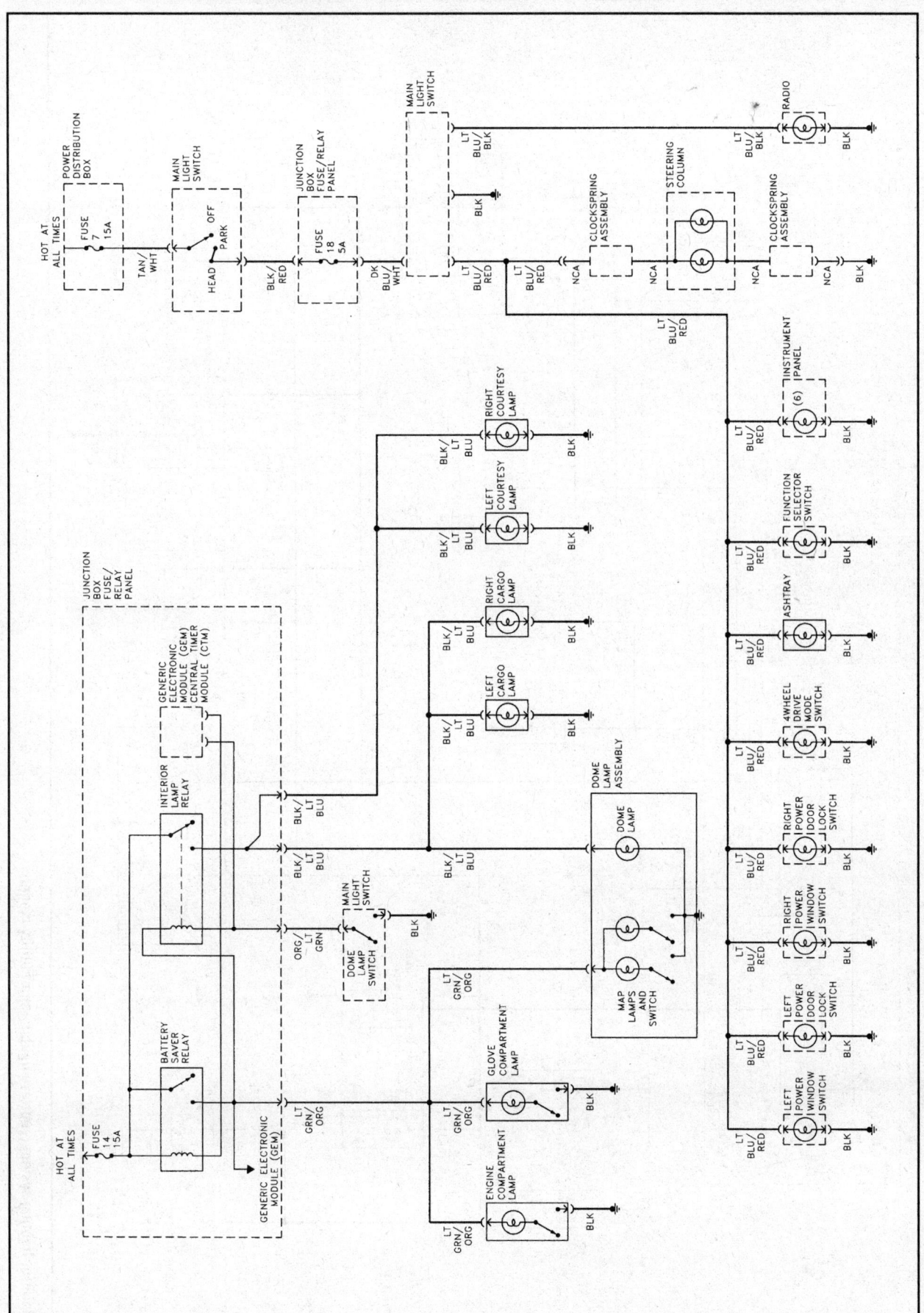

Interior lighting system - pick-up models

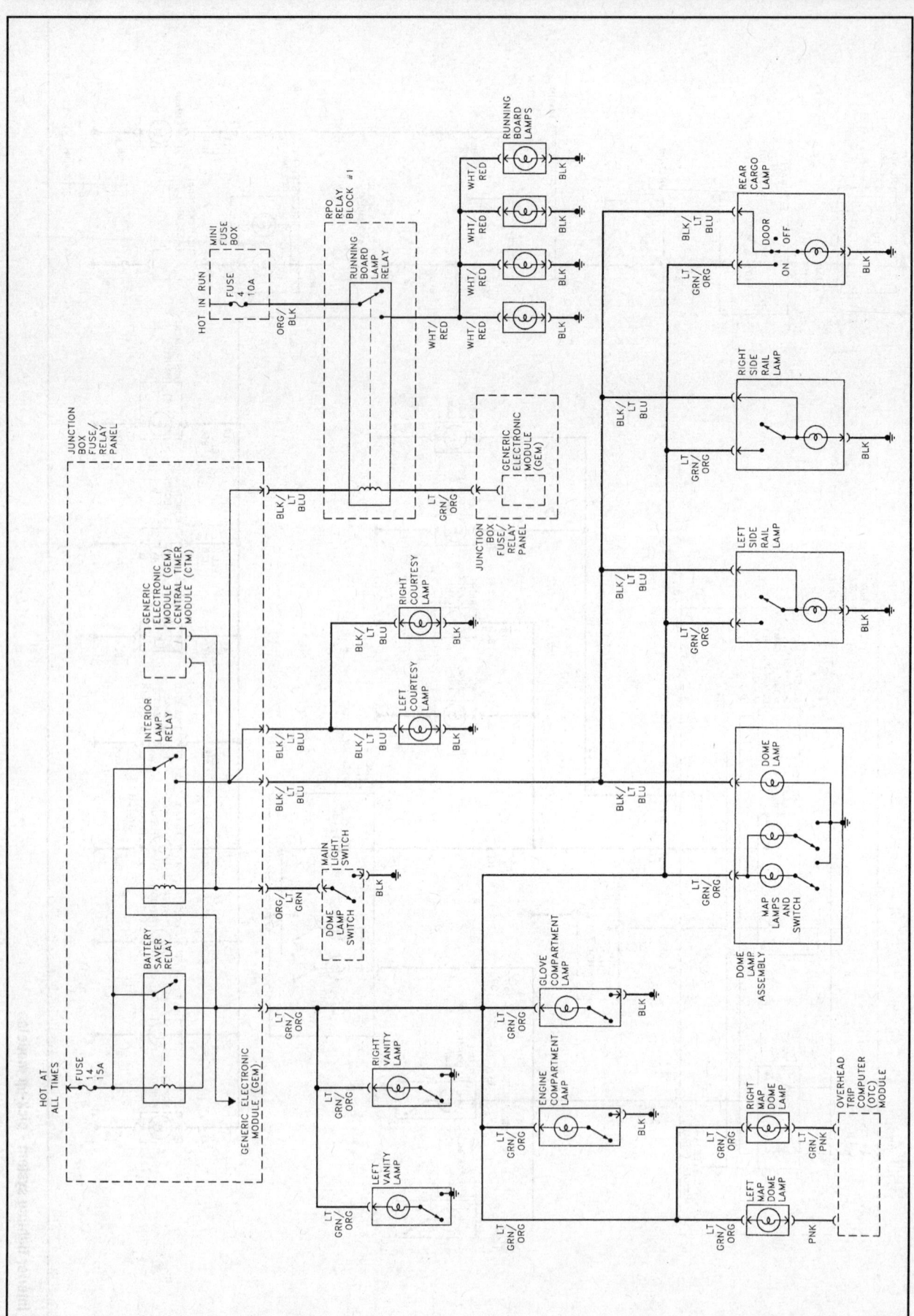

Interior lighting system (courtesy lights) - sport utility models

CHASSIS ELECTRICAL SYSTEM 12-37

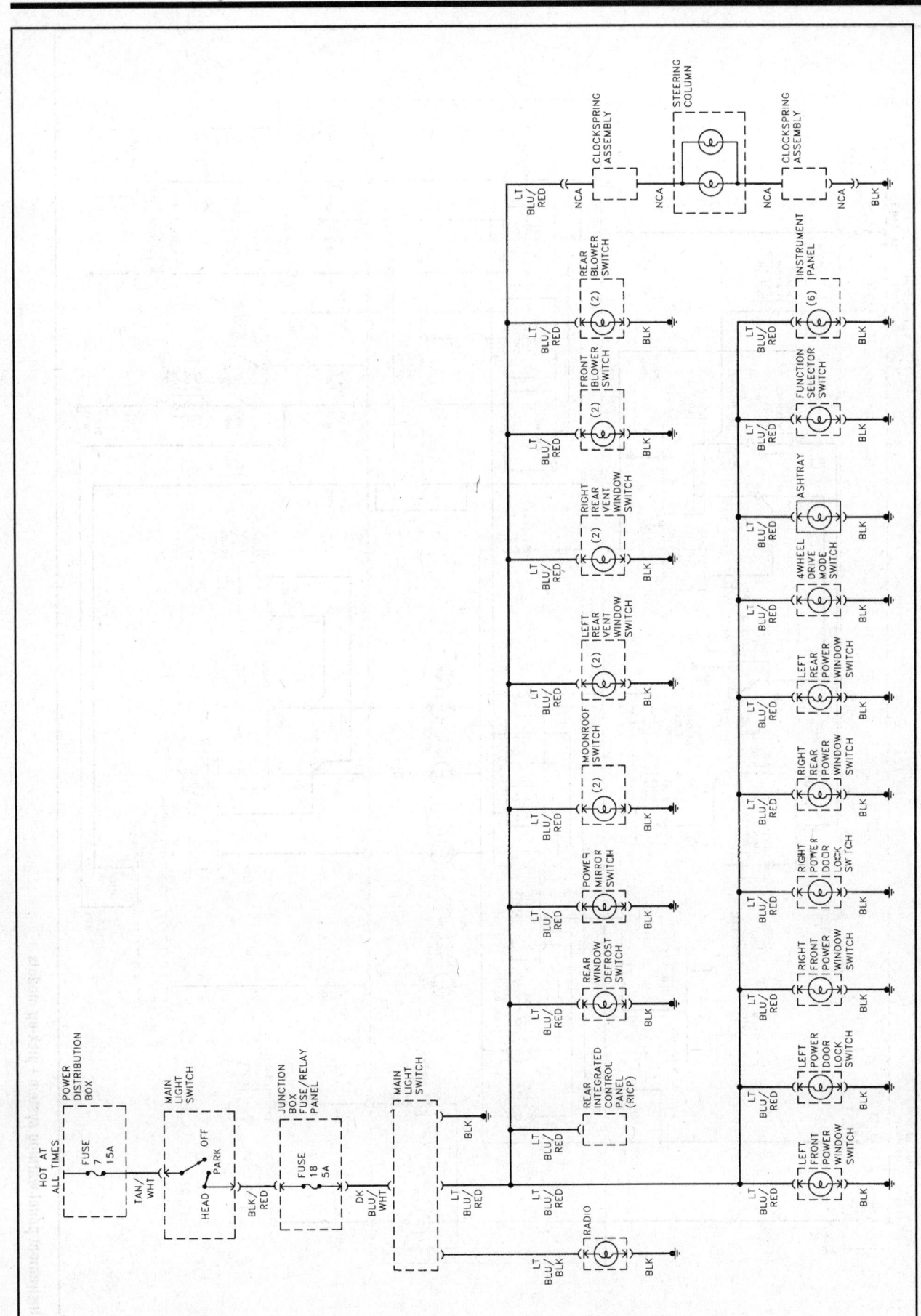

Interior lighting system (interior lights) - sport utility models

12-38 CHASSIS ELECTRICAL SYSTEM

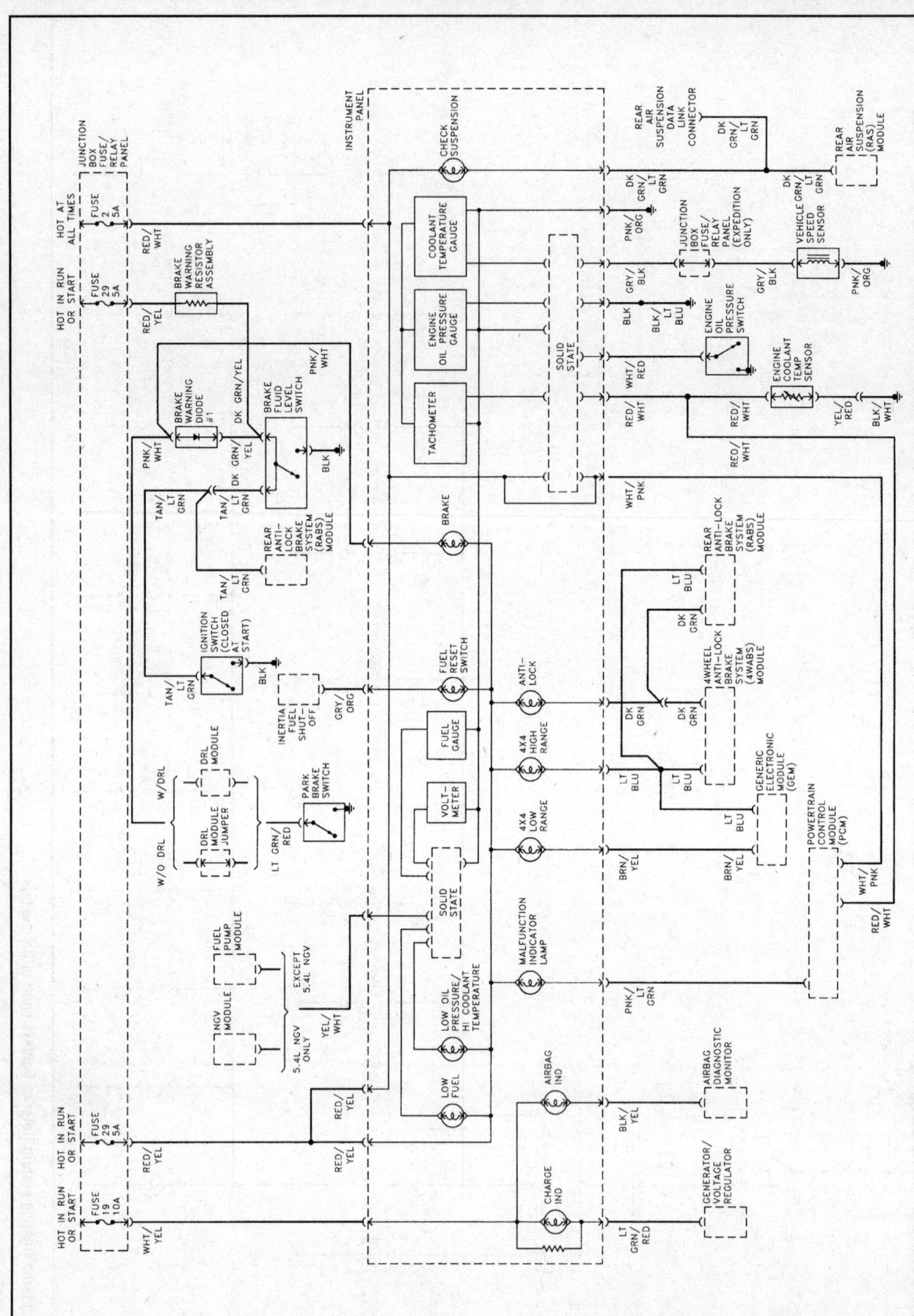

Instrument panel warning system - pick-up models

CHASSIS ELECTRICAL SYSTEM 12-39

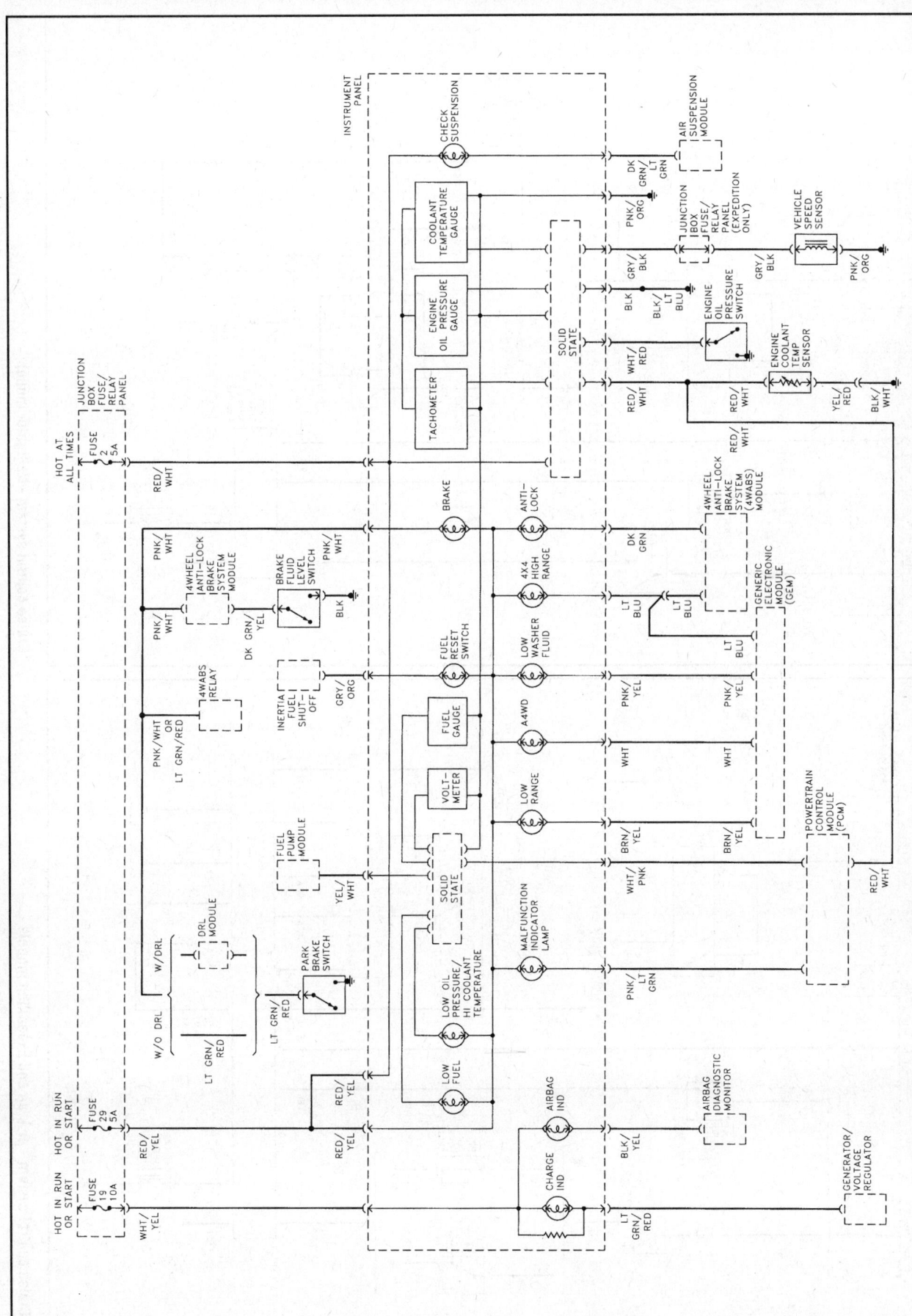

Instrument panel warning system - sport-utility models

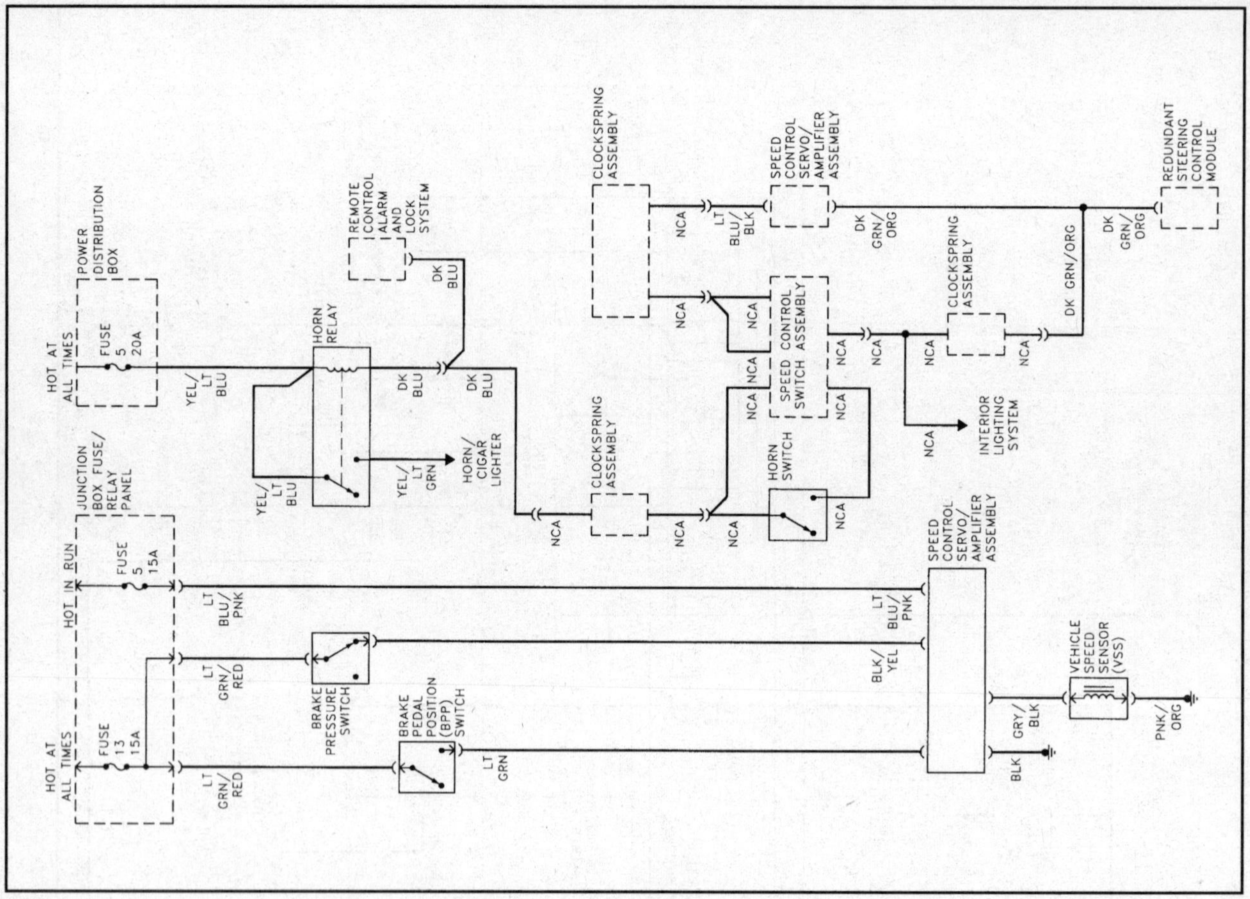

12-40 CHASSIS ELECTRICAL SYSTEM

Cruise control system - Navigator model

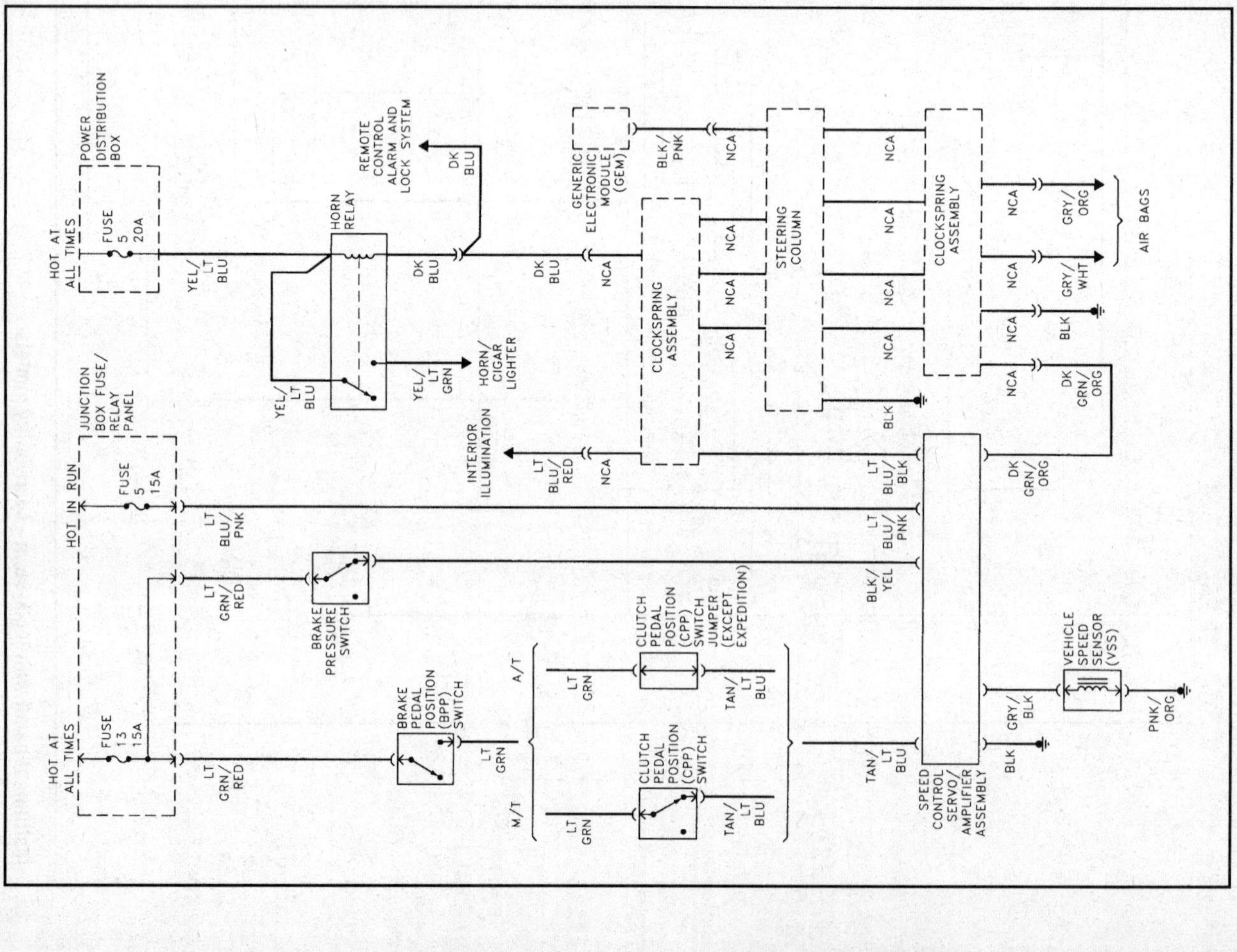

Cruise control system - pick-up and Expedition models

CHASSIS ELECTRICAL SYSTEM 12-41

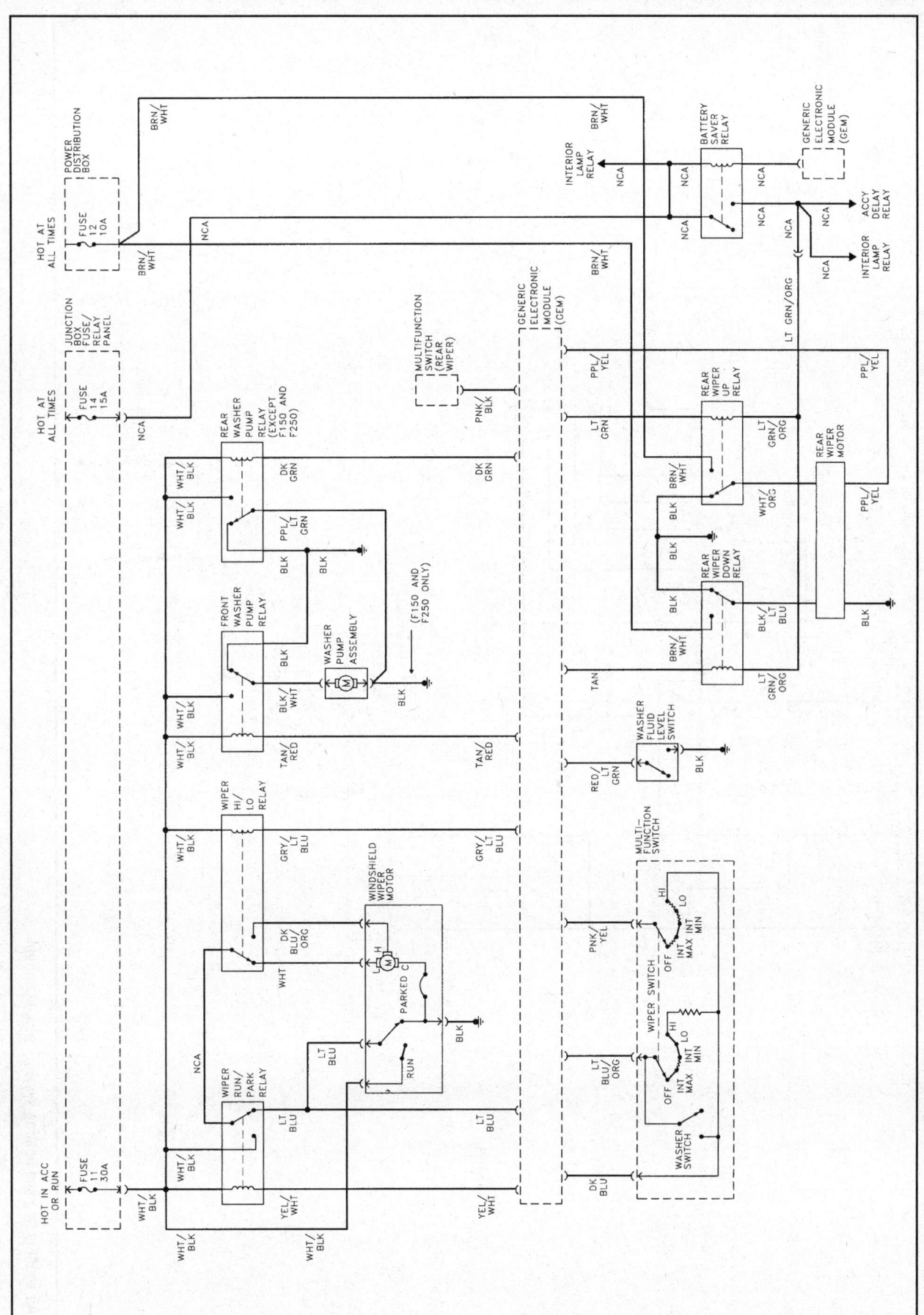

Windshield wiper and washer system

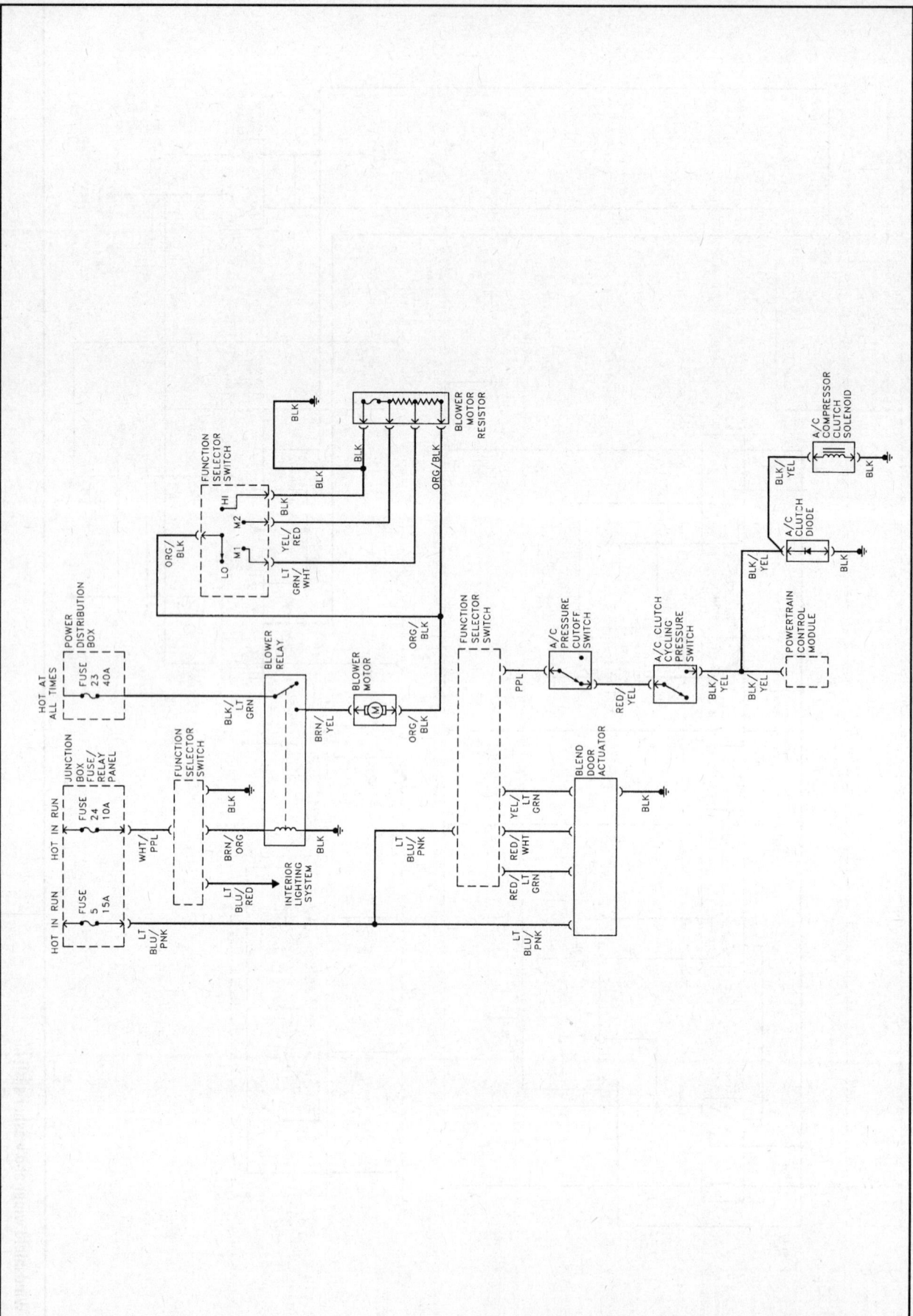

Air conditioning and heating system - pick-up models

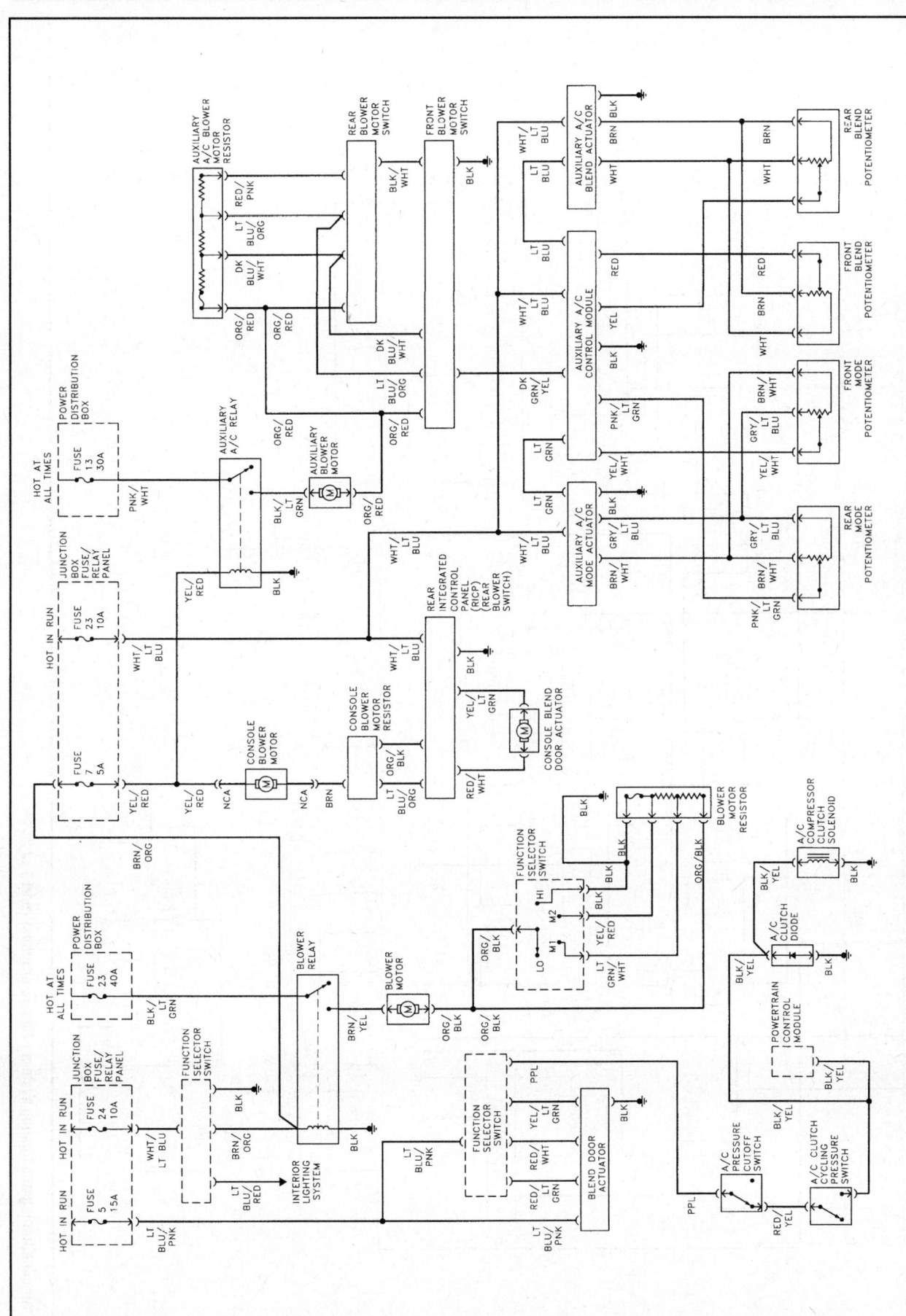

Air conditioning and heating system - Expedition model

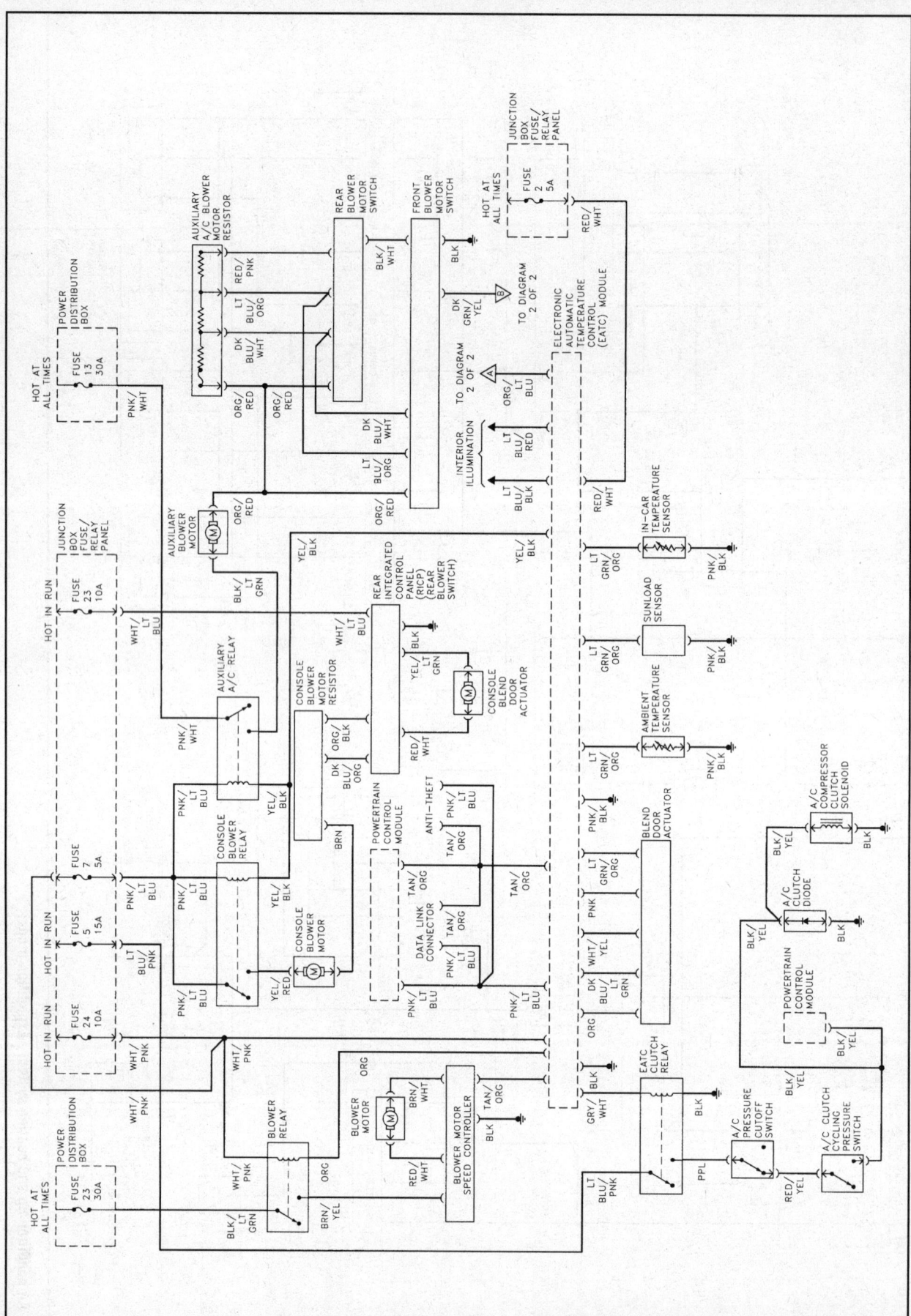

Air conditioning and heating system - Navigator model (part 1 of 2)

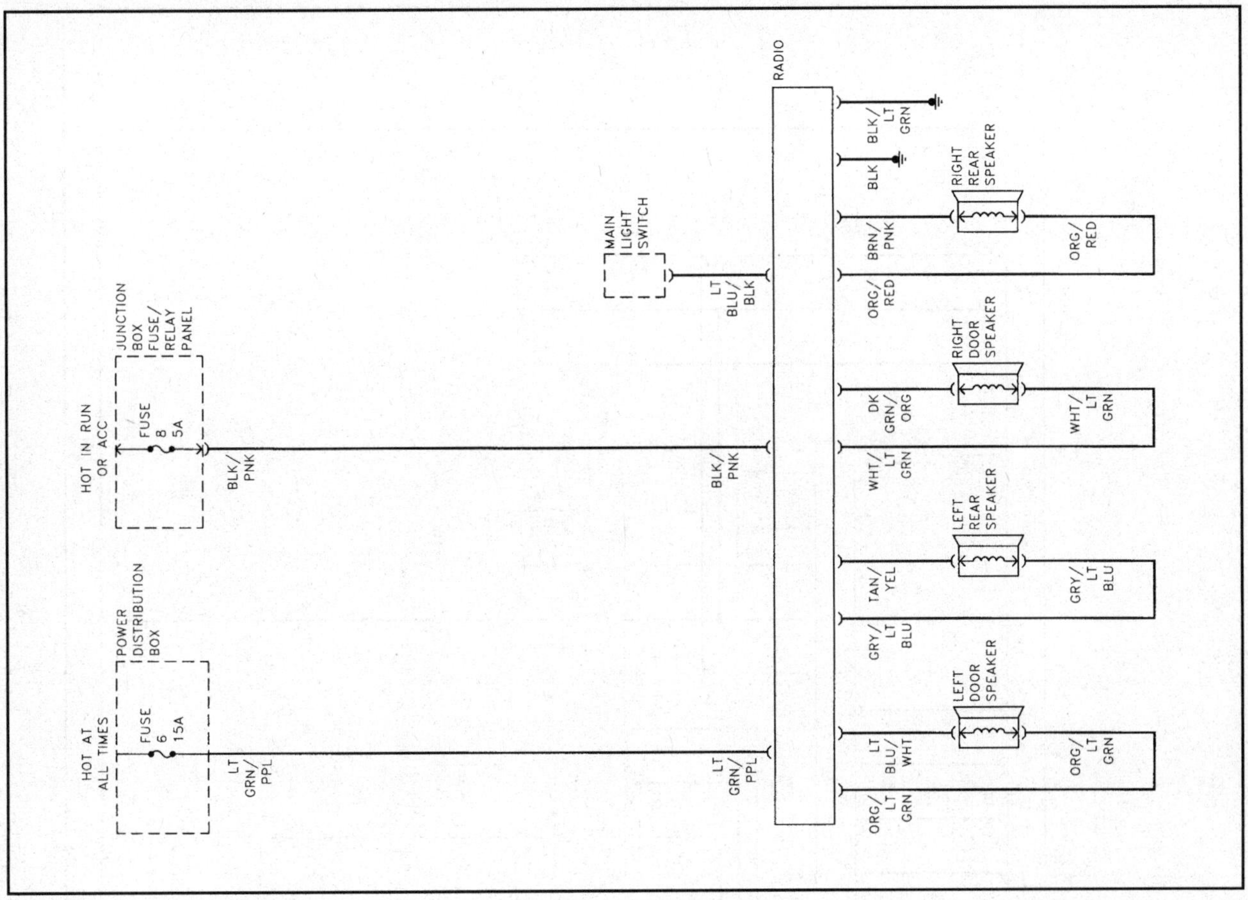

CHASSIS ELECTRICAL SYSTEM 12-45

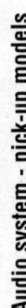

Audio system - pick-up models

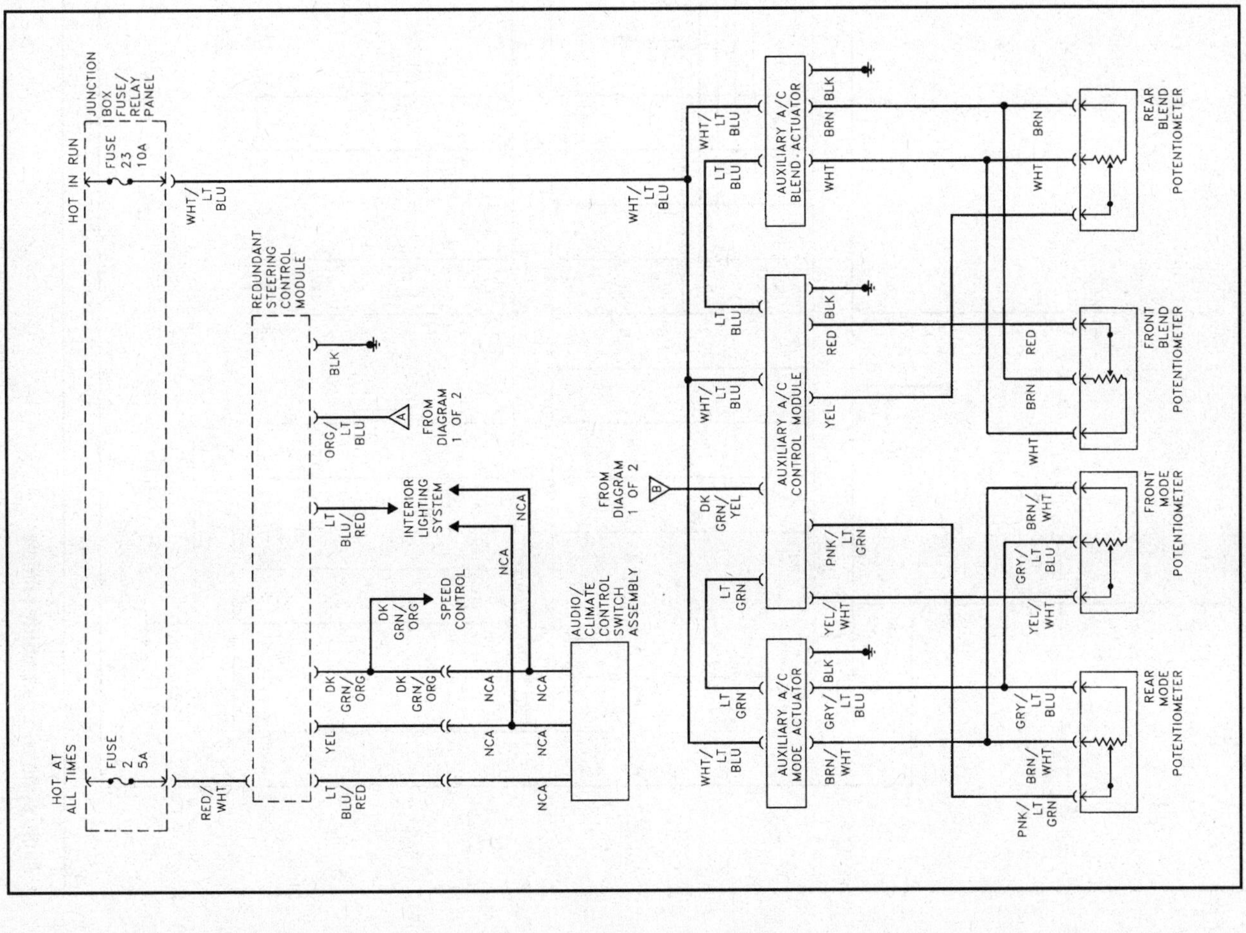

Air conditioning and heating system - Navigator model (part 2 of 2)

12-46 CHASSIS ELECTRICAL SYSTEM

Audio system - sport-utility models

A

ABOUT THIS MANUAL, 0-5
ACCELERATOR CABLE - REMOVAL, INSTALLATION AND ADJUSTMENT, 4-12
ACCELERATOR PEDAL POSITION (APP) SENSOR (2005 AND LATER MODELS) - REPLACEMENT AND ADJUSTMENT, 6-28
ADJUSTABLE BRAKE PEDAL AND BRACKET - REMOVAL, INSTALLATION AND INDEXING, 9-30
AIR CONDITIONING
    accumulator/drier - removal and installation, 3-19
    and heating system - check and maintenance, 3-16
    compressor - removal and installation, 3-21
    condenser - removal and installation, 3-22
    evaporator - removal and installation, 3-23
AIR FILTER CHECK AND REPLACEMENT, 1-26
AIR FILTER HOUSING - REMOVAL AND INSTALLATION, 4-12
AIRBAG - GENERAL INFORMATION, 12-22
ALTERNATOR
    components - replacement, 5-11
    removal and installation, 5-10
ANTENNA - REMOVAL AND INSTALLATION, 12-11
ANTIFREEZE - GENERAL INFORMATION, 3-2
ANTI-LOCK BRAKE SYSTEM - GENERAL INFORMATION, 9-2
AUTOMATIC TRANSMISSION FLUID
    and filter change, 1-28
    fluid level check, 1-10
AUTOMATIC TRANSMISSION, 7B-1
    Automatic transmission - removal and installation, 7B-6
    Diagnosis - general, 7B-2
    General information, 7B-2
    Shift cable - removal, installation and adjustment, 7B-4
    Shift indicator cable adjustment, 7B-3
    Shift interlock system - description, check and actuator replacement, 7B-5
    Shift lever - removal and installation, 7B-3
    Transmission control switch - description, check and replacement, 7B-3
    Transmission Range (TR) sensor - description, adjustment and replacement, 7B-4
AUTOMOTIVE CHEMICALS AND LUBRICANTS, 0-30
AXLE (FRONT) (4WD MODELS) - REMOVAL AND INSTALLATION, 8-29
AXLE/DIFFERENTIAL (REAR) - REMOVAL AND INSTALLATION, 8-23
AXLES - DESCRIPTION AND CHECK, 8-19
AXLESHAFT
    bearing (rear) - replacement, 8-21
    oil seal (rear) - replacement, 8-20
    rear - removal and installation, 8-19

B

BALLJOINTS - CHECK AND REPLACEMENT, 10-8
BATTERY
    cables - check and replacement, 5-3
    check, maintenance and charging, 1-15
    emergency jump starting, 5-3
    removal and installation, 5-2
BODY, 11-1
    Body - maintenance, 11-2
    Body repair - major damage, 11-6

MASTER INDEX

BODY, 11-1 (CONTINUED)
- Body repair - minor damage, 11-2
- Bumpers - removal and installation, 11-9
- Center console (Expedition and Navigator) - removal and installation, 11-18
- Cowl cover - removal and installation, 11-23
- Dashboard trim panels - removal and installation, 11-19
- Door - removal, installation and adjustment, 11-12
- Door latch, lock cylinder and handles - removal and installation, 11-13
- Door trim panels - removal and installation, 11-11
- Door window glass - removal and installation, 11-14
- Door window glass regulator - removal and installation, 11-14
- Front fender - removal and installation, 11-10
- General information, 11-2
- Hinges and locks - maintenance, 11-6
- Hood - removal, installation and adjustment, 11-7
- Hood and rear liftgate support struts - removal and installation, 11-6
- Hood latch and release cable - removal and installation, 11-8
- Instrument cluster bezel - removal and installation, 11-18
- Instrument panel - removal and installation, 11-21
- Liftgate - removal, installation and adjustment, 11-16
- Radiator grille - removal and installation, 11-8
- Seats - removal and installation, 11-23
- Sideview mirrors - removal and installation, 11-15
- Steering column cover - removal and installation, 11-20
- Tailgate - removal, installation and adjustment, 11-16
- Tailgate latch, handle and lock cylinder - removal and installation, 11-17
- Upholstery and carpets - maintenance, 11-2
- Vinyl trim - maintenance, 11-2
- Windshield and fixed glass - replacement, 11-6

BOOSTER BATTERY (JUMP) STARTING, 0-27
BRAKE
- check, 1-23
- fluid change, 1-31

BRAKES, 9-1
- Adjustable brake pedal and bracket - removal, installation and indexing, 9-30
- Anti-lock brake system - general information, 9-2
- Brake disc - inspection, removal and installation, 9-10
- Brake hoses and lines - check and replacement, 9-21
- Brake hydraulic system - bleeding, 9-22
- Brake light switch - check and replacement, 9-23
- Disc brake caliper - removal, overhaul and installation, 9-9
- Disc brake pads - replacement, 9-3
- Drum brake shoes - replacement, 9-11
- General information, 9-2
- Master cylinder - removal, overhaul and installation, 9-18
- Parking brake cables - replacement, 9-25
- Parking brake shoes (models with rear disc brakes) - replacement, 9-28
- Power brake booster - check, removal and installation, 9-24
- Wheel cylinder - removal, overhaul and installation, 9-17

BULB REPLACEMENT, 12-14
BUMPERS - REMOVAL AND INSTALLATION, 11-9
BUYING PARTS, 0-17

C
CAMSHAFT(S) - REMOVAL, INSPECTION AND INSTALLATION, 2C-16
CAMSHAFT, BALANCE SHAFT AND BEARINGS - REMOVAL, INSPECTION AND INSTALLATION, 2A-17
CAMSHAFTS AND TAPPETS - REMOVAL, INSPECTION AND INSTALLATION, 2B-11
CATALYTIC CONVERTER, 6-26
CENTER CONSOLE (EXPEDITION AND NAVIGATOR) - REMOVAL AND INSTALLATION, 11-18
CHARGING SYSTEM
- check, 5-8
- general information and precautions, 5-8

CHASSIS ELECTRICAL SYSTEM, 12-1
- Airbag - general information, 12-22
- Antenna - removal and installation, 12-11
- Bulb replacement, 12-14
- Circuit breakers - general information, 12-4
- Cruise control system - description and check, 12-19
- Daytime Running Lights (DRL) - general information, 12-16
- Electric side view mirrors - description and check, 12-22
- Electrical troubleshooting - general information, 12-2
- Fuses - general information, 12-3
- Fusible links - general information, 12-4
- General information, 12-2
- Headlight switch - replacement, 12-8
- Headlights - adjustment, 12-13
- Headlights - replacement, 12-12
- Horn - check and replacement, 12-18
- Ignition lock cylinder - removal and installation, 12-7
- Ignition switch - replacement, 12-7
- Instrument cluster - removal and installation, 12-10
- Instrument panel gauges - check, 12-9
- Power door lock and keyless entry system - description and check, 12-21
- Power window system - description and check, 12-20
- Radio and speakers - removal and installation, 12-10
- Rear window defogger - check and repair, 12-18
- Relays - general information and testing, 12-4
- Steering column switches - replacement, 12-5
- Turn signal/hazard flasher - check and replacement, 12-5
- Wiper motor - check and replacement, 12-16
- Wiring diagrams - general information, 12-24

CHASSIS LUBRICATION, 1-13
CIRCUIT BREAKERS - GENERAL INFORMATION, 12-4
CLUTCH AND DRIVELINE, 8-1
- Axle (front) (4WD models) - removal and installation, 8-29
- Axle/differential (rear) - removal and installation, 8-23
- Axles - description and check, 8-19
- Axleshaft (rear) - removal and installation, 8-19
- Axleshaft bearing (rear) - replacement, 8-21
- Axleshaft oil seal (rear) - replacement, 8-20
- Clutch - description and check, 8-2
- Clutch components - removal, inspection and installation, 8-6
- Clutch hydraulic system - bleeding, 8-10
- Clutch master cylinder - removal, inspection and installation, 8-2
- Clutch pedal position switch - check and replacement, 8-11
- Clutch release bearing - removal, inspection and installation, 8-4

MASTER INDEX IND-3

Clutch release cylinder - removal, inspection and installation, 8-5
Differential pinion seal - replacement, 8-21
Driveaxle boot replacement (4WD models), 8-25
Driveaxles - removal and installation, 8-24
Driveline inspection, 8-12
Driveshaft - removal and installation, 8-12
Driveshaft slip yoke boot and center bearing (two-piece driveshafts) - inspection and replacement, 8-18
Driveshafts and universal joints - general information, 8-11
General information, 8-2
Pilot bearing - inspection and replacement, 8-9
Universal joints - replacement, 8-14
Vacuum shift system (4WD models) - description, check and component replacement, 8-29

COIL SPRING (2WD MODELS) - REMOVAL AND INSTALLATION, 10-12
CONVERSION FACTORS, 0-28
COOLING SYSTEM
check, 1-19
servicing (draining, flushing and refilling), 1-26

COOLING, HEATING AND AIR CONDITIONING SYSTEMS, 3-1
Air conditioning accumulator/drier - removal and installation, 3-19
Air conditioning and heating system - check and maintenance, 3-16
Air conditioning compressor - removal and installation, 3-21
Air conditioning condenser - removal and installation, 3-22
Air conditioning evaporator - removal and installation, 3-23
Antifreeze - general information, 3-2
Coolant temperature sending unit - check and replacement, 3-7
Engine cooling fan and fan clutch - check, removal and installation, 3-4
Engine oil cooler - removal and installation, 3-8
General Information, 3-2
Heater and air conditioning blower motor – removal and installation, 3-13
Heater and air conditioning blower motor and circuit - check, 3-11
Heater and air conditioning control assembly - removal and installation, 3-14
Heater core - removal and installation, 3-15
Radiator and degas bottle - removal and installation, 3-5
Thermostat - check and replacement, 3-3
Water pump - check, 3-9
Water pump - removal and installation, 3-9

COWL COVER - REMOVAL AND INSTALLATION, 11-23
CRANKSHAFT
inspection, 2D-21
installation and main bearing oil clearance check, 2D-24
removal, 2D-15
oil seals - replacement
4.2L V6 engine, 2A-21
V8 engines, 2C-28

CRANKSHAFT PULLEY - REMOVAL AND INSTALLATION, 2C-4
CRANKSHAFT PULLEY AND CRANKSHAFT FRONT OIL SEAL - REPLACEMENT, 2B-6
CRUISE CONTROL SYSTEM - DESCRIPTION AND CHECK, 12-19
CYLINDER COMPRESSION CHECK, 2D-3
CYLINDER HEAD
cleaning and inspection, 2D-9
disassembly, 2D-8
reassembly, 2D-12

CYLINDER HEADS - REMOVAL AND INSTALLATION
3.5L V6 engine, 2B-6
4.2L V6 engine, 2A-10
V8 engines, 2C-23

CYLINDER HONING, 2D-18

D

DASHBOARD TRIM PANELS - REMOVAL AND INSTALLATION, 11-19
DAYTIME RUNNING LIGHTS (DRL) - GENERAL INFORMATION, 12-16
DIAGNOSIS - GENERAL, 7B-2
DIFFERENTIAL LUBRICANT
change, 1-38
level check, 1-25

DIFFERENTIAL PINION SEAL - REPLACEMENT, 8-21
DISC BRAKE
caliper - removal, overhaul and installation, 9-9
pads - replacement, 9-3

DOOR
removal, installation and adjustment, 11-12
latch, lock cylinder and handles - removal and installation, 11-13
trim panels - removal and installation, 11-11
window glass - removal and installation, 11-14
window glass regulator - removal and installation, 11-14

DRIVEAXLE BOOT REPLACEMENT (4WD MODELS), 8-25
DRIVEAXLES - REMOVAL AND INSTALLATION, 8-24
DRIVEBELT CHECK AND REPLACEMENT, 1-31
DRIVELINE INSPECTION, 8-12
DRIVEPLATE - REMOVAL AND INSTALLATION, 2B-14
DRIVESHAFT - REMOVAL AND INSTALLATION, 8-12
DRIVESHAFT SLIP YOKE BOOT AND CENTER BEARING (TWO-PIECE DRIVESHAFTS) - INSPECTION AND REPLACEMENT, 8-18
DRIVESHAFTS AND UNIVERSAL JOINTS - GENERAL INFORMATION, 8-11
DRUM BRAKE SHOES - REPLACEMENT, 9-11

E

ELECTRIC SHIFT MOTOR (ELECTRIC-SHIFT MODELS) - REPLACEMENT, 7C-3
ELECTRIC SIDE VIEW MIRRORS - DESCRIPTION AND CHECK, 12-22
ELECTRICAL TROUBLESHOOTING - GENERAL INFORMATION, 12-2
EMISSIONS AND ENGINE CONTROL SYSTEMS, 6-1
Accelerator Pedal Position (APP) sensor (2005 and later models) - replacement and adjustment, 6-28
Catalytic converter, 6-26
Evaporative Emissions Control System (EVAP), 6-24
Exhaust Gas Recirculation (EGR) system, 6-22
General information, 6-2
Information sensors, 6-11
Oil pressure control solenoid (3.5L V6 engine) – replacement, 6-27
On Board Diagnosis (OBD) system and trouble codes, 6-4
Positive Crankcase Ventilation (PCV) system, 6-26

IND-4 MASTER INDEX

EMISSIONS AND ENGINE CONTROL SYSTEMS, 6-1 (CONT'D)
  Powertrain Control Module (PCM) - replacement, 6-10
  Turbocharger bypass valve (3.5L V6 engine) – replacement, 6-27
  Turbocharger wastegate vacuum sensor - replacement, 6-27
  Engine - removal and installation, 2D-5
ENGINE, IN-VEHICLE REPAIR PROCEDURES
  3.5L V6 engine, 2B-1
    Camshafts and tappets - removal, inspection and installation, 2B-11
    Crankshaft pulley and crankshaft front oil seal - replacement, 2B-6
    Cylinder heads - removal and installation, 2B-6
    Driveplate - removal and installation, 2B-14
    Engine front cover - removal and installation, 2B-7
    Engine mounts - check and replacement, 2B-16
    Exhaust manifolds - removal and installation, 2B-5
    General Information, 2B-2
    Intake manifold - removal and installation, 2B-5
    Oil pan - removal and installation, 2B-13
    Oil pump - removal and installation, 2B-14
    Rear main oil seal - replacement, 2B-15
    Repair operations possible with the engine in the vehicle, 2B-2
    Timing chains and sprockets - removal and installation, 2B-9
    Valve clearance - check and adjustment, 2B-2
    Valve covers - removal and installation, 2B-3
  4.2L V6 engine, 2A-1
    Camshaft, balance shaft and bearings - removal, inspection and installation, 2A-17
    Crankshaft oil seals - replacement, 2A-21
    Cylinder heads - removal and installation, 2A-10
    Engine mounts - check and replacement, 2A-24
    Exhaust manifolds - removal and installation, 2A-9
    Flywheel/driveplate - removal and installation, 2A-23
    General information, 2A-2
    Intake manifold - removal and installation, 2A-6
    Oil pan - removal and installation, 2A-19
    Oil pump - removal and installation, 2A-20
    Repair operations possible with the engine in the vehicle, 2A-2
    Rocker arms and pushrods - removal, inspection and installation, 2A-4
    Timing chain and sprockets - inspection, removal and installation, 2A-14
    Timing chain cover - removal and installation, 2A-12
    Top Dead Center (TDC) for number one piston - locating, 2A-2
    Valve covers - removal and installation, 2A-3
    Valve lifters - removal, inspection and installation, 2A-16
    Valve springs, retainers and seals - replacement, 2A-5
  V8 engines, 2C-1
    Camshaft(s) - removal, inspection and installation, 2C-16
    Crankshaft oil seals - replacement, 2C-28
    Crankshaft pulley - removal and installation, 2C-4
    Cylinder heads - removal and installation, 2C-23
    Engine mounts - check and replacement, 2C-30
    Exhaust manifolds - removal and installation, 2C-22
    Flywheel/driveplate - removal and installation, 2C-27
    General information, 2C-2
    Intake manifold - removal and installation, 2C-21
    Oil pan - removal and installation, 2C-25
    Oil pump - removal and installation, 2C-26
    Repair operations possible with the engine in the vehicle, 2C-2
    Rocker arms and valve lash adjusters - removal, inspection and installation, 2C-14
    Timing chain cover - removal and installation, 2C-5
    Timing chains, tensioners and sprockets - removal, inspection and installation, 2C-7
    Top Dead Center (TDC) for number one piston - locating, 2C-2
    Valve covers - removal and installation, 2C-3
    Valve springs, retainers and seals - removal and installation, 2C-19
ENGINE BLOCK
  cleaning, 2D-16
  inspection, 2D-18
ENGINE COOLING FAN AND FAN CLUTCH - CHECK, REMOVAL AND INSTALLATION, 3-4
ENGINE ELECTRICAL SYSTEMS, 5-1
  Alternator - removal and installation, 5-10
  Alternator components - replacement, 5-11
  Battery - emergency jump starting, 5-3
  Battery - removal and installation, 5-2
  Battery cables - check and replacement, 5-3
  Charging system - check, 5-8
  Charging system - general information and precautions, 5-8
  General information, 5-2
  Ignition coils - check and replacement, 5-6
  Ignition system - check, 5-4
  Ignition system - general information, 5-3
  Ignition timing - check, 5-7
  Starter motor - removal and installation, 5-13
  Starter motor and circuit - in-vehicle check, 5-12
  Starter solenoid - replacement, 5-13
  Starting system - general information and precautions, 5-12
ENGINE FRONT COVER - REMOVAL AND INSTALLATION, 2B-7
ENGINE MOUNTS - CHECK AND REPLACEMENT
  3.5L V6 engine, 2B-16
  4.2L V6 engine, 2A-24
  V8 engines, 2C-30
ENGINE OIL AND FILTER CHANGE, 1-12
ENGINE OIL COOLER - REMOVAL AND INSTALLATION, 3-8
ENGINE OVERHAUL
  disassembly sequence, 2D-7
  general information, 2D-2
  reassembly sequence, 2D-22
ENGINE
  rebuilding alternatives, 2D-7
  removal - methods and precautions, 2D-4
EVAPORATIVE EMISSIONS CONTROL SYSTEM (EVAP), 6-24
EXHAUST GAS RECIRCULATION (EGR) SYSTEM, 6-22
EXHAUST MANIFOLDS - REMOVAL AND INSTALLATION
  3.5L V6 engine, 2B-5
  4.2L V6 engine, 2A-9
  V8 engines, 2C-22
EXHAUST SYSTEM
  check, 1-18
  servicing - general information, 4-22

F

FLUID LEVEL CHECKS, 1-6
FLYWHEEL/DRIVEPLATE - REMOVAL AND INSTALLATION
  4.2L V6 engine, 2A-23
  V8 engines, 2C-27
FRACTION/DECIMAL/MILLIMETER EQUIVALENTS, 0-29

MASTER INDEX IND-5

FRONT END ALIGNMENT - GENERAL INFORMATION, 10-30
FRONT FENDER - REMOVAL AND INSTALLATION, 11-10
FRONT WHEEL BEARING CHECK, REPACK AND ADJUSTMENT (2WD MODELS), 1-29
FUEL AND EXHAUST SYSTEMS, 4-1
    Accelerator cable - removal, installation and adjustment, 4-12
    Air filter housing - removal and installation, 4-12
    Exhaust system servicing - general information, 4-22
    Fuel injection system - check, 4-14
    Fuel injection system - general information, 4-13
    Fuel level sending unit - check and replacement, 4-11
    Fuel lines and fittings - general information, 4-6
    Fuel pressure relief procedure, 4-3
    Fuel pump - removal and installation, 4-9
    Fuel Pump Driver Module (FPDM) - replacement, 4-23
    Fuel pump/fuel pressure - check, 4-4
    Fuel rails and injectors (3.5L V6 engine) - removal and installation, 4-24
    Fuel tank - cleaning and repair, 4-9
    Fuel tank - removal and installation, 4-8
    General information, 4-2
    High pressure fuel pump (3.6L V6 engine) removal and installation, 4-23
    Idle Air Control (IAC) valve (2004 and earlier models) - check, removal and adjustment, 4-20
    Intake Air Systems, 4-21
    Intercooler (3.5L V6 engine) - removal and installation, 4-26
    Sequential Electronic Fuel Injection (SEFI) system (2014 and earlier models) - component check and replacement, 4-15
    Turbocharger(s) (3.5L V6 engine) - removal and installation, 4-25
    Wastegate control actuator (3.5L V6 engine) - replacement and adjustment, 4-26
FUEL FILTER REPLACEMENT, 1-21
FUEL SYSTEM CHECK, 1-20
FUSES - GENERAL INFORMATION, 12-3
FUSIBLE LINKS - GENERAL INFORMATION, 12-4

G

GENERAL ENGINE OVERHAUL PROCEDURES, 2D-1
    Crankshaft - inspection, 2D-21
    Crankshaft - installation and main bearing oil clearance check, 2D-24
    Crankshaft - removal, 2D-15
    Cylinder compression check, 2D-3
    Cylinder head - cleaning and inspection, 2D-9
    Cylinder head - disassembly, 2D-8
    Cylinder head - reassembly, 2D-12
    Cylinder honing, 2D-18
    Engine - removal and installation, 2D-5
    Engine block - cleaning, 2D-16
    Engine block - inspection, 2D-18
    Engine overhaul - disassembly sequence, 2D-7
    Engine overhaul - general information, 2D-2
    Engine overhaul - reassembly sequence, 2D-22
    Engine rebuilding alternatives, 2D-7
    Engine removal - methods and precautions, 2D-4
    General information, 2D-2
    Initial start-up and break-in after overhaul, 2D-33
    Main and connecting rod bearings - inspection, 2D-22
    Piston rings - installation, 2D-23
    Pistons/connecting rods - inspection, 2D-19
    Pistons/connecting rods - installation and rod bearing oil clearance check, 2D-30
    Pistons/connecting rods - removal, 2D-13
    Vacuum gauge diagnostic checks, 2D-3
    Valves - servicing, 2D-11

H

HEADLIGHT SWITCH - REPLACEMENT, 12-8
HEADLIGHTS
    adjustment, 12-13
    replacement, 12-12
HEATER AND AIR CONDITIONING
    blower motor – removal and installation, 3-13
    blower motor and circuit - check, 3-11
    control assembly - removal and installation, 3-14
HEATER CORE - REMOVAL AND INSTALLATION, 3-15
HIGH PRESSURE FUEL PUMP (3.6L V6 ENGINE) REMOVAL AND INSTALLATION, 4-23
HINGES AND LOCKS - MAINTENANCE, 11-6
HOOD
    and rear liftgate support struts - removal and installation, 11-6
    latch and release cable - removal and installation, 11-8
    removal, installation and adjustment, 11-7
HORN - CHECK AND REPLACEMENT, 12-18
HUB AND BEARING ASSEMBLY (2003 AND LATER 2WD AND ALL 4WD MODELS) - REPLACEMENT, 10-13

I

IDLE AIR CONTROL (IAC) VALVE (2004 AND EARLIER MODELS) - CHECK, REMOVAL AND ADJUSTMENT, 4-20
IGNITION
    coils - check and replacement, 5-6
    lock cylinder - removal and installation, 12-7
    switch - replacement, 12-7
    system
        check, 5-4
        component check and replacement, 1-37
        general information, 5-3
        timing - check, 5-7
INDEPENDENT REAR SUSPENSION (2003 AND LATER EXPEDITION/NAVIGATOR) - COMPONENT REPLACEMENT, 10-28
INFORMATION SENSORS, 6-11
INITIAL START-UP AND BREAK-IN AFTER OVERHAUL, 2D-33
INSTRUMENT CLUSTER
    bezel - removal and installation, 11-18
    removal and installation, 12-10
INSTRUMENT PANEL
    gauges - check, 12-9
    removal and installation, 11-21
INTAKE AIR SYSTEMS, 4-21

IND-6 MASTER INDEX

INTAKE MANIFOLD - REMOVAL AND INSTALLATION
3.5L V6 engine, 2B-5
4.2L V6 engine, 2A-6
V8 engines, 2C-21
INTERCOOLER (3.5L V6 ENGINE) - REMOVAL AND INSTALLATION, 4-26
INTRODUCTION TO THE F-150, F-250, EXPEDITION AND NAVIGATOR, 0-5

J

JACKING AND TOWING, 0-26

L

LEAF SPRING/AIR SPRING - REMOVAL AND INSTALLATION, 10-17
LIFTGATE - REMOVAL, INSTALLATION AND ADJUSTMENT, 11-16
LOWER CONTROL ARM - REMOVAL AND INSTALLATION, 10-10

M

MAIN AND CONNECTING ROD BEARINGS - INSPECTION, 2D-22
MAINTENANCE SCHEDULE, 1-2
MAINTENANCE TECHNIQUES, TOOLS AND WORKING FACILITIES, 0-18
MANUAL TRANSMISSION, 7A-1
General information, 7A-2
Manual transmission - removal and installation, 7A-5
Manual transmission overhaul - general information, 7A-8
Oil seal - replacement, 7A-2
Shift lever - removal and installation, 7A-2
Transmission mount - check and replacement, 7A-4
MANUAL TRANSMISSION LUBRICANT
change, 1-37
level check, 1-25
MASTER CYLINDER - REMOVAL, OVERHAUL AND INSTALLATION, 9-18

O

OIL PAN - REMOVAL AND INSTALLATION
3.5L V6 engine, 2B-13
4.2L V6 engine, 2A-19
V8 engines, 2C-25
OIL PRESSURE CONTROL SOLENOID (3.5L V6 ENGINE) - REPLACEMENT, 6-27
OIL PUMP - REMOVAL AND INSTALLATION
3.5L V6 engine, 2B-14
4.2L V6 engine, 2A-20
V8 engines, 2C-26

OIL SEAL – REPLACEMENT
manual transmission, 7A-2
transfer case, 7C-3
ON BOARD DIAGNOSIS (OBD) SYSTEM AND TROUBLE CODES, 6-4

P

PARKING BRAKE
cables - replacement, 9-25
shoes (models with rear disc brakes) - replacement, 9-28
PILOT BEARING - INSPECTION AND REPLACEMENT, 8-9
PISTON RINGS - INSTALLATION, 2D-23
PISTONS/CONNECTING RODS
inspection, 2D-19
installation and rod bearing oil clearance check, 2D-30
removal, 2D-13
POSITIVE CRANKCASE VENTILATION (PCV)
system, 6-26
valve check, 1-34
POWER BRAKE BOOSTER - CHECK, REMOVAL AND INSTALLATION, 9-24
POWER DOOR LOCK AND KEYLESS ENTRY SYSTEM - DESCRIPTION AND CHECK, 12-21
POWER STEERING
fluid level check, 1-10
pump - removal and installation, 10-27
system - bleeding, 10-28
POWER WINDOW SYSTEM - DESCRIPTION AND CHECK, 12-20
POWERTRAIN CONTROL MODULE (PCM) - REPLACEMENT, 6-10

R

RADIATOR AND DEGAS BOTTLE - REMOVAL AND INSTALLATION, 3-5
RADIATOR GRILLE - REMOVAL AND INSTALLATION, 11-8
RADIO AND SPEAKERS - REMOVAL AND INSTALLATION, 12-10
REAR COIL SPRING/AIR SPRING (2002 AND EARLIER EXPEDITION/NAVIGATOR) - REMOVAL AND INSTALLATION, 10-20
REAR MAIN OIL SEAL - REPLACEMENT, 2B-15
REAR STABILIZER BAR (2002 AND EARLIER EXPEDITION/NAVIGATOR) - REMOVAL AND INSTALLATION, 10-19
REAR SUSPENSION ARMS (2002 AND EARLIER EXPEDITION/NAVIGATOR) - REMOVAL AND INSTALLATION, 10-19
REAR WINDOW DEFOGGER - CHECK AND REPAIR, 12-18
RECALL INFORMATION, 0-8
RELAYS - GENERAL INFORMATION AND TESTING, 12-4

MASTER INDEX IND-7

REPAIR OPERATIONS POSSIBLE WITH THE ENGINE IN THE VEHICLE
    3.5L V6 engine, 2B-2
    4.2L V6 engine, 2A-2
    V8 engines, 2C-2
ROCKER ARMS AND PUSHRODS - REMOVAL, INSPECTION AND INSTALLATION, 2A-4
ROCKER ARMS AND VALVE LASH ADJUSTERS - REMOVAL, INSPECTION AND INSTALLATION, 2C-14

S

SAFETY FIRST!, 0-31
SEAT BELT CHECK, 1-18
SEATS - REMOVAL AND INSTALLATION, 11-23
SEQUENTIAL ELECTRONIC FUEL INJECTION (SEFI) SYSTEM (2014 AND EARLIER MODELS) - COMPONENT CHECK AND REPLACEMENT, 4-15
SHIFT
    cable - removal, installation and adjustment, 7B-4
    indicator cable adjustment, 7B-3
    interlock system - description, check and actuator replacement, 7B-5
SHIFT LEVER - REMOVAL AND INSTALLATION
    automatic transmission, 7B-3
    manual transmission, 7A-2
    transfer case (manual-shift models), 7C-2
SHIFT RANGE SELECTOR SWITCH (ELECTRIC-SHIFT MODELS) - REPLACEMENT, 7C-3
SHOCK ABSORBER (1997 THROUGH 2002 SUV MODELS/2003 PICK-UP MODELS) OR SHOCK ABSORBER/COIL SPRING ASSEMBLY (2003 AND LATER SUV MODELS) (FRONT) - REMOVAL AND INSTALLATION, 10-6
SHOCK ABSORBER (REAR) - REMOVAL AND INSTALLATION, 10-16
SIDEVIEW MIRRORS - REMOVAL AND INSTALLATION, 11-15
SPARK PLUG CHECK AND REPLACEMENT, 1-34
STABILIZER BAR (FRONT) - REMOVAL AND INSTALLATION, 10-8
STARTER MOTOR - REMOVAL AND INSTALLATION, 5-13
STARTER MOTOR AND CIRCUIT - IN-VEHICLE CHECK, 5-12
STARTER SOLENOID - REPLACEMENT, 5-13
STARTING SYSTEM - GENERAL INFORMATION AND PRECAUTIONS, 5-12
STEERING
    and suspension check, 1-22
    column cover - removal and installation, 11-20
    column switches - replacement, 12-5
SUSPENSION AND STEERING SYSTEMS, 10-1
    Balljoints - check and replacement, 10-8
    Coil spring (2WD models) - removal and installation, 10-12
    Front end alignment - general information, 10-30
    General information, 10-2
    Hub and bearing assembly (2003 and later 2WD and all 4WD models) - replacement, 10-13
    Independent rear suspension (2003 and later Expedition/Navigator) - component replacement, 10-28
    Leaf spring/air spring - removal and installation, 10-17
    Lower control arm - removal and installation, 10-10
    Power steering pump - removal and installation, 10-27
    Power steering system - bleeding, 10-28
    Rear coil spring/air spring (2002 and earlier Expedition/Navigator) - removal and installation, 10-20
    Rear stabilizer bar (2002 and earlier Expedition/Navigator) - removal and installation, 10-19
    Rear suspension arms (2002 and earlier Expedition/Navigator) - removal and installation, 10-19
    Shock absorber (1997 through 2002 SUV models/2003 pick-up models) or shock absorber/coil spring assembly (2003 and later SUV models) (front) - removal and installation, 10-6
    Shock absorber (rear) - removal and installation, 10-16
    Stabilizer bar (front) - removal and installation, 10-8
    Steering gear - removal and installation, 10-25
    Steering knuckle (2WD models) - removal and installation, 10-12
    Steering knuckle (4WD models) - removal and installation, 10-13
    Steering linkage (2002 and earlier Pick-ups and Expedition/Navigator) - inspection, removal and installation, 10-23
    Steering wheel and clockspring - removal and installation, 10-21
    Suspension Load Leveling Control/Vehicle Dynamic Suspension systems - general information, 10-15
    Torsion bar (4WD models) - removal and installation, 10-14
    Track bar - removal and installation, 10-18
    Upper control arm - removal and installation, 10-9
    Wheels and tires - general information, 10-30

T

TAILGATE - REMOVAL, INSTALLATION AND ADJUSTMENT, 11-16
TAILGATE LATCH, HANDLE AND LOCK CYLINDER - REMOVAL AND INSTALLATION, 11-17
THERMOSTAT - CHECK AND REPLACEMENT, 3-3
TIMING CHAIN AND SPROCKETS - INSPECTION, REMOVAL AND INSTALLATION, 2A-14
TIMING CHAIN COVER - REMOVAL AND INSTALLATION
    4.2L V6 engine, 2A-12
    V8 engines, 2C-5
TIMING CHAINS AND SPROCKETS - REMOVAL AND INSTALLATION, 2B-9
TIMING CHAINS, TENSIONERS AND SPROCKETS - REMOVAL, INSPECTION AND INSTALLATION, 2C-7
TIRE AND TIRE PRESSURE CHECKS, 1-8
TIRE ROTATION, 1-17
TOP DEAD CENTER (TDC) FOR NUMBER ONE PISTON – LOCATING
    4.2L V6 engine, 2A-2
    V8 engines, 2C-2
TORSION BAR (4WD MODELS) - REMOVAL AND INSTALLATION, 10-14
TRACK BAR - REMOVAL AND INSTALLATION, 10-18
TRANSFER CASE, 7C-1
    4WD indicator switch (manual-shift models) - replacement, 7C-2
    Electric shift motor (electric-shift models) - replacement, 7C-3
    General information, 7C-2
    Oil seal - replacement, 7C-3
    Shift lever (manual-shift models) - removal and installation, 7C-2
    Shift range selector switch (electric-shift models) - replacement, 7C-3
    Transfer case - removal and installation, 7C-4

MASTER INDEX

TRANSFER CASE LUBRICANT
  change (4WD models), 1-38
  level check, 1-25
TRANSMISSION CONTROL SWITCH - DESCRIPTION, CHECK AND REPLACEMENT, 7B-3
TRANSMISSION MOUNT - CHECK AND REPLACEMENT, 7A-4
TRANSMISSION RANGE (TR) SENSOR - DESCRIPTION, ADJUSTMENT AND REPLACEMENT, 7B-4
TROUBLESHOOTING, 0-32
TUNE-UP AND ROUTINE MAINTENANCE, 1-1
  Air filter check and replacement, 1-26
  Automatic transmission fluid and filter change, 1-28
  Automatic transmission fluid level check, 1-10
  Battery check, maintenance and charging, 1-15
  Brake check, 1-23
  Brake fluid change, 1-31
  Chassis lubrication, 1-13
  Cooling system check, 1-19
  Cooling system servicing (draining, flushing and refilling), 1-26
  Differential lubricant change, 1-38
  Differential lubricant level check, 1-25
  Drivebelt check and replacement, 1-31
  Engine oil and filter change, 1-12
  Exhaust system check, 1-18
  Fluid level checks, 1-6
  Front wheel bearing check, repack and adjustment (2WD models), 1-29
  Fuel filter replacement, 1-21
  Fuel system check, 1-20
  Ignition system component check and replacement, 1-37
  Introduction, 1-4
  Maintenance schedule, 1-2
  Manual transmission lubricant change, 1-37
  Manual transmission lubricant level check, 1-25
  Positive Crankcase Ventilation (PCV) valve check, 1-34
  Power steering fluid level check, 1-10
  Seat belt check, 1-18
  Spark plug check and replacement, 1-34
  Steering and suspension check, 1-22
  Tire and tire pressure checks, 1-8
  Tire rotation, 1-17
  Transfer case lubricant change (4WD models), 1-38
  Transfer case lubricant level check, 1-25
  Tune-up general information, 1-5
  Underhood hose check and replacement, 1-18
  Windshield wiper blade inspection and replacement, 1-16
TURBOCHARGER
  bypass valve (3.5L V6 engine) – replacement, 6-27
  wastegate vacuum sensor - replacement, 6-27
TURBOCHARGER(S) (3.5L V6 ENGINE) - REMOVAL AND INSTALLATION, 4-25
TURN SIGNAL/HAZARD FLASHER - CHECK AND REPLACEMENT, 12-5

U

UNDERHOOD HOSE CHECK AND REPLACEMENT, 1-18
UNIVERSAL JOINTS - REPLACEMENT, 8-14
UPHOLSTERY AND CARPETS - MAINTENANCE, 11-2
UPPER CONTROL ARM - REMOVAL AND INSTALLATION, 10-9

V

V8 ENGINES, 2C-1
  Camshaft(s) - removal, inspection and installation, 2C-16
  Crankshaft oil seals - replacement, 2C-28
  Crankshaft pulley - removal and installation, 2C-4
  Cylinder heads - removal and installation, 2C-23
  Engine mounts - check and replacement, 2C-30
  Exhaust manifolds - removal and installation, 2C-22
  Flywheel/driveplate - removal and installation, 2C-27
  General information, 2C-2
  Intake manifold - removal and installation, 2C-21
  Oil pan - removal and installation, 2C-25
  Oil pump - removal and installation, 2C-26
  Repair operations possible with the engine in the vehicle, 2C-2
  Rocker arms and valve lash adjusters - removal, inspection and installation, 2C-14
  Timing chain cover - removal and installation, 2C-5
  Timing chains, tensioners and sprockets - removal, inspection and installation, 2C-7
  Top Dead Center (TDC) for number one piston - locating, 2C-2
  Valve covers - removal and installation, 2C-3
  Valve springs, retainers and seals - removal and installation, 2C-19
VACUUM GAUGE DIAGNOSTIC CHECKS, 2D-3
VACUUM SHIFT SYSTEM (4WD MODELS) - DESCRIPTION, CHECK AND COMPONENT REPLACEMENT, 8-29
VALVE CLEARANCE - CHECK AND ADJUSTMENT, 2B-2
VALVE COVERS - REMOVAL AND INSTALLATION
  3.5L V6 engine, 2B-3
  4.2L V6 engine, 2A-3
  V8 engines, 2C-3
VALVE LIFTERS - REMOVAL, INSPECTION AND INSTALLATION, 2A-16
VALVE SPRINGS, RETAINERS AND SEALS
  removal and installation, V8 engines, 2C-19
  replacement, 4.2L V6 engine, 2A-5
VALVES - SERVICING, 2D-11
VEHICLE IDENTIFICATION NUMBERS, 0-6
VINYL TRIM - MAINTENANCE, 11-2

W

WASTEGATE CONTROL ACTUATOR (3.5L V6 ENGINE) - REPLACEMENT AND ADJUSTMENT, 4-26
WATER PUMP
  check, 3-9
  removal and installation, 3-9
WHEEL CYLINDER - REMOVAL, OVERHAUL AND INSTALLATION, 9-17
WHEELS AND TIRES - GENERAL INFORMATION, 10-30
WINDSHIELD AND FIXED GLASS - REPLACEMENT, 11-6
WINDSHIELD WIPER BLADE INSPECTION AND REPLACEMENT, 1-16
WIPER MOTOR - CHECK AND REPLACEMENT, 12-16
WIRING DIAGRAMS - GENERAL INFORMATION, 12-24